SILICON PROCESSING

FOR

THE VLSI ERA

VOLUME 2:

PROCESS INTEGRATION

SILICON PROCESSING

FOR

THE VLSI ERA

VOLUME 2:

PROCESS INTEGRATION

STANLEY WOLF Ph.D.
Professor, Department of Electrical Engineering
California State University, Long Beach
Long Beach, California

LATTICE PRESS

Sunset Beach, California

Published by:

Lattice Press,
Post Office Box 340
Sunset Beach, California 90742, U.S.A.

Cover design by Roy Montibon, Visionary Art Resources, Inc., Santa Ana, CA.

Library of Congress Cataloging in Publication Data
Wolf, Stanley

Silicon Processing for the VLSI Era
 Volume 2 : Process Integration

Includes Index
1. Integrated circuits-Very large scale
 integration. 2. Silicon. I. Title

ISBN 0-961672-4-5

9 8 7 6 5 4 3 2 1

PRINTED IN THE UNITED STATES OF AMERICA

To my wife, Carrol Ann,
and my children, Jennifer Laura and Stanley Charles Ross

———————

CONTENTS

CHAP. 5 - *MOS DEVICES AND NMOS PROCESS INTEGRATION*   298

LIST OF TECHNICAL REVIEWERS

Each of the chapters was reviewed for technical correctness. The following persons graciously undertook the review task for the chapters indicated:

Chapter 2 Dr. Joseph R. Monkowski Dr. Haiping Dun
Lam Research Corp. Intel Corp.
 - CVD Division Santa Clara, CA
Fremont, CA

Chapter 3 Dr. Robert S. Blewer Dr. Stan Swirhun
Sandia National Laboratories Honeywell SSPL
Albuquerque, NM Bloomington, MN

Chapter 4 Dr. Farhad K. Moghadam Dr. Terry Herndon
Intel Corp. MIT - Lincoln Laboratory
Santa Clara, CA Lexington, MA

Chapter 5 Mr. Andrew R. Coulson
TRW Electronic Systems Group
Redondo Beach, CA

Chapter 6 Dr. John Y. Chen Dr. Samuel T. Wang
Boeing Electronics International CMOS Technology, Inc.
Seattle, WA San Jose, CA

Chapter 8 Professor Al F. Tasch, Jr.
University of Texas
Austin, TX

Chapter 9 Dr. Michael Kump
Technology Modeling Associates, Inc.
Palo Alto, CA

PREFACE

SILICON PROCESSING FOR THE VLSI ERA is a text designed to provide a comprehensive and up-to-date treatment of this important and rapidly changing field. The text will consist of three volumes, of which this book is the second, subtitled, *Process Integration*. Volume 1, subtitled *Process Technology*, was published in 1986. Volume 3, to be subtitled *Assembly, Packaging and Manufacturing Technology* is scheduled for publication in 1993. In Volume 1, the individual processes utilized in the fabrication of silicon VLSI circuits are covered in depth (e.g., epitaxial growth, chemical vapor and physical vapor deposition of amorphous and polycrystalline films, thermal oxidation of silicon, diffusion, ion implantation, microlithography, and etching processes).

In this volume, we undertake to explain how the individual processes described in Volume 1 are combined in various ways to produce silicon integrated circuits. This task is referred to as *process integration*. The first part of the book deals with *sub-process integration;* that is, the effort involved in forming circuit structures that can be implemented into a variety of circuit types. These structures include *isolation structures* (Chap. 2), *metal-silicon contacts* (Chap. 3), and *device-interconnect structures* (Chap. 4).

The second part of the book covers the process integration tasks of full-device-type technologies, including *NMOS* (Chap. 5), *CMOS* (Chap. 6), *bipolar* and *BiCMOS* (Chap. 7), and *semiconductor memories* (Chap. 8). Chapter 9 describes the process simulation tools that are available for aiding in the process integration and development efforts.

Volume 3 will cover process control, VLSI manufacturing issues and facilities, contamination and yield, automation, assembly, packaging, and parametric testing.

The purpose of writing this text was to provide professionals involved in the microelectronics industry with a single source that offers a complete overview of the technology associated with the manufacture of silicon integrated circuits. Other texts on the subject are available only in the form of specialized books (i.e., that treat just a small subset of all of the processes), or in the form of edited volumes (i.e., books in which a group of authors each contributes a small portion of the contents).

Such edited volumes typically suffer from a lack of unity in the presented material from chapter-to-chapter, as well as an unevenness in writing style and level of presentation. In addition, in multi-disciplinary fields, such as microelectronic fabrication, it is difficult for most readers to follow technical arguments in such books, especially if the information is presented without defining each technical "buzzword" as it is first introduced.

In our books such drawbacks are avoided by treating the subject of VLSI fabrication from a unified and more pedagogical viewpoint, and by carefully defining technical terms when they are first introduced. The result is intended to be a user friendly book for workers who have come to the semiconductor industry after having been trained in but one of the many traditional technical disciplines.

An important technical breakthrough has occurred in publishing that the author also felt could be exploited in creating a unique book on silicon processing. That is, revolutionary electronic publishing techniques have recently become available, which can cut the time required to produce a published book from a finished manuscript. This task traditionally took 15-18 months, but can be now reduced to less than 3 months. If traditional techniques are used to produce books in such fast-breaking fields as VLSI fabrication, these books automatically possess a built-in obsolescence, even upon being first published.

We have taken advantage of these rapid production techniques, and have been able to successfully meet the reduced production-time schedule. As a result, information contained in technical journals and conferences which was available within three months of the book's publication date has been included.

Written for the professional, the book belongs on the bookshelf of workers in several microelectronic disciplines. Microelectronic fabrication engineers who seek to develop a more complete perspective of the subject, or who are new to the field, will find it invaluable. Integrated circuit designers, test engineers, and integrated circuit equipment designers, who must understand VLSI processing issues to effectively interface with the fabrication environment, will also find it a uniquely useful reference. The book should also be very suitable as a text for graduate-level courses on silicon processing techniques, offered to students of electrical engineering, applied physics, and materials science. It is assumed that such students already possess a basic familiarity with semiconductor device physics. Problems are included at the end of each chapter to assist readers in gauging how well they have assimilated the material in the text.

The book is an outgrowth of several intensive seminars conducted by the author through the Engineering Extension of the University of California, Irvine. Over one thousand engineers and managers from more than 75 companies and government agencies have enrolled in these short courses since they were first presented in 1984.

A book of this length and diversity would not have been possible without the indirect and direct assistance of many other workers. To begin, virtually all of the information presented in this text is based on the research efforts of a countless number of scientists and engineers. Their contributions are recognized to a small degree by citing some of their articles in the references given at the end of each chapter. The direct help came in a variety of forms, and was generously provided by many people. The text is a much better work as a result of this aid, and the authors express heartfelt thanks to those who gave of their time, energy, and intellect.

Each of the chapters was reviewed after the writing was completed. The engineers and scientists who participated in this review were numerous. The main reviewers are listed on the page before the preface, and we would like to thank them once again at this point for their contributions. That is, the following professionals (each one an expert in the topic covered by the chapter they reviewed), read an entire chapter for technical correctness, and provided appropriate comments and corrections: Robert S. Blewer, John Y. Chen, Andrew R. Coulson, Haiping Dun, Terry Herndon, Michael Kump, Farhad K. Moghadam, Joseph R. Monkowski, Stan Swirhun, Al F. Tasch, Jr., and Samuel T. Wang. In addition, J. B. Price, of Spectrum CVD, Chris A. Mack, of the National Security Agency, and Sidney Marshall, Editor of Solid State Technology, provided other valuable technical and editorial input. Robert Shier, of VTC Corp., and Jim Cable of TRW also kindly provided the author with timely and valuable technical literature.

The copy editing of the book was undertaken by Mary Nadler, and the clarity and grammatical correctness of the prose owes a great deal to her professional efforts. The aesthetically pleasing graphics of the cover were designed by Roy Montibon of Visionary Art Resources, Inc., Santa Ana, CA.

Stanley Wolf

P.S. Additional copies of the books can be obtained from:

Lattice Press
P.O. Box 340-W
Sunset Beach, CA, 90742

An order form, for your convenience, is provided on the final leaf of the book.

CHAPTER 1

PROCESS INTEGRATION

FOR VLSI and ULSI

Since the creation of the first integrated circuit in 1960, the density of devices that can be fabricated on semiconductor substrates has steadily increased. In the late 1970s the number of devices manufactured on a chip exceeded the generally accepted definition of *very large scale integration,* or *VLSI* (i.e., more than 100,000 devices per chip) (Fig. 1-1a). By 1990 this number had grown to more than 32 million devices per chip (16-Mbit DRAMs), and it is generally acknowledged that the era of *ultra-large-scale integration (ULSI)* has begun. The increasing device count has been accompanied by a shrinking minimum feature size (Fig. 1-1b), which as of 1990 had decreased to ~0.5 μm in the most advanced commercially available chips.

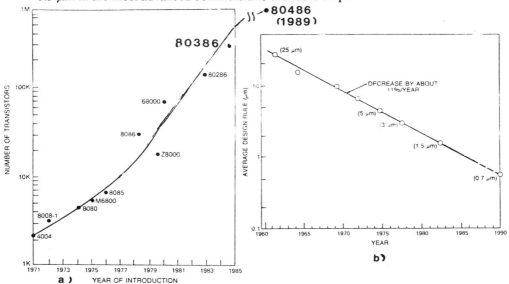

Fig. 1-1 (a) Increase in the number of transistors per microprocessor chip versus year of introduction, for a variety of 8-bit and 16-bit microprocessors, and (b) The decrease in minimum device feature size versus time on integrated circuits.

1

Progress seems likely to continue at a rapid pace, with even further reductions in the unit cost per function and in the power-delay product of integrated circuits projected. Silicon processing has been the dominant technology of IC fabrication and is likely to retain this position for the foreseeable future. The entire adventure of silicon device manufacturing represents a remarkable application of scientific knowledge to the requirements of technology. Our books* are intended to serve as a comprehensive and cohesive report on the state of the art of this technology, as practiced at the time of publication.

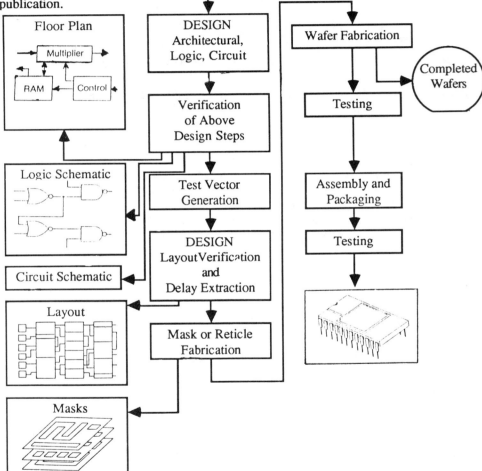

Fig. 1-2 Steps required for the manufacture of very large scale integrated circuits (VLSI).

* SILICON PROCESSING FOR THE VLSI ERA - Volume 1: Process Technology; Volume 2 - Process Integration; and Volume 3 - Assembly, Packaging, and Manufacturing Technology (the latter is scheduled to to be published *approximately* in 1992)

Figure 1-2 illustrates the sequence of tasks followed in the manufacture of an integrated circuit. These tasks can be grouped into three phases: *design, fabrication*, and *testing*. Our books are concerned primarily with the fabrication phase. We describe the IC manufacturing steps that occur from the point at which the circuit design has been completed and the necessary design information has been rendered into the form of a circuit layout. In this form, the layout information is ready to be used to generate a set of masks (or reticles) that will serve as the tools for specifying the circuit patterns on silicon wafers. For VLSI and ULSI circuits, the layout information is stored in a computer.

The details concerning the individual fabrication processes (e.g., those associated with creating patterns, introducing dopants, and depositing films on silicon substrates to form integrated circuits) are the subjects of Volume 1. In this second volume, we

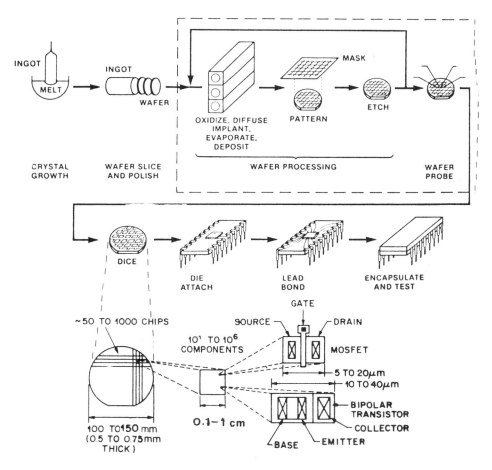

Fig. 1-3 The fabrication process sequence of integrated circuits.

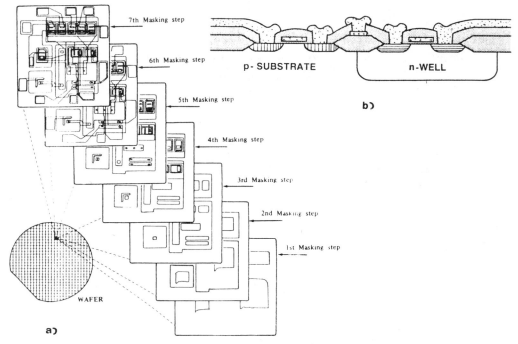

Fig. 1-4 (a) Example of the patterns transferred to a wafer during a seven-mask process sequence, and (b) Cross section of completed devices in a basic CMOS process.

undertake a discussion of the *sequences* of the steps performed to produce the device structures of the circuits. The text describes how the various individual processes are integrated together so that the end result is a completely realized integrated circuit.

One perspective of such a *process-integration* effort is illustrated in Fig. 1-4. It can be seen that a series of masking steps must be sequentially performed for the desired patterns to be created on the wafer surface. The other processing procedures performed between the masking steps serve to create the desired device structures. An example of the end result of a complete process sequence is shown in Fig. 1-4b as an IC cross-section.

This book is roughly divided into two parts. The first part deals with what we refer to as *subprocess integration* (Chapters 2, 3, and 4), while the second covers *complete process integration* (Chapters 5, 6, 7, and 8). Chapter 9 covers the subject of *process simulation*, an approach applicable to both the subprocess and complete process-integration development efforts.

The function of subprocess integration is to produce device structures that can be used in a variety of integrated-circuit technologies (i.e., isolation, contact, and interconnect structures). The purpose of the process-integration task is to design the process sequences used to manufacture complete IC technologies (i.e., NMOS, CMOS, bipolar, BICMOS, and IC memory devices).

1.1 PROCESS INTEGRATION

The approach used in building integrated circuits on monolithic pieces of silicon involves the fabrication of successive layers of insulating, conducting, and semiconducting materials. Each layer is patterned to form a structure that performs a specific function, usually linked with surrounding areas and subsequent layers. The fabrication steps used to manufacture an integrated circuit must therefore be executed in a specific sequence, which constitutes an IC *process flow* (or *process*).

1.1.1 Process Sequence Used to Fabricate an Integrated-Circuit MOS Capacitor Structure.
A simple example of the sequence of steps that must be used to form an integrated circuit MOS capacitor is shown in Fig. 1-5.[3] First, a relatively thick SiO_2 layer is grown on a bare *n*-type Si substrate (using oxide-growth techniques, as described in Vol. 1, chap. 7). This oxide layer (Fig. 1-5a) serves to isolate the metallization-interconnect film deposited at the end of the process from the Si substrate.

An opening in the SiO_2 layer is then formed using a photolithographic process (Vol. 1, chaps. 12 and 13) and an etch process (see Vol. 1, chaps. 15 and 16). Boron atoms are ion-implanted into the wafer surface (Fig. 1-5b). Wherever an opening exists in the SiO_2 layer, the boron atoms are implanted into the Si substrate. A sufficient

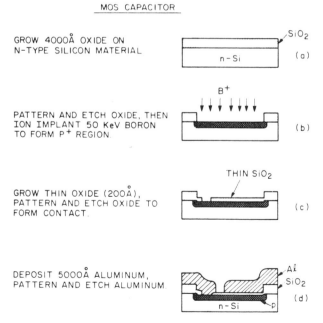

MOS CAPACITOR

GROW 4000Å OXIDE ON N-TYPE SILICON MATERIAL   — SiO_2 / n–Si (a)

PATTERN AND ETCH OXIDE, THEN ION IMPLANT 50 KeV BORON TO FORM P$^+$ REGION.   — B^+ (b)

GROW THIN OXIDE (200Å), PATTERN AND ETCH OXIDE TO FORM CONTACT.   — THIN SiO_2 (c)

DEPOSIT 5000Å ALUMINUM, PATTERN AND ETCH ALUMINUM.   — Aℓ / SiO_2 / n–Si / p (d)

Fig. 1-5 Process sequence for fabrication an MOS capacitor. From S. M. Sze, Ed., *VLSI Technology*, 2nd Ed., Copyright 1988 McGraw-Hill. Reprinted with permission.

implant dose is used to cause these regions to become *p*-type. The energy of the implant and the SiO_2 thickness are also selected so that the boron atoms do not penetrate to the Si wherever the SiO_2 layer remains. In this manner, *p*-regions (which will serve as the lower plate of the MOS capacitor) are selectively formed in the Si.*
In addition, if the voltage between the *p* and *n* regions of the Si substrate keeps the *pn* junctions reverse-biased, the lower plates of any capacitors built on the same Si substrate will be *junction-isolated* from one another (see chap. 2).

The remaining steps are shown in Figs. 1-5c and 1-5d. A thin SiO_2 film is grown over the exposed *p*-type Si region, and a *contact hole* is etched in the film to allow an Al film to make contact to the lower plate of the capacitor. The Al film, deposited next, also serves as the top plate of the capacitor. The formation of the contact holes and Al patterns involves alignment, another important process that must be successfully accomplished when integrating a process sequence. The capacitor structure will function properly only if the two plates of the capacitor are isolated from each other, except where an intentional contact must be made. Therefore, if the Al pattern that serves as the top plate of the capacitor is misaligned (resulting in an accidental contact between the Al and the *p*-type Si region), the two plates of the capacitor will be shorted. This alignment restriction is one of the limitations faced in device miniaturization, as discussed in Volume 1, chapter 13.

1.1.2 Specifying a Process Sequence. Once a process has been completely developed, a set of instructions is produced for fabrication of the specific technology. These instructions include the appropriate times, temperatures, gas flows, power settings, etc. needed for each process step in order for the desired structures to be produced. The instructions also list the specifications of the device structures that should be achieved at the conclusion of each of the process steps (e.g., line width, sheet resistance, film thickness, and step height), as well as the inspections and measurements that must be made to ensure that these specifications have been met. This set of instructions accompanies the group of wafers (or *lot*) as it moves through the process, and is therefore known as a *lot traveler*, or *run sheet*.

Since integrated circuits are normally processed in *batches* (or *lots*), each of the steps is usually performed on all of the wafers in a lot before that lot is moved to the next processing step. As each step is completed, the fabrication worker (*operator*) responsible for that step indicates that the process step has been accomplished and records on the *lot traveler* the time of completion, as well as any required data concerning the process conditions, resultant measured parametric data, and any pertinent comments. When a lot has been completed, the data gathered about all of the fabrication details of that *process run* is thus contained on the lot traveller. Lot travellers initially consisted of printed

---

* Note that this region is also *self-aligned* to the edges of the SiO_2 layer. This is an example of the self-alignment process, a fabrication approach that is critical for allowing progress to be made in device miniaturization techniques.

sheets of paper, but they are now stored on computers. Data concerning the manufacturing details of each lot-fabrication sequence is entered directly into the computer.

1.1.3 Levels of Process Integration Tasks.

From a general perspective, process integration involves the combining of various individual processes (described in Vol. 1) to produce IC structures (or even an entire IC). Process integration tasks can thus involve varying degrees of complexity. Examples of the levels at which such tasks can be performed can be grouped according to increasing complexity, as follows:

1. *Development of a process sequence to modify an existing structure that is only one part of an integrated circuit.* Two examples of such process-integration tasks are: (a) the development of a modified local-oxidation isolation structure with reduced bird's beak or boron channel-stop encroachment (see chap. 2), and (b) the improvement of the characteristics of a barrier-layer material used in metal-Si contacts (e.g., by replacing Ti with Ti:W; see chap. 3).

2. *Development of a process sequence to produce a new device structure.* Three examples of this are: (a) development of a selective epitaxial isolation structure to replace LOCOS (see chap. 2); (b) implementation of a double-level-metal (DLM) structure to replace a single-level-metal interconnect approach (see chap. 4); and (c) creation of a TiN local-interconnect structure for CMOS (chap. 3).

3. *Enhancement of a complete process sequence by shrinking the minimum feature size used in the technology.* While such "technology-shrinks" are generally less complex than the tasks involved in the development of an entirely-new complete-process sequence, they usually entail more than just the implementation of higher-resolution lithography processes. That is, many of the other processes must also be modified, and entirely new ones may need to be developed. Two examples of such technology evolution that have been detailed in the literature are the shrinking of Intel's NMOS process from NMOS I (L = 6 μm) to HMOS III (L = 1.5 μm), and the shrinking of of AT&T's Twin-Tub CMOS process from CMOS I (L = 2 μm) to CMOS VI (L = 0.5 μm).

4. *Implementation of a new technology through modification or enhancement of an older technology.* Two examples of this type of process-integration effort are:

    a. Development of an *n*-well CMOS process based on the experience gained from prior NMOS technology development. In this case, NMOS devices are fabricated in the *p*-substrate, using the same process sequence as in an NMOS process developed earlier, while the fabrication of PMOS devices in the *n*-well (and the integration of these process steps into the NMOS sequence) represent new process integration tasks.

    b. Development of a BiCMOS process through integration of the steps needed to fabricate bipolar *npn* transistors into an existing CMOS process sequence (see chap. 7).

5. *Development of a new technology from scratch.* Examples of this effort are the development of new CMOS and BiCMOS processes without reliance on the enhancement of older process sequences. This kind of effort is normally the most difficult, and so it is rarely attempted. When a new process is to be developed from scratch (especially where new materials are to be used), the effort required to establish the interrelationships among the various process steps can be extremely expensive and time consuming. A more conservative approach is to modify or enhance a previously developed process. Consequently, when a company decides to launch a line of products based on a new process sequence, it often purchases the technology details from a company that has already developed such a process. The acquired process technology can either be used as is, or enhanced.

1.2 PROCESS-DEVELOPMENT AND PROCESS-INTEGRATION ISSUES

When a new process or device structure is proposed for development, the intent is to improve the circuit performance in some way or to build new products that could not be fabricated with existing technology. It is very important that the circuit designers, product engineers, device designers, and process engineers interact closely in such a development effort. The circuit designers and product engineers identify the performance requirements and package sizes needed for the applications to be served (e.g., SRAMs, ASICs, microprocessors, or logic circuits). From these specifications, the device designers can target candidate device structures, as well as the various structural options and process flows that can be potentially investigated for achieving these goals. The process engineers can then work on finding new materials and innovative processes to implement the advanced device structures. The odds of developing new device structures that offer improved performance (while still meeting reliability and manufacturability constraints) are greatly enhanced when each of the members of a technology development team understands the goals and limitations of the other disciplines involved.

Several key issues must be considered when a process-integration task is to be undertaken. One of the most important is that virtually all of the steps in a process sequence are significantly interrelated. For example, each of the thermal cycles in a process sequence contributes to the total vertical and lateral diffusion effects in a process. To end up with device structures that have the desired doping profiles at the end of a process sequence, it is necessary to take into account *all* of the thermal processing steps that a device undergoes. Furthermore, each new thermal step may impact the formation of defects in the silicon or may play a role in the chemical or morphological characteristics of the structures created in the IC.* Since the fabrication steps are so strongly interrelated, once a process has been well established, a cardinal rule − which must be rigidly followed − is that *no step in the process may be arbitrarily changed.*

Nevertheless, when a new process is developed, it is obvious that at least some changes must be made to an existing process. The ability to predict the effect of a process change on the circuit or device electrical parameters is an important part of the process-integration effort. In the early days of integrated-circuit manufacturing, hand calculations of the doping profiles in the silicon and SiO_2 based on simple one-dimensional analytical models were used to estimate the electrical characteristics of device structures fabricated using a particular process sequence. These predicted estimates were then compared to the electrical characteristics measured following the completion of process runs which were designed according to the analytically determined process steps (trial-and error approach).

As device sizes grew smaller, however, such an approach proved to be inadequate. That is, two-dimensional effects within the device structures made the measured electrical characteristics deviate significantly from the characteristics predicted with one-dimensional models. To produce small-geometry devices with the desired electrical characteristics, device and process engineers had two options: (1) to perform more experiments, or (2) to use computer-aided simulation of the fabrication process to model the device structures that are obtained from a specific process sequence (Fig. 1-6).[1]

The first option would mean a drastic increase in the costs and cycle time of new process integration. Furthermore, even if the experiments did yield adequately performing devices, if complicated processes and structures were involved, the result would still be poor physical insight and inadequate quantitative analysis of the factors governing device operation.

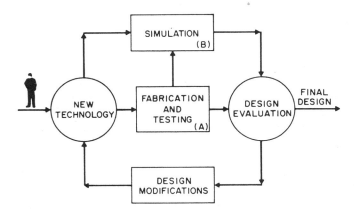

Fig. 1-6 Simulation and experiment are approaches to IC process development. Path A depicts the experimental approach. Path B is the approach that exploits computer-aided simulation.[1] (Copyright IEEE, 1984).

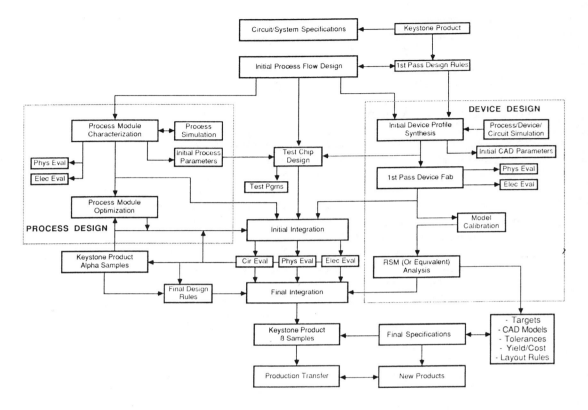

Fig. 1-7 Technology development flowchart, integrating process, device, and design considerations.[2] (Copyright IEEE, 1987).

The second approach has therefore become more widely adopted. The information about the device structures calculated from the process simulation (in the form of impurity profiles, electrical characteristics, and layer and pattern dimensions) can be used with circuit models of the devices to calculate the device's terminal characteristics. The circuit parameters (e.g., SPICE parameters) can then be extracted from the terminal characteristics and the device layouts. These parameters, together with the information calculated on the connectivity characteristics of the circuit (extracted from circuit-layout and process-simulation data on the interconnect structures of the circuit), can yield the frequency-response or switching characteristics of the resultant circuits. The simulated design can be verified by building circuits based on the results once a satisfactory set of circuit-performance parameters has been obtained. In addition to being far less costly and time consuming than laboratory experiments, this design approach produces detailed information on device operation in a well-controlled environment. The simulation tools available for process integration as described in chapter 9.

A comprehensive approach for developing new IC technologies has been outlined by Alvarez et al.[2] and is summarized in the flow-chart shown in Fig. 1.7. It can be seen that the process integration task is but one part of this larger technology development effort. Other important tasks that must also be successfully performed include device design, module development (i.e., development of each of the individual steps in the process sequence), specification of the circuit/system characteristics that the technology will be serving, and test-chip design.

Reference 2 also addresses the impact of the conflicting requirements that arise in the attempt to optimize of all of the device characteristics (here, for the specific case of a BiCMOS technology). Statistical design of experiments (see Vol. 1, chap. 18) and response surface methodology (RSM) are used to identify the effect of tradeoffs between the various device characteristics on device structures. Process and device simulators (described in chap. 9) are used to conduct the analyses that identify the optimum device structures for the applications to be served. Process conditions can then be selected to allow such optimum device structures to be fabricated.

REFERENCES

1. W. Fichtner et al., *Proceedings IEEE,* **72**, 96, (1984).
2. A. R. Alvarez, "BiCMOS Technology," 1987 IEDM Short Course on BiCMOS Technology, December 1987.
3. S. J. Hillenius, "VLSI Process Integration," Chap. 11, in *VLSI Technology*, 2nd. Ed., S.M. Sze, Editor, McGraw-Hill Book Company, New York, 1988.

CHAPTER 2

ISOLATION TECHNOLOGIES
FOR INTEGRATED CIRCUITS

Implementing electric circuits involves connecting *isolated* devices through specific electrical paths. When fabricating silicon integrated circuits it must therefore be possible to isolate devices built into the silicon from one another. These devices can subsequently be interconnected to create the specific circuit configurations desired. From this perspective, we can see that isolation technology is one of the critical aspects of fabricating integrated circuits.

A variety of techniques have been developed to isolate devices in integrated circuits. One reason is that different IC types (e.g., NMOS, CMOS, and bipolar) have somewhat different isolation requirements. Furthermore, the various isolation technologies exhibit differing attributes, with respect to minimum isolation spacing, surface planarity, process complexity, and density of defects generated during fabrication of the isolation structure. Tradeoffs can be made among these characteristics when selecting an isolation technology for a particular circuit application. This chapter surveys the various isolation technologies, including their evolution up to those being evaluated for submicron devices.

Before the invention of integrated circuits, only discrete diodes, bipolar transistors and field-effect transistors (FETs) could be fabricated. In the early 1950s, these discrete devices exhibited relatively high reverse-bias junction leakages and low breakdown voltages (caused by the large density of traps at the surface of single-crystal silicon).

In 1958, a group of workers at Bell Telephone Laboratories, led by Atalla, found that when a thin layer of SiO_2 was grown on the surface of silicon where a *pn* junction intercepts the surface, the leakage current of the junction was reduced by a factor from 10 to 100. It was later understood that the oxide reduces and stabilizes many of the interface and oxide traps. Not only did such oxide-passivation of the silicon surfaces allow diodes and transistors to be fabricated with significantly improved device characteristics, but the leakage path along the surface of the silicon was also effectively shut off. Thus, one of the fundamental isolation capabilities needed for planar devices and integrated circuits had also been developed. In his report on the evolution of the MOS transistor, C. T. Sah remarks that the successful effort by the Bell Labs group to stabilize Si surfaces was the most important technological advance in microelectronics during the 1950s, and that it blazed the trail that led to the development of the silicon integrated circuit.[90]

With the advent of integrated circuits, it became necessary to provide electrical isolation between devices fabricated on the same piece of silicon. Since bipolar ICs were the first to be developed, a technology for isolating the collector regions of the bipolar devices was also the first to be invented (termed *junction isolation*).

PMOS and NMOS ICs did not need junction isolation, but it was still necessary to provide an isolation structure that would prevent the establishment of parasitic channels between adjacent devices. The most important technique developed was termed *LOCOS isolation* (for LOCal Oxidation of Silicon), which involved the formation of a semirecessed oxide in the nonactive (or *field*) areas of the substrate.

Eventually, bipolar ICs adopted a similar LOCOS-isolation technology (with the oxide performing a somewhat different function than in MOS circuits). The nature of CMOS also required that isolation exist between devices in adjacent tubs as well as between devices within each tub.

As device geometries reached submicron size, conventional LOCOS isolation technologies reached the limits of their effectiveness, and alternative isolation processes for CMOS and bipolar technologies were needed. Modified LOCOS processes (which overcome some drawbacks of conventional LOCOS for small-geometry devices), trench isolation, and selective-epitaxial isolation, were among the newer approaches adopted.

Devices that must function under high voltages and in harsh radiation environments require even more stringent isolation techniques. These techniques are generally referred to as *silicon-on-insulator* (SOI) isolation methods. They include such older approaches as *dielectric isolation* (DI) and *silicon-on-sapphire* (SOS), as well as more recently developed technologies, such as *separation by implanted oxygen* (SIMOX), *zone-melting-recrystallization* (ZMR), *full isolation by porous-oxidized silicon* (FIPOS), and *wafer bonding*. These technologies will be described in the final sections of the chapter.

2.1 BASIC ISOLATION PROCESSES FOR BIPOLAR ICs

2.1.1 Junction Isolation

As mentioned in the introduction, isolation techniques for bipolar ICs were the first to be developed. To see why isolation is needed in bipolar ICs consider Fig. 2-1a, which shows several non-isolated bipolar transistors on a monolithic substrate. It can be seen that all the collector regions are electrically connected through the substrate - obviously, an unacceptable circumstance for all bipolar IC structures (except those few in which common collectors are desired). To circumvent this problem, several approaches to isolating the devices have been developed.

The most important isolation technique used in early bipolar ICs is *junction isolation*, and it has been incorporated as part of the *standard buried collector* (SBC) process (described in this section as well as in chap. 7), the *triple diffused (3D)* process

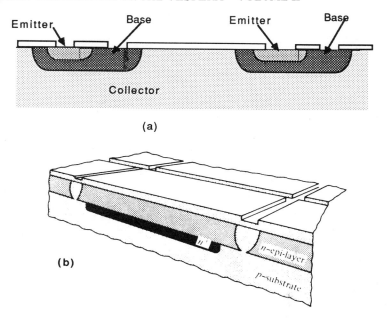

Fig. 2-1 (a) Two non-isolated bipolar transistors on a monolithic substrate. It can be seen that all the collector regions are electrically connected through the substrate - obviously, an unacceptable circumstance for all bipolar IC structures (except those few in which common collectors are desired). (b) A p-type "moat" is diffused into an n-type epitaxial layer on a p-type substrate The p diffusion extends through the epitaxial layer, joining with the p-type substrate to completely surround the n-type epi islands.

(described further in chap. 7), and the *collector diffused* process (described in this section).

2.1.1.1 Junction Isolation in the SBC Process.
The SBC process is still widely implemented in bipolar analog-circuit and power-circuit applications where power-supply voltages are in the 15-30 V range. Most digital bipolar technologies (which operate with supply voltages smaller than 10 V), however, have come to adopt oxide-isolated structures.

When junction isolation is used for *npn* devices, a p-type "moat" is diffused into an n-type epitaxial layer on a p-type substrate (Fig. 2-1b). The p diffusion extends through the epitaxial layer, joining with the p-type substrate to completely surround the n-type epi islands. (The n-type epitaxial layer is typically less than 10-μm thick, which is thick enough to accommodate the subsequent base and emitter regions of the bipolar transistors.) Hence, each device is contained in an island of n-type material surrounded by a *pn* junction; good isolation exists if the substrate p material is held at the most negative potential of the circuit. In effect, each component is surrounded by a reverse-biased *pn* junction, and room-temperature leakage current for reverse-biased silicon junctions can be held within the nanoampere range. With the exception of capacitance effects at high frequencies, excellent isolation between circuit elements can be provided.

The isolation structure is formed after the n^+ buried layer and the epitaxial layer have been grown in the p-type substrate (see chap. 7 for more details on bipolar device structures). An oxide is grown on the epi layer and is then patterned and etched. Alignment of the isolation diffusion now becomes an issue. That is, the isolation structure should not touch the buried layer because not only would the high doping on both sides of the resulting pn junction drastically decrease the breakdown voltage, but the capacitance of the collector/substrate junction would also increase. Thus, a minimum spacing between the n^+ buried layer and the isolation diffusion must be specified among the layout design rules. This dimension is determined by taking into account the lateral diffusion distances of both the isolation diffusion and buried layer, with an allowance made for alignment tolerance (Fig. 2-2).

Because of the depth to which the p-region must be driven, the boron isolation diffusion typically takes the longest of any furnace step in the bipolar process. During this step, the lateral diffusion of the boron atoms is about 75 to 80% of the vertical diffusion, and thus the isolation region width is approximately twice the thickness of the epi-layer. Following diffusion, a low-temperature oxidation-and-etch sequence is performed to remove the boron-rich layer on the silicon surface, and a new oxide layer is grown over the isolation diffusion.

Successful fabrication of the isolation structures can be confirmed by stripping the oxide over the isolation diffusion and probing two adjacent structures. If complete isolation has been achieved, no current will flow unless the applied voltage exceeds the breakdown voltage of the isolation junction. If a small gap exists at the bottom of the isolation, leakage current will occur. Poor isolation results in catastrophic device failure since all transistor collector regions are shorted together.

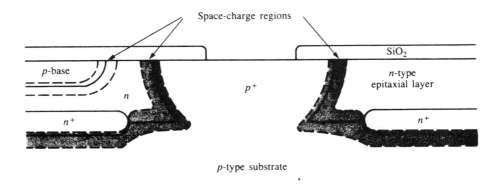

Fig. 2-2 Isolation region between two bipolar transistors. The spacing must be large enough to ensure that the two space-charge regions do not merge together. From R. C. Jaeger, *Introduction to Microelectronic Fabrication*. Copyright, 1988, Addison-Wesley. Reprinted with permission.

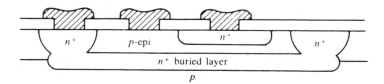

Fig. 2-3 Cross section of a transistor fabricated in the CDI process.[129] (© 1969 IEEE).

Junction-isolation, while quite simple, exhibits two major drawbacks. First, the wide isolation regions represent large inactive areas on the silicon surface, and this adversely impacts IC packing-densities. Second, the isolation diffusions cause large parasitic collector-to-substrate and collector-to-base capacitance values, which degrade the speed performance of the circuits (see chap. 7).

2.1.1.2 Collector-Diffusion Isolation. The collector-diffusion isolation (CDI) structure[1] allows isolation between adjacent bipolar devices to be achieved without the use of a *p*-type isolation diffusion (Fig. 2-3). This reduces device area as well as process complexity, as compared to the SBC process. The CDI process is used to best advantage in digital applications in which high speed is an important priority and low power-supply voltages are utilized (e.g., +5 V, as used in digital circuits).

The process is begun by selectively diffusing low-sheet-resistance (15-25 Ω/sq) arsenic buried layers 3-5 μm deep, onto a lightly doped *p*-type wafer. These buried layers act as the collectors of the transistor and as isolation regions for the diffused resistors. This is followed by formation of a thin *p*-type (0.2 Ω-cm) epitaxial layer (1-2 μm thick) which serves as a part of the base. Device isolation and definition of the base are accomplished through an n^+ phosphorus diffusion, which surrounds the transistor peripherally and also provides contact to the collector layer. A *p*-diffusion is then carried out to form the remainder of the base region. Since the surface of the n^+ regions are heavily doped, this *p*-diffusion can be done non-selectively, without the need for another mask. This diffusion prevents surface inversion, minimizes edge injection from the emitter, provides a built-in field in the base region, and provides control of the composite base-layer resistance (see chap. 7 for more details in bipolar device operation). A shallow emitter is next implanted or diffused; this is followed by contacts, metallization.

The process in this basic form utilizes only five masks (or six, if the pad mask is counted), and the transistor is considerably smaller than a junction-isolated bipolar transistor (though not as small as an oxide-isolated device). Two-level metallization adds two additional masking steps. At the 2.5-μm geometry level, no plasma etching or ion implantation is required, and no LPCVD of polysilicon or nitride is needed. The flat surface topography of CDI is also well suited to further device scaling.

The primary disadvantages of the CDI process are the large collector-base and collector-substrate capacitances (as well as relatively low collector-base junction breakdown voltages, which limits them to applications which use small power-supply voltages). Consequently, advanced oxide-isolated structures that minimize these capacitances have largely replaced CDI. However, the benefits of oxide isolation are gained only at the cost of significantly increased process complexity. Work is therefore continuing to increase the performance of CDI so that its compact structure and process simplicity can be exploited for additional applications.[2]

2.2 BASIC ISOLATION PROCESS FOR MOS ICs (*LOCOS ISOLATION*)

The isolation requirements of MOS ICs are somewhat different than those of bipolar ICs. Because MOS transistors are self-isolated, as long as the source-substrate and drain-substrate *pn* junctions are held at reverse bias, the drain current (I_D) should only be due to current flow from source to drain through a channel under the gate. This also implies that no significant current between adjacent MOS devices should exist if no channel exists between them (Fig. 2-4). The buried n^+-p junction needed to isolate bipolar transistors is thus not necessary for MOS circuits.

The self-isolation property of MOS devices represents a substantial area savings for NMOS (and PMOS) circuits as compared to junction-isolated bipolar circuits. Hence, ICs having the highest component densities are fabricated with MOS technologies.

The method by which the components of an integrated circuit are interconnected involves the fabrication of metal stripes that run across the oxide in the regions between the transistors (the field regions, Fig. 2-4). However, these metal stripes form the gates of parasitic MOS transistors, with the oxide beneath them forming a gate oxide and the diffused regions (2) and (3) acting as the source and drain, respectively. The threshold voltage of such parasitic transistors must be kept higher than any possible operating voltage so that spurious channels will not be inadvertently formed between devices (see chap. 5 for a discussion of inversion in MOS transistors).

In order to isolate MOS transistors, then, it is necessary to prevent the formation of channels in the field regions, implying that a large value of V_T is needed in the field regions. In practice, the V_T of the field regions needs to be 3-4 V above the supply voltage to ensure that less than 1 pA of current flows between isolated MOS devices[108] (e.g., for 5 V circuit operation, the minimum field-region V_T must be 8-9 V). This condition must be maintained at the minimum isolation spacing used in a given technology, since decreases in field-region V_T arise for same reasons as in short-channel MOS devices (see section 5.5.1.1). Furthermore, V_T also decreases as the temperature rises; reductions of as much as 2 V in field-V_T values have been observed as the temperature rises from 25°C to 125°C.[109]

As will be discussed in chapter 5, several methods can be used to raise the threshold voltage. The two utilized to increase field-region V_T's involve increasing the field-

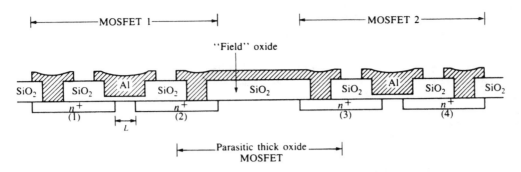

Fig. 2-4 No current should flow between source and drain regions of adjacent MOS transistors. This figure shows the parasitic field transistor with a possible channel under the field oxide if the substrate under this oxide is inverted.[128] From R. C. Jaeger, *Introduction to Microelectronic Fabrication.* Copyright, 1988, Addison-Wesley. Reprinted with permission.

oxide thickness and raising the doping beneath the field oxide. If the field oxide were made sufficiently thick, it alone could cause a high enough threshold voltage in the field parasitic device. Unfortunately, the large oxide steps produced by this approach would give rise to step coverage problems, and reduced field-oxide thicknesses are thus preferred. To achieve a sufficiently large field threshold voltage with such thinner oxides, the doping under the field oxide must be increased (see chap. 5, Example 5-5).

The field oxide is nevertheless typically made seven to ten times thicker than the gate oxide in the active regions. (This thick oxide also reduces the parasitic capacitance between the interconnect runners and the substrate, improving the speed characteristics of the circuits.) Normally, ion implantation is used to increase the doping under the field oxide. This step is called a *channel-stop implant*. The combination of thick oxide and channel-stop implant can provide adequate isolation for PMOS and NMOS (and as we shall see, for oxide-isolated bipolar) ICs. Additional isolation considerations that arise in the case of CMOS circuits will be described in chapter 6.

The thick field oxide can be fabricated in a number of ways. In one approach (used until about 1970), the oxide is grown to the desired thickness on a flat silicon surface and then etched in the active regions (leaving thick oxide in the field regions, Fig. 2-5b). Although this *grow-oxide-and-etch* approach also allows the isolation regions to be most sharply defined, two disadvantages have prevented it from being utilized in VLSI applications: (1) the field-oxide steps are high and have sharp upper corners, making them very difficult to cover with subsequent metal-interconnect lines; and (2) the channel-stop implant must be performed before the oxide is grown, and thus the active regions must subsequently be aligned to the channel-stop regions in the silicon with a lithography step. This imposes a severe packing density penalty.

In another approach, the oxide is selectively grown over the desired field regions. This is done by covering the active regions with a thin layer of silicon nitride that

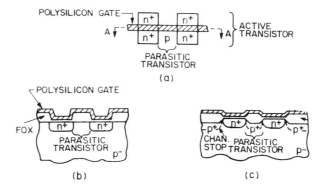

Fig. 2-5 MOSFET isolation. (a) Top view of adjacent MOS transistors with common polysilicon gate illustrating active and parasitic transistor conduction paths. Cross section through A - A for (b) Grow-oxide-and-etch isolation, and (c) LOCOS isolation structures.[130] From S. M. Sze, Ed., *VLSI Technology*, 2nd Ed., Copyright 1988 McGraw-Hill. Reprinted with permission.

prevents oxidation from occurring beneath them. After the nitride layer has been etched away in the field regions (and prior to field-oxide growth), the silicon in those regions can also be selectively implanted with the channel-stop dopant. Thus, the channel-stop region becomes self-aligned to the field oxide, overcoming one of the drawbacks of the grow-oxide-and-etch method.

The more serious step-coverage limitation of the grow-oxide-and-etch isolation method is also overcome to some degree by the selective-oxidation approach. If the silicon is etched (to a depth of about half the desired field oxide thickness) after the oxide-preventing layer is patterned, the field oxide can be grown until it forms a planar surface with the silicon substrate. (Note: this occurs because the growing oxide film is about twice as thick as the silicon layer it consumes; see Vol. 1, chap. 6). This is known as a *fully recessed isolation oxide process.*

If the field oxide is selectively grown without etching the silicon (Fig. 2-5c) the resulting field oxide will be *partially (or semi) recessed.* In the semirecessed process, the oxide step height is larger than in the fully recessed process, but it is smaller than in the grow-oxide-and-etch process. In addition, the step has a gentle slope that is more easily covered by subsequent polysilicon and metal layers. Semirecessed LOCOS is also less complex than fully recessed LOCOS and results in fewer process-induced defects in the silicon substrate. The semirecessed oxide has become the workhorse isolation technology for MOS devices down to about 1.5-μm geometries, while the fully recessed oxide has primarily been used in bipolar circuits.

The selective oxidation process described above was introduced by Appels and Kooi in 1970,[3] and it has become the most widely used approach for forming the thick field oxide. We will first describe this process in detail, discussing its limitations for submicron devices. Subsequently, alternative isolation schemes that have been proposed to overcome these limitations will be considered.

2.2.1 Punchthrough Prevention between Adjacent Devices in MOS Circuits

Parasitic conduction between adjacent devices due to *punchthrough* (Fig. 2-4c) must also be prevented (see chap. 5 for additional information on punchthrough). In MOS circuits, the source and drain regions of each transistor must be kept far enough from the source and drain regions of any neighboring devices so that the depletion regions do not merge together (i.e., this distance must be greater than twice the maximum depletion-region width). Substrate doping must also be considered, because lighter doping allows wider depletion-region widths, and punchthrough is more likely to occur at lower voltages in lightly doped substrates.

2.2.2 Details of the Semirecessed-Oxide LOCOS Process

Each of the steps used to fabricate conventional semirecessed LOCOS structures will now be discussed in additional detail (Fig. 2-6).

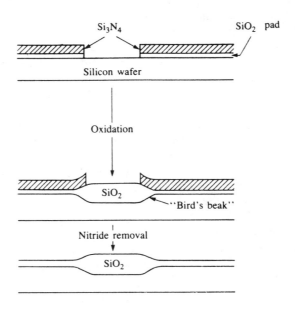

Fig. 2-6 Cross section depicting process sequence for semirecessed oxidations of silicon.[128] From R. C. Jaeger, *Introduction to Microelectronic Fabrication.* Copyright, 1988, Addison-Wesley. Reprinted with permission.

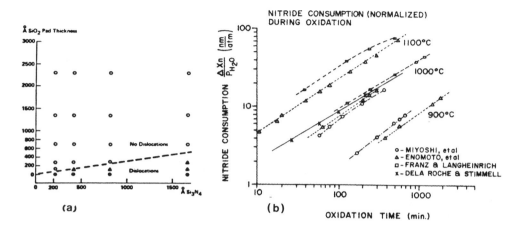

Fig. 2-7 (a) Dislocation generation at Si_3N_4 film edges versus CVD pad-oxide thickness, after 6 h of dry-wet-dry thermal oxidation at 950°C.[7] Copyright 1978, reprinted with permission of the AIOP. (b) Nitride consumption during oxidation.[131] Reprinted with permission of Semiconductor International.

2.2.2.1 Pad-Oxide Layer. A wafer with a bare silicon surface is cleaned, and a 20-60 nm SiO_2 layer is thermally grown on the surface. The function of this layer, called a *pad* or *buffer* oxide, is to cushion the transition of stresses between the silicon substrate and the subsequently deposited nitride. In general, the thicker the pad oxide, the less edge force is transmitted to the silicon from the nitride (Fig. 2-7a).[110] On the other hand, a thick pad-oxide layer will render the nitride layer ineffective as an oxidation mask by allowing lateral oxidation to take place. Therefore, the minimum pad-oxide thickness that will avoid the formation of dislocations should be used. Figure 2-7b indicates that the minimum thickness of a thermally-grown pad oxide should be at least one third the thickness of the nitride layer.

Alternative techniques to allow the use of thinner pad oxides have also been reported. One method involves the use of CVD SiO_2 in place of thermal SiO_2.[4] Because CVD SiO_2 is more effective than thermal SiO_2 for avoiding edge defects, such pad layers can be about 25% as thick as thermal-oxide pad layers. The second method utilizes a pad layer consisting of a thin thermal SiO_2 layer and a buffer polysilicon layer.[5,6,34,35]

2.2.2.2 CVD of Silicon Nitride Layer. Next, a 100-200 nm thick layer of CVD silicon nitride which functions as an oxidation mask is deposited. Silicon nitride is effective in this role because oxygen and water vapor diffuse very slowly through it, preventing oxidizing species from reaching the silicon surface under the nitride. In addition, the nitride itself oxidizes very slowly as the field oxide is grown (typically, only a few tens of nm of nitride are converted to SiO_2 during the field-oxide growth process). Thus, the nitride should remain as an integral oxidation-barrier layer during the entire field-oxide-growth step. Figure 2-7c compares the oxidation rate of silicon and silicon nitride, and shows that silicon oxidizes approximately 25 times faster than silicon nitride. One criterion for selecting the nitride thickness is that it should be greater than the thickness that will be converted to SiO_2 during the subsequent field oxidation step.

Silicon-nitride films, however, have the well-known drawback of exhibiting a very high tensile stress when deposited by CVD on silicon (on the order of 10^{10} dynes/cm^2). The termination of intrinsic stresses at the edge of a nitride film gives rise to a horizontal force that acts on the substrate. Under some circumstances, this stress can exceed the critical stress for dislocation generation in silicon, and will thus become a source of fabrication-induced defects. For example, at 1000°C in oxidizing ambients, defects can be generated at the edges of nitride films as thin as 21 nm if they are deposited *directly* on silicon.[7] Pad oxides are used to combat these stresses and avoid dislocation generation. The effect of the pad layer is to reduce the force transmitted to the silicon at the nitride edge, and to relieve the stress of the nitride via the viscous flow of the pad oxide.

2.2.2.3 Mask and Etch Pad-Oxide/Nitride Layer to Define Active Regions.
The active regions are now defined with a photolithographic step. A resist pattern is normally used to protect all of the areas where active devices will be formed. The nitride layer is then dry etched, and the pad oxide is etched by means of either a dry- or wet-chemical process. After the pad oxide has been etched, the resist is not removed but instead is left in place to serve as a masking layer during the channel-stop implant step.

2.2.2.4 Channel-Stop Implant.
An implant is next performed in the field regions to create a channel-stop doping layer under the field oxide. In NMOS circuits, a p^+ implant of boron is used, while in PMOS (and in the *n*-tubs of CMOS circuits) an n^+ implant of arsenic is utilized. Although this normally requires two masking steps in CMOS circuits, a single mask process for implanting both *p* and *n* channel stops has been reported.[8] After the implant has been completed, the masking resist is stripped.

2.2.2.5 Problems Arising from the Channel-Stop Implants.
During field oxidation, the channel-stop boron experiences both segregation and oxidation-enhanced diffusion. Thus, relatively high boron doses are needed (mid 10^{12}-10^{13} atoms/cm^2) in order for acceptable field threshold voltages to be achieved. This also implies that the peak of the boron implant must be deep enough that it is not absorbed by the growing field-oxide interface (implant energies in the 60-100 keV range are used). If the channel-stop doping is too heavy, it will cause high source/drain-to-substrate capacitances and will reduce source/drain-to-substrate *pn* junction breakdown voltages.

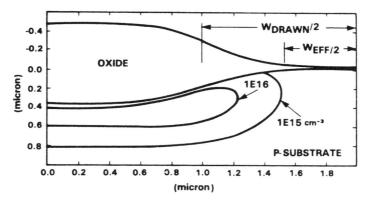

Fig. 2-8 Field oxide and boron encroachment in LOCOS.[96] From K. M. Cham et al., *Computer-Aided Design and VLSI Device Development.* Copyright 1986, Kluwer Academic Publishers. Reprinted with permission.

The lateral diffusion of the boron also causes it to encroach into the NMOS active areas (Fig. 2-8). Such redistribution raises the boron surface concentration near the edge of the field oxide, causing the threshold voltage to increase in that region of the active device. As a result, the edge of the device will not conduct as much current as the interior portion, and the transistor will behave as if it were a narrower device. (This effect is also enhanced as the dose of the channel-stop implant is increased.)

Finally, dislocations generated during the channel-stop implant step (which would ordinarily climb out to the surface during appropriate annealing conditions) can be caused to glide under the nitride-covered regions by the stresses at the nitride edges. The penetration of such dislocations through nitride-edge-defined junctions can cause increased emitter-base leakage in bipolar devices.[10]

The extent of the channel-stop dopant diffusion can be reduced by using high-pressure oxidation (HIPOX) to grow the field oxide,[8] by use of a germanium-boron co-implant,[9] or by use of a chlorine implant.[51] HIPOX allows the oxide-growth temperature to be reduced, which reduces the diffusion length of the boron. The germanium-boron co-implant exploits the fact that boron diffuses with a lower diffusivity in the presence of implanted germanium, and boron segregation effects are also reduced (see Vol. 1, chap. 7). The result is a 40% increase in field threshold voltage for the same dose of implanted boron, and a corresponding decrease in lateral-boron encroachment. The chlorine implant is performed in the field regions prior to field oxidation, and this causes the oxide to grow at a faster rate. Consequently, the field oxide can be grown in less time at the same temperature.

2.2.2.6 Grow Field Oxide. After the channel-stop implant has been performed, the field oxide is thermally grown by means of wet oxidation, at temperatures of around 1000°C for 2-4 hours (to thicknesses of 0.3-1.0 μm). The oxide grows where there is

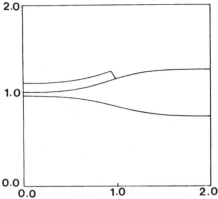

Fig. 2-9 Schematic of the bird's beak that occurs during semirecessed LOCOS.

no masking nitride, but at the edges of the nitride, some oxidant also diffuses laterally. This causes the oxide to grow under and lift the nitride edges. Because the shape of the oxide at the nitride edges is that of a slowly tapering oxide wedge that merges into the pad oxide, it has been named a *bird's beak* (Fig. 2-9). The bird's beak is a lateral extension of the field oxide into the active area of the devices. Although the length of the bird's beak depends upon a number of parameters – including the thicknesses of the buffer oxide, nitride, and isolation oxide (as well as the oxidation temperature and oxygen partial pressure) – the length for a typical 0.5-0.6-μm field oxide is ~0.5 μm/side. This would make a lithographically defined 1-μm feature disappear on the chip following the field oxidation step (Fig. 2-10). Although this effect led to predictions that LOCOS isolation would have to be replaced for device dimensions smaller than 2 μm, optimization of process steps has allowed conventional LOCOS to continue to be used for device-isolation spacings as narrow as 1.25-1.5 μm.[11,14]

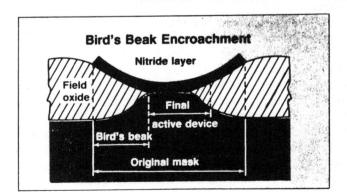

Fig. 2-10 Bird's beak encroachment limits the scaling of channel widths to about 1.2-1.5 μm.[11] Reprinted with permission of Semiconductor International.

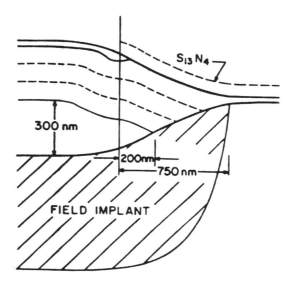

Fig. 2-11 Eventual exposure of the *p*-substrate region when etching back the LOCOS field oxide in an NMOS device (in an attempt to regain some of the active device area lost by bird's beak encroachment).

The LOCOS bird's beak also creates another problem during the later process step of contacting metal to the source and drain regions of an MOS device. Because of the shape of the bird's beak, any overetching that must be performed during the contact-window-opening step will etch away part of the bird's-beak oxide. This may expose the substrate region under the source or drain region (Fig. 2-11). If this occurs, the source will become shorted to the well region when the metal interconnect film is deposited, impairing or destroying device operation.

The problem worsens in CMOS as shallower junctions are used, due to exposure of the well region. (If deeper junctions are used, the lateral diffusion distance is also longer and etching back part of the bird's beak may not expose the well region.) In NMOS processes, this problem was counteracted by redoping of the contacts with phosphorus after they were opened (and before the metal was deposited). This created a deep junction in the contact areas only. The source/drain junction near the gate oxide was kept shallow in order for good device performance to be to retained. In CMOS processing, not only are shallow junctions used, but redoping of the contacts presents an additional problem because both n^+ and p^+ contact openings are defined at the same masking step. As a result, in CMOS processes enclosed contacts are commonly employed. This solution, however, reduces packing density, as area must be provided in the source and drain regions to keep the contact from overlapping the bird's beak. Another solution would be to use an alternative isolation process in which the contacts are able to coincide with (or overlap) the active areas.

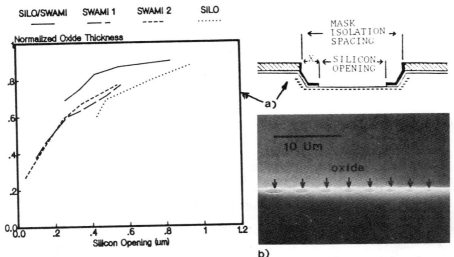

Fig. 2-12 (a) Normalized field oxide thickness versus silicon opening for various advanced LOCOS processes.[14] (© 1985 IEEE). (b) SEM photo of the thinning of the field oxide as the dimension of the exposed silicon gets smaller.[121] This paper was originally presented at the Spring 1989 Electrochemical Society Meeting held in Los Angeles, CA.

Modifications to the basic process have allowed the bird's beak to be reduced in length. These include etching back a portion of the field oxide after it is grown; using silicon nitride without a pad oxide; and using a thin pad oxide covered with polysilicon under the masking nitride layer. These advances will be described in more detail in a later section.   One last limitation of LOCOS-based isolation schemes for submicron structures is the oxide field-thinning effect.[14] The field-oxide thickness in submicron-isolation spacings is significantly less than the thickness of field-oxides grown in wider spacings (Fig. 2-12).[14,121] The narrower the width of the exposed substrate silicon region, the thinner the field oxide. For example, a field oxide that is grown to a thickness of 400 nm above a region of exposed Si that is 1.5-μm wide would be only about 290 nm thick if grown above a 0.8-μm-wide region of Si.

This effect is believed to be caused by the reduction in the oxidants available in the submicron opening compared to those available in wider openings.[15] Thin field oxides that result from this effect can have a great impact on field-threshold voltages and on the interconnect capacitances to substrate. It is predicted that as a result of this field-thinning effect and the need to maintain defect-free isolation structures (see next section), the minimum space that must be allowed for a LOCOS-type isolation structure with a field-oxide thickness of 550 nm will be 0.75 μm.

2.2.2.7 Strip the Masking Nitride/Pad-Oxide Layer. Following field oxidation, the masking layer is removed. Since 20-30 nm of the top of the nitride is converted to SiO_2 during the field oxidation, this layer must be etched off first. The

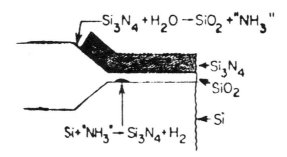

Fig. 2-13 Kooi's model for explaining nitride growth under the gate oxide.[16] Reprinted by permission of the publisher, The Electrochemical Society, Inc.

remaining nitride and pad oxide are then etched. These steps can be carried out by means of wet-chemical etching, since there is no need to maintain dimensions or to otherwise tightly control the etching. Since chemical methods that remove nitride offer excellent selectivity with respect to the underlying oxide, considerable overetching can be used.

2.2.2.8   Regrow Sacrificial Pad Oxide and Strip (Kooi Effect).
During the growth of the field oxide, another phenomenon occurs that causes defects in when the gate oxide is grown. Kooi et al., discovered that a thin layer of silicon nitride can form on the silicon surface (i.e., at the pad-oxide/silicon interface) as a result of the reaction of NH_3 and silicon at that interface (Fig. 2-13).[16] The NH_3 is generated from the reaction between H_2O and the masking nitride during the field-oxidation step. This NH_3 then diffuses through the pad oxide and reacts with the silicon substrate to form silicon-nitride spots or ribbon (these regions are sometimes called *white ribbon)*. When the gate oxide is grown, the growth rate becomes impeded at the locations where the silicon nitride has formed. The gate oxide is thus thinner at these locations than elsewhere, causing low-voltage breakdown of the gate oxide. One way to eliminate this problem is to grow a "sacrificial" gate oxide after stripping the masking nitride and pad oxide, and removing it before growing the final gate oxide.[17,18]

2.2.3   Limitations of Conventional Semi-Recessed Oxide LOCOS for Small-Geometry ICs

The limitations of conventional LOCOS for submicron technologies can be summarized as follows:

  • The bird's-beak structure causes unacceptably large encroachment of the field oxide into the device active regions (Fig. 2-10).

• Boron from the channel-stop implant of *n*-channel MOSFETs is excessively redistributed during the field-oxide growth and other high-temperature steps, leading to unacceptable narrow-width effects (see chap. 5).

• The planarity of the surface topography is inadequate for submicron lithography needs (see section 2.9.2).

• The thickness of the field oxide in submicron regions of exposed silicon is significantly thinner than that grown in wider spacings (Fig. 2-12). This field-oxide-thinning effect can produce problems with respect to field threshold voltages, interconnect-to-substrate capacitance, and field-edge leakage. Finally, it introduces additional nonplanarity across the surface of the wafer.

2.3 FULLY RECESSED OXIDE LOCOS PROCESSES

Several conventional, fully recessed oxide LOCOS methods have been developed for bipolar integrated-circuit applications, including the ISOPLANAR,[19] ROI,[20] and OXIS[21] processes. Such fully oxide-isolation processes allow higher performance bipolar ICs to be fabricated than if junction isolation is used. The collar of SiO_2 that forms the sidewall isolation eliminates the sidewall contribution to collector-to-

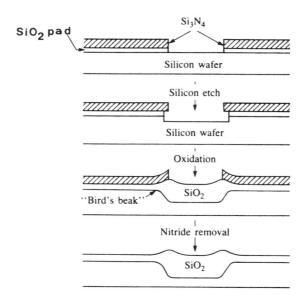

Fig. 2-14 Cross section depicting process sequence for fully-recessed oxidation of silicon.[128] From R. C. Jaeger, *Introduction to Microelectronic Fabrication*. Copyright, 1988, Addison-Wesley. Reprinted with permission.

substrate and base-to-collector capacitances, thereby increasing the cutoff frequency of oxide-isolated transistors to 3-5 GHz – well above the 1-GHz limit of junction-isolated devices. In addition, the area of the transistor is reduced by about 70%.

In all of these approaches, the basic semirecessed LOCOS process is still followed, with the following modification: After the nitride and pad oxide have been defined, the silicon-epitaxial layer of the standard buried-collector (SBC) *npn* bipolar process is etched (Fig. 2-14). About one-half to two-thirds of the epi layer is removed, which creates grooves in the silicon. This step can be performed using wet-isotropic etching (using an HNO_3/HF solution); wet-anisotropic etching, to yield slope-walled grooves (using a KOH solution – see Vol. 1, chap. 15); or dry etching. Boron is diffused or implanted into the bottom of the etched areas as a channel stop. The patterned nitride/oxide regions again mask the implant so that the active areas are not implanted.

Next the field oxide is grown (typically to a thickness of 1-2 μm). About 45% of the oxide growth is downward (see Vol. 1, chap. 6) and extends completely through the epi layer. Since 55% of the oxide growth is upward, it expands to fill the etched space, and a relatively planar device surface can be produced. (We will see in the next section that the surface of the fully-recessed-oxide process is actually not completely planar.)

This field oxide performs a somewhat different isolation function in bipolar ICs than in MOS circuits. That is, it primarily serves to replace the *p*-diffusion at the device sidewalls, thereby preventing adjacent collector regions from being directly electrically connected. Whereas in MOSFETs the minimum oxide thickness is determined by the required V_T in the field regions, in the bipolar device the oxide must penetrate all the way through the epi layer, and so must normally be thicker. To achieve this without producing intolerably high steps at the surface, it is necessary to etch grooves grooves into the silicon initially. Since the growth of field oxides with thicknesses greater than 2 μm is impractical (even when grooves are etched in the Si), oxide isolation is not used in bipolar ICs that require power supply voltages exceeding 15 V (i.e., such technologies require epi layers greater than 2-μm thick).

The oxide-isolated bipolar transistor is smaller than the junction-isolated transistor because the lateral dimension of the oxide region is somewhat smaller than that of the *p*-diffusion region, but its size is also reduced for several other reasons (see chap. 7, Fig. for a comparison of bipolar device sizes with various isolation processes). First, the SiO_2 can be butted directly against the buried-collector n^+ region since the problem of depletion-region merging (inherent in junction isolation) does not exist. Second, the collector-metal stripe and the base-metal stripe can overlap the sidewall oxide. Finally,

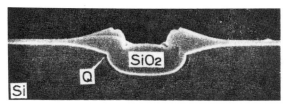

Fig. 2-15 SEM micrograph showing bird's head and beak.

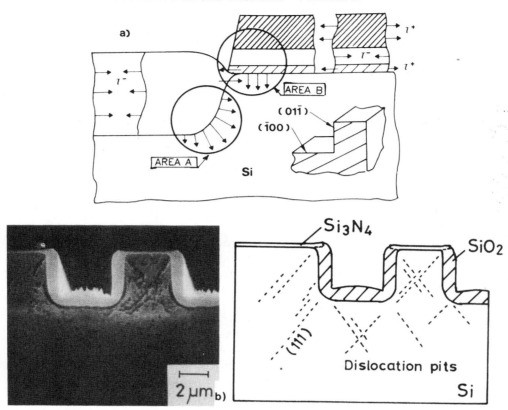

Fig. 2-16 (a) Schematic representation of the stress field induced in the silicon substrate by the SILO process.[32] (© 1985 IEEE). (b) Dislocations caused by stress in growing oxide in trench structures as revealed by Secco etching.[49] Reprinted by permission of the publisher, The Electrochemical Society, Inc.

the collector n^+ contact, the p-type base, and even the emitter region can extend to the isolation region. Since such smaller-area devices further reduce parasitic capacitances, oxide isolation has become the technology of choice for advanced, high-performance, 5-V digital circuits.

There are, however, several problems that arise from the etching of the grooves in the silicon. First, the fully recessed field oxide is not completely planar - that is, a bulge (or *bird's head*) occurs in the oxide surrounding the nitride pads (Fig. 2-15) which does not occur in the semirecessed LOCOS field oxide. The bird's head can cause step-coverage problems in the metallization layers due to uneven surface topology. Etchback procedures have been proposed to minimize the problem in multimetal processes.[22] Second, as described in the next paragraphs, the growth of the field oxide in the grooves causes stresses that can lead to defects in the silicon.

The most important defect-generating stresses arise at the bottom corners of the etched grooves during field oxide growth and are due to the volume misfit of the growing oxide. First, as the oxide grows, it pushes against the sidewalls of the groove. In addition, as the oxide in the corner grows, its top surface is pushed outward by the newly forming oxide layer, and thus becomes tangentially stressed. Finally, as the laterally encroaching oxide grows under the nitride, it is restrained from growing upward by the stiffness of the nitride layer, causing further stress downward against the silicon in the corner of the recess. These stresses, if not relieved, will generate dislocations in the silicon (Fig. 2-16).

The two mechanisms that can prevent these stresses from producing defects as the oxide is grown in the silicon recesses are *stress reduction* and *stress relief*. The stress produced at the corners of the recessed silicon can be reduced by ensuring that the corners of the etched structures are rounded, and by using thinner silicon-nitride masking layers. Stress can be relieved through the viscoelastic flow of the oxide using growth above 960°C.[23] Enhanced viscous flow also occurs under high-pressure oxidation at reduced temperatures. It has been reported that high-pressure oxidation can be used to drastically reduce field-oxidation time, and thereby to also reduce the lateral diffusion of the channel-stop implants and the updiffusion of the buried-collector layer in bipolar ICs.[24] Fewer defects occur if the sidewalls of the recessed regions are inclined away from the vertical at angles of 75° or more. This type of structure appears to facilitate stress relief by enhancing the viscoelastic-flow effect.

2.3.1 Modeling the LOCOS Process

Among the many aspects of the LOCOS process that would be useful to simulate are the oxide thickness and morphology, the shape and dimensions of the bird's beak and bird's head, the redistribution of the channel-stop dopant, and the stresses induced in the silicon during the LOCOS processing. Models for simulating these features have been developed, and they are described more fully in chapter 9.[26,27,28]

2.4 ADVANCED SEMIRECESSED-OXIDE LOCOS ISOLATION PROCESSES

The conventional LOCOS process described earlier is limited by oxide and boron encroachment, as well as by nonplanar surface topographies and the field-thinning effect. These limitations must be overcome to allow the LOCOS process to be used for submicron IC designs.

2.4.1 Etched-Back LOCOS

The simplest way to reduce the bird's beak and to obtain a more planar surface is to etch back a portion of the field oxide after it has been grown.[12, 13] In this procedure, the upper region of the field oxide is removed to recover the active-area loss and to thin the

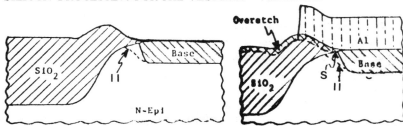

Fig. 2-17 Implanting the base of a bipolar transistor allows the bird's head and beak to be etched back to some degree without exposing the substrate region of opposite doping polarity.

field oxide so that a more planar topography is produced (Fig. 2-11). What keeps this from from being useful for submicron processes is that in order for the bird's beak to be significantly reduced, the p-substrate must become exposed, which will cause the drain-to-substrate regions of a MOSFET to be shorted. Implantation of source and drain regions (or base regions of a bipolar transistor), instead of forming them by diffusion, extended the use of this approach, as shown in Fig. 2-17. However, the problem of boron encroachment is not alleviated by this approach.

2.4.2 Polybuffered LOCOS

Another semirecessed-LOCOS bird's-beak compaction scheme exploits the fact that a thinner pad oxide results in a shorter bird's beak. Therefore, the usual pad-oxide layer is replaced with a polybuffered pad layer (poly [50-nm]/oxide [5-10-nm]) and a thicker-than-usual nitride (100-240 nm).[5,6,34] A bird's beak of only 0.1-0.2 μm/side is produced for a field oxide of 400-nm thickness. Reported leakage currents of the n^+-p junction were as low as with a conventional LOCOS process. The basic polybuffered LOCOS process reduces bird's beak length but does not solve the boron-encroachment or planarity difficulties of conventional LOCOS. Reference 125 provides an extensive characterization of the polybuffered LOCOS method, including the effects of varying the pad oxide, polysilicon, and nitride layer thicknesses on the bird's beak length and defect formation.

An more successful advanced polybuffered LOCOS process has recently been reported (Fig. 2-18).[35] In this approach, a thinner field oxide is used, and the channel-stop implant is performed through this 220-nm-thick field oxide. The 100-keV boron implant that forms the channel stop also serves as the punchthrough implant in the active regions. Since the channel-stop implant is done after the field-oxide growth, much less boron encroachment occurs. In addition, a smaller field-oxide step is produced, thereby helping the planarity requirement. A thinner field-oxide region can be used because the process is designed for 0.7-μm (and smaller) device applications. At these short channel lengths, the doping of the substrate must be increased to prevent punchthrough (see chap. 5). As a result, the dopant concentration of the channel stop also ends up being high enough to prevent field inversion, even with the use of such a thin field oxide.

2.4.3 SILO (Sealed-Interface Local Oxidation)

The SILO isolation process is another modification of conventional LOCOS. It reduces bird's-beak length to about 0.2 μm by forming a layer of silicon nitride directly on the silicon surface before depositing an oxide pad layer over it.[32] In this way the interface between the silicon surface and the nitride is sealed, and lateral diffusion of oxidants is prevented (Fig. 2-19). As a result, a shorter bird's beak is produced (Fig. 2-20).[123]

In SILO, a thin (10-20-nm) nitride layer is formed on the silicon surface by thermal nitridation of the silicon, or CVD. (The use of such a very thin reduces the defect-causing nitride-edge stress.) Next, a 25-30-nm CVD SiO_2 film and a 150-200-nm nitride film are sequentially deposited. After the three-layer-masking film has been patterned by RIE, the field oxide can be grown on the unetched silicon surface to create a semirecessed field oxide, or the silicon surface can be etched so that a fully recessed field oxide can be grown. The channel stop is implanted before the field oxide is grown. After field oxidation, the masking layer is etched away.

There are some problems with SILO. If the silicon is not etched, the semirecessed field oxide will cause steps on the wafer surface. While these are no higher than those

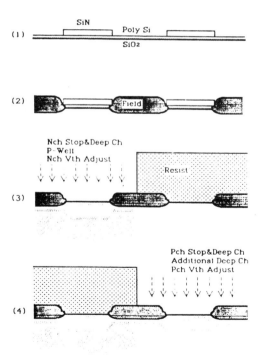

Fig. 2-18 Process sequence of APPL isolation structure.[35] (© 1988 IEEE).

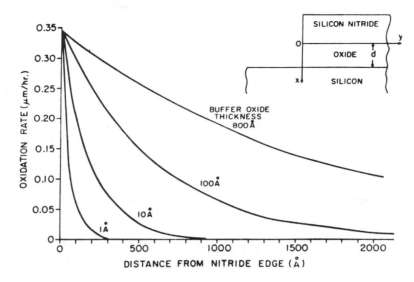

Fig. 2-19 Oxidation rates versus distance from the edge of the nitride edge with the buffer (pad) oxide thickness as a parameter.[110] (© 1982 IEEE).

of conventional LOCOS, they are steeper, since the bird's beak has been nearly eliminated (Fig. 2-21).[33] If the silicon is etched prior to oxide growth, many dislocations will be produced if the etched-silicon sidewalls are vertical. In order to

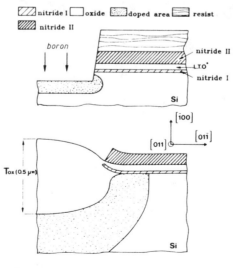

*LTO≡ low temperature oxide

Fig. 2-20 Example of a shortened bird's beak achievable with a SILO process.[32] (© 1985 IEEE).

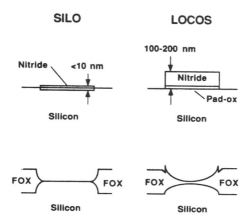

Fig. 2-21 Schematic of SILO and LOCOS structures before and after field oxidation, showing the steeper field oxide sidewalls with SILO.[33] (© 1988 IEEE).

avoid this, the bottom corner of the recess should have a gentle, rounded slope, best produced through isotropic etching. Unfortunately, this leads to lateral oxide encroachment, as illustrated by a 1.2-μm CMOS process described by Motorola in 1986. The SILO process oxide encroachment was 0.2 μm/side.[36] Finally, it has been reported that the stress associated with the abrupt transition from field oxide to thin gate oxide leads to increased sensitivity with respect to localized interface-trap generation.[37]

2.4.4 Laterally Sealed LOCOS Isolation

Ghezzo et al. developed another approach for sealing the silicon interface during field-oxide growth, called *laterally-sealed-LOCOS isolation*.[31] This process seeks to maintain the bird's-beak reduction of SILO while reducing the defects that are induced when a nitride is in direct contact with the silicon. The active-device area is covered with a pad-oxide layer, while the silicon interface is sealed with nitride. This achieves the low-defect densities provided by the pad oxide, together with the reduced bird's beak of SILO.

This method employs an initial masking layer (called the *active-area stack*) consisting of a 20-nm pad oxide, an 80-nm CVD nitride, and a 200-nm plasma-enhanced (PE) CVD oxide. The purpose of the top PECVD oxide is to provide vertical support for the subsequently deposited second nitride layer which will serve to seal the sidewalls of the active-area stack. In the next stack patterning step, overetching ensures

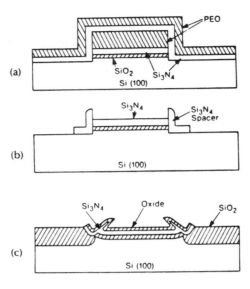

Fig. 2-22 Process sequence for the laterally sealed LOCOS isolation. (a) Active area stack consisting of thin SiO_2 layer covered with Si_3N_4 and a thicker SiO_2 layer (PEO) and "sealed" with a double layer of Si_3N_4 and SiO_2. (b) After RIE to produce a Si_3N_4 spacer around the edges of active area. (c) Active area and field structure after thermal oxidation to form thick field oxide.[31] Reprinted by permission of the publisher, The Electrochemical Society, Inc.

complete film removal in the field regions. It also removes ~10 nm of Si, but this extra silicon etch actually helps to seal the edge of the original pad oxide by forming a thin inlay for the nitride spacer that is subsequently deposited (see Fig. 2-22a).

After the resist has been stripped, a second 80-nm CVD-nitride film and a second 220-nm PECVD-oxide film is deposited to provide conformal coverage of the active-area stack. An anisotropic RIE step is then carried out to remove these second films from the field, along with the first PECVD oxide. This forms an L-shaped nitride spacer that seals the sidewalls of the stack (Fig. 2-22b). The field oxide can now be grown. The foot of the L-shaped nitride spacer is quite thin and easily bends back toward the sidewall under the volumetric expansion of the growing field oxide (Fig. 2-22c).

A thin layer of nitride is converted to oxide during the field oxidation. The process is completed by removing this oxide and the residual nitride. Figure 2-23a shows an active-area stack and the nitride spacer following spacer formation and sacrificial-oxide removal, while Figure 2-23b shows a cross-section after a 800-nm-thick field oxide has been grown and the nitride is still in place (the bird's beak is 0.2 μm long). Diode leakage similar to that seen in conventional LOCOS (with a bird's-beak length of 0.6 μm) has been observed in this structure, which has a bird's-beak length of 0.45 μm.

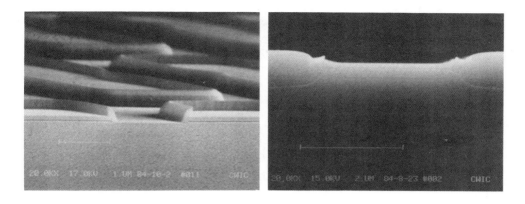

Fig. 2-23 SEM micrographs of isolation structures with laterally sealed LOCOS.[31] Reprinted by permission of the publisher, The Electrochemical Society, Inc.

In a variation of the laterally sealed LOCOS process (Fig. 2-24) that incorporates a protective-oxide pad and nitride spacer that seals the masking-layer sidewall (named POP-SILO),[38] a 12.5-nm pad oxide, an 80-nm CVD nitride, and a 150-nm LPCVD oxide are sequentially deposited. The thickness of the oxide layer is chosen to provide

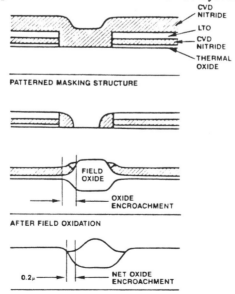

Fig. 2-24 POP-SILO process flow.[38] This paper was originally presented at the Spring 1989 Meeting of The Electrochemical Society, Inc. held in Los Angeles, CA.

the necessary step height for the spacer formation. After the previously described layers have been etched to the silicon substrate to define the active areas, a second CVD-nitride film is deposited to achieve a spacer length of 250 nm. The base of the spacer defines the length of the sealed interface. Field oxide is grown under and beyond the spacer so that the area of silicon on which subsequent gate oxide will be grown is not exposed to silicon nitride. After field oxide growth, the masking layer is removed. The typical isolation encroachment for this process is 0.2 μm larger than the originally masked pattern. However, the problems of planarity, boron encroachment, and thinning of the field oxide (when the windows to the exposed silicon are of submicron dimensions) do not appear to be addressed by these two approaches.

2.4.5 Bird's Beak Suppression in LOCOS by Mask-Stack Engineering

When isolation spacings approach 1 μm in dimension some other effects can be exploited to suppress lateral-oxide encroachment in a semirecessed LOCOS process. First, the stiffness of the mask stack (which consists of the pad-oxide and nitride films) is dramatically increased, and as a result it exhibits much more resistance to bending as the oxide film grows laterally under it, than when the mask is wider. If the deformation of the mask stack can be decreased, lateral oxidation is also reduced since the oxide growth requires volumetric expansion. The magnitude of the lateral encroachment can be reduced by as much as 30% for a 0.5 μm mask length, compared to a "long" mask length. Through optimization of the relative film thicknesses of the mask stack, along with the mask length and the degree of interface sealing (i.e., by using nitride, oxynitride, or polybuffered pad layers), an acceptable degree of suppression of the bird's beak can be produced for a given technology. This process-development task has been referred to as *mask-stack engineering.*[121]

2.4.6 Planarized SILO with High-Energy Channel-Stop Implant

An advanced SILO approach has also been reported that attacks the problems of bird's-beak encroachment, boron channel-stop encroachment, and planarity.[39,111,112] In this process (Fig. 2-25), a sandwich film consisting of a silicon-oxynitride pad layer and a silicon-nitride masking layer is used to suppress bird's-beak formation. The semirecessed field oxide is then planarized with an etchback process (see chap. 4 for more information on etchback processes). Finally, the channel-stop implants are implanted through the field oxide: hence, very little boron encroachment occurs. With the exception of the field-thinning problem, this approach is predicted to allow such modified LOCOS techniques to be extended to 0.5-μm CMOS technology.

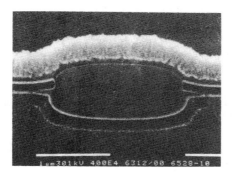

Fig. 2-25 SEM micrograph of 700 nm SLOCOS.[111] This paper was originally presented at the Spring 1989 Meeting of The Electrochemical Society, Inc. held in Los Angeles, CA.

2.5 ADVANCED FULLY RECESSED OXIDE LOCOS ISOLATION PROCESSES

Work has also been done to overcome the limitations exhibited by fully recessed oxide LOCOS isolation technologies. Although the conventional, fully recessed LOCOS offers the benefit of a planarized surface, it also possesses all of the drawbacks of the semirecessed LOCOS approach (as well as those of increased process complexity and stress-induced defect generation). The advanced fully recessed LOCOS isolation techniques were developed to overcome these drawbacks.

2.5.1 SWAMI (Sidewall-Masked Isolation Technique)

The *sidewall-masked isolation* (SWAMI) technique has received much attention since it was introduced as a bird's-beak-free isolation scheme by Chiu et al. in 1982.[29] Its main advantages over conventional LOCOS are an increase in packing density (due to limited lateral encroachment), no severe restrictions on the field-oxide thickness, improvement of the surface planarity, and complete elimination of the gate-oxide-thinning phenomenon (Kooi effect).

   SWAMI pad-oxide and CVD-nitride layers are formed and etched in the same manner as in conventional LOCOS. Grooves are then etched in the silicon to a depth of approximately half the desired field-oxide thickness. This is most often carried out by means of an orientation-dependent silicon etch, such as KOH. This produces recesses having sidewalls that are inclined about 60° when formed on a <100> surface. During field oxidation these sloping sidewalls help to reduce the stresses that contribute to the generation of edge defects.

A second stress-relief oxide layer is then grown, followed by the deposition of a second CVD nitride (which provides conformal coverage of the entire surface, including the sidewalls of the silicon recesses) and a CVD oxide (Fig. 2-26a). This composite CVD SiO_2-CVD Si_3N_4-thermal SiO_2 layer is anisotropically etched in the field region so that it remains only on the sidewalls of the recessed silicon. Because the CVD oxide forms a spacer that protects part of the second nitride layer, this nitride forms a structure with a foot that extends partway into the exposed silicon at the bottom of the recess. After the oxide spacer has been etched away, the final structure is a silicon mesa whose sidewalls are surrounded by the second nitride and oxide. At this point the channel-stop implant is performed, and the field oxide is grown.

In the latter step, the thin sidewall nitride is bent upward due to the expansion of the converted SiO_2. The length of the foot of the nitride on the floor of the recessed silicon is selected to minimize the growth of the bird's beak into the active area (the active-area boundary is defined by the edge of the first nitride). Finally the masking nitride/oxide layers are removed.

A perspective view of a SWAMI isolation structure is shown in Fig. 2-26b. The bird's-beak is reduced, and a relatively planar surface topography is achieved.

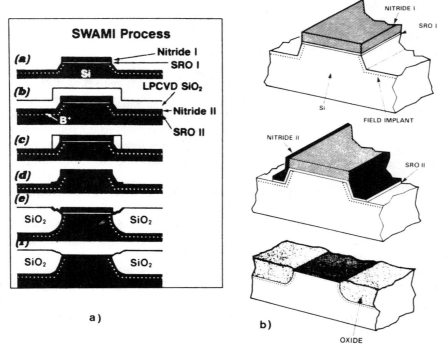

Fig. 2-26 (a) The SWAMI process developed by Hewlett-Packard.[29] (© 1982 IEEE). (b) Perspective views of the SWAMI process.[96] From K. M. Cham et al., *Computer-Aided Design and VLSI Device Development*. Copyright 1986, Kluwer Academic Publishers. Reprinted with permission.

Several problems, however, have been reported with the SWAMI method. First, this process is more complex than conventional LOCOS. Second, the depth of the KOH-formed silicon recesses may vary depending on the position on the wafer, especially since no etch stops are employed. This can lead to differences in the corner structure at the field-oxide/active-area border, which in turn might widen the distribution of electrical-device parameters (such as threshold voltage and punchthrough leakage current). Third, the benefits of reduced edge defects obtained by using a KOH etch step (i.e., due to the 60°-inclined sidewalls that the etch produces) are offset by the characteristic property of this type of etch to delineate surface-crystallographic imperfections. If such imperfections are present, the field-oxide quality and uniformity are affected.

A double-silicon plasma etch that still produces sloped sidewalls has therefore been proposed as a replacement for the KOH etch process.[30,124] The electrical characteristics of SWAMI structures fabricated by this dry-etch process have been reported to be as good as those obtained with standard LOCOS. Although the double-silicon etch SWAMI method involves a more complex manufacturing sequence than LOCOS or single-etch SWAMI, it also provides a greater degree of planarization and a reduction in the lateral field oxide encroachment than the former, and significantly decreased leakage current levels than the latter.

2.5.2 SPOT (Self-Aligned Planar-Oxidation Technology)

Another modified SWAMI-like fully recessed LOCOS process that eliminates the bird's head and produces a highly planar surface with no observed dislocation generation has been reported by Sakuma et al.[25] In this isolation process, called *self-aligned planar-oxidation technology* (SPOT), a conventional semirecessed field oxide is first grown using high-pressure oxidation. This field oxide is then removed with a buffered HF solution. Next, a second pad oxide (100 nm thick) is grown on the exposed recessed silicon, followed by a 100-nm CVD nitride that conformally covers all surfaces. These two layers are then anisotropically etched, which removes them from the bottom of the recessed silicon areas but not from under the overhanging nitride film (Fig. 2-27a). The second field oxide (2 μm thick) is grown using high-pressure oxidation at 900°C.

Figure 2-27b shows the SPOT isolation structure compared to the structures obtained with semirecessed and fully recessed LOCOS. Because the shape of the exposed silicon following removal of the first field oxide is very smooth, only small stresses are created between the silicon and SiO_2 as the second field oxide is grown. Hence, dislocation generation is avoided.

2.5.3 FUROX (Fully Recessed Oxide)

The fully recessed-oxide (FUROX)[39] isolation technology is very similar to the SPOT process just described, except that a nitridized oxide is used instead of using pad oxide to

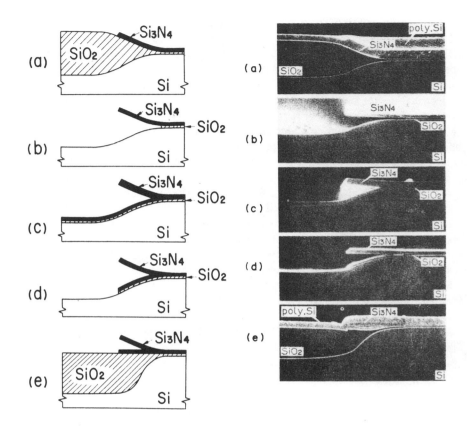

Fig. 2-27 Schematic of the SPOT isolation process.[25] Reprinted by permission of the publisher, The Electrochemical Society, Inc.

seal the silicon surface during growth of the oxide that forms the recess in the silicon substrate. Thus, there is less oxide encroachment into the active-device regions.

A thermal-oxide layer of ~20-nm thickness is grown and is then nitridized in an NH_3 ambient at 1200°C for seven hours. This provides a pad layer consisting of nitridized oxide that is capable of shutting off the diffusion path along the Si_3N_4-SiO_2-Si interface, with only a slight loss of the pad-oxide viscosity. The first LPCVD-nitride layer of 80-nm thickness and an undoped CVD oxide of 20-nm thickness are then deposited and patterned by means of plasma etching (Fig. 2-28a). The exposed nitridized oxide under the CVD nitride is etched away in a hot phosphoric bath. At this point the first field oxide is grown (450 nm thick), with the bird's beak largely suppressed by the nitridation-enhanced sealing ability. This field oxide is then chemically etched away to provide a moat of well-controlled depth and sidewall angle. The silicon surface is free of dry-etch damage and has a short nitride overhang (Fig. 2-28b).

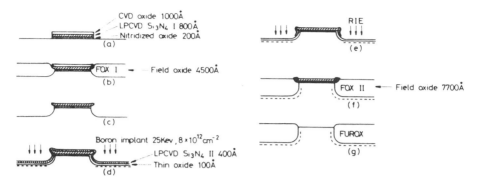

Fig. 2-28 Major process steps of the FUROX isolation technology.[39] (© 1988 IEEE).

A second pad-oxide layer (10 nm thick) is then grown and covered with a second CVD nitride (40 nm thick). After the self-aligned boron channel-stop implant is performed, the second nitride layer is etched anisotropically, so that it still remains on the sidewall for oxidation masking (Fig. 2-28c). After the second field oxide has been grown, a defect-free, near-zero bird's-beak, fully recessed oxide is produced with fairly good planarity. The narrow-width effects observed in conventional LOCOS are also reported to be significantly reduced. Figure 2-29 shows a successfully fabricated 770-nm-thick field oxide using FUROX, with a 1.1-μm isolation width and a bird's beak of about a 0.15 μm. A similar process, named PROXIS, has also been reported.[127]

2.5.4   OSELO II

Another recessed selective-oxidation-based isolation process that reduces the bird's-beak length through lateral sealing of the silicon surface with a nitride layer is called OSELO II.[40] This process is designed to achieve isolation at spacings as small as 0.7 μm with low parasitic-junction capacitance. Conventional patterning of the masking layer, which here consists of a 50-nm pad oxide and a 240-nm CVD-nitride film, is followed by a *surface* boron implant (Fig. 2-30a), and a second CVD nitride layer (30 nm thick)

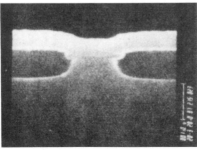

Fig. 2-29 SEM cross sectional photograph of the FUROX structure.[39] (© 1988 IEEE).

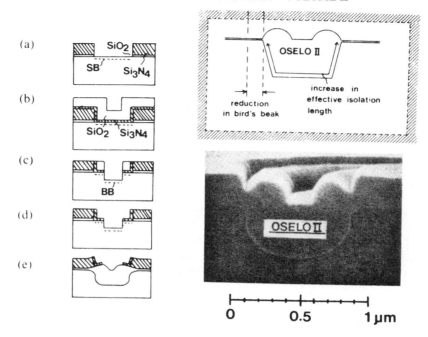

Fig. 2-30 (a) Fabrication process steps for OSELO II. (b) SEM photograph of a cross section of OSELO II.[40] (© 1988 IEEE).

- which prevents bird's-beak intrusion into the active region - and a 200-nm CVD SiO_2 film (Fig. 2-30b).

Dry etching removes this second nitride/oxide film everywhere except along the sidewalls of the first nitride layer, creating an offset structure of sidewall spacers. With the spacers used as a mask, 0.2 μm deep grooves are etched into the exposed Si with a dry etch that minimizes etching of the spacer materials (Fig. 2-30c). This is then followed by a second channel-stop boron implant that penetrates the silicon at the bottom of the groove (Fig. 2-30d), but is prevented from reaching the area that received the surface implant by the nitride/oxide sidewall spacer. Following the *bottom* implant, the CVD oxide remaining on the spacer is removed. Finally, a 550-nm-thick oxide is grown over the field region by means of wet-oxidation step for 120 minutes at 1000°C (Fig. 2-30e).

Since dry etching is used to form the grooves in the OSELO II process, depth control is better than in the SWAMI process, which uses KOH etching. The surface implant keeps the sidewalls of the etched groove from being inverted, while the bottom implant prevents inversion of the bottom of the groove. The effective isolation length of the OSELO II structure (defined as the length along the SiO_2-Si interface under the thick oxide) is 1.3 μm, even though the isolation-mask dimension is only 0.8 μm. Thus, a long, effective-isolation length is achieved. However, OSELO II was also shown to generate dislocation defects during oxide growth.

2.6 NON-LOCOS ISOLATION TECHNOLOGIES, I: (TRENCH ETCH AND REFILL)

Refilled trench structures have been developed for use in four primary VLSI and ULSI applications:

- Replacement of LOCOS as the isolation technology for like devices within the same tub in CMOS.

- Isolation of bipolar devices.

- Isolation of n-channel from p-channel devices and prevention of latchup in CMOS.

- Use as trench-capacitor structures in DRAMs.

Trench/refill approaches for isolation applications fall into the following three categories: shallow trenches (<1 μm), moderate depth trenches (1-3 μm), and deep, narrow trenches (>3 μm deep, <2 μm wide). *Trench-capacitor* structures are normally fabricated with narrow, deep trenches. The next sections will describe the fabrication details of the three trench types, as well as the chief uses of each type. (Trench-capacitor structures will be considered in detail in chap. 8.)

2.6.1 Shallow Trench and Refill Isolation

Shallow, refilled trenches are used primarily for isolating devices of the same type, and hence they can be considered as replacements for LOCOS isolation.

2.6.1.1 BOX Isolation. The *buried-oxide* (or BOX) isolation technology uses shallow trenches refilled with a CVD-SiO_2 layer that is then etched back to yield a planar surface. While the use of shallow trenches makes this process less difficult to develop than the deep-trench etch processes, the shallow trenches are also much less effective in preventing latchup and isolating n-channel from p-channel devices in CMOS. BOX isolation can be considered as a replacement isolation technology for LOCOS, but one that must be supplemented with another isolation method in CMOS for latchup protection.

In the basic BOX technique (Fig. 2-31),[41] shallow trenches (0.5-0.8 μm deep) are anisotropically etched into the silicon substrate through dry etching. Next, a CVD oxide is deposited onto the wafer surface and is then etched back so that it remains only in the recesses, its top surface level with the original silicon surface. Etchback is performed by etching photoresist and SiO_2 at the same rate. The top surface of the resist layer is highly planarized prior to etchback through the application of two layers of resist, and flowing the first of these layers. Active regions are those that were protected from etch when the trenches were created. This technique has the advantages of eliminating the bird's beak of LOCOS and of providing a planar surface.

The basic BOX process has several drawbacks, including the following:

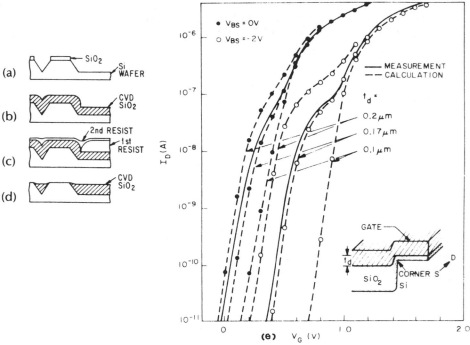

Fig. 2-31. The BOX isolation technique. (a) Si is etched using an oxide transfer mask. (b) Oxide is deposited by CVD. (c) Two layers of resist are applied, with the first being flowed. (d) The oxide is planarized using an RIE etchback technique which etches oxide and resist at the same rate.[46] (© 1983 IEEE). (e) Measured and simulated drain current versus gate potential for devices having a downward step in the field oxide (t_d).[43] (© 1983 IEEE).

• Void formation can occur if the CVD oxide is deposited into trenches that are narrower than about 2 μm (see chap. 4 for additional details about this CVD-deposition effect). A high-temperature-SiO_2 CVD-deposition step that gives conformal coverage without void formation has been reported to prevent this problem.[42]

• Inversion of the silicon at the sidewalls of p-type active regions may occur (see the following section which defines isolation in CMOS and gives further details of this problem).

• The field-oxide profile following the planarization etch varies among different areas. (Various approaches to eliminate this problem have been reported, see the following discussion on improvements to the basic BOX approach).

• In practice, the resist layer ends up being not perfectly planarized, even when a two-layer-resist technique is used. This is especially true if there is a wide range of active-area sizes and spacings. To ensure that the CVD SiO_2 is removed from all of the active areas at the conclusion of the etchback step, one must overetch of the field oxide by about 100 nm. The field oxide is thus etched below the active-area surface, exposing a portion of the active-area sidewall (Fig. 2-31b). This can lead to sidewall and edge-parasitic conduction (observable in the subthreshold-device characteristics), as well as to a higher electric field in the gate oxide at the active-area edge.

The electric-field concentration at the sharp-corner region of the active transistor reduces the threshold voltage of the corner region, and this part of the device turns on at a lower voltage than does the interior portion of the device.[43] The problem becomes worse if there is a downward step in the field oxide, as can occur in BOX isolation. The larger the step, the lower the threshold voltage, and unwanted subthreshold conduction begins at progressively lower values of V_{GS}.

2.6.1.2 Modifications to Improve BOX Isolation.
Modifications to improve the basic BOX process include eliminating sidewall inversion of p-active-device regions through a shallow-angle implant ($7°$ to the sidewall) of boron (see Fig. 2-32a).[44] This increases the threshold voltage of the parasitic n-channel that could occur along the sidewalls. Alternatively, a CVD borosilicate glass (BSG) layer may be deposited into the trench as a boron-diffusion source.[45] The BSG is etched away from the regions of the trench where the boron doping is not desired (Fig. 2-32b).

Several methods have been proposed to solve the problem of varying field-oxide depths. The first involves the use of an additional masking step.[46] The second employs LOCOS in the large-area field regions together with BOX-isolation between closely spaced devices.[42] The third uses a thin, sputtered SiO_2 layer between double photoresists (as an etchback layer) and a polysilicon film above active regions (as an etchback buffer). This process is called *planarization with resist/oxide and polysilicon* (PROPS). This process offers a uniform field-oxide thickness in both narrow and wide field regions.[45]

To avoid exposing the sidewalls of the active region, a *buried-oxide with etch-stop process* (BOXES) has been proposed (Fig. 2-33). The process uses a Mo/nitride etch-stop layer and passivates the recessed-silicon surfaces with a thin thermally-grown oxide before the CVD oxide is deposited. The oxide is grown by means of wet-hydrogen oxidation (i.e., oxidation in an ambient of a small percentage of O_2 in H_2) since Mo is not oxidized in this ambient.[47]

Advanced BOX processes are reported to be capable of isolating MOS and bipolar transistors of the same type (i.e., n-channel devices) with submicron spacings. It appears, however, that their use will still require either a large spacing, a deeper trench, or an SEG-based technique between n^+ and p^+ regions in CMOS circuits to avoid latchup and the inversion of parasitic field devices between n^+ and p^+ structures (see the following sections on deep-trench isolation and SEG, and see chap. 6 [the material on CMOS isolation issues] for reasons that support this latter statement).

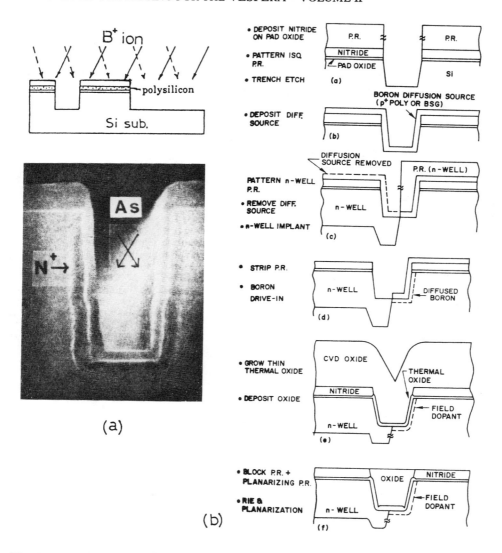

Fig. 2-32 (a) Schematic of oblique ion implantation and SEM cross section result of this procedure.[44] (© 1987 IEEE). (b) Shallow trench isolation process flow using a doped oxide as diffusion source of trench sidewall doping.[45] (© 1988 IEEE).

2.6.2   Moderate-Depth Trench and Refill Isolation

Two moderate-depth trench and refill processes will be described: *U-groove isolation* for bipolar ICs, and a *moderate-depth trench process to prevent latchup in CMOS*. Moderate-depth trenches are still easier to fabricate and refill than deep, narrow trenches.

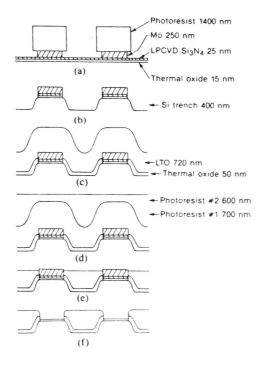

Photoresist 1400 nm
Mo 250 nm
LPCVD Si₃N₄ 25 nm

(a)

Thermal oxide 15 nm

Si trench 400 nm

(b)

LTO 720 nm
Thermal oxide 50 nm

(c)

Photoresist #2 600 nm
Photoresist #1 700 nm

(d)

(e)

(f)

Fig. 2-33 Schematic process flow of BOXES process.[47] (© 1988 IEEE).

2.6.2.1 U-Groove Isolation.[48]

U-groove isolation is a method of isolating bipolar devices that was introduced in 1981, and has has reportedly found use in bipolar SRAMs. It actually combines the techniques of selective oxidation and trench refill of a unique, moderately deep groove to create the isolation structure. It requires less isolation spacing than fully recessed LOCOS and generates fewer dislocations under certain processing conditions.[49]

A U-shaped groove is etched in the silicon field regions. The upper 0.5-μm is etched anisotropically with a KOH/isopropyl-alcohol solution to yield sidewalls angled at 60°, while the remaining trench depth is dry etched to yield less sloped sidewalls and rounded bottom corners (Fig. 2-34a). With the nitride mask that was used to define the trenched areas still in place, 0.4 μm of SiO_2 is thermally grown in the silicon trench. The trench is refilled with polysilicon, and the structure is etched back and thermally oxidized to yield a planar surface.

When the field oxidation is conducted at 1100°C, fewer dislocations are generated than in a comparable fully recessed LOCOS structure.[49] The shape of the U groove, which has rounded upper and lower corners and sloped sidewalls is credited with reducing the dislocation generation. Reduced bird's-beak and thinner SiO_2 field oxide allow this structure to be fabricated with smaller bipolar isolation spacing than does

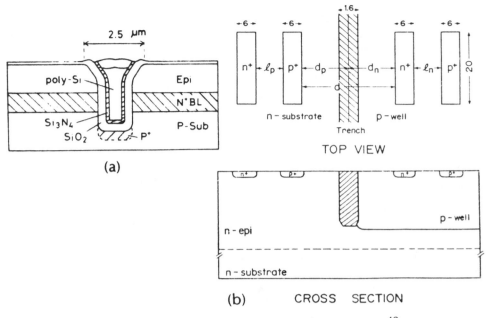

(a)

TOP VIEW

(b) CROSS SECTION

Fig. 2-34 (a) Cross sectional view of U-groove isolation structure.[48] Reprinted by permission of the publisher, The Electrochemical Society, Inc. (b) Schematic view of moderate-depth Toshiba trench process.[50] (© 1985 IEEE).

fully recessed LOCOS - for example, 3 μm-deep U grooves can provide a 2.5-μm-wide isolation spacing.

2.6.2.2 Toshiba Moderate-Depth Trench Isolation for CMOS.

A relatively simple trench technology for preventing latchup and isolating n^+ and p^+ regions in CMOS circuits was described by Nitsu et al. in 1985.[50] It utilizes a shallow epi layer and shallow trenches (both can be less than 3 μm deep). This reduces the complexity of trench etch-and-refill.

The process begins with the formation of p-well regions in a thin (2-3 μm thick) n-epi layer on a heavily doped n^+ substrate. Moderately deep trenches at the borders of the n and p regions are created through dry-etching, refilled with polysilicon, and are etched back to give a planar surface (Fig. 2-34b).

Trenches are 2.5 μm deep and 1.6 μm wide, and have n^+ to p^+ spacings of 2 μm. Shallower trenches may be used (e.g., 1.4 μm deep) if the n^+ to p^+ spacings are increased to 5.8 μm. CMOS circuits having this isolation approach exhibit a holding voltage > 10 V, ensuring latchup-free operation in circuits operated with a 5-V power supply. Because the poly trench-refill process cannot be used to fill trenches of varying widths, CMOS circuits using this latchup-prevention isolation technology must still utilize another method (e.g., LOCOS) for isolating like devices in a common well.

2.6.3 Deep, Narrow Trench and Refill

Deep (greater than 3 μm) and narrow (less than 2 μm) trench structures have been utilized for three major purposes in advanced-IC fabrication: (1) to prevent latchup and to isolate n-channel from p-channel devices in CMOS circuits; (2) to isolate the transistors of bipolar circuits; and (3) to serve as storage-capacitor structures in DRAMs. Such trench structures are attractive for several reasons. First, trench isolation, because of its vertical wall, exhibits a significantly smaller width than a LOCOS isolation field oxide of the same depth. Second, if the trench is used together with a lightly doped epi layer on a heavily doped substrate, latchup in CMOS will be eliminated if the trench extends completely through the epi. Finally, without a deep trench, n^+ to p^+ spacing must be ≥ 8 μm to avoid latchup in most CMOS technologies, but since the trench width can be made very small (less than 1 μm), the packing density of the CMOS devices is significantly increased.

Despite its advantages, deep-trench technology has been hindered by fabrication complexity. Deep trenches are not only difficult to reliably manufacture from run-to-run, and over an entire wafer, but they also makes it difficult to meet the throughput demands of full production. In addition, inversion is possible because of positive charges at the trench surface or in the trench-refill material.[93] The problem is more severe for n-well CMOS because the n-well is biased at a positive voltage (see chap. 6). In this case the n-well acts like an electrode and tends to invert the p-type substrate opposite the n-well. If the trench surface is inverted, a conducting path will connect the n-well and the n^+ region, causing leakage between isolated n^+ regions and between the source and drain of the n-channel transistor. Latchup sensitivity will also be increased. In order for this problem to be avoided, the n^+ regions must be kept about 1 μm away from the trench (which reduces the packing density), or the substrate must be doped more heavily (which increases the junction capacitance and the body-effect factor, see chap. 5).

In this section we describe the sequence of process steps used to fabricate and refill such deep and narrow trenches. In subsequent sections we will discuss how these trenches are applied for isolation purposes. Finally, in chapter 8 (which deals with memory-device fabrication), we will examine trench storage-capacitor structures.

There are three steps in the trench fabrication sequence: reactive ion etching of the substrate, refilling with dielectric materials, and planarization (Fig. 2-35).

2.6.3.1 Reactive Ion Etching of the Substrate. The deep, narrow trenches that are etched into the silicon should have the following characteristics:

• The sidewalls should be smooth and slightly tapered, with an angle of ~87°. There should be no undercutting of the mask used to define the trench width, no bowing, or kinks along the trench walls, nor should the etch process damage the sidewalls. Control of the sidewall profile is needed so that the trench can be refilled by CVD without the formation of voids. Undamaged sidewalls are necessary to avoid leakage along the trench. The angle of the sidewall cannot be

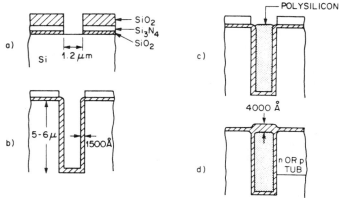

Fig. 2-35 Process sequence for forming a deep, narrow trench isolation structure. (a) Trench mask patterning. (b) Trench etching and oxide growth. (c) Polysilicon deposition to fill trench.[62] (© 1984 IEEE).

too great, since this would enlarge the dimension of the trench width at the surface, negating the benefit of increased packing density.

• The bottom of the trench should be smooth, with rounded corners (ideally U-shaped, Fig. 2-36a), in order to maintain the integrity of the oxide that is grown on the trench surface after etch, and for keeping stress-induced defects from being generated during this oxide growth.[52]

• The trench depth should be uniform both across the wafer and from wafer to wafer, especially in capacitor structures where capacitance values must exhibit a tight distribution.

A material must first be deposited on the silicon to act as the masking material that will defines the trench width. This layer must remain intact during the dry etching of the trench. Various masking layers have been reported, but a combination layer of thermal oxide, CVD silicon nitride, and CVD oxide seems to be the most popular.[53]

After patterning of the mask layer and removal of the resist, the exposed silicon-substrate regions must be cleaned to remove any native oxide that has formed during the resist-removal process. This is a key step, since any remaining native oxide will result in "grass" or "black silicon" at the bottom of the trench (Fig. 2-36b). A separate, preliminary dry-etch step (performed *in situ* with the trench-etch step) can be used to effect this cleaning.[53] It should also be noted that two other causes of "black silicon" have been identified: (1) sputtered material from the mask layer can redeposit on the Si and interfere with the Si etching at local spots, thereby causing the towers seen in Fig. 2-36b; and (2) the presence of oxygen in the Si can produce micromasks that lead to this condition.[105]

Taper of the trench sidewalls is controlled by the deposition of a carbonaceous polymer layer during the etch process. This protects the sidewalls from the etchant species, thus preventing lateral etching. The polymer deposit is continually removed

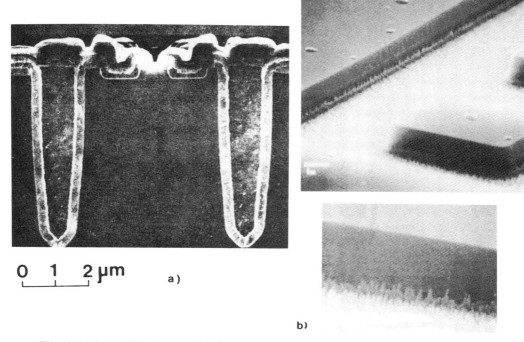

0 1 2 μm   a)

b)

Fig. 2-36 (a) SEM photograph of a deep trench with smooth sidewalls and a rounded bottom. (b) SEM photograph of the bottom of a trench that has "black grass" or "towers", resulting from a residual oxide on the silicon surface before trench etching is begun.[53] Reprinted with permission of Solid State Technology.

from the bottom of the trench by energetic ions so that the etching reaction can continue downward. In one process, the polymer layer on the sidewalls builds up in a uniform fashion during the etching, causing the pattern geometry to shrink slowly (Fig. 2-37).[54] This results in a linear, positive slope on the sidewall without enlarging of the pattern geometry. Another report indicates that the temperature during the etch process must be kept constant to maintain smooth sidewalls with a uniform slope (implying that the polymer deposition rate may be strongly dependent on temperature).[56] Nevertheless, the deposited polymer layer must be removed prior to refilling of the trench, and this may be done using a wet-chemical procedure.[53,54]

Different Cl-based etch gases have reportedly produced good trench-etch profiles. These include BCl_3/Cl_2,[53] $H_2/Cl_2/SiCl_4$,[55] and $CHCl_3/O_2/N_2$.[56] Good trench-wall profiles have also been obtained through the use of a triode-type dry-etch reactor and $Cl_2/O_2/SiCl_4$[116] or $CBrF_3N_2$[117] gas mixtures.

The etch rate must be uniform throughout the entire etching step so that the same depth of trench can be created from wafer to wafer and from run to run. This is apparently possible, provided only one width of trench is being etched. Because etch

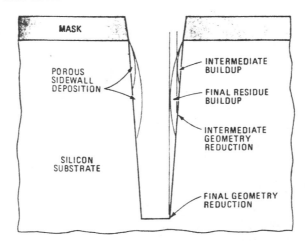

Fig. 2-37 The etch products in one trench etch technique are selectively deposited on the trench sidewalls to protect them during deep etching and to produce a controlled slope on the trench sidewalls.[54] Reprinted with permission of Electronics.

rate varies depending on trench width, only one width can thus be used in a given trench process. This may be a limitation from the point of view of circuit or device design.

In order for a rounded-bottom profile to be produced, the dry-etch process conditions must change to isotropic near the end of the etch step.[53] Increasing Cl_2 flow is one way to do this. No barreling of the trench profile occurs because the sidewalls are still protected by the polymer film that was deposited during the previous part of the etch process. Two other reports that review all of the details of trench etching are found in references 106 and 107.

2.6.3.2 Refilling the Trench. After trench etching, refilling of the trench begins by removing the carbonaceous sidewall polymer, as well as some of the silicon from the surface of the trench.[52] S. Samukawa et al. have reported that the leakage current of bipolar transistors isolated by trenches is significantly reduced when this

(a) 900°C   (b) 1100°C   200nm

Fig. 2-38. Cross-sectional SEM photograph of a convex corner after oxidation in dry O_2 ambient.[60] (© 1987 IEEE).

deposited layer is removed with a combination of a wet HF solution etch and a dry O_2 etch.[57]

Next, a sacrificial-oxide layer (e.g., 50 nm thick) may be thermally grown on the sidewalls. The growth and etching of this oxide can remove silicon damage caused by high-energy-ion bombardment during the trench-etch step.[58] In addition, the growth and removal of the oxide can round any sharp, concave corners of the bottom of the trench, as well as the sharp convex corner at the trench-oxide/gate-oxide point of connection (Fig. 2-38).[59,60,61] Trench capacitors were reported to exhibit much lower leakage currents when fabricated with such a rounded-off oxidation. The removal of the sacrificial oxide is normally done with a dilute HF solution, but removal in a gas-phase HF ambient has also been suggested.

If the trench is to act as an isolation structure, a thicker thermal oxide (~100-150 nm) is usually used than if a storage capacitor is to be formed.[62] In trench capacitors, a 15-25-nm thin thermal oxide,[54] a combination thermal-oxide/CVD-nitride,[53,63] or even a combination thermal-oxide/nitride/CVD-oxide layer[64] of about the same thickness is used. Herb et al. reported that the combination oxide/CVD-nitride layer produced capacitor structures with a higher breakdown voltage, as well as a tighter breakdown-voltage distribution than those produced by the oxide layer alone.[53] Use of silicon nitride is advantageous because its higher dielectric constant (6-7) produces a trench structure with increased capacitance.

The thermal-oxide growth conditions that generate the fewest number of defects in fully recessed LOCOS isolation structures also generally produce oxide films in deep trenches with fewest defects. Trenches with rounded bottom corners, and higher-temperature oxidation conditions (>1050°C) have given the best results.[52] The growth rate is reduced on concave surfaces (Fig. 2-39), an effect that becomes more pronounced as the degree of concavity is increased and the radius decreased.[65] This accounts for the greatly reduced oxide growth at the sharp corners of a trench bottom.

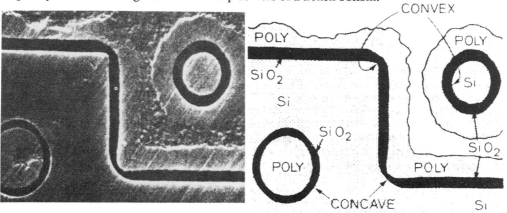

Fig. 2-39 A thermally grown oxide on a concave surface is thinner than that on a convex surface. Similarly, the concave corner also grows thinner oxide than the convex corner.[65] (© 1985 IEEE).

For trench-refill, the next step, various materials have been used, depending on the application. In *trench-isolation structures*, atmospheric-CVD polysilicon,[66] LPCVD polysilicon, CVD SiO_2, and selective, epitaxially-grown (SEG) silicon[67] have reportedly been used. In *trench-capacitor* applications, the charge is stored either on the edge of the trench facing the substrate or on the edge of the material inside the trench, and thus the one of the capacitor plates must reside within the trench. Consequently, polysilicon (typically-heavily doped with phosphorus) is used to refill trench capacitors. (Additional information on trench-capacitor structures is presented in chap. 8.)

With the exception of the SEG process, the refill process proceeds by means of a conformal covering of the trench surfaces and the original substrate surface (see Vol. 1, chap. 6 for a discussion of conformal-CVD deposition). In deep, narrow trenches, the deposition occurs in four distinct stages (Fig. 2-40):[66]

    1. Initial conformal stage.
    2. Seam-formation stage.
    3. Seam-closure stage.
    4. Cusp-planarization stage.

At the initial cusp-formation stage, voids can begin to form as the surfaces of the two opposite sidewalls meet centrally to close and fill the trench. Voids that form are of the order of the roughness of the two meeting surfaces. In some cases there can be "healing" or refill of these small voids with continued trench fill, since reactant gas apparently has access through the seams. This helps to explain why slightly angled trench sidewalls aid in filling high-aspect-ratio trenches in a void-free manner. The cusp that forms at the central regions of the trench is associated with the closure of the seam. Self-planarization of the cusp occurs with extended overfill. If the refill step is ended at the point of exact fill, the thickness of the film on the top surface will be about half that of the trench width.

If trenches of widely varying widths are filled, the narrow ones must be well overfilled in order for the wider ones to be filled completely. Thus, the thickness of the top-surface-deposited film will vary, making planarization very difficult. This is one reason why deep trenches refilled with CVD polysilicon or SiO_2 cannot replace LOCOS as an isolation technology for field regions of varying widths. SEG refill, however, does not suffer this limitation.[67]

2.6.3.3 Planarization after Refill. After trench refill, the dielectric material deposited on the horizontal surface of the wafer must be etched back until the nitride/oxide mask film is exposed. This can be done in one of two ways: by directly removing the layer with a fully anisotropic etch step, or by applying a resist film over the wafer surface, and etching to remove the resist and the dielectric film at the same rate. In either case, the etch step must be carefully controlled to prevent the notch (associated with the trench at the top of the filled layer) from being greatly increased or exaggerated due to removal of some of the material from the filled trench.

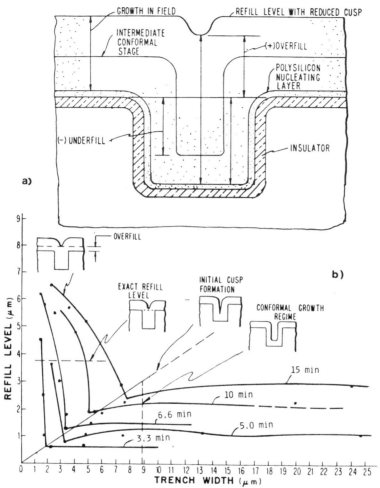

Fig. 2-40 Refilling a trench with polysilicon. (a) Schematic diagram of trench cross section showing experimental quantities measured. (b) Refill level versus trench width plotted for experimental refill times. This figure also shows the four phases of the deposition process.[66] Reprinted by permission of the publisher, The Electrochemical Society, Inc.

One novel way to improve the control of planarization is to deposit a buffer layer of CVD SiO_2 on top of the nitride mask before etching the trench. Later, during poly etchback, this CVD SiO_2 acts as an etch-stop. In addition, if a buffered-HF solution is used to strip the CVD SiO_2, the nitride mask is completely preserved. Thus, it can be ensured that all poly residue is removed and an undamaged nitride surface remains. Note that if polysilicon is used for refill, its top surface must be oxidized to complete the

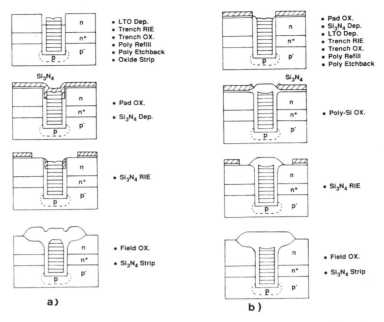

Fig. 2-41 Trench isolation process with LOCOS isolation. (a) Original deep trench process. (b) New planarized deep-trench process.[68] This paper was originally presented at the Fall 1988 Meeting of The Electrochemical Society, Inc. held in Chicago, IL.

process sequence. An example of a deep-trench isolation process combined with LOCOS for bipolar circuits is shown in Fig. 2-41.[68]

2.7 NON-LOCOS ISOLATION TECHNOLOGIES, II: SELECTIVE EPITAXIAL GROWTH (SEG)

The use of *selective epitaxial growth* (SEG) for isolating bipolar, CMOS, and BiCMOS devices is a relatively new development. The basic concepts of SEG (see Vol. 1, chap. 5) were established in 1962, but the full potential of the concept could not be implemented until reduced-pressure epitaxial reactors were developed in the late 1970s. The effort to apply reduced-pressure SEG to isolation structures began shortly thereafter.[69,70]

Low-pressure epitaxy has overcome some of the problems of atmospheric SEG, such as:

• Attack of the sidewalls of oxide openings, which takes place as these openings are refilled with SEG.

• Local loading effects that produce poor thickness uniformity in the SEG layers.

• Loss of selectivity, which deposits epitaxial Si on single-crystal Si surfaces, and polysilicon on other surfaces, such as oxide.

But remaining problems with SEG still include junction leakage, sidewall inversion, and corner faceting. *Junction leakage* is caused by sidewall-interface defects. Their generation is reduced by a lowering of the deposition temperature.[70] *Sidewall inversion* can be avoided by proper doping control.[71] It is also argued that the doping levels in the channel regions of submicron MOS devices will be so high that sidewall inversion will disappear or be minimal. *Corner faceting* along the <100> direction (Fig. 2-42) can be minimized by optimizing the SEG growth conditions and device-layout geometry.[70]

A number of different methods have been proposed for accomplishing isolation by SEG. These may be categorized as follows.

2.7.1 Refill by SEG in Windows Cut into Surface Oxide

Openings are etched in a surface oxide down to the silicon substrate and then filled by means of SEG, until a planar surface is achieved (Fig. 2-43).[71,72,73] This produces an isolation structure with no bird's beak, and with a field-oxide thickness that does not depend on the width of the oxide space.[74] In addition SEG can provide a planar surface (except for facets), and the channel-stop region is removed from the source and drain regions (so that boron encroachment into the active regions is avoided). Furthermore, isolation structures with various widths can be planarized. Electrical isolation exists not only between wells, but also between devices (i.e., there is no need to use LOCOS to provide device-to-device isolation). Another bonus of this method is that it allows oversize contacts to be made to the source and drain regions of CMOS devices. It was noted in an earlier section that such contracts cannot be made when conventional

Fig. 2-42 Corner faceting problem that occurs during selective epitaxial growth in oxide windows.[70] Reprinted with permission of Solid State Technology.

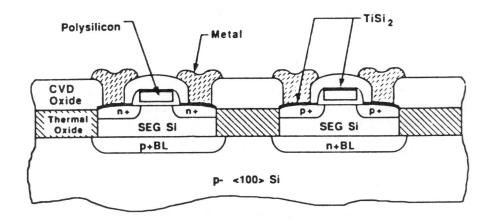

Fig. 2-43 Schematic cross section of a twin-well SEG CMOS technology.[71] (© 1987 IEEE).

LOCOS is used with CMOS (because of the bird's-beak structure), and that this restriction imposes a packing-density penalty.

The SEG isolation process has several several drawbacks:[74] (1) facets still occur (even under conditions that minimize their depth), which produce depressions in the silicon surface that may be very hard to clean out when filled with overlying film material. Chemical-mechanical polishing has been proposed for facet removal from the top silicon layer; (2) the sharp transition between the field oxide and the gate oxide at the edges of the device causes a penetration of the electric field from the field region to the active-channel area (Fig. 2-44). This lowers the V_T of the device (a phenomenon known as the *reverse narrow-width effect*). The problem may also be reduced as the doping is increased in the channel regions in submicron devices; (3) channel stops along the oxide sidewalls are needed to prevent device leakage caused by inversion along the sidewalls.

2.7.2 Simultaneous Single-Crystal/Poly Deposition (SSPD)

Here again, openings in the surface oxide are etched down to the silicon substrate. Silicon is then deposited non-selectively in an epi reactor.[75,76] This deposits polysilicon over the oxide regions, but produces single-crystal silicon in the exposed single-crystal areas. This approach is known as *simultaneous single-crystal/poly deposition* (or SSPD).

2.7.3 Etching of Silicon Trenches and Refilling with SEG to Form Active Device Regions

Recesses with vertical sides are etched into the silicon substrate by means of an anisotropic dry etch and an oxide transfer mask. An insulating layer is then formed on the recessed-silicon surfaces (e.g., through thermal oxidation[77] or through a thermal oxide and CVD nitride[78]), and the insulating layer is removed from the bottom of the recessed silicon by means of another anisotropic dry etch (Fig. 2-45). This provides a nucleation site for epitaxial growth, while simultaneously allowing the insulating layer to remain on the sidewalls. The recesses are then refilled with SEG until they are level with the original silicon surface.

2.7.4 Selective-Epitaxial-Layer Field Oxidation (SELFOX)

With this approach, SEG refills trenches (~1 μm deep) etched in silicon.[79] Yu et al. reported that the trench sidewalls and the surfaces of the active-device regions are masked with a CVD-nitride layer before the SEG step. SEG fills the trenches about halfway full and selectively oxidation forms 1 μm-thick field-oxide regions. No bird's beak is formed, no defect generation during field-oxide growth is observed, and isolation spacings as small as 1 μm wide are fabricated. Device regions are formed only in areas of the original single-crystal substrate (not in any epi-layer regions), and a planar surface is produced. No mention has been made of possible sidewall-inversion problems in n-channel active-device regions, nor has the structure been evaluated for its

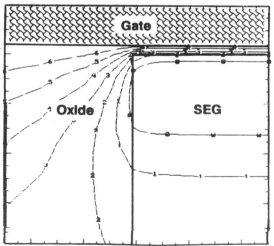

Fig. 2-44 The sharp transition between the field oxide and the gate oxide at the edges of the device causes a penetration of the electric field from the field region to the active-channel area. This lowers the V_T of the device (a phenomenon known as the *reverse narrow-width effect*).

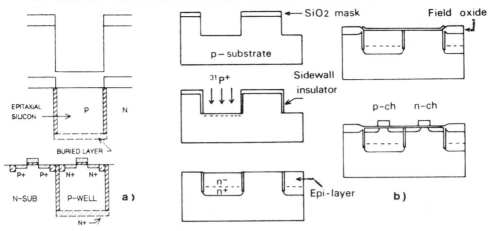

Fig. 2-45 (a) Schematic diagram showing SEREPI fabrication process, culminating in a SEREPI CMOS structure.[77] (© 1987 IEEE). (b) Selective epi and LOCOS isolation.[132] (© 1985 IEEE).

effectiveness in preventing latchup in CMOS circuits. This approach appears to be another possible LOCOS replacement for isolating like-type MOS devices in a common tub.

2.7.5 SEG Refill of Trenches (as an Alternative to Poly Refill)

Trenches are again etched into the silicon substrate and are refilled by means of SEG.[67] In this case, however, the SEG material is not used to form device regions, but instead acts merely as filler for the trenches that are the isolation structures between active regions. Note that the sidewall oxide of the trench must again be removed from the trench bottom to provide a site for epitaxial nucleation. Epitaxial growth is terminated when planarity is achieved.

For refill of trenches the advantages of SEG over CVD SiO_2 or CVD poly are that trenches of widely varying widths can be filled simultaneously, there is no void formation, and the refill process is insensitive to trench profile. Trenches can also be refilled to the top without the need for overfill and etchback, which must be done with polysilicon refill.

2.7.6 Epitaxial Lateral Overgrowth (ELO)

In this approach, totally isolated silicon-on-insulator structures are fabricated by means of epitaxial-silicon overgrowth (Fig. 2-46).[80]

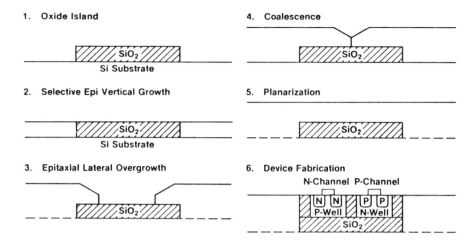

Fig. 2-46 Silicon on insulator by epitaxial lateral overgrowth.

2.8   MISCELLANEOUS NON-LOCOS ISOLATION TECHNOLOGIES

2.8.1   Field-Shield Isolation

An alternative approach for forming submicron isolation structures that does not require the growth of a thick field oxide is the *field-shield* technology.[101] A thin (50-nm-thick) thermal oxide known as the *field-shield gate oxide* is grown and a doped-polysilicon film (200 nm thick) called the *field-shield plate,* is deposited on it and is patterned to remain over the field regions. A CVD oxide is deposited over the field-shield plate (Fig. 2-47). With a 5-V gate voltage applied to the field shield, the silicon surface in the field regions does not invert.

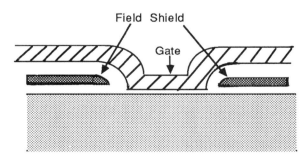

Fig. 2-47 Cross sectional view of a field-shield isolation structure.[101] (© 1988 IEEE).

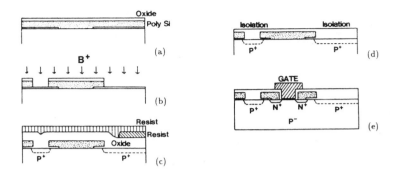

Fig. 2-48 Process sequence of BIPS isolation.[102] (© 1988 IEEE).

Since there is no channel-stop implant and no thick field oxide, there is no lateral bird's beak encroachment or redistributed boron. The threshold voltage of the narrow-width active devices is not degraded, and in addition, this isolation structure exhibits good resistance to punchthrough, even for field-shield-transistor gate lengths of 0.6 μm. The method's drawbacks are the nonplanarity of the isolation structure, and the need to provide a contact for each isolated field-shield plate. The latter problem could reduce the packing density of circuits and provide additional interconnection difficulties.

Another report which provides more extensive details concerning the fabrication and the isolation characteristics of field-shield structures for VLSI has recently been published.[126] The structure is self-aligned without any alteration in circuit design, and is also a self-aligned directly insertable structure. There is no penalty in the layout area and junction capacitance. In addition, the field shield protects the silicon surface from process-induced radiation during the active device fabrication and multilevel metal processing.

2.8.2   Buried Insulator between Source/Drain Polysilicon (BIPS)

Another novel non-LOCOS isolation structure is the so-called *buried insulator between source/drain polysilicon* (BIPS).[102] The isolation structure is created by the refilling of CVD oxides in openings between source/drain polysilicon patterns through double-photoresist etchback (Fig. 2-48). In this process, 0.5-μm isolation spaces are achieved with no bird's beak, and narrow-width effects are almost entirely eliminated.

There are two concerns associated with this approach: first, when the polysilicon is etched to define the channel region, the etch must be stopped at the polysilicon/single-crystal interface; and second, there is no apparent way to easily create an LDD structure in the polysilicon source/drain transistors.

2.9 SUMMARY: CANDIDATE ISOLATION TECHNOLOGIES FOR SUBMICRON DEVICES

2.9.1 Basic Requirements of VLSI and ULSI Isolation Technologies

Isolation technologies for submicron CMOS and BiCMOS ICs will be selected based on how well they can meet the following requirements:

1. *Device-to-device isolation within a well.* The threshold and punch-through voltages of the field parasitic transistors must be large enough to prevent leakage between active-device regions through the parasitic field devices.

2. *Well-to-well isolation.* This type of isolation is needed to prevent leakage between parasitic MOS devices (see section 6.5 on CMOS isolation), as well as to prevent latchup of the parasitic bipolar-devices that exist in CMOS structures (see section 6.4 on latchup in CMOS).

3. *Minimum area.* Isolation structures need to be as small as possible, and they must be potentially scalable to even smaller sizes.

4. *Process simplicity.* Processes should be simple, transparent to the existing device-process flow, and resistant to contamination. Further, it should require a minimal number of additional masking steps.

5. *Active-device characteristics should not be degraded by parasitic effects.* For example, the isolation structure should not increase the device junction capacitance and body-effect sensitivity, nor should it decrease the drain breakdown voltage. Narrow-width effects caused by boron encroachment should be alleviated.

2.9.2 The Need for Planarity

A major limiting factor in the processing of submicron-feature-sized ICs will be the very small depth of focus of the optical steppers used to pattern the circuit features.[83,104] In order to be able to obtain the maximum resolution from these steppers, everything within the field of view will have to be kept within a depth of focus of about 0.5 μm. Accordingly, the imaging surface will need to be very flat. Loss of flatness in the field of view arises from several factors: resist-thickness differences at different locations, residual lens aberrations, autofocus errors, wafer nonflatness, and wafer-surface topography. To overcome non-planarity due to the last group of effects, use of device structures that result in a planar topography will be mandatory.

2.9.3 How the Various Isolation Technologies Meet the Requirements

• *LOCOS-based processes.* If a planarized semirecessed SILO or polysilicon-buffered LOCOS process is used, along with a channel-stop implant performed after the field-oxide growth, then most of the requirements can be met. Only the problem of field thinning remains, and the same limitation applies to fully recessed LOCOS processes. Advanced LOCOS-based processes, however, involve increased process complexity.

• *Trench-based isolation processes.* The modified BOX technique with selectively-doped sidewalls,[42,45] has the potential of providing excellent planarity, as well as freedom from encroachment and field thinning. However, it requires two more masking steps than does a conventional LOCOS process, as well as three or four dry-etch steps (including one etchback step). Moderate- and deep-trench processes offer the benefit of producing latchup-immune or even latchup-free CMOS circuits, as well as dense isolation structures for bipolar devices. Deep-trench structures generally require significantly increased process complexity, and consequently result in potentially lower productivity (i.e., lower throughput and yield).

• *Selective-epitaxial-growth isolation processes.* SEG processes appear to have the potential of providing planar topology and small isolation spacing. In addition, compared to the other isolation approaches, SEG processes offer the benefits of process simplicity. However, they require the use of a relatively new technology, which still has some problems; further, a history of use in a production-line environment has not yet been established.

• *Miscellaneous processes.* The field-shield process creates a nonplanar surface, and every field-area polysilicon must be electrically connected to V_{DD} (which requires an area for contact, as well as an additional interconnection). The BIPs process looks interesting, but requires a polysilicon etch stop at the single-crystal silicon surface. This appears to be challenging for production applications.

2.10 SILICON-ON-INSULATOR (SOI) ISOLATION TECHNOLOGIES

Junction isolation is not suitable for high-voltage applications* because at supply voltages of ± 30 V junction breakdown occurs under reasonable doping levels and device-structure dimensions. Junction isolation is also ineffective in high-radiation environments due to transient photocurrents produced in *pn* junctions by gamma rays. For these applications, a preferred isolation is one that depends on completely surrounding devices with an insulator, rather than with a *pn* junction. Technologies that

---

* For this discussion the term *junction isolation* includes structures that are isolated by an oxide along the sidewalls and by a junction at the bottom.

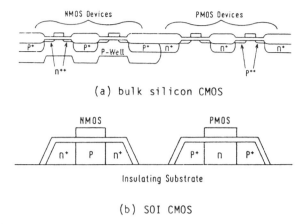

(a) bulk silicon CMOS

(b) SOI CMOS

Fig. 2-49 Comparison of (a) structure cross sections and (b) process flows of bulk and SOI CMOS.[122] (© 1986 IEEE).

provide this full isolation are collectively referred to as *silicon-on-insulator* (SOI) technologies.

SOI offers many additional advantages. In some cases it uses a simpler fabrication sequence and resultant cross-section compared to circuits fabricated on bulk silicon. This is clearly demonstrated in Fig. 2-49,[122] which compares a mesa-isolated SOI CMOS process with a *p*-well bulk CMOS process. SOI also provides reduced capacitive coupling between various circuit elements over the entire IC chip, and in CMOS circuits latchup is eliminated. SOI reduces chip size and/or increases packing density, and minimum device separation is determined only by the limitations of lithography. Finally, it provides increased circuit speed, due to reductions in parasitic capacitance and chip size.

When an SOI technology based on a thin silicon film is used, two other important advantages can also be gained. A relatively benign surface topography (for step-coverage) is produced if device isolation can be achieved by a complete island, sloped-wall etch process of the thin silicon film. Second, since SOI isolation eliminates the parasitic field FET between adjacent devices, LOCOS processes are not needed.

As with all technologies SOI has its own disadvantages. To date, active-device regions in SOI technologies are poorer in crystalline quality than their counterparts in bulk silicon. In addition, the presence of an insulating substrate, or insulating layer, may complicate or prevent the adoption of effective defect- and impurity-gettering processes. Nevertheless, SOI's potential advantages are sufficiently attractive that development work in this area remains quite active.

2.10.1 Dielectric Isolation

The first technique developed for fully isolating the collectors of bipolar transistors with a layer of SiO_2 rather than with a *pn* junction was given the name *dielectric isolation* (DI). This has been a production process for at least ten years, and circuits are available on DI wafers in both CMOS and bipolar technologies.

The fabrication steps used to produce DI wafers are shown in Fig. 2-50. The starting material is a bare *n*-type silicon wafer with a resistivity appropriate for the collector region of the transistor. The first step etches V-grooves in the back side of the wafer; these will become the isolation regions in the finished circuit. Using oxide as an etching mask, an orientation-dependent etching step is used to delineate these grooves (~20 μm deep for typical analog-circuit applications). Next, an oxide is grown on the grooved surface (to a thickness of ~2-4 μm), and this is followed by a layer of thick CVD polysilicon. This poly layer will become the mechanical support for the finished DI wafer and must therefore be ≥200 μm thick.

The starting wafer is then lapped from the top side until the silicon dioxide in the V-grooves is exposed. The only remaining single-crystal silicon material is that which is left in the isolated active-device islands. The active-device surface is then mechanically and chemically polished using a polishing procedure very similar to the final step used by vendors of silicon wafers. The wafer is then turned over, yielding regions of single-crystal silicon ~8-10 μm thick that are completely isolated from one another by an oxide layer. Standard processing can then be used to fabricate bipolar transistors in these islands.

Although the DI process provides structures that will allow high-speed, high-voltage, and/or radiation-hardened circuits to be built on them, the process has several drawbacks. The fabrication sequence used to produce DI wafers is expensive. For example, even in 1984 the cost of the largest DI wafers was ~$60-$100; high compared to the cost of processing an entire CMOS wafer on a conventional wafer which then was ~$150-$250. Next, precise mechanical alignment of the wafer during the lapping

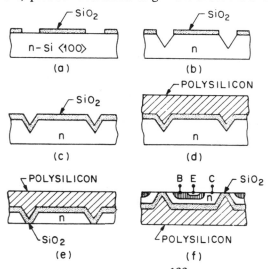

Fig. 2-50 Process sequence for dielectric isolation.[133] Reprinted by permission of the publisher, The Electrochemical Society, Inc.

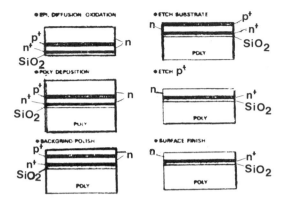

Fig. 2-51 FELT/SOI process sequence.[120] This paper was originally presented at the Spring 1989 Meeting of The Electrochemical Society, Inc. held in Los Angeles, CA.

process must be maintained to ensure that this step removes exactly the correct thickness of silicon everywhere on the wafer. Third, the polysilicon-deposition processes cause warping of the wafer due to the thermal-expansion coefficient differences between the polysilicon and the single-crystal silicon. Thus, as the wafer cools down from the polysilicon-deposition temperature, bowing can occur. During lapping, this produces some isolation islands in which the silicon is too thin, or some "islands" that are not completely isolated. Finally, the formation of the V-grooves at the front end of the process places limitations on the density achievable with DI technology.

A recent improvement on the conventional dielectric-isolation process is called *full-etchstop layer transfer/silicon-on-insulator* (or FELT/SOI).[120] A p^+ etchstop layer is formed on a lightly doped wafer, and a layer of lightly doped n-type material is epitaxially deposited to the desired thickness and doping concentration (Fig. 2-51). This n-epi layer will be the device layer when the substrate processing is completed. A CVD insulator is then applied and a thick polysilicon layer is deposited (as in the conventional DI process).

The original single-crystal silicon wafer is then lapped back to within a few mils of the p^+ etchstop layer. A doping-concentration-dependent etch removes the remaining silicon but stops at the p^+ etchstop layer (which is subsequently removed with another doping-concentration-dependent etch). The end result is a layer of silicon of known thickness and doping concentration fabricated on an insulating layer. In addition, since device isolation is not built into the material, this method offers more design flexibility than does the DI process – that is, device isolation does not have to be predesigned into the substrate, allowing reduction in the overall cycle time for new circuit designs.

2.10.2   Wafer Bonding

A planar process similar to the conventional DI process can be implemented with wafer bonding. In this method, two oxidized silicon wafers are fused together through a high-temperature furnace step. The following approaches have been developed to consumate the bonding:

• Two oxidized-silicon wafers are pressed together and subjected to an oxidizing ambient at 700°C.[81] Bonding under these conditions is believed to occur as a result of the conversion of the gaseous oxygen between the wafers to SiO_2. This conversion is thought to create a partial vacuum that forces the wafers into intimate contact, bringing about the bonding chemical reaction. (This is believed to be a polymerization of silanol bonds to form a siloxane network).

• When a moderate voltage is applied between two silicon wafers, bonding is induced at temperatures of 1100-1200°C (Fig. 2-52).[82] One advantage of electrostatic bonding is that since no mechanical force needs to be applied to press the wafers together initially, it does not introduce crystal defects into the silicon.

• One oxidized silicon wafer and one bare silicon wafer are cleaned with an H_2O_2-H_2SO_4 mixture. Treatment with an acid solution is then used to form an O-H group on the wafer surfaces. After drying, the wafers are placed face-to-face at room temperature in clean air. A self-adhesive contact is formed, and thereafter no weight or pressure need be applied to the contacted wafers. The bonding is then carried out by a heat treatment of 1100°C in a nitrogen ambient for four hours.[83]

The SOI structure is also formed in a number of different ways. With one method, a lightly doped epitaxial layer is grown on a heavily doped substrate, and this layer is then thermally oxidized (i.e., to a thickness of 1-2 μm). Following bonding, a preferential etch is used to remove the heavily doped substrate, leaving a thin, lightly doped epitaxial layer above the thermally grown oxide.[81] The active-device layer is

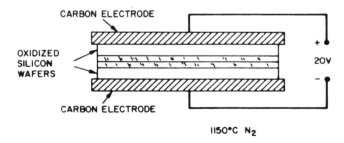

Fig. 2-52 Electrostatic technique for bonding silicon wafers.[82] Reprinted by permission of the publisher, The Electrochemical Society, Inc.

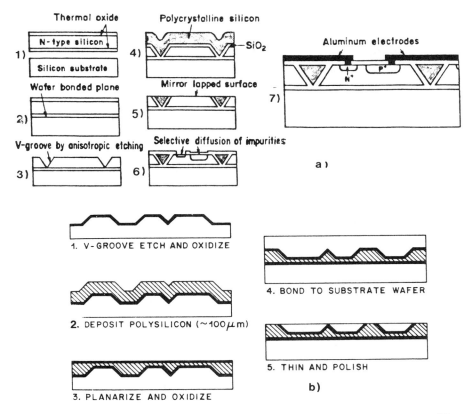

Fig. 2-53 Isolation structure formation in bonded silicon wafers. (a) SDB technique,[83] (© 1986 IEEE). (b) Electrostatically bonded technique.[82] Reprinted by permission of the publisher, The Electrochemical Society, Inc.

determined by the thickness of the epitaxial layer. Alternatively, the heavily doped substrate is removed by lapping, and the remaining epi layer is polished to the desired thickness.[84]

In a second method, the surface of the bonded silicon wafer is V-groove etched and oxidized. A polysilicon film, much thinner than that used in conventional DI, is then deposited, planarized, and oxidized (Fig. 2-53a).[83] This wafer is then bonded to another wafer. Finally, the single-crystal silicon of the wafer with the deposited polysilicon is thinned and polished to form dielectrically-isolated active regions.

In yet another approach, the upper part of one of the wafers is chemically polished until a 70-μm-thick Si layer remains (Fig. 2-53b). V-grooves are formed, and SiO_2 is regrown over the surface. Polysilicon is deposited to a thickness of 100 μm and the wafer is again polished until most of the polysilicon is removed and the 70-μm-thick region has been reduced to a thickness of 50 μm. The remaining poly is that which remains in the grooves. An SiO_2 layer is then grown on the surface, and the wafers are ready for device processing.

In a report comparing the various SOI approaches, it was stated that bonded-wafer technologies exhibit very low leakage currents.[84] On the other hand, they are limited by the minimum thickness that can be achieved in the silicon and the minimum variation of that thickness across the wafer. This characteristic might still make SOI-by-wafer-bonding an attractive choice for advanced bipolar applications, as these require somewhat thicker silicon-device regions. In addition, it could be useful in some CMOS applications, such as those used for very low-power or high-voltage circuits, in which film thickness is not required to be minimized. A 1-kbit ECL SRAM fabricated on a bonded SOI wafer is described in reference 85.

A recently reported problem associated with the wafer-bonding process is that voids can occur at the interface between the wafers if a particle exists on the surface of either wafer (Fig. 2-53b). Such particles can keep two mechanically rigid surfaces from mating and bonding in the vicinity of the particle. The acute susceptibility of this bonding process to particles is reported to cause low production yield.[120]

2.10.3   Silicon-on-Sapphire (SOS)

Another early SOI technology, begun in the 1960s, uses sapphire for both substrate and insulator (hence the name *silicon-on-sapphire*, or SOS). Sapphire has the advantage of being fairly well lattice-matched to silicon. Although there have been notable successes with SOS over the years, technology and material issues have prevented it from becoming a mainstream circuit technology. Recent advances in SOS, however, continue to preserve the viability of this material, and in the late 1980s it remained the only commercial, non-DI, SOI technology.[86] (More information about SOS can also be found in Vol. 1, chap. 5.)

2.10.4   Separation by Implanted Oxygen (SIMOX)

The creation of a buried layer of SiO_2 by implanting oxygen into silicon, a process known as SIMOX (for *separation by implanted oxygen*), is one of the major contenders for creating SOI structures.[87] The technique requires a high dose (~$2x10^{18}$ cm^{-2}) of oxygen (O^+) ions, as this provides the minimum concentration necessary to ensure that a continuous layer of stoichiometric SiO_2 will be formed by reaction of the oxygen with the silicon during the annealing process (Fig. 2-54).[88] The energy of the implant must also be high enough (150-180 keV) so that the peak of the implant is sufficiently deep within the silicon (0.3-0.5 μm). The wafer is normally heated to more than 400°C during the implantation process to ensure that the surface maintains its crystallinity during the high-dose implantation step (see Vol. 1, chap. 9 for more details on this topic).

A post-implant anneal is performed in a neutral ambient (N_2) for a sufficient time (3-5 hours) and at a high enough temperature (1100-1175°C) to form a buried layer of

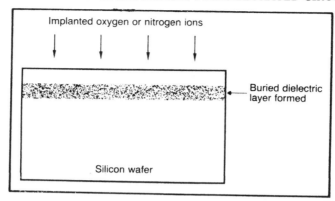

Fig. 2-54 Implanted oxygen ion process for the formation of a SIMOX wafer.[87] (© 1987 IEEE).

SiO_2. The anneal step also allows excess oxygen in the surface silicon to out-diffuse, thereby increasing the dielectric strength of the buried-oxide layer.

After the anneal step, the crystalline-silicon surface layer is typically only 100-300 nm thick. As a result, an additional layer of epitaxial silicon is usually deposited so that single-crystal device regions ≥0.5 μm thick are available for fabricating devices. CMOS devices produced in such regions on SIMOX substrates have exhibited satisfactory operating characteristics, especially in cases in which submicron-CMOS transistors were evaluated.[89,90]

SIMOX exhibits some advantages over other SOI technologies. The most important is that the technology is transparent to the manufacturing line since the fabrication of SIMOX-based circuits uses processing steps similar to those used in conventional IC manufacturing. Even the novel high dose implant of oxygen is a procedure that is understood by process engineers.

SIMOX does, however, have a number of disadvantages, including the following:

▪ The process requires the availability of special oxygen implanters. Beam accelerators with the ability to implant oxygen at 150 keV with a 10-mA beam current first became commercially available with the introduction of the NV-10 implanter by the Eaton Corp. The beam current of this machine puts it into the medium-current-implanter category. However, with such an implanter, an entire day is needed for implanting of a 50-wafer SIMOX lot. High-beam-current machines (100 mA) have recently been introduced (e.g., the Eaton/Nova-100), making the high-volume production of SIMOX wafers more feasible.

▪ The thickness of the buried-oxide layer is limited to about 0.5 μm. For some high-voltage applications, a greater thickness may be required. Furthermore, the relatively thin buried-oxide layer will increase parasitic capacitance, and can make SIMOX-based circuits somewhat slower than those based on an SOI process with a thicker oxide.

- The thickened device-silicon layer may not be free of defects, since the SOI epitaxial-seeding layer may still contain some residual damage from the implant step. (SIMOX substrates have been found to contain larger densities of threading dislocations than conventional bulk wafers. These defects considerably reduce the minority-carrier lifetime and increase the surface-recombination velocity.)[118]

Defect generation in the surface-silicon film and the deposited epi layer can occur from metal contamination sputtered from the beam-line and end-station by the primary oxygen ion-beam. The contamination can be as high as 1% of the primary dose. To combat this problem, some metallic parts in oxygen implanters have been replaced with graphite parts; however, this results in carbon contamination. During the anneal step, carbon precipitates out of solid solution, forming microdefects that degrade the subsequently deposited epi layer. For removal of the carbon precipitates, a shallow plasma etch step is suggested.[91]

- Implantation parameters and anneal schedules must also be appropriately chosen to provide optimum IC performance because the microstructure of the surface-silicon film is very sensitive to the oxygen dose and the post-oxygen-implant annealing temperature.[89] For example, a lower dose of oxygen results in a higher oxygen content in the silicon film and a higher density of oxygen precipitates at the silicon-film/buried-oxide interface following an 1150°C anneal. For oxygen doses of 2.25×10^{18} cm^{-2}, thermal annealing at 1275°C annihilates the oxygen precipitates in the silicon film.

SIMOX is attractive for military electronic systems requiring high-speed, low-power, radiation-hardened devices that can operate reliably over a wide temperature range as well as for power IC and telecommunications applications which must integrate low-voltage control circuits with high-voltage devices. SIMOX is also a candidate for CMOS circuits with feature sizes below 0.8 μm. Finally, SIMOX is well-suited for integrating bipolar and CMOS devices on the same chip. The manufacturability of SIMOX has been demonstrated, but although commercially available, the 1989 cost of SIMOX wafers is still quite high.

Recently, use of wafers on which no epitaxial layer is deposited following oxygen implantation and annealing (so-called *thin-film* SIMOX wafers) has been studied.[113,114,115] Fabrication of MOS devices in such wafers offers several important advantages. Long-channel behavior (see chap. 5) can be preserved at gate lengths below half a micron because of the limited vertical extension of the depletion region into the thin film, and a reduced electric field at the drain junction in such *fully depleted SOI devices* is produced (eliminating the kink effect observed in thin-film transistors and lowering hot-carrier degradation). In addition, thin-film SIMOX processing is simpler than a bulk process because there is no need to form doped wells, and shallow junctions are automatically formed when the doping impurity reaches the underlying buried oxide.

Implanting a high dose of nitrogen and annealing at 1200°C to form SOI structures has also been investigated. Potentially, this has several advantages over the implanting of oxygen. Hot-filament, high-current nitrogen-implant sources are easier to construct than are similar oxygen sources. The defect-laden layer that exists at the top Si/insulator interface for implanted oxygen is absent in the case of implanted nitrogen. Annealing of the as-implanted nitrogen atoms results in the formation of a top silicon layer, which is sufficiently thick that a subsequent epitaxial-growth step may not be needed, and a continuous buried silicon-nitride layer may be formed at about 40-50% of the dose necessary to form a buried SiO_2 layer with an abrupt top Si/SiO_2 interface.[92] However, the higher dielectric constant of silicon nitride may reduce the speed performance of ICs compared with those fabricated on SIMOX wafers.

2.10.5 Zone-Melting Recrystallization (ZMR)

In the ZMR process, an active-device layer is formed by recrystallizing a polysilicon layer that has been deposited on a thermally oxidized silicon wafer,[94] i.e., the grain sizes of the as-deposited CVD-poly layer are enlarged by the movement of a narrow molten zone. The most promising method for creating the molten zone is heating the substrate close to the melting point of silicon by an auxiliary means, and then using a graphite heating strip (Fig. 2-55),[95] or focused lamp[96] to supply the additional heat necessary to achieve melting. In a typical process, 0.5-2 μm of thermal oxide is grown, and 0.5 μm of polysilicon film is deposited over it. A 2 μm layer of SiO_2 is deposited to cover the poly. Next, the wafer is heated to 1100-1300°C by a lower heater so that radiative heating from a strip above the wafer can then produce a molten

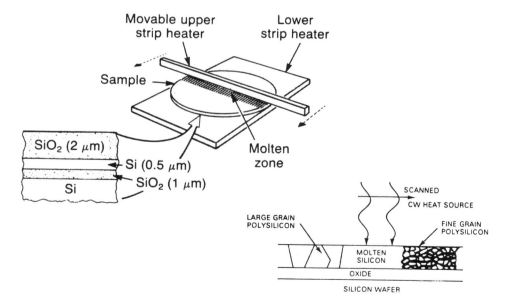

Fig. 2-55 Schematic diagram of sample and graphite strip heaters used in zone melting recrystallization of SOI films.[94] (© 1987 IEEE).

zone in the silicon film directly below it. The zone is scanned across the sample by moving the upper strip-heater, typically at speeds of ~1 mm/sec. The recrystallization process is performed in a flowing Ar gas ambient. Several types of CMOS circuits have been fabricated in ZMR SOI films, including 1-kbit static RAMs and 1.2-k gate arrays.

ZMR can implement greater buried-oxide layer thicknesses than SIMOX (in which the range of buried oxide is only 0.3-0.5 μm). With a thicker oxide, of course, there is lower parasitic capacitance and therefore, higher circuit speed. Thicker buried oxides are also needed for power ICs.ZMR's chief disadvantage is that defects remain in the recrystallized film. The predominant ones are low-angle grain boundaries (see Vol. 1, chap. 2), spaced 20-50 μm apart. Between these grain boundaries, however, the material is free of crystallographic defects. The grain boundaries in ZMR material have very little effect on majority-carrier transport, and thus there is little mobility degradation. However, they do provide paths for enhanced dopant diffusion. Such diffusion in short-channel MOSFETs can result in excess transistor leakage and even in shorting between source and drain. The threshold-voltage variation is greater in ZMR SOI devices than in bulk MOS devices. Furthermore, the minority-carrier lifetimes in ZMR films are reduced from those in conventional bulk wafers (2.5 μs vs. 240 μs).[118]

The ZMR process requires high operating temperatures, which make it sensitive to small perturbations. For example, particulate contamination of the starting wafers can result in macroscopic defects in the recrystallized films, and temperature nonuniformity can lead to wafer distortion. Nevertheless, the overall quality of ZMR wafers is good enough for some VLSI applications and efforts are being made to establish a commercial supply of this material.

2.10.6 Full Isolation by Porous Oxidized Silicon (FIPOS)

The SOI technology known as FIPOS is more complicated than other SOI technologies, but offers the potential for essentially defect-free active silicon.[97] This full-isolation technology is implemented by depositing an epitaxial film on a silicon substrate that has a heavily doped surface layer. Openings are cut through the epi layer, and the entire heavily doped silicon region under the epi layer between the etched recesses is converted to porous silicon (Fig. 2-56). The original process employs anodization of a heavily p-doped layer with an HF solution to preferentially render the silicon porous. More recently an n^+ layer between an n^- island and an n^- substrate has been selectively converted to porous silicon.[98] This allows the thicknesses of the porous silicon layer and the island to be easily controlled and to be made quite uniform. In addition, it provides an automatic end point for the island formation: as soon as all of the easily-anodized n^+ layer is converted to porous silicon, the anodization current drops, and the conversion reaction stops. Finally, the formation of porous silicon occurs only laterally, and hence large silicon islands can be isolated with this technique.

Porous silicon oxidizes much more readily than single-crystal silicon, and this characteristic is exploited to form silicon islands completely surrounded by SiO_2. The

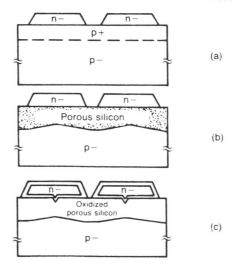

Fig. 2-56 Schematic of the buried p^+ anodization process for forming SIPOS isolation structures.[97] (© 1987 IEEE).

thermal-oxidation process yields material that has electrical and physical properties similar to thermal oxide. High-pressure oxidation reputedly eliminates the formation of defects in the silicon islands during the oxide growth and significantly reduces wafer warpage (which is one of the drawbacks of the FIPOS process). Oxidized porous silicon etches about an order of magnitude more rapidly than thermal oxide. The etch rate can be slowed by densifying the porous material in steam at about 1100°C. During the oxidation step, a layer of SiO_2 also grows on the surface and sidewalls of the single-crystal islands.

The FIPOS advantages are that the buried-oxide layer can be made much thicker than with SIMOX, and that it can provide arbitrarily thick island regions, without the need to resort to an extra epi step, as in SIMOX or ZMR. As a result, the islands on FIPOS substrates can have far fewer defects than those of SIMOX or ZMR. High-speed, 64-kbit SRAMs have been among the devices reported to have been built on FIPOS substrates. One of the main problems with FIPOS, however, is that a variety of nonconventional silicon-processing steps must be developed in order to implement it. These include the anodization process to form the porous silicon layer and an oxidation and anneal sequence for this layer.

2.10.7 Novel SOI CMOS Processes with Selective Oxidation and Selective Epitaxial Growth

Two novel SOI processes were recently reported that do not need a special oxygen implanter (as does SIMOX) nor a special anodization process (as does FIPOS), and yet

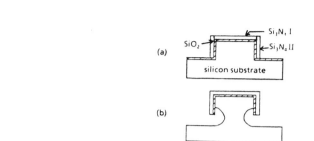

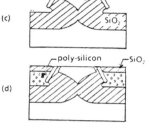

Fig. 2-57 Essential process steps of SOI method described in section 2.10.5.[100] (© 1986 IEEE).

produce active regions without high-angle grain boundaries (as occur in ZMR active regions).

In the first process, a selective oxidation beneath a top silicon layer is used to form the SOI structure.[100] A double silicon-nitride layer (similar to the structure used in SWAMI) is first created (Fig. 2-57a). The bottom of the trenches that were etched in the course of producing the structure just described are isotropically etched to create the structure shown in Fig. 2-57b. This structure is then selectively oxidized until the silicon islands are electrically isolated from the substrate. The trenches are finally filled to form a planar-surface topography with polysilicon and CVD SiO_2. The process was used to form islands with widths of 1.1-1.7 μm and minimum separations of 1.5 μm. One obvious difficulty with this method is that the width of the islands created is quite small. The possible generation of defects that might result from the growth of the oxide in a confined area was also not discussed in the report.

In the second process, wide trenches are etched 15 μm deep into a Si substrate (15-30 μm wide), and a 2-μm-thick CVD oxide is deposited into the trenches. Next, a seed hole of ~3-μm width is etched into the CVD oxide at the bottom of the trenches. A selective epitaxial silicon region is then grown through the seed hole to fill the trenches. After the epi is deposited, the wafers are oxidized to grow a 2-μm-thick oxide film to fill the gap between the epi islands and the substrate and to remove the seed hole to the silicon substrate.[119]

REFERENCES

1. B. T. Murphy, V. J. Glinski, P. A. Gary, and R. A. Pederson, "Collector Diffusion Isolated Integrated Circuits," *Proceedings of the IEEE,* **57,** 1523 (September, 1969).
2. J. A. Bruchez and P. Pollok "The Philosophy of a Simple Collector Diffusion Isolation Bipolar Process" *Solid State Technology,* p. 93, Aug. 1987.
3. E. Kooi and J. A. Appels, in "Semiconductor Silicon 1973", H. R. Huff and R. Burgess, Eds., p. 860, *The Electrochemical Symp. Series,* Princeton, N.J. (1973).
4. I. Magdo and A. Bogh, *J. Electrochem. Soc.,* **125,** 932 (1987).
5. R. Chapman et al., "0.8 μm CMOS Technology" *IEDM Tech. Dig.* 1987, p. 362.
6. Y. Han and B. Ma "Poly Buffer Layer for Scaled MOS", *VLSI Sci &Tech, Electrochem. Soc.,* 1984, p. 334.
7. A. Bogh and A. K. Gaind, *Appl. Phys. Lett.* **33,** 895 (1978).
8. L. C. Parrillo, *Semiconductor Intl.,* April, 1988, p. 64.
9. J. Pfiester and J. Alvis, "Improved CMOS Field Isolation Using Ge/B Ion Implantation", *IEEE Elect. Dev. Lett.,* Aug. 1988, 391.
10. S. M. Hu, "Defects and Device Processing: Achievements and Limitations", *VLSI Sci. & Tech., Electrochem. Soc.,* 1987, p. 722.
11. J. Manoliu, *Semiconductor Intl.,* April, 1988, p. 90.
12. T. Mizuno et al., "Oxidation Rate Reduction in the Submicron LOCOS Process" *IEEE Trans. Elect. Dev.,* Nov. 1987, p. 2255.
13. W. R. Hunter et al., *IEEE Trans. Electron Dev.,* **ED-26,** 353, (1979).
14. J. Hui, P. Vande Voorde, and J. Moll, "Scaling Limitations of Submicron LOCOS Technology", *Tech. Dig. IEDM,* 1985, p. 392.
15. T. Mizuno et al., *IEEE Trans. Electron Dev.,* Nov. 1987, 2255.
16. E. Kooi et al. , *J. Electrochem. Soc.* **123,** 1117 (1976).
17. T. A. Shankoff et al., *J. Electrochem. Soc.,* **127,** 216 (1980)
18. C. A. Goodwin and J. W. Brossman, *J. Electrochem. Soc.,* **129,** 1066 (1982).
19. V.A. Dhaka et al., "Subnanosecond ECL using Isoplanar II", *IEEE J. Solid State Ckts.,* **SC-8,** Oct. 1973, p. 368.
20. E. Bassous et al., *J. Electrochem. Soc.* **123,** 1729, (1976).
21. H. Murrmann, *Siemens Forsch. - Entwicklungsber.* **5,** 353, (1976).
22. R. C. Ellwanger et al., *Ext. Abs, Fall Meeting ECS,* 1984, No. 418, p. 603.
23. A. Cosand and S. Prussin, *ECS Meeting,* May, 1976. Ext Abs. No. 109, p. 292.
24. B. O. Kolbesen and H.P. Strunk, "Analysis, Electrical Effects, and Prevention of Defects", *Solid-State Devices, Inst. Phys. Conf. Series,* **57,** 1980, p. 21.
25. K. Sakuma et al., "A New Self-Aligned Planar Oxidation Technology", *J. Electrochem. Soc.,* June, 1987, p. 1503.
26. N. Guillemot et al., "A New Analtytical Model of the Birds Beak", *IEEE Trans. Elect. Dev.* **ED-34,** May 1987, p. 1033.
27. P. Sutardja et al., *IEDM Tech. Dig.,* 1986, p. 526.
28. A. Lewis et al., "Device Isolation in High-Density LOCOS CMOS", *IEEE Trans Elect. Dev.,* June 1987, 1337.
29. K. Y. Chiu et al., *IEEE Trans Electron Dev.,* **ED-29,** p. 536, (1982)

30. C. Claeys and J. Vanhellemont "Structural and Electrical Characterization of SWAMI for Submicron Technologies" *Ext. Abs. of Spring Meeting of the ECS*, 1988, Abs. No. 140, p. 211.

31. M. Ghezzo et al., "Laterally Sealed LOCOS Isolation", *J. Electrochem. Soc.*, June 1987, p. 1475.

32. P. Deroux-Dauphin & J.P. Gonchond, *IEEE Trans. Electron Dev.*, Nov. 1985, p. 2392.

33. M. P. Brassington, R. Razouk, and C. Hu, *IEEE Trans. Electron Dev.*, Jan 1988, p. 96.

34. N. Hoshi et al., *Tech. Dig. IEDM*, 1986, p. 300.

35. T. Nishihara et al., *Tech. Dig. IEDM*, 1988, p. 100.

36. L. C. Parrillo et al., *Tech Dig. IEDM*, 1986, p. 244.

37. M. Brassington et al., *IEEE Trans. Electron Dev.*, Jan 1988, p. 96.

38. F. Bryant et al., *Ext. Abs. Spring Meeting Electrochem. Soc.* 1988, Abs. No. 137, p. 205.

39. H. H. Tsai et al., *IEEE Trans. Electron Dev.*, March 1988, p. 275.

40. T. Kaga et al., "Advanced OSELO Isolation with Shallow Grooves" *IEEE Trans. Electron Dev.*, July 1988, p. 893.

41. M. Mikoshiba et al., *IEDM Tech. Dig.*, 1984, p. 578.

42. G. Fuse et al., *IEEE Trans. Electr. Dev.*, Feb. 87, p. 356.

43. T. Iizuka et al., *Tech. Dig. IEEE IEDM*, 1981, p. 384.

44. G. Fuse et al., *IEDM Tech. Dig.*, 1987, p. 732.

45. B. Davari et al., *Tech. Dig. IEDM*, 1988, p. 92.

46. T. Shibata et al., *IEDM Tech. Dig.*, 1983, p. 27.

47. R. Kwasnick et al., "Buried Oxide Isolation with Etch-Stop", *IEEE Elect. Dev. Lett.*, Feb. 1988, p. 62.

48. Y. Tamaki et al., *Jap. J. Appl. Phys.*, **21**, Supplement 21-1, p. 37 (1982).

49. Y. Tamaki et al., *J. Electrochem. Soc.*, March 1988, p. 726.

50. Y. Niitsu et al., "Latchup-Free CMOS Structure Using Shallow Trench Isolation",*Tech. Dig. IEDM*, 1985, p. 509.

51. J. R. Pfiester and J. R. Alvis, "A Novel CMOS VLSI Isolation Technology Using Selective Cl Implantation", *IEEE Electron Dev. Letts,* Nov. 1988, p. 561.

52. M. J. Kim et al., *Ext. Abs. Fall Mting Electrochem. Soc.*, 1988, Abs. No. 241, p. 338.

53. G. K. Herb, D. J. Rieger, and K. Shields, "Silicon Trench Etch in a Hex Reactor", *Solid-State Technol.*, Oct., 1987, p. 109.

54. "TI's Trench Technology Moves into the Factory", *Electronics*, July 9, '87, p. 75.

55. M. Sato and Y. Arita, *J. Electrochem. Soc.*, Nov. 1987, p. 2856.

56. R. N. Carlile et al., "Trench Etches in Silicon with Controllable Sidewall Angles", *J. Electrochem. Soc.*, Aug., 1988, p. 2058.

57. S. Samukawa et al., *Ext. Abs. Fall Meeting Electrochem. Soc.*, 1988, Abs. No. 110, p. 163.

58. R. E. Nowak, *Solid State Tech.*, March 1988, p. 39.

59. K. Yamada et al., *Tech. Dig. IEDM,* 1985, p. 702.

60. K. Yamabe and K. Imai, *IEEE Trans. Electron Dev.*, Aug. 87, p. 1681.

61. K.V. Rao et al., *Tech. Dig. IEDM,* 1986, p. 140.

62. R. D. Rung, "Trench Isolation Prospects for Application in CMOS VLSI", *Tech. Dig. IEDM,* 1984, p. 574.

63. K. Tsukamoto et al., *IEDM Tech. Dig.,* 1987, p. 328.

64. F. Horiguchi et al., *IEDM Tech. Dig.* 1987, p. 324.

65. D.-B. Kao et al., "Two-Dimensional Silicon Oxidation Experiments and Theory", *Tech. Dig. IEDM,* 1985, p. 388.

66. V. J. Silvestri, "Growth Kinematics of a Polysilicon Trench Refill Process", *J. Electrochem. Soc.,* Nov. 1986, p. 2374.

67. V. J. Silvestri, "Selective Epitaxy Trench", *J. Electrochem. Soc.,* July, 1988, p. 1808.

68. Y.-C. S. Yu, "Planarized Deep-Trench Process for Bipolar Isolation", *Ext. Abs. Fall Meeting 1988, Electrochem. Soc.* Abs. No. 234, p. 326.

69. J. Borland and C. Drowley, *Solid State Tech.,* Aug. 1985, p. 141.

70. J. Borland et al., "Silicon Epitaxial Growth for Advanced Structures", *Solid State Tech.,* Jan 1988, p. 111.

71. J. Manoliu and J. O. Borland, "A Submicron Dual Buried Layer Twin Well CMOS SEG Process", *IEDM Tech. Dig.,* 1987, p. 20.

72. S. Nagao et al., *IEEE Trans. Electron Dev.,* Nov. 1986, p. 1738.

73. N. Endo et al., *IEEE Trans. Electron Dev.,* Nov. 1986, p. 1659.

74. A. Stivers and C.H. Ting, "Selective Epitaxial Growth for CMOS Isolation", *Presented at the Third Annual Applied Materials Inc.,"Innovations in Epitaxial Technology for Advanced Device Structures"* Dec. 1988, Palo Alto, CA.

75. J. Borland et al., *Ext. Abs. of Intl. Conf. on Solid State Dev. and Matls.,* 1986, Tokyo, Japan, p. 53.

76. F. Mieno et al., *Tech. Dig. IEDM,* 1987, p. 16.

77. T. I. Kamins and S. Shiang, *Electron Dev. Letts.,* Dec 1985, p. 617.

78. N. Endo et al., *IEEE Trans. Electron Dev.,* Nov. 1986, p. 1659.

79. Y.-C. Yu and A. Witkowski, *J. Electrochem. Soc.,* October 1988, p. 2562.

80. J. O. Borland, D. Schmidt and A. Stivers, *Ext. Abs. of 18th Conf. on Solid-State Devices and Materials,* 1986, Tokyo, Japan, p. 1659.

81. J.B. Lasky et al, "SOI by Bonding and Etchback"*Tech. Dig. IEDM,* 1985, p. 684.

82. R. C. Frye et al., *J. Electrochem. Soc.,* Aug 1986, p. 1673.

83. H. Ohashi et al., *Tech. Dig. IEDM,* 1986, p. 210.

84. W. A. Krull et al., *IEEE Circuits and Systems Mfg.,* July 1987, p. 20.

85. K. Ueno et al., *Tech. Dig. IEDM,* 1988, p. 870.

86. P. K. Vasudev "Recent Advances in Solid-Phase Epitaxial Recrystallization of SOS for CMOS and Bipolar Devices", *IEEE Circuits and Devices Magazine,* July 1987, p. 17.

87. H. W. Lam, "SIMOX SOI for IC Fabrication", *IEEE Circuits and Sys. Mag.,* July 1987, p. 6.

88. D. J. Foster et al., *IEEE Trans. Electron Dev.,* March 1986, p. 354.

89. B.-Y. Mao et al., *IEEE Electron Dev. Letts.,* July, 1987, p. 306.

90. J. R. Davis et al., *IEEE Electron Dev. Letts.,* July, 1987, p. 291.

91. L. Jastrzebski et al., *J. Electrochem. Soc.,* July, 1988, p. 1746.

92. L. Nesbit et al., *J. Electrochem. Soc.,* Nov. 1985, p. 2713.

93. K. M. Cham, S. Oh, D. Chin, and J. L. Moll, *Computer-Aided Design and VLSI Device Development"*, Boston, Klewer, 1986, Chap. 9.

94. B.-Y. Tsaur, "ZMR SOI Technology", *IEEE Circuits and Sys. Mag.*, July, 1987, p. 12.

95. J. C. C. Fan et al., *J. Cryst. Growth*, **63**, No. 3, 1983.

96. T. Stultz et al., in *Laser-Solid Interactions and Transient Thermal Processing of Materials*, Ed. by J. Narayan, North-Holland, New York, p. 463, 1983.

97. S. S. Tsao, "Porous Silicon Techniques for SOI Structures", *IEEE Circuits & Sys. Mag.*, November 1987, p. 3.

98. E. J. Zorinsky et al., *Tech. Dig. IEDM*, 1986, p. 431.

99. C. T. Sah, "The Evolution of the MOS Transistor", *Proceedings of the IEEE,* October 1988, p. 1280.

100. M. Kubota et al., *Tech. Dig. IEDM*, 1986, p. 814

101. W. Wakayima et al., *Tech. Dig. IEDM*, 1988, p. 246.

102. M. Shimizu et al., *Tech. Dig. IEDM*, 1988, p. 96.

103. M. H. El-Diwany et al., *Tech. Dig. IEDM*, 1987, p. 917.

104. H. J. Levinson and W. H. Arnold, *J. Vac. Sci. Tech.* B5(1), p. 293 (1987).

105. D. W. Hess, *Solid State Technol.*, Apr. 1981, p. 192.

106. M. Ameen et al., *Semiconductor Internatl.*, September, 1988, p. 122.

107. R. N. Carlile et al., "Trench Etches in Silicon with Controllable Sidewall Angles", *J. Electrochem. Soc.*, Aug. 1988, p. 2058.

108. J. Y. Chen, *CMOS Science and Technology*, Prentice-Hall, Englewood Cliffs, N. J., 1990, p. 237.

109. K. L. Wang et al., *IEEE Trans. Electron Dev.* **ED-29**, p. 541, 1982.

110. J. Hui et al., *IEEE Trans. Electron Dev.*, **ED-29**, p. 554, 1982.

111. R. P. Kramer, *Ext. Abs. Spring Meeting Electrochem. Soc.*, 1989, Abs. No. 131, p. 182.

112. P. A. van der Plas et al., *Dig. of 1987 Symp. on VLSI Technology*, III-4, p. 19.

113. J. C. Sturm, *Materials Research Soc. Symp. Proc.*, **107**, p. 295, (1988).

114. T. W. MacElwee, *Ext. Abs. Spring Meeting Electrochem. Soc.*, 1989, Abs. No. 192, p. 280.

115. J. -P. Colinge, *Ext. Abs. Spring Meeting Electrochem. Soc.*, 1989, Abs. No. 193, p. 282.

116. G. R. Powell, S. V. Davies, and A. H. Chambers, *Ext. Abs. Electrochem. Soc. Conf.*, Spring, 1989, p. 252.

117. M. Englehardt and S. Schwarzl, *Ext. Abs. Electrochem. Soc. Conf.*, Spring, 1989, p. 254.

118. G. A. Rozgonyi et al., *Ext. Abs. Electrochem. Soc. Conf.*, Spring, 1989, p. 339.

119. D. M. Boisvert and J. D. Plummer, *Ext. Abs. Electrochem. Soc. Conf.*, Spring, 1989, p. 431.

120. S. Mahli et al., *Ext. Abs. Electrochem. Soc. Conf.*, Spring, 1989, p. 427.

121. B. Coulman et al., *Ext. Abs. Electrochem. Soc. Conf.*, Spring, 1989, p. 293.

122. S. L. Partridge, *Tech. Dig. IEDM*, 1986, p. 428.

123. J. Hui et al., *IEEE Trans. Electron Dev.*, vol. ED-29, p. 554, 1982.

124. C. Claeys et al., *J. Electrochemical Society,* September 1989, p. 2619.

125. M. Ghezzo et al., *J. Electrochemical Soc.*, July 1989, p. 1992.

126. l. Manchanda et al., "A High Performance, Self-Aligned, Ultra-Rad-Hard Field Oxide Technology for NMOS/CMOS VLSI'", *IEEE Trans. Electron Dev.,* April 1989, p. 651.

127. H. Kabza et al., *IEEE Electron. Dev. Lett.,* August 1989, p. 344.

128. R. C. Jaeger, *"Introduction to Microelectronic Fabrication,* Addison-Wesley, Reading MA, 1988.

129. B. T. Murphy et al., *Proc. of the IEEE,* **57,** 1523 (1969).

130. S. M. Sze, Ed. *VLSI Technology,* 2nd Ed., New York, McGraw-Hill, 1988, p. 476.

131. D. R. Craven and J. B. Stimmell, *Semiconductor International,* June 1981, p. 59.

132. N. Kasai et al., *Tech. Dig. IEDM,* 1985, p. 419.

133. K. E. Bean and W. R. Runyan, *J. Electrochem. Soc.,* **124,** 5C (1977).

CHAPTER 3

CONTACT TECHNOLOGY AND
LOCAL INTERCONNECTS FOR VLSI

When silicon integrated circuits are fabricated, isolated active-device regions are created within the single-crystal substrate (see chap. 2). The technology used to connect these isolated devices through specific electrical paths employs high-conductivity, thin-film structures, fabricated above the SiO_2 insulator that covers the silicon surface. Wherever a connection is needed between a conductor film and the Si substrate, an opening in the SiO_2 must be provided to allow such *contacts* to occur.

Interconnections in *ideal* electric circuits are assumed to exhibit zero impedance to the flow of electric current. Although the finite impedance of *real* conductor structures can be ignored in some applications, in others such parasitic effects can significantly impact circuit performance, and must therefore be considered during circuit analysis or design. The same is true with integrated circuits. The electrical characteristics of interconnects and contacts must be studied in order to determine under what circumstances their parasitic behavior can significantly impact circuit performance.

In this chapter we will focus primarily on *ohmic contacts* between metal or metal-silicide thin films and single-crystal silicon substrates. We will also briefly describe metal-Si Schottky contacts, polysilicon-to-single-crystal Si contacts, and local-interconnect technology. The topics of metal-to-metal and metal-to-metal-silicide contact properties will be covered in chapter 4.

3.1 THE ROLE OF CONTACT STRUCTURES IN DEVICE AND CIRCUIT BEHAVIOR

Parasitic resistances exist in the path between the metal-to-Si substrate interface and the region in the device where the actual transistor action begins. In this chapter we are concerned with the processing of metal-silicon contact structures, and thus are interested in the fraction of this device parasitic resistance due to the contacts. In order to see how

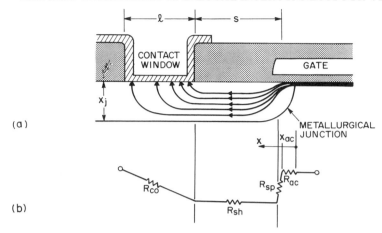

Fig. 3-1 Schematic drawing showing (a) current paths within the source and drain regions, of a planar MOSFET and (b) their corresponding representative series resistance components.[84] (© 1979 IEEE).

this part of the parasitic resistance is extracted from the total value, the physical structure of planar MOSFETs and bipolar transistors must be examined.

3.1.1 Contact Structures in Planar MOSFETs and Bipolar Transistors

In the planar MOSFET, the device region in which transistor action occurs is the *channel region under the gate*. The current flowing from the metal-interconnect lines to these channel regions must be transported from the metal-Si contact to the edge of the channel through the source or drain regions (Figs. 3-1 and 5-1). In the planar bipolar transistor, the region of transistor action is the *intrinsic-base region*. Base current from the external circuit must be transported from the base contact through the extrinsic-base region to reach the intrinsic base (Fig. 3-2).

In the MOSFET, current enters the contact perpendicular to the wafer surface, and then travels parallel to the surface in order to reach the channel. The parasitic series resistance, R_S, of the current path from the contact to the edge of the channel can be modeled by the sum of four components,

$$R_S = R_{co} + R_{sh} + R_{sp} + R_{ac} \qquad (3\text{-}1)$$

where R_{co} is the contact resistance between the metal and the source/drain region, R_{sh} is the sheet resistance of the bulk region of the source and drain, R_{sp} describes the resistance of the current lines crowding near the channel end of the source, and R_{ac} is the accumulation-layer resistance. Qualitatively, for a nonabrupt junction, R_{ac} and R_{sp}

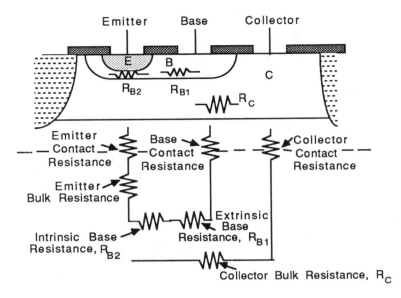

Fig. 3-2 Schematic drawing of the parasitic series-resistance components in planar bipolar transistors.

arise because, after leaving the channel, the current does not immediately spread into the bulk of the source-drain regions (the parasitic resistance effect in these bulk regions is R_{sh}). Thus R_{ac} and R_{sp} exist because the accumulation-layer conductance is much higher than the semiconductor conductance in the vicinity of the junction.

In the *planar bipolar transistor*, the series resistance of the base region, R_B, is the sum of the contact resistance, R_{co}, and the resistance of the current path of the extrinsic base, R_{B1}, and the intrinsic base, R_{B2} (Fig. 3-2). Again, the base current enters the base contact perpendicular to the wafer surface but must travel parallel to the wafer surface in order to reach the intrinsic-base region. The emitter current experiences a series resistance that is basically due only to the contact resistance of the emitter contact because the emitters are so shallow. In this case the current flow is perpendicular to the wafer surface.

We will quantitatively describe the impact of R_S on the device and circuit behavior of MOS devices in detail later in this chapter. (The effect on bipolar devices will be described in chap. 7.) At this point we merely want to make the point that R_{co} is always one of the parasitic-series-resistance components whenever metal-to-semiconductor contacts are made. In addition, as semiconductor device dimensions shrink both vertically and laterally, device currents and current densities increase. Hence, if the metal-to-semiconductor contact area is also scaled, R_{co} also increases. Since this may significantly degrade device performance, techniques for accurately determining R_{co} need to be available (both from measurements and by simulation).

3.2 THEORY OF METAL-SEMICONDUCTOR CONTACTS

Ideal nonrectifying contacts would exhibit no resistance to the flow of current in either direction through the contact, and their I-V characteristic would appear as shown in Fig. 3-3a. In general, however, when metal-to-semiconductor contacts are fabricated, they possess *non-ohmic* I-V characteristics (Fig. 3-3b). Nevertheless, it is possible to fabricate contacts with electrical characteristics that approach those of the ideal. We will refer to real metal-semiconductor contacts as low-resistance *ohmic contacts* if they exhibit a near-linear current-voltage characteristic in both directions of current flow, and negligible resistance when compared to the bulk resistance (Fig. 3-3c). It is the ohmic contact that is normally fabricated for the purpose of interconnecting devices in an IC. Non-ohmic, rectifying metal-semiconductor contacts (*Schottky contacts*) find some application in ICs, and they are described in section 3.7.

The reason that metal-semiconductor contacts generally exhibit non-ohmic I-V characteristics is that the work functions of the metal and the semiconductor (φ_m and φ_s) are not equal, and in most metal-silicon contacts φ_m is greater than φ_s (Fig. 3-4a). When such metal and semiconductor materials are in contact at equilibrium (i.e., no current flows), the Fermi level must be constant throughout the system (which in this case includes both materials). If φ_m is not equal to φ_s, a potential-energy barrier must exist between the metal and the semiconductor at equilibrium, and the height of this barrier, $q\varphi_b$ (assuming $\varphi_m > \varphi_s$) is given by:

$$q\varphi_b = q(\varphi_m - \chi_s) \qquad\qquad (3\text{-}2)$$

where χ_s is the electron affinity of the semiconductor (Fig. 3-4b). The effect of the barrier on current flow under bias is similar to that of the barrier that exists in a *pn* junction; hence the I-V characteristic of a metal-semiconductor contact, in general, exhibits rectifying properties that are essentially like those of a *pn* junction.

For an ohmic-contact structure to be created, the effect of the barrier on carrier flow must be made negligible. In theory, this could be achieved by making φ_b very small

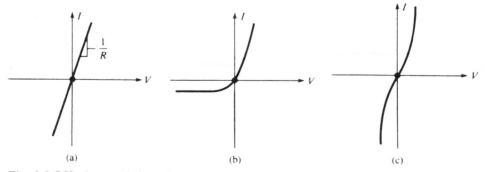

(a)   (b)   (c)

Fig. 3-3 I-V characteristics of contacts between metal and semiconductor in integrated circuits. (a) Ideal ohmic contact. (b) Rectifying contact. (c) Practical nonlinear "ohmic" contact.

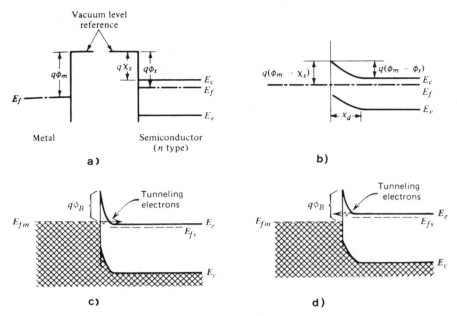

Fig. 3-4 Energy band diagram of a metal-semiconductor contact with $\phi_m > \phi_s$: (a) before contact and (b) after contact and at thermal equilibrium.[130] From E. S. Yang, *Microelectronic Devices*. Copyright 1988 McGraw-Hill. Reprinted with permission. Metal-semiconductor barrier with a thin space-charge region through which tunneling can take place. (c) tunneling from metal to semiconductor, (d) tunneling from semiconductor to metal. From R. S. Muller and T. I. Kamins, 2nd Ed., *Device Electronics for Integrated Circuits*. Copyright 1986, John Wiley & Sons. Reprinted with permission.

(i.e., by using a metal that would give a low ϕ_b when in contact with Si). Most carriers would then have enough energy to overcome the barrier, and current would flow easily when a bias was applied. Unfortunately, this is not a viable option in Si devices, because the smallest value of ϕ_b for a metal-to-n-type Si contact is about 0.5 V. Such large values of ϕ_b end up yielding contact-resistance values on lightly doped n-type Si that are far too large to be of practical use for ohmic contacts to MOS or bipolar devices.

Another approach would be to use a heavily doped n^+ or p^+ surface in the silicon. This technique is effective since the charge transport across a metal-semiconductor contact can be indirectly influenced by the doping concentration in the semiconductor. That is, when the doping concentration is low ($N_D < 10^{19}$ cm^{-3}), only carriers that have energies greater than the barrier height can overcome the barrier (i.e., through the phenomenon of thermionic emission). If the doping concentration exceeds this value, carrier transport becomes dominated by *quantum-mechanical tunneling* (i.e., the carriers go *through* the potential barrier).

Significant tunneling between two regions can occur if the width of the potential-energy barrier that separates them is small. At a metal-silicon interface, the width of the potential energy barrier is essentially equal to that of the depletion-region width in the semiconductor, W. When the silicon is heavily doped, W is very thin - only a few nanometers thick. (The following relation is used to compute W: $W = \sqrt{2\,\varepsilon_s\,\varphi_i\,/qN_D}$, where φ_i is the voltage across the space charge region of the semiconductor, and N_D is the dopant concentration.) For such small values of W, carriers with less energy than the barrier height are still able to pass through the barrier in either direction by the mechanism of tunneling.

As shown in Fig. 3-4c, if the metal is biased negatively with respect to the semiconductor, electrons in the metal can tunnel through the barrier to the lower-energy conduction-band states in the semiconductor. Similarly, if the metal is biased positively with respect to the semiconductor, the electrons from the semiconductor can tunnel into the lower-energy electronic states in the metal (Fig. 3-4d). Thus, an essentially symmetrical I-V curve for both forward and reverse bias is realized. In addition, since there is a high electron concentration in both the metal and the conduction band of a heavily doped semiconductor, many electrons are available to take part in these processes. Consequently, currents increase very steeply as bias is applied, and low-resistance contacts can be achieved.

The physical parameter that characterizes the incremental resistance of a metal-to-semiconductor interface is the *specific contact resistivity*, ρ_c. That is, ρ_c is a measurement-independent quantity that describes the incremental resistance of an infinitely small area of such an interface. The units of ρ_c are therefore V-cm^2/A or Ω-cm^2. The values of ρ_c are commonly expressed in these units, or their equivalent, Ω-μm^2. Since ρ_c of the ohmic contacts used in IC technology is typically in the 10^{-6}-10^{-8} Ω-cm^2 range (and 1 μm = 10^{-4} cm), the latter units are often considered more convenient. (Of course, the goal is to fabricate ohmic contacts with as small a value of ρ_c as possible.)

The contact resistance, R_{co}, on the other hand, is a macroscopic quantity that depends not only on ρ_c, but also on the contact size, the semiconductor sheet resistance, and the contact geometry. Since ρ_c is the single measurement-independent parameter that describes the quality of the interface, it is the proper parameter on which to focus attention when seeking to optimize a contact structure. The following section describes techniques that allow ρ_c to be extracted from measurements of special contact test structures. In the present section we consider how material parameters can be quantitatively varied to produce contacts with minimum values of ρ_c.

The values of ρ_c derived from experimental measurements are found to be well modeled by the theoretical expression

$$\rho_c \approx \exp\left(\frac{4\pi\,\sqrt{\varepsilon_{si}m_e\varphi_b}}{h\,\sqrt{N_D}}\right) \qquad (3\text{-}3)$$

where m_e is the *tunneling effective mass*. In Figure 3-5, this equation is plotted for φ_b = 0.85, 0.6, and 0.5 eV and for various doping levels in the silicon substrate.[1] In addition, experimental data points for Al-nSi and for PtSi-nSi contacts (where φ_b = 0.72

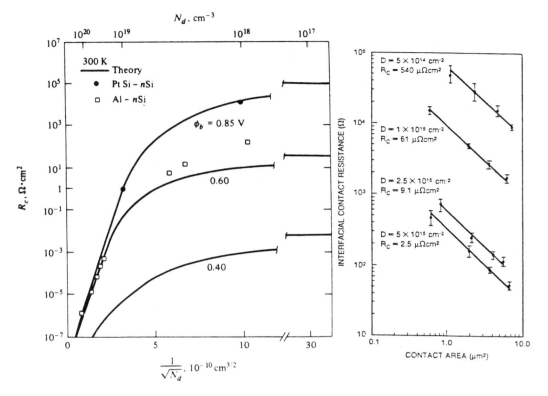

Fig. 3-5 (a) Theoretical and experimental values of specific contact resistance.[1] (© 1971 IEEE). (b) Specific contact resistivity versus implant dose.[2] (© 1985 IEEE).

eV and 0.84 eV, respectively) are also plotted. It can be seen that Eq. 3-3 quite accurately predicts the measured values of ρ_c as a function of doping. For doping concentrations of $N_D > 10^{20}$ cm^{-3}, it is predicted that ρ_c values as low as $\approx 10^{-7}$ Ω-cm^2 can be obtained.

In practice, the doping concentration in the silicon can be increased through diffusion and/or ion-implantation methods. Figure 3-5b plots the value of ρ_c versus increases in implant dose into *n*-type Si.[2] It can be seen that ρ_c decreases inversely with the square root of the implant dose, again fitting well with Eq. 3-3's premise that $\rho_c \sim 1/\sqrt{N_D}$.

When the doping concentration in the semiconductor is decreased, the depletion region is wider, and this reduces current due to tunneling. Eventually, as doping further decreases, the current-flow mechanism due to thermionic emission over the barrier becomes dominant. Hence, a rectifying I-V curve is generally exhibited by contacts fabricated between metal and lightly doped Si. The I-V curve of a rectifying contact is similar to that of a *pn* junction, which is used as a diode. Hence, the contact structure of a metal to lightly doped Si can be used as a diode as well; such devices are known as *Schottky diodes*.

Aluminum is the most common and important metal used in IC interconnections. "Ohmic" contacts between Al and n-type Si can be made if the surface concentration of dopant in the Si is high (e.g., $>10^{19}$ cm^{-3}). However, when Al is deposited directly onto less heavily doped p-Si and is subsequently sintered (i.e., heated to ~450°C following deposition to ensure good contact formation), a low value of ρ_c is still observed, even for Si as lightly doped as 10^{16} cm^{-3}. This occurs because Al is a p-type dopant, and some of the Al diffuses into the Si during the sinter step; a heavily doped p-region is thus established under the Al-Si contact region.

In the fabrication of actual contacts, the values of ρ_c that are obtained are also strongly dependent on the condition of the Si surface. That is, if the surface is not clean (e.g., if there is a thin polymer or native SiO$_2$ film on it), the value of ρ_c that is obtained can be much higher than that predicted by Eq. 3-3. Such phenomena will be described in more detail in the section dealing with the practical issues of fabricating metal-semiconductor contacts.

Note that metal-to-metal contacts must also be made when fabricating ICs. The contact resistance of such contacts is usually negligibly small compared to that of metal-Si contacts, and hence is normally ignored. In some cases, however, metal-to-metal contact resistance may be significantly large (e.g., if contact is made to an Al film that has a native Al$_2$O$_3$ layer on the surface, or when contact is made between CVD W and Al films), and measures must be taken to reduce it. Issues dealing with metal-to-metal contact resistance are discussed in chapter 4.

3.3 EXTRACTING VALUES OF SPECIFIC CONTACT RESISTIVITY FROM MEASUREMENTS

In order to analyze the effects of R_{co} on the device characteristics of bipolar and MOS transistors, the value of R_{co} in specific device structures must be known. However, the measured value of R_{co} depends not only on ρ_c but also on the contact size, the semiconductor sheet resistance, and the contact geometry. As a result, the measured value of R_{co} in a test structure may not by itself provide an accurate estimate of R_{co} in real FET and bipolar transistor contacts which have different dimensions than those in the test structure (especially in the smaller contacts of future generations of devices).

The general procedure used to determine R_{co} instead involves the extraction of the contact-size-independent parameter ρ_c from test-structure measurements. This procedure is tedious, but once an accurate value of ρ_c has been obtained, it can be used to calculate the R_{co} in actual devices with contacts of various sizes (based on the specific contact size, semiconductor sheet resistivity, and contact geometry).

The flow chart in Fig. 3-6 shows that a measured value of V/I for a given test structure is first obtained; next the extraction procedure is used to get ρ_c; finally the value of ρ_c is used together with the other factors to get R_{co} in an actual device. A 2-D device simulator, such as PISCES (see chap. 9), can be used to accurately calculate R_{co} for a specific contact structure once ρ_c is available.

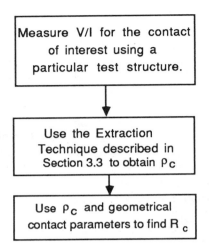

Fig. 3-6 Flow chart that illustrates how to obtain accurate values of R_{co} in an actual contact.

Since each metal-semiconductor contact (uniquely defined by the metal and semiconductor type, as well as by the doping concentration) exhibits its own value of ρ_c, values of R_{co} for different contact structures can also be compared if they are based on accurate values of ρ_c.

3.3.1 Extraction of the Specific Contact Resistivity from an Ideal Contact Structure

To understand how values of ρ_c can be extracted, consider once again the current that flows into the contacts of MOSFETs and bipolar transistors. If the current density over the entire area, A, of the contact were uniform (an ideal case), the value of ρ_c could be extracted from a measurement of the voltage across the contact interface, V, and the current through the contact, I. The measured value of V/I would then yield a value, R_k, and ρ_c would be found from the following expression:

$$\rho_c = R_k / A \qquad\qquad (3\text{-}4)$$

However, in most cases the current density is not uniform across the entire area of the contact, and hence this simple expression is usually not valid for extracting ρ_c from R_k. Nevertheless, many reports in the literature have used this equation to extract values of ρ_c from test-structure measurements; unfortunately such reported values of ρ_c are usually in error. In general, Eq. 3-4 is useful for providing an estimate of the *upper bounds* to the value of ρ_c. Hence, when Eq. 3-4 is used to determine ρ_c, the reported value is actually greater than the correct value. The more recently reported values of ρ_c

for various contact structures (for which the more correct models may have been used to extract ρ_c) have usually been lower than the earlier reported values.

3.3.2 Current Flow in Actual Contact Structures

The reason that current flow in actual MOS (and bipolar base) contact structures is not uniform over the entire contact area can be explained through a simple, one-dimensional perspective (Fig. 3-1). In such contacts the laterally moving current in the semiconductor flows in a region that has a finite resistance (characterized by the value of its sheet resistance, ρ_{sh} [in Ω/sq]), and the much lower-resistivity metal is assumed to be an equipotential. The parasitic resistance offered to lateral current flow is therefore smallest for current entering the metal at the leading edge of the contact, and this resistance effectively increases as the current flows laterally under the contact before entering the metal. The density of current entering the contact is thus greatest at the leading edge and smallest at the trailing edge. This phenomenon is termed *current crowding*.

Such current crowding is increased as ρ_{sh} is increased (or ρ_c decreased), making it useful to define a quantity known as the *transfer length*, l_t, in terms of these parameters. l_t is defined as the distance from the edge of the contact at which the value of the current density has dropped to 1/e of its value at the leading edge. The value of l_t is calculated from

$$l_t = \sqrt{\frac{\rho_c}{\rho_{sh}}} \qquad (3\text{-}5)$$

This equation implies that if the length of the contact were made much longer than l_t, most of the current would flow into the contact near the leading edge, and there would be almost none in the rest of the contact. In such a contact, only a small fraction of the area of the contact would be used to carry the current. If a value of ρ_c was extracted from the measurement of R_k in such a long contact using Eq. 3-4, and this value of ρ_c was used to estimate the contact resistance, R_{co}, of a much smaller (and therefore shorter contact), the resulting estimate of R_{co} in the smaller contact would not be accurate.

The value of l_t in typical device contacts can be estimated by considering a contact in which $\rho_c = 10^{-6}$ Ω-cm^2 and $\rho_{sh} = 100$ Ω/sq. Equation 3-5 predicts the value of l_t of such a contact to be 1 μm. (If ρ_c is even smaller, l_t will also be smaller.) Thus, in actual tests, the use Eq. 3-4 to extract values of ρ_c and R_{co} from contact structures whose lengths are 3 μm or longer is likely to produce erroneous values, since l_t values of 1 μm or less would cause significant current-crowding effects in such larger contacts.

One of the earliest papers that reported this effect was published in 1983.[3] In it, the measured values of R_k versus contact size (using the cross-bridge Kelvin resistor described in the next section) was compared to the values of R_{co} predicted by the value of ρ_c extracted from a large contact (e.g., 5 μm x 5 μm) using Eq. 3-4. With Eq. 3-4 used to extract ρ_c from the R_k measured in a large contact, the dotted line in Fig. 3-7 gives the estimated value of R_{co} of smaller contacts, based on the extracted value of ρ_c.

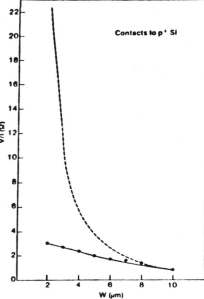

Fig. 3-7 Values of R_{co} for small contacts predicted by Eq. 3-4 compared to measured values.[3] Reprinted by permission of the publisher, The Electrochemical Society, Inc.

The *observed* value of R_k in the actual smaller contacts, however, was much smaller, as shown by the solid line.

Thus, the value of ρ_c extracted from the large contact using Eq. 3-4 does not allow accurate values of R_{co} to be estimated in smaller contacts. The reason for the discrepancy is that since the current density over the area of the contact is not uniform, it yields an erroneous value of ρ_c if Eq. 3-4 is used to extract it. A more accurate procedure for extracting ρ_c from contact-test-structure measurements must therefore be provided.

3.3.3 Contact Structures Used to Extract ρ_c

Several test structures have been developed for extracting ρ_c from electrical measurements. The three most commonly used are: the *cross-bridge Kelvin resistor* (CBKR, Fig. 3-8a);[4] the *contact end-resistor* (CER, Fig. 3-8b);[5] and the *transmission-line tap resistor* (TLTR, Fig. 3-8c).[6] In all of these structures, a specific current is sourced from the diffusion level up into the metal level through the contact window. A voltage is measured between the two levels using two other terminals. The contact resistance measured for each structure is this voltage divided by the source current.

Since each of the test structures measures the voltage at a different position along the contact, as shown in Fig. 3-8, the contact resistance values measured are different, and they must therefore be distinguished from one another. The resistances measured by the

CBKR, the CER, and the TLTR structures are, respectively, R_k (Kelvin), R_e (end), and R_f (front). The 1-D equations used to extract ρ_c from the measured resistance values are derived from the distributed-resistor network that is used to model the current-crowding effects described in the previous section. In addition, in deriving these equations it is assumed that the diffusion width, w is equal to the contact width. (In the absence of a "diffusion overlap" of the contact, the current flow is uniform across the diffusion area width; that is the current flow is one-dimensional). The 1-D model equations used to extract ρ_c from each of the test structures are as follows:

$$R_k = V_{14} / I_{32} = \rho_c / l^2 \qquad\qquad (3\text{-}6a)$$

$$R_e = \frac{V_{43}}{I_{12}} = \frac{\sqrt{\rho_{sh}\rho_c}}{w \, \sinh\left(l \sqrt{\rho_{sh}\,\rho_c}\right)} \qquad\qquad (3\text{-}6b)$$

$$R_f = \frac{\left(\frac{V_{56}}{I_{23}}\right) l_1 - \left(\frac{V_{45}}{I_{12}}\right)}{2 \, (l_1 - l_2)}$$

$$= \frac{V_{43}}{I_{12}} = \frac{\sqrt{\rho_{sh}\rho_c}}{w \, \tanh\left(l \sqrt{\rho_{sh}\,\rho_c}\right)} \qquad\qquad (3\text{-}6c)$$

where the subscripts in V and I denote the pad numbers shown in the diagrams, and w and l are the width and length of the contact window.

Several problems arise, however, if these equations are used to extract a value of ρ_c from practical test structures in which the contact is smaller than the diffusion area. First, if a value of ρ_c is extracted from same-type contacts with identical dimensions using two different types of structures, the extracted values often yield disagreeing estimates of ρ_c. Second, when the CBKR structure is used and contact resistance is plotted against area, a sublinear behavior is observed, instead of the expected inverse linear behavior. Third, the extracted values of ρ_c from experimental data based on the 1-D models appear to be not only contact-size dependent, but also appear to be a function of diffusion sheet resistance, ρ_{sh} (even if the active surface-dopant concentrations are identical). This is a serious problem, since the variations are often more than an order of magnitude!

To account for and explain these results, the current flow *around* the contact must be included, a 2-D model must be employed instead of the 1-D model that results in Eqs. 3-6. Such a 2-D model has been rigorously derived and is described in reference 7. This model provides a basis for extracting the value of ρ_c from any of the three contact test structures; the value of ρ_c that is obtained can then be used to calculate the value of R_{co} in contact structures of any size. When the validity of this model was being checked, it was found to yield identical values of ρ_c, even when contacts of different dimensions and from different test structures on the same wafer were used. Thus, it is claimed that when

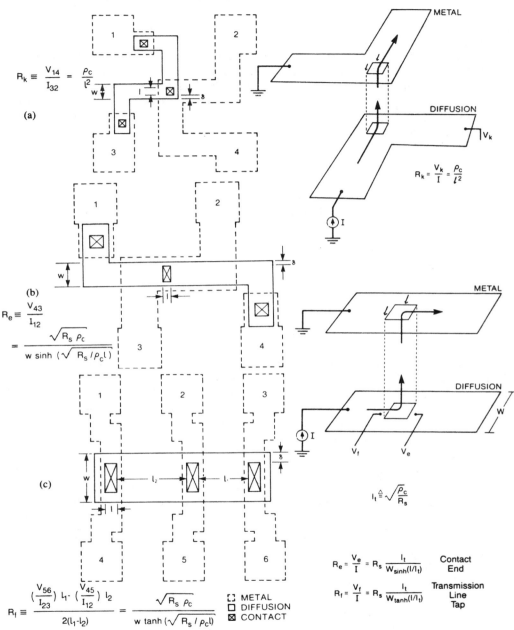

Fig. 3-8 The layouts and 3-D depictions of the three test structures used for measuring contact resistance. (a) Cross bridge Kelvin resistor (CBKR). (b) Contact end resistor (CER). (c) Transmission line tap resistor (TLTR).[7] (© 1987 IEEE).

this model is used to extract ρ_c, the values obtained are accurate, self-consistent, and independent of geometry and test structure type.

The results of the 2-D model are presented in a universal form so that values of ρ_c can be extracted without the need to perform any additional computer simulations. The method for using the results of this model to extract values of ρ_c from data obtained from the CBKR test structure is presented in the following section.

3.3.4 Procedure for Accurately Extracting ρ_c from CBKR Test Structures

The procedure for accurately extracting ρ_c using the CBKR method and the universal CBKR curves of Loh et al.[7] is described in this section. The CBKR structure has been selected to illustrate the procedure because it is the structure recommended in reference 7; similar procedures to extract ρ_c using the other test structures can also be used.

1. Needed are two sets of CBKR test structures of varying contact sizes, l (e.g., two sets of 10 contacts each, varying in length from 1 to 25 μm), with at least two diffusion region overlaps, δ, for each set of structures (Fig. 3-9). (Note that in reference 1 δ values of 1.25 μm and 2.5 μm are used to generate an exhaustive set of curves for extracting ρ_c values. Thus, if this reference is to be closely followed, these same δ values might suffice.) The diffused region under the contacts should be fabricated to closely emulate the junctions that will be built in the devices for which the contact structures are being evaluated. If contacts to both p^+ and n^+ Si will be studied, both p^+ on n and n^+ on p junctions will need to be produced. The value of the sheet resistance of the diffused layers, ρ_{sh}, is first measured.

2. After the contacts to the Si have been fabricated, the value of the Kelvin contact resistance, $R_k = V/I$, of each contact is measured (on an appropriate number of samples). Then the value of $\log_{10} (R_k/\rho_{sh})$ is calculated for each contact.

3. The value of $\log_{10} (l/\delta)$ is calculated for each contact.

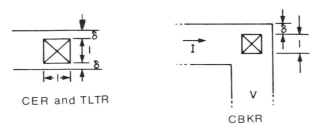

CER and TLTR

CBKR

Fig. 3-9 Diffusion region overlap in the three test structures used for measuring contact resistance.[7] (© 1987 IEEE).

4. The values of $\log_{10} (R_k/\rho_{sh})$ versus $\log_{10} (l/\delta)$ are plotted on the graph containing the universal CBKR curves (Fig. 3-10a). (In reference 1, this is done first for the set of contacts in which $\delta = 1.25 \ \mu m$, and then for the set of contacts in which $\delta = 2.5 \ \mu m$.)

5. Each of these two sets of data points should lie along one of the universal curves. Two values of $y = l_t /\delta$ should be extractable from the curves in this way.

6. Since the values of δ are known, the value of l_t can be found from $l_t = y \ \delta$.

7. Since $l_t = \sqrt{\rho_c/\rho_{sh}}$, the value of ρ_c is found from $\rho_c = l_t^2 \ \rho_{sh}$.

As an example, if R_k is obtained for two sets of contacts, and these data are used to plot $\log_{10} (R_k/\rho_{sh})$ on the universal CBKR curves (as shown in Fig. 3-10b), it is found that

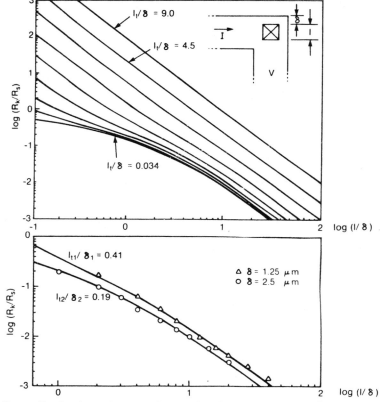

Fig. 3-10 (a) Generalized universal curves for the CBKR. Curves show l_t/δ approximately in octave steps. (b) Illustration of the use of the CBKR universal curve to extract l_t from devices with a different overlap δ on the same wafer, $l_t = 0.5 \ \mu m$ for all points.[7] (© 1987 IEEE).

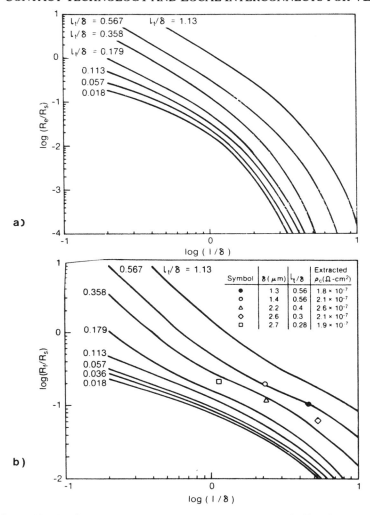

Fig. 3-11 Generalized universal curves for (a) the CER. (b) the TLTR. (Curves show l_t/δ approximately in octave steps).[7] (© 1987 IEEE).

these values of R_k lie on the curves for which y = 0.41 and y = 0.19. (Note that in this figure, only the two curves from Fig. 3-10a for the particular y values 0.41 and 0.19 are shown). If in this example the values of the diffusion-region overlaps, δ, are 1.25 μm and 2.5 μm, and ρ_{sh} = 44.4 Ω/sq, we determine that l_t = 0.5 μm for both sets of contact structures. As a result, ρ_c is found to be (0.5 μm)^2 44.4 Ω/sq = 11.1 Ω-μm^2.

The sets of universal curves from reference 7 that allow ρ_c to be extracted from TLTR and CER test structures (in the same manner as outlined above) are presented in Figs. 3-11a and 3-11b. The values of ρ_c extracted from measurements using these structures should be the same as those extracted from CBKR test structure measurements.

Therefore, if a set of TLTR or CER test structures is available, they can be used instead of the CBKR structures.

Loh, however, recommends the use of the CBKR test structures over the others. In the TLTR method, linear extrapolation is needed to eliminate the diffusion potential drop; this can result in large errors when ρ_c is extracted. In the CER method, parasitic components usually dominate the potential contribution from the contact end, again making the extraction of ρ_c difficult. The CBKR therefore emerges as the best compromise in almost all cases between extraction ease and sensitivity to errors in the diffused layer and overlap-width dimensions, especially when it is aided by 2-D simulations.

3.3.5 Reported Values of ρ_c for Various Contact Structures

Table 3-1 presents the values of ρ_c for various contact structures using the procedure outlined in the previous section. It should be noted that these values generally represent the minimum values of ρ_c that have been reported in the literature; higher values of ρ_c have been reported by other workers. In some cases, such previously reported values overstated ρ_c because the values were extracted using the simpler models that did not account for parasitic current-flow effects in the contact structures. In others, however, a lower value may also have been due to alternative process-fabrication procedures (e.g., variations in the annealing temperature, contact cleaning procedure, and doping of the Si under the contact).

Thus, while the values of ρ_c in Table 3-1 may represent the "best case" that can be achieved under well-controlled laboratory conditions, the values obtained by means of a specific contact-fabrication sequence in a production environment may be significantly higher. Therefore, if the value of R_{co} is considered to be one of the critical parameters that may limit device performance, it is important that each contact-fabrication process be characterized using the ρ_c-extraction procedure presented earlier. In particular, the

Table 3.1 Specific Contact Resistivities of Various Metal-Si Contacts

Metal-Si	ρ_c $(\Omega\text{-}\mu m^2)$	Reference
Al:Si to n^+ Si	15	6
Al:Si - TiN - n^+Si	1.0	62
Al:Si - TiN - p^+Si	20	62
PtSi to n^+Si	5	7
CVD W to n^+ Si	11	7
Al - Ti:W - TiSi$_2$ - p^+ Si	60-80	153
Al - Ti:W - TiSi$_2$ - n^+ Si	13-25	153

effect of changing (especially reducing) the doping concentration at the surface of the Si (e.g., by varying the implant doses, the implanted species, and the dopant-activation-annealing steps following implant) has been reported to have a potentially significant negative impact on the value of ρ_c exhibited by a particular contact structure.

Finally, it should also be noted that if ρ_c values of $10\ \Omega\text{-}\mu m^2$ or less can be obtained, then the value of R_{co} in devices with such contacts will probably not have much adverse effect on performance behavior, regardless of how small they are scaled. This important conclusion will be considered in more detail in section 3.8.

3.3.6 Use of a Simple Contact-Chain Structure to Monitor Contact Resistance

It should be briefly noted that an accurate value of R_{co} generally cannot be extracted from resistance data obtained from a simple *contact-chain structure*. That is, to determine R_{co} from the measurement of the total resistance of the contact chain shown in Fig. 3-12, the resistances of the metal-interconnect lines and the diffused region of the semiconductor must be subtracted from the measured reistance value. Since these resistances amount to a large fraction of the total measured value, the total contact resistance of the chain of contacts is obtained by subtracting two large numbers to obtain a smaller one. Such experimental data are therefore subject to large experimental errors.

Nevertheless, these kinds of contact chains are useful as test structures that can provide rapid monitoring of the contact-fabrication process. As long as the total measured-resistance value of such a contact chain is less than the specified value following a contact anneal step, the contact-fabrication process has probably been successfully performed.

3.4 THE EVOLUTION OF CONVENTIONAL METAL-TO-SILICON CONTACTS

Like other aspects of IC processing, the technology for fabricating metal-to-silicon contacts has had to evolve in order to keep pace with other advances in the process sequence. The technology we refer to as *conventional contact fabrication* involves the

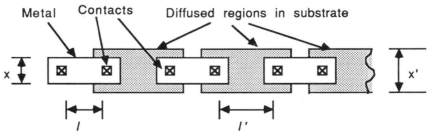

Fig. 3-12 Contact-chain. A test vehicle for monitoring contact resistance.

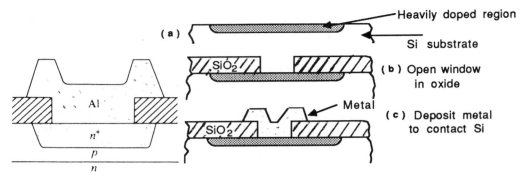

Fig. 3-13 Process sequence for forming conventional metal-semiconductor contact structure, in which the metal is deposited over a window in the passivating surface oxide.

fabrication of a contact to silicon at locations where a window has been etched in the oxide that covers and passivates the silicon surface (Fig. 3-13). While this type of contact has been used since integrated circuits were first invented, alternative contact structures have also been implemented recently. In this section we describe the evolution of the conventional contact structure, and show the reasons why alternative technologies were developed.

3.4.1 The Basic Process Sequence of Conventional Ohmic-Contact Structures to Silicon

Following is the basic sequence of steps involved in the formation of a metal-to-silicon contact structure for an integrated circuit:

1. Heavily doped regions, which extend relatively deeply into the silicon, are created in the silicon substrate in the locations where contacts are to be established.

2. A window (or *contact hole*) is etched in the oxide that covers the silicon wafer surface (typically, this is a thermal oxide or a CVD oxide/thermal oxide layer). The etching can be done using either wet or dry etching techniques.

3. Prior to deposition of the metal, the surface of the silicon is cleaned to remove the thin native-oxide layer that rapidly forms on a silicon surface whenever it is exposed to an oxygen-containing ambient.

4. The metal film is deposited onto the wafer surface and makes contact with the silicon wherever contact holes have been created in the oxide. In the simplest contact structures, the deposited metal is Al (or an Al:Si alloy). A variety of procedures have been developed to ensure that the deposited-metal films adequately cover the sidewalls of the contact windows without severe thinning (i.e., thereby achieving good *step coverage* into the contact windows).

5. After deposition, the contact structure is subjected to a thermal cycle known as *sintering* or *annealing*. The purpose of this step is to bring the metal and the silicon surfaces into intimate contact. In the case of Al contacts, during sintering the Al film will consume any residual native oxide by chemically reacting with it (assuming that this residual film is not too thick).

3.4.2 Additional Details Concerning the Processing Steps

3.4.2.1 Formation of the Heavily Doped Regions in the Silicon.
As described earlier, in order for an ohmic contact to be formed at a metal-semiconductor interface, the surface of the semiconductor must be heavily doped, with the dopants having been usually selectively introduced through ion implantation or diffusion. A masking layer is used to restrict the introduction of dopants to the desired locations on the silicon surface. Such heavily doped regions may also form part of the transistor device structure (e.g., the source and drain regions in MOSFETs and the emitter regions in bipolar transistors).

The contact resistance is inversely proportional to the surface concentration of the dopant, and thus as heavy a doping as possible would seem desirable. However, the maximum doping concentration is limited by the solid solubility at the temperature at which the dopant is introduced. Clustering effects may also reduce the concentration of the electrically active impurities below the concentration of deposited dopant atoms. For example,[8] arsenic atoms undergo such clustering if their concentration is above 10^{20} cm^{-3}. A model for As clustering is described in reference 9.

In advanced IC processing, it is necessary to form heavily doped regions that extend into the silicon to very shallow depths only. Since higher implant doses result in deeper junction depths, the dose cannot be chosen independently of the device design. The topic of shallow-junction formation and the problems associated with the fabrication of contacts to shallow junctions is explored in section 3.10.

3.4.2.2 Formation of Contact Openings (Etching).
The formation of contact holes in the oxide that covers the wafer surface is also a key step in the fabrication of contact structures. The minimum size of the contact holes is usually determined by the minimum-resolution capability of the patterning technology, because this results in the smallest device sizes. As a result the patterning technology is generally pushed to its limits in the patterning of contact holes.

When the contact holes are larger than ~2.0 μm, wet etching has often been used to open them. Even in such contact holes, however, wetting and reactant product removal can be a problem. To partially overcome the wetting problem and to ensure better material flow into the vicinity of the reacting surface, mechanical or ultrasonic agitation was introduced. Such wet-etching technology of SiO_2 films is described in more detail in Volume 1, chapter 15.

The isotropic nature of wet etching, however, made it ineffective for the patterning of smaller-sized contact holes. As a result, development of SiO_2 dry-etching processes had to be pursued (see Vol. 1, chap. 16). Dry etching, however, introduced a new set of problems, including polymer contamination, damage of the Si surface, and decrease of oxide etch rate with decrease in contact size. One consequence of the former effect was that either gases with less propensity for forming polymeric products would have to be found, or more complex processes to remove the polymers from the contact holes after etch, would have to be developed. Dry etching also exhibited selectivity problems, which became more significant as junctions grew shallower.

Several approaches were also proposed to alleviate the problems of damage to the silicon surface and redeposition of nonvolatile by-products from the chamber surfaces and wafer holders. One was to use isotropic (ion-bombardment-free) dry etching to remove surface damage and contaminants.[10] Another combined a dry-etch step (to remove the first 80% of the contact oxide layer) and a buffered HF wet etch (to etch the remainder of the oxide film). This method, however, still presents the problem of wetting in small contact holes. One proposed solution to the wetting difficulty is to use an anhydrous-HF gas to remove the final oxide layer.[11,134]

A process has been reported that uses a CHF_3/CO_2 dry-etch process to open the contact holes, followed by a two-step dry-etch cleaning procedure to remove the residual damage.[12,13] Finally, an RTP process that repairs the silicon-crystal damage in the contact hole has also been described.[14]

As the contact hole size decreases below 1 μm there is a strong decrease in etch rate with contact size dimension. To avoid long overetch times, contact holes of identical dimension would need to be used if this problem was not solved. Reference 156 describes a dry-etch process which overcomes this limitation. By adding 10% Cl_2 to a CHF_3 plasma the effective fluorine to carbon ratio in small holes is kept constant, maintaining uniform etch rates, independent of contact hole size. Nevertheless, it is very common that only one contact size is used. Larger contacts are implemented with multiple contact windows.

3.4.2.3 Sidewall Contouring of the Contact Holes by Reflow. In addition to the need to ensure that the contact holes are opened and that silicon-surface damage and contamination are minimized, it is also important to give the contact hole a shape that will result in good step coverage by the metal that is deposited into it. In general, better step coverage will be obtained if the walls of the contact opening are sloped and the top corners are rounded, and several different approaches have been pursued to achieve these desired sidewall profiles.

One of the most popular involves *reflow of the contact hole dielectric layer*. Wafers are exposed to a high-temperature step after the contact holes have been opened. This causes the CVD doped-SiO_2 layer to flow slightly, producing rounded corners and sloped sidewalls in the contact holes (see Vol. 1, chap. 6). *Borophosphosilicate glass* (BPSG) flows at the lowest temperatures (800-850°C at atmospheric pressures), and even lower-temperature reflow cycles (using high-pressure conditions) have reportedly been developed.[15] Reflow using RTP has also been investigated.[16] A recently introduced LPCVD reactor allows BPSG film deposition to be followed by an in situ flow step, which eliminates a process step and reduces the overall cycle time (although reflow must still be performed after contact holes have been opened).[141] A study of the reflow mechanisms in BPSG films presented in reference 138. It indicates that reflow in BPSG film will cause sidewall overhang and reentrant angles at the base of the contact hole if the aspect ratio of the contact hole is larger than 0.393.

It has also been mentioned that it can be useful to grow a thin-oxide layer over the contact hole following patterning, but prior to reflow. The function of this oxide is to prevent autodoping of the contact regions during reflow by the BPSG film. This oxide would be grown at a lower temperature (e.g., 8 minutes in steam at 800°C) than the reflow temperature (e.g., 900°C).[135] Alternatively, the reflow can be done in a dry O_2 ambient to simultaneously grow the thin protective-oxide layer.[140]

H. Ozaki et al., however, have described how the B_2O_3 in the BPSG film can react with the PH_3 (used as part of the gas ambient during reflow to redope the contacts in NMOS technology) to form a volatile compound of boron. This boron is then taken up by the oxide film that is grown in the contact openings during the reflow, resulting in a much thicker oxide in the contact hole following reflow.[17] Consequently, another suggested approach for preventing contact-degradation effects during the reflow step is the deposition of diffusion-barrier materials (e.g., silicon nitride or TiN) over the opened contacts prior to the reflow step.[15]

3.4.2.4 Sidewall Contouring by Etching.

Other sidewall-contouring processes involve the etching procedure used to pattern the contact holes. The first of these uses a wet etch (which is isotropic) to partially etch the oxide, and follows this with an anisotropic dry etch.[18] The method yields a contact hole whose profile is sloped at the top but is vertical at the bottom. While good step coverage can be achieved in some applications, difficulty may be encountered in obtaining good wetting, especially for very small contacts. In addition, the sidewall profile may still have a sharp corner at the upper edge, which gives rise to step-coverage problems. A variation of this method is to use a triple layer (oxide/nitride/oxide). The top oxide is wet etched to provide a sloped contact-hole sidewall, and the nitride serves as an etch stop. The remaining nitride and oxide are then etched with a vertical dry-etch step.[19]

Another general approach involves the controlled erosion of photoresist (PR) that has been baked to produce a sloped PR wall.[20,21] In this method, PR images of the contacts are exposed and developed using standard lithographic techniques. Following develop, the resist images are subjected to a postdevelop bake of ~150°C (Fig. 3-14a). The resist flows during the bake, relaxing the vertical resist profile. Etching the resist and oxide at approximately the same rate replicates the tapered-resist profile into the contact sidewall. While the bake-to-slope process is quite adequate for large contacts or vias (>2 μm), it is not easily scalable to smaller geometries. For such smaller contacts resist baking step becomes critical. Too little baking results in vertical contact profiles, while excessive baking can result in closed contacts. Furthermore, continued etching of

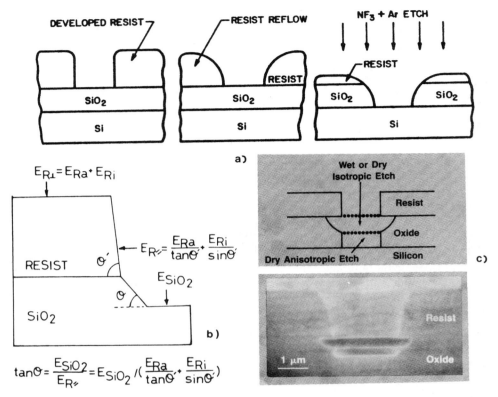

Fig. 3-14 (a) Tapered contact-hole etching by reflowing resist profile with a resist bake, followed by a dry etch to transfer the resist profile into the oxide. (b) Tapering of contact-window sidewall by optimizing the lateral to vertical etch rates of the resist mask.[25] Reprinted by permission of the publisher, The Electrochemical Society, Inc. (c) Sloping oxide sidewalls using a triode dry etcher.[152] Reprinted with permission of Solid State Technology.

the contact results in continued growth in the contact diameter (although this effect can be alleviated somewhat by using a two-step etch process).[22]

Several other, more controllable resist-erosion techniques have subsequently been described. In one, the effect of resist faceting (see Vol. 1, chap. 10) caused by the preferential sputtering of the resist corners was used to produce tapered-contact sidewalls.[23] Another method, developed by Jillie et al.,[24] uses a single-wafer etcher, and a two-step etch process to etch a multilevel-resist layer (i.e., consisting of a thin positive resist, an antireflection coating, and a thick PMMA). Good control over the sidewall slope has been reported, and a reflow step has also been used to round the top corners of the sidewall. The drawback to this approach is increased process complexity.

A third resist-erosion approach was reported by Kudoh et al.[25] In this procedure, etching of the contact oxide is carried out with a PR mask that has vertical sidewalls.

Oxygen is added to the CHF_3 gases being used to etch the oxide. The oxygen attacks the PR at a controlled rate, thereby producing lateral as well as vertical etching of the PR mask. More of the top SiO_2 is slowly exposed as vertical etching of the SiO_2 proceeds, and a sloped oxide sidewall is produced (Fig. 3-14b). Taper angles of 40-85° were obtained by varying the oxygen concentration.

Another technique for tapering contact-hole sidewalls without using resist erosion involves implantation of the surface of the oxide following deposition. This damages the top layer of the oxide and causes it to etch more rapidly, even when dry etching is used. The faster- etching top-surface layer causes a slope in the sidewalls.

A final group of processes utilizes two or more dry-etch steps to obtain tapered-sidewall profiles. In one example, a high-rate isotropic SiO_2 etch (i.e., with a lateral-to-vertical etch-rate ratio of 0.9 to 1 being exhibited for doped oxides) is used to etch the top portion of the contact-oxide layer, and an anisotropic-etch process is used to remove the oxide from the bottom of the contact hole.[26] In another, a downflow etcher operated at 2.45 GHz is used to etch part of a doped SiO_2 layer in an isotropic manner using a $CF_4 + O_2$ mixture; this is followed by an RIE step to give a vertical profile for the bottom portion of the layer.[27] A method that uses a tri-electrode dry-etch chamber for this type of process is described in reference 152 (Fig. 3-14c).

The problem of end-point detection for contact holes can also be difficult. That is, if a timed etch is used, a sufficient overetch must be allowed to ensure that all the contacts are opened. However, this demands a high selectivity to the Si to prevent too much Si from being consumed during the overetch. End-point detection is difficult, because the total area of the contacts being etched is significantly smaller compared to other layers. A technique for electronically enhancing the weak end-point signal produced when the contacts are finally opened is described in reference 144.

3.4.2.5 Removal of the Native-Oxide Layer Prior to Metal Deposition. Before the metal is deposited into the contact openings, it is necessary to ensure that the exposed silicon surface is as free as possible of contamination or of a native-oxide layer. Silicon will grow a thin native oxide within a matter of seconds upon exposure to an oxygen-containing ambient or an oxidizing solution. This native oxide can represent an impediment to current flow through the contact interface, result-

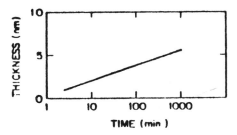

Fig. 3-15 Native oxide growth rate on Si exposed to room air.[28] Copyright 1984. Reprinted with permission of the AIOP.

ing in high resistance for ohmic contacts. Oxide films of minimum thickness   (2–5Å) do not present a problem, since many contact metals (e.g., Al, Ti, Pt) can consume such thin oxide layers during the sinter step. Nevertheless, as shown in Fig. 3-15, the native-oxide layer can grow thicker than this value in 30 to 60 minutes* if the Si is exposed to atmosphere at room temperature.[28,139] Consequently, techniques must normally be employed to remove native oxide layers on Si, and the metal must be deposited quickly enough after such premetal cleaning that the film does not have enough time to regrow.

A variety of techniques have been suggested for cleaning the Si prior to metal deposition. Polymers deposited during a dry-etch step used to etch the contact oxide are typically removed through a final oxygen-ashing step. This step may, however, may create a thin oxide layer on the exposed Si. The most widely used method for removing native-oxide layers on the Si involves dipping the wafer in a dilute H_2O:HF (100:1) solution for ~1 minute, followed by rinsing and drying and immediate insertion into the sputter or evaporation chamber. (The HF dip also serves to remove heavy-metal contamination that might occur as a result of the dry-etch process.)

It is acknowledged that this method is not perfect, since some SiO_2 will regrow on the Si during the rinse-and-dry step, and inadequate wetting may prevent complete removal of the native oxide in small-geometry contacts. Therefore, a technique of sputter-etching the contacts in the sputter chamber and depositing the metal before reexposing the contacts to atmospheric conditions is also frequently used in addition to the HF dip. Most sputtering systems currently provide the capability for performing such sputter etching.

However, some concerns have also been raised about sputter etching. For small, high-aspect-ratio contact holes, sputtering and redeposition of material from the contact sidewalls and the wafer surface may cause more oxide to be deposited onto the Si than is removed through the sputter-etch process (Fig. 3-16).[29] Two alternative procedures to overcome these concerns have therefore been suggested. The first involves an in situ, chemically driven dry-etch step to remove the native oxide. While this would eliminate the resputtering effect, it would require that sputtering equipment be designed with this extra capability; no such products have yet been commercially introduced. In the second approach, an anhydrous-HF gas could be used to remove the native oxide, as described in a previous section.

The effect of "dirty" Si surfaces prior to Al deposition has been studied by Faith et al.[30] These researchers showed that if the Si surface is not successfully cleaned prior to Al deposition, the post-sinter contact resistance will be an order of magnitude higher than if an effective pre-metal clean had been used. They also noted that clean contacts

---

*A recent report indicates that both moisture and oxygen must be present in air in order for the oxide to grow. Furthermore, if oxygen is present in DI water, a native oxide will also grow when Si wafers are immersed in it. Since the growth rate of native oxide decreases with the concentration of moisture, this may provide an approach to the control of native-oxide films on Si.[147] For example, virtually no native oxide was observed to grow during a seven day period of time when Si wafers were stored in air containing < 0.1 ppm moisture.

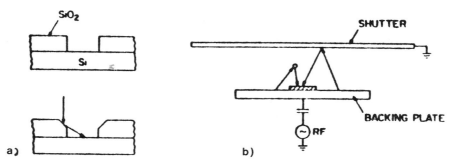

Fig. 3-16 Sources of contamination of Si surface at the bottom of the contact hole as a result of RIE. (a) Faceting occurs during anisotropic etching of contact openings. Material sputtered from the facet deposits in the contact opening at a rate that may exceed the rate of removal of the material from the bottom of the opening. (b) Backscattering of material sputtered from the backing plate may occur due to collisions with gas molecules of the glow discharge or from reflection from the shutter.[28] Copyright 1984. Reprinted with permission of the AIOP.

exhibit a lower contact resistance than dirty ones *prior* to the sinter step, and that this effect could be exploited to test the contacts prior to sintering. If a dirty interface were detected at that point, a decision to strip and rework the metal might still be possible, whereas this option would no longer be available once sintering had been performed. As described in the following section on contact sintering, ion-beam mixing has also been proposed as a way to disperse any native-oxide layers at the interface.

3.4.2.6 Metal Deposition and Patterning. The major issue in the deposition of metal for fabricating contacts is ensuring that adequate step coverage is obtained into the contact holes. When contact-hole sizes are comparable to the oxide thickness (i.e., when the holes have high aspect ratios), good step coverage can be difficult to achieve. The deposition process, as well as the profile of the contact-hole sidewalls, can significantly impact the quality of the step coverage. Several aspects of the metal- deposition procedure can also play a role in this issue.

First, the type of process selected for deposition is significant. Some CVD processes can completely fill high-aspect-ratio contact holes, even those with nearly vertical sidewalls, while physical vapor deposition (PVD) methods are not apt to fill the holes so well. This advantage has been exploited in filling contact holes with CVD W (selectively and through blanket deposition). Polysilicon and selectively grown epitaxial layers of Si are other CVD processes that have been reported for such contact-filling applications.

If films are needed that cannot be deposited by CVD, the conditions of the PVD process can be selected to give improved step coverage (see chap. 4 in this volume, and Vol. 1, chap. 10). For example, heating of the wafers to ~300-350°C has been shown to significantly improve the step coverage of sputtered Al films into contact holes. Full planarization (i.e., complete filling of the holes) has also been demonstrated by applying even higher temperatures during sputter deposition (400-500°C). Laser planarization of Al films following deposition has also been developed to improve step coverage into contacts (see chap. 4).

In addition, sputtered Ti:W films have also been reported to give excellent step coverage. This method can be used to enhance the overall step coverage of contact structures – for example, if such a Ti:W film is used under a layer of Al. On the other hand, the problem of film thinning at the bottom corners of the contact window can produce reliability problems in contacts. Even if the top layer of Al appears to have been deposited with adequate step coverage, the diffusion-barrier layer, for instance might have been deposited with thin spots at these edges (Fig. 3-17). As a result, the contact could fail because the barrier layer was not thick enough at these locations to survive the sintering steps following deposition of the metal (see section 3.5).

The issues dealing with the patterning of the metallization layers used in contact structures are described in further detail in Volume 1, chapter 16.

3.4.2.7 Sintering the Contacts. As described earlier, the sintering or annealing of contacts (i.e., the alloying of contacts through treatment at an elevated temperature) is performed to allow any interface layer that exists between the metal and the Si to be consumed by a chemical reaction, and to allow the metal and Si to come into intimate contact through interdiffusion. The details of the various reactions that occur at specific metal-Si interfaces during the sinter cycle are described in the following sections examining the properties of each contact type. In this section we merely mention the several methods by which such intermixing of materials at the interface is performed.

The traditional method carries out the sinter step in a diffusion furnace, usually at 400-500°C for 10 to 30 minutes in the presence of H_2 or a *forming gas* {a mixture of H_2 (5-10 at%) and N_2 (95-90 at%)}. Such gases are used because many metals are

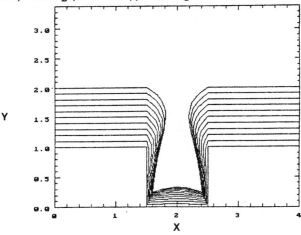

Fig. 3-17 Simulated metal coverage of a contact hole with vertical sidewalls, showing thinning at the bottom corners.

sensitive to oxygen at elevated temperatures, and because this step is also used to anneal out the surface states (i.e., interface traps at the oxide-Si interface) by tying up the dangling Si and oxygen bonds. The hydrogen is responsible for passivating and deactivating the interface traps.

More recently, alternative methods – including RTP, laser annealing, isothermal e-beam sintering, and ion-beam mixing – have also been investigated for performing this step. Although ion-beam mixing does not subject the contacts to elevated temperature, the chemical outcome is very similar to that in an alloying process. The technique involves the implantation of a relatively high dose of ions at an energy that is high enough to allow full penetration through the metal layer and into the Si substrate. If the species that is implanted is the same as the dopant of the junction to be formed, the junction and the contact can be fabricated simultaneously.[31] Normally, the implant is followed by a furnace step (conventional or RTP) to repair the implant damage, and possibly to form a silicide or to distribute the implanted dopant.

3.4.3 Aluminum-Silicon Contact Characteristics

The contact structure consisting of pure Al deposited directly onto Si was adopted in the earliest stages of silicon technology primarily because of its simplicity, but also because it exhibits some other advantages. Al and Al alloy thin films are still the most common interconnect materials in silicon ICs today. Al-to-Si ohmic contacts can be fabricated with low values of contact resistance to both n^+ and p^+ Si, and both the contact and the interconnect structure can be fabricated through a single deposition step. Al is desirable as an interconnect material because it exhibits low resistivity ($\rho_{Al} = 2.7$ $\mu\Omega$-cm) and offers excellent compatibility with SiO_2 (i.e., Al will readily react with SiO_2 to form a thin layer of Al_2O_3 at the interface, thus promoting strong adhesion between the Al and the SiO_2). The Al thin films used for interconnects are typically 0.5-1.5 μm thick.

However, the Al-to-Si contact exhibits some poor contact characteristics and also introduces some limitations into the processing sequence. Because the melting point of Al (660°C) and the eutectic temperature of Al and Si mixtures (577°C) are so low, Al is a problematic material with respect to the maintenance of contact stability. Thus, the simple contact structure of pure Al-to-Si has had to be modified, especially for VLSI applications. In addition, Al must be introduced into the IC process sequence after all high-temperature processing has been completed.

3.4.3.1 The Kinetics of the Al-Si Interface During Sintering.
We noted earlier that a sinter step is performed after a contact metal film has been deposited and patterned. In the case of Al-Si contacts, such sintering causes the Al to react with the native-oxide layer that forms on the silicon surface. In the case of e-beam-evaporated Al films on MOS devices, sintering also anneals out the radiation damage at the SiO_2/Si (gate oxide/Si) interface from the x-rays generated by the evaporation process; see Vol. 1, chap. 9. As the Al reacts with the thin SiO_2 layer, Al_2O_3 is formed, and in a good ohmic contact, the native oxide is eventually completely

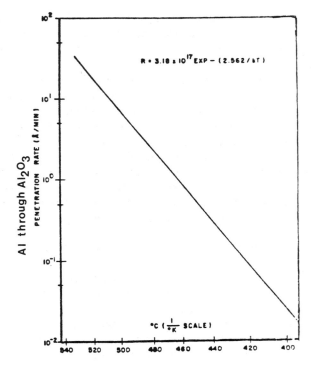

Fig. 3-18 Rate of reaction for Al + SiO$_2$ --> Al$_2$O$_3$ + Si.

consumed. Thereafter, Al diffuses through the resultant Al$_2$O$_3$ layer to reach the Si surface, forming an intimate metal-Si contact.

Note that Al must diffuse through the Al$_2$O$_3$ layer to reach the remaining SiO$_2$. As the Al$_2$O$_3$ layer increases in thickness, it takes longer for the Al to penetrate it. Thus, if the native-oxide layer is too thick, the Al$_2$O$_3$ layer eventually also becomes too thick for Al to diffuse through it. In this case, not all of the SiO$_2$ will be consumed, and a poor ohmic contact will result (i.e., one that exhibits a high specific contact resistivity).

Figure 3-18 shows the penetration rate of Al through Al$_2$O$_3$ as a function of temperature. For acceptable sinter temperatures and sinter times (e.g., 450°C and 30 min.) the thickness of the Al$_2$O$_3$ should be in the 5-10 Å thickness range. Since the maximum Al$_2$O$_3$ thickness is of the order of the thickness of the native oxide that is consumed, an approximate upper limit to the allowable thickness of the native-oxide layer is set. The longer the silicon surface is exposed to an oxygen-containing ambient the thicker the native oxide will be (Fig. 3-15). This explains why surface-cleaning procedures in most contact processes are performed just prior to loading of the wafers into the deposition chamber for metal deposition.

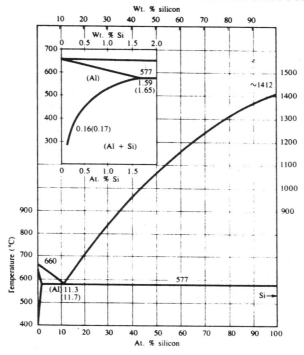

Fig. 3-19 Phase diagram of the aluminum-silicon system. The silicon-aluminum eutectic point occurs at a temperature of 577 °C. At contact-alloying temperatures between 450 and 500 °C, aluminum will absorb from 0.5 to 1% silicon. Copyright 1958, McGraw-Hill Book Company, with permission from ref. [161].

Since the use of a sinter step is inevitable when Al-Si contacts are fabricated, the other effects of this thermal cycle on the contact structure must also be considered. The phase diagram of the aluminum-silicon system (Fig. 3-19), indicates that the solubility of silicon in aluminum rises as the temperature increases (e.g., it is ~0.5 at% at 450°C, and ~1 at% at 500°C). This means that if a pure Al film were heated to 450°C and a source of silicon were provided, the Al would take up the silicon in solution until a concentration of 0.5 at% of Si was reached. The substrate serves as such a source, and and the silicon will enter the Al by diffusion at such elevated temperatures. If silicon diffusion were slow and the contact was maintained at the elevated temperature for only a short time, not much silicon would enter the Al, so there would be no problem.

In the actual case of Al thin films (which are polycrystalline in structure), the grain boundaries of the polycrystalline Al film provide very fast diffusion paths for the Si at temperatures above 400°C. Figure 3-20 shows the diffusivity of Si in Al as a function of temperature, for both bulk and thin-film Al.[32] The diffusivity in the thin-film Al is seen to be much faster. As a result, if a large volume of Al is available, a significant quantity of the Si from below the Al-silicon interface can diffuse into the Al film. Simultaneously, the Al from the film will move rapidly to fill the voids created by the departing Si (Fig. 3-21). If the penetration of the Al is deeper than the *pn*-junction depth below the contact, the junction will exhibit large leakage currents or even become electrically shorted. This effect is referred to as *junction spiking*.

The depth to which Si is consumed and Al penetrates into the Si substrate can be calculated as a function of the thickness and width of a long Al line, the contact area, and the temperature and time of the sintering. As shown in Fig. 3-20, at temperatures in excess of 400°C the diffusion length in one minute exceeds 10 μm. The effect of the diffusion with respect to an actual contact structure is illustrated in Fig. 3-21. As a result of such diffusion, the silicon in a 16 μm^2 Al-Si contact would be consumed to a depth of ~0.3 μm if the contact was exposed to a temperature of 500°C for 30 min (assuming uniform Si consumption across the contact area).[33] Consequently, junctions with depths of less than 0.3 μm would be shorted, rendering the devices inoperable.

In fact, since the Si consumption is not uniform across the entire contact area, the situation is even more severe than is portrayed by the above example. The native oxide that exists at the Si surface is not uniform in thickness but has local thin spots, or defects, as shown in Fig. 3-21. Thus, when the native oxide is consumed during sintering, it will first be removed at these thin spots, and Si will therefore be able to dissolve into the Al at these points. As a result, Al penetration, or *spikes* can occur to a depth of more than 1 μm. In (100)-oriented Si substrates, such spikes tend to move perpendicularly to the surface and are bounded by (111) planes. In addition, Al spiking occurs preferentially at the contact-hole edges, where stresses present in the Si lead to enhanced dissolution of the Si during contact sintering. If the Al is stripped after such sintering, the Si pits appear as shown in Fig. 3-22.

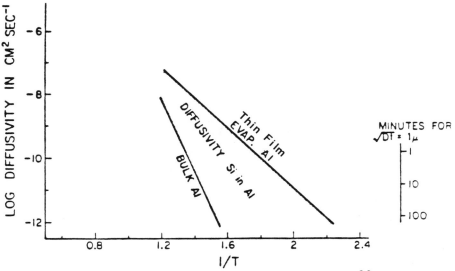

Fig. 3-20 Diffusivity of silicon in bulk and thin film aluminum.[32] Copyright 1974. Reprinted with permission of the AIOP.

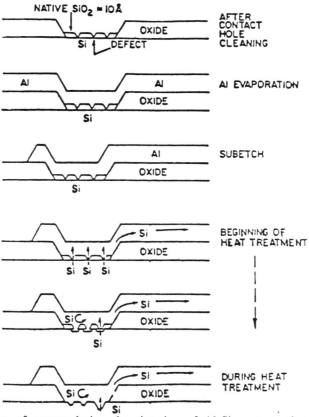

Fig. 3-21 Sequence of events during the sintering of Al-Si contacts that leads to Al junction spiking.

The pit dimensions is also observed to increase as the area of the contact hole is decreased.[34] This effect imposes a severe limitation on the usable combination of

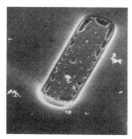

Fig. 3-22 SEM micrograph of Si pitting that exists after sintering of an Al-covered contact. (Al has been stripped after the sinter step.)[165]

junction depth, x_j, and contact-hole size, d, or $x_j \gg 4.7d^{-0.8}$. More concisely, since junction depths of less than 200 nm are being utilized in advanced ICs, this type of contact cannot be used.

3.4.4 Use of Aluminum-Silicon Alloys to Reduce Junction Spiking

To alleviate the problem of junction spiking at contacts without abandoning the advantages of the simple Al-to-Si contact structure, Si can be added to the Al film as it is deposited. This is typically accomplished by sputter depositing the film from a single target containing both Al and Si, although coevaporation of Si and Al has also been used. If enough Si is added to the Al so that the concentration exceeds the Si solubility at the maximum process temperature, diffusion of Si into the Al will no longer occur, and junction spiking can be prevented. The amount of Si that must be added to accomplish such *presaturation* of Al with Si therefore depends on the maximum process temperature to which the contact structure will be exposed. Since this maximum temperature is normally less than 500°C, in most cases slightly more than 1 wt% Si is added. The change in resistivity of the Al, $\Delta\rho$, caused by such additions of Si is very small (i.e., $\Delta\rho$ for each percentage of Si added is only 0.07 $\mu\Omega$-cm, and $\rho_{\text{pure Al}} = 2.7$ $\mu\Omega$-cm).

Such Al:Si alloys were widely adopted for fabricating the contacts and interconnects of NMOS integrated circuits. To improve the reliability of such contact structures, deeper junctions immediately below the contact opening were also employed in some processes. (These deep junctions were created by redoping the Si with phosphorus after the contacts were opened. Such redoping can be accomplished during the reflow step, by performing the reflow in an $N_2/PH_3/O_2$ ambient.[17] Doing so allows the source/drain regions near the channel to be kept shallow for good device performance, while providing a deep junction under the contact to prevent junction spiking. This method cannot of course be used in the shallow source/drain regions of devices fabricated in the wells of CMOS structures (see chap. 6 on CMOS device structures).

Although the use of Al:Si instead of pure Al prevented junction spiking, it also introduced other problems. During the cooling cycle of a thermal anneal, the solid solubility of silicon in Al decreases with decreasing temperature. The aluminum thus becomes supersaturated with Si, which causes nucleation and growth of Si precipitates out of the Al:Si solution. Such precipitation occurs both at the Al-SiO$_2$ interface and the Al-Si interface in the contacts (preferentially around the contact hole edges where solid-phase epitaxial growth occurs easily). These precipitates consist of p-type silicon (since Al is a p-type dopant in Si), doped with Al to a concentration of ~5×10^{17} cm^{-3}.

If such precipitates form at a p^+Si-Al contact interface, they do not degrade the contact properties. On the other hand, if they form at the contact interface to n^+Si, an undesirable increase in contact resistance results. In practice, for contacts sizes of 3 x 3 μm or greater, such effects are not significant, and Al:Si alloys can be used effectively. For smaller contacts, however, the increase in resistivity can be excessive.

In addition, the Si precipitates formed within the Al interconnect lines can increase the susceptibility of the lines to electromigration failure. That is, in narrow Al lines the precipitates can be large enough to obstruct a large fraction of the cross-sectional area of the metal line at random locations.[35,36] The size of the Si precipitates in Al:1% Si alloys is about 0.4 μm; if the films are cooled slowly, the precipitates can grow to be as large as 1.5 μm. At locations where such precipitates are formed, a large flux-divergence in the current is produced. This can lead to early failure of the conductor by an electromigration-induced open circuit.

Al-Si contacts exhibit two other important limitations that would prevent their use in advanced VLSI applications, even if the problems above could be overcome. First, in IC technologies in which small-sized contacts and shallow junctions must both be used, Al:Si alloys cease to be completely effective in preventing junction spiking. The precipitation that occurs during the cooldown from the first thermal cycle (e.g., the sinter step) can cause the Al near the contact interface to contain an Si concentration that is lower than that needed for solid solubility during subsequent processing steps. When the wafer is heated again (e.g., during the deposition of the passivation film or during the die-attach or packaging process), Si from the substrate can diffuse into the Al.

Second, when high current densities are passed through Al-Si contacts, the electron current leads to the transport of Si from the contacts into the Al. This leads to spiking of the junction at the negative terminal and deposition of Si at the positive terminal. Such susceptibility to *contact electromigration failures* also increases as the contact area is reduced.

In summary, the simple Al-Si contact structure that was used extensively for NMOS IC fabrication has become increasingly less suitable for advanced CMOS and bipolar integrated circuits, even if an Al:Si alloy is used for the metallization. As a result, more elaborate contact structures that can overcome the limitations of the Al-Si contact have had to be developed.

3.4.5   Platinum Silicide-to-Silicon Contacts

An early approach to overcoming the junction-spiking problem of Al-Si contacts used a platinum-silicide layer (PtSi) between the Si and Al in the contact structure. Not only could such contacts withstand temperatures of 350-400°C without junction spiking, but they also exhibited two other advantages over Al-Si contacts: (1) they offer a self-alignment capability, which restricts the formation of the PtSi layer to the region in the contact hole without there being a need for additional masking steps; and (2) they exhibit excellent electrical characteristics, due to the mechanism by which the silicide is formed. That is, the PtSi layer is formed by reacting a Pt layer in contact with Si at low temperatures (250-400°C). Rapid diffusion and reaction of the Pt occur at such low temperatures, allowing the silicide to be produced. During the growth process, the original contaminants from the Pt-Si interface are swept to the surface of the newly formed silicide, and the PtSi-Si interface is buried beneath the original Si surface. As a result, an absolutely clean and intimate silicide-silicon contact is formed.[39]

Platinum-silicide layers are able to provide low contact resistivity for ohmic contacts to both n^+ and p^+ Si (approximately a factor of two lower than the Al-to-Si contacts described in an earlier section), and they provide a film that is stable with respect to the silicon substrate for process temperatures in excess of 500°C. They also exhibit relatively high conductivities. For example, PtSi films of 60-nm thickness exhibit a sheet resistivity of ~34 $\mu\Omega$-cm.

The first application of PtSi contacts to silicon technology was reported by Hosack in 1972.[37] Since that time, numerous other noble and near-noble silicides (e.g., Pd_2Si, $CoSi_2$, and NiSi), and refractory metal silicides ($TiSi_2$, WSi_2, $MoSi_2$, and $TaSi_2$) have also been investigated for this purpose. The electronics industry, by and large, early on adopted PtSi-Si for Schottky diodes and ohmic contacts to bipolar transistors. (PtSi Schottky contacts will be described in section 3.7.) The properties and fabrication technology of $CoSi_2$ and $TiSi_2$ will also be described later, as these two silicides have been incorporated into advanced contact structures. A recent study on the applicability of Pd_2Si ohmic contacts has indicated that they may also offer many attractive benefits for VLSI applications.[38]

3.4.5.1 Process Sequence Used to Form PtSi-Si Contacts. The process sequence used to form the PtSi layer in the contact hole (Fig. 3-23) includes the following steps:

1. After an opening is etched in the SiO_2 layer, the surface of the silicon is subjected to an additional cleaning step, such as a dip in dilute HF just prior to loading of the wafers into the sputter chamber, or preferably, an in situ sputter etch. This step is important, because a smooth PtSi-Si interface of uniform penetration requires a Si surface that is free of contaminants and has a minimal native oxide. (i.e., planar, very shallow PtSi-Si interfaces are needed to prevent junction penetration of shallow junctions by the growing silicide layer.) Figure 3-24 shows the effect of the surface quality on the smoothness of the PtSi-Si interface; it can be seen that a contaminated Si surface leads to a PtSi layer of irregular depth.[40,41] It has also been reported that as long as the native oxide on the Si surface is less than 20 Å thick, the PtSi layer will form and, in doing so, will break up the oxide layer and sweep the oxygen to the PtSi surface. On the other hand, thicker native-oxide layers will inhibit the formation of PtSi.

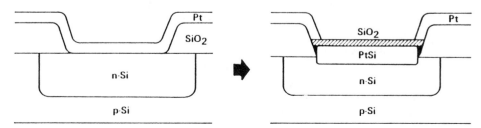

Fig. 3-23 Formation of self-aligned PtSi-Si contacts.

Fig. 3-24 The effect of the surface quality on the smoothness of the PtSi-Si interface. It can be seen that a contaminated Si surface leads to a PtSi layer of irregular depth. (a) shows the rough PtSi -Si interface grown from a severely contaminated surface. (c) shows a smooth, uniform PtSi-Si interface that results when the PtSi is formed on a clean Si film.[40] Copyright 1984. Reprinted with permission of the AIOP.

2. A thin Pt film is then sputter deposited onto the wafer surface to a thickness of 30-60 nm. This thickness range produces silicide layers thick enough to provide good contact properties but without consuming an excess amount of Si below the contact. In forming the silicide layer, Si is consumed to about the same depth as the thickness of the Pt film (e.g., 50 nm of Pt will react with ~60 nm of Si to form a 100-nm-thick PtSi film). A slow sputter-deposition technique (such as *rf sputtering*) is commonly employed in order to maintain adequate thickness control for such thin Pt layers. A clean Si surface (i.e., with a native oxide no thicker than ~10 Å) and a deposition process that does not allow oxygen to be incorporated into the Pt film are both critically important, since the growth of the PtSi layer can be inhibited by the presence of oxygen.

3. The silicide is then formed by heating the wafers in an N_2 ambient. During the annealing process, the Pt reacts with the Si in the contact hole to form

PtSi, but it does not react wherever it is in contact with the SiO_2. Toward the end of this annealing step, some oxygen is introduced into the furnace. This causes a thin film of SiO_2 to grow on the surface of the just formed PtSi, but no such oxide is grown on the unreacted Pt regions. Various temperatures have reported for the annealing process, but 550°C for 10 minutes is typical. Chang et al., however, have reported that a three-temperature sequence at 200°-350°-550°C yields PtSi films with improved properties. (The formation and properties of the passivating-oxide film on top of the PtSi under various annealing conditions are also described by Chang.)[42] An RTP process for reacting the Pt has also been published.[41]

4. The unreacted Pt is then stripped from the wafer surface by etching in a dilute aqua regia solution at 85°C. The aqua regia, however, does not attack the PtSi because of the protective layer of SiO_2 that was grown on it during the last part of the annealing process. Note that this SiO_2 layer must be removed prior to deposition of the next layer of metal onto the (sputter etching, e.g., has been effectively used for removal of this film).[43]

3.4.5.2 Limitations of the PtSi-Si Contact Structure.

Junctions under PtSi-Si interfaces exhibit no degradation (e.g., leakage or shorting) up to temperatures of 800°C. Unfortunately, when an Al layer is deposited onto the PtSi, such junctions undergo junction spiking failures if temperatures exceed 350°C. This occurs because above this temperature, the Al begins to react with the underlying PtSi, causing it to decompose and form various Pt-Al-Si intermetallic compounds. Once the aluminum-silicide reaction front reaches the silicide-Si interface, silicon pits and aluminum spikes are created almost immediately, and shallow junctions rapidly become shorted.

In the early days of IC processing, when device junctions were 0.5 μm in depth (or deeper) and the contact areas were >10 μm^2, such (Al-PtSi-Si) contacts were nevertheless able to maintain adequate junction stability if the annealing temperatures following metallization were limited to 400°C. When the device junctions became shallower than 0.25 μm, however, the Al-PtSi-Si contact (and all other near-noble silicide contacts) could no longer prevent junction spiking at annealing temperatures of 400°C. As a result, it became necessary to introduce yet another barrier layer between the PtSi and the Al to allow the contacts to survive temperatures exceeding 400°C (see Fig. 3-26a). Such so-called *diffusion barriers* will be described in the next section.

PtSi also has two other characteristics that also need to be mentioned. First, because PtSi layers agglomerate at temperatures above 800°C, post-Pt deposition temperatures are limited. Second, the formation of the PtSi occurs in a lateral, as well as a vertical direction. Thus, a 100-nm-thick PtSi layer also penetrates ~100 nm under the edge of the contact oxide window. In the case of a MOSFET with a gate length of 1.0 μm, this would reduce the nominal gate length by 20%. Several approaches have been investigated to minimize this effect, including the deposition of Si and Pt together and the deposition of an alloy of Pt and W (Fig. 3-25). These approaches are reviewed by Tu in reference 44.

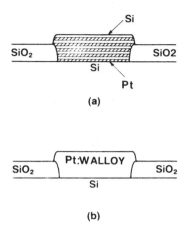

(a)

(b)

Fig. 3-25 Two approaches investigated to minimize the effect of excess Si consumption during PtSi formation: (a) the deposition of Si and Pt together, and (b) the deposition of an alloy of Pt and W.[44] Copyright 1981. Reprinted with permission of the AIOP.

3.5 DIFFUSION BARRIERS

3.5.1 Theory of Diffusion Barrier Layers

The intermixing of materials from two layers in contact (such as Al and Si) can be prevented by sandwiching another material between them (Fig. 3-26). The role of this third material is to prevent (or ar least retard) the diffusion of the two original materials into each other, or to resist the tendency of a chemical reaction to form a new phase between the adjoining materials. In practice, the available diffusion barrier materials are not perfect: they are capable of extending the life of devices only to some degree, not indefinitely. The efficiency of such a diffusion barrier is therefore determined by how long it can extend the lifetime of the contact structure under various thermal treatments, compared to its lifetime without a diffusion barrier. End-of-contact-life (or *contact failure*) occurs when the junction under the contact is short-circuited (e.g., by junction spiking) or when the contact exhibits open-circuit or high-resistance behavior.*

A diffusion barrier used in IC fabrication is a thin film (i.e., material X between materials A and B in Fig. 3-26b) inserted between an overlying metal and an underlying

---

* Note that diffusion barriers are also used in many other applications besides microelectronics. For example, the paint on wood, the galvanizing on metal, and the wax coating on cardboard containers all serve the same purpose – that is, such barriers are used to protect the underlying materials by preventing diffusion of substances, such as moisture, that would degrade their properties.

semiconductor material, or between two metals in multilevel metal systems. Such diffusion barriers should have the following characteristics:

• The diffusion of A and B through the barrier should be low.

• The barrier layer should be stable in the presence of A and B, so that very little of the barrier itself is lost through such mechanisms as diffusion into A or B or reaction with A or B.

• The barrier material should adhere well to A and B and should have a low contact resistivity to A and B.

• The barrier should either be compatible insofar as the coefficients of thermal expansion of A and B are concerned, or it should be very resistant to thermal and mechanical stresses.

• Good electrical conductivity (maximum allowable resistivity ~200 $\mu\Omega$-cm).

In practice, the contact-metallization systems of mainstream IC technologies are subjected to maximum temperatures of ~500°C in the course of the final IC fabrication and packaging processes. As a result, some diffusion barriers need to maintain the integrity of contacts at 500°C for 30-60 minutes in order to be useful for such IC technologies. It should be noted that it is difficult to get all of the above listed characteristics in a single material. Since the ideal material, X, is not yet available, compromises must be sought among these characteristics when a diffusion-barrier material is selected.

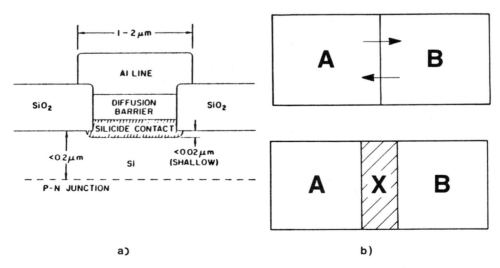

Fig. 3-26 (a) Addition of another layer of material (the diffusion barrier) between the Al and the silicide (or just the Si) to prevent degradation of the contact properties under high temperature processing. (b) Diffusion barrier (material X) reduces intermixing of A and B.

Nicolet has classified diffusion barriers into three types, according to the mechanisms by which they prevent or retard the diffusion process: *passive barriers, sacrificial barriers,* and *stuffed barriers.*[45]

A *passive barrier* is chemically inert with respect to A and B and has a low solubility for A and B. Materials that generally make good passive diffusion barriers are those that have strong bonds and therefore possess good chemical stability while in contact with other materials. The passive barrier is considered to be probably the ideal diffusion barrier. Nitrides (e.g., TiN), carbides, and borides are examples of materials that could act as such passive barriers, since they exhibit good electrical conductivity, high melting points, chemical inertness, and strong atomic bonds.

A *sacrificial barrier*, on the other hand, is not inert, but it reacts with either A or B. As a result, it can maintain a separation between A and B, but only for a predictable amount of time at a given temperature of operation. The materials A and/or B diffuse into X and form compounds (or material X may diffuse into A or B). If the reaction or diffusion rate is sufficiently slow at the temperature of operation, the expected lifetime of the sacrificial barrier can exceed the predicted lifetime of the device. Once the barrier layer has been totally consumed by the formation of compounds with A and B (or has been dissipated by diffusion into A or B), the barrier characteristic may no longer exist. Polysilicon and titanium films are two examples of sacrificial barriers that have been used in IC fabrication.

A *stuffed barrier* exists when the grain boundaries of the diffusion-barrier films are filled with some other material that blocks diffusion. These grain boundaries would normally be fast diffusion paths, as described earlier in the section on the diffusion of Si in Al polycrystalline films. However, if certain materials are introduced into the film (e.g., during deposition), these materials can become located at the grain boundaries. When the grain boundaries are thereby filled (or "stuffed") with a material, the fast diffusion paths are rendered inoperable. Typically, thin-film impurity concentrations on the order of 1,000 ppm are required to significantly reduce the diffusion rate. An example of a stuffed-diffusion-barrier material is the alloy Ti (10wt%):W, which has been sputter deposited in an ambient containing nitrogen. Nitrogen is thereby incorporated into the film and is thought to stuff the grain boundaries in the manner described.

It is frequently easier to evaluate the efficiency of a diffusion barrier than to establish unequivocally the mechanism by which it operates.[46] Often more than one mechanism may contribute to the effectiveness of a particular barrier material, as is likely the case for the Ti:W diffusion barrier films cited above. That is, once the fast diffusion paths have been preempted (such as by being stuffed with nitrogen in the case of Ti:W), the barrier properties may be more of a function of the inherently high chemical stability of the material. Then the material the material behaves as a highly effective passive barrier.

It should be noted that care must be taken when a barrier layer is deposited into deep, high-aspect-ratio contact windows, to ensure that the minimum required barrier-layer thickness is maintained at the window edges. Thinning at such locations can result in contacts that will fail at lower temperatures than expected. Ultimately, CVD may be needed to provide adequate barrier-layer deposition processes for such applications.

3.5.2   Materials Used as Diffusion Barriers

Of the various materials investigated as diffusion barriers in ICs, those that have achieved the widest adoption have been Ti:W (sputter deposited), Ti (sputter deposited), polysilicon, TiN, and CVD W.[47] Their properties will be described in more detail here; a brief mention of some other experimental diffusion-barrier materials will also be made at the close of this section.

3.5.2.1   Sputter-Deposited Titanium-Tungsten (Stuffed Barrier).

Titanium tungsten (Ti:W) was among the first materials to be employed as a diffusion barrier, although Ti:W thin films were in fact first introduced by Cunningham et al. in 1970 to improve the thermal stability and corrosion resistance of contacts with gold wire bonding and thermal packaging.[48] In 1978 Ghate et al. described the use of Ti:W as a contact barrier for the ohmic contact structures of bipolar integrated circuits. The shallow emitter-base junctions of bipolar devices were more susceptible to Al spiking than were the NMOS source/drain junctions of that time. Ti:W diffusion barriers were eventually considered for CMOS use when the source/drain junctions of the well devices approached the same shallow depths as those of the bipolar emitter-base junctions.[49]

A typical ohmic contact structure that uses a Ti:W diffusion barrier is shown in Fig. 3-27a. A PtSi layer (50-100 nm thick) is in direct contact with the heavily doped Si regions and is covered with a sputter-deposited Ti:W layer (100-200 nm thick). Finally, a layer of Al (or an Al alloy) is deposited on the Ti:W. When the metal is patterned, the Ti:W remains under the Al layer and thus becomes part of the interconnect structure as well.

While tungsten is by itself a fairly good diffusion barrier,* the Ti is added for several reasons. First, Ti improves the adhesion of the tungsten to SiO_2. Second, it protects the tungsten from corrosion by forming a thin layer of titanium oxide on the surface, making the tungsten an even better diffusion barrier. Finally, the maximum temperature that the contacts can withstand is increased to ~500°C.[47]

Ti:W diffusion-barrier layers are deposited by sputtering, normally from a single target. Such targets are now available in MOS grade – that is, with an alkali metal content as low as that available in Al (1%Si) targets. The nominal composition of the target used in the sputter deposition of Ti:W films is $Ti_{0.3}W_{0.7}$ (10:90wt%). Films deposited from such targets exhibit resistivities of 60-100 $\mu\Omega$-cm, depending on the deposition conditions.

The sputter-deposition process is commonly carried out in an Ar-N_2 ambient, so that nitrogen is incorporated into the Ti:W film.[50] The nitrogen is believed to improve the barrier properties of the film by "stuffing" the grain boundaries, thereby substantially reducing the rate of interdiffusion. In addition, it may decrease the reactivity of the titanium in the film through the formation of TiN. Each of these effects acts to extend

---

* Reference 47 presents data that contacts made with W as the barrier layer can withstand temperatures of up to 450°C.

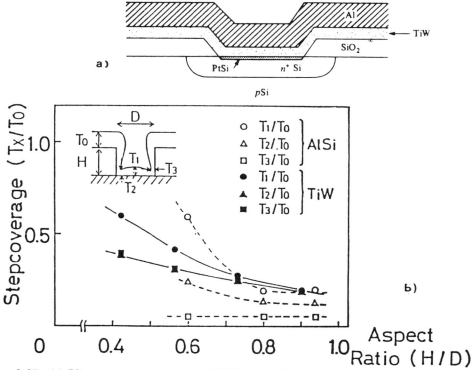

Fig. 3-27 (a) IC contact structure using Ti:W as a diffusion barrier to prevent reaction between PtSi and Al. (b) Comparison of the step coverage of Ti:W (0.6 μm thick) and Al:Si (0.9 μm thick) deposited into a 1.2-μm-wide vertically sided, 0.9-μm-deep contact hole.[143]

the lifetime of the layer. The resistivity of Ti:W is found to increase from 75 $\mu\Omega$-cm in a pure film to 200 $\mu\Omega$-cm when the nitrogen content in the film is 25 at%.[50] This is not of consequence, however, since the resistance of a 100-nm-thick film with this resistivity in a 1-μm^2 contact hole would still be only 0.2 Ω.

The sputter-deposited Ti:W films also exhibit another important advantage for contact structure fabrication. Good step coverage results when these films are sputter-deposited into high-aspect-ratio contact holes. This improves the reliability of the metal interconnects fabricated with Ti:W underlayers.[54] Figure 3-27b illustrates the step coverage of sputtered Ti:W as a function of the aspect ratio of the contact hole opening and shows its superiority with respect to step coverage by sputtered Al.[143]

In the contact-metallization system just described, the Ti:W does not act as a perfect barrier. During annealing, the following reactions with the Al occur:

$$Ti + 3Al \rightarrow TiAl_3 \ (400°C) \qquad (3-7a)$$
$$W + 12Al \rightarrow WAl_{12} \ (450-500°C) \qquad (3-7b)$$

Interdiffusion is the dominant process that destroys these contact structures. A detailed failure analysis performed by Canali et al.[53] shows that Al begins to diffuse through the Ti:W barrier at 500°C and decomposes the underlying PtSi layer to form the intermetallic compound Al_2Pt with the platinum. At the same time, WAl_{12} forms at the Al/Ti:W interface. Volume expansion due to the Al_2Pt formation causes the Ti:W layer to break up, leading to the dissolution of Si and the growth of WSi_2 and $Ti_xW_{1-x}Si_2$ ternary compounds. It has been observed, however, that little degradation of the contacts occurs when a nitrogen-stuffed Ti:W diffusion layer is subjected to up to a 525°C anneal for 30 minutes.

If the Ti:W is used as a contact layer (i.e., without an underlying PtSi layer) with an overlying Al layer, the incorporation of nitrogen is also found to reduce the contact resistance to n^+ Si but to increase the contact resistance to p^+ Si. In one report, a 550°C anneal was needed before a Ti:W layer deposited directly on Si could dissolve the native-oxide layer.[51] Alloying at a temperature of 625°C was necessary to obtain the lowest value of specific contact resistivity to p^+ Si. This annealing temperature would be problematic if an Al layer were present on the Ti:W. It was also noted that low contact resistivity by Ti:W to n^+ Si was nevertheless achievable at the lower annealing temperature of 400°C.

In another report, a multistep surface-preparation routine (after contact holes have been dry-etched in SiO_2) was used to reduce the contact resistance of Ti:W in direct contact with Si by a factor of 4. The specific contact resistivity of Ti:W to p^+ Si was reduced to ~$1x10^{-6}$ $\Omega-cm^2$ following a 450°C anneal.[14] The silicon surface at the bottom of the contact holes was first cleansed of any polymers (that may have been deposited during the contact opening process through an oxygen plasma), and then was subjected to a 1000°C RTP anneal for 10 seconds in nitrogen. This served to repair any surface damage that might have been produced during dry etching. Next, the wafer was dipped in a 1% HF solution for 1 minute just prior to being loaded into the sputter chamber for metal deposition. Another paper reported low contact resistance between Ti:W and n^+ and p^+ Si following furnace anneals of 440°C in N_2 for 30 minutes.[52]

However, the use of Ti:W also has a major drawback for VLSI and ULSI processing, in that the film is quite brittle and highly stressed upon being deposited. Although these stresses can be sufficiently reduced so that problems on the wafer can be avoided, the film that is deposited on the walls of the sputtering chamber eventually reaches a thickness at which the stress causes it to begin to flake off (or *spall*). Such particulates can significantly reduce yield; for large chips this effect may produce unacceptably large yield loss.

3.5.2.2 Polysilicon (Sacrificial Barrier).

A thin layer of polysilicon can be used to separate the Al and single-crystal Si substrate (Fig. 3-28).[55] The polysilicon layer in such contact structures serves as a sacrificial diffusion barrier that protects junctions beneath the contact from undergoing contact-electromigration failure. Under high current stresses, Si from the polysilicon (rather than from the substrate) is transported into the Al. Consequently, void formation in the substrate (which would lead to junction spiking) is prevented. This contact structure is attractive because it exploits a technology already developed for buried contacts, and because it avoids the use of materials other than Al and Si. The phosphorus-doped polysilicon film (200-300 nm thick) is deposited following the opening of the contact holes, but prior to the deposition of Al:Cu (0.5 wt%). After the Al:Cu-polysilicon film has been patterned, the contact structure is annealed.

Although this structure alleviates the problem of contact-electromigration failure of Al:Si-Si contacts to some degree, such failures still occur at rates higher than those due to interconnect electromigration failures. In addition, the contact resistance of the n^+-doped polysilicon to Si is ~50 times as large as that exhibited by Al:Si-to-Si after a 450°C anneal,[56] and the interconnect structures are still prone to the problems caused by silicon precipitates in the Al lines. Furthermore, while this contact structure is easy to integrate into NMOS technology, it is not as compatible with CMOS. In order for it to be used it with CMOS, the polysilicon underlayer must be doped p-type wherever contacts to p-channel devices are needed. This requirement introduces all of the problems associated with the use of dual-doped polysilicon, as described in chapter 6.

3.5.2.3 Titanium (Sacrificial Barrier).

Titanium films have also been used as diffusion barrier. Ti is an oxygen-gettering material and oxide-reducing agent, which causes it to dissolve the native oxide layer on the Si surface during annealing and to adhere well to both Si and SiO_2. In addition, because the Schottky-barrier height of Ti on n-Si is about half the silicon band gap, Ti can form good ohmic contacts to both types of heavily doped Si. In fact, the contact resistivities of Al-Ti-Si contacts are comparable to those of PtSi-Si contacts.

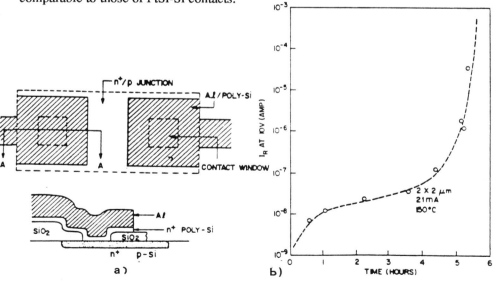

Fig. 3-28 Al-polysilicon-Si contact structure. (a) Layout and cross section. (b) Junction leakage after passing 21 mA at 150 °C through a 2 x 2 μm contact hole versus time.[55] Copyright 1981. Reprinted with permission of the AIOP.

The Ti layer between Si and Al behaves as a sacrificial barrier because it reacts with Al to form TiAl$_3$ at temperatures above 400°C. Ti is a reasonably good diffusion barrier for Si below 500°C, and as long as the Ti is not completely consumed, this barrier property is maintained. Although Ti is not as good a diffusion barrier for Al, the diffusion of Al through the Ti into the Si does not lead to the destruction of the contact since the diffusivity and solid solubility of Al in Si are quite low. Once the Ti has completely reacted to form TiAl$_3$, however, its diffusion barrier properties are lost, and the contacts rapidly fail. The reaction kinetics of Al and Ti were first described by Bower in 1973.[57]

If Ti is selected as a diffusion barrier, a thick enough layer must be used to ensure an adequate lifetime at given fabrication and operating conditions. It has been determined that a 100-nm-thick Ti film can withstand 500°C for 15 minutes and 477°C for 30 minutes;[58] at 425°C, contact failure was observed after 6 hours.[59]

3.5.2.4 Titanium Nitride (Passive Barrier).

The use of the TiN layer as a diffusion barrier was first proposed by Nelson in 1969.[60] The first successful application in devices was reported in 1979 by Garceau et al., who used TiN between Ti and Pt in the Au-Pt-Ti-Si beam-lead metal system.[61] Wittmer reported on the diffusion-barrier properties of TiN for silicon devices in 1980.[62]

TiN is an attractive material as a contact diffusion barrier in silicon ICs because it behaves as an impermeable barrier to silicon, and because the activation energy for the diffusion of other impurities is high (e.g., the activation energy for Cu diffusion into TiN thin films is 4.3 eV, whereas the normal value for diffusion of Cu into metals is only 1 to 2 eV). TiN is also chemically and thermodynamically very stable, and it exhibits one of the lowest electrical resistivities of the transition metal carbides, borides, or nitrides (all of which are chemically and thermally stable compounds).

The specific contact resistivity of TiN films to Si is somewhat higher than that of Ti or PtSi ($\sim 10^{-5}$ Ω-cm^2), and as a result it is ordinarily not used to make direct contact to Si. Instead, it has most commonly been used in a contact structure consisting of Al-TiN-Ti-(or TiSi$_2$)-Si. Such contact structures exhibit very low specific contact resistivities to Si and remarkably high thermal stability, with the ability to withstand temperatures up to 550°C without contact failure. The methods used to form TiN are described in section 3.6.2, which details the characteristics of Al-TiN-Ti-Si contacts. A report on the barrier properties of reactively sputter-deposited TiN as a function of deposition conditions, however, is given in reference 63. Another report indicates that the incorporation of oxygen into TiN films increases the barrier properties of these films, at the expense of a somewhat increased contact resistivity.[64] Al-TiN-Ti-Si contacts have also been reported to exhibit much less susceptibility than Al:Si-Si contacts to contact electromigration failure.

3.5.2.5 CVD Tungsten.

Chemically vapor-deposited tungsten (CVD W) has been investigated for a variety of applications in both metal-Si contacts and interconnect structures. In this section we consider both its applicability as a diffusion barrier and its

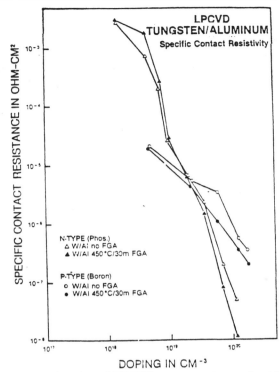

Fig. 3-29 Specific contact resistivity of LPCVD tungsten (covered with an Al layer) to *n*-type and *p*-type Si.[162] (© 1984 IEEE).

usefulness as a material for making direct contact to Si. The details of CVD W deposition and etching, as well as the application of CVD W as an interconnect material and as a material for filling contact holes and vias, will be considered in chapter 4.

Selective CVD W has been investigated for direct contact formation to Si (Fig. 3-29). This process appears attractive for many reasons. First, specific contact resistivity can be very low to n^+ Si, and reasonably low to p^+ Si.*

* Such low contact-resistivity with selective CVD W is achieved only if the silicon surface is clean and reasonably free of a native-oxide layer. If the native oxide is less than 10°A thick, low-resistance contacts will be reliably fabricated, and the Si consumed during the deposition process is self-limited to ~100 °A. (Note that if the native oxide is removed from a Si surface by an HF dip, it takes ~2 hours for it to regrow to a thickness of 10 °A. Thus, if the wafers are loaded into the CVD chamber and W is selectively deposited within that time, good contacts can be repeatably produced.) However, if the native oxide is >10 °A thick, not only will the selective CVD W process produce higher resistance contacts, but the W film will be rough and more Si will be consumed from the substrate (up to 300 Å).

The polymer film that is produced when contact holes are opened by dry etching must also be removed prior to CVD W deposition. If the polymer film is allowed to remain in place,

Second, the W can simultaneously serve as a barrier layer. CVD W layers covered with Al have been found to be able to withstand temperatures of up to 500°C without undergoing significant reaction with the Al.[65] Third, selectively deposited W can partially or completely fill contact holes, thereby easing the problem of step coverage by the first-level metal layer into the contact hole. Finally, as mentioned in the footnote, the Si consumption is self-limiting at ~100 °A if the native oxide layer is less than 10 °A thick. This is an advantage over $TiSi_2$ and $CoSi_2$ contacts (see section 3.9.1). That is, when such silicide contacts are formed, the Si consumption is not self-limiting, but depends on the thickness of the Ti or Co that is deposited before the silicide reaction is carried out.

In spite of the significant benefits offered by the selective CVD W process, concerns have been expressed about some of the side effects observed under certain deposition conditions. Specifically, when W is selectively deposited by the hydrogen reduction of WF_6, lateral encroachment of the W under the Si-SiO_2 interface can occur, and wormholes are occasionally observed at the Si surface (see Vol. 1, chap. 11). Both of

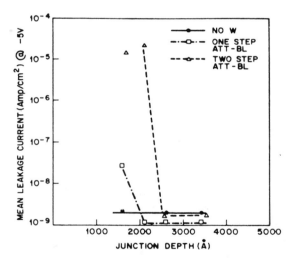

Fig. 3-30 Comparison of mean leakage current density vs. n^+ (As/P) junction depth in Si for contacts fabricated with no W and for W deposited with the hydrogen reduction process.[70] This paper was originally presented at the Spring 1988 Meeting of The Electrochemical Society, Inc. held in Atlanta, GA.

poor contacts are likely to be formed. To ensure polymer layer removal, a slight silicon etch that removes ~100 °A of Si is recommended. (It is not clear whether the etch undercuts the polymer layer, or removes it directly, but at the end of this so-called *silicon polishing procedure*, the polymer layer is gone.) The silicon polishing procedure should be performed in situ as the last step of the dry etch process used to etch the contact holes. This will also ensure good CVD W-to-polysilicon contacts when such contacts are opened at the same time as those to the Si substrate.[157]

these effects can produce junction leakage (Fig. 3-30).

A variety of techniques for reducing wormhole formation,[66] and encroachment have been studied.[67] One of the most promising approaches involves formation of selectively deposited W films using SiH_4 reduction of WF_6 (see chap. 4, section 4.5.4.1).[68,69] This process results in minimal erosion of the Si and minimal encroachment of the Si contacts, as well as the formation of W films with smoother surfaces than those obtainable by the hydrogen reduction process. Although it has been reported that leakage may still occur when junction depths are less than 200 nm deep,[70] more recently, other evidence has been presented that this problem can be eliminated in such shallow junctions using silane reduction of WF_6. Thus, selective CVD W appears to be a very promising technique for forming contacts directly to Si in submicron IC technologies.

Two other aspects of selective CVD W should also be mentioned. First, commercial equipment must be available in which the silane reduction of WF_6 process can be performed. Several equipment vendors have introduced new models with this capability, and others are sure to follow suit. Second, the value of ρ_c of CVD W to p^+Si is still significantly higher than that of Al:Si or PtSi to p^+Si; a lower value would improve the characteristics of CVD W as a contact material.

3.5.2.6 Experimental Diffusion Barrier Materials.

Among other experimental materials that have demonstrated diffusion barrier capabilities are Mo_xNi_{1-x}[71] and $Fe_{0.45}W_{0.55}$.[72] Note that these materials may be good for high-temperature applications, together with the transition-metal carbides (e.g., Ti_xC, x = 3.1)[73] and the transition-metal borides (e.g., TiB_2).[46]

An additional category of materials being investigated for use as diffusion barriers is that of *amorphous thin-film* materials.[74] These are unlike the other diffusion-barrier materials thus far considered, which are polycrystalline in structure (and hence contain grain boundaries, which represent potential paths for rapid diffusion). Amorphous materials can exhibit smaller diffusion coefficients than comparable polycrystalline materials. For example, at 400°C, the diffusion coefficient of Au in amorphous $Ni_{0.45}Nb_{0.55}$ is reduced to less than 10^{-21} cm^2/sec, from 1.6×10^{-15} cm^2/sec in a poly-crystalline alloy of $Ni_{0.45}Nb_{0.55}$. The absence of grain boundaries is also expected to reduce the susceptibility of the contact structures to failure due to electromigration effects.

3.6 MULTILAYERED OHMIC-CONTACT STRUCTURES TO SILICON

Earlier, we showed why the simple Al-to-Si contact structure had to evolve into a variety of more complex contact structures. In some such structures, the material used as the diffusion barrier is also utilized to form a direct contact to Si (e.g., Al-polySi-Si, Al-Ti-Si, Al-Ti:W-Si, and Al-CVD W-Si). These have been discussed previously. In the most complex contact structures, more than one material is used to separate the Al and Si. This section will describe those of the latter type that have been widely adopted.

3.6.1   Al-Ti:W-PtSi-Si Contacts

The Al-Ti:W-PtSi-Si system (Fig. 3-27) has been a workhorse contact structure for bipolar ICs since the late 1970s. It exhibits low contact resistance to n^+ and p^+ Si, and it can withstand temperatures of up to 500°C. It also provides excellent protection against contact electromigration failure. The main reasons that MOS technologies did not adopt such contact structures were the higher defect density associated with Ti:W deposition, the increased process complexity needed to form the contact structures, and the nonavailability of low-Na Ti:W sputtering targets (although such targets now exist).

3.6.2   Al-TiN-Ti-Si Contacts

This contact structure (with the Ti layer ranging from 30-80 nm in thickness, and the TiN from 100-200 nm) has recently been studied with a great deal of interest. When TiN is used as a diffusion barrier between Ti and Al, the thermal stability of the contacts is excellent (annealing temperatures of 500°C or more can be tolerated), and the contact resistivity is good (10^{-6} Ω-cm^2), although not quite as low as with contacts that use PtSi.

The Ti promotes good adhesion to SiO$_2$, and it forms TiSi$_2$ by reacting with the Si substrate during the anneal step. During this step, however, large quantities of dopant from the Si substrate (particularly boron) can be absorbed by the TiSi$_2$ as it is being formed. If this dopant loss is not controlled, high contact resistance can result. (This is an especially serious problem if TiSi$_2$ is used to connect n-type to p-type polysilicon.) It is possible to limit the dopant loss by minimizing the the amount of TiSi$_2$ allowed to form. This can be achieved by using a very thin layer of Ti covered with the TiN diffusion layer.[137]

The TiN layer can be formed in one of five ways: by evaporating the Ti in an N$_2$ ambient;[47] by reactively sputtering the Ti in an Ar + N$_2$ mixture;[75] by sputtering from a TiN target in an inert (Ar) ambient;[76] by sputter depositing Ti in an Ar ambient and converting it to TiN in a separate plasma nitridation step;[77] or by CVD.[126] The simultaneous formation of TiSi$_2$ and TiN following a deposition of Ti is also described in several reports.[78]

In another variation of this contact structure, a layer of Ti (75 nm) and a layer of W (100 nm) are sequentially sputter deposited, followed by an anneal in N$_2$ at 650°C for 20 minutes. The Ti at the Si interface is converted to TiSi$_y$, and the Ti near the W interface to TiN$_x$. Such contacts exhibit good stability even after an eight-hour anneal at 400°C.[79]

In some other processes, improved filling of the contact holes is achieved by coating the wafers with a nonselective CVD W layer that results in good step coverage in the contact holes.[19] An etchback step is performed to remove this W film in the SiO$_2$ field regions and to achieve a planar surface at the contact holes. When the CVD W plug process is not used, the Ti-TiN is patterned and becomes a part of the Metal 1 film as well. A report on the behavior of such contacts (both with and without the additional W film), under high current stressing (to induce contact-electromigration failures) is given by Fu and Pyle.[80]

In another article, it was reported that if the Al-TiN-Ti-Si contact is made to a shallow BF_2^+-implanted region, high leakage currents could result as a result of the thinning effect of the TiN at the contact-hole edges. That is, Al can react with the Si through pinholes in the TiN at these locations. If the Al is replaced with Al:Si, however, excellent leakage characteristics are observed.[81] In addition, it has been observed that the Ti layer in such contacts should be between 50 and 80 nm thick when used with a 0.2-μm-deep junction. In this case, the $TiSi_2$ that is formed will be thick enough to entirely consume the damaged Si layer caused by the BF_2 implant, but not so thick as to penetrate beyond the 0.2-μm-thick junction.

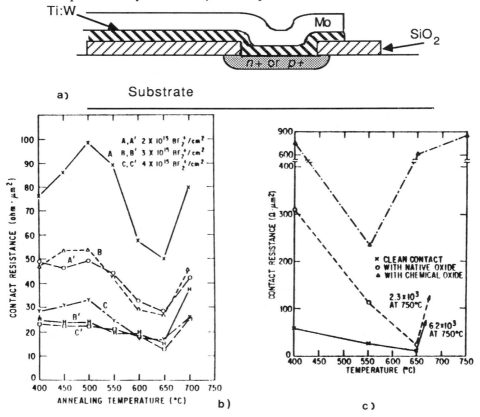

Fig. 3-31 Mo-Ti:W-Si contacts. (a) Structure cross section. (b) Effect of isochronal annealing on the p^+ contact resistance versus BF_2^+ implant dose. (c) Contact resistance change to n^+ Si as a function of isochronal annealing temperature with and without interface oxide.[82] Reprinted by permission of the publisher, The Electrochemical Society, Inc.

3.6.3 Mo-Ti:W-Si and Mo-Ti-Si Contacts

The contact structures described in sections 3.6.1 and 3.6.2 exhibit satisfactory contact properties for VLSI applications, but at the price of increased process complexity. Somewhat simpler contact structures that exhibit equally good characteristics would therefore be attractive. The Mo-Ti:W-Si[82] and Mo-Ti-Si[83] systems have been suggested as two alternative candidates (Fig. 3-31), since such structures exhibit good stability when exposed to temperatures exceeding 500°C (i.e., up to 650°C), and both are able to form low-resistance ohmic contacts to Si. The double-layer film that constitutes the contact structure can also serve as an interconnect layer, provided that it is restricted to the first level of a multilevel-metal scheme. The difficulty of bonding to Mo thin films requires that Al be used for the final metal (i.e., bonding-pad) layer.

In the Mo-Ti:W-Si contacts, an 80-nm-thick Ti:W film is sputter deposited; this is followed by the deposition of a 300-nm-thick Mo layer, in one pump-down. The wafers are allowed to cool before being exposed to atmosphere in order to prevent formation of an oxide on the Mo. After the patterning of the bilayer, the contacts are sintered. It is reported that the ρ_c of Ti:W to p^+Si reaches a minimum after a 625°C anneal (50 Ω– μm^2). Since such a high-temperature-anneal step must be used, a metal other than Al must be selected as the top layer of this structure (Fig. 3-31b). Since Mo has a melting point of 2620°C, it can successfully replace Al for this application. Both the leakage current of the contact structure and the sheet resistance of the bilayer remain low after the 625°C anneal. The Ti:W appears to be capable of dissolving up to 1.5 nm of native oxide during the contact-annealing step, allowing the low-resistance ohmic contacts to Si to be formed.

In the more recently reported Mo-Ti-Si contacts, a 40-80-nm-thick sputtered Ti layer is used instead of Ti:W and is covered with a sputtered Mo layer. This bilayer is etched by means of an RIE process using a mixture of SF_6 and O_2 gas. After being subjected to a 600°C anneal, these contacts exhibit lower ρ_c values than do the Mo-Ti:W-Si contacts ($\rho_c = 10$ Ω–μm^2 to n^+Si and 20 μm^2 to p^+Si). In addition, the Ti layer dissolves native-oxide layers more reliably than does Ti:W. Finally, a pure Ti sputter target can be used instead of a Ti:W target. Ti:W targets are generally fabricated from a hot-pressed mixed powder, which has a greater probability of being impure than a single-element Ti target: the Ti:W target can also occasionally produce particles.[83]

3.7 SCHOTTKY-BARRIER CONTACTS

Schottky barriers are used in a variety of device and circuit applications. One of the most important utilizes the *Schottky-barrier diode* to increase the switching speed of bipolar transistors. Figure 3-32 shows how it is connected in parallel with the collector-base junction of the transistor for this purpose. In this configuration, the *turn-on voltage*, V_F, of the Schottky diode is lower than that of a silicon *pn* diode (i.e., V_F is the voltage at which a particular current, I_F – for example 1 mA – flows through the diode). The lower value of V_F and the charge-storage properties of the metal-semicon-

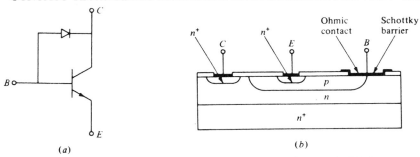

Fig. 3-32 Schottky clamped transistor: (a) circuit representation and (b) integrated structure.

ductor junction are exploited to reduce the transistor switching speed. (Additional information on Schottky-barrier applications in bipolar integrated circuits is given in chap. 7.)

The forward I-V characteristics of a Schottky-barrier diode are given by:

$$I = I_s \{ [\exp (qV/n\,kT)] - 1 \} \qquad (3-8)$$

where

$$I_s = RT^2A \, \exp \{ -q \, \varphi_b/kT \} \qquad (3-9)$$

and I is the current through the Schottky diode, V is the external voltage applied across it, A is the area of the diode, T is the temperature (in kelvins), R is Richardson's constant (120 $A/cm^2\text{-}K^2$), φ_b is the Schottky-barrier potential difference, and n is the ideality factor.

For the case of V > 3kT/q

$$I \approx I_s \exp \{qV/nkT\} = RT^2A \exp \{ (q/kT) (V - \varphi_b) \} \qquad (3-10)$$

or

$$(\varphi_b - V) = (kT/q) \ln \{ (RT^2A)/I \} \qquad (3-11)$$

For design purposes, it is necessary that Schottky diodes exhibit a stable value of V_F. Therefore, from Eq. 3-11 we can see that if φ_b is not stable (i.e., if it varies for some reason, possibly due to a dependence on the maximum thermal anneal temperature, on the Si surface-contamination condition, or on time during normal operating conditions), the value of V_F for a fixed value of I_F would have to vary as well. As a result, it must be possible to produce a constant barrier height for a Schottky-barrier diode in a production environment, and that barrier height value must remain stable under further device fabrication and operating conditions.

PtSi is the material most commonly used for fabricating Schottky-barrier diodes to lightly doped *n*-Si (i.e., when $N_c \leq 10^{17}$ cm^{-3}). This is due to the overall reliability with which PtSi-Si contacts can be fabricated, compared to the other candidate structures. Since any contaminants that are present at the original Si interface are swept up to the silicide surface during the formation reactions, a very clean silicide-Si interface is formed, ensuring the reproducibility of the Schottky-barrier height. In addition, the

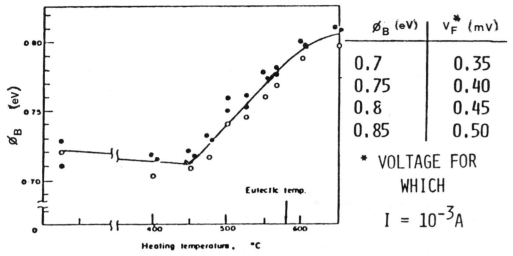

ϕ_B (eV)	V_F^* (mV)
0.7	0.35
0.75	0.40
0.8	0.45
0.85	0.50

* VOLTAGE FOR WHICH

$I = 10^{-3}A$

Fig. 3-33 (a) Barrier heights of Al-Si Schottky barrier diodes as a function of the heating temperature. (b) V_F (voltage at which the forward bias current in a Schottky diode = 1 mA) as a function of ϕ_b.

agglomeration of PtSi layers occurs at ~825°C, whereas, for example, Pd_2Si layers agglomerate at lower temperatures (700°C).

The Al-Si system forms a poor Schottky-barrier contact because its Schottky-barrier height varies with heating temperature (as shown in Fig. 3-33). This is due to the interdiffusion of Al and Si, as described in an earlier section of this chapter. Figure 3-33 also shows how V_F would change as the barrier height of the Al-Si Schottky contact varied from 0.7 to 0.85 V, eliminating such structures from being considered for Schottky diode applications in IC production.

The Schottky-barrier height of PtSi to n-Si has a value of 0.84 eV. This value will remain stable to temperatures of 500°C or more if a suitable diffusion barrier (such as Ti:W or TiN) is interposed between the PtSi and Al. In a bipolar npn transistor, a PtSi layer can therefore be used to form an excellent Schottky barrier to the lightly n-doped collector region, as well as to form an exemplary ohmic contact to the heavily p-doped base region (as shown in Fig. 3-32). Both of these contacts can be fabricated simultaneously by means of a single, maskless metallization procedure (akin to the one described in the earlier section on PtSi-Si ohmic contacts). It should also be noted, however, that the low Schottky-barrier height of PtSi to p-type Si (0.25 eV) causes such contacts to exhibit ohmic behavior even when the Si is relatively lightly p-doped. This has prevented Schottky contacts to p-type Si from being implemented in IC technologies.

Three examples of Schottky-barrier device structures are shown in Fig. 3-34. In the first (Fig. 3-34a), ideal Schottky-barrier characteristics are not obtained because of the sharp edge and the positive fixed charges that exist at the Si-SiO₂ interface. These conditions create a high electric field in the depletion region in the semiconductor near

Fig. 3-34 Special processing techniques improve the performance of Schottky diodes. (a) The diffused guard ring leads to a uniform electric field and eliminates breakdown at the junction edge. (b) The field plate is an alternative means for achieving the same effect. From R. S. Muller and T. I. Kamins, 2nd Ed., *Device Electronics for Integrated Circuits*. Copyright 1986, John Wiley & Sons. Reprinted with permission.

the periphery, which leads to excess current at the corners, a soft reverse-bias characteristic, and a low breakdown voltage. Such problems can be eliminated by allowing the metal to overlap the oxide (Fig. 3-34b) or by diffusing a p^+ guard ring around the periphery of the Schottky structure (Fig. 3-34c). The overlapped metal structure is preferred in IC processes because it is less complex to fabricate.

3.8 THE IMPACT OF THE INTRINSIC SERIES RESISTANCE ON MOS TRANSISTOR PERFORMANCE

The intrinsic series-resistance components of metal-to-Si contact structures were introduced earlier in this chapter, and a discussion followed on how to establish the value of R_{co}. In this section we undertake a discussion of the remaining intrinsic series components of the MOSFET (R_{sh}, R_{ac}, and R_{sp}) and describe under what circumstances these effects (when combined with R_{co}), significantly impact device performance. The impact of the parasitic series-resistance effects on the bipolar transistor is presented in chapter 7.

3.8.1 The Impact of R_S on MOSFET Performance

The series resistance, R_S, of the MOS source and drain regions should be small compared to the resistance of the channel, R_{ch}, in order for degradation of the device characteristics to be avoided. In the past, when larger design rules were used, R_S was a minor component of the total MOS resistance of a turned-on device. As devices got smaller, the channel impedance became smaller, while R_S grew larger (because of shrinking contact sizes and shallower source/drain junction depths). It therefore became necessary to determine whether R_S might become a significant fraction of the total MOS resistance. Under such circumstances, R_S would degrade the device performance in a number of ways. For example, I_{Dsat}, g_{msat}, V_{GS}, and V_T would all be reduced (see chap. 5).

Since the benefits of making devices with smaller channel lengths are so vital, it is important to find ways to design devices so that these smaller lengths can be obtained without degrading device performance through an increase in the value of R_S. The maximum value of R_S (relative to R_{ch}) that can be tolerated in MOS devices should thus be identified, and then process and device technology developed to achieve such allowable R_S values. One proposed constraint is that R_S be kept at less than 10% of R_{ch}. Although this 10% percent limit may seem arbitrary, it is approximately equal to the ratio of R_S to R_{ch} in MOS devices fabricated with a typical 1.0-μm process. Therefore, maintaining this ratio as devices shrink is a reasonable goal. We will report here on how the values of the component resistances of R_S are estimated to change as device dimensions are scaled.

To begin, we present a modified form of the relation used to calculate the *MOS channel resistance*, R_{ch} (normalized to unit width, and therefore expressed in units of Ω-μm), when the device is in the linear region of the MOS characteristics. The value of R_{ch} in the linear region is chosen because it is the worst case comparison with R_S (i.e., the channel resistance of an MOS device in saturation is larger than when the device is in the linear region; see chap. 5).

$$R_{ch \text{ (linear region)}} = [L_{eff} + V_{DS}]/[\mu_o C_{ox} (V_{GS} - V_T - 0.5\ V_{DS})] \qquad (3\text{-}12)$$

The terms in this equation are defined in chapter 5, but for the case at hand we merely graph the values of R_{ch} versus channel length for NMOS devices and PMOS devices (Fig. 3-35), assuming $V_{GS} = 5$ V. We also show the 10% value of R_{ch} in these figures. We note that as the devices get smaller, the value of R_{ch} also decreases. These graphs indicate that the maximum allowed values for R_S for submicron devices will be in the 300-500 Ω-μm range.

3.8.2 Estimates of R_{sh}, R_{sp}, R_{ac}, and R_{co}

Let us examine the components of R_S in Eq. 3-1 (R_{sh}, R_{ac}, R_{sp}, and R_{co}), and see how their estimated values compare with the maximum allowed values of R_S. The information used to estimate the values of these parasitic resistances is based on the analysis presented by Ng and Lynch in references 84 and 85.* Other analyses predicting R_S have also been performed, but an earlier model was used to estimate the values of R_{sp} and R_{ac} leading to somewhat different results from those given here.[86,87] Because it is believed that Ng and Lynch's model treats the effects of R_{sp} and R_{ac} more correctly, it is their conclusions that are presented here. The basis of their analysis is that as the devices are scaled, the physical and electrical parameters of the device will also be scaled. Figure 3-36 gives the values of these scaled parameters as the device channel length is reduced from 0.7 to 0.15 μm. This analysis also assumes that the power-supply voltage will scale as the square root of the channel length.

The *diffusion sheet resistance*, R_{sh}, is calculated using the formulas

$$R_{sh} = \rho_{sh}\ S/W \qquad\qquad (3\text{-}13)$$

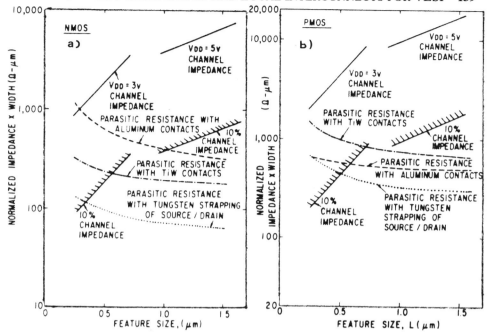

Fig. 3-35 Plot of minimum channel resistance and parasitic series resistance versus design rule feature for (a) NMOS, and (b) PMOS. The parasitic resistances are calculated for three different contact technologies: 1) Al - - - ; 2) Ti:W alloy - - · · ; and 3) self-aligned W strapping of source/drain · · · · ·).[86] (© 1986 IEEE).

and

$$\rho_{sh} = \rho/X_j \qquad\qquad (3\text{-}14)$$

where ρ_{sh} is the sheet resistance in ohms per square, ρ is average bulk resistivity in the n^+ or p^+ layer, W is the device width, and S is the distance from the edge of the channel to the leading edge of the contact window (see Fig. 3-1). For example, in a PMOS device, R_{sh} is calculated to be ~65 Ω-μm if the device has an effective channel length of 0.7 μm, a minimum drawn channel length of 1.01 μm (which dictates that a value of $W = 1.01$ μm be used in Eq. 3-13), $\rho_{sh} = 132$ Ω/sq, and $S = 0.5$ μm.

The analysis of Ng and Lynch estimates that R_{sh} will not significantly contribute to R_S in submicron MOSFETs with gate lengths down to 0.15 μm because they predict that the bulk resistivities used in Eq. 3-13 will remain essentially constant, at the values of 0.001 and 0.0033 Ω-cm for n^+ and p^+ materials, respectively.[85] As a result, R_{sh} values are also predicted to remain essentially constant, at ~20 Ω-μm and ~60 Ω-μm in n- and p-channel MOSFETs, respectively, as the effective channel length is

---

* Measured values of the extrinsic drain resistance of MOSFETs with silicided p^+ and n^+ source/drain regions have recently been published.[166]

SCALED PHYSICAL AND ELECTRICAL PARAMETERS USED IN THIS STUDY

L_c	V_{dd}	V_t	N_c	T_{ox}	X_j	L_{eh}*	K(scaled)	K(const)	S	$\rho_c(n^+)$	$\rho_c(p^+)$
(μm)	(V)	(V)	(cm^{-3})	(Å)	(μm)	(μm)	(cm^{-1})	(cm^{-1})	(μm)	($\Omega/\square$)	($\Omega/\square$)
0.70	3.00	0.75	5×10^{16}	276	0.25	1.01	2.0×10^6	2.0×10^6	0.5	40	132
0.60	2.78	0.70	6.5×10^{16}	225	0.22	0.82	2.3×10^6	2.0×10^6	0.4	45	150
0.50	2.54	0.64	8.9×10^{16}	175	0.20	0.65	2.5×10^6	2.0×10^6	0.35	50	165
0.40	2.27	0.57	1.3×10^{17}	129	0.17	0.48	2.9×10^6	2.0×10^6	0.3	59	194
0.30	2.00	0.50	2.2×10^{17}	86	0.14	0.32	3.6×10^6	2.0×10^6	0.25	71	236
0.20	1.60	0.40	4.5×10^{17}	48	0.10	0.18	5.0×10^6	2.0×10^6	0.2	100	330
0.15	1.40	0.35	7.8×10^{17}	31	0.07	0.12	7.1×10^6	2.0×10^6	0.15	143	471

Fig. 3-36 Scaled physical and electrical parameters in the study of parasitic series resistance components on MOS transistor performance.[85] (© 1987 IEEE).

scaled from 0.7 μm down to 0.15 μm. Note that since X_j and S are both scaled to the same extent, R_{sh} remains constant if ρ is assumed to stay constant. This result is graphed in Figs. 3-37a and 3-37b.

The values of the *accumulation-layer resistance*, R_{ac}, and *spreading resistance*, R_{sp}, were calculated together as a single quantity by Ng and Lynch,[84] using a model they derived. This model incorporates the doping gradient near the junction, whereas an earlier model assumed an abrupt doping profile.[88] Thus, their estimated values were significantly different than those obtained with the previous model. In particular, their results show that the values of $(R_{sp} + R_{ac})$ depend critically on the steepness of the doping profile, K. Specifically, in order to obtain minimum values of $(R_{ac} + R_{sp})$, the junction profile should be as steep as possible.

The model also estimates that $(R_{ac} + R_{sp})$ will decrease with shrinking channel length. The predicted values of $(R_{ac} + R_{sp})$ versus channel length are given in Fig. 3-37, both for the case in which K is maintained at the values that exist in the 0.7-μm technology (i.e., K is not scaled, but remains constant), and for the case in which future technological breakthroughs will permit the fabrication of a higher K (i.e., K is scaled). The advantage to scaling K – if this is in fact possible – is obvious from these results. It is also apparent that if the value of R_{co} is optimized, $(R_{ac} + R_{sp})$ can be the dominant component of R_S. In 0.5-0.7-μm devices the estimated values of $(R_{ac} + R_{sp})$ is >200 Ω-μm for NMOS devices and >300 Ω-μm for PMOS devices (assuming that K is not scaled).

The value of the *contact resistance*, R_{co}, can be accurately calculated using a 2-D device simulator, such as PISCES (see chap. 9). It can also be calculated by hand using the 1-D TLTR equation (Eq. 3-6c), with the value of ρ_{sh} obtained from measured data and the value of ρ_c extracted from test structure measurements (as described in section 3.3.4). Ng and Lynch's analysis assumes values for ρ_{sh} and ρ_c and calculates the range of values of R_{co} for the most probable assumed values. The minimum value of

$R_{co(min)}$ is obtained if the contact length, l, is greater than 1.5 $[\sqrt{\rho_c/\rho_{sh}}]$. For a specific set of values of ρ_{sh}, ρ_c, and contact width (w), this value can be calculated from

$$R_{co(min)} \approx \frac{\sqrt{\rho_{sh}\,\rho_c}}{w} \qquad\qquad (3\text{-}15)$$

In such cases, the coth term in Eq. 3-6c will approach 1, and Eq. 3-6c will approach Eq. 3-15. In typical devices, this means that the contact length, l, will need to be one to four times the channel length in order to produce a metal-Si contact with the minimum value of R_{co} (for a given ρ_c and ρ_{sh}). This result indicates that the use of *self-aligned contacts* (as will be described in section 3.9) will be an important approach for minimizing R_{co}, since such contacts allow the contact length and width to be increased without there also being an increase in the size of the device that is fabricated using a conventional contact structure.

Figures 3-37a and 3-37b show the estimated values of $R_{co(min)}$ that will be attain-

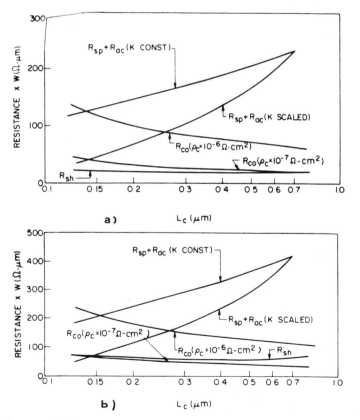

Fig. 3-37 The contributions of each resistance component as a function of channel length, L_c, for (a) *n*-channel and (b) *p*-channel MOSFETs.[85] (© 1987 IEEE).

able for values of ρ_c between 10 and 100 Ω-μm^2 for NMOS and PMOS devices, respectively. This indicates that the values of $R_{co(min)}$ are smaller than $(R_{ac} + R_{sp})$ for devices with 0.5-0.7-μm channel lengths if such ρ_c values are used. Thus, by using sufficiently long contacts, R_{co} will not be the dominant component of R_S.

3.8.3 Impact of R_S on Device Characteristics

It was earlier proposed that R_S should be no larger than 10% of R_{ch}. Ng and Lynch, on the other hand, performed a more sophisticated analysis in which they estimated the reduction in switching speed as R_S was increased and L_{eff} decreased. They found that the circuit-speed degradation is technology dependent. That is, since the channel resistance in the linear region of operation of a MOSFET is smaller than the channel resistance in the saturation region of operation (see chap. 5), the impact of R_S will be different on enhancement-depletion (E/D) NMOS logic (i.e., more detrimental) than on CMOS logic. As we have moved into an era in which CMOS technology predominates, this helps solve the problem. The analysis of Ng and Lynch also indicates that as device size continues to shrink, performance will continue to improve, but that it will not be as good as it would be if the values of R_S obtained had been negligibly small.

3.8.4 Summary of the Impact of Intrinsic Series-Resistance Effects on MOSFET Performance

The following points summarize the results obtained by Ng and Lynch:

• Compared to the other components, R_{sh} contributes only a small value; for silicided source/drain regions, its effect should be negligible.

• R_{co} can be important in degrading MOS device performance in some cases, while in other cases it may not be significant. If the value of R_{co} is small relative to $(R_{sp} + R_{ac})$, it will not significantly impact device performance. R_{co} is essentially determined by the contact length l, and the value of ρ_c. The minimum value of R_{co} for a given ρ_c will be obtained for a contact length that is greater than the transfer length, l_t (Eq. 3-5). Therefore the main advantage of the self-aligned silicided source/drain regions will be an increase in the contact length that is sufficient to minimize R_{co}.

• An upper bound on a value of ρ_c that will still allow contacts with sufficiently small values of R_{co} is probably 100 Ω-μm^2, and this value appears to be attainable for the contact metallization systems in current use or under development. If ρ_c is smaller than 10 $\mu\Omega$-cm^2, the effect of R_{co} becomes insignificant. If ρ_c is between 10 and 100 Ω-μm^2, R_{co} is likely to be less important than $(R_{sp} + R_{ac})$. Nevertheless, the ρ_c value of a contact structure being used in production should be well characterized, especially as it can vary with implant dose, annealing conditions, contact premetal cleaning, etc.

- The value of $(R_{ac} + R_{sp})$ is likely to dominate the value of R_S for MOS devices in which channel lengths are smaller than 0.5 μm. Minimum values of $(R_{ac} + R_{sp})$ are achieved by fabricating source/drain junctions with as steep a doping profile as possible.

Overall, Ng and Lynch predict that the intrinsic series-resistance effects will negatively impact the speed performance of MOS devices as these devices are scaled. Nevertheless, they foresee that it will be possible to build devices that will still demonstrate improvement of speed performance as they are scaled to 0.15 μm. The use of self-aligned, silicided source/drain regions to reduce R_{co} will be important, as will the development of processes that produce junctions with more steeply graded doping profiles. In addition, contact technologies that provide smaller ρ_c and lower values of ρ_{sh} of the diffusion layer under the contact window (or the silicided region) will need to be developed.

3.9   ALTERNATIVE (SELF-ALIGNED) CONTACT STRUCTURES FOR ULSI MOS DEVICES

As transistor dimensions approached 1 μm, the conventional contact structures used up to that point began to limit device performance in several ways. First, it was not possible to minimize the contact resistance if the contact hole was also of minimum size (as explained in section 3.8.2), and problems with cleaning the small contact holes became a concern. In addition, the area of the source/drain regions could not be minimized because the contact hole had be aligned to these regions with a separate masking step, and extra area had to be allocated for misalignment. (The larger area also resulted in increased source/drain-to-substrate junction capacitance, which slowed down device speed). Finally, when nonminimum-width MOSFETs were fabricated with conventional contacts, several small, uniform-sized contact holes were usually used rather than one wider contact hole. (The reason for this is that if all the contact holes are of identical size, they are more likely to clear simultaneously during the etching process.) The problem with using several small, equally sized contact holes rather than one wider one, was that the full width of the source/drain region was thus not available for the contact structure (see Fig. 3-38). As a result the device contact resistance was proportionally larger than it would have been in a device having minimum width.

A variety of alternative contact structures have been investigated in an effort to alleviate these problems. The most important that have emerged are the following:

- self-aligned silicides on the source/drain regions (when these silicides are formed at the same time as the polycide structure, the approach referred to as a *salicide process*);

- elevated source/drain or emitter regions (formed by Si deposition onto the exposed source/drain and emitter regions); and

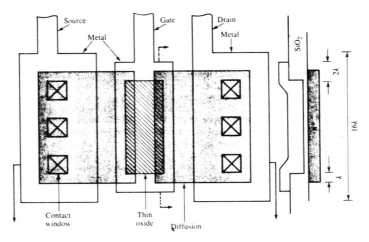

Fig. 3-38 In transistors that are wider than minimum width, conventional contacts may still be made using several contact holes of minimum size, rather than one wider contact hole.

- buried-oxide MOS (*BOMOS*) contacts

- selectively deposited layers of metal in the contact holes (either by CVD, as described in sections 3.5.2.5 and 4.5.4.3, or by electroless deposition, as discussed in section 4.5.6.3).

In this section we will describe the salicide and BOMOS contact approaches. In the last two sections of the chapter we will consider elevated source/drain contacts and the advantages these offer with regard to making contacts to shallow junctions. Mention will also be made of the important topics of *unframed contacts* and *vias*. The filling of contact holes and vias will be addressed in chapter 4, as part of the multilevel-interconnect discussion.

3.9.1 Self-Aligned Silicide Contacts

One alternative to using the conventional contact structure is to contact the entire source and drain regions of a MOSFET with a conductor film (Fig. 3-39). This approach becomes even more attractive if such a film can be formed using a self-aligned process that does not entail any additional masking steps, in a process sequence as follows:

1. After the source and drain regions have been implanted to form the source/drain junctions (Fig. 3-39a), the poly-Si sidewall spacers are formed (Fig. 3-39b).

2. The metal metal used to form the silicide is deposited.

3. The wafer is heated, which causes the silicide reaction to occur wherever the metal is in contact with the silicon. Everywhere else, the metal remains unreacted (Fig. 3-39c).

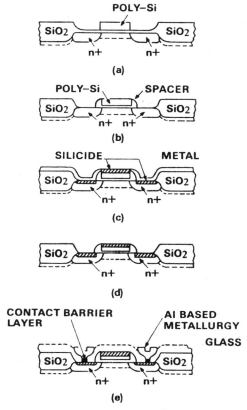

Fig. 3-39 Salicide processing steps and final structure.[163] (© 1984 IEEE).

4. The unreacted metal is selectively removed through the use of an etchant that does not attack the silicide, the silicon substrate, or the SiO_2 (Fig. 3-39d). As a result, each exposed source and drain region is now completely covered by a silicide film, but there is no film elsewhere.

5. A dielectric layer is deposited onto the silicide, and contact holes are opened in it down to the silicide layer.

6. Metal is deposited into the contact holes to make contact with the silicide (Fig. 3-39e).

The above contact-formation method offers several advantages over the conventional contact structure:

- The value of R_{sh} of Eq. 3-1 becomes negligibly small because the ρ_{sh} value of the silicide is typically 1-2 Ω/sq, whereas that of the diffused junction region alone is 40-120 Ω/sq.

• The contact area of the silicide and the Si is much larger than that of the metal-Si area in a conventional contact in a device of the same total area. Thus, the value of R_{co} for the same ρ_c value is significantly lower.

• The contact hole is now used to make contact between the metal and the silicide, but the value of ρ_c exhibited by silicide-metal interfaces is typically $\leq 10^{-9}$ Ω-cm^2. This is two orders of magnitude lower than nominal metal-to-silicon specific contact resistivities, $\sim 10^{-7}$ Ω-cm^2, obtained at or near solid-solubility silicon doping limits. Hence, the contribution to R_{co} due to the metal-silicide contact resistance at the contact hole is usually negligibly small.

• In some processes the deposition of the metal and the formation of the silicide can occur simultaneously with the silicide formation process of the gate; hence, this step can be integrated into the process essentially for "free."

As it became apparent that a self-aligned contact structure would be beneficial for reducing the values of the parasitic series resistances of the conventional contact structure, some other processing trends also allowed a new group of materials to be considered as contact metals to Si. These materials were the group-VIII silicides (PtSi, Pd$_2$Si, CoSi$_2$, and NiSi$_2$) as well as TiSi$_2$. As it became necessary to fabricate shallow junctions, processing technology had to be performed at lower temperatures to avoid the formation of deep junctions. When it became possible to restrict the maximum temperature after the silicide formation to less than 900°C, this also opened the door to these silicides, since all of them are unstable at temperatures higher than 900°C.

There are several important reasons why TiSi$_2$ and the group-VIII metal silicides are especially attractive as candidates for self-aligned ohmic contacts and local interconnects to silicon. These are as follows:

• Such silicides exhibit lower resistivities than the refractory-metal silicides of W, Ta, and Mo (e.g., the respective resistivities of TiSi$_2$, CoSi$_2$, PtSi, Pd$_2$Si, and NiSi$_2$ are 13-20, 16-18, 28-30, 30-35, and 50 $\mu\Omega$-cm).

• All of the group VIII metals react with Si at 600°C or less, and, with the exception of Ti, there is no reaction of the metal with the masking oxide during the silicide formation. Even with Ti, the reaction is minimal, and the reaction products appear to be soluble in the Ti etch.[89]

• When the silicides are formed at such low temperatures, the metal atoms diffuse into the Si, and react with it. As the metal atoms diffuse into the SiO$_2$, no Si is available for reaction, and hence, the metal over device areas covered with SiO$_2$ remains unreacted. This also prevents lateral formation of the silicide over the narrow oxide-spacer region during the silicide formation step.*

*There is an exception to this statement. During TiSi$_2$ formation at temperatures above 600°C, Si diffuses into the Ti. The diffused Si then reacts to form TiSi$_2$ over the oxide-spacer regions, resulting in shorting of the gate and source/drain. Procedures for avoiding this effect are discussed in section 3.9.1.1.

• Dopant atoms in the silicon are "snowplowed" into the silicon substrate as group-VIII-metal silicides are formed[138] (but not necessarily as $TiSi_2$ is formed).[98] Such snowplowing maintains a high doping concentration at the silicide-Si interface, which is beneficial for producing low resistance contacts.

• In all cases, after the silicide has been formed, the unreacted metal can be etched off from the oxide surface using chemical etchants that do not attack the oxide, the silicide, or the substrate silicon. This last characteristic allows self-aligned contacts to be formed through the use of these silicides.

Of this group of silicides, the two that have been the focus of most development for the salicide process are $TiSi_2$ and $CoSi_2$. These two silicides exhibit the lowest resistivities of the group and can withstand process temperatures in excess of 800°C (such temperatures are frequently used to reflow the doped glasses deposited over the salicided source/drain regions). Pd_2Si and $PtSi$, on the other hand, agglomerate when in contact with Si at high process temperatures (Pd_2Si at 700°C, and $PtSi$ at ~800°C). The use of $PtSi$ was not limited in conventional contacts because the self-aligned $PtSi$ contacts were formed after the reflow step, but in the salicide process, reflow is done after silicide formation, making $PtSi$ incompatible with reflow. $NiSi_2$ exhibits significantly higher resistivity, and the stresses formed in $NiSi_2$ films at low temperatures (400-600°C) seem high enough to cause mechanical instability.[90]

The salicide process exhibits a limitation related to the fact that the gate and the source/drain silicides are formed at the same time. On the gate, it is desirable for the silicide to have the lowest possible sheet resistance (so that the gate electrode will also possess a low interconnect resistance). To achieve this, a thick silicide layer is needed. Over the source/drain regions, however, the silicide can be only of limited thickness, in order to prevent excess consumption of the substrate silicon by silicide formation. Thus, a thicker silicide, though favorable at the gate level, is detrimental to contact formation, and vice versa. One solution is to use a two-step process[91] in which the silicide is formed on the gate level first, and on the contact regions at a later stage (and at a different thickness).

Another problem is that it is difficult to metallize shallow junctions (≤ 200 nm), because of the junction leakage that is observed when the silicidation process is used. The implant damage (caused during the source/drain junction formation) may not be sufficiently annealed-out by the lower-temperature silicide-formation step. This damaged region may remain within the depletion region of the shallow junction, giving rise to higher leakage. In addition, the RIE process used to open the contact holes in the doped oxide (deposited after silicide formation), may damage the silicide.[96] Techniques to reduce or eliminate this last set of problems are covered in section 3.10. A final problem of salicide technology is that it degrades the effectiveness of some electrostatic-discharge protection devices used with CMOS circuits (see chap. 6, section 6.7.1.4).

3.9.1.1 Self-Aligned Titanium Silicide Contacts. $TiSi_2$ is attractive for the salicide application because it exhibits low resistivity, and because it can reduce

native-oxide layers (making it the only known refractory metal that can reliably form a silicide on both polycrystalline and single-crystal silicon through a thermal reaction). Furthermore, devices fabricated with titanium silicide on the gate electrode are more resistant to high-field-induced hot-electron degradation than are conventional poly-Si gate devices.[95]   It is conjectured that the $TiSi_2$ is an effective getter for the hydrogen atoms introduced during the hydrogen anneal. Less hydrogen is thus incorporated into the gate oxide, and this improves the hot-electron reliability (see chap. 5, section 5.6.6).

Some aspects of the $TiSi_2$ process are not as favorable, including: (a) the reactivity of Ti with SiO_2 can cause unwanted reactions with the oxide spacers during the silicide-formation process; (b) $TiSi_2$ is less stable than WSi_2 or $MoSi_2$; and (c) because Ti films have a high propensity to oxidize, the silicide must be formed in ambients that are free of oxygen.

Oxide spacers at the edges of the polysilicon are formed to separate the silicided gate and source/drain regions, but these spacers are typically only 200-300 nm wide. Thus, any lateral formation of silicide can easily bridge the separation and cause the gate to be come shorted to the source/drain (this effect is referred to as *bridging*).

It has been observed that if the $TiSi_2$ is formed by conventional furnace annealing in an inert-gas atmosphere (e.g., Ar for ~30 min), lateral $TiSi_2$ formation occurs rapidly. This occurs because silicon diffuses into the Ti regions that cover the spacer oxide and subsequently reacts with the Ti. When annealing is performed in an N_2 ambient, the Ti absorbs a significant amount of nitrogen (e.g., more than 20 at%). The N_2 is absorbed preferentially at the Ti grain boundaries, which "stuffs" the grain-boundary diffusion paths. This reduces the diffusivity of Si in the Ti, and essentially suppresses the lateral silicide reaction. Annealing in pure N_2 or pure forming gas (95% N_2 + 5% H_2) thus results in $TiSi_2$ formation without bridging. It is also important that the N_2 ambient contains fewer than 5 ppma of oxygen or water to avoid unwanted oxidation of the Ti film.[130] The absorbed N_2 also reacts with the Ti to form TiN at the Ti surface.[89,130]

If the temperatures exceed 700°C during $TiSi_2$ formation, the Ti and the spacer SiO_2 can also react to form titanium oxides. Any residues of this reaction can degrade device performance by compromising the oxide integrity or producing bridging. To avoid such effects, it is recommended that the $TiSi_2$-formation temperature be held to <700°C, and that a minimum field-oxide thickness of 100 nm be utilized. In practice, a two-step formation process is typically used.[136] During the first step, the temperature is kept at ~650°C. Following selective etching and removal of the unreacted Ti in a room-temperature mixture of DI H_2O, 30% H_2O_2, and NH_4OH (5:1:1), a second temperature step of ~800°C in Ar is used to lower the $TiSi_2$ sheet resistance and to stabilize the $TiSi_2$ phase.[92]

Rapid thermal processing (RTP) at 600-800°C in Ar (with the specific reaction time depending on the temperature selected), has also been used to effect the silicide formation. Following selective removal of the unreacted Ti, a stabilization anneal of 1000°C for 30 seconds in Ar is conducted to reduce the $TiSi_2$ resistivity.

$TiSi_2$ in contact with Si was originally thought to be capable of being subjected to temperatures up to ~900°C ($TiSi_2$ agglomerates at higher temperatures). It has been observed, however, that if the temperature exceeds 800°C, the value of ρ_c to p^+ Si

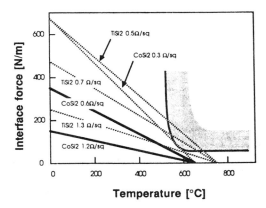

Fig. 3-40 Force at the silicide edge (silicide stress x silicide thickness) versus temperature acting during cooling after silicide formation for the various samples. (Dotted lines correspond with samples in which defects are observed.)[100] This paper was originally presented at the Spring 1988 Meeting of The Electrochemical Society, Inc., in Atlanta, GA.

becomes too large.[93] (However, the ρ_c of $TiSi_2$ to n^+Si can be less than 10^{-6} Ω-cm^2 for anneal temperatures of 800°C.) Even at anneal temperatures of 800°C, the value of ρ_c to p^+ Si is 10^{-5} Ω-cm^2, while at temperatures of 900°C, it increases to 10^{-3} Ω-cm^2. The cause of this problem is the rapid diffusion of the boron from the PMOS source/drain regions into the overlying silicide, causing dopant depletion at the silicide/junction interface (and consequently, high contact resistance between the junction and the silicide layer). Temperatures in excess of 800°C must thus be avoided following $TiSi_2$ formation in CMOS technologies. This may represent a problem if doped-glass reflow is used in a CMOS process.[94]

It has been reported that As from the Si substrate will also diffuse into the silicide during the silicide formation, leading to increased contact resistance. To minimize such As dopant redistribution, an RTP process was successfully used to form the silicide (700°C in N_2 for 20 seconds, followed by 10 seconds at 850°C after the unreacted Ti is removed, for a 60-nm-thick Ti film).[97] An extensive study of the effects of $TiSi_2$ formation on dopant redistribution is presented in reference 98.

Another reported method for enhancing the $TiSi_2$ formation process uses *ion-beam mixing*. In this method, Si atoms are implanted into the Ti film before it is annealed. The surface morphology of the silicide film is improved, and its sheet resistance is lowered by ~20%.[149]

Al-TiN-$TiSi_2$-Si contact structures were observed to undergo early electromigration failure if the current passing through the contact caused local heating such that the temperature exceeded 465°C. It was also found that thermal stresses due to even lower temperatures produced damage to the contacts, which also led to contact-electromigration failure.[99]

Another problem associated with $TiSi_2$ contacts involves the generation of defects at the edge of the $TiSi_2$ film due to stresses in the film (Fig. 3-40).[100] Such defects are reported to begin occurring once the thickness of the $TiSi_2$ film exceeds 100 nm. Another report suggests that a Ti film no thicker than 70 nm should be used when $TiSi_2$ contacts to shallow junctions (0.2 μm) are being implemented.[142] Thicker Ti films would cause excessive silicon consumption and rough interfaces, and these in turn would produce increased junction-leakage current and high contact resistance.

The formation of CVD W plugs in contact with $TiSi_2$ salicide layers has also been investigated. Early reports indicated that an interfacial TiF_3 layer is formed between the W and the $TiSi_2$ if the W deposition is done in a hotwall process at 300-350°C, and that this layer inhibits good contact between the plug and the underlying silicide. Higher-temperature W depositions (600-700°C) were found to reduce the contact resistance;[101] nevertheless, random contacts exhibited a "crusting" or overgrowth problem when this process was used. An alternative method for overcoming the high contact resistance between $TiSi_2$ and CVD W is to deposit a thin bilayer of Ti and W before doing the CVD W deposition (with the Ti and W layers each being 10-30 nm thick). The underlying Ti forms a low resistance contact to the $TiSi_2$ (since any native SiO_2 on the silicide is consumed by reaction with the Ti), and the sputtered-W layer prevents the formation of the high-resistivity TiF_3 layer on the $TiSi_2$ during the CVD W process.[158]

3.9.1.2 Self-Aligned Cobalt Silicide Contacts.   Although $TiSi_2$ has been the most widely implemented material for the salicide process, $CoSi_2$ has also received significant attention as an alternative candidate material. $CoSi_2$ exhibits the following attractive characteristics for this application:

• It offers the benefits of low resistivity (16-18 $\mu\Omega$-cm) and high temperature stability. For example, $CoSi_2$ films of 300-nm thickness exhibit a ρ_{sh} of 0.5 Ω/sq and remain stable when in contact with Si up to 900°C.[102,103] This demonstrates the compatibility of $CoSi_2$ with a 900°C glass-reflow step.

• It can be formed by the reaction of Co with Si in a single annealing step, with no lateral formation of the silicide or encroachment under the oxide (i.e., bridging is not a problem).

• The interface of the silicide and silicon is smooth, and low-resistance contacts to the silicon can be formed (as low as 15 Ω-μm^2 to both n^+ and p^+Si).

• Contacts to shallow As and B junctions can be successfully fabricated, and the dopant profiles in such junctions appear to be essentially unchanged after the silicide has been formed.

• A selective chemical etch exists that makes it possible to remove the Co without affecting the $CoSi_2$ (e.g., $HCl:H_2O_2$ [30%] in a volume ratio of 3:1 at room temperature).

- $CoSi_2$ is far less susceptible to removal by plasma etching than $TiSi_2$. Hence there is negligible silicide loss during the overetch time that may be needed when dry-etching the contact openings through the glass overlayer.

- No competing reaction to the $CoSi_2$ formation reaction occurs, whereas when $TiSi_2$ is formed in nitrogen, TiN is simultaneously formed. In addition, lower shear forces (important to device integrity) are present in $CoSi_2$ than in $TiSi_2$ of the same thickness.[104]

The process sequence for forming the $CoSi_2$ polycide and the self-aligned contacts to Si is as follows:

1. The silicon and polysilicon surfaces are chemically cleaned by dipping the wafers in a dilute (100:1) solution of H_2O:HF for 2 minutes just prior to loading them into the sputtering chamber. Although in one report this step is followed with an rf backsputtering procedure to remove ~10 nm of the silicon surface,[103] in another the rf-backsputter step was observed to produce a rough interface following silicidation.[102] Nevertheless, the cleaning procedure is important, because cobalt is not able to reduce SiO_2 and because the presence of a native oxide can inhibit the silicidation reaction.

2. The cobalt is sputter deposited to the desired thickness. The thickness is selected based on how much of the silicon in the source/drain regions can be consumed, as well as what silicide sheet resistance value is needed. For example, 35.9 nm of Si is consumed by 10 nm of Co to form 35.9 nm of $CoSi_2$. To form a $CoSi_2$ layer with ρ_{sh} = 1.5 Ω/sq, a 30 nm thick layer of Co must be deposited. It should be noted that the Si consumption during $CoSi_2$ formation is about 25% greater than that during $TiSi_2$ formation. This constitutes one of the disadvantages of $CoSi_2$ compared to $TiSi_2$.

3. The $CoSi_2$ is formed by reacting the Co with the Si and polysilicon. The reaction kinetics are described in reference 103. Although the reaction can be accomplished through either a furnace anneal or RTP, the latter has emerged as the more attractive approach.[129] An 80-nm-thick Co film will be completely converted to $CoSi_2$ at 700°C for 30 seconds. The gas ambient must be kept free of oxygen to ensure that oxide-free silicide films are formed.

4. The unreacted Co is etched away using the etchant mentioned earlier.

5. Finally, the wafer is covered with a CVD glass layer. Note that when Al is contacted to the $CoSi_2$, the maximum anneal temperature must thereafter be limited to 400°C (since Al will react with $CoSi_2$ at temperatures in excess of 400°C). To allow a higher anneal temperature to be used, a diffusion barrier layer (such as Ti:W), must be used between the $CoSi_2$ and the Al layers.[105]

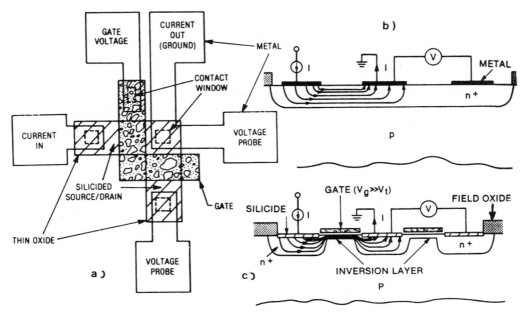

Fig. 3-41 Layout of the silicided test structure described in ref. 106. (b) Cross sections showing the current pattern in (a) the conventional Kelvin test structure, and the new silicided test structure.[106] (© 1988 IEEE).

A report describing the operation of 0.5-μm MOS devices built with $CoSi_2$ source/drains and $CoSi_2$/poly gates is presented in reference 148. Two studies comparing the properties of $TiSi_2$ and $CoSi_2$ for MOS salicide applications are given in references 150 and 151. Both reports identify $CoSi_2$ as a better candidate than $TiSi_2$ for use in the salicide process for the reasons listed above.

3.9.1.3 Measuring ρ_c of Self-Aligned Silicide Contacts. The standard CBKR test structure used to extract ρ_c of a conventional contact is not suitable for obtaining ρ_c for a self-aligned silicide contact. In the CBKR structure, three contacts are required over the same diffusion region (Fig. 3-41a). Therefore, it is possible to measure only the metal-to-silicide contact resistance, but not the silicide-to-Si contact resistance. To use the CBKR structure to measure the latter contact resistance value, the silicide formation must be restricted to the contact hole; however, this condition does not exist in the actual self-aligned contact. A test structure has therefore been suggested to overcome this difficulty.[106] A cross-section of this structure is shown in Fig. 3-41b. In addition to providing the ρ_c value, this structure can provide the value of ρ_{sh} beneath the silicide. Such structures can be directly integrated into production wafers with no modification of the process sequence. Additional details of how this structure is fabricated can be found in reference 106.

3.9.2   Buried-Oxide MOS Contact Structure (BOMOS)

In the buried-oxide MOS (or BOMOS) method, the contact structure is formed after the field oxide, but before the gate oxide. Figure 3-42 shows the key steps of the BOMOS process sequence. The process is begun by growing and patterning the *first oxide film* (FOX 1), which underlies the polysilicon interconnect runners and the contact areas (Fig. 3-42b). This is followed by the simultaneous deposition of epitaxial single-crystal silicon in the exposed silicon areas and of polysilicon over the oxide.

The active areas on both the polysilicon and the epi regions are then defined by growing a second field-oxide layer. As this oxide is grown, it selectively consumes the polysilicon layer. To insure good isolation, it is necessary to completely oxidize this poly layer on top of FOX 1. The process is then completed with gate oxidation, gate patterning, and source/drain formation.

The BOMOS structure allows larger contact areas to polysilicon, since the contact hole is opened on top of the oxide (Fig. 3-43). In addition, p^+ and n^+ poly can abut on top of the oxide, and a shared contact can be made to this joined area. Finally, the source/drain areas can be made smaller than in MOSFETs with conventional contacts, which results in devices with smaller source/drain-to-substrate junction capacitances. Although the BOMOS structure is formed by means of a self-aligned contact process, it still suffers from the narrow-width effects associated with the use of LOCOS. In addition, there are alignment and topography limitations in the source/drain areas.

The original BOMOS process was proposed in 1977.[107] Since then, several improvements to it have been suggested, including the LID-MOS[108] and the COO-MOS[109] processes. The process sequence shown in Fig. 3-42 incorporates the improvements described in these two reports. An additional enhancement would involve siliciding the poly extensions to reduce the high sheet resistance of the poly.

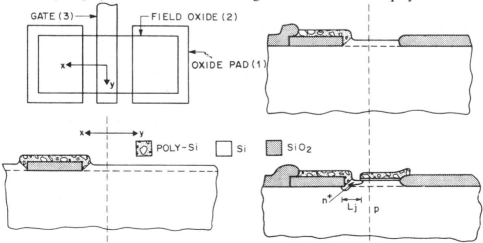

Fig. 3-42 The key process steps of the BOMOS self-aligned contact approach.[164] (© 1987 IEEE).

3.10   FORMATION OF SHALLOW JUNCTIONS AND THEIR IMPACT ON CONTACT FABRICATION

As VLSI device structures shrink laterally in dimension, it becomes necessary to scale the vertical dimensions of the devices as well. One important vertical dimension is the depth of *pn* junctions. For example, in MOS devices with channel lengths of less than 0.8 μm, source/drain junction depths smaller than 250 nm are required to maintain adequate device performance.[110] In this section we will describe the technological approaches to forming shallow junctions.

3.10.1   Conventional Shallow-Junction Formation

The conventional approaches to forming shallow junctions involve ion implantation and annealing. Shallow n^+ layers for NMOS source/drain regions are relatively easy to form in this way. Arsenic is typically used to form such layers, since the projected range of As at implant energies of 75 keV is only 50 nm. In addition, the heavy As atoms tend to create amorphous layers at the Si surface during the implant step. Such amorphous layers prevent As from undergoing significant channeling during implant (see Vol. 1, chap. 9). The amorphous layers can also be annealed through solid-phase epitaxy regrowth at relatively low temperatures. Finally, since As diffuses relatively slowly, such layers remain relatively shallow during the high-temperature anneal steps that must be performed following the source/drain implantation. Sheet resistances as low as 26 Ω/sq in As implanted layers 200 nm thick have been reported.

The formation of shallow p^+ junctions, however, is more difficult. Boron is a light atom (B^{11}), and when B implants are carried out at room temperature, an amorphous surface layer is not formed. As a result, boron implants exhibit channeling tails that penetrate relatively deeply into the Si, producing deep, as-implanted junctions. While attempts to use very low implant energies (e.g., 1-2 keV) might be effective,[111] such efforts would require the availability of low-energy ion implanters and perhaps even new ion sources. In addition, sputtering of the surface by the ion beam would have to be considered. Finally, the crystalline-defect damage caused by boron implantation must be annealed at temperatures >900°C, leading to significant diffusion of the implanted atoms, and thus driving the junctions even deeper.

Several approaches have been developed to minimize the problems associated with implanting shallow boron-doped junctions. First, the implanted dopant profile can be moved closer to the Si surface by implanting through a thin (e.g., 20-nm-thick) screen oxide. While this shifts the peak of the B film closer to the surface by roughly the oxide thickness, this approach introduces the problem of recoiled oxygen atoms and also worsens the straggle of the implanted doping distribution. Furthermore, if the oxide film thickness is close to the projected range of the ions, it is difficult to accurately control the dose penetrating through to the Si.

Second, BF_2^+ is implanted instead of B, in which case the B^+ ion acquires 22% of the BF_2^+ energy. This reduces the depth to which the B atoms are implanted (e.g., the projected range of B atoms from a 50 keV BF_2^+ implantation is ~30 nm). While the fluorine ions damage the surface and thus reduce B channeling, significant B channeling may still occur. (Fluorine, however, does cause unwanted defects when BF_2^+ is implanted, which is a drawback to this technique.)[131] Consequently, a prior implantation step may be performed using Si^+, Sn^+, or Ge^+ to preamorphize the Si surface[112] as a way to minimize B channeling.* Although tilting of the wafers by 7 or 8° with respect to the ion beam has also been used to reduce channeling, this procedure can cause asymmetric implant effects in the source/drain regions. Such effects can significantly impact MOS device behavior, especially for PMOS devices in which lightly doped drains are used (see chap. 5, section 5.6.5.2). Hence, this procedure may no longer be feasible for deep-submicron (< 0.7 μm) PMOS device processes.

The redistribution of B atoms following implantation may still cause undesirably deep junctions, even when a BF_2^+ implant is performed. The use of *rapid thermal processing* to replace furnace annealing addresses this problem. In addition, the solubility of the dopant, and hence the conductivity of the junction region, depends rather strongly on the temperature of the anneal for high dose implants. The data presented by Ryssel[132] indicate an increase by a factor of two in B solubility at equilibrium for every increase of 100°C above 900°C. Finally, it has been observed that when B implants are annealed at temperatures close to 800°C, they give rise to very large diffusive motions.[133] Thus, simply lowering the temperature in a furnace anneal in an attempt to reduce profile motion could instead lead to a much deeper junction, depending on specific experimental conditions.

It was also observed in Ryssel's report that a pre-anneal at a higher temperature using RTP diminishes this transient annealing effect. Junction depths following an RTP anneal have been shown to be 25-35% as deep as those performed with a furnace anneal. Boron-doped junctions with depths of 0.2 μm and surface concentrations of $N_A = 10^{20}/cm^3$ have been reported (Fig. 3-43).[113] The modeling of diffusion in shallow junctions under RTP has been discussed by Fair.[114,155]

3.10.2 Alternative Approaches to Forming Shallow Junctions

Conventional techniques for forming B-doped shallow junctions require complex multistep processes with high-temperature annealing steps to remove the implant damage and activate the dopant. A variety of alternative techniques for forming shallow junctions in Si are being pursued in an attempt to overcome these limitations, includ-

---

* Note that pre-amorphizing of the surface with a vertical ion beam will not prevent lateral channeling of the boron, since the Si under the mask edge will not be amorphized. Hence, even though the pre-amorphization technique is effective in reducing the vertical junction depth, it does not help reduce the lateral junction depth. Such lateral channeling presents a serious problem in MOS devices with submicron gate lengths, since it causes a significant reduction in channel length. A reported technique for tackling this problem is a preamorphization implant carried out at a 30° tilt angle. This causes amorphization under the mask and thus significantly suppresses lateral channeling.[147]

ing: implantation into silicide contact layers and diffusion of the dopant into the substrate; selective growth of Si over source/drain regions, followed by implantation into the epi Si and diffusion of the implanted dopants to form the junctions; depositionof doped layers by CVD, followed by diffusion to form the junctions; and gas immersion laser doping (GILD).

Implantation into already formed silicide layers has been described for a number of different silicides that are used to make contact to Si. These include WSi_2,[115] $CoSi_2$,[96] and $TiSi_2$.[116,149] In this method, virtually all of the implanted ions stop within the silicide layer and do not penetrate into the Si. The shallow junctions are formed by outdiffusion of the implanted dopant into the underlying Si. This approach has the following advantages:

• Since all the implanted ions stop in the silicide, no implant damage is created in the Si. As a result, leakage currents in the junction (caused by damage within the depletion region of the junction) are eliminated (Fig. 3-44).

• The processing temperature following implantation is no longer determined by the need to anneal out the implant damage (for B-doped junctions, this temperature would ordinarily be ≥900°C). Instead, very shallow (40-70-nm), uniform junctions can be fabricated by outdiffusing the junctions from the silicide at lower temperatures (e.g., 800°C).

• Diffusion results in steep dopant profiles without damage to the silicon lattice.

• Uniform dopant profiles can be achieved through outdiffusion of the dopants

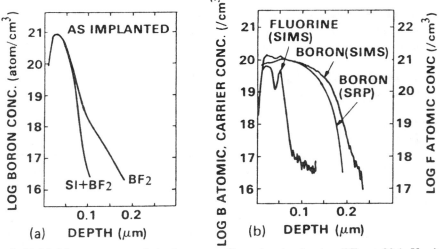

Fig. 3-43 (a) Measurements of the boron profiles after implanting BF_2 at 30 keV with a dose of $3x10^{15}$ cm^{-2} into monocrystalline and preamorphised silicon. Silicon implant parameters: 50 keV, $1x10^{15}$ cm^{-2}. (b) SIMS and spreading resistance (SRP) profiles for a BF_2 implant at 30 keV with a dose of $3x10^{15}$ cm^{-2} after rapid thermal annealing at 1150 °C for 5 sec.[113] Reprinted by permission of the publisher, The Electrochemical Society, Inc.

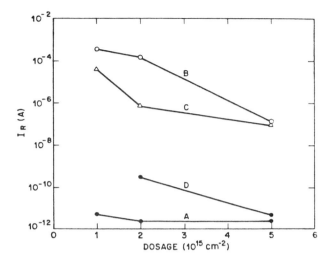

Fig. 3-44 Silicide junction leakage: (a) good n^+/p diode with no silicide, junction depth of 0.15 μm;   (b) after forming 0.07 μm of CoSi$_2$, Si consumed 0.07 μm; (c) after CoSi$_2$ formation at 700 °C, Si consumed 0.11 μm; (d) after a second 900°C/30 min drive.[96] (© 1986 IEEE).

from the silicide, since anisotropy and shadowing effects are eliminated. That is, the dopant diffuses rapidly in the silicide, and this wipes out any nonuniform dopant concentrations produced in the silicide as a result of shadowing during implantation. For more details on these shadowing effects in short-channel MOSFETs see chap. 5, section 5.6.5.2.

• Silicide formation occurs before junction formation, and as a result the same silicide thickness is produced on both p and n channel devices (i.e., Si consumption during silicide formation is reportedly dependent on the Si doping type and concentration).[149]

Although a high surface concentration of dopants can be achieved, a sufficiently high implant dose must be used, as the silicide may retain a considerable amount of the dopant within its grains and grain boundaries to satisfy the solubility of the dopant in the silicide. In addition, it must be verified that the dopant drive-in step does not degrade the stability of the silicide. It has also been reported that the dopant can be implanted into the metal prior to silicide formation, and that the diffusion that forms the shallow junction occurs simultaneously with the silicide formation in a single subsequent anneal step.[117]

Selective growth of Si and diffusion of the implanted dopants to form the junctions has also been investigated by several groups.[118,119] In this approach (Fig. 3-45 and Fig. 6-43, chap. 6), silicon is selectively grown (SSG) over the source/drain regions of MOS devices to a depth of 200-400 nm, following the completion of oxide-spacer

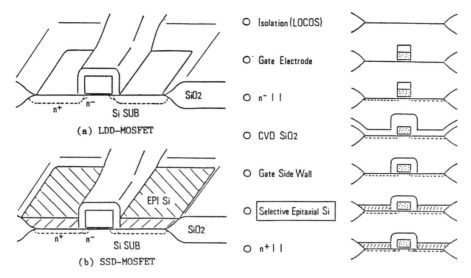

Fig. 3-45 Schematic diagram of (a) an LDD-MOSFET. (b) a stacked-source-drain (SSD) MOSFET using selective epitaxy. (c) Fabrication process flow of an SSD-MOSFET.[118] This paper was originally presented at the Spring 1988 Meeting of The Electrochemical Society, Inc. held in Atlanta, GA.

formation. A reduced-pressure, 900°C process using SiH_2Cl_2-HCl gas is employed for the SSG step. This produces raised source/drain regions. In one report, a BF_2^+ implant into the SSG film is followed by a 1050-1075°C RTP step to form a low resistance (60 Ω/sq) shallow junction (100 nm deep).[119] The RTP step is performed after a BPSG film has been deposited and contact holes in it have been opened. Thus, the RTP step also serves to reflow the contact holes. PMOS devices with gate lengths of 0.5 μm fabricated with this approach exhibited good device characteristics.

In the other report, a phosphorus implant into the SSG layer is performed so that a gradual-drain n^+ junction can be formed to reduce hot-carrier degradation. Non-selectively deposited polysilicon has also been used as a dopant source for shallow-junction formation: such processes are described in sections 3.11.2.4 and 3.11.2.5. Deposition of a *dopant source by CVD* and diffusion of dopant from it to form the junction has also been investigated.[120]

Gas-immersion laser doping (GILD) is another candidate process being evaluated for shallow-junction formation. In this process, the desired dopant species is incorporated into the Si during a melt/regrowth step that is initiated by a homogenized $\lambda = 308$ nm XeCl pulsed excimer laser beam (Fig. 3-46). A significant feature of this approach is that no high-temperature anneals are required following the source/drain doping step. Boron-doped junctions with depths of 25 - 150 nm and sheet-resistance values down to 20 Ω/sq (at a junction depth of 110 nm) have reportedly been fabricated using the GILD process (Fig. 3-47).[121] PMOS devices with gate lengths of 0.8 μm exhibiting good

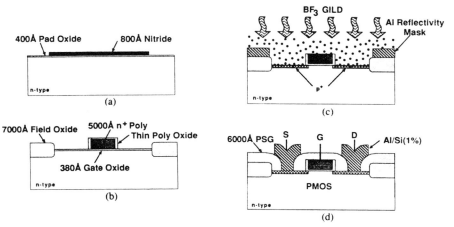

Fig. 3-46 Key steps in the GILD PMOS process: (a) active area definition, (b) after poly deposition, n^+ doping, patterning, and oxidation, (c) GILD laser doping step for p^+ source/drain formation, and (d) after PSG and Al/Si (1 %) deposition and etching.[121] (© 1988 IEEE).

performance have also been demonstrated using the GILD method.

A final novel technique for forming shallow junctions is *photo-enhanced epitaxy*. In this approach, a silicon epitaxial film is grown at low temperatures (<650°C) by the photo-enhanced dissociation of disilane under UV irradiation.[146] The UV radiation also

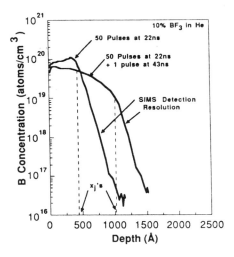

Fig. 3-47 SIMS profiles of boron (^{10}B plus ^{11}B) laser-doped junctions. The shallow profile received 50 laser pulses at 22-ns silicon melt, while the deeper profile received one additional pulse at 43-ns melt time. The actual junction depths are estimated to be 450 and 1000 Å, respectively.[121] (© 1988 IEEE).

activates the boron that is incorporated in the epitaxial film during growth. Ultra-shallow (< 0.1 μm) p^+n junctions were reported to have been formed with photo-enhanced epitaxy.

3.10.3   Impact of Shallow Junctions on Contact Formation

The use of shallow junctions implies the need for planar contact interfaces that penetrate the original Si surface only slightly (or better, not at all), as well as a barrier layer to prevent Si from diffusing into any Al layers (which would thereby cause junction spiking). If a contact layer is used that consumes some Si during the contact formation (e.g., as occurs when PtSi, TiSi$_2$, CoSi$_2$, and CVD W layers are used to contact Si), measures must be taken to prevent excess Si from from being consumed. The considerations necessary for selecting the correct thickness of Ti in an Al-TiN-Ti-Si contact structure to 0.2-μm deep junctions are presented in reference 81. As noted earlier, CVD W seems to be adequate for making direct contact to junctions greater than 250 nm deep, if the silane reduction process is used.

The approach of diffusing the dopant from a polysilicon layer or a selectively grown epitaxial layer on the Si (described in the section on self-aligned contacts) offers another solution to the problem of how to form a reliable contact to shallow junctions. In such cases the metal makes contact to the deposited Si far from the original Si interface (and therefore away from the shallow junction as well).

3.11 BURIED CONTACTS AND LOCAL INTERCONNECTS

3.11.1 Butted Contacts and Buried Contacts

When silicon-gate technology was developed, a means had to be provided for making contact between the polysilicon layer and the single-crystal substrate. In early silicon-gate MOS circuits, such contacts were made by using either a metal link to interconnect the polysilicon and the substrate (Fig. 3-48), or by so-called *butted contacts*. The latter were subsequently replaced by *buried contacts*, which became a standard structure in NMOS ICs.

With the *butted contact*, polysilicon is aligned with the edge of the active-device area to which the contact will be established. This is done by patterning the polysilicon film after it has been deposited. After an insulating layer has been deposited to cover the poly, a contact window that overlaps both the poly and the substrate is opened, exposing both the poly layer and the substrate. Metal is deposited to fill the contact, thereby electrically strapping the two regions together (Fig. 3-49). The butted contact conserves area by eliminating the space required between the separate contact windows when the approach of Fig. 3-48 is used.

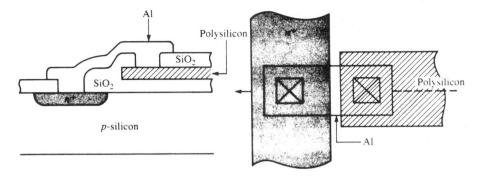

Fig. 3-48 Aluminum film link for connecting polysilicon and the silicon substrate with an intervening space. From R. C. Jaeger, *Introduction to Microelectronic Fabrication.* Copyright, 1988, Addison-Wesley. Reprinted with permission.

The term "butted contact" derives from the fact that the poly and the substrate edges butt up against one another, but do not actually make electrical contact. Since the poly is separated from the substrate by an oxide layer even when the contact window is opened, a linking metal film must still be deposited to electrically connect the poly and the substrate.

With the *buried contact*, direct contact is made between polysilicon and the substrate, eliminating the need for a metal link to form the contact (Fig. 3-50). In this structure, a window is opened in the thin gate oxide over the substrate area at which the contact is to be established. When the polysilicon is subsequently deposited, it is in direct contact with the substrate in these openings but is isolated from the substrate by the gate and field oxides everywhere else. An ohmic contact is formed at the poly-substrate Si interface by the diffusion into the substrate of dopant present in the polysilicon. This

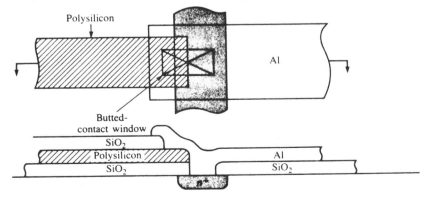

Fig. 3-49 Cross section of a butted-contact structure. From R. C. Jaeger, *Introduction to Microelectronic Fabrication.* Copyright, 1988, Addison-Wesley. Reprinted with permission.

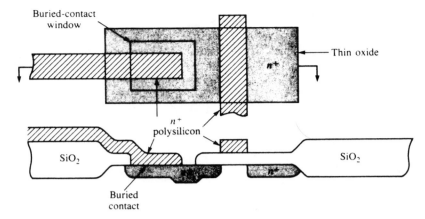

Fig. 3-50 Cross section of a buried-contact structure. From R. C. Jaeger, *Introduction to Microelectronic Fabrication*. Copyright, 1988, Addison-Wesley. Reprinted with permission.

in effect merges the two silicon regions. A CVD layer of insulating film is then deposited to cover the contact. The structure is called a "buried contact" because a metal layer can cross over the area of the substrate where a contact has been established without making an electrical connection to it.

The use of buried contacts in silicon-gate technology provides an important benefit in that it makes available an additional level of interconnect on the integrated circuit. This extra layer allows some of the circuit connections to be formed before the metal layer is deposited, and its availability adds significant interconnect routing flexibility to the design task. However, this additional interconnect level does not exist in technologies using butted contacts, since a metal link is needed to establish each poly/substrate contact.

3.11.2 Local Interconnects

Even when used with buried contacts, the polysilicon interconnect layer is not as universal as the metal interconnect layer because the sheet resistance of the polysilicon layer is much higher than that of Al. Furthermore, the poly layer is used both as a *gate material* and an *interconnect material*. Consequently, when the poly film acts as an interconnect structure, it cannot cross over regions where a transistor gate exists without making a contact to the gate. Therefore, unless such contacts are desired, gate locations represent areas on the wafer that cannot be crossed by poly films when these are being used as interconnect structures.

The metal layer, on the other hand, is allowed to pass over polysilicon gates (since these have been covered with SiO_2 before the deposition of the metal). That is to say, there are no areas of the chip surface over which the metal line is restricted from being routed. Such unrestricted interconnect levels are therefore referred to as *global interconnect levels* to distinguish them from such routing-restricted interconnect levels as the polysilicon with buried contacts. The routing-restricted levels have been termed *local interconnect (LI) levels*, because their higher resistivities make them best used for such short metallization runs as those that locally interconnect gates and drains in NMOS circuits.

Other materials besides polysilicon have been used to establish such local interconnects. For example, in NMOS, local interconnect structures were implemented using bilayer conductor structures, with a refractory metal-silicide layer (i.e., WSi_2, $TaSi_2$, $MoSi_2$, or $TiSi_2$) on top of a polysilicon layer. Such structures, referred to as *polycides*, were adopted as a means of reducing the relatively high resistivity of heavily doped polysilicon.

Establishing an LI level is more complex in CMOS than in NMOS, and most early CMOS processes did not implement a local interconnect level. The difficulties were these:

- The polysilicon layer in CMOS was normally doped *n*-type. Even when used just as an underlayer in a polycide structure, such poly could not be used to make buried contacts to the source/drain regions of PMOS devices.

- Phosphorus outdiffusion limits device isolation and buried-contact design rule scaling (Fig. 3-51).

- In the conventional buried-contact process, the gate oxide must be patterned before the poly is deposited. This patterning step can adversely impact yield if the oxide is scaled to 25 nm and below.

The fact that buried contacts were not implemented was one of the factors that historically kept the packing density of CMOS circuits less than that of NMOS circuits. Recently, however, several LI technologies for advanced CMOS have been reported, including selective formation of $TiSi_2$, Ti:W over $CoSi_2$, TiN over $TiSi_2$, and dual-doped polysilicon.

Despite the fact that LI levels cannot be used as universally as metal layers, their

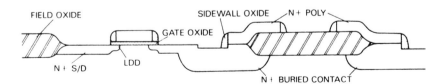

FIELD OXIDE   SIDEWALL OXIDE   N+ POLY
                 GATE OXIDE
N+ S/D   LDD
                 N+ BURIED CONTACT

Fig. 3-51 Limitations of buried contacts: (1) Phosphorus outdiffusion limits device isolation and buried-contact-to-gate design-rule scaling. (2) The use of *n*-type polysilicon limits buried contacts in CMOS to *n*-channel transistors only.[124] (© 1987 IEEE).

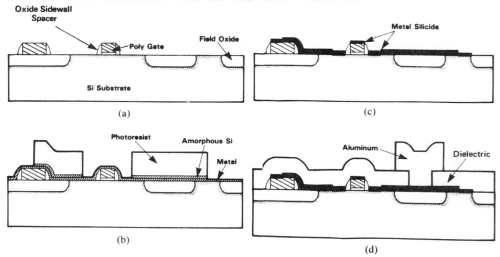

Fig. 3-52 Fabrication of HPSAC technology. Cross-sectional view after (a) sidewall spacer formation; (b) patterning and etching of amorphous silicon layer; (c) silicide formation; and (d) aluminum alloying.[122] (© 1984 IEEE).

implementation still allows significant chip-area savings for VLSI logic and SRAM designs. According to one report, SRAM cell layouts were reduced by up to 25% when an LI level was made available.[124] The reduction of cell size in such regular structures as SRAMs by use of local interconnects is considered in further detail in a more recent reference as well.[159]

3.11.2.1 Selectively Formed TiSi$_2$.

The selective formation of TiSi$_2$ to form an LI level (Fig. 3-52) was first described in 1984, and the process is referred to as *high-performance self-aligned contact and local interconnection* (HPSAC).[122] In this approach, conventional MOS processing is carried out up to the formation of oxide sidewall spacers and the definition of source/drain regions (Fig. 3-52a). A layer of Ti is then deposited, followed by a layer of amorphous Si (e.g., 40-50 nm and 80-100 nm thick, respectively). The thickness of the Si layer is chosen so that the Si will react fully with the underlying Ti layer.

Both layers are sequentially sputter-deposited in one pumpdown to minimize the formation of an oxide on the Ti layer (which could retard silicide formation). The amorphous Si layer is then patterned, by means of a plasma-etching process that has high selectivity to the underlying Ti layer (so that as little of the Ti is removed during any overetch of the Si; see Fig. 3-52c).

Next, the silicide is formed in a 600°C anneal in N$_2$. The Ti in contact with the substrate reacts with that Si to form TiSi$_2$. The Ti on SiO$_2$ does not react unless it is covered by amorphous Si, in which case TiSi$_2$ is also formed. The unreacted Ti (on which a thin layer of TiN actually forms) is then selectively removed with an H$_2$SO$_4$ + H$_2$O$_2$ solution. A final 800°C anneal is used to reduce the resistivity of the TiSi$_2$.

Since this process produces an LI layer of $TiSi_2$, as well as $TiSi_2$-Si contacts, it can be used to establish contacts between gate and source/drain regions (in addition to contacts between neighboring source/drain regions). Because the $TiSi_2$ layer in this structure has a sheet resistance of ~3 Ω/sq and is thin (<150 nm), metal-layer step-coverage problems are not significantly exacerbated. In addition, the ρ_c of $TiSi_2$ to Si is 20 Ω-μm^2 to n^+Si and 100 Ω-μm^2 to p^+Si.

A CVD layer of BPSG is then deposited, and openings are created in it so that an Al overlayer can make contact to the $TiSi_2$ LI layer. Such Al-$TiSi_2$ contacts can be made over field-oxide region because the $TiSi_2$ extends over these regions. However, if the Al-$TiSi_2$ contacts overlap the field oxide and source/drain region (as that shown in Fig. 3-53), a barrier metal such as Ti:W, needs to be inserted to prevent Al from diffusing through the silicide into the Si substrate.

Several processing challenges must nevertheless be overcome in order for the HPSAC process to be successfully implemented. First, the Si-to-Ti etch selectivity must be quite high so that Ti is not excessively removed over contact regions during the overetch of the amorphous Si. Second, the gate edges must be cleared of Si during the Si-removal step, and of Ti during the Ti-removal step. Third, low-leakage, shallow-junction processes must be incorporated (since the $TiSi_2$ consumes some Si during the Si contact formation). Fourth, as the amorphous Si that covers the Ti prevents nitrogen stuffing of the Ti grain boundaries, Si can rapidly diffuse along these grain-boundary paths at the same time that the Ti is reacting with the Si to form $TiSi_2$. This effect becomes a problem when a small area of the Si substrate is in contact with a larger area of Ti (>5x the area of the exposed Si substrate), because sufficient Si from the substrate can be "sucked out" by the Ti to cause pitting of the Si. Thus, the area of the Ti interconnect runner above the field oxide connected to a silicon contact must be restricted by design.[154] Finally, phosphorus has a high diffusivity in $TiSi_2$. Hence, under some conditions phosphorus from the polysilicon gate regions in contact with the $TiSi_2$ could diffuse through the $TiSi_2$ and counterdope p^+ junctions. To prevent this

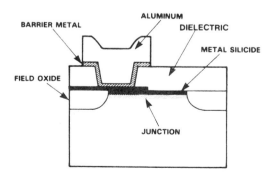

Fig. 3-53 HPSAC contact overlapping source-drain region and field oxide.[122] (© 1984 IEEE).

occurrence, either the thermal budget following silicide formation must be appropriately restricted, or an alternative polysilicon electrode structure must be employed (see chap. 6, section 6.3).

3.11.2.2 Ti:W over CoSi$_2$.

In this approach, a Ti:W layer is sputter-deposited over the entire wafer surface, including regions in which self-aligned contacts to Si by CoSi$_2$ are exposed. The sputter deposition is carried out in an ambient containing Ar + N$_2$, so that nitrogen is incorporated into the Ti:W film to improve its diffusion-barrier properties. The Ti:W film is then patterned to form the LI structure. The advantage of this method is that the Ti:W should be a better barrier to the diffusion of phosphorus than is the TiSi$_2$ film of the HPSAC process. Regions of n^+ and p^+ Si can therefore be connected with the Ti:W film with less concern about counterdoping of one type of poly by the other.[123] The implementation of a CoSi$_2$ layer as an LI was also studied.[123] However, because void formation at the CoSi$_2$/SiO$_2$ interface was observed, concerns were raised about possible adhesion problems between the CoSi$_2$ and SiO$_2$ materials.

3.11.2.3 TiN Formed over TiSi$_2$.

In this process of LI formation for CMOS, first reported in 1985,[124] a TiN layer is formed on a layer of Ti as the reaction to form TiSi$_2$ is occurring. The process steps are the same as those used in the HPSAC process through the deposition of the Ti layer (Fig. 3-54a). At that point, instead of depositing an amorphous Si layer, the Ti (e.g., 100-nm thick) is annealed in an ambient containing

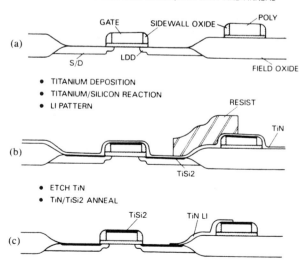

Fig. 3-54 Self-aligned TiSi$_2$ process showing the TiN LI patterning step.[124] (© 1987 IEEE).

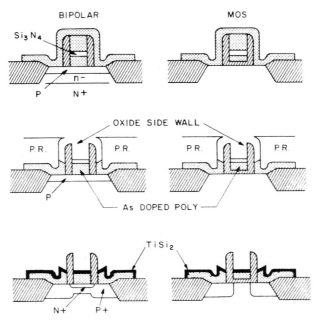

Fig. 3-55 (a) SDB poly is deposited over the wafer. (b) Planarization resist is etched back, exposing top of the gate/emitter stack. This figure depicts the structure after isotropic polysilicon etch. (c) This figure shows the devices after salicide process.[127] (© 1987 IEEE).

N_2 at 675°C. The Ti in contact with the SiO_2 reacts with the nitrogen to form a film of TiN ~110 nm thick. The Ti in contact with the Si forms a top layer of TiN that is ~40-nm thick: beneath this surface layer of TiN, a 150-nm-thick layer of $TiSi_2$ is formed. By varying the temperature and ambient gas during the anneal, it is possible to obtain various TiN-$TiSi_2$ thickness combinations.

Following this anneal step, the TiN is patterned with a two-step, dry/wet etch process (Fig. 3-54b). The fluorine-based dry-etch step removes the TiN from the unmasked SiO_2 regions, as well as from on top of the unmasked $TiSi_2$ regions (Fig. 3-54c). The wet portion of the etch process is used to remove any stringers of TiN that might remain at the base of the transistor sidewall oxide. Following the etch-and-resist strip, an 800°C anneal is performed to reduce the $TiSi_2$ and the TiN sheet resistivities to 1.0-1.5 Ω/sq and 10-20 Ω/sq, respectively.

This process has several advantages over the HPSAC process. First, the TiN LI layer can be formed without the need for any layers other than Ti to be deposited. Second, the TiN is an excellent diffusion barrier, and prevents phosphorus from n^+ poly from counterdoping p^+ junctions. Similarly, the TiN can serve as a diffusion-barrier material when contacts such as those shown in Fig. 3-53 are used.

3.11.2.4 Dual-Doped Polysilicon LI with Diffused Source/Drain Junctions. In the second dual-doped polysilicon LI process, some of the limitations

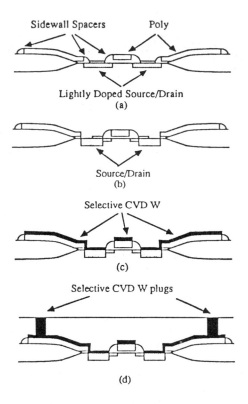

Fig. 3-56 Processing sequence of selective CVD W local interconnect technology. Cross sectional view after (a) sidewall spacer formation., (b) selective removal of sidewall spacer of local interconnect and source/drain formation, (c) selective CVD W deposition, and (d) selective CVD W plug filling.[128] (© 1988 IEEE).

of the first poly LI process are overcome, but at the cost of introducing a more complex process for formation of the gate structure.[124,127] In this approach, three separate polysilicon layers are deposited (Figure 3-55). The final poly layer makes direct contact with the exposed source/drain regions and is selectively doped by means of ion implantation after it is deposited.

An etch step is then used to remove part of poly 3 and all of poly 2, creating the gate structure shown in Fig. 3-55b. This structure has oxide sidewalls protruding 600 nm above the top of poly 1 (gate poly) and the remaining parts of poly 3 (source/drain dopant and LI poly). The dopants implanted in the poly are then driven into the Si substrate by means of a diffusion step, to simultaneously form the buried contacts and the source/drain junctions. All of the polysilicon (poly 3 and the gate poly) is then selectively clad with a $TiSi_2$ layer. The protruding oxide sidewalls prevent any bridging of TiSi$_2$ between gate and LI poly. It should be noted, however, that this approach requires increased process complexity, and the problem of dual-doped poly counterdoping through the silicide can still occur.

3.11.2.5 CVD W-Clad Polysilicon LI.

Another LI process in which W is selectively deposited onto patterned polysilicon in the field regions and onto the exposed source/drain regions was recently described.[128] In this method, isolated polysilicon runners in the field regions are patterned at the same time as the polysilicon gate structures (Fig. 3-56). Following LDD and source/drain-region formation, a thin layer of CVD W is selectively deposited onto the gate poly, the exposed source/drain regions, and the LI poly runners in the field regions.

With this approach, it is not necessary to perform selective etching of the LI layer. In addition, redistribution of the dopants through the poly into the source/drain regions will not occur, since the LI poly is not in contact with the source/drain regions (instead, the LI poly and the source/drain regions are connected by the CVD W). Since the CVD W also straps the entire source/drain region, care must be taken to ensure that increased diode leakage currents are not produced. This is believed to be feasible if the CVD W is performed using the silane (SiH$_4$) reduction of WF$_6$, which has been shown to suppress encroachment and wormholes (two of the main sources of such leakage currents).

REFERENCES

1. C. Y. Chang, Y. Y. Fang, and S. M. Sze, *Solid State Electron.*, **14**, p. 541 (1971).
2. R. L. Maddox, *IEEE Trans. Electron Dev.*, **ED-32**, p. 682 (1985).
3. S. S. Cohen et al., *J. Electrochem. Soc.*, **130**, April 1983, p. 979.
4. S. J. Proctor, L. W. Linholm, and J. A. Mazer, *IEEE Trans. Electron Dev.*, **ED-30**, 1983, p. 1535.
5. J. Chern and W. G. Oldham, *IEEE Electron Dev. Letts.*, **EDL-5**, May 1984, p. 178.
6. H. H. Berger, *J. Electrochem. Soc.*, April 1972, p. 507.
7. W. M. Loh et al., "Modeling and Measurement of Contact Resistances," *IEEE Trans. Electron Dev.*, March 1987, p. 512.
8. R. B. Fair, in "Semiconductor Silicon 1981", p. 963, Electrochem Soc., Princeton, N. J., 1982.
9. E. Guerrero et al., *J. Electrochem. Soc.*, **129**, (1982), p. 1826).
10. J. Dieleman et al., *Solid State Technology*, April 1984, p. 191.
11. C. R. Cleavlin and G. T. Duranko, *Semiconductor Internat.* Nov 1987, p. 94.
12. M. J. Kim et al., *IEDM Tech. Dig.*, 1984, p. 137.
13. K. Shenai et al., *Ext Abs. Electrochem Soc. Meeting,* Spring, 1988, p. 169.
14. R. A. M. Wolters and A. J. M. Nellissen, *Proceedings 5th Intl.VMIC Conf.*, Santa Clara, CA, 1988, p. 351.
15. R. Kopp, *Semiconductor Intl.*, Jan 1989, p. 54.
16. R. I. Baker and S. R. Jennings, *Proc. 3rd Intl IEEE VMIC Conf.*, Santa Clara, CA, 1986, p. 484.

17. H. Ozaki et al., "Contact Resistance Behavior in BPSG Glass", *Proc. 4th Intll. IEEE VMIC Conf.*, Santa Clara, CA, 1987, p. 323.

18. J. S. Chang, *Solid State Tech.*, April, 1984, p. 214.

19. G. Higelin et al., *Proc. 5th Annual IEEE VMIC Conf.*, Santa Clara, CA, 1988, p. 29.

20. J. A. Bondur and R. G. Frieser, in "Plasma Etching and Deposition" Electrochemical Society Proceedings, Pennington, N.J., 1981, p. 180.

21. R. Catellano, "Profile Control in Plasma Etching of SiO_2", *Solid State Tech.*, May, 1984, p. 203.

22. L. Pechaud et al., Abs. 396, *Ext. Abs. of Electrochem. Soc. Meeting,* Oct. 1984, p. 554.

23. S. Roth, W. Ray, and G. Wissen, *Semicond. Intl.*, May, 1988, p. 138.

24. D. Jillie et al., *J. Electrochem. Soc.*, Aug. 1987, p. 1988.

25. H. Kudoh et al., *J. Electrochem. Soc.*, Aug. 1986, p. 1666.

26. V. Grewal, H.P. Erb, and P. Mokrisch, *Proc. 4th Intl IEEE VMIC Conf.*, Santa Clara, CA, 1987, p. 298.

27. T. Tsuchiya et al., Abs. 261, *Ext. Abs. Electrochem. Soc. Meeting,* Oct.1988, p. 373.

28. J. L. Vossen et al., *J. Vac. Sci Tech.* A2 (2), 1984, p. 212.

29. H. Tomioka, S.-I. Tanabe, and K. Mizukami, "A New Reliability Problem Associated with Ar Ion Sputter Cleaning of Interconnect Vias", *Internat. Reliability Sympos.*, 1989.

30. T. J. Faith, et al., "Contact Resistance Monitor for Si ICs", *J.Vac. Sci. Tech.,* **82**, Jan-Mar. 1984, p. 54.

31. E. Nagasawa et al., *IEEE Trans. Electron Dev.*, March 1987, p. 581.

32. J. O. McCaldin, *J. Vac. Sci. Tech.*, November 1974, p. 990.

33. Y. Pauleau, "Interconnect Materials for VLSI Circuits, Part II", *Solid-State Tech.,* April, 1987, p. 155.

34. A. K. Sinha, *Thin Solid Films,* **90** (3), p. 271, 1982.

35. J. Curry et al., *Proc. Internat. Reliability Symp.*, 1984, p. 6.

36. D. Flowers, "Processing Requirements for Multilevel Interconnect Fabrication", *Proceedings of the 3rd IEEE VMIC Conference,* Santa Clara, CA, 1986, p. 78.

37. H. H. Hosack, *Appl. Phys. Lett.,* **21**, (1972), p. 256.

38. R. N. Singh, D. W. Skelly, and D. M. Brown, *J. Electrochem. Soc.,* Nov. 1986 p. 2390.

39. C. Canali et al., *J. Appl. Phys. Letts.*, **31**, (1977), p. 43.

40. C. A. Crider et al., *J. Appl. Phys.*, **52** (4), April 1981, p. 2860.

41. K. E. Broadbent et al., *IEEE Elect. Dev. Letts.*, July 87, p. 318.

42. C.-A. Chang et al., *J. Electrochem. Soc.* June 1986, p. 1256.

43. P. Merchant and J. Amano, "Thermal Stability of Diffusion Barriers for Alloy/PtSi Contacts", *J. Vac. Sci. Technology,* **A1(2)**, Apr.-June, 1983, p. 459.

44. K. N. Tu, "Shallow and Parallel Silicide Contacts", *J. Vac. Sci. Technology,* **19** (3), Sept./Oct. 1981, p. 766.

45. M. A. Nicolet and M. Bartur, *J. Vac. Sci. Tech.*, 19 (3), Sept/Oct. 1981, p. 786.

46. J. R. Shappirio, "Diffusion Barriers in Advanced Semiconductor Device Technology", *Solid State Technology,* Oct. 1985, p. 161.

47. T. C. Y. Ting and M. Wittmer, *Thin Solid Films,* **96**, 1982, p. 327.

48. J. A. Cunningham et al., *IEEE Trans. Reliab.*, **19**, (1970), p. 182.

49. P. B. Ghate et al., *Thin Solid Films,* **53**, (1978), p. 117.

50. R. A. M. Wolters and A. J. M. Nellisen,"Properties of Reactive Sputtered Ti:W," *Solid State Technol.,* Feb. 1986, p. 131.

51. M. J. Kimet al., *VLSI Science and Technology, 1985,* Electrochemical Society, Pennington, N.J., p. 213.

52. T. Hara et al., *IEEE Trans. Electron Dev.,* March 1987, p. 593.

53. C. Canali, F. Fantini, and E. Zanoni, *Thin Solid Films,* **97,** (1982), p. 325.

54. J. Nulty, G. Spadini, and D. Pramanik, *Proceedings of the 5th IEEE VMIC Conference,* Santa Clara, CA, 1988, p. 453.

55. S. Vaidya, *Appl. Phys. Lett.,* **39,** p. 900, (1981).

56. M. Finetti et al., *Solid State Electronics,* **23,** 1980, p. 255.

57. R. W. Bower, *Appl. Phys. Lett.,* **23,** (1973), p. 99.

58. C. Y. Ting and B. L. Crowder, *J. Electrochem. Soc.,* **129,** (1982), p. 2590.

59. M. M. Farhani, T. E. Turner, and J. J. Barnes, *J. Electrochem. Soc.,* Nov. 1987, p. 2835.

60. C. W. Nelson, *Proc. Internatl. Symp. on Hybrid Microelectronics,* Dallas, TX, 1969, p. 413.

61. W. J. Garceau et al., *Thin Solid Films,* **53,** (1978), p. 195.

62. M. Wittmer, *Appl. Phys. Letts,* **37,** (1980), p. 540.

63. M. Inoue et al., *Proc. 5th Intl IEEE VMIC Conf.,* Santa Clara, CA, 1988, p. 205.

64. B. Lee et al., *Proc. 4th Intl IEEE VMIC Conf.,* Santa Clara, CA, 1987, p. 344.

65. T. Hara et al., *Japan J. Appl. Phys.,* **24(7),** (1985), p. 828.

66. C. Yang et al., *Proc. 4th Intll IEEE VMIS Conf.,* Santa Clara, CA, 1987, p. 200.

67. R. A. Levy et al., *J. Electrochem. Soc.,* Sept, 1986, p. 1905.

68. T. Ohba, S. Inoue, and M. Maeda, *Tech. Dig. IEDM,* 1987, p. 213.

69. H. Kotani et al., *Tech. Dig. IEDM,* 1987, p. 217.

70. M. L. Green, *Ext. Abs. of Electrochem. Soc. Meeting,* Spring, 1988, Abs. No. 208, p. 322.

71. K. T.-Y. Kung et al., *Appl. Phys. Lett.* **42,** 987, (1983).

72. M. Finetti et al., *J. Appl. Phys.* **55,** 3882, (1984).

73. M. Eisenberg et al., *J. Appl. Phys.,* **54,** p. 3799, (1984).

74. J. D. Wiley et al., *Proc. High-Temp. Electronics Conf., 1st, Tuscon, Ariz.* p. 35, (1985).

75. N. Circelli and J. Hems, *Solid State Technol,* Feb 1988, p. 75.

76. T. Brat et al., *Ext. Abs. Electrochem. Soc. Meeting, Fall, 88,* Abs. No. 250, p. 354, 1988.

77. S. Murakami et al., *Proceedings of the 4th Intl VMIC Conf.,* Santa Clara, CA, 1987. p. 148.

78. M. Delfino et al., *IEEE Electron Dev. Letts.,* Nov. 1985, p. 591.

79. S. W. Sun et al., *IEEE Electron Dev. Letts.,* Feb 1988, p. 68.

80. K.-Y. Fu and R.E. Pyle, *IEEE Trans. Electron Dev.,* Dec. 1988, p. 2151.

81. T. Maeda, et al., *IEEE Trans. Electron Dev.,* Mar. 1987, p. 599.

82. M. J. Kim et al., *IEEE Trans. Electron Dev.,* July, 1985, p. 1328.

83. M. J. Kim et al., *J. Electrochem. Soc.,* Oct. 1987, p. 2603.

84. K. K. Ng and W. T. Lynch, *IEEE Trans. Electron. Dev.,* July, 1986, p. 965.

85. K. K. Ng and W. T. Lynch, *IEEE Trans. Electron. Dev.*, March, 1987, p. 503.
86. D. M. Brown, M. Ghezzo, and J. M. Pimbley, *Proceedings of the IEEE*, Dec., 1986, p. 1678.
87. G. Sh. Gildenblatt and S.S. Cohen, Chapter 6 "Contact Metallization", in *VLSI Metallization*, VLSI Electronics, Vol. 15, Academic Press, 1987, Orlando, FL.
88. G. Baccarani and G.A. Sai-Halasz, *IEEE Electron Dev. Letts.*, Feb. 1983, p. 27.
89. R. A. Haken, "Applications of the Self-Aligned TiSi$_2$ Process to VLSI MOS CMOS Technologies", *Workshop on Refractory Metals and Silicides for VLSI-IIII*, San Juan Batista, CA, May, 1985.
90. S. P. Murarka. et al., *J. Electrochem. Soc.*, **129**, (1982), p. 293.
91. S. P. Murarka, *J. Vac. Sci. Tech.*, Nov. 1986.
92. C. Y. Ting et al., Interaction Between Ti and SiO$_2$", *J. Electrochem. Soc.* **131**, p. 2934, 1984.
93. D. B. Scott et al., *IEEE Trans. Electron Dev.*, March 1987, p. 562.
94. R. K. Shukla and J. S. Multani, *Proceedings of the 4th Intl VMIC Conf.*, Santa Clara, CA, 1987, p. 471.
95. S.-T. Chang and K. Y. Chiu, *IEEE Electron Dev. Letts.*, May, 1986, p. 244.
96. R. Liu, D. S. Williams, and W. T. Lynch, *Tech. Dig. IEDM*, 1986, p. 58.
97. C. Mallardeau, Y. Morand, and E. Abonneau, *J. Electrochem. Soc.*, January 1989, p. 238.
98. C. M. Osburn et al., *J. Electrochem. Soc.*, June 1988, p. 1490.
99. K.-Y. Fu and R. E. Pyle, *Proceedings 5th Intl VMIC Conf.*, Santa Clara, CA, 1988, p. 469.
100. L. Van den Hove et al., *Ext. Abs. of the Electrochem. Soc. Meeting*, Spring, 1988, p. 312.
101. G. C. Smith et al., *Proceedings of the 4th Intl VMIC Conf.*, Santa Clara, CA, 1987, p. 155.
102. L. C. Van de Hove et al., *IEEE Trans. Electron Dev.*, March, 1987, p. 554.
103. S. P. Murarka et al., *IEEE Trans. Electron Dev.*, Oct. 1987, p. 2108.
104. L. Van den Hove et al., *Electrochem. Soc. Abstracts*, Vol. 88-1, Atlanta, GA, May 15-20, 1988, p. 312.
105. K. E. Broadbent et al., *Proceedings of the 5th IEEE VMIC Conf.*, Santa Clara, CA, 1988, p. 175.
106. W. T. Lynch and K.K. Ng, *Tech. Dig. IEDM*, 1988, p. 352.
107. J. Sakuai, *IEDM Tech. Dig.*, 1977, p. 388.
108. H. Inokawa, T. Kobayashi, and K. Kiuchi, *IEEE Electron Dev. Letts*, **EDL-8**, Mar. 1987, p. 98.
109. C.H. Dennison et al., *IEDM Tech. Dig.*,1985, p. 204.
110. J. W. Brews et al., "Generalized Guide for MOSFET Miniaturization", *IEEE Electron Dev. Lett.*, **EDL-1**, p. 2, 1980.
111. T. E. Seidel et al., *Nucl. Instr. and Meth. in Phys. Res.*, B7/8 (1985), p. 251.
112. M. C. Ozturk et al., *IEEE Trans. Electron Dev.*, May 1988, p. 659.
113. M.E. Lunnon, J. T. Chen, and J. E. Baker, *J. Electrochem. Soc.*, **132**, 2473 (1985).

114. R. B. Fair, *Ext. Abs. Electrochem. Soc. Meeting, Spring, 1988*, Abs. No. 194, p. 303, 1988.

115. F. C. Shone, K. C. Saraswat, and J. D. Plummer, *Tech. Dig. IEDM*, 1985, p. 407

116. B. Dance, *Semicond. Internal.*, January 1989, p. 22.

117. M. Horiuchi and K. Yamaguchi, *IEEE Trans. Electron Dev.*, 33, (1986), p. 260.

118. T. Makino et al, *Ext. Abs. Electrochem. Soc. Meeting, Spring, 1988*, Abs. No. 193, p. 301, 1988.

119. H. Shibata et al., *Tech. Dig. IEDM*, 1987, p. 590.

120. J. F. Gibbons et al., *Matls. Res. Soc. Proc.* Vol. 92, p. 281, (1987).

121. G. Carey, K. H. Weiner, and T. W. Sigmon, *IEEE Electron Dev. Letts,* Oct. 1988, p. 542.

122. D. C. Chen *et al., Tech. Dig. IEDM*, 1984, p. 118.

123. R. Wolters and L. Van den Hove, *Proc. 5th IEEE VMIC Conf.,* Santa Clara, CA, 1988, p. 149.

124. T. Tang et al., *IEEE Trans. Electron Dev.*, March, 1987, p. 682.

125. M. H. El-Diwany et al., *IEEE Trans. Electron Dev.*, , Sept 1988, p. 1556.

126. N. Yokoyama, K. Hinode, and Y. Homma, *J. Electrochemical Soc.*, Mar. 1989, p. 882.

127. T. Y. Chiu et al., *Tech. Dig. IEDM*, 1987, p. 24.

128. V.V. Lee, S. Veronckt-Vandebroek, and S.S. Wong, *Tech. Dig. IEDM*, 1988, p. 450.

129. A. R. Sitaram and S. P. Murarka, *Ext. Abs. Electrochem. Soc. Meeting,* Fall 1988, Abs. No. 263, p. 376, 1988.

130. R. M. Vadjikar and R. P. Roberge, *Ext. Abs. Electrochem. Soc. Meeting,* Fall 1988, Abs. No. 467, p. 681, 1988.

131. T. O. Sedgwick, *Tech. Proceedings Semicon/East 1986, p. 2.*

132. H. Ryssel et al., *Appl. Phys.*, 22, p. 35, (1980).

133. A. E. Michel, *Mater. Res. Soc. Symp. Proc.*, T. E. Seidel and B. Y. Tsaur, Eds. Vol. 52, (1986), p. 3.

134. R. E. Novak, *Solid State Tech.*, March, 1988, p. 39.

135. P. B. Johnson and P. Sethna, *Semicond. Internat.*, October, 1987, p. 80.

136. C. Y. Ting et al., *J. Electrochem. Soc.*, December, 1986, p. 2631.

137. J. K. Elliott, *Semiconductor Internatl.*, March, 1988, p. 46.

138. M. Wittmer, *J. Electrochem. Soc.*, Aug. 1988, p. 2049.

139. E. A. Taft, "Growth of Native Oxide on Si, *J. Electrochem. Soc.*, Apr. 88, p. 1022.

140. W. Kern and D. Freeman, *Ext. Abs. Electrochem. Soc. Meeting,* Fall 1988, Abs. No. 238, p. 333, 1988.

141. D. Freeman et al., *Ext. Abs. Electrochem. Soc. Meeting,* Fall 1988, Abs. No. 240, p. 337, 1988.

142. D. C. Chen et al., *IEEE Trans. Electron Dev.*, Oct. 1986, p. 1463.

143. S. Saito et al., *Tungsten and Other Refractory Metals for VLSI Applications II*, 1984, San Juan Batista, CA (R. Blewer Ed., p. 69).

144. F. Moghadam and D. Ranadive, *Ext. Abs. Electrochem. Soc. Meeting,* Fall1986, Abs. No. 297.

145. T. Ohmi et al., *Ext. Abs. of the Electrochemical Soc. Meeting,* Spring, 1989, Abs. No. 160, p. 227.

146. T. Yamazaki et al, *Tech. Dig. IEDM,* 1987, p. 586.

147. M. Nakano, *Ext. Abs. of the Electrochemical Soc. Meeting,* Spring, 1989, Abs. No. 195, p. 140.

148. S. J. Hillenius et al., *Ext. Abs. Electrochem. Soc. Conf.,* Spring, 1989, p. 184, Abs. No. 132.

149. Y. H. Hu, S. K. Lee, and D. L. Kwong, *Ext. Abs. Electrochem. Soc. Conf.,* Spring, 1989, p. 199, Abs. No. 142.

150. K. H. Jung et al., *Ext. Abs. Electrochem. Soc. Conf.,* Spring, 1989, p. 201, Abs. No. 143.

151. C. Wei et al., *Proceedings 6th Internatl. IEEE VMIC Conf.,* Santa Clara, CA, p. 129, 1989.

152. L. Giffen et al., *Solid State Technology,* April, 1989, p. 55.

153. J. Hui, S. Wong, and J. Moll, *IEEE Electron Device Letters,* Sept. 1985, p. 479.

154. R. P. Kramer, *Ext. Abs. Electrochem. Soc. Conf.,* Spring, 1989, p. 182, Abs. No. 131.

155. R. B. Fair, "Low-Thermal Budget Process Modeling with the PREDICT Computer Program", *IEEE Trans. Electron Devices,* March, 1988, p. 285.

156. M. Sato and Y. Arita, *Ext. Abs. Electrochem. Soc. Conf.,* Fall 1989, p. 350, Abs. No. 247.

157. R. S. Blewer, private communication.

158. E. K. Broadbent et al., *IEEE Trans. Electron Dev.,* July 1988, p. 952.

159. C. G. Sodini, S. S. Wong, and P. -K. Ho, *IEEE J. Solid State Circuits,* February, 1989, p. 118.

160. E. S. Yang, *Micro-electronic Devices,* New York, McGraw-Hill, 1988.

161. M. Hansen and A. Anderko, *Constitution of Binary Alloys,* McGraw-Hill, New York, 1958.

162. S. Swirhun et al., *IEEE Electron Dev. Letts.,* **5.** p. 209, (1984).

163. C. Y. Ting, "Silicide for Contacts and Interconnects," *Tech. Dig. IEDM,* 1984, p. 110.

164. W. Lynch, *Tech. Dig. IEDM, 1987,* p. 354.

165. S. Vaidya, *J. Electron. Mater.,* **10,** p. 337, (1981).

166. A. H. Perera and J. P. Krusius, *Tech, Dig. IEDM,* 1989, p. 625.

1

CHAPTER 4

MULTILEVEL-INTERCONNECT TECHNOLOGY
FOR VLSI AND ULSI

In order to build an integrated circuit, its necessary to fabricate many active devices on a single substrate. Initially, each of the devices must be electrically isolated from the others, but later in the fabrication sequence specific devices must be electrically interconnected so as to implement the desired circuit function. In chapter 2 we considered the technologies used to isolate the devices, and in chapter 3 we described the fabrication of *contacts* between the interconnect materials and the Si substrate. This chapter is concerned with the fabrication technology of the interconnect structures themselves.

Since both MOS and bipolar VLSI and ULSI devices invariably require more than one level of interconnect, much of this chapter deals with the issues involved in creating such *multilevel-interconnect* structures. The three main challenges in implementing such structures for submicron devices are the following:

1. Planarization of the intermetal dielectric layers,

2. Filling of high-aspect-ratio (and varying-depth) contact holes and vias, and

3. Integration of new conductor materials that exhibit low resistance, high reliability, and process compatibility.

4.1 EARLY DEVELOPMENT OF INTERCONNECT TECHNOLOGY FOR INTEGRATED CIRCUITS

4.1.1 Interconnects for Early Bipolar ICs

Robert N. Noyce invented the monolithic integrated-circuit concept while at Fairchild. His patent application in 1959 described how planar silicon bipolar transistors and resistors could be interconnected with thin and narrow aluminum lines over the surface-passivation oxide. The interconnects in the early generations of bipolar ICs typically consisted of a single level of metal (Al) and heavily doped diffused regions in the silicon substrate (although single-level-metal layers consisting of other materials – including the multilayer Ti/Pt/Au film – were also developed).* By the late 1960s, however, interconnects with two levels of metal were needed for bipolar SSI and MSI circuits, and these were implemented using Al for both levels of metal.

The *diffused regions in the silicon substrate* used in early ICs as interconnect paths (Fig. 4-1) were fabricated by selectively diffusing dopant impurities of the type opposite to those present in the substrate. In order for isolation to the substrate to be maintained, any voltage applied to such regions had to be of a polarity that kept the junctions under reverse bias. The relatively high sheet resistance and capacitance of such diffused regions, however, limited them to short-distance interconnection-path applications; the relatively large RC product of long diffused lines causes unacceptably large propagation delays for signals transmitted along them.

Such diffused regions have a minimum resistivity of ~1000 $\mu\Omega$-cm; much higher than that of Al (2.7 $\mu\Omega$-cm). In addition, the capacitance of the structure is that of a reverse-biased *pn* junction formed between the diffused region and the substrate. The capacitance can be calculated by treating the structure as a one-sided step junction in which the depletion layer extends primarily into the lighter doped substrate. The capacitance, C, per unit area is then given by

$$ C = \sqrt{\frac{q\, N_s K_{si} \varepsilon_o}{2\,(\varphi_{bi} + V_R)}} \qquad (4-1) $$

$$ \varphi_{bi} = \left[\frac{kT}{q} \ln\left(\frac{N_s}{n_i}\right) \right] + 0.56 \ \text{V} \qquad (4-2) $$

where N_s is the substrate doping, φ_{bi} is the built-in potential of the junction, n_i is the intrinsic carrier concentration, and V_R is the reverse bias applied to the junction. In practice, the capacitance of the diffused regions is about 2.5 times as large as the capacitance of the polysilicon interconnects (~1x10^{-4} pf/μm^2 vs. ~0.4x10^{-4} pf/μm^2).

The term *metal pitch* is widely used to describe the dimensions of a metal system. The pitch of a metallization system is the minimum centerline-to-centerline dimension of two adjacent metal lines (Fig. 4-2a); consequently, it is also the sum of the minimum metal-line width and minimum spacing of a metal system. Figure 4-2b shows how the metal pitch has decreased as IC technology has evolved.[218]

---

* Jack S. Kilby of Texas Instruments made a working phase-shift oscillator on a single chip of Ge in September 1958 and filed a patent application for the integrated circuit in February 1959. He is generally regarded as an independent co-inventor of the integrated circuit. Because Kilby used gold wires to interconnect the devices in the Ge substrate, a U.S court of appeals ruled that Noyce was the inventor of the *monolithic* technique. It was recognized that Noyce's invention implemented the integrated circuit through the use of an adherent oxide, junction isolation, and Al-film interconnection lines – the latter which were also adherent to the oxide and were formed by deposition and photolithographic etching.

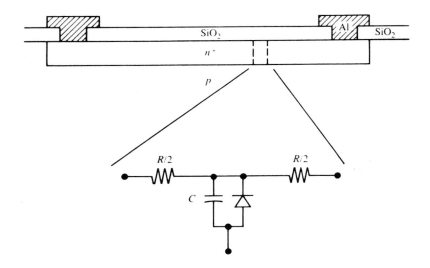

Fig. 4-1 A lumped circuit model for a small section of an n^+ diffusion. The RC delay limits the use of diffusions for high-speed signal distribution. From R. C. Jaeger, *Introduction to Microelectronic Fabrication.* Copyright, 1988, Addison-Wesley. Reprinted with permission.

4.1.2   Interconnects in Silicon-Gate NMOS ICs

Diffused regions and Al thin films were used as interconnection structures in early aluminum-gate MOS ICs. This did not significantly degrade their performance, since they were already relatively slower than their bipolar counterparts. With the advent of silicon-gate MOS technology, MOS ICs gained another level of interconnect – namely, heavily doped polysilicon (see chap. 5). With the implementation of buried contacts (see chap. 3), the polysilicon level could function as a local-interconnect level. This additional interconnect layer gave MOS devices an advantage in routing flexibility over bipolar ICs fabricated with only a single-metal interconnect level.

The major drawback of polysilicon is that its resistivity is comparable to that of the diffused regions, and is thus much higher than that of Al (i.e., 1000 $\mu\Omega$-cm vs. 2.7 $\mu\Omega$-cm). Polysilicon lines were nevertheless better suited than diffused regions to serve as interconnect paths because of their smaller capacitance per unit area. The development of double-polysilicon technologies (in which two levels of polysilicon are used) extended the use of polysilicon interconnects in ICs to the mid 1980s. For example, the double-poly process is needed to implement compact SRAM-cells that use polysilicon loads, as well as compact DRAM cells (see chap. 8). Such double-poly processes were used until the advent of the 64-kbit DRAM generation.

Eventually, however, chip sizes became so large and MOS-device performance so much improved (primarily as a result of shrinking device dimensions) that the resistance of the polysilicon interconnect structures began to significantly degrade the speed of the circuits. As a result, *polycides* were developed as a means of retaining the advantages of polysilicon while reducing the impact of the large polysilicon resistivity (see Vol. 1, chap. 11). Through the formation a refractory-metal silicide on top of the polysilicon runners, a polycide structure (in which the resistivity was reduced to ~50-100 $\mu\Omega$-cm) became available as an interconnect and gate layer. Interconnect technologies that used one level of metal, one level of polycide, and one level of polysilicon were therefore common in NMOS ICs used to fabricate 16-bit microprocessors, as well as in MOS DRAMs of the 256-kbit and 1-Mbit generations.

4.1.3 Evolution of Interconnects for Bipolar ICs

As bipolar technology advanced, smaller device sizes and circuits with higher functional

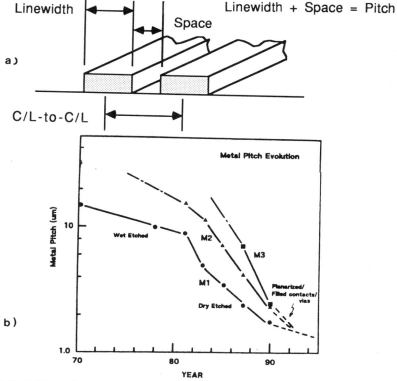

Fig. 4-2 (a) Definition of *metal pitch*. (b) Advanced process integration techniques have enabled the development of multilayer interconnects with tight pitches.[218] This paper was originally presented at the Spring 1989 Meeting of The Electrochemical Society, Inc. held in Los Angeles, CA.

densities were implemented. In order to obtain the flexibility needed to interconnect larger numbers of devices, it became necessary to develop a second interconnect level. Since no equivalent to Si-gate technology evolved in bipolar IC processing, polysilicon did not have to be automatically accepted as the second interconnect level. In fact the speed penalty imposed by the use of polysilicon would have eliminated the chief advantage of bipolar technology over MOS ICs (higher speed); thus, two-level *metal* interconnect technologies had to be developed. These emerged during the late 1960s and the 1970s. The shallow emitter-base junctions of bipolar devices also forced manufacturers of bipolar ICs to develop diffusion barrier technology for contacts before barrier layers were needed for MOS contacts.

4.1.4 Evolution of Interconnects for CMOS ICs

When CMOS replaced NMOS as the leading MOS IC technology in the mid 1980s, the interconnect technology also had to be modified to fit the needs of CMOS. For example, the local interconnect level of polysilicon could not serve CMOS as effectively as NMOS; since polysilicon in CMOS is normally *n*-doped, it cannot be used to make ohmic contacts to the source and drain regions of PMOS devices. As a result, it became necessary to develop two-level-metal technologies to provide CMOS VLSI circuits with sufficient routing flexibility. The shallow junctions needed in   2 μm and smaller MOS technologies also forced the use of diffusion-barrier technology in the contacts of CMOS ICs.

Implementing a two-level-metal process is more difficult in CMOS than in bipolar technologies because the underlying wafer topography is more rugged in CMOS (due primarily to the field-oxide steps arising from the semirecessed LOCOS process, and to the steps caused by the polysilicon layer). Severe topography conditions give rise to step-coverage problems. Nevertheless, the significant benefits offered by multilevel metal-interconnect technology eventually forced the challenges of integrating it into CMOS processes to be undertaken.

4.2 THE NEED FOR MULTILEVEL-INTERCONNECT TECHNOLOGY

Throughout the evolution of integrated circuits, the aim of device scaling has been twofold: (1) to increase circuit performance (mainly by increasing circuit speed), and (2) to increase the functional complexity of the circuits. At the outset, scaling down of active device sizes was a very effective means of achieving these goals. Eventually, the scaling of active devices became less profitable, as the limitations of the circuit speed and maximum functional density came to depend more on the characteristics of the interconnects than on the scaled devices. In addition, the aspects of silicon utilization, chip cost, and ease and flexibility of IC design were also adversely affected by the interconnect-technology restrictions. The approaches to lifting these limitations have predominantly involved the implementation of multilevel-interconnect schemes.

This section will first explain how the speed and functional density of ICs can be limited by the interconnect technology, and it will then indicate why multilevel interconnects are effective in counteracting this problem. Also, notation that is used throughout the chapter in descriptions of various features of a multilevel-interconnect structure will also be introduced.

4.2.1 Interconnect Limitations of VLSI

4.2.1.1 Functional Density.

In the course of integrated-circuit evolution, the maximum number of devices per chip has steadily increased, mainly as a result of the increase in functional density (although the growth in chip area has also played a role). *Functional density* is defined as the number of interconnected devices per chip area, while the number of devices per chip area is referred to as the *active-device density*. As the minimum feature size on an IC decreases, the active device density increases. The functional density, however, also depends on how effectively these devices can be interconnected. Unless a larger number of the devices can be interconnected as the active-device density is increased, there will be no gain in functional density.

When integrated circuits contained only a relatively small number of devices per chip, the devices could be more easily interconnected. This situation persisted for some time, even as the active-device density steadily increased. Nevertheless, the area occupied by the interconnection lines on the chip surface grew more quickly than the area needed to accommodate the active devices. Eventually, the condition was reached in which the minimum chip area became interconnect-limited - that is, the area needed to route the interconnect lines between the active devices exceeded the area occupied by the active devices. At this point, continued shrinking of active devices produced less circuit-performance benefits.

One way to overcome such a limitation is to implement a multilevel-interconnection system in which the area needed by the interconnect lines is shared among two or more levels. This allows the fractional area of the chip occupied by active devices to be increased, leading to an increase in the functional density. The following is a simple example illustrating why the chip surface becomes interconnect-area-limited as the active-device density is increased.

Assume that devices are fabricated using a *single-level interconnect technology* on a chip of constant area, and that the active-device density is increased by a factor of 10 (e.g., the number of logic gates per chip is increased from 100 to 1,000). Furthermore, assume that the minimum line width of the metal lines is scaled by the same factor used to scale the size of the logic gates (i.e., by the square root of the area decrease). If the minimum metal line width is initially w, and B is the number of squares of metal in the total line length, the total line length is Bw, and the area of the chip covered by the metal will be Bw^2 (Fig. 4-3a).

When the condition of increased active-device density is considered, the scaled area of one minimum square of metal film is seen to be be $w^2/10$. In addition, the total length of the interconnection lines will also be increased, due to two factors: (1) since there are more gates, a larger number of connections between gates must be made, and (2) the

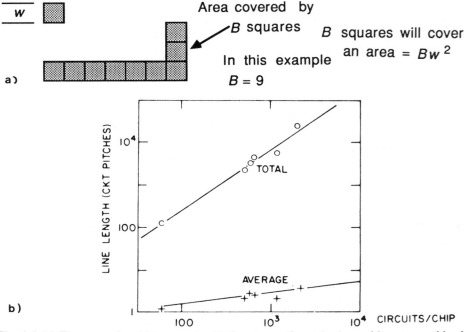

Fig. 4-3 (a) The area of a chip covered with B squares of metal whose sides are w wide, is Bw^2. (b) The average length of interconnect and the total wire length on a logic chip as a function of device density. The lengths are measured in circuit pitches, i.e., the square root of the area per circuit.[1] (© 1981 IEEE).

average length of the interconnection lines will increase. This latter effect is plotted in Fig. 4-3b.[1] The graph in Fig. 4-3 also shows the total line length as the number of logic gates is increased from 100 to 1000. (Note that the lengths are expressed in *circuit pitches* – that is, the graph actually indicates the number of squares that exist in the total line length.)   Thus, the graph shows that the total number of squares in the interconnect lines for the condition of higher device density has increased by a factor of 100. The *absolute* total line length is therefore $100B(w/\sqrt{10})$. The area of the chip covered by the metal lines for the higher logic device density will then be:

$$100\,B\left(\frac{w}{\sqrt{10}}\right)\left(\frac{w}{\sqrt{10}}\right) = 10\,Bw^2 \qquad (4-3)$$

This illustrates that as the active-device density is increased by a factor of 10, *the area of the chip covered by the metal also increases by a factor of 10*. As a result, a condition is soon reached in which the metal must occupy more area than that occupied by the active devices, and the chip area becomes interconnect limited. Keyes cites the example of a bipolar chip with a gate count of 1500 gates, and a chip area of 0.29 cm^2, fabricated using a single-level metal with a pitch of 6.5 μm.[1] The total wiring area occupied by the metal is 0.26 cm[2], which is about 90% of the surface area of the chip.

4.2.1.2 Propagation Delay. As the minimum line width on an IC shrinks, the active-device density increases. For example, in 256-kbit DRAMs 2-μm lines are used, while in 1-Mbit, 4-Mbit, and 16-Mbit DRAMs, 1.2-μm, 0.8-μm, and 0.5-μm lines are used, respectively. At these dimensions, the MOS transistor switching speed itself no longer limits the logic delay or access time of the IC. Instead, the time required for the transistor to charge capacitive loads becomes the limit to the performance of the IC. That is, as the devices shrink, the device contribution to the propagation delay of a digital signal also decreases. On the other hand, the scaling of the interconnect line widths does not bring about a corresponding decrease in the propagation delay time through the interconnect lines. As the chip sizes increase, the interconnect-path lengths also increase, and in fact most large VLSI circuits have become interconnect propagation-delay-time limited.

For example, as pointed out by Sah,[2] the intrinsic gate delay caused by electrons drifting through a 2-μm channel at their phonon-scattering-limited velocity of 10^7 cm/sec is 20 ps,[3] while the delay of of a 256-kbit DRAM is ~100 ns. This is 5,000 times longer than the intrinsic gate delay. Thus, a chip with devices whose gate lengths are 2 μm or less cannot be limited by the intrinsic gate delay unless 5,000 devices are connected in series. This is certainly not the case in a memory chip, and it is highly unlikely that it would occur even in complex logic circuits.

The propagation delay exhibited by VLSI circuits is therefore almost always limited by the large RC delay of the interconnection lines. The delay resulting from various

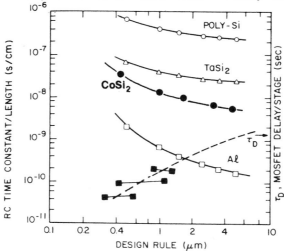

Fig. 4-4 RC time constant per unit length for several conductive materials as a function of feature size. Also shown is delay per stage of NMOS ring oscillators as a function of feature size. The RC time constant is calculated assuming a field oxide thickness of 1 μm.[258] Copyright 1981 AIOP. Reprinted with permission.

interconnect materials per millimeter of length is given in Table 4-1, and is graphed for some of these materials (assuming a 1-cm-long interconnect line) in Fig. 4-4.* For example, if a polysilicon line having a 20 Ω/sq sheet resistance is used to connect one corner of a 0.7-x-0.7 cm chip to the diagonal corner, the data from Table 4-1 predict that the propagation delay will be 1.4 μs. Even if Al is used as the interconnect layer, the delay caused by 1-cm-long interconnection lines will limit the circuit speeds once device dimensions fall below ~2 μm.

The above indicates that in order for circuit performance to be increased as device dimensions shrink, two goals must be met. First, the materials used for transmitting signals over long distances on a chip must have the lowest possible resistance values. This has been the key motivation in the drive to replace polysilicon interconnects (20 Ω/sq) with polycide interconnects (1.7 Ω/sq for TiSi$_2$). It may eventually cause refractory metals (0.5 Ω/sq for W) to be used as the first-level interconnect layer in multilevel-interconnect systems. In addition, it also points outs why Al will continue to be used as an interconnect material, even as devices shrink further.

Second, the length of interconnect lines on a chip must be made as short as possible. The RC delay can be shown to be proportional to the *square* of the length of the interconnect line. That is,

$$R = (\rho \, l)/(w \, t_m) \qquad (4\text{-}4)$$

and

$$C = \varepsilon w \, l \, / \, t_{ox}, \qquad (4\text{-}5)$$

and therefore

$$RC = \frac{\rho \, \varepsilon l^2}{t_m \, t_{ox}} \qquad (4\text{-}6)$$

where ρ is the resistivity, l is the interconnect line length, w is the line width, ε is the permittivity, t_m is the thickness of the metal, and t_{ox} is the thickness of the oxide. A multilevel-interconnect structure is an effective way to allow the longest lines to be reduced in length.

There are several other propagation-delay-related considerations that make multilevel interconnects attractive. First, they make it more feasible to lay out all interconnect lines with as close to the average length as possible. This is important because if some lines are much longer than the rest, the overall delay time of the circuit will be increased. The length variation also introduces nonuniform on-chip RC response times, which can disturb the switch synchronization of the circuit.

Figure 4-5a shows a histogram of the line lengths of a typical single-level-metal IC.[4] Whereas most lines are short, a few are much longer (e.g., in this example ~6% of them are >7 mm long). Multilevel interconnects can help eliminate the incidence of the few long lines.

Second, as shown by McGreivy in Fig. 4-5b,[5] if the line spaces become smaller than ~0.6 μm the total parasitic capacitance of the interconnect lines will increase, because

---

* Note that in Fig. 4-4 a somewhat thicker field oxide is used, yielding a smaller value of C.

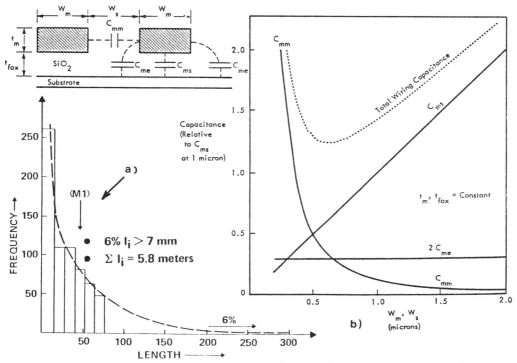

Fig. 4-5 (a) Example of lead length distribution of metal interconnects on an IC having a single level of metal.[4] Reprinted with permission of Electronics. (b) Variation of components of interconnect capacitance with design rules. The various components are defined in the adjacent figure.[5] (© 1982 IEEE).

the side-by-side capacitance increases with decreasing line pitch. The end result is an overall increase in the system cycle time. Again, the use of a multilevel-metal interconnect system, in which the wires run orthogonally in adjacent layers can reduce this problem.

Third, the layout should minimize cross-talk in order to reduce the level of inductance noise coming from the random switching of individual circuits. This goal can be accomplished through the use of a multilevel-interconnect structure in which all the wire tracks on one level run in one direction, while those on the adjacent level run in the perpendicular direction.

Finally, because the functional density of the chips can be increased, less chip-to-chip signal transmission is required. Since such off-chip driving is much slower than on-chip signal transmission, the overall system speed will be enhanced.

The above discussion implies that the lower levels of metal or polysilicon (which generally have a finer pitch, and therefore higher resistance and capacitance) should be reserved for interconnecting neighboring devices with short interconnection lines. The

Table 4.1 Interconnection Delay in Silicon VLSI Chips[1]

Conductor Materials	Resistivity ($\mu\Omega$-cm)	Thickness (nm)	Sheet Resistance (Ω/square)	Delay (psec/mm^2)
Polysilicon	1,000	500	20	7,000
TaSi$_2$	46	100	4.6	1,587
Ti	42.7	100	4.3	1,484
PdSi	32	100	3.2	1,104
MoSi$_2$	22	100	2.2	759
TiSi$_2$	17	100	1.7	586
Ta	13.1	100	1.3	448
WSi$_2$	12.5	100	1.3	448
Pd	10.5	100	1.1	380
W	5.3	100	0.53	183
Mo	5.3	100	0.53	183
Al	2.6	100	0.26	90

upper interconnect levels, which are usually designed with a larger pitch and lower resistance, should be used to transmit signals across the entire chip.

4.2.1.3 Ease of Design and Gate Utilization for ASICs and Wafer-Scale Integration. In VLSI manufacturing, an interrelationship exists between the design and processing tasks, particularly when gate-array and standard-cell circuits are being produced. Such *application-specific ICs* (ASICs) are widely used because they allow new circuits to be quickly designed and rapidly fabricated. Gate-array wafers can be processed up to the metallization layers and then stockpiled. Users can implement their circuit designs by interconnecting the uncommitted logic gates through the use of custom metal masks. The design effort of interconnecting the uncommitted logic gates, however, must be minimal, and it should be possible to utilize a large fraction (if not all) of the gates in the uncommitted array.

Multilevel-interconnect technology allows both of these goals to be achieved much more easily. For example, a two- to fourfold decrease in design time was realized when a bipolar gate array was implemented with a three-level-metal process rather than with a two-level technology.[6] In another example, the ratio of the chip area required for power buses as a function of the number of gates in a gate array is plotted for two-level-metal and three-level-metal processes (Fig. 4-6).[7] The extra level of metal significantly decreases the area of the chip that must be used for the power buses. Hence, this area can be used to interconnect the devices – which reduces the difficulty of the design task.

Wafer-scale integration is a another concept in which a large number of chips from a single wafer are interconnected to create large digital systems without the chips being separated and mounted in separate packages. Since good die are interconnected after having been fabricated and tested, a multilevel-interconnection technology is essential for implementing this approach.

4.2.1.4 Cost. If multilevel-interconnect processes are used to fabricate integrated circuits, the die size should decrease. Thus, more die per wafer can be manufactured. If the manufacturing cost per wafer remains the same and the yield is not impacted by the implementation of a multilevel-interconnect process, the cost per chip will decrease. In fact, smaller die sizes should imply higher yields, and enhancing the benefit of chip-size reduction. In addition, improved device performance may allow the circuit to command a higher market price.

However, the implementation of a multilevel-interconnect system requires that at least two additional masking steps be used for each additional level of interconnect. The extra process steps add to the manufacturing cost of each wafer, and the number of defects/cm^2 is also generally proportional to the number of masking steps. In addition, the manufacturing yield and long-term reliability for a multilevel metal process are typically lower, since the process becomes more technically demanding. As a result, it must be determined whether the chip-size reduction and enhanced chip value will produce a margin of profit that is greater than the amount lost due to additional incurred process costs and yield and reliability loss.

4.2.2 Problems Associated with Multimetal-Interconnect Processes

As alluded to in the previous section, adding a multilevel interconnect process to a fabrication sequence introduces a new set of difficulties. In addition to added process

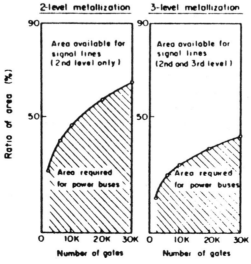

Fig. 4-6 Ratio of area required for power buses to the total chip area for ICs using two- and three-level-metal interconnect technologies.[7] (© 1983 IEEE).

complexity and loss of topological planarity, there are several other concerns.

First, new materials must be used, which necessarily involves an extensive characterization of their properties to ensure that they are compatible with all other aspects of the process technology.

Second, new process-related manufacturing difficulties may be encountered that can adversely impact manufacturing yield (e.g., interlevel shorts due to pinholes; stringers due to incomplete etching over severe steps; failure to open vias due to difficulty in implementing reliable endpoint-detection techniques in the dry-etch process; film delamination due to poor adhesion or high stress; and difficulty in bonding to some metal alloys).

Third, new failure modes may be encountered – for example, electromigration, corrosion, and hillock formation – and these must also characterized to determine whether they will significantly compromise circuit reliability.

The problems related to multilevel interconnects is listed at this point in order to show that the benefits can be gained only by successfully pursuing a considerable technical-development effort. More specific details on these problems (and on how they can be overcome) will be provided throughout the chapter. Section 4.7 also gives an overview of the yield and reliability problems that occur when multilevel-interconnect technologies are implemented.

4.2.3 Terminology of Multilevel-Interconnect Structures

Figure 4-7 shows the terminology associated with a double-level-metal structure for MOS technologies. The MOS structure has a dielectric layer between the polysilicon gate/interconnect level and Metal 1, which we refer to as the *polysilicon/Metal 1*

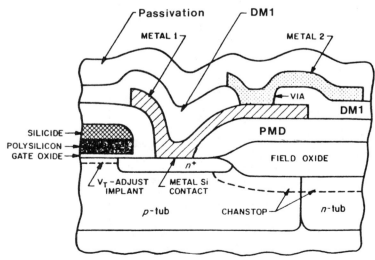

Fig. 4-7 Terminology of double-level-metal interconnects.

dielectric (or PMD). The dielectric layers between metal levels are called *intermetal dielectrics*. The intermetal dielectric between Metal 1 and Metal 2 is designated as DM1, and dielectric layers between other levels of metal (e.g., Metals 2 and 3, or Metals 3 and 4) are DM2, DM3, etc. The openings in PMD are referred to as *contact holes*. Contact through them is established between Metal 1 and polysilicon, as well as between Metal 1 and the Si substrate. Openings in the intermetal dielectric layers are known as *vias:* these allow contact to be made between Metals 1 and 2, Metals 2 and 3, etc.

In bipolar technology, the dielectric layer between Metal 1 and the substrate is still referred to as PMD, despite the fact that it may not isolate Metal 1 from polysilicon. The openings in PMD are again called contact holes, although they are only used to allow contact to be established between Metal 1 and the substrate (i.e., not poly). The notation for the other metal and dielectric layers is otherwise identical to that used in MOS technologies.

A distinction exists between our use of the terms *multi-level* and *multilayer*. A *multilayer-interconnect structure* is a thin film consisting of more than one layer of material, but existing at just one level of the interconnect system. Hence, a multilayer film can serve as the conductor (or dielectric) at each level of a multilevel-interconnect system.

4.3 MATERIALS FOR MULTILEVEL INTERCONNECT TECHNOLOGIES

The two groups of materials employed in multilevel-interconnect technologies are *thin-film conductors* and *thin-film insulators*. In this section we will describe the properties of such materials that have been adopted for use in VLSI applications.

4.3.1 Conductor Materials for Multilevel Interconnects

4.3.1.1 Requirements of VLSI Conductor Materials. Before describing properties of the specific conductor materials that have been considered for VLSI interconnect technologies, it's useful to examine the general requirements needed by such conductors. The most important of these are listed in Table 4-2. The list is long and many of the requirements are quite stringent. Nevertheless, unless a conductor structure can satisfy virtually satisfy all of them, it is unlikely to find use in VLSI applications. As a result, the number of materials that have been found suitable for VLSI interconnects is rather small. Table 4-3 summarizes the important properties of this group. The resistivities given in the table are typical for polycrystalline films with thicknesses of 10 - 1000 nm. The lower-resistivity values are exhibited by films that are purer and thicker, and that are large-grained. Deviations from stoichiometry for alloy conductors usually also lead to increased resistivity. As a result, the resistivity of a given type of film may vary from one deposition run to another, unless all deposition parameters are controlled to yield identical films.

Another of the key properties of VLSI thin-film conductors is their ability to adhere to Si and to SiO_2. In this respect, Al, Ti, Ti:W, and TiN films are the best. Al and Ti reduce SiO_2 to form interfacial metal-oxide bonds that promote adhesion and stability. WSi_2, $MoSi_2$, and $TiSi_2$ also adhere well to SiO_2, but not as well as Al and Ti. On the other hand, W and Mo do not reduce SiO_2, and therefore they exhibit poor adhesion to SiO_2 surfaces. As a result, W and Mo are not used as stand-alone conductor materials. Instead, a layer such as Ti, Ti:W, or TiN_x is required under W or Mo films to promote adequate adhesion to Si and SiO_2.

Table 4.2   Desired Properties of Conductor Structures for VLSI

---

a. Low resistivity (has to be $< 4 \, \mu\Omega$-cm).

b. Surface smoothness (and hillock resistant).

c. Resistance to electromigration failure effects.

d. Bondable (final level of a multilevel system).

e. Adhesion to underlying substrate materials must be excellent.

f. Stable - the mechanical and electrical properties, remain constant with time under the following conditions:

   1. throughout processing, including sintering, interlevel, and passivation dielectric deposition;

   2. in an oxidizing ambient;

   3. under normal operating conditions; and

   4. under ordinary storage (stress voids).

g. Corrosion resistant.

h. Should not contaminate devices, wafers, or processing equipment.

i. Depositable to controlled thickness and uniformity, and free of deposited particulates.

j. Anisotropically etchable with high selectivity with respect to substrate and mask material (see Vol. 1, chaps. 15 and 16).

k. Depositable over vertical walls with conformal coverage of steps ("good step coverage").

l. Film reflectance should be controllable to allow effective photolithographic processing.

m. If necessary, the metallization film should be depositable as a combination metal film (multilayer) (see Vol. 1, chap. 10).

n. Each required layer should be depositable in alloy form, with alloy composition tightly controllable (see Vol. 1, chap. 10).

o. Depositable pure -- i.e., without reaction with the gases present in the deposition chamber, and without incorporation of residual gases into the deposited films) (see Vol. 1, chap. 10).

p. Low film stresses.

q. Deposition and patterning processing i. should be economically viable, through:

   1. high throughput;

   2. reasonable processing-equipment purchase, maintenance, and operating costs;

   3. high reliability and up time of processing equipment;

   4. low enough in complexity that highly skilled operators not required.

4.3.1.2 Local-Interconnect Conductor Materials (Polysilicon, Metal Silicides, and Polycides).

The *local-interconnection level* is defined in chapter 3, section 3.11.2; various schemes to implement local interconnects are also described there. Polysilicon and polycides (i.e., structures in which a refractory-metal layer is formed on top of a polysilicon layer) have been used as local interconnect structures in NMOS technologies. (More information on the properties, formation, and patterning of polysilicon and polycides can be found in Vol. 1, chaps. 6 and 11, respectively.)

Various conductors have been proposed as local interconnect materials in CMOS technologies, including: dual-doped polysilicon, $TiSi_2$, Ti:W, TiN, and CVD W. Information on the characteristics of CMOS local interconnect materials is presented in chapter 3, section 3.11.2, as well as elsewhere in this volume.

4.3.1.3 Aluminum Metallization. Patterned aluminum thin films have been the most widely used interconnect structures in the manufacture of silicon ICs. As described in detail in chapter 3 of this volume and in Volume 1, chapter 10, the main reasons for the pervasiveness of Al are its low resistivity (2.7 $\mu\Omega$-cm) and its good adhesion to SiO_2 and Si. In addition, Al is generally used as the top level of metal in multilevel-metal systems because wire-bonding technology to Al thin films is a well-characterized process.

The main limitations of Al are its low melting temperature (660°C) and its low eutectic temperature with Si (577°C). In addition, Al thin films begin to form hillocks at relatively low processing temperatures (i.e., above 300°C), and they offer relatively poor resistance to electromigration effects and corrosion. (The electromigration and hillock problems are described in more detail in section 4.7.) The problems that arise when Al contacts are made to Si (i.e., related to the eutectic temperature of Al and Si) are considered in chapter 3, section 3.4.3.

Attempts to alleviate the above problems have generally involved the addition of alloy materials to Al or the formation of multilayer-Al conductor structures. The addition of Si to Al was discussed in chapter 3, section 3.4.3 as a technique for preventing junction spiking in Al contacts to Si. Electromigration and hillock resistance have been improved through the addition of such elements as Cu, Ti, Pd and Si to Al to form alloys. Another approach to overcoming the latter problems has involved the fabrication of multilayer-metal structures consisting of Al and other metal layers, such as the following: (1) Ti, Cr, and Ta layers sandwiched between Al layers; (2) Ti:W layers deposited below or on top of the Al film; and (3) W selectively deposited over the top and sidewalls of patterned Al lines. Such multilayer structures are also described in section 4.7. The addition of other metals to form Al alloys, however, generally degrades one or more of the Al characteristics (e.g., resistivity, corrosion resistance, etchability, or bondability).

Thin films of Al for VLSI interconnects are commonly deposited by dc magnetron sputtering (see Vol. 1, chap. 10). Although evaporation of Al films was carried out in the early days of IC fabrication, the need for Al-alloy films with tightly controlled alloy composition was the primary reason for the displacement of evaporation by sputtering. Sputtering generally allows more control over the alloy composition than evaporation (see Vol. 1, chap. 10). Some recent advances in sputter deposition are reviewed in reference 8. Research is also being conducted on CVD Al, as described in section 4.5.5.2.

4.3.1.4 Tungsten and Other Conductor Materials for VLSI Interconnects.
Tungsten has been intensively investigated for a number of roles in multilevel-interconnect systems for the following reasons:

- It exhibits excellent resistance to electromigration effects, hillock formation, and humidity-induced corrosion.

- It can be deposited by means of CVD, and thus allows much better step coverage than can be obtained by sputter-deposited or evaporated films (e.g., Al films).

- The use of selective CVD allows W to fill contact holes and vias that have very high aspect ratios, and it also allows W to be selectively deposited onto patterned Al lines.

- Finally, CVD W allows unframed vias and contact holes to be implemented, thereby increasing circuit packing density.

The main disadvantages of CVD W compared to Al are its higher resistivity (6-15 $\mu\Omega$-cm), its rough surface after deposition, and the difficulties involved with etching it. The details of CVD-W deposition, which involves planarization of W conductor structures and the filling of contact holes and vias with W, will be described in section 4.5.4. The issues involved with W etching will be considered at this point.

The etching of CVD W presents several problems. First, the surfaces of most CVD W films are very rough (usually 25% or more of the film thickness in the hydrogen reduction process – but much improved via SiH_4 reduction). Second, W and SiO_2 both form volatile fluoride by-products. Thus, it is difficult to obtain high selectivity with respect to the underlying oxides if a fluorine-based chemistry is used for dry-etching W. Such selectivity is necessary during the overetch time that must be used to clear away any stringers of W that remain at steps on the wafer surface (see Vol. 1, chap. 15). On the other hand, when chlorine chemistries (which allow higher selectivity to the oxide, and are therefore most commonly used to etch W) are employed, the erosion rate of photoresist is very high. As a result, an inorganic mask or a composite resist/SiO_2 mask has usually been used when W is dry etched.[110, 111] Such approaches, however, represent an undesirable process complexity.

One approach that allows a resist mask to be utilized makes use of a two-step etch process.[113] In the first step, a high-power RIE mode is used to rapidly etch 0.8 μm of a 1.0-μm-thick W film. A gas mixture of SF_6, HBr, and CH_4 is employed for this step. A selectivity to the resist of 4:1 is obtained by using this gas mixture together with a thermal UV treatment of the resist mask prior to etching. The second step uses a low-power rf mode, coupled with a remote microwave plasma to etch the remaining W. A gas mixture of SF_6, HBr, and CCl_4 is used, and selectivity of 4:1 to the oxide is obtained.

Interconnect films using sputter-deposited Mo (together with an underlayer of Ti:W or Ti) have also been implemented as the Metal 1 layer of a multilevel-interconnect system.[114] The Mo offers the advantages of excellent electromigration and hillock resistance. More information on this conductor structure is found in chapter 3, section 3.6.3.

Interconnect films using Au were used in some early-generation integrated circuit technologies. Au offers low resistivity (2.2 $\mu\Omega$-cm), resistance against corrosion, and good electromigration properties. However, it does not adhere well to SiO_2, and it is "poison" to devices (i.e., it causes deep-level traps in the forbidden gap, degrading the minority carrier lifetime). These drawbacks make the fabrication of Au metallization systems quite complex, and hence Au has not found much use in VLSI applications. However, reference 9 describes an example of a recent two-level gold metallization process that uses Ti:W as a diffusion barrier and adhesion layer under the Au, and in which the Au is deposited by electroless plating.

Copper thin films offer even lower resistivity (1.7 $\mu\Omega$-cm) than Au films, and they are also expected to show good electromigration resistance. They would thus appear to be attractive as VLSI conductor materials, especially as device dimensions approach deep-submicron sizes. A dry-etching process for Cu films has not been successfully developed (see Vol. 1, chap. 16), however, and this is one of the chief reasons that Cu has not been widely considered for such applications.

A low-temperature Cu-deposition process that exhibits good adhesion to SiO_2 is given in reference 10. Since the presence of Cu in the Si also poisons devices, it would be necessary to prevent Cu migration into the Si substrate through the use of diffusion barriers. Such barriers are described in reference 99. A recent report on using Cu interconnects for ICs with device dimensions smaller than 0.5 μm is given in reference 233. Possible approaches to overcoming the problems of deposition, etching, device contamination, and corrosion associated with Cu interconnects are discussed. The

Table 4.3 Properties of Various Thin-Film Conductors Used in VLSI Multilevel Interconnects

Metal or Alloy	ρ ($\mu\Omega$-cm)	Melting Point (°C)	Reaction with Si (°C)	Stable on Si up to (°C)
Al	2.7-3.0	660	~250	~250
Mo	6-15	2,620	400-700	~400
W	6-15	3,410	600-700	~600
$MoSi_2$	40-100	1,980	-	≤1,000
$TaSi_2$	38-50	~2,200	-	≤1,000
$TiSi_2$	13-16	1,540	-	≤ 900
WSi_2	30-70	2,165	-	≤1,000
$CoSi_2$	10-18	1,326	-	≤950
PtSi	28-35	1,229	-	≤750
Ti:W (10 wt% Ti)	75-200	-	600-700	~700
TiN	25-200	~2880		

etching problem to be circumvented by selectively depositing the Cu; device contamination can be controlled by utilizing silicon nitride under the Cu lines; and corrosion can be stopped by selectively covering the Cu lines with a layer of nickel.

4.3.2 Dielectric Materials for Multilevel Interconnects

4.3.2.1 Requirements of Dielectric Layers in Multilevel Interconnects.
Dielectric layers must be used to electrically isolate one level of conductor from another in multilevel-interconnect systems. The list of properties that must be possessed by such dielectric layers is given in Table 4-4. As is the case for conductor materials used in such applications, the list of requirements is long and stringent. When we describe the materials that have been developed for this role, Table 4-4 will serve as our point of reference.

It should also be mentioned that there can be a significant difference between the dielectric film referred to in Fig. 4-7 as PMD (used between polysilicon, or other local-interconnect level material, and Metal 1) and the dielectric films that are employed between the metal layers (*intermetal dielectrics* – e.g., DM1 in Fig. 4-7). PMD films can be deposited (and densified if necessary) at a higher temperature than is possible for the intermetal dielectric layers. Furthermore, PMD films can be *flowed* and *reflowed* at temperatures in excess of 800°C. On the other hand, when Al is present on the wafer surface, the maximum temperature of the intermetal dielectric layers is limited to ~450°C. We will therefore discuss these two dielectric types separately, even though some dielectric films can be used for both applications.

4.3.2.2 Poly-Metal Interlevel Dielectric (PMD) Materials.
Doped CVD SiO_2 films have found the widest application as PMD layers in MOS ICs. Silicon-nitride films have generally not been used as stand-alone PMD layers because they possess a much higher dielectric constant than SiO_2 films and because they cannot be flowed or reflowed. High-temperature CVD-oxide films (deposited at 900°C by the reaction of dichlorosilane and nitrous oxide) were among the first materials used for this purpose. Such films provide excellent step coverage, as well as dielectric properties almost identical to those of thermally grown SiO_2 films (see Vol. 1, chap. 6). Unfortunately, these films cannot be doped because of the high temperature at which they are deposited; as a result, they cannot be flowed or reflowed at temperatures lower than 1100°C.

Undoped TEOS films (deposited by the decomposition of *tetraethyl orthosilicate*, [TEOS] at 600-650°C) were also used as PMD layers in some early IC processes because of their excellent step coverage. In order to allow such TEOS films to be reflowed, phosphorus- and boron-doped TEOS deposition processes were subsequently developed. Recent efforts with doped-TEOS processes have been aimed at replacing the relatively toxic phosphine and diborane dopant gases with less toxic liquid sources.[12] In addition, PECVD TEOS processes have recently been developed in which TEOS films can be deposited at lower temperatures (e.g., 375°C; see section 4.3.2.4).

Table 4.4   Desired Properties of Interlevel Dielectrics for VLSI[11]

---

1. Low *dielectric constant* for frequencies up to ~20 MHz, in order to keep capacitance between metal lines low.

2. High breakdown field strength (>5 MV/cm).

3. Low leakage, even under electric fields close to the breakdown field strength. Bulk resistivity should exceed 10^{15} Ω-cm.

4. Low surface conductance. Surface resistivity should be $>10^{15}$.

5. No moisture absorption or permeability to moisture should occur.

6. The films should exhibit low stress, and the preferred stress is compressive (~5×10^8 dynes/cm^2), since dielectric films under tensile stress exhibit more of a tendency to crack.

7. Good adhesion to aluminum, and of aluminum to the dielectric. (Good adhesion is also needed to the other conductors used in VLSI, such as doped polysilicon and silicides). In cases of poor adhesion (such as with gold or CVD W), a glue layer (such as Ti or Ti:W) may need to be applied between the conductor and the dielectric.

8. Good adhesion to *dielectric layers* above or below. Such dielectric layers could be thermal oxides, doped-CVD oxides, nitrides, oxynitrides, polyimides, or spin-on glasses.

9. Stable up to temperatures of 500°C.

10. Easily etched (by wet or dry processing).

11. Permeable to hydrogen. This is important for IC processing, in which an anneal in a hydrogen containing ambient must be used to reduce the concentration of interface states between Si and the gate oxides of MOS devices (see Vol. 1, chap. 8).

12. No incorporated electrical charge or dipoles. Some polyimides in particular contain polar molecules that can orient themselves in an electric field and give rise to an electric field even when the externally applied field is removed.

13. Contains no metallic impurities.

14. Step coverage that does not produce reentrant angles.

15. Good thickness uniformity across the wafer, and from wafer to wafer.

16. In the case of doped oxides, good dopant uniformity across the wafer, and from wafer to wafer.

17. Low defect density (pinholes and particles).

18. Contains no residual constituents that outgas during later processing to the degree that they degrade the properties of other layers of the interconnect system (e.g., outgassing from some polyimide films, SOG films, or low-temperature TEOS films).

Silane-based phosphorus-doped SiO_2 films deposited at low temperatures (350-450°C) have also been used as PMD layers. The addition of phosphorus to the films allows reflow to be performed at ~1000°C in steam. However, the need for lower flow and reflow temperatures has made silane-based BPSG and boron/phosphorus-doped TEOS films more attractive.[13] Reflow temperatures of less than 850°C can be achieved with BPSG.

Silane-based BPSG films (3-5 wt% each of B and P) are deposited at low temperatures (400-450°C) and are then immediately densified at ~800°C for one hour. The purpose of this step is to completely stabilize the BPSG films, which would otherwise be prone to blistering during subsequent processing. The higher the boron concentration in such films, the lower the flow temperature; however, BPSG films with high boron concentrations (i.e., greater than 5%) are not stable, and hence are not used.

A report has been made of a BPSG film with 4.8 wt% B and 4.6 wt% P that was flowed at 900°C for 30 minutes; this film exhibited a nearly planar surface over patterned polysilicon.[14] A two-layer film was actually used for the PMD – that is, a thin (120-nm-thick) silicon-nitride layer was deposited before the BPSG. The nitride prevents dopants from the BPSG from diffusing into the poly or substrate device regions during the flow and reflow thermal cycles. Note, however, that the temperature used to flow the film in this example may be too high for processes that employ shallow junctions or $TiSi_2$ self-aligned contact structures (as described in chap. 3, section 3.9.1.1).

The phosphorus content of the BPSG films can be measured easily and quickly by using energy-dispersive X-rays (see Vol. 1, chap. 17). The boron content is more difficult to measure, but two methods for doing this are *wavelength dispersive X-ray analysis* (see Vol. 1, chap. 17), and a wet-chemical technique known as *colorimetry* (see ref. 15).

4.3.2.3 CVD SiO_2 Films as Intermetal Dielectrics.

Until recently, the materials most widely used for intermetal dielectrics have been doped, silane-based CVD SiO_2 films. There are two main reasons for the widespread adoption of such films for this application: first, CVD SiO_2 films with good electrical and physical properties can be deposited at temperatures compatible with the presence of Al on the wafer (≤ 450°C); and second, when CVD SiO_2 is doped with phosphorus, it serves as an excellent gettering layer for sodium ions and other lifetime-killing metallic impurities.

Good-quality SiO_2 films for intermetal applications are characterized by the following measures:

- The P-etch rates are on the order of 30 nm/min (*P-etch* is a mixture of 300 parts H_2O, 10 parts HNO_3 [70%], and 15 parts HF [49%]).

- The index of refraction is 1.46 (this value increases with oxygen deficiency).

- The dielectric strength is 4×10^6 V/cm.

- There is an absence of nodular growths and pebble-like surface defects.

- The films are pinhole-free.

- Good edge coverage of underlying metal steps is provided.

Intermetal SiO_2 films can be deposited by means of a number of CVD processes, including *atmospheric CVD, low-pressure CVD, and plasma-enhanced CVD* (see Vol. 1, chap. 6). While the CVD SiO_2 films deposited by these methods can exhibit good

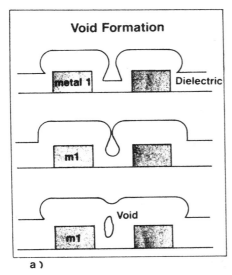

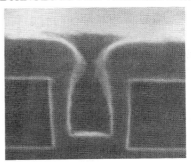

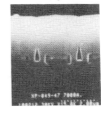

Fig. 4-8 (a) Diagram of void formation due to cusping of the dielectric. (b) SEM of voids left between 1.5-μm-spaced Metal 1 lines in a 3.0-μm-thick DM1 PECVD oxide.

electrical and physical properties, all silane-based CVD methods share a number of problems. First, coverage of these films over closely spaced steps is inadequate, primarily as a result of geometrical shadowing; as a result, the film thickness at the top of the step exceeds that on the bottom and sidewalls. In addition, the angle of the SiO_2 film at the base of the step can be *reentrant*, making step coverage and etching conditions more difficult. Related to this problem is the formation of voids when the thicker sidewalls of the oxide touch, leaving a void below (see Fig. 4-8).

The addition of phosphorus to silane-formed SiO_2 films (known as *phosphosilicate glass, PSG,* or *P glass*) helps to reduce stress in the SiO_2 films, makes them more resistant to moisture, and improves their gettering of sodium. However, it also increases the film etch rate. This can impact dielectric etchback planarization processes (see further sections 4.4.3 and 4.4.8). That is, if the incorporation of phosphorus is not uniform across the wafer (and from run to run), the etchback rate will not be uniform across the wafer. Such variations in etch rate can produce devices with very thin dielectric layers on some locations on a wafer.[16]

Phosphorus doping also degrades the step coverage of the CVD SiO_2 films by enhancing their cusping tendency. In addition, when PECVD SiO_2 is used to fill high-aspect-ratio spaces, a seam forms where the oxide sidewalls meet. This seam region may exhibit a much higher dry-etch rate than the bulk regions of the film. As a result, in an etchback planarization process, the faster-etching seam regions can cause the formation of deep grooves on the oxide surface.[16]

A third problem, gas-phase nucleation, is associated with silane-based plasma-enhanced CVD (PECVD) SiO_2 films. If the rf field becomes too strong, the dissociated gas molecules may react in the gas stream and form SiO_2 particles, which fall into the growing film.[17]

A PECVD borosilicate glass (BSG) has also been reported on as an intermetal dielectric material.[18] Its advantages include good-quality dielectric properties for deposition temperatures as low as 300°C, and a dry-etch rate higher than that of PSG, which makes it attractive for use in etchback SOG processes (see section 4.5.9). In addition, BSG provides excellent step coverage and can reportedly fill narrower spaces between metal lines than can silane-based PSG films.[11] Therefore, SiO_2 layers containing both B and P (to exploit the advantages that are obtained when each of these dopants is present in the film) should, in principle, produce intermetal-dielectric films with the best overall characteristics.

4.3.2.4 Low-Temperature-TEOS Films as Intermetal Dielectrics.

Processes that allow both pyrolytic and plasma-enhanced deposition of SiO_2 from TEOS (PETEOS) at temperatures of 375°C (rather than from silane) have recently become commercially available.[16,19] Such low-temperature PETEOS oxides to some degree alleviate the problems of silane-based CVD SiO_2 films. First, the drawback of cusping is reduced in TEOS-based films, because when the oxide is deposited from organic silicon compounds, the adatoms have a higher surface mobility. The larger the mean free path for surface migration, the more local averaging of the local solid angle there is, improving step coverage and conformality. Second, this effect allows filling of spaces between adjacent lines with aspect ratios as high as 0.8, while silane-based PECVD SiO_2 films become ineffective when aspect ratios exceed 0.5. (PETEOS SiO_2 films showed no cusping, nor did etching in dilute HF delineate any seams when 0.8 aspect-ratio spaces were filled.) Third, the problem of non-uniform phosphorus incorporation (which might impact the etch rate) is alleviated, since the film does not include any phosphorus during deposition. Finally, the problem of gas-phase nucleation is eliminated, since TEOS is not *pyrophoric* (i.e., it will not spontaneously ignite in the presence of oxygen).

A report on the properties of boron-doped plasma CVD TEOS films for intermetal dielectric applications is given in reference 20. In another report, both the low-temperature pyrolytic TEOS film and the plasma-enhanced TEOS films were used, together with an etchback process, to produce a planar surface over topographies with steps ≤10 μm apart.[21]

Another PETEOS process that produces directional deposition (thus allowing void-free filling of high-aspect-ratio spaces) has recently been reported.[239] Step coverages with 80% bottom- to top-surface ratios and only 60% sidewall to top-surface ratios were observed on features with aspect ratios of ~1.0. Such directional deposition was observed in PETEOS films with low O_2:TEOS feed-gas ratios. The novel step coverage is attributed to the directionally impinging ions which induce anisotropic step coverage profiles and enhance the crosslinking in the formation of the silicate network.

An alternative source compound to TEOS or silane has also been investigated for low-temperature plasma-enhanced CVD SiO_2 film deposition. This compound is TMCTS (1,3,5,7-tetramethylcyclo-tetrasiloxane). It reportedly produces films of comparable quality to those of PECVD TEOS, but at a faster deposition rate. The addition of NF_3 to TEOS and TMCTS plasmas produces void-free filling of holes with aspect ratios as high as 1.0 (compared to void-free filling of holes with 0.7 aspect ratios when PECVD TEOS and TMCTS are used without NF_3).

4.3.2.5 Other Materials and Deposition Processes Used to Form Intermetal Dielectrics.

In addition to doped CVD SiO_2 and low-temperature TEOS-based SiO_2 films, a variety of other materials have been implemented as intermetal dielectrics. These include PECVD silicon nitrides, bias-sputtered SiO_2, polyimides, and spin-on glasses. The PECVD nitrides have rarely been used as stand-alone inter-metal dielectric layers, because they posses a higher dielectric constant than CVD SiO_2 films. Instead, they are usually employed as one layer of a multilayer intermetal dielectric film; some examples will be given in later sections. The dielectric properties of bias-sputtered SiO_2 films, polyimide films, and SOG films will also be described in subsequent sections (4.4.5, 4.4.6, and 4.4.9, respectively). Research on other intermetal dielectric materials and alternative deposition processes for VLSI applications has also been reported, including photo-CVD,[22] laser activated CVD dielectric films,[23] CVD SiO_2 films deposited in a downstream afterglow reactor,[24] and evaporated Al_2O_3 layers.[25]

4.4   PLANARIZATION OF INTERLEVEL DIELECTRIC LAYERS

As pointed out in the introduction to this chapter, the planarization of interlevel dielectrics is one of the three critical issues that must be addressed when implementing a multilevel interconnect system for VLSI applications. We will now justify this assertion and describe the various approaches developed for dealing with this issue.

4.4.1   Terminology of Planarization in Multilevel Interconnects

As additional levels are added to multilevel-interconnection schemes and circuit features are scaled to submicron dimensions, the required degree of planarization is increased. Such planarization can be implemented in either the conductor or the dielectric layers. In this section we will consider processes developed to planarize dielectric layers; in later sections, techniques for planarizing conductor layers and vias will be considered.

The term *planarization* is employed quite frequently (and loosely) both here and in other technical literature. At this point, therefore, it is useful to define this term more carefully and completely, especially as it applies to the planarization of dielectric layers in multilevel-interconnect technology. The example case we will use for this discussion is that of a dielectric layer (DM1) that is deposited after Metal 1 is patterned. In the case where *no planarization exists* (Fig. 4-9a), the step heights on the DM1 surface closely approximate the step heights of the Metal 1 layer and the underlying topography.

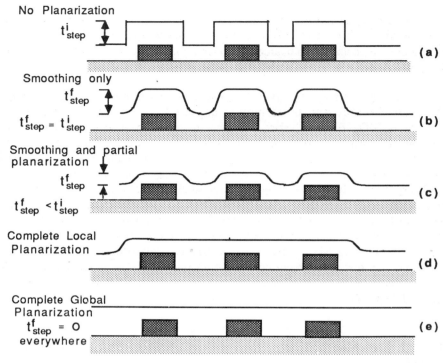

Fig. 4-9 Degrees of dielectric planarization

Furthermore, in this case the steps on the DM1 surface also have steep slopes (i.e., vertical, or even reentrant).

4.4.1.1 Degree of Planarization. The steps on the surface of DM1 can be made less severe through various planarization processes. The degree to which this can be successfully accomplished differs according to the planarization technique used. We classify the degree of planarization according to the following *qualitative planarization criteria:*

1. The first degree of planarization (*smoothing*) involves a lessening of the step slopes at the DM1 surface (Fig. 4-9b). The step heights in this case, however, are not significantly reduced in magnitude.
2. In the second degree of planarization (*partial* or *semi-planarization*) the step heights are reduced (but not eliminated), and the slopes of the steps are also smoothed (Fig. 4-9c).
3. In the third degree of planarization (*complete local planarization*), the steps at the surface of DM1 are completely eliminated wherever the spaces in the underlying topography are relatively close together (e.g., <10 μm apart), but the steps at isolated, wide features still exhibit a step (Fig. 4-9d).

4. In the fourth degree of planarization (*complete global planarization*), the surface of DM1 is completely planarized over arbitrary topography (Fig. 4-9e).

A *quantitative measure* of the step-height reduction, referred to as the *planarization factor*, β, is given by[26]

$$\beta = 1 - (t^f_{step}/t^i_{step}) \qquad (4-7)$$

where t^i_{step} and t^f_{step} are the initial and final step heights, respectively. In complete-planarization cases, $\beta = 1$; if no planarization (or only smoothing) exists, then $\beta = 0$.

4.4.1.2 The Need for Dielectric Planarization.
As the number of levels in an interconnect technology is increased, the stacking of additional layers on top of one another produces a more and more rugged topography. Let us consider, for example, a two-level metal, single-poly CMOS process. Assume that the step height of the semi-recessed field oxide is 0.3 μm, and the thicknesses of the poly and the first and second metals are 0.4, 0.5 and 1.0 μm, respectively. The maximum height of the steps on the wafer surface after each of these processes will correspondingly be 0.3, 0.7, 1.2, and 2.2 μm (Fig. 4-10). It is apparent that the surface of the wafer must be planarized

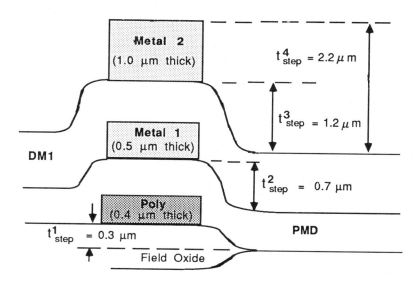

Fig. 4-10 As the number of levels in an interconnect technology is increased, the stacking of additional layers on top of one another produces steps of progressively greater heights.

in some fashion to prevent the topography roughness from growing with each level. Without such planarization, the microscopic canyons that result on the wafer surface from stacking of device features can lead to topography conditions that would eventually reduce the yield of circuits to unacceptably small values.

A nonplanar wafer topography results in three conditions that prevent the fabrication of reliable electrical connections in multilevel interconnect systems. These conditions are as follows:

1. *Poor step coverage of the metal lines as they cross over the high and steep steps*. A measure of how well a film maintains its nominal thickness is expressed by the ratio of the minimum thickness of a film as it crosses a step, t_s, to the nominal thickness of the film on horizontal regions, t_n (Fig. 4-11a). This property is referred to as the *step coverage* of the film, and it is expressed as the percentage of the nominal thickness that occurs at the step:

$$\text{Step coverage (\%)} = (t_s/t_n) \times 100\% \qquad (4-8)$$

Step coverage of 100% is ideal for a conductor, but in general the height of the step and the aspect ratio of the features being covered impact the expected step coverage. (The *aspect ratio* is defined as the height-to-spacing ratio of two adjacent steps.) The greater the step height or the aspect ratio, the more difficult it is attain coverage of the step without a corresponding thinning of the film that overlies it (Fig. 4-11b shows an example of poor step coverage.). Hence, worse step coverage is expected under these conditions. In addition to the step height and aspect ratio, step coverage depends on two other topological factors: the *contour* and the *slope* of the step. In general, the smoother the steps contour and the less vertical its slope, the better will be the step coverage of any overlying films.

2. *Metal stringers remain behind at the foot or sides of a steep step when anisotropic etching is used.* This condition is shown in Fig. 4-11c (and see Vol. 1, chap. 15). In addition, the resist is typically eroded by the same gases that etch the metal. The resist must therefore be thick enough in all locations that it will not be completely removed before the etch process is completed. If the dielectric surface has steps, in certain locations (i.e., over the upper corners of the steps in the dielectric, see Fig. 4-11d) there may not be sufficient resist thickness to survive during the time of overetch that must be used for removal of the stringers. The resist-erosion rate during the overetch time is also generally enhanced: because most of the exposed metal has been etched away the reactant gases that were being consumed during etching of the metal are now present in greater concentrations. In cases where the resist is completely eroded at thin spots, the metal beneath the resist will be exposed to the etch gases and may be etched away to an unacceptable degree.

Even if the above two limitations could be overcome, the following obstacle would eventually force planarization to be implemented:

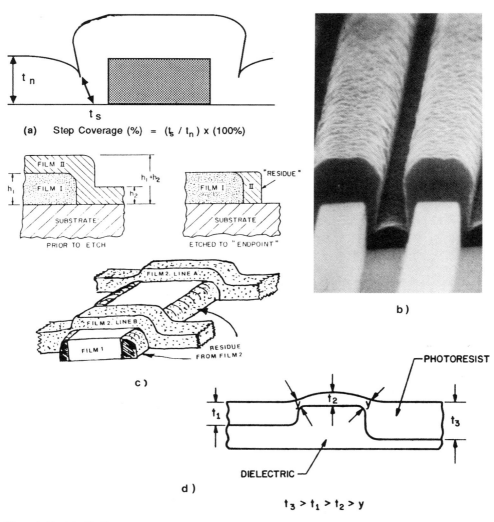

(a) Step Coverage (%) = $(t_s / t_n) \times (100\%)$

b)

c)

$t_3 > t_1 > t_2 > y$

d)

Fig. 4-11 (a) Definition of step coverage. (b) SEM micrograph showing poor step coverage. (c) Metal stringers remain behind at the foot or sides of a steep step when anisotropic etching is used. (d) Problem of resist thinning as film crosses underlying steps.

3. *The depth-of-field limitations of submicron optical-lithography tools will require surfaces to be planar within ±0.5 μm.* As a result, if steps larger than 0.5 μm exist on the surface of DM1 (or any other layer on the wafer), it will not be possible to pattern features in Metal 1 to the maximum resolution of the stepper. Thus, planarization will be mandatory if optical lithography is to be usable for fabricating ICs with submicron feature sizes. Note that this topic is discussed in more detail in chapter 2, section 2.9.2.

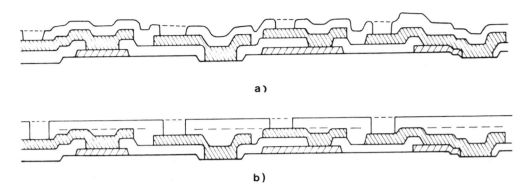

a)

b)

Fig. 4-12 (a) When no planarization is used, the via and contact holes between adjacent conductor levels all have approximately the same depth. (b) If planarization of the intermetal-dielectric layer, DM1, is achieved, the via depths can vary very widely in depth.

Planarization of the intermetal dielectric layers is one of the chief approaches that have been taken to alleviate the problem of rough surface topography.

4.4.1.3 The Price That Must Be Paid as the Degree of Dielectric Planarization Is Increased.
In the sections that follow, we will describe a progression of planarization techniques, beginning with the smoothing of dielectric films, and going through complete global planarization. However, at least two significant penalties are encountered as the degree of planarization is increased, since the maximum variation in the thickness of the dielectric layer at different locations on the circuit will also increase. This is illustrated in Figs. 4-12a and b, in which the variations in the thickness of the dielectric in a nonplanarized topography are compared to the variations that occur in a fully planarized interconnect structure.

The first penalty associated with this effect is that the the dielectric layer may end up being too thin over some of the underlying conductors after the dielectric has been planarized (Fig. 4-12b). If this occurs, it may be necessary to deposit an additional dielectric layer to increase the thickness of DM1 before depositing Metal 2.

The second, and more serious, penalty involves the process of creating openings in the dielectric layers to allow selective contacting between conductors on different interconnect levels. As the degree of planarization is increased, such openings (referred to as *vias* when they are established in intermetal dielectric layers) will have different depths. For example, consider the two-level-metal structure shown in Fig. 4-13a. In this structure, the LOCOS field-oxide step is 200 nm, the polysilicon thickness is 400 nm, and the PMD layer thickness is 400 nm. It can be seen that if a completely planarized DM1 layer is used, via depths will vary from 0.3 to 1.3 μm. Such a large variation can lead to insurmountable via-filling problems.

If the technology for adequately covering vias with metal requires that the via sidewalls be sloped, an etch process for forming such sloped sidewalls will have to be

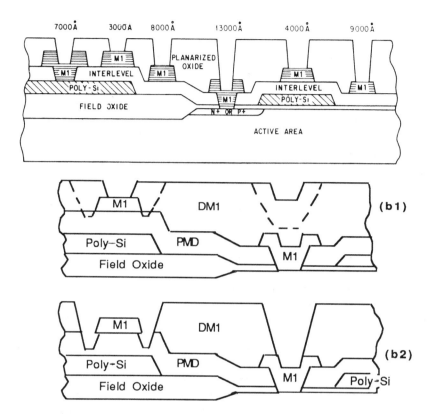

Fig. 4-13 (a) Example which shows how much the via depth can vary in a DLM process if full planarization is achieved.[259] (© 1986 IEEE). (b) If a sloped-sidewall etch is used to etch the vias shown in part (a), the dimensions of the shallow vias will continue to increase during the time needed to completely etch the deep vias. If the lateral dimensions of the shallow vias exceed the width of the Metal 1 pattern below these vias, the etch process will begin to erode the PMD layer (b2), even if a nested via is used.

used.* However, if the vias have significantly different depths, it may not be possible to implement a sloped-sidewall etch process. If a sloped-sidewall etch is used in the case shown in Fig. 4-12b, the dimensions of the shallow vias will continue to increase during the time needed to completely etch the deep vias. If the lateral dimensions of the

---

* Via sidewall sloping is usually needed when metal films are deposited by sputtering. In such cases, if the via sidewalls are too steep, metal step coverage becomes inadequate, and metal thinning (or even metal-film discontinuity) occurs. Therefore, it becomes necessary to slope the sidewalls as much as feasible.

shallow vias exceed the width of the Metal 1 pattern below these vias, the etch process will begin to erode the PMD layer (Fig. 4-13b2), even if a nested via is used (see section 4.5).

In such cases, it may be necessary to avoid the use of a complete planarization process and to resort instead to a smoothing or semi-planarization procedure. With these methods it is possible to select a small enough maximum variation in via depths that a sloped-sidewall etch process can be successfully implemented.[27] In general, however, the smoothing and semi-planarization techniques lose their effectiveness when three or more levels of metal are required and when device dimensions are smaller than 1 μm. Here, the use of complete planarization processes becomes mandatory, and technologies must be employed that allow vertically sided contact holes and vias of varying depths to be completely filled. A variety of candidate approaches for satisfying this need are described in section 4.5.

4.4.1.4 Design Rules Related to Intermetal Dielectric-Formation and Planarization Processes.

The type of intermetal dielectric material, its thickness, and the method of its deposition and planarization can all have an impact on the design rules that must be applied when laying out an integrated circuit with a particular multilevel-metal system. We will consider here a two-level-metal system with an underlying polysilicon level, and we'll assume that the topography beneath the polysilicon is completely planar. In this system, design rules dealing with the following conditions must be specified: (1) minimum polysilicon-to-Metal 1 spacing, assuming no metal overlap of the poly, (2) minimum distance between coincident edges when Metal 1 overlaps polysilicon, and (3) minimum spacing of Metal 1 to Metal 1.

If polysilicon and Metal 1 come too close together, a gap will exist between them that will lead either to void formation in the CVD oxide DM1 layer following deposition (Fig. 4-14a) or to the formation of a deep narrow crevice that cannot be covered by the next level of metal, even after some planarization procedures have been implemented (Fig. 4-14b).

In the second case, if polysilicon and metal edges are coincident, the total step height may be too high for adequate coverage by Metal 2 to be achieved (Fig. 4-14c). Figure 4-14d illustrates the case in which the problems of coincident edges and a narrow crevice are exacerbated by their simultaneous occurrence.

In the third case, the spacing between adjacent Metal 1 lines must be large enough to prevent void or crevice formation in the deposited dielectric film. The step that exists in the LOCOS field oxide must also be considered in real situations when such design rules are being formulated.

One possible way to prevent the formation of voids, crevices, and excessively high steps, is to allow the spacings between Metal 1 and underlying features to be sufficiently large that any possible combination of misalignment cannot lead to the problems listed. However, this alternative is impractical, because it increases chip size and causes excessive complexity in layout design.

The trend in advanced IC designs today is toward the elimination of all such restrictive dielectric-layer-related design rules, with the development focus on dielectric

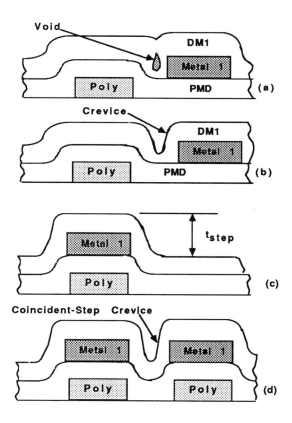

Fig. 4-14 Conditions associated with the relative spacing of polysilicon and Metal 1 that lead to layout design rules of these layers.

planarization processes that allow *all spaces between any adjacent or overlapping features (and all possible step heights), to be covered with a dielectric layer whose surface does not exhibit any voids, crevices, or excessively high steps.* Even though this is an actively pursued goal, it has not yet been attained in all processes and technologies. We will describe the various planarization approaches that have been developed along the path toward the condition in which no restrictive design rules need be invoked.

4.4.2 Step-Height Reduction of Underlying Topography as a Technique to Alleviate the Need for Planarization

In two-level-metal processes, it may be possible to avoid having to perform *any* planarization procedures with respect to DM1 if the topography underlying DM1 is planarized by the process steps that precede DM1 deposition. Although this approach may appear to be obsolete for VLSI and ULSI circuits, several of the step-height reduction techniques also turn out to be useful for providing a planar surface under Metal 1. Such a condition will ultimately be required for three- and four-level-metal processes. We will therefore begin the survey of planarization techniques of dielectric layers by considering this approach.

4.4.2.1 Provide a Completely Planar Substrate Topography. The methods used to provide a completely planar substrate surface at the conclusion of isolation-structure formation in the substrate are described in chapter 2. In general, all of these methods rely on the formation of a fully-recessed field oxide layer in the substrate. The semirecessed LOCOS technology that has served MOS ICs for a number of generations, will ultimately have to be abandoned or modified in order for this goal to be achieved.

4.4.2.2   Provide a Planar Surface over Local-Interconnect Levels. Once a completely planar surface has been established following formation of the isolation structures, a polysilicon gate or emitter layer is fabricated. This layer can also serve as a *local-interconnect layer* (silicides and refractory metals have also been developed for this function). The oxide spacers that are formed at the edges of the polysilicon during the fabrication of lightly doped drain structures (see chap. 5, section 5.6.5) and salicides (see chap. 3, section 3.9.1) result in some smoothing of the steps in the dielectric layer between the poly and Metal 1.

In addition, since this local-interconnect level is made of materials that can normally tolerate relatively high temperatures (e.g., >800°C), the CVD glass that covers it can be flowed to smooth the steps at its surface. For example, as shown in Fig. 4-15, a 30-minute 900°C anneal in N_2 of a BPSG film (4.8 wt% B and 4.6 wt% P) over patterned polysilicon can result in an almost completely planar surface.[28] If a 900°C flow temperature is too high, a sacrificial etchback step can be used to planarize the surface of the glass layer to the same degree. In one report, a spin-on glass film is used as the sacrificial layer in such a process.[29]

Recent advancements in the glass-flow process have included the use of RTP (so that shallow junctions can be maintained)[30] and the design of a CVD-SiO_2 reactor in which simultaneous deposition and flow can be accomplished.[31] Mention of a reduced-temperature (<800°C), high-pressure reflow process has also been made, although not much information on its details has been published as of this writing.[32]

4.4.2.3.   Minimize the Thickness of the Metal 1 Layer. The thinner the Metal 1 layer, the better will be the step coverage of Metal 2 as it crosses the steps

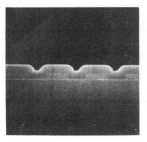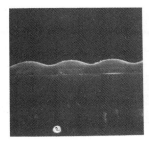

Fig. 4-15 Planarization of the PMD layer (BPSG) by thermal flow.[28] Reprinted with permission of Semiconductor International.

caused by Metal 1. A minimum thickness of ~500 nm has been proposed for Metal 1.[33] To ensure that such a thin Metal 1 layer will itself provide adequate thickness as it crosses over steps in the underlying topography, the planarization techniques described in the previous section will have to be employed.   In addition, a useful approach in double-level-metal technologies is to utilize Metal 1 to interconnect the gates, while reserving Metal 2 for the power (V_{DD}) and ground (V_{SS}) buses. This also allows a relatively thin first-level metal to be used.

4.4.2.4   Achieve Smoothing of Steps in DM1 by Sloping the Sidewalls of Metal 1 Lines.

As shown in Fig. 4-16, the reentrant angles of the steps in DM1 depend on the sidewall profile of Metal 1. If the sidewalls of Metal 1 are vertical (or even reentrant), the steps in atmospherically deposited CVD-SiO_2 DM1 layers will also be reentrant. This will reduce the step coverage of Metal 2 and will give rise to stringers if an anisotropic Metal 2 etch is used. If the sidewalls are sloped, as shown in Figs. 4-16c and 4-16d, the DM1 steps will be smoothed (the more slope, the better the smoothing). The most widely used technique for obtaining sloped sidewalls in metal lines with dry etching is to controllably erode the masking-resist layer during the metal etch. This is done by isotropically etching a sloped-resist mask edge at the same time that the metal is being etched. The slope in the resist sidewall is achieved either by flowing the resist at a high temperature or by reactively facet-etching the resist corner.[34]

As dimensions decrease, however, the latitude for sloping decreases. Another way to produce smoothing is to modify the metal-sidewall profile without sloping the sidewalls. One such approach involves cutting the sharp top corner of the metal line, as shown in Fig. 4-16e.[35] In this technique, a polymer is formed on the sidewalls of the Al lines following dry etching (but preceding removal of the resist). Next, a slight etch of the resist is performed so that the top corners of the Al lines are exposed. A wet etch of the Al is then used to cut the top corners. Finally, the resist is removed. Figure 4-16f shows an SEM photographs of an Al line with such cut top corners.

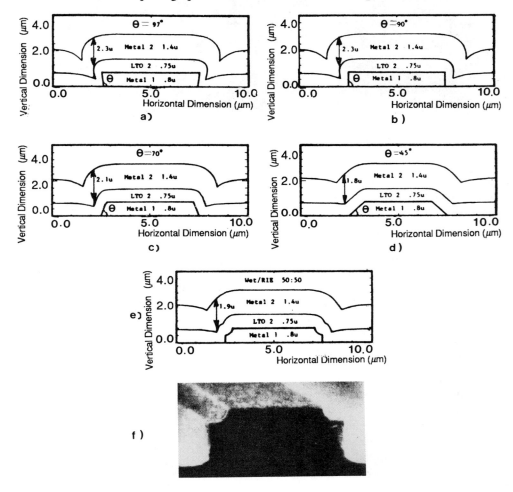

Fig. 4-16 Dependence of the angle of the DM1 sidewall on the profile of Metal 1. (a) Reentrant M1 sidewalls produce reentrant angles. (b) Vertical Metal 1 sidewalls still lead to DM1 reentrant angle sidewalls. (c) and (d) Sloped Metal 1 sidewalls produce DM1 sidewalls that are properly sloped and smooth, for ease of Metal 2 step coverage.[33] (© 1985 IEEE). (e) Etching away the top corner of the metal line also helps smooth the DM1 sidewall profile. (f) SEM micrograph of Metal 1 lines with the top corners cut off.[35] (© 1987 IEEE).

4.4.3   Deposition of Thick CVD-SiO$_2$ Layers, and Etching Back without a Sacrificial Layer

As described in an earlier section, one of the simplest methods available for smoothing steps in DM1 is to deposit a CVD-glass layer that is significantly thicker than the step height it must cover. That is, it has been determined that in low-temperature CVD dielectric layers, the slope of the DM1 step decreases as the DM1 thickness increases. For example, it has been shown that if a PECVD oxide is deposited to 1.5X the underlying metal thickness, the steps of DM1 no longer exhibit reentrant angles.[36] From this effect, it would appear that the greater the ratio of the DM1 thickness to the underlying step height, the better the profile of DM1 would be (from the point of view of step coverage by an overlying metal film).

In practice, however, this is unrealistic, because increasing the thickness of DM1 also increases the via depth between Metal 1 and Metal 2. Furthermore, as the Metal 1 lines become more closely spaced, voids will form in the dielectric (see Fig. 4-8) if conventional atmospheric, low-pressure, or plasma enhanced low-temperature CVD-SiO$_2$ processes are used (ACVD, LPCVD or PECVD SiO$_2$, respectively). In order for void formation to be prevented, the thickness of DM1 must not be greater than half the minimum spacing between Metal 1 features. For example, for a 2-μm metal spacing, the maximum thickness of DM1 would have to be less than 1 μm. For a 1-μm space, no more than 0.5 μm of DM1 thickness could be deposited – and unfortunately, a film of this thickness (using the 1.5X guideline) would not prevent reentrant angles over 0.5-μm-thick metal lines.

A recently introduced process that utilizes a dielectric layer of plasma-enhanced low-temperature CVD TEOS together with a low-temperature thermal CVD TEOS (as described in section 4.3.2) allows narrow, high-aspect-ratio spaces between metal lines to be filled without the formation of voids. (As described earlier, plasma TEOS allows spaces with aspect ratios as high as 0.85 to be filled without void formation, whereas PECVD silane-based SiO$_2$ films exhibit voids when the aspect ratio of a space exceeds 0.5.)[37] This process will make the use of a nonsacrificial-etchback technique applicable to smaller-sized devices. However, it can be seen from Fig. 4-17a that this process does not provide global planarization.

If the spacing between metal lines is large enough to allow this technique to be employed (or if the low-temperature CVD TEOS process mentioned above is used), but the thickness of DM1 is too great after deposition, the DM1 can be etched back until a desired thickness is obtained. If an etch process that has a sputtering component is used, the etch will also round off the corners (see Vol. 1, chaps. 9 and 16), enhancing the smoothing effect (Fig. 4-17b).[38] One advantage of this type of process over a sacrificial-etchback process (see section 4.4.8) is that it is much simpler and does not cause contamination of the reactor walls by polymer deposition. A report detailing such a plasma TEOS nonsacrificial-layer etchback approach for a double-level-metal CMOS process is found in reference 37.

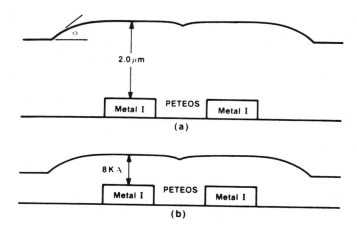

Fig. 4-17 (a) Thick PETEOS deposition (2.0 μm) over 0.5-μm Metal 1 creates a "smoothed" topography for Metal 2 at the edge of an isolated Metal 1 feature while planarizing closely spaced Metal 1 features. (b) An anisotropic etchback of the interlevel dielectric preserves the "smoothed" surface.[37] (© 1987 IEEE).

4.4.4 Oxide Spacers

We have seen that the step coverage of metal lines is dramatically improved if the steps have gentle slopes. Flowing of doped CVD glass layers is used to achieve such slopes when the PMD layer covers polysilicon. Unfortunately, when metal is deposited on the wafer, the high temperatures needed to cause glass flow cannot be tolerated, and hence this solution cannot be applied. Another method of producing such smoothing is to taper the sidewalls of the underlying metal lines. Techniques to achieve such tapering were described in an earlier section. Tapered metal lines, however, are difficult to form, and they require tight process control. In addition, as the line widths get narrower (without a corresponding decrease in line thickness), the latitude for sloping decreases. The use of oxide spacers at the sidewalls of the metal lines allows shallower-sloped steps to be created without tapering of the metal lines.[39,40,41]

The oxide-spacer approach begins with the deposition of a doped PECVD-SiO$_2$ layer over Metal 1. This layer exhibits conformal coverage over the Metal 1 lines and other steps (see Vol. 1, chap. 6). The PECVD SiO$_2$ layer is then anisotropically etched back until Metal 1 is exposed. Unetched portions of SiO$_2$ at the vertical sides of metal lines and other steps remain following this etch; such residual structures are known as *oxide spacers* (Fig. 4-18). The slope of the spacer sidewalls is less severe than the slope of the as-deposited conformal PECVD oxide. The smoothing occurs as a result of faceting during the anisotropic dry-etch step (see Vol. 1, chap 9). Following spacer formation, a second layer of CVD SiO_2 is again conformally deposited, and the gentler slopes of the spacers are thus replicated in the steps at the surface of the dielectric layer.

The oxide-spacer technology is seen to be a semi-planarization technique. As a result, it offers the following advantages:

- It is a simple approach for achieving adequate Metal 2 step coverage in that it requires no photolithography and only one extra SiO_2 deposition and its etchback.

- It is not sensitive to the oxide-thickness uniformity or to the degree of overetch used in the etchback step.

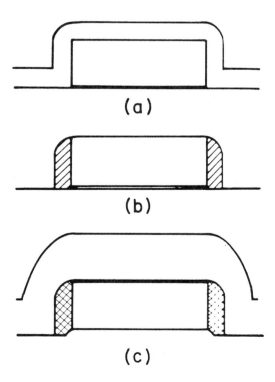

(a)

(b)

(c)

Fig. 4-18 Spacer formation: (a) CVD deposition over a polysilicon line; (b) spacer formation by RIE; (c) spacer structure covered with CVD oxide.

- It avoids the deposition and etching of a sacrificial resist layer, steps that can cause contamination of the etch chamber through the deposition of polymer by-products arising from the resist etching.

- All of the via holes are of the same depth, since the final SiO_2 layer is equally thick in all places (making the etching of sloped via holes much easier).

Oxide spacers also have several important limitations, including the following:

- Since the step heights themselves are not reduced by this technique, as each layer of a multilevel structure is added the steps additively grow larger. This causes severe lithography problems for small depth-of-field steppers (see chap. 2, section 2.9.2); step-coverage problems would become severe for third and fourth levels of metal.

- If adjacent Metal 1 lines get too close (e.g., when the aspect ratio of the spaces between adjacent Metal 1 lines approaches 0.6), voids will form when the first PECVD-SiO_2 layer is deposited (Fig. 4-8). The narrow spacing leads to shadowing effects, and when combined with the lower surface mobility of reactive species at temperatures below 600°C, these effects can result in deposition rates that are higher on the upper surfaces than in the valleys between structures. Such voids may open up during the etchback step and subsequently trap moisture, photoresist, or metal from the next deposition.

- Long Metal 2 lines will have measurably larger resistances than those deposited on a completely planar surface (since the line must travel up and down the steps of the semiplanarized surface).

Because of these limitations, the oxide spacer approach is viewed as a stopgap solution, one that can be used only in two-level metal processes with Metal 1 spacings that are no closer than ~2 μm.

4.4.5 Polyimides as Intermetal Dielectrics

The use of polyimide films as interlevel dielectrics has been pursued as another technique for providing partial planarization of the dielectric surface. Polyimides offer the following attractive properties for such applications:

- They produce surfaces in which the step heights of underlying features are reduced, and step slopes are gentle and smooth.

- They are able to fill small, high-aspect-ratio openings without producing the voids that occur when low-temperature CVD-oxide films are deposited.

- The cured polyimide films can tolerate temperatures of up to 500°C without degradation of their dielectric film characteristics.

• Polyimide films have a dielectric-breakdown strength that is only slightly lower than that of SiO_2;

• The dielectric constant of polyimides (3.2-3.4) is smaller than that of SiO_2 (3.5-4.0) and of silicon nitride (6.0-9.0);

• The films are free of pinholes and cracks;

• The process used to deposit and pattern the polyimide films is relatively simple, and it is significantly less expensive to carry out than are the non-spin-on inorganic planarized-dielectric alternatives. That is, polyimide film is spun on in the form of a liquid (polyamic-acid precursor), much as in the process used to deposit photoresist films. During a high-temperature-cure step (e.g., at 150°C for 30 min and 300°C for 60 min), the polyamic acid undergoes a chemical change (imidization) that causes it to become the solid polyimide film. Vias are normally dry etched in the cured film using oxygen- or fluorine-based plasmas.

The partial planarization produced by polyimide films occurs because the film is deposited as a liquid. Just as the surface of a lake is flat because of the surface tension of the liquid, the polyamic acid in liquid form fills the crevices on the wafer surface and produces a flat surface. During the bake step (in which the polyamic acid is imidized), the solvent in which the polyamic acid is dissolved is also driven off, along with OH as H_2O. Since only a relatively small percentage of the film consists of solids (typically, 15-30%), the final surface height of the polyimide in areas of the wafer over which the liquid depth is larger will be at a lower level than in areas over which the liquid depth is smaller (Fig. 4-19). In general, the larger the percentage of solids in the original liquid material, the higher the degree of planarization that is produced.

The planarization of actual polyimide films ranges from 49% to 69% if a single-coat application is used. Planarization can be further increased through the application of a second coat.[42] In many processes, a thin, so-called *hard-mask layer* has been used between the polyimide and the resist etch mask. This hard-mask layer is often needed because photoresist and polyimide both etch at about the same rate in the O_2 plasmas used to etch vias in the polyimide. Various films have reportedly been used for this purpose, including PECVD SiO_2, Al, Ti, and spin-on glass. After via etching is completed, the hard mask is removed.

Some impressive processes using polyimide as an interlevel dielectric have been reported, including a three-level-metal process for bipolar analog circuits from IBM (in which the polyimide acts as the interlevel dielectric between Metal 1 and Metal 2, and between Metal 2 and Metal 3, as well as the passivation layer);[43] a three-level-metal process for a bipolar gate array by Siemens;[44] and a multilevel-tungsten process from Hewlett-Packard.[45] Despite such successes, polyimide films have not gained widespread acceptance as interlevel dielectrics. The reasons for this have to do with the following manufacturing-compatibility and reliability concerns:

• Polyimide is hygroscopic (i.e., it readily absorbs water). When the polyimide films are rapidly heated (e.g., during a final sinter step or during the die-attach

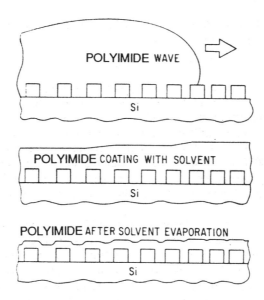

Fig. 4-19 Although the surface of polyimide film is planar when first spun on in liquid form (a) and (b), as the film is cured, solidified and shrunk, the steps on the surface reappear (c), especially at widely spaced features.

step), any absorbed water will be desorbed by the polyimide film. If a wide metal line covers a portion of the polyimide film, the moisture under the metal film will attempt to escape at the surface covered by the metal, which can lead to bubbling of the metal film. To prevent such occurrences, a slow, ramped dehydration bake must be used whenever the wafers with polyimide films are heated.[46] Some of these problems may be alleviated through the use of new polyimide materials that exhibit much less moisture absorption than those of earlier generations.[47]

• The absorbed moisture can potentially cause corrosion of underlying metal lines. Although a number of studies have indicated that this may not be a problem, preventive measures have been included in some polyimide processes to forestall such possibilities. For example, a coating of silicon nitride was added over each layer of polyimide in the processes of references 44 and 45. In the IBM process,[43] a silicon-nitride layer was deposited *under* the first polyimide intermetal-dielectric film to prevent moisture and ionic penetration to the devices.

• Reliable adhesion of metal films to polyimide underlayers has also been a concern. Although an adhesion-promoting layer is deposited onto SiO_2 layers prior to the spinning on of the polyimide (which is quite effective in promoting good adhesion between the polyimide and the SiO_2), there is as yet no well-understood model that explains how to produce good adhesion between the overlying metal and the polyimide. The concern is magnified because the thermal-expansion coefficients of metals and polyimides are generally not equal, and residual stresses caused by such mismatches may lead to delamination failures. This problem, however, can be circumvented by coating the polyimide with a nitride film[44] or an oxide film.[47] A report dealing with adhesion of metal to polyimide is given in reference 48.

• Because polyimide films are easily etched in oxygen- and fluorine-based plasmas, a suitable etch process must be developed when metal films are patterned over polyimide layers. Since the underlying polyimide can be easily eroded during the overetch part of the metal-film-etching process, in some cases it may be necessary to provide a special etch-stop material under the main metal layer to prevent such polyimide erosion.

4.4.6   Planarization with Bias-Sputtered SiO$_2$

Deposition of SiO_2 films through rf sputtering of a silica-glass target is another technique used to achieve partially or completely planarized intermetal dielectric layers. Because the wafers themselves are also subjected to a negative rf bias during the sputter-deposition process, the process is termed *bias-sputtered SiO$_2$*.*

The application of a bias to the wafers serves two key purposes in this process. First, if SiO_2 films are sputter deposited without bias, they are porous, they etch very rapidly in HF solutions, and they have a columnar structure with scattered nodular defects. A dielectric film with such a structure is obviously undesirable from the point of view of pinhole density, dielectric strength, and the ability to provide corrosion protection to underlying metal films. Ion bombardment during film growth (resulting from biasing of the wafers) can modify the microstructure of the film so that dense SiO_2 films with properties very similar to those of thermally grown SiO_2 can be obtained.

The second role of the applied bias is to establish deposition conditions that bring about planarization of the SiO_2 film. The planarization mechanism in bias-sputtered SiO_2 was analyzed in a paper by C. Y. Ting et al.[49] These researchers determined that planarization occurs due to the interaction of two factors: (1) the film buildup on the substrate resulting from sputtering at the target is independent of the geometry of the features of the substrate; and (2) resputtering at the substrate (which occurs as a result of the bias applied to the wafers) is a strong function of the geometry of the substrate features (i.e., sputtering yield is a function of the ion angle of incidence, and sloped

---

* The glass-sputtering target is commonly referred to as a *quartz target;* as a result, this process is also frequently called *bias-sputtered quartz,* or *BSQ.*

surfaces on the substrate thus resputter faster than flat areas; Fig. 4-20a). In general, the sputtered SiO_2 process can proceed in one of three modes: (a) the nonplanarization mode; (b) the planarization mode; or (c) the erosive mode.

If the film is deposited without biasing of the wafer, the deposition will proceed in a *nonplanarization mode*. In such cases, the deposition rate will be greater than the resputtering rate, regardless of the slope of the substrate feature. Film growth as a function of time in this mode is shown in Fig. 4-20b; no planarization is seen to occur.

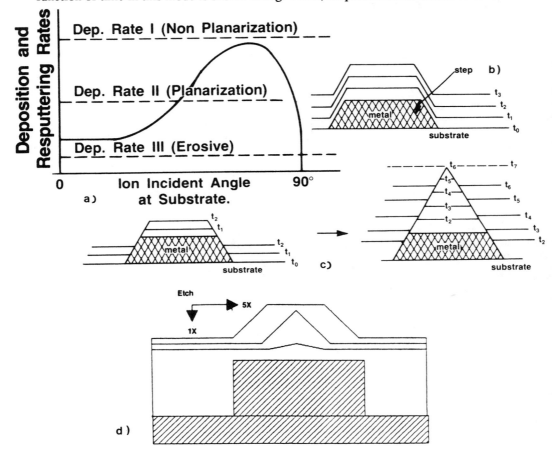

Fig. 4-20 (a) Sputtering (and hence also resputtering) yield is higher on surfaces whose angle toward the incident ions is less than 90°. (b) Nonplanarization mode - deposition rate is greater than resputtering rate (no bias is applied to the wafer). (c) Planarize mode - net deposition on the sloped areas is much lower than that on the flat areas. (d) Erosive mode - no net deposition anywhere, but material is sputtered away more rapidly on the sloped areas than on the flat areas.   Although the pyramid structure on the surface eventually disappears first over narrow lines, it remains much longer over wider lines.

If a bias is applied to the wafer, the deposition can be made to proceed in either a *planarized* mode or an *erosive* mode. In the *planarized mode,* the net deposition on the sloped areas is much lower than it is on the flat areas (Fig. 4-20c). In the *erosive mode,* there is no net deposition anywhere, but the sloped regions will be etched more rapidly than the flat regions. Under conditions of *planarized deposition,* as the SiO_2 film grows thicker, the surface of the SiO_2 over a narrow metal line will change, as shown in Fig. 4-20d. Eventually the pyramid structure on the surface will disappear, and complete planarization will be achieved.

For several reasons, the process is rarely allowed to proceed to this point in practical circumstances. First, the planarized SiO_2 film thickness would end up being much too thick for most applications. Second, complete planarization would have occurred only over the narrow lines – that is, over wide lines no reduction in the step height would have occurred, and over medium-width lines a pyramid would remain.

As a result of the second factor, in some bias-sputtered SiO_2 processes the maximum width of the metal lines to be covered is restricted, and the film is deposited to a thickness in which pyramids remain over these lines. The *erosive mode* is then used to sputter away the pyramids so that a planarized surface is achieved. This erosive step, however, causes some of the oxide in the flat areas to also be etched away. To end up with a sufficiently thick dielectric film in such processes, it might be necessary to carry out another dielectric deposition step.[229] If such multiple deposition/etch/deposition steps are performed, it may take several hours to produce an SiO_2 layer of adequate thickness and planarity. (In fact, to produce pyramid structures with less steep slopes, a greater resputtering rate during the planarization mode is needed, which further reduces the overall growth rate.) Thus, one limitation of the planarized, bias-sputtered SiO_2 process is low throughput due to low net-deposition rates. This limitation is emphasized by the high capital-equipment cost needed to implement the process.[50]

The bias-sputtered SiO_2 approach for planarizing films offers the advantages of being able to fill very narrow spaces and of doing so at relatively low temperatures (e.g., the deposition is typically carried out at <400°C). Nevertheless, even though IBM has reportedly used bias sputtered SiO_2 for many generations of IC products,[51] and other companies have also implemented it for some special applications,[217] this approach has not gained acceptance as a mainstream dielectric planarization technology in IC production. In addition to the low-throughput drawback already mentioned, there are a number of factors responsible for this. Among them are the following:

- The sputter-deposited SiO_2 films exhibit high compressive stresses (~100 MPa, compressive), which contribute to the problem of stress-induced cracks in the metal films underlying such dielectrics.

- The sputtered-SiO_2 material also deposits on the shields, shutters, and other surfaces of the sputter chamber. Since these films are also highly stressed, they have a strong tendency to flake off in the form of small particulates. As a result, particulate levels in the bias-sputtered SiO_2 process can be much higher than those observed in conventional CVD-oxide processes, and they can represent a serious yield limiter in manufacturing environments. Special rigorous cleaning procedures of the sputtering equipment must therefore be followed if bias-sputtered SiO_2 is to be successfully implemented.

• If the pyramid structures of Fig. 4-20d are not removed, they can cause cracking of metal lines deposited over them.

• Metallic contamination arising from resputtering of the sputter-chamber surfaces (e.g., iron contamination caused by resputtering of the stainless-steel fixtures inside the chamber) is another possible source of manufacturing yield loss.

• The purity of the SiO_2 targets has been a concern, and radiation damage associated with the sputter-deposition and etch phenomena has been reported to produce reliability degradation in MOS devices.

Progress has reportedly been made on several of the above problems. The pyramid structures on the SiO_2 surface have been eroded through the use of a photoresist sacrificial-etchback step (see section 4.4.8) after the SiO_2 film has been initially deposited by means of a planarization-mode deposition step.[52] In a more recent report, a double photoresist layer was used to increase the planarization over wide metal features.[53] In addition, the purity of the SiO_2 targets has been improved. Finally, a process has been described in which the radiation damage is reduced to an acceptable degree for practical use.[52] In this report, device damage that could not be annealed out by a 450°C step was identified as gate-oxide damage caused by bremsstrahlung X-rays (which are apparently produced when high-energy secondary electrons, originating from the sputter target, strike the wafer). The suggested way to decrease this damage was to prevent these high-energy electrons from reaching the wafers. This was achieved by placing a stainless-steel grid between the target and the wafers in the sputter chamber. The presence of the grid, which was biased to intercept the high-energy target electrons was reported to not measurably contaminate the devices. It did, however, result in an even lower deposition rate. A high-deposition-rate bias-sputtered SiO_2 process is described in reference 57.

4.4.7 CVD SiO2 and Bias-Sputter Etchback

An alternative method similar to the bias-sputtered SiO_2 approach for forming partially planarized dielectrics is the *CVD/bias-sputter etchback* technique (CVD/Etch). In this process, the SiO_2 layer is formed through the use of plasma-enhanced CVD (PECVD) rather than through the sputtering of a glass target (Figs. 4-21a and b. The advantages are that the slow sputter-deposition process is replaced by a faster CVD process, and that the CVD process eliminates the problem of dielectric-film contamination due to impurities in the sputter target. In addition, the deposited films are not as highly stressed and particulates not as severe as with the bias-sputter approach. Finally, the high-energy electron damage caused by the sputter deposition is eliminated. The only bombardment of the wafer surface is by the argon ions. Since these penetrate no more than a few atomic layers, they affect only the interlayer dielectric, not the devices underneath.

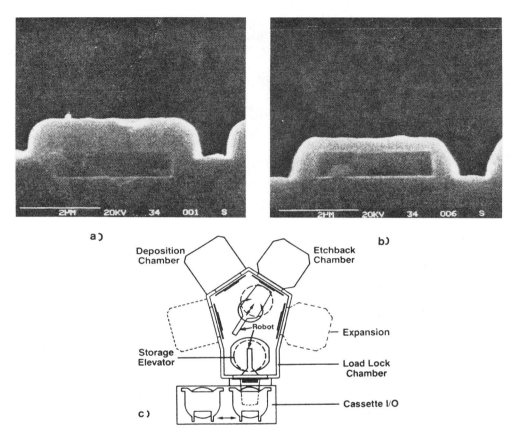

a)   b)

c)

Fig. 4-21 CVD SiO_2 and in situ etchback. (a) Photograph of an as deposited PECVD oxide over vertical sidewalled Metal 1 lines. (b) Shape of PECVD oxide after reactive ion etching using energetic ion bombardment.[54] (© 1987 IEEE). (c) Example of a multi-chamber tool that can perform deposition in one chamber an etching in another. Courtesy of Applied Materials.

Early attempts to implement this technique involved deposition of the film in one machine, followed through sputter etching of the deposited layer in another. An enhancement of this technique involved reactive facet tapering of the oxide by the use of a reactive gas (CHF_3) in conjunction with Ar.[54,56] This produced steps on the surface of the oxide with slopes that were not as gentle as they would have been if the steps had been created by purely physical sputtering. However, the erosion rate was much faster, which significantly increased the throughput. Subsequent work integrated the process into a single reactor, first with a parallel-plate reactor,[55,243] and later with a multi-chamber tool that can perform deposition and etching without exposing the wafers to

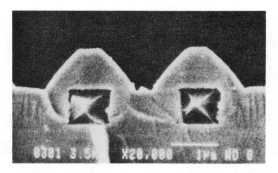

Fig. 4-22 At the conclusion of a CVD/Etch process, the surface topography over narrow lines will have a pyramid structure.[58] (© 1987 IEEE).

atmosphere between process steps (Fig. 4-21c).[58,244]

One disadvantage of the CVD/Etch method is that at low temperatures the CVD film does not fill high aspect ratio openings (e.g., a 0.5-μm space between 0.7-μm-thick metal lines) as effectively as does bias-sputtered SiO_2. Thus, for such high-aspect-ratio applications, the deposition and etchback procedure must be performed in several sequential deposition/etch sequences to prevent the formation of voids in such spaces.[58] First, an oxide layer is deposited that is not thick enough to cause seam closure in the opening. Sputter etching is then used to form sloped surfaces in the small openings. This keeps the small trenches open during deposition, allowing them to eventually be completely filled without the formation of keyholes (voids). This multistep approach also tends to stop the propagation of pinholes.

At the end of the multistep process, the surface topography will have pyramid structures over narrow lines, as shown in Fig. 4-22. To eradicate these pyramids, an erosive-mode step is carried out so that a planar surface is finally established. The pyramid structures can also be removed using a resist-etchback technique.[59] A report has been made of a two-level-metal wiring technology with metal pitches of 1.9 μm for Metal 1 and 2.4 μm for Metal 2 that uses this method to form the interlevel dielectric.[60]

There are two main drawbacks to this multiple CVD/Etch process: (1) the throughput is low, even in a batch reactor (e.g., eight wafers in a batch); and (2) complete planarization is not achieved over wide steps.

4.4.8 Planarization through Sacrificial-Layer Etchback

Planarization of CVD interlevel-dielectric films can also be achieved using the *sacrificial-layer etchback* technique. This method has gained the most widespread acceptance in two-level-metal processes down to ~1 μm device technologies (and even in some reported three-level-metal bipolar processes).[61] Using this approach, it is possible to achieve a high degree of planarization between steps that are ~2-10 μm apart. With more closely spaced features, the technique runs into problems, and for more widely spaced steps, planarization is less complete. The process is carried out by first depositing the CVD film that will serve as the interlevel dielectric. This layer is then coated with a film that will later be etched off (sacrificed). In most cases, such sacrificial layers are deposited as low-viscosity liquids. Upon being baked, these liquids become solid thin films that are thick enough to produce an essentially flat surface. A few other reports have described the use of conformally deposited silicon nitrides over CVD oxides or oxynitrides. These processes rely on controlling the difference in plasma etch rates between the two materials to obtain semi-planarization of the dielectric surface.

When sacrificial liquid layers are used, the material is spun on over the underlying dielectric film (which, at best, may exhibit conformal coverage of the wafer surface, but essentially no planarization). The sacrificial layer is relatively thick (typically, 1-3 μm) for improved surface planarity. The wafer is then baked until the spun-on film becomes a solid, and the surface (above features that are less than ~10 μm apart) becomes reasonably well planarized, as shown in Fig. 4-23a.

Photoresists have been the most widely used sacrificial liquid layers, although polyimide and spin-on-glass (SOG) layers have also been occasionally used for this role. Resists are popular primarily because of their purity, their well-characterized process history, and their lower cost.[62,63] High-temperature baking (>150°C) may be used to flow the resist to improve planarization even further. However, the effect of such high bake temperatures on the plasma-etch rate of the resist must be taken into account in order for good process control to be achieved. Figure 4-24 illustrates the degree of planarization that is obtained as a function of the width of an isolated feature for several photoresists (and the AZ protective-coating material).[64] As long as the feature size is smaller than 10 μm, the degree of planarization can exceed 70% with most photoresists.

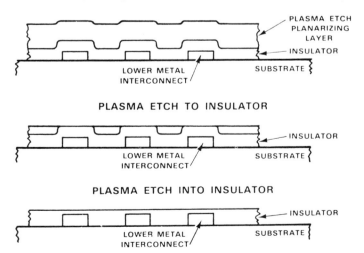

Fig. 4-23 Sequence of steps used in a sacrificial-layer-etchback process for planarization.

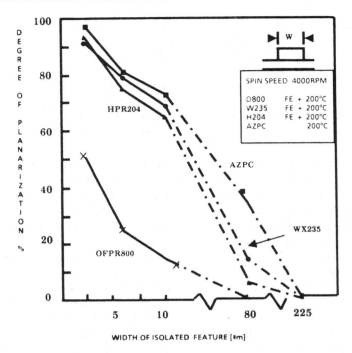

Fig. 4-24 Degree of planarization versus various feature sizes for different polymer films.[64] (© 1987 IEEE).

In the next step, the sacrificial layer is first rapidly etched back in a plasma (typically, O_2 or O_2 mixed with CF_4)[65] until the topmost regions of the dielectric layer are just exposed (Fig. 4-23b). The etch chemistry is then modified so that the sacrificial-layer material and the dielectric are etched at approximately the same rate. The etch is continued under these conditions until all of the sacrificial layer has been etched away. At this point, the surface of the dielectric film is highly planarized since the profile of the sacrificial layer is transferred to the dielectric layer by the etchback procedure. The thickness of the dielectric film over underlying features, such as metal lines, may be thinner than desired after the etchback is completed. In some processes, in fact, the etchback step is allowed to proceed until the Metal 1 lines are exposed. In any case, an additional layer of CVD dielectric is generally deposited in order to establish a minimum adequate thickness everywhere on the wafer surface.

4.4.8.1 Degree of Planarization Achieved by Sacrificial Etchback.
The sacrificial-etchback process planarizes the final dielectric-layer surface by reducing the step height and smoothing the steep steps of the as-deposited dielectric layer. Although the step height can be reduced to zero (complete planarization) over closely spaced features, this may not always be the desired result. Such complete planarization will lead to widely varying via depths, which may be difficult to adequately fill with the

next level of metal. As a result, a trade-off of β (defined by Eq. 4-7) against *maximum via depth variation* can be made.   The final step height is related to the initial step height and the ratio of the etch rates of the resist, E_r, and the oxide-etch rate, E_o, by[66]

$$t^f_{step} = t^i_{step} - [(E_o / E_r)(t^i_{step} - t_r)] \qquad (4-9)$$

where t_r is the step height in the planarizing layer. Combining Eqs. 4-7 and 4-9 yields the following simple relation between the process parameters and β:[65]

$$\beta = (E_o / E_r)[1 - (t_r / t^i_{step})] \qquad (4-10)$$

The linear relationship between β and the etch rate ratio (E_o/E_r) indicates the need for maintaining precise control over both the deposition processes and the planarization etch.

The maximum degree of planarization, β_1, based on the maximum via depth variation and the minimum acceptable post-etchback thickness of oxide, t_{ox}, covering Metal 1 (or polysilicon), is given by

$$\beta_1 = (t_{ox}' - t_{ox}) / t'_{step} \qquad (4-11)$$

where t'_{step} is the step height of the underlying topography and t_{ox}' is the maximum depth of the vias following etchback. Thus, the appropriate degree of planarization is $\beta \leq \beta_1$. A model for predicting the planarization that will be produced by an etchback process is described in reference 68. The desired etch-rate ratio is typically obtained by varying the oxygen-flow rate in a CF_4/O_2, CHF_3/O_2, or C_2F_6/O_2 plasma, as illustrated in Fig. 4-25.

The above discussion is generally valid only for the planarization of topographies in which features are less than 10 μm apart. For wide, isolated features, the step height will not be reduced, since the resist thickness on top of such features will be the same as the thickness over adjacent substrate regions (Fig. 4-26a). If it is necessary for t_r to be zero everywhere on the wafer surface – regardless of the underlying feature widths, spacings and density – it is possible to use a dual photoresist process (which also involves an additional resist-patterning step, as well as a *planarization block mask* [PBM] to pattern the inverse of the undesired topography), as described in references 67 and 70 (Fig. 4-26b). In this procedure, the first organic film is spun on, and the PBM mask is used expose and develop this film. As a result, this layer remains in the wide "low" regions, which are thereby effectively filled. Consequently, the final planarizing-resist film only has to fill small crevices (Fig. 4-26b).

It has been pointed out, however, that creation of the PBM mask cannot be implemented by merely reversing the mask that produced the underlying topography. That is, this simple approach results in errors due to changes in the resist thickness caused by the underlying pattern, misalignment, and "overfilling", as the resist competes with (partial) planarization accomplished by the underlying deposited films (Fig.

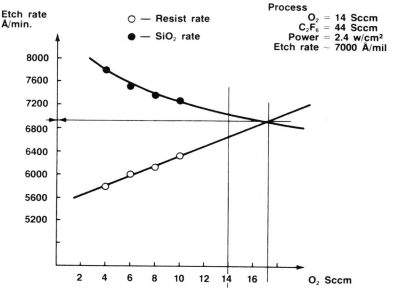

Fig. 4-25 Example of how the etch rates of resist and SiO$_2$ can be equalized by varying the gas mixture used in the plasma-etching process. Courtesy MRC, Inc.

4-26c).[228] The PBM mask must therefore be designed to compensate for these errors.

A novel thermal flow/thermal setting polymer (aromatic polyalkylenes) has recently been described that planarizes even the largest geometries, such as 100 μm bonding pads, after a reflow step at 200°C (Fig. 4-26d).[249] In this figure the results obtained using 1.2 μm of AZ1370 photoresist coating are also included. The new polymer provides a 60% planarization even at feature sizes of 250 μm.

Another etchback technique for obtaining complete global planarization that does not rely on dry etching is *chemical-mechanical lapping*.[69] This technique has been considered for removing the top layer of the Si substrate after a selective epitaxial process has been performed to remove facet indentations (see chap. 2), as well as for intermetal-dielectric planarization. The technique is still being investigated in a research mode, and is discussed further in section 4.4.11.

4.4.8.2 Advantages of the Sacrificial Etchback Process.
In addition to being a low-temperature process, the sacrificial etchback method uses familiar processing technology, materials, and equipment (e.g. spinners, dry etchers, and CVD dielectric-deposition systems). It is also relatively straight-forward, and the final interlevel-dielectric film is a material (CVD SiO$_2$) for which there is already a long process history. Finally, most competing processes exhibit more severe drawbacks than this approach.

4.4.8.3 Sacrificial-Etchback Process Problems.
To achieve the desired degree of planarity and final dielectric film thickness, tight process control is necessary

for such parameters as the magnitude of the etch rate, the etch-rate uniformity across a wafer, end-point detection, and oxide-thickness uniformity under the resist layer. The resist etch rate is sensitive to the cure cycle (especially if a high-temperature bake step [>150°C] is used to improve surface planarity), to chamber preconditioning, and in batch processes, to the number of wafers present in the chamber. Using SOG as a sacrificial layer in lieu of resist has been suggested as a way to alleviate the problem of resist etch-rate variability, since the etch rate of SOG is found to be less variable with bake temperature and processing ambient conditions (e.g., humidity).[71]

The process requires a multitude of steps, which implies a lengthy process, and thus a low throughput. In general, etchback is a difficult process to successfully implement in a batch-etch mode, resulting in an even lower throughput.

While the thickness of the resist is a function of many variables, in general, it will be thickest over narrow spaces and thinnest over high, narrow, isolated steps. Thus, the dielectric film thickness may become unacceptably thin over the latter after etchback. In addition, the small amount of resist remaining between narrowly spaced features makes end-point detection difficult.

If an insufficient cure cycle is used, solvent remaining in the resist may bubble out

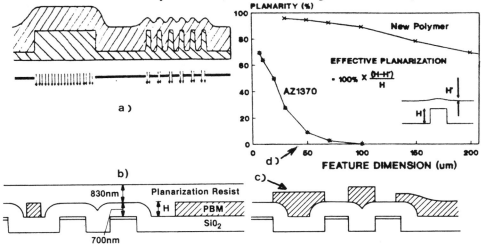

Fig. 4-26 (a) Step height at wide isolated features is not reduced by spinning on a photoresist because the resist thickness over the wide feature is the same as that over adjacent substrate areas. (b) A planarization block mask (PBM) together with an additional resist deposition have been used to overcome the problem of part (a).[67] (c) If the PBM mask is not properly compensated for underlying surface topography, a non-planarized final surface is still obtained.[228] This paper was originally presented at the Spring 1989 Meeting of The Electrochemical Society, Inc. held in Los Angeles, CA. (d) The effective planarization for the new polymer and AZ1370 versus width of the feature being covered.[249] This paper was originally presented at the Fall 1989 Meeting of The Electrochemical Society, Inc.

during the etchback step, causing craters in the resist surface. These defects may be transferred to the dielectric surface by the etchback step, causing thin spots in the dielectric film.

When a heavy resist-erosion process is performed, the chamber may become contaminated with a polymer coating. As a result, the two-step process described earlier is typically used; in the first step, an oxygen plasma (which does not cause polymer formation) is used to erode the resist until the SiO_2 is exposed. Etch-gas mixtures that produce less polymerization are also employed to reduce this problem.

If the aspect ratio between adjacent Metal 1 lines with vertical sidewalls exceeds 0.4, voids will form when most CVD-SiO_2 layers are deposited. These voids cause resist and moisture trapping (and subsequent release when the voids are opened during the etchback process, as shown in Fig. 4-26b). Opened voids will also trap metal during the Metal 2 deposition step, which will remain in the form of metal stringers following Metal 2 patterning. Thus, this process is not feasible for Metal 1 spacings of less than ~1.25 μm unless a process is available for the deposition of SiO_2 layers at low temperatures without the formation of voids in high-aspect ratio spaces (e.g., ECR deposition of oxides, or PECVD TEOS films).

The plasma-etch chemistry must be adjusted during the etchback step, since the etch-rate ratio established from etch-rate measurements of unpatterned wafers will be different from that observed when oxide and resist are etched at the same time.[59] There are two reasons for the etch rate increase when oxide is simultaneously etched:[72] first, since less resist is exposed, the concentration of the reactive species (oxygen) is higher; second, once oxide etching has begun, the oxygen that is liberated as a byproduct ($SiO_2 + 4F$ --> $SiF_4 + O_2$) may react with the resist, leading to enhanced resist etching. The magnitude of the increase, which can be up to threefold, depends on the percentage of oxide exposed surface. These effects are known as *global loading effects* (see Vol. 1, chap. 16). Typically, the etch chemistry is adjusted by reducing O_2 flow.

The etch-rate ratio is also sensitive to the pattern density on the wafer surface (this is known as a *local-loading*, or *micro-loading effect*). Etch-rate-ratio data should therefore be obtained from monitor wafers with a pattern density similar to that of product wafers. Reference 73 describes a multiple etchback process that takes the loading effects (both global and local) into account, thereby reducing planarity variations across the wafer.

4.4.8.4 Alternative Sacrificial Etchback Processes.
Several alternative sacrificial-etchback processes have also been reported. One uses a spun-on AZ Novolac-resin protective coating (essentially positive photoresist without the photoactive compound) as a sacrificial layer, because it is less expensive and exhibits somewhat-better planarizing and dry-etching properties than do resists (Fig. 4-24). Such material is normally used to protect the wafer surface during back lapping and etching procedures.[64] A 200°C bake step after spin-on is recommended to improve the degree of planarization. Integration of a sacrificial-etchback process using such an AZ protective coating was reported for a 1.2-μm DLM-CMOS process.

Another method uses a dual CVD-oxynitride/nitride dielectric layer (deposited sequentially in a single chamber) plus resist.[74] Partial planarization is sought so that the via-depth variation is less than 0.25 μm. This is accomplished by selecting the appropriate oxynitride and nitride film thicknesses (0.55 μm and 0.85 μm, respectively) together with suitable etchback process conditions. The resist acts as a *complete* sacrificial layer, and the nitride as a *partial* sacrificial layer (which also fills the gaps between steps). Following the etchback process, all of the nitride is removed over the Metal-1 regions, but remains in the troughs between the steps. The oxynitride layer still remains over the Metal-1 steps following etchback, at essentially its originally deposited thickness.

The etchback step in this method is carried out in two parts, beginning with a gas mixture of $SF_6 + O_2$ that etches resist and nitride at the same rate. Only resist is removed initially, but eventually nitride over the narrow spaces is exposed. The second part of the etchback step is initiated once this occurs. The plasma chemistry is altered by deleting the O_2 from the feed gas because the only resist remaining on the wafer is in the narrow spaces between Metal-1 lines. It therefore etches more rapidly than when large regions of resist are being etched (microloading effect). By deleting O_2 from the plasma, the etch-rate ratio of resist to nitride is essentially returned to unity for this condition. The oxynitride layer is hardly etched by the etchback step, since the selectivity over nitride in SF_6 is ~5:1, allowing the nitride over Metal-1 regions to be removed without etching the CVD oxide. As a result, no redeposition after etchback is needed.

In the next example process, a CVD nitride/CVD oxide film with no covering resist layer is used as the sacrificial layer (Fig. 4-27a). A plasma-etch step which removes the nitride more rapidly than the SiO_2 and partially-planarizes the surface by leaving some unetched nitride in the grooves, is used. Once the oxide over the Metal-1 steps is exposed, the etch process is changed so that the nitride-to-oxide etch-ratio becomes unity (Fig. 4-27b). Etchback is continued until all the nitride is removed (Fig. 4-27c). Finally, an additional layer of CVD SiO_2 is deposited (Fig. 4-27d).[75]

In the last alternative method, the etchback process is integrated with a sloped-via-etch process, reducing the overall number of process steps.[76] A 2-μm CVD oxide is deposited, followed by resist deposition and via pattern exposure (Fig. 4-28a). The via pattern is anisotropically transferred into the SiO_2 by means of an RIE step (Fig. 4-28b) and an etchback planarization step is performed. The remainder of the via is etched during the etchback step, including tapering of its sidewalls.

4.4.9 Spin-On Glass (SOG)

Spin-on glass (SOG) is another interlevel-dielectric material that is applied in liquid form, and therefore exhibits planarization capabilities similar to those of polyimide films. SOG and polyimide films can both fill narrower spaces without causing voids than can CVD intermetal dielectric films. Even such crevices as those caused by closely spaced Metal-1 and polysilicon edges (Fig. 4-14b) can be planarized by SOG to a degree that allows for adequate metal step coverage. Other advantages of SOG films include: (a) simpler processing; (b) lower defect density; (c) higher throughput; (d) relatively low cost; and (e) no handling of hazardous gases.

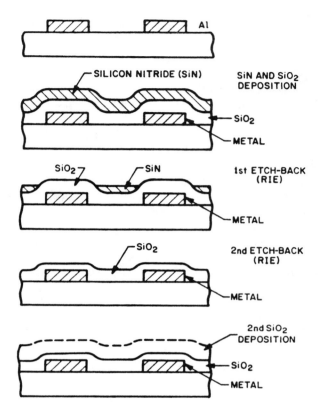

Fig. 4-27 Example of a sacrificial-etchback process that uses SiN as the sacrificial film.[75] (© 1983 IEEE).

SOG materials are *siloxanes* or *silicates* mixed in alcohol-based solvents, the primary difference between them being that a small percentage of Si-C bonds remains in the siloxane-based SOGs following the final cure cycle. Upon baking, the solvents are driven off and the remaining solid film exhibits properties similar to those of SiO_2 (as opposed to the organic film, in the case of polyimides). Silicate SOGs can also be doped with such compounds as P_2O_5 to improve the dielectric film properties. Case studies on the use of SOG materials can be found in references 77 and 78 (silicate based) and in references 79 and 80 (siloxane based).

Desirable properties of an SOG material are good consistency and shelf life, a simple cure cycle, excellent thermal stability, low stress and good crack resistance, spin-on uniformity, good adhesion, and an appropriate film-thickness range. Because the early SOG products, however, were proprietary mixes of organic monomers and attached organic groups mixed in solvents, it was difficult to predict their properties based on fundamental ideas of chemistry and materials science. The material properties of

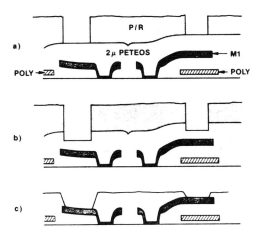

Fig. 4-28 Sequence showing how the etchback and via-sloping processes can be integrated to reduce the overall number of process steps.This paper was originally presented at the Spring 1988 Meeting of The Electrochemical Society, Inc. held in Atlanta, GA.

commercially available SOGs, however, have recently been described in more detail in references 81, 82, 83, 225, and 226 (siloxane based), and in references 84 and 225 (silicate based).

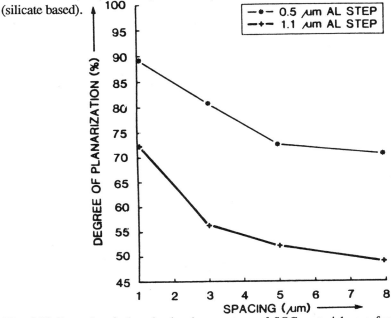

Fig. 4-29 Example of planarization by one type of SOG material as a function of spacing between steps. The underlying oxide step heights are 0. 5 and 1.0 μm.[77] (© 1988 IEEE).

SOG films provide somewhat lower degrees of planarization than do photoresists. However, their smoothing effect on sharp, vertical edges is satisfactory; enabling good step coverage of Metal 2 lines without line width variation. Figure 4-29 shows the planarization capabilities of SOG films as a function of metal pitch. The planarization can be as high as 80% for very closely spaced features, while for larger spacings it drops to less than 50%.[85]

4.4.9.1 SOG Process Integration. SOG processing involves some special considerations. Since the solvents evaporate from the SOG very quickly after being dispensed, an optimum dispense/spin cycle must be developed so that adequate thickness uniformity can be obtained. The exhaust system is also very important for controlling film uniformity. In addition, the dispense nozzle must be cleaned frequently to prevent clogging, and the exhaust bowl must be rinsed so that any particles spun off the wafer will be washed away. Specially designed equipment is commercially available to meet these requirements (Fig. 4-30). A static dispense is recommended for coatings with thicknesses of 200-400 nm.

After being spun on, the SOG is baked first at a low temperature (e.g., 150-250°C for 1-15 min in air), and then at a higher temperature (e.g., 400-425°C for 30-60 min in air). The solvent is first driven off, and water is evolved from the film (due to polymerization of the silanol [SiOH] groups). The loss of considerable mass together with material shrinkage creates a tensile stress in the film. If too thick a layer is applied, the stress can lead to film cracking. Organic groups, such as methyl (CH_3) or phenyl (C_6H_5) have been added to siloxane films (from 1 to 12 wt% carbon) to improve

Fig. 4-30 Photograph of an SOG dispense nozzle. Courtesy of SEMIX, Inc.

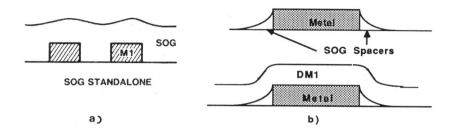

Fig. 4-31 (a) SOG standalone film. (b) Spin on SOG over underlying metal; etchback to leave SOG spacers; deposit DM1 oxide over this structure.

cracking resistance. (While a higher carbon content would be more effective for this purpose, it would also tends to degrade adhesion and steam resistance. Hence, a tradeoff is usually made in selecting an optimum carbon content.)[100] The simplest technique for integrating an SOG film into a multilevel interconnect process would be to use it as a stand-alone film. Unfortunately, when an SOG film is formed to the thickness required of an interlevel dielectric, it exhibits an intolerable degree of cracking. SOG can also be spun on over a CVD-oxide layer (Fig. 4-31a), or CVD SiO_2 can be applied over a thin spun-on layer of SOG; however, such approaches have also proved to be unworkable, since frequent adhesion failures are observed when SOG is in contact with metal or resist layers. In addition, resist stripping procedures may attack the SOG film and degrade its stability.[86] Still another technique involves deposition of an SOG film over a CVD oxide to smooth the surface. The SOG and oxide are then sacrificially etched back to leave gradually sloped CVD SiO_2 spacers, and a second CVD-SiO_2 layer is deposited over the metal/spacer structure (Fig. 4-31b).[87] This technique offers the benefits and drawbacks of the spacer semi-planarization method described in section 4.4.4.

The most widely reported implementations of SOG material involve *sandwich-type structures*, in which SOG is deposited over an initial CVD dielectric layer (*interlevel dielectric layer 1,* or ILD1), and a second CVD dielectric layer (ILD2) is then applied. There are two variations of this sandwich: *etchback SOG*, and *non-etchback SOG*. In the first technique, the SOG is partially etched back after being cured so that it remains only in the troughs between metal lines (Fig. 4-32a).[88,89,97,100] No SOG remains over locations at which vias will be etched. After the etchback step, ILD2 is deposited. In the *non-etchback SOG structures*, ILD2 is deposited over the SOG following curing. The SOG remains permanently as a continuous thin film between the two layers of CVD SiO_2 (Fig. 4-33).[90,91,92,93]

4.4.9.2 The Etchback SOG Process.

Despite its greater complexity, the *etchback* approach has been more widely adopted, primarily because it avoids the so-called *poisoned-via problem* that can be encountered with the non-etchback method. The etchback method is normally implemented with the less crack-susceptible siloxane-based

SOG films (the process sequence is illustrated in Fig. 4-32). Since sufficient planarization is normally not obtained with a single coating of SOG, two or more layers are often spun on before etchback is performed. A detailed description of an etchback SOG process is given in reference 255.

One drawback to the etchback process is that a thicker ILD1 film must be deposited than will ultimately remain, since some is removed during the etchback step. For a semi-recessed field-oxide thickness of 0.6 μm, a polysilicon layer 0.4 μm thick, and a 0.2-μm-thick SOG film, a 1.0-μm-thick ILD1 layer is needed. A minimum spacing of 1.2 μm is required between Metal-1 steps to prevent the formation of voids with a conventional-PECVD SiO_2 ILD1 layer. For smaller Metal-1 spacings, the etchback process becomes impractically complex with conventional PECVD ILD1 layers.

A plasma-TEOS/SOG process has been reported to allow spaces as small as 0.8 μm to be planarized by means of this etchback-SOG method,[94] with a 300-nm PECVD layer and a plasma-TEOS film used as a bilayer ILD1 film (fig. 4-32b). The plasma TEOS allows the small spacings to be filled without the formation of voids. The SOG is next applied to fill the "intermediate-width" gaps (i.e., near 2.5 μm) that are not sufficiently filled by the bilayer ILD1 film. Following an SOG etchback step, a thin ILD2 layer of PECVD SiO_2 is used to complete the sandwich structure. A second

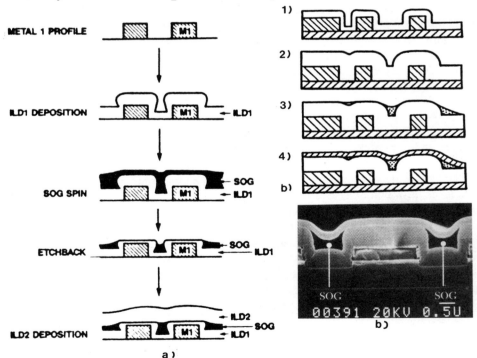

Fig. 4-32 (a) Sequence of steps used in the etchback SOG process. (b) Etchback SOG process which uses plasma TEOS as part of the CVD film that is first deposited. This allows smaller spaces between adjacent steps to be filled without void formation.[94] (© 1988 IEEE).

method for increasing the step coverage of Metal 2 over these "intermediate gaps" is to cut the top edges of the Metal-1 Al lines, as described in section 4.4.2.4.[35]

Other problems of the etchback technique include:

  •SOG-to-SOG adhesion problems are sometimes observed when more than one layer of SOG is used to improve the degree of planarization. The cause of SOG-to-SOG delamination is thought to be the slow out-gassing of the first SOG layer during the cure of the second layer, which pushes up the top layer.[95]

  • Lifting of the SOG from small gaps, occurs due to stress caused by volume shrinkage of the SOG and poor adhesion to the ILD1 film.[95]

  • Etch-rate differences exist between the SOG and the ILD1 layer (this problem is compounded by the added complication that the SOG etch rate is apparently also stress dependent);

  • Microloading effects occur during the etchback step (similar to those described in the sacrificial etchback section).[96]

  • The etchback process has a very small process margin for submicron technology. The CVD layer under the SOG must be thick enough to prevent Metal 1 from being exposed during the etchback step. The SOG thickness varies with topography and the amount of SOG to be removed has to include the thickest area where vias are allowed to be opened. A thick CVD layer will reduce the spacing between Metal 1 lines such that the SOG cannot flow in, or the CVD layer pinches off, leaving voids between the metal lines.

The use of polyimide instead of SOG in an etchback sandwich type structure has also been reported,[97] with good filling of small spaces (~1 μm) between Metal 1 lines observed. In this approach, an LPCVD oxide of 600-700-nm thickness is first deposited. At this thickness, the 1-μm spaces between metal lines are not yet pinched off by the deposited SiO_2 layer. Next, a 1.2-μm polyimide film is spun on and cured. This composite film is then etched back, using a process that compensates for the variation in oxide etch rate as the amount of oxide area is exposed.

A two-step etch process was developed for this application: (1) O_2 is used to etch the majority of the polyimide thickness (to reduce polymer-residue formation), and (2) simultaneous etchback of polyimide and SiO_2 is performed using a $C_2F_6 + O_2$ mixture. Following the etch process, a bakeout step is used to drive out any moisture from the polyimide remaining in the crevices. The bake step is followed immediately by the ILD2 deposition. Two and 3 levels of metal have been fabricated with this process.

4.4.9.3 The Nonetchback SOG Process. There are fewer steps with this process, and the difficult etchback step is eliminated. However, as noted, the non-etchback process suffers from the *poisoned-via* problem if siloxane-type SOG materials are used. During Metal 2 deposition, this type of SOG outgasses, causing higher contact resistance (e.g., 300-400 mΩ/via versus 60 mΩ/via for non-SOG controls) between

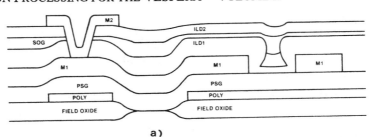

a)

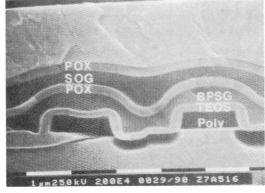

b)

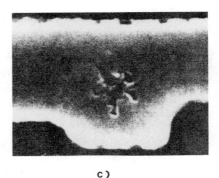

c)

Fig. 4-33 (a) Drawing of a cross-section of structure formed using the non-etchback SOG process. (b) SEM micrograph of a structure formed with the non-etchback SOG process. Courtesy of Allied-Signal, Inc. (c) SEM micrograph of a cross section of a "poisoned via."[93] (© 1987 IEEE).

Metal 2 and Metal 1, or even an electrical "open" (Fig. 4-33c). The source of the gases has been found to be primarily moisture[253] that remains within the SOG (due to insufficient curing), or that has been reabsorbed during the wet process steps performed following spin-on and curing of the SOG film (e.g., during resist removal after the via etch or during pre-metal-deposition cleaning steps).

Although it has been suggested that the problem could be overcome through the use of an all-dry process sequence following SOG spin-on coupled with complete curing of the SOG film, most non-etchback approaches instead use a silicate-type SOG. Special crack-resistant silicate SOG materials (phosphorus doped to increase crack resistance) have been developed, since silicate SOG is inorganic. When such phosphosilicate glasses (containing ~9 wt% P_2O_5) are properly cured, the poisoned via effect is apparently eliminated.[91,92,93] Such nonetchback processes (with multiple coatings of SOG applied for adequate planarization) have been reported for Metal 1 spaces as small as 1 μm.[101] Two detailed reports of how a non-etchback phosphosilicate SOG process was integrated into a 1-μm CMOS processes for both polysilicon and first metal interconnect planarization is given in references 245 and 246.

4.4.10   Electron-Cyclotron-Resonance Plasma CVD

Electron-cyclotron-resonance (ECR) plasma CVD SiO_2 and silicon-nitride films are being investigated as another method for planarizing intermetal-dielectric layers.[102,231] The most significant advantages of ECR-deposited dielectric films are their low deposition temperatures (i.e., room temperature to 150°C) and their apparent capability to fill higher aspect-ratio spaces than either PECVD or bias-sputtered CVD. This latter characteristic allows ECR oxides to fill spaces as small as 0.5 μm wide.* In addition, it has been observed that ECR oxide and nitride films contain lower concentrations of hydrogen and exhibit other promising physical properties.

ECR plasmas are capable of producing such films because they can operate at pressures an order of magnitude lower than conventional rf plasmas (i.e., at 1-2 mtorr pressure). The ECR plasma is generated at microwave frequencies of 2.45 GHz. For SiO_2 deposition a gas mixture of SiH_4, O_2, and Ar is metered into the reaction chamber (Fig. 4-34). Magnetic fields surrounding the microwave cavity produce a field gradient that directs the ions of the plasma to the surfaces of the wafers (which can either be maintained at room temperature or heated). The densities of the extracted ions are an order of magnitude higher than those observed in conventional tubular or parallel-plate PECVD reactors.

Bias is also applied to the wafers if a planarized dielectric layer is desired. Planarization is achieved during deposition through essentially the same mechanisms that are operative in bias-sputtered SiO_2 (i.e., CVD-SiO_2 deposition occurs simultaneously with sputter etching of the growing film by argon ions). This, combined with the low operating pressure of the plasma, allows small, high-aspect-ratio spaces to be filled. The high ion densities allow greater throughputs than are possible with bias-sputtered SiO_2 processes, but they are still rather low. Reports on the dielectric

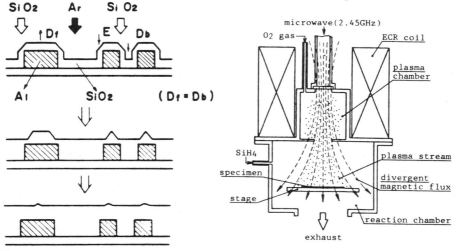

Fig. 4-34 (a) Principle of bias-ECR plasma CVD; (b) ECR plasma CVD apparatus.

properties of ECR silicon-oxide and silicon-nitride films are given in references 103, 104, and 105.

It has been observed that the physical properties of the ECR films are better than those obtained with conventional PECVD processes. However, the step coverage can be very nonconformal, and the sidewall films can be quite porous if the film is deposited without rf bias.

Films deposited in conventional ECR machines exhibit non-uniform thickness, especially on large wafers ($\geq$150 mm), due to the divergent magnetic field that extracts the ions from the microwave cavity. The modified magnetic coils of more recently designed machines make it possible to produce films with $\pm$5% thickness uniformities across 150-mm wafers.[106]

A system that uses ECR to generate a plasma with oxygen gas (instead of Ar) has been reported for sputtering layers of SiO_2 by formed CVD in the same chamber.[261] The oxygen ions sputter the SiO_2 (formed earlier by reaction of SiH_4 and O_2) to achieve planarization. This can be achieved without applying a dc bias the substrate or by having self-bias develop between the wafer and the plasma. That is, an rf signal at 400 kHz is applied to the substrate. The impinging oxygen ions from the plasma can follow this electric field, and thus bombard the surface without having a self-bias build up on the substrate. Electrostatically-caused damage (due to the buildup of charge on the wafer surface as a result of self-bias) is thus prevented, while the sputtering needed to achieve planarization still occurs. High sputtering rates of SiO_2 films with low damage have been reported.

4.4.11 Chemical Mechanical Polishing

The sacrificial-resist etchback process, even with compensated PBM resist still, still has severe problems for multilevel metal applications. The deposition and etchback tolerances associated with large film thicknesses are cumulative, and any non-planarity of the resist is replicated in the final planarized oxide surface. Under RIE etchback conditions designed to reduce such non-planarities (i.e., by increasing the oxide/resist etchrate ratio to more than 1.0), oxide spikes are caused (Fig. 4-35a). A chemical-mechanical polishing (CMP) process has recently been reported for removing such spikes.[265] The CMP process can rapidly remove such small elevated features without significantly thinning the oxide on the flat areas (Fig. 4-35b). It is pointed out that the CMP process alone is not viable for planarization because residual oxide will be left in the middle of large active areas or arrays after polishing (Fig. 4-35c). By combining CMP with RIE, however, an effective planarization method can be achieved.

Not many details about CMP technology have been published as of this writing, but references to silicon wafer-polishing techniques have been cited.[266] To prevent mechanical work damage from remaining on the polished film, the chemical component of the polishing should probably dominate. Since the wafers will likely be polished one at a time, the throughput is also likely to be low (e.g., on the order of 5-10 wafers per hour). Other predicted problems include the clean-up of the polishing-slurry particles from the wafer surface (slurry-particle size <0.1 μm), the compatibility of such a dirty

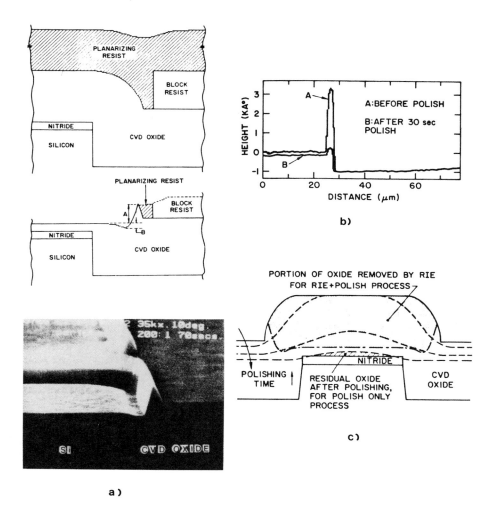

Fig. 4-35 (a) Cross section after selective RIE etchback, showing local oxide spikes that remain after the etchback when the oxide/resist etch-rate ratio > 1. (b) Measured oxide step heights before and after CMP, showing the fast removal rate of such spikes. (c) Schematic drawing showing the fundamental problem of "CMP-only" planarization.[264] (© 1989 IEEE).

process with the clean room environment (i.e., reducing the density of the 50,000 to 100,000 particles/ft^3 generated to less than 50/ft^3), the lack of an end point detection method, and maintaining sufficient polishing-rate uniformity across the wafer, and from wafer to wafer. Commercially available equipment for performing this process is reportedly being developed.

4.5. METAL DEPOSITION AND VIA FILLING

4.5.1 Conventional Approach to Via Fabrication and Formation of Metal-to-Metal Contacts through Vias

One aspect of minimum feature size that applies to via fabrication is the issue of the via sidewalls: for the same contact-opening sizes at the bottom of vias, straight sidewalled vias would require less area than would those with sloped sidewalls. Furthermore, when dry etching is used to open the vias, it is often easier to produce straight rather than sloped sidewalls.

Unfortunately, when physical vapor deposition is used to deposit metal over the vias, straight sidewalls result in worse step coverage by the metal than if the sidewalls were sloped. This has been shown both theoretically[117] and experimentally (Fig. 4-36a).[118] While adequate step coverage of the metal is critical, there is no single value of minimum step coverage that is considered adequate for all conditions. This is because the minimum value depends on many factors (including the current density passing through the metal, its absolute film thickness at the thinnest spot, the type of metal, etc.). In general, a range of 20-50% minimum step coverage is probably acceptable, depending on the application.

The step coverage of a sputtered metal film can be improved in a number of ways. The first is to optimize the deposition conditions by increasing the surface-migration ability of the atoms that have already arrived on the surface. These atoms will then

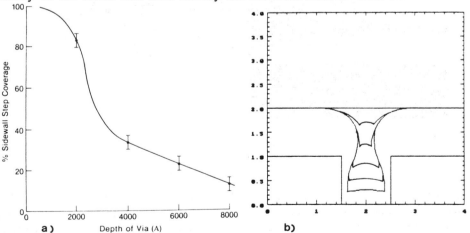

Fig. 4-36 (a) Percent sidewall step coverage of a standard dc sputtered aluminum film (0.8-μm thick) into vertically-sided 1.2-μm-diameter vias of different depths.[118] Copyright 1986. Reprinted with permission of the AIOP. (b) Simulation of a sputtered Al film into a via with an aspect ratio of 1.0, with the via being partially filled and unfilled. The profiles form the top represent the metal film deposited into a 100% filled, 80% filled, 60% filled, 40% filled, 20% filled and unfilled via.[249] This paper was originally presented at the Fall 1988 Meeting of The Electrochemical Society, Inc. held in Chicago, IL.

rearrange themselves until the film has a uniform thickness over all topographical features. Surface mobility can be increased by raising the substrate temperature and by increasing ion bombardment (which also causes heating, since a large fraction of the incident energy flux is transferred to the film/substrate as heat). Sputtering processes that are capable of heating the wafers during deposition and causing them to be bombarded by ions (i.e., through bias sputtering; see Vol. 1, chap. 10) have been shown to produce improved step coverage. *Assuming that a sputtering process is developed that optimizes step coverage, however, the coverage will still depend on the sidewall slope and aspect ratio of the via.*

The aspect-ratio dependence of the step coverage into contact holes and vias is critical as feature sizes are scaled into the submicron regime. In order for the interlevel capacitance of the interconnections to the underlying and overlying metallization lines to be maintained (or even reduced), the thickness of the dielectric layers in which the contact holes and vias are formed may not be decreased at all. Thus, the aspect ratio will grow larger. For example, when an Al film is sputter deposited without bias and at <300°C, the step coverage into a vertically-sided via whose aspect ratio is 0.5 is typically ~20%. As the aspect ratio increases to 1.0 (so that the via depth equals the width), the step coverage decreases to less than 5%.

An example of a simulation of a sputtered Al film into a via with an aspect ratio of 1.0 is shown in Fig. 4-36b. (Another example of step coverage that has been reported as a function of aspect ratio is given in Fig. 3-27b.) As shown in Fig. 4-36, sputtered-Al depositions used in production processes in 1988 provide more than 50% step coverage into vertically sided vias only if the aspect ratio is smaller than 0.25 (e.g., for 1.2-μm-wide vias, the depth can be no more than 0.3 μm). Since the thicknesses of the PMD and DM1 layers generally exceed 0.5 μm each, adequate step coverage may not be achieved for small vertically-sided vias.

The conclusion to be drawn from the above discussion is that in *conventional via processing, sloped vias are necessary to ensure adequate step coverage in most applications when contact holes and vias are less than 1.5 μm wide.*

Two other important aspects of conventional via processing are the planarization of the intermetal-dielectric films and the decision of whether the design rules are to allow *stacked vias* (i.e., vias located on top of silicon and polysilicon contacts) or only *staggered vias* (Fig. 4-37a). As shown in Fig. 4-37b, if complete planarization of DM1 is achieved and if stacked vias are also permissible, via depths can vary by as much as 1 μm. Such large variations have two significant impacts on the via fabrication process. First, the deepest vias will have poor (and generally inadequate) step coverage unless sloped sidewalls are formed. Second, since a sloped-via process must be used, the Metal-1 patterns under the shallowest vias must be wide enough to prevent the etching of grooves along the pattern sidewalls when the deep vias are being opened (Fig. 4-38a). This increase in pattern size results in an area penalty that can significantly decrease packing density.

In general, in the earliest double-level-metal (DLM) processes for CMOS, conventional via processing was employed. The rules that had to be followed because of the difficulty in filling vertical vias included these:

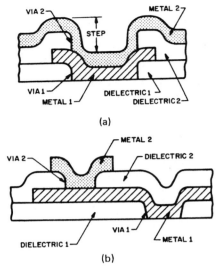

(a)

(b)

Fig. 4-37 Contact schemes: (a) stacked vias; and (b) staggered vias. From A. G. Sabnis, in G. S. Gildenblat and S. S. Cohen, Eds., *VLSI Electronics: Microstructure Science, Vol. 15, Metallization,* Chap. 7. Copyright 1987, Academic Press. Reprinted with permission.

- No step-height reduction planarization process could be used with respect to PMD and DM1 layers (i.e., only the first degree of planarization – *smoothing* – was employed), so that all vias could be maintained at the same depth.

- Only staggered vias were allowed.

- Contact to Metal 2 was allowed only through Metal 1.

- The vias had to be sloped to allow adequate step coverage by sputter-deposited metal.

However, vias with sloped sidewalls give rise to a set of design rules that must be followed when laying out patterns of the metal and via structures in a multilevel metal system.

4.5.1.1 Design Rules of Multilevel Metal Systems which are Impacted by Conventional Via-Processing Limitations.   In the ideal case, metal pitch is determined by the minimum line and space dimensions that can be patterned using the most recent advances in lithographic techniques. In practice, metal pitch is also limited by the via size and the underlying metal pad size. The following design rules must thus be obeyed:

- The underlying metal pad must be larger than the via opening. This condition is referred to as a *framed* or *nested* via. If the via were larger than the metal pad, grooves would be etched alongside the pad during the via-etch step (Fig. 4-38a)

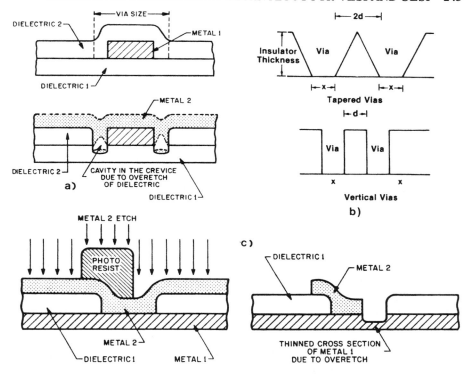

Fig. 4-38 (a) If the via dimension in the direction of the Metal-2 current flow is larger than the Metal-1 width, then crevices may be formed into the dielectric below Metal 1, and cavities can result in Metal 2. (b) If sloped vias are used, the minimum spacing between vias must be larger than if vertical vias are used. (c) If the via dimension in the direction of Metal 1 current is larger than the Metal-2 width, then Metal 1 can get overetched, thus the overlying metal must be larger than the top of the via opening. From A. G. Sabnis, in G. S. Gildenblat and S. S. Cohen, Eds., *VLSI Electronics: Microstructure Science, Vol. 15, Metallization,* Chap. 7. Copyright 1987, Academic Press. Reprinted with permission.

and could cause microcracking or thinning of Metal 2 when it is deposited over the via to contact Metal 1. The minimum dimension by which the metal pad must frame the via is dependent on the misalignment tolerances of the lithography step. If vias with varying depths must be etched, the enlargement that will occur when shallow-sloped vias continue to be etched during etching of the deep vias must also be taken into consideration.

• The slope of the vias must be taken into account when deciding on the minimum spacing between vias (Fig. 4-38b). A minimum space is needed between the tops of the etched via openings so that metal space on this horizontal surface can be reproducibly patterned.

• The overlying metal pad must be larger than the top of the via opening to ensure full coverage of the via (Fig. 4-38c).

4.5.2 Advanced Via Processing (Vertical Vias and Complete Filling of Vias by Metal)

As long as an adequate sidewall taper of contact windows and vias can be implemented (and the aspect ratios do not become too great), sputter deposition can produce metal films with adequate step coverage into vias. However, design rules are rapidly evolving to a point at which sputtered metals will no longer be able to acceptably fill contact and via holes. In addition, in submicron technologies the extra area needed to accommodate sloped vias severely limits the maximum packing density (particularly since it is also difficult to tightly control the slope of the sidewalls in a production environment).

Finally, if sputter etching is used to clean the native oxide from the Al at the bottom of the vias prior to deposition of the next Al film, SiO_2 from the via sidewalls will be redeposited in the bottom, causing contact resistance problems. The problem is minimized through the use of vertically sided vias (or vias with a slight sidewall slope, e.g., 85°). Once the limits that mandate the use of such vertically-sided vias have been reached, *advanced via processing* (in which it is possible to completely fill vertical vias with metal) must be used.

4.5.2.1 Increases in Packing Density Resulting from Advanced Via Process Technology. Several space saving benefits are achieved through the use of this technology. First, the minimum distance between adjacent metal lines at the via is reduced, since the via has no slope (Fig. 4-38b). Second, the pad area of Metal 2 can be decreased, since the plug in the completely filled via provides ample overetch protection to underlying metal structures without mask coverage (Fig. 4-39a).

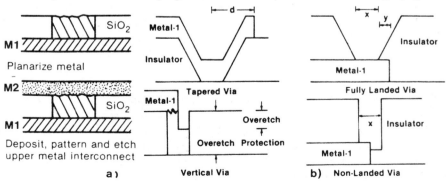

Fig. 4-39 Metal 1 and Metal 2 pad areas can be smaller when vertical vias and fully filled vias are used because: (a) such filled openings provide protection during any overetch that may occur during the etching of Metal 1 or Metal 2; (b) any grooves that may be created during the overetching that occurs when the via openings are etched are easily filled by the via-filling process.[128] (© 1987 IEEE).

Third, the minimum size of the Metal 1 pad areas can be reduced, since any small grooves that might be etched alongside the pad during the via etch will be easily filled during the plug formation process (Fig. 4-39b). As a result, the pitches of both Metal 1 and Metal 2 can be reduced, allowing significantly increased packing density.

4.5.3 Processing Techniques that Allow for Vertical Vias

Two general fabrication approaches make the implementation of vertical vias possible:

1. *Filling of vias through deposition of metal into the opened via to form a plug in the opening.* In theory, this can be accomplished either independently of the metal-runner formation process, or through simultaneous fabrication of the plugs and metal runner. An example of the latter is the deposition and patterning of a blanket CVD W layer.

On the other hand, when the plugs are formed separately, it is not necessary to replace the Al interconnect runners with those of a higher resistivity metal (W). Conversely, W contact plugs do not significantly increase the total resistance of interconnect lines because their length is so short (and because they maintain a large cross-sectional area): hence, they are appropriate for this use.

2. *Creation of a post, or pillar, which is then surrounded with a dielectric.*

Although the post approach offers several layout advantages, the plug approach is much easier to implement in a production-worthy process. Nevertheless, we will discuss both approaches.

4.5.3.1 Required Degree of Via Filling by Plugs.   Although the plug structures described in the previous paragraphs will completely span the cross-sectional area of the contact holes or vias, they may not fill the openings up to the top. However, as noted earlier, adequate step coverage (e.g., >50%) by PVD metal into a vertically sided opening can still be achieved if the aspect ratio of the opening is less than 0.25. Step coverage by means of sputtered Al can thus be adequate under some circumstances, with more than 50% coverage possible in 1.2-μm wide, vertically sided, 600-nm-deep vias with 300-nm plugs. Figure 4-36b shows the step coverage of a 1.0 μm metal film into partially filled vias.[249]

4.5.4 CVD W Techniques for Filling Vertical Vias and Contact Holes

Two CVD W methods that have been developed for this application are *blanket CVD W and (etchback)* and *selective CVD W.*

4.5.4.1 General Information on the CVD Tungsten Process.   The chemical vapor deposition (CVD) of tungsten is generally performed in either a hot-

wall, low-pressure system or a cold-wall, low-temperature system. Although tungsten can be selectively deposited either from WF_6 or WCl_6, the former (tungsten hexafluoride) is better suited as the W source gas, since it is a liquid that boils at room temperature (while WCl_6 is a solid that melts at 275°C). WF_6 can also be reduced by silicon, hydrogen, or silane, as shown by the following equations:

Silicon reduction

$$2WF_6 \text{ (vapor)} + 3Si \text{ (solid)} \longrightarrow 2W \text{ (solid)} + 3SiF_4 \text{ (vapor)} \qquad (4-12)$$

Hydrogen reduction

$$WF_6 \text{ (vapor)} + 3H_2 \text{ (vapor)} \longrightarrow W \text{ (solid)} + 6HF \text{ (vapor)} \qquad (4-13)$$

Silane reduction

$$2WF_6 \text{ (vapor)} + 3SiH_4 \text{ (vapor)} \longrightarrow 2W \text{ (solid)} + 3SiF_4 \text{ (vapor)} + 6H_2 \text{ (vapor)} \quad (4-14)$$

When the hydrogen reduction is used, the rate-limiting step is the dissociation of H_2 into atomic hydrogen on the reaction surface. Selectivity is therefore achieved through deposition at a temperature below which SiO_2 will not catalyze the H_2 dissociation (~500°C), but at which other surfaces (silicon, metal, and silicide) will do so. If the entire surface is covered with a material that forms a good nucleating surface, the W will be deposited everywhere even when temperatures are below 500°C (*blanket W deposition*). If the nucleating surface exists only in certain locations (e.g., at the bottom of contact holes or vias), W will be selectively deposited in those locations. Above these temperatures, however, W will deposit on the SiO_2 as well. If the silane reduction is used, selectivity can be reproducibly achieved if the temperature is kept below 325°C, but it will be lost if the SiH_4/WF_6 flow-rate ratio is greater than 1.6.

On a silicon surface, the deposition starts with the silicon-reduction reaction (Eq. 4-12) even when hydrogen is present, and it will occur exclusively if hydrogen is absent. This reaction is self-limiting, however, since once a W layer is deposited it serves as a diffusion barrier between the Si and WF_6; the silicon reduction essentially stops when the W film reaches a thickness of 10-15 nm. For every two atoms of W deposited, three atoms of Si are consumed and volatilized as SiF_4. Typically, 10 to 15 nm of Si are consumed. No deposit occurs on the SiO_2 during the reaction. The silicon reduction of WF_6 can be used to produce thin films of selectively deposited W, with excellent surfaces and with sheet resistances of 10-15 Ω/sq.

The *hydrogen reduction process* can be used to selectively deposit thicker CVD W films, as was first reported by Blewer and Wells.[120] Normally, the silicon reduction reaction deposits the initial thin layer on a Si surface, and the hydrogen reduction then reaction takes over. As a result, the carrier gas at the outset the of this type of deposition is normally Ar. After the Si reduces the WF_6, either H_2 or SiH_4 is added to the gas flow, and the Ar flow is stopped.

When used to deposit W into contact and via holes through the blanket deposition method, hydrogen reduction gives excellent conformal coverage of the topography (which results in near-complete contact hole and via planarization). Broadbent and Ramiller studied the deposition kinetics of the hydrogen reduction of WF_6.[121] In this study, which spanned the temperature range of 250-500°C and the pressure range of 0.1-5 torr, an activation energy of 0.71 eV was obtained for all pressures. The growth rate was found to be independent of WF_6 partial pressure, but was observed to increase as the square root of the H_2 partial pressure. These dependencies suggest that *surface-adsorbed H_2 dissociation* is the rate-limiting reaction mechanism. The activation energy of 0.71 eV is in close agreement with the value of 0.69 eV for the H_2 surface diffusion of W. A more recent report of the kinetics of the hydrogen reaction is given in references 122 and 123.

The hydrogen reduction process, however, has several drawbacks. First, the deposition rate is relatively low (e.g., ~8 nm/min at 350°C), especially at the low deposition temperatures necessary when the CVD W process is used to fill via holes in which Al is the underlying metal layer. Second, the surface of such thicker deposited films is quite rough, with surface features 25% taller than the film thickness. (One report indicates that the surface roughness can be reduced through the use of a clean reactor, if a hot-wall furnace is used for the deposition process.)[110] Finally, the HF gas by-product of the reaction is believed to be responsible for the problems of encroachment under the oxide, as well as for wormholes which lead to junction leakage.

The *silane reduction process* was developed as a way to overcome the problems of the hydrogen reduction process.[124,125] High deposition rates can be achieved at low temperatures (up to 600 nm per minute at 250°C has been observed, with no loss of selectivity). In addition, the grain size of the deposited films is much smaller than that obtained with the hydrogen reduction process, resulting in much smoother W-film surfaces. Finally, since no HF gas is produced as the reaction by-product, the problems of encroachment and wormholes are also apparently eliminated. Note that the deposition temperature must be kept below 325°C so that selectivity can be maintained. The resistivity of silane W films produced by silane reduction is comparable to the 13 $\mu\Omega$-cm of W films produced by hydrogen reduction. Low and stable contact resistances to both *p* and *n* type silicon are also formed. A process for selective W CVD in a hot-wall reactor by silane reduction of WF_6 has recently been described.[254] The advantage of using a hot-wall reactor is that it can accommodate a larger number of wafers than can cold wall reactors.

4.5.4.2 Blanket CVD W and Etchback.

Due to lateral encroachment and wormholes - problems that exist with selective CVD W deposited with the hydrogen reduction - blanket CVD W and etchback has been more widely adopted for contact hole and via filling (at the penalty of using a higher cost process).

The process is normally begun by depositing a 100-nm-thick adhesion layer on the wafer surface (e.g., Ti, Ti:W, TiN, WSi_x, or Ti/TiN).[126,127,128,129,243] Such adhesion layers are needed because of the extremely poor adhesion of CVD W on such insulators as BPSG, thermal oxide, and plasma-enhanced oxide and silicon nitride.

Tungsten, however, will adhere well to these materials, and they in turn will adhere well to the insulators listed. Thus, a method which allows good adhesion of CVD W to the substrate is achieved.* Next, a layer of CVD W is blanket deposited. This can be accomplished using silane reduction. (A WSi_2 layer can be deposited as part of the silane-reduction deposition process. Since this is done in situ, the W can be deposited after the WSi_2 layer without the need to break vacuum.) As the deposition proceeds, the sidewalls of the vias are covered by the conformal CVD W film and they eventually become thick enough that they come into contact with one another. In the ideal case, the holes are thereby filled (Fig. 4-40a); in practice, however, the voids created when PVD processes are used to fill vias can still occur under some CVD W conditions (Fig. 4-40b). Process optimization of deposition temperature, pressure, and relative gas flows, however, can minimize or eliminate void formation.

The silane reduction process is more prone to void formation, probably because of its faster deposition rate. A combined hydrogen and silane reduction process has been reported as a means of eliminating void formation. The process has a deposition rate faster than that of the hydrogen reduction process alone, but still allows the establishment of a set of deposition conditions in which no voids form.[130] In another report, this combined process was used to fill vias through deposition of W at 400°C. This resulted in sufficiently high W deposition rates, and limited the growth of hillocks in the underlying Al films.[131]

Since a blanket W film is deposited everywhere, it must be etched back so that it remains only in the vias. The simplest way to do this is to etch back the W directly, without using a sacrificial resist layer. When vias of varying widths must be filled, however, the "dimples" that occur at the center of the W-filled vias (Fig. 4-40c) will be deeper in those that are wider and will also be replicated into the final film. In some cases, faster etching of W will occur along the seam. When a metal layer is deposited over these via plugs, poor step coverage may result.

Use of a planarizing resist layer has been suggested to overcome this limitation. If a 1:1 resist-to-W etch-rate ratio can be achieved – not an easy thing to do – such an etchback process could be used to planarize both wide and narrow vias. One study abandoned this approach in favor of the direct etchback process.[132] Another group reported success[119] by using an RIE etch-gas mixture of $CCl_2F_2 + O_2$ to obtain a 1:1 selectivity between W and resist (Fig. 4-41). Nevertheless, it was indicated that both high W film-thickness uniformity and high etch-rate uniformity are necessary for successful implementation of this process.

In another report, a polyimide film was used instead of resist as the sacrificial layer.[131] Still another group reported a two-step etchback process.[224] In the first step,

---

* TiN has been reported to have the best set of properties for this adhesion-layer application, including the lowest contact-resistance values – two orders of magnitude lower than when Ti is used – and the best resistance to the etch gases used to etch the CVD W.[130] In another report, WSi_x was found to yield poor W adhesion in submicron contacts, so Ti:W was used instead.[230] The Ti/TiN layer was used in a two-level metal process in which it also served as a contact to regions of silicon and polysilicon covered with $TiSi_2$.

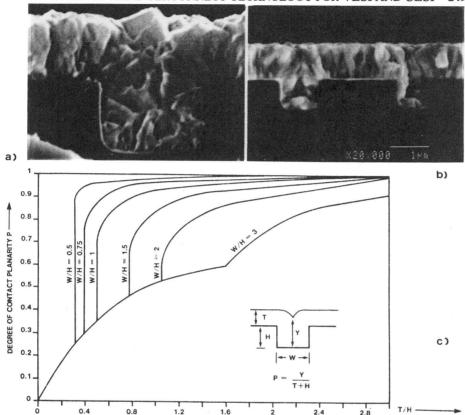

a)

b)

c)

Fig. 4-40 (a) CVD W filling of contact hole without void formation. (b) SEM micrograph of voids that may form when CVD W is deposited into contact openings. (c) "Dimples" that are formed when CVD W is deposited into contact holes and etched back.[128] (© 1987 IEEE).

an SF_6 and C_2F_6 mixture was used to etch the W and the resist at similar rates (400 nm/min). The C_2F_6 acts to reduce the microloading effect in which the W etch rate increases as the amount of W exposed decreases during etching decreases. In the second step, residual W and the TiN adhesion layer are removed with a Cl_2-based plasma. The TiN is removed preferentially with respect to W and the underlying SiO_2. To eliminate metallic residue, an appreciable overetch is carried out, without excessive loss of W in the vias or of interlevel SiO_2.

In both types of etchback processes the dielectric surface must be highly planarized; otherwise, the overetch required to remove W stringers at steps will lower the height of the W plugs in the vias. The surface roughness of the CVD W film (Fig. 4-42a) worsens this problem, since a longer overetch is needed to clear all of the W residue from the field. In addition, the roughness of the tungsten surface can be transferred to the surface of the field oxide (which will produce other problems in further processing),

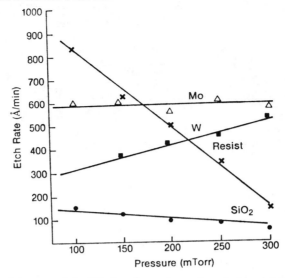

Fig. 4-41 Etch conditions used for W plug formation using photoresist planarization. Etch parameters are: flow – 60 sccm, 75% O_2 in CCl_2F_2/O_2; and power – 300 W (0.25 W cm^2); and a carbon cathode cover in the reactive ion etcher.[137] (© 1986 IEEE).

unless either an etchback process that planarizes the W surface, or a deposition process that produces smoother W films, is used. A technique to produce smoother films through the deposition of W in several layers (interspersed with a thin polysilicon layer between W layers) has been described (Fig. 4-42b).[127]

The erosion of W in the vias is also enhanced by a microloading effect that increases the etch rate when only a small area of the wafer contains W (i.e., when the only W exposed is in the vias or in the form of stringers). A three-step etchback step that minimizes this microloading effect has been reported.[133]

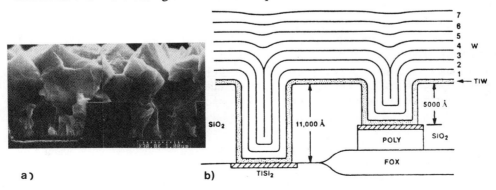

Fig. 4-42 (a) Surface roughness of conventional CVD-W films formed using the hydrogen-reduction process. (b) Smoother CVD-W films can be formed using a multi-layer-deposition technique in which a thin silicon layer is deposited between the W layers.[127] (© 1988 IEEE).

Several other problems can be encountered with the blanket etchback process. First thick-blanket W films exhibit high stress, and such stressed films may cause wafer warpage. In addition, if W is deposited on the back of the wafer – where no adhesion layer may have been applied – the film will delaminate, producing unwanted particulates in the deposition chamber. Because W exhibits such good conformal coverage, backside deposition can occur unless the wafer is very well clamped to the plate on which it sits in the reaction chamber. Thus, the wafer clamping design is important. Furthermore, appropriate wafer clamping during the adhesion layer deposition is also important. That is, if the adhesion layer is not deposited on areas beneath the clips that hold the wafer, W can delaminate from these areas, again producing particles.[264]

In addition, The W deposited on the chamber walls can also build up until it spalls, producing particulates in the chamber. However, a recently designed single-wafer W-deposition chamber provides for in situ cleaning of the back of the wafer and the chamber (after each wafer is removed following film deposition).[252]

The second problem occurs during overetch, as a result of the etch gases attacking the adhesion layers. This can cause the formation of a groove between the W plug and the SiO_2 surrounding it, leading to step coverage problems for overlying Al films. TiN is reported to resist attack by the etch gases better than Ti, WSi_2, or Ti:W.[130]

A novel W etchback approach that uses a PECVD nitride sacrificial layer is described in reference 232. In this scheme, the problems of SiO_2 surface roughness and micro-loading effects during etchback are reported to be essentially eliminated. Overetching thus becomes less critical and can be used to compensate for nonuniformities in W deposition and etching. The elimination of loading effects allows coplanar W plugs to be easily achieved.

4.5.4.3 Selective CVD W. This method allows vias of varying widths to be filled without the problem of dimple occurrence in the wider vias. However, if deep vias are completely filled, then shallower ones will be overfilled (Fig. 4-43), necessitating an etchback step to planarize the surface. An alternative approach that does not re-quire etchback is to partially fill the deeper vias and completely fill the shallower ones.

The main issues of importance for selective CVD W filling of vias are deposition selectivity, deposition rate, and the possibility of selectively depositing W on substrate materials other than silicon.[134] A review of the physical and electrical properties of selectively deposited W films produced in cold-wall systems through the hydrogen and silane reduction process is presented in reference 135.

The selectivity in deposition is dependent on various factors, including the type of insulator and the condition of the insulator surface. In general, selectivity tends to degrade as the W gets thicker. Ion bombardment or contamination of the insulator surface, however, causes the insulator surfaces to exhibit selectivity breakdown at an earlier stage in the deposition process. PSG demonstrates better selectivity than undoped SiO_2. Lowering of the WF_6 partial pressure, a decrease in the deposition temperature, and rapid scavenging the HF and SiF_4 reaction products will improve selectivity. A recent model for selectivity loss is described in reference 136.

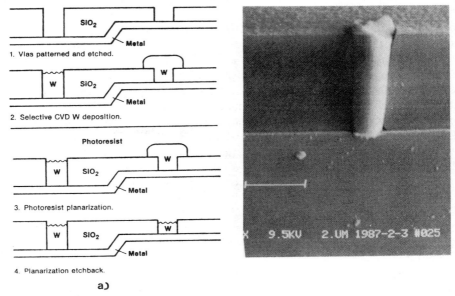

Fig. 4-43 Method of forming metal plugs using selective W deposition, resist planarization, and etchback. The shallow openings are overfilled and must be etched back to obtain a planar surface. (b) An SEM photo of a via plug 1.4 μm in diameter and 2.2 μm deep formed by selective-W deposition.[137] (© 1986 IEEE).

A *high deposition rate* is desirable from a manufacturing viewpoint. While the conditions that provide such high deposition rates in the hydrogen reduction process also degrade selectivity, the silane reduction process is able to provide high deposition rates (600-1000 nm/min) at lower temperatures (250-300°C, vs. 400-500°C for the hydrogen reduction process).

Since W will deposit selectively with either the hydrogen or the silane reduction of WF_6 as long as a good nucleating surface is exposed and the proper process conditions are met, the selective CVD W process can be used to deposit W into vias in which nucleating surface exists at the bottom. Vias with exposed Mo were used in one selective via-fill process (i.e., Metal 1 consisted of Ti:W overlaid with Mo).[137] In this process, a cold-wall reactor is used to deposit W at 50 - 500 nm/min. This process is compatible with the metal system used, since the contacts to Si using such a Metal 1 layer are stable to 650°C (see chap. 3, section 3.6.3). Other nucleating surfaces included $MoSi_2$,[138] and Al. The $MoSi_2$ was deposited over an Al film because a high contact resistance was observed when W was selectively deposited on Al. It was believed that this was due to nonvolatile reaction products (e.g., AlF_3) produced during the initial stages of W deposition. A later report indicated that this problem could be circumvented by a higher deposition temperature (>375°C) which would volatilize the AlF_3.[139]

4.5.5   Other CVD Via-Filling Processes

4.5.5.1   Blanket CVD Polysilicon and Etchback for Contact Hole Filling.

The use of a blanket-CVD polysilicon film together with etchback has been proposed as an alternative to the blanket-CVD-W filling of contact holes.[140,141] This approach circumvents the problems of film stress, backside deposition, deposition uniformity, and difficulty in etchback. In addition, the poly films can be deposited through the use of conventional hot-wall furnaces, rather than the special developmental CVD reactors needed for the deposition of W. Finally, because no adhesion layer is needed, a simple resistless etchback technique can be used.

An in situ heavily phosphorus-doped polysilicon film was used as the filling material in one report.[140] This procedure used an RTP anneal at 1050°C for 20 sec in an N_2 ambient following the poly deposition to produce films with the lowest sheet-resistance values (~9 Ω/sq). In another report,[141] the polysilicon plugs were doped with phosphorus following the etchback step. An undoped layer of SiO_2 on the BPSG prevented dopants from interacting with the diffusion ambient. Void formation in the center of the poly plug was eliminated through the use of a sloped-contact-hole etch procedure. In both reports, the polysilicon was also etched back without the use of a sacrificial resist layer, and a $Cl_2 + SF_6$ gas mixture was employed in the etchback step to give good selectivity to the underlying BPSG film and to ensure that no residue remained on the surface.

For 1.5 x 1.5 μm^2 contact holes, a contact resistance of 25±5 Ω per contact was measured for the in situ doped poly plugs (compared to 50 Ω per contact for Al:Si). Low junction breakdown-voltages were observed, however, and were determined to have been caused by junction punchthrough during the etchback process. In addition, it was noted that in order for this process to be used in CMOS, it would be necessary to use either two different types of poly doping, or a diffusion barrier (such as TiN) between the poly plugs and the Si substrate.

4.5.5.2 Selective Deposition of Poly.

Contact-hole filling through selective deposition of polysilicon has also been reported.[221] In this approach, the native surface of the Si is purged by $SiCl_4$ and a carrier gas of hydrogen. This step deposits polysilicon seed crystals on the Si exposed at the bottoms of the contact openings in SiO_2. However, no polysilicon is deposited on the SiO_2 surfaces. Following the seeding, the gas mixture is switched to dichlorosilane, HCl, and H_2, and the process is continued as for selective epitaxial growth of Si (see chap. 2, section 2.7.5). Note that polysilicon rather than single-crystal silicon is chosen for selective deposition, since the grain boundaries of the poly enhance the diffusion of impurities. As a result, a lower-temperature process can be used to uniformly distribute dopants throughout the entire plug volume; ensuring fabrication of low-resistance plugs. However, the series resistance of such polysilicon plugs is relatively high.

4.5.5.3 Selectively Formed Silicide Contact Plugs.

Intrinsic epitaxial Si is selectively deposited to approximately half the depth of contact holes in SiO_2,[222] and a film of Co or Ti is deposited to a thickness that will react the Si. A subsequent heating step is used to effect this reaction, and silicide plugs are thereby formed in the contact holes. The unreacted metal is then etched away with a selective metal etch. This technique has the advantage of not requiring a heavily doped SEG process for contact filling – a process that would be incompatible with CMOS technology.[223]

4.5.5.4 CVD Aluminum. While most of the work on CVD of metal films has involved tungsten, some research efforts have been directed at developing a CVD process for Al (although this technique is not yet sufficiently mature to be commercially available). The most successful CVD Al process involves the pyrolysis of triisobutyl aluminum (or TIBA).[142] Such films exhibit excellent conformal coverage, as well as resistivities of ~3.4 $\mu\Omega$-cm; in addition, they adhere well to Si and SiO_2. However, the surfaces of these blanket-deposited films are rough, which can cause lithography problems. Depositions are typically run at ~250°C and 100-200 torr, producing deposition rates of 10-20 nm/min. A review of this process and of the properties of CVD Al films is given by Levy and Green.[123]

It has recently been demonstrated that Al films can also be selectively deposited through the thermal decomposition of TIBA. Via holes as small as 0.4 μm have been successfully filled in this way.[144] In addition, whereas WF_6 can react extensively with Si during the selective CVD process, TIBA does not react with Si, and so leaves it undamaged. Finally, the surface-roughness drawback is apparently alleviated through this process.

Another recent paper described an LPCVD process for Al in a new batch-type, load-locked multichamber reactor, with smooth, highly-reflective Al films of high purity produced. Either blanket or selective Al deposition can be accomplished using this machine.[234]

There are two major problems associated with CVD Al. First, because TIBA is pyrophoric, explosive in contact with water, and toxic, it must be cold-trapped during processing rather than exhausted through the vacuum pump, and then disposed of after processing. In response to these difficulties, a process for forming CVD Al from a chloride source has also been studied.[145] The second problem exhibited is poor electromigration resistance. Although CVD Al has been shown to be as good as pure sputtered Al, it is not as good as Al:Cu, and a process for forming alloys must therefore be developed.

4.5.6 Alternatives to CVD for Filling of Vias

4.5.6.1 Bias Sputtering of Al for Complete Filling of Via Holes. If sufficient heat and bias are applied, complete filling of via holes can be achieved through sputter deposition of Al (Fig. 4-44).[107,108,115,116] Temperatures of between 400 and 500°C are necessary to sufficiently enhance the surface mobility is sufficiently enhanced so that complete filling is promoted as the film is deposited. However, such high temperatures, may cause unwanted effects in the film – for example, at 475°C and above, Cu in Al films tends to form large grains of Al-Cu intermetallics, which become

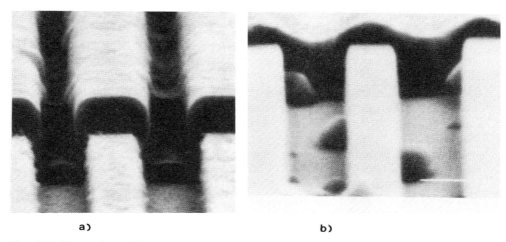

a)   b)

Fig. 4-44 Comparison of step coverage of: (a) conventionally sputter-deposited Al, and (b) high-temperature, high-bias sputterdeposition process that yields highly planarized Al films, even over high-aspect ratio contact openings with vertical sidewalls. Courtesy of MRC, Inc.

a serious problem for dry etching. At lower temperatures (200-350°C), Cu tends to segregate at the film/substrate interface. As a result, it has been reported that these films exhibit poor electromigration resistance, unless deposition temperatures in excess of 500°C are used.[262] It has also been reported that completely planarized Al causes heavy damage to the Si that cannot be fully annealed out. Work is being done to understand the nature of the damage, and to develop ways to minimize it.[109]

4.5.6.2 Laser Planarization of Al Films.

Another technique for improving the step coverage of Al lines over vias involves melting the Al film through irradiation with a pulsed laser prior to patterning. The mass transport that occurs while the Al is molten arises from high surface tension (which is about 50 times that of water). Such flow can completely fill and planarize the surface over the vias (Fig. 4-45). All hillocks

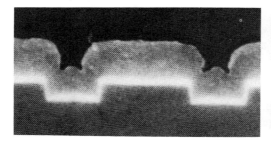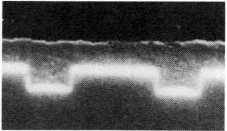

Fig. 4-45 SEM photos of an Al film before and after laser planarization.[210] Reprinted with permission of Solid State Technology.

and surface irregularities are also removed. In addition, laser-planarized films show a reduced tendency to regrow hillocks.[146,220]

A dye laser (wavelength 504 nm) was used in the original report, and a laser pulse of ~1 μs duration was estimated to be adequate to allow the surface to reach a sufficient level of smoothness.[147] A thin (20 nm) layer of Si was in situ sputter deposited onto the Al film to passivate it against native aluminum-oxide formation (which would remain as a solid skin and impede planarization), and also to act as an antireflection coating (since ~92% of the light would otherwise be reflected from the uncovered Al film surface).

In later work, an excimer-laser process was developed, and the laser-pulse duration was reduced to 15-30 ns.[148,240] The shorter pulse and ultrarapid heating prevent undesirable metallurgical reactions. A thin Cu film (<100 nm) was sputter deposited onto the Al layer as an antireflection coating (since the reflectivity of Cu is only ~20%). The excimer laser can planarize the surface of a 150-mm wafer in about 40 sec, with either a step-and-repeat or a raster-scan motion employed. Data showed that no junction leakage was observed in Al:Si excimer-laser planarized films, and that contact resistance was comparable to that obtained with conventional furnace annealing. Contact openings 0.75 μm wide and 1.5 μm deep have reportedly been filled using this technique. The redistribution of alloy constituents in the film as a result of this procedure may play a role in the electromigration behavior of these films. Another report indicated that laser-planarized Al:Cu films exhibited no hillock growth following 450°C furnace annealing, and initial data also suggest that an improvement in electromigration resistance may be possible under specific process conditions.[241]

Recent work indicates that the method shows promising results in the research laboratory, but that it has a narrow process window of only 6-8% (e.g., the energy applied to the Al film must be carefully controlled to provide adequate melting without producing ablation).[242] The dominant variable is the laser energy which can vary by ±5%, but Al film thickness variations and substrate heating uniformity are other variables which can impact the local quantity of energy applied to the Al. Methods to widen this process window will need to be developed to make it a viable manufacturing option.

In one recent study an attempt was made to optimize the wafer temperature and laser power to provide the widest possible process window for the laser power. Good via filling and coverage of the Al over steps was achieved. The laser-flowed Al:Cu lines also showed electromigration resistance comparable to those of non-flowed Al:Cu lines.[263]

4.5.6.3 Contact-Hole and Via Filling with Selective Electroless Metal Deposition.

Various metal films (including Ni, Cu, Au, and Pd) can be deposited through electroless deposition of metal films from aqueous solutions consisting of metal ions and reducing agents.[149] Reports on via filling have focused on Ni and Pd because of the abundant information processes, and because of the wide choice of reducing agents.

In one process, the Ni was able to fill the vias to the top, producing a uniform and planar surface (Fig. 4-46). Vias with several different diameters were filled simulta-

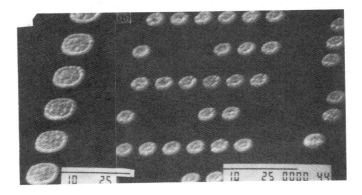

Fig. 4-46 SEM photos of 2.5-μm TiSi$_2$-based contacts filled with Ni.[149] Reprinted by permission of the publisher, The Electrochemical Society, Inc.

neously. An Al film was then deposited. Low contact resistance in the vias was exhibited following annealing. Unframed vias can be used, since the vias are filled with a different material than the main conductor, and since good selectivity can be obtained between the Al and the Ni during the Al etch process.

Three critical aspects of the Ni process must be controlled in order to successfully implement via filling with this technique.[150] First, the surface of the Al at the bottom of the via must be free of native oxide; this can be accomplished by performing a dilute HF dip. Second, the Al surface must be activated (e.g., by first depositing Pd), since Al is not catalytic for Ni deposition. This is achieved by using a displacement reaction ($3Pd^{+2} + 2Al \rightarrow 3Pd + 2Al^{+3}$) that deposits a high density of discrete Pd particles on the Al surface. The third critical aspect is the Ni deposition itself. A solution containing Ni ions and a reducing agent (e.g., hypophosphite, at a concentration of 0.1 mole/L, or dimethylamine borane) at ~70°C gives a deposition rate of ~3 μm/hr. Following the deposition, an anneal step of 400°C in N$_2$ is used to remove any traces of the solution in the vias and to improve the adhesion of Ni to the substrate.

In the selective electroless process for contact-hole filling, either Ni or Pd is deposited onto CoSi$_2$ or TiSi$_2$ at the bottom of contact holes in BPSG.[151] The conditions for Ni deposition are like those just described for via filling. Pd can be deposited without activation of the surface; however, it has been observed that both diodes and transistors fabricated with this process show degraded device characteristics following the metal deposition and anneal, due to interaction between the Ni/silicide or Pd/silicide and the underlying n^+ Si. A diffusion barrier will therefore be needed. For example, a 150-nm selective CVD W layer can be deposited onto the TiSi$_2$ layer in the contact holes. This has been observed to prevent the reaction between the Pd plugs and the underlying substrate for anneal temperatures < 450°C.[152]

4.5.7 Pillar Formation as an Alternative to Contact-Hole and Via Filling

Pillars (also called *studs* or *posts*) may be formed prior to deposition of the dielectric. In contrast to conventional contact holes and vias, pillars allow the use of non-nested design rules, and device areas can thus be reduced. In addition, stacking of contact holes and vias is possible. Pillar-formation methods include:

• *Pillar formation by lift-off.* A lift-off technique (see Vol. 1, chap. 15) has been used as the method used to form pillars in a four-level metal interconnect system for bipolar gate arrays.[153,154]

• *Simultaneous deposition of the Metal-1 and pillar films, followed by two-mask patterning of pillars and Metal-1 lines.* This approach also includes a technique for self-aligning the pillars to the Metal-1 lines, even though two masking steps are used.[155] The Metal-1 patterns are etched, and then the pillars (Fig. 4-47).[251] To obtain a highly planar surface, a resist etchback technique is used that includes an additional masking step. This technique was designed for use with a metallization system having a 2.5-μm pitch.

• *Etching the pillars from a thin metal film using contrast-enhancement lithography.* In this approach, submicron-wide pillars are defined and etched using a contrast-enhancement lithography (CEL) technique.[156] The first pillar, which consists of a Ti/W/Al trilayer film, is deposited directly onto $TiSi_2$. The film is imaged using a CEL technique to form 1 x 1 μm pillars. The dielectric-

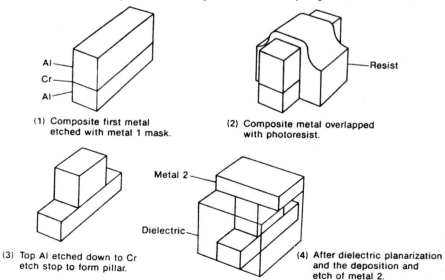

(1) Composite first metal etched with metal 1 mask.

(2) Composite metal overlapped with photoresist.

(3) Top Al etched down to Cr etch stop to form pillar.

(4) After dielectric planarization and the deposition and etch of metal 2.

Fig. 4-47 Sequence used to produce Al/Cr/Al pillars using photoresist overlap method.[251] (© 1984 IEEE).

layer formation is accomplished by etching back a CVD SiO_2 film. The tops of both high pillars (located on field-oxide and poly steps) and low pillars (located on substrate contact regions) must be exposed following the etchback procedure. Metal 1, which makes contact to the posts, is then deposited and patterned.

4.6 FILLED GROOVES IN A DIELECTRIC LAYER

Assume that a planarized dielectric surface exists between polysilicon and Metal 1 and that a process is available that can completely fill the contact holes etched in this layer. The fabrication of a multilevel interconnect system can now follow the process sequence that has been described, or an alternative process sequence can be carried out. In this sequence, part of the DM1 dielectric layer is deposited, and grooves are patterned in it. Metal 1 is then selectively formed in the grooves. Finally, the entire surface is covered with an additional dielectric layer.

This alternative filled groove approach has the important advantage of being able to produce a completely planarized dielectric surface *without the need of a planarizing procedure for DM1*. Because this approach is inherently planar and self-aligned, and also allows stacked vias, it can be used to fabricate three or more levels of metal. It also allows thicker metal layers to be deposited without the creation of topographical conditions that would make planarization even more difficult. This capability is necessary, for example, to produce W and Al lines of the same line resistance and equal widths. However, since the vertical elongation of the W lines increases the intra-level capacitance, this approach may not be suitable for all circuit designs and types.

The unique process step in this method is the deposition of metal in the dielectric grooves. Techniques to implement the process include the following:

- *Blanket CVD deposition and etchback of W*. This technique, dubbed the *filled-interconnect-groove* (or FIG) method, was described by Broadbent et al. (Fig. 4-48).[157] A thin Ti:W adhesion layer is first sputter deposited into the grooves. The CVD-W process is then carried out by means of hydrogen reduction of WF_6 so that a void-free filling of the grooves with W is achieved. This W film is then etched back until W remains only in the grooves of the SiO_2 film.

- *Selective CVD W deposition in the grooves*. This technique was presented by D. C. Thomas et al., who called it the *tungsten (W) interconnect technology*, WIT.[158] Si is implanted into the bottom of the grooves etched in the SiO_2 (with an implant dose of $\sim10^{17}$ cm^{-2}), where it serves to selectively initiate the deposition of W on the SiO_2 surface. A silicon-nitride layer prevents the Si from penetrating the nonetched SiO_2 surface and the sidewalls of the groove (Fig. 4-49). The nitride layer is removed by means of a hot phosphoric etch after the Si implant and before the W deposition.

- *Lift-off using edge-detection (LOPED)*. This technique, illustrated in Fig. 4-50,[159] can be used to fill narrower spaces than would be possible with conventional lift-off methods (see Vol. 1, chap. 15).

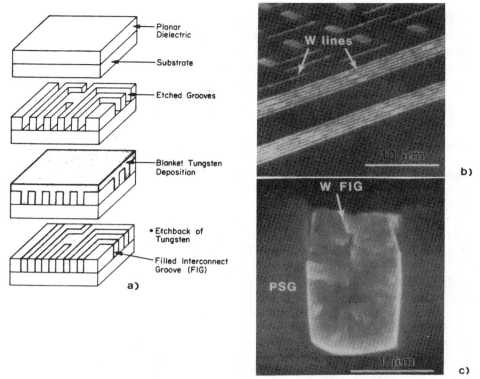

Fig. 4-48 (a) Representation of process sequence used to construct a level of filled interconnect groove (FIG) metallization. (b) Perspective top view of SEM micrograph of FIG metallization after etchback of the CVD W. (c) Cross section SEM micrograph of a FIG conductor (~1.1-μm wide) after CVD deposition and etchback.[157] (© 1988 IEEE).

• *Electroless selective deposition of metal.* Here the process technology employed to fill via holes is the same as that described in section 4.5.6.3, except that a thin Al layer is deposited into the grooves prior to the starting of the process sequence.[160]

4.7 MANUFACTURING YIELD AND RELIABILITY ISSUES OF VLSI INTERCONNECTS

Yield-limiting factors are flaws in the manufacturing process that result in the production of parts that fail to operate properly immediately following fabrication. Other flaws, known as *reliability factors* result in parts that operate correctly immediately following manufacture, but that fail later during normal operation (or even during storage). Defect-related random failures tend to follow yield trends, while wearout-related reliability failures may be totally independent of yield. (Information of a more detailed nature dealing with these issues will be presented in Vol. 3 of this text.)

4.7.1 Factors That Impact Manufacturing Yield

Many of the yield-limiting factors associated with multilevel-interconnect fabrication are the same as those that afflict single-level interconnect processes:

- Junction spiking (see chap. 3, section 3.4.3),

- High contact resistance between Metal 1 and Si (see chap. 3, section 3.4.2.3),

- Gate-to-source shorts due to lateral silicidation over oxide spacer (see chap. 3, section 3.9.1.1),

- Contamination of Al interconnects by incorporation of residual gases present during sputter deposition.[161] For example, oxygen incorporation can cause an increase in the film resistance and hardness, which may lead to bonding problems. Nitrogen incorporation increases stress and thus the tendency to crack.[162]

- Misalignment of metal lines to contact holes (see Vol. 1, chap. 12).

4.7.2   Multilevel Interconnect-Related Yield Issues

Following are several of the factors that adversely impact yield when multilevel interconnects are used:

- *Pinholes and weakspots in the PMD and intermetal-dielectric layers*, which result in shorts or unacceptably low breakdown voltages between different conductor levels. These have several causes:

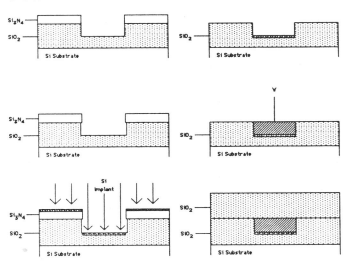

Fig. 4-49 Process flow for planar *W interconnect technology* (WIT).[158] (© 1988 IEEE).

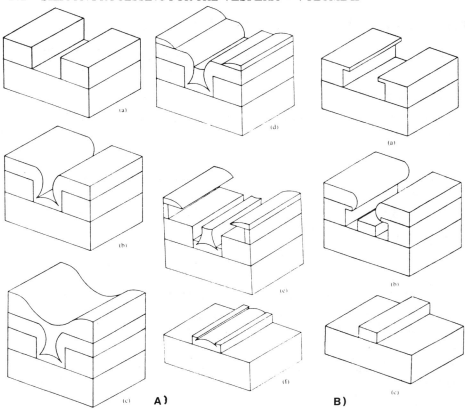

Fig. 4-50 (A) Outline of the LOPED process: (a) A layer of positive resist is patterned by a dark field mask and flood exposed after development; (b) The desired thin film is deposited with good step coverage; (c) A second resist layer is applied; (d) The second resist layer is etched; (e) The deposited film is etched; (f) The liftoff process is completed. **(B)** Process flow of direct liftoff: (a) A layer of photoresist is patterned by a dark-field mask and processed to form a reentrant angle; (b) The desired film is deposited with poor step coverage; (c) The liftoff step is completed by dissolving the lifting medium in a solvent. The excess film is removed at the same time.[159] (© 1989 IEEE).

- *Particulates.* These may be generated in three ways: (1) they may originate in the process environment; (2) they may be formed by gas-phase nucleation during CVD of the dielectric (see section 4.3.2.3); or (3) they may arise as a result of dielectric-film deposition on moving parts in the CVD chamber, later flaking off.
- *Hillock growth* in the underlying metal layer.
- *Thinning of the photoresist* over the corners of steep steps in the dielectric. This may may lead to complete resist erosion during via etching and subsequent attack of the dielectric (which can give rise to thin spots).

• *Misalignment during via patterning process*, leading to open-circuit conditions.

• *Topography-related issues*. These include step-coverage problems that cause opens in interconnect lines or thinning of the metal lines to such a degree circuit performance becomes unacceptable; and stringers that short adjacent metal lines.

• *Open circuit or high contact resistance between conductor materials of different levels*, due to:

  - Incomplete etching of all via holes (see section 4.6),
  - Failure to remove the native oxide from the conductor (e.g., Al_2O_3 on Al) prior to the deposition of the next level of metal (see chap. 3, section 3.4.2.3),
  - "Poisoned vias," which occur when the constituents of an SOG film outgas during the metal deposition, producing high via resistances (see section 4.4.9).
  - High resistance between two conductor layers (e.g., high via resistance).

Since the very last effect listed above has not been discussed elsewhere, we will consider it here.

4.7.2.1 High Resistance between Two Conductor Layers. Generally speaking, the specific contact resistivity of metal to metal is very low, and does not represent a problem when contact between metal layers is made through a via. For example, the specific contact resistivity of Al-to-Al contacts is $<10^{-8}$ Ω-cm^2. However, relatively high contact resistance has been observed when W via plugs are used to make contact to underlying Al layers.[138] This is believed to be caused by fluorine (from the WF_6) reacting with the Al to form a nonvolatile AlF_3 layer at the   Al-W interface.

Procedures for obtaining lower contact-resistance values between selective CVD W and Al have been reported. One report found that at temperature and deposition rates of $\geq 375°C$ and ≥ 20 nm/min, respectively, specific contact resistivities of $\sim 10^{-8}$ Ω-cm^2 are obtained, (possibly because at such temperatures the AlF_3 is volatilized as soon as it forms).[139] In the second report, it was also observed that the contact resistance of selective-W-to-Al interfaces in submicron vias is deposition-temperature dependent, and that it continues to decrease up to temperatures of 550°C.[247] Another approach that produces low contact resistance between Al and CVD W plugs involves capping the Al film with a layer of sputtered Ti:W during the same pumpdown as the Al deposition. Following via patterning of the intermetal dielectric, CVD W is selectively deposited on the Ti:W. Since no Al is exposed, the possibility of forming AlF_3 is avoided.[257]

Relatively high contact resistance between selectively deposited W and $TiSi_2$ has also been observed.[163] This finding was confirmed by Ng and Merchant, who reported that the use of a higher deposition temperature could also reduce the contact resistance value.[164]

4.7.3 General Reliability Issues Associated with IC Interconnects

4.7.3.1 Electromigration. Electromigration is the motion of the ions of a conductor (such as Al) in response to the passage of current through it. These ions are moved "downstream" by the force of the "electron wind." A positive divergence of the ionic flux leads to an accumulation of vacancies, forming a void in the metal. Such voids may ultimately grow to a size that results in an open-circuit failure of the conductor line. Electromigration is, therefore, a wearout mechanism.

In general, the failure rate is increased when the current density in the conductor line is increased, or the operating temperature is raised. Thinning of the conductor lines as they cross steep steps will also accelerate electromigration failure rates, because the current density at such locations along a line are increased. It has likewise been found that Si precipitates (or nodules) which occur in Al:Si films, can grow large enough to constrict Al cross-sections in narrow lines.

Techniques for increasing the resistance of an interconnect process to electromigration failure include the following:

- Adding Cu (0.5-4 %) to the Al film.
- Adding Ti (0.1-0.5 %) to the Al film.
- Using a layered Al film structure, with a highly electromigration resistant metal (such as Ti, W, or Mo) as the central layer of a trilayer film.
- Planarizing the intermetal dielectric, to eliminate thinning of the conductor lines as they cross steps.
- Selectively depositing a layer of CVD W over the Al lines.
- Avoiding the use of Al:Si when fabricating narrow, multilevel-metal structures.
- Replacing the Al metallization with a more electromigration-resistant metal, such as W or Mo.

4.7.3.2 Electromigration at the Contacts. Electromigration can also occur as a result of current flow through the contact interface. The "electron wind" drives the Si atoms along the grain boundaries of the Al film, causing the formation of voids in the Si. These voids result in either open circuit failures, or — if they are back-filled by Al - result in junction spiking failures. Such failures can be prevented through the use of a diffusion barrier material between the Si and the Al or of a metallization in which Si does not rapidly diffuse (e.g., W).

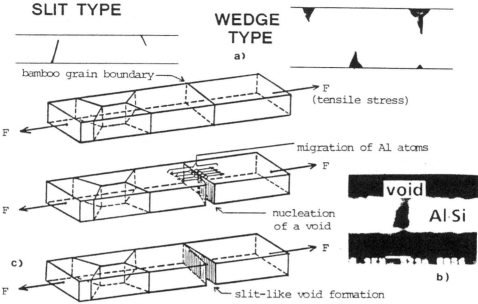

SLIT TYPE

WEDGE TYPE

a)

b)

c)

bamboo grain boundary

F (tensile stress)

F

migration of Al atoms

F

nucleation of a void

void

Al·Si

F

slit-like void formation

Fig. 4-51 (a) Schematic drawing of stress-induced voids in Al interconnect lines. (b) SEM photo of such a void. (c) Schematics of one proposed formation-mechanism of slit-like voids.[260] (© 1985 IEEE).

4.7.3.3 Stress-Induced Metal Cracks and Voids.
Metal cracks and voids have been observed in Al interconnect lines (Fig. 4-51a and b) after the lines have been stored at elevated temperatures (even when no current has been passed through them).[165] Parts have also been found to fail after doing no more than sitting on a shelf for one or two years, even when they had passed earlier tests. In some cases the voids extend all the way across a line, causing an open-circuit failure. In others, the notches at the edges of the metal lines cause partial loss of a cross-sectional area, resulting in early electromigration failure. The problem is so severe that industry observers have claimed it contributed to putting at least one company out of business (Mostek).[166]

Two stress-related failure mechanisms have been identified: (1) high compressive stress in overlying dielectric films, which causes tensile stresses in the metal film; and (2) tensile stress in the Al films during heating, caused by the mismatch in the coefficients of thermal expansion of the Al lines and the silicon substrate (and the consequent inability of the Al lines to relieve this stress by differential movement, since it is then tightly bonded to the deposited dielectric). It is believed that the stress must be relieved by mass transport through the diffusion of Al atoms into the the grain boundaries from neighboring grains, with void formation resulting (Fig. 4-51c). In one report, the first mechanism is held primarily responsible for void formation in wide Al lines, while the second mechanism is thought to cause voids in narrow Al lines.[167]

The problem is expected to become even more severe as devices shrink, for several reasons. First, the narrower the line width becomes, the greater the stress will be (Fig. 4-52a).[168] Second, the metal lines in lowest level of multimetal interconnect structures experience the greatest stress. Finally, the lines of the lowest-level metal are often the narrowest.

Several measures can be taken to reduce the severity of these effects in Al lines:

1. The stresses in the overlying films should be tensile or at least have a low compressive value. For example, the failure rate was far lower for silicon-oxynitride passivation layers that exhibited low compressive stress, than for highly compressively stressed PECVD nitride passivation layers.

2. Since the incorporation of N_2 into Al films makes the films very hard and

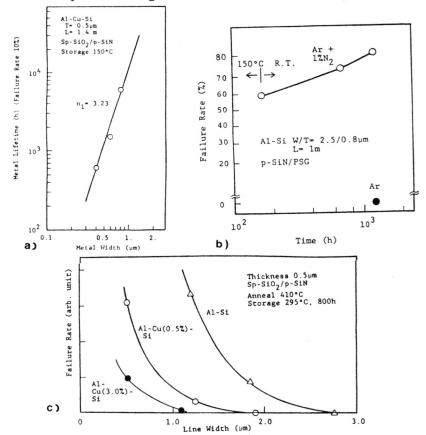

Fig. 4-52 (a) Line width dependence of stress-induced migration of Al:Si:Cu. (b) Stress-induced migration enhanced by N_2 contamination. (c) Dependence of stress-induced migration on the Cu content of the Al films.[168] This paper was originally presented at the Fall 1988 Meeting of The Electrochemical Society, Inc. held in Chicago, IL.

and brittle, and thus dramatically increases the stress-induced void formation rate (Fig. 4-52b), care must be taken to prevent N_2 from being incorporated during sputter deposition.[162]

3. Adding Cu to the films reduces the failure rate (Fig. 4-52c). A recent report indicates that adding Pd to the Al films seems to be even more effective than adding Cu.[238]

4. A layered structure (with such materials as $MoSi_2$) produces lines with much longer life. These structures help prevent the formation of voids across the entire cross-sectional area of the interconnect line.

5. Because steep steps on the wafer surface enhance stress and increase failures, the interlevel dielectrics should be planarized.

6. A slow cooling time from 400°C to room temperature increases the failure rate, while rapid-quench cooling results in much lower failure rates.

7. Silicon nodules in the Al cause local stresses in the lines, which also increase the tendency for void formation.[169] See section 3.4.4, chap. 3 for a discussion on silicon-nodule formation in Al lines.

Other suggested approaches for alleviating the problem include the use of higher-melting-temperature metals, such as W (which exhibit lower diffusivities); the development of a more pliable passivation layer; the use of dielectric-film deposition processes that require lower temperatures; and intentional debonding of the interface between the metallization line and the dielectric layers that cover it.

4.7.3.4 Corrosion. Corrosion of the metal lines can be caused by the following:

• Transport of moisture and such contaminants as Cl through the passivation layer, and subsequent reaction of these with the metal lines (see section 4.8). The moisture and contaminants may either exist in the package material itself or may arrive through cracks in the package.

• Leaching of phosphorus from phosphorus-doped SiO_2 intermetal or passivation dielectric layers, followed by reaction of the phosphorus with absorbed moisture to form phosphoric acid (which then attacks the Al - see Vol. 1, chap. 10).

• Residual Cl (which remains on the wafer surface following a Cl-based dry-etch step of Al) reacts with moisture to form HCl, which then attacks the Al (see Vol. 1, chap. 10). The standard way of dealing with this problem has been to expose the wafer to a short, in situ CHF_3 or CF_4 etch step after the Al has been patterned and the resist has been dry etched. A recently reported alternative technique uses a reactor that contains an integrated spin-rinse system in the exit loadlock, which sequentially dispenses an organic solvent and water onto the wafer. This technique is believed to remove the carbonaceous chlorine-containing polymer that remains on the wafer following the etch process.[219]

4.7.4 Reliability Issues Associated with Multilevel Interconnects

4.7.4.1 Hillock Formation and Prevention Measures.
Hillocks are spike-like projections that erupt in response to a state of compressive stress in metal films and consequently protrude from the film's surface (4-53a). There are two reasons why hillocks are an especially severe problem in Al thin films (in which they may become twice as large as the film thickness), for two reasons. First, the thermal coefficient of expansion of Al ($23.5 \times 10^{-6}/°C$) is almost ten times as large as that of Si ($2.5 \times 10^{-6}/°C$). When a silicon wafer is heated, the thin Al film (which is tightly

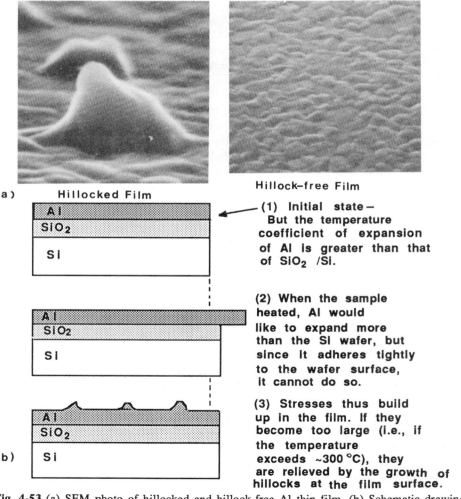

a) Hillocked Film   Hillock–free Film

(1) Initial state –
But the temperature coefficient of expansion of Al is greater than that of SiO_2 /Si.

(2) When the sample heated, Al would like to expand more than the Si wafer, but since it adheres tightly to the wafer surface, it cannot do so.

(3) Stresses thus build up in the film. If they become too large (i.e., if the temperature exceeds ~300 °C), they are relieved by the growth of hillocks at the film surface.

Fig. 4-53 (a) SEM photo of hillocked and hillock-free Al thin film. (b) Schematic drawing of how thermal stress causes hillock growth in Al thin films on a Si wafer.

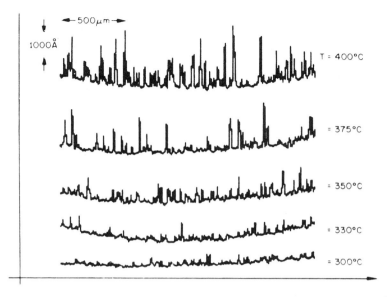

Fig. 4-53 (c) Hillock formation as a function of deposition temperature.[172] (© 1987 IEEE).

adherent to the much more massive wafer) "wants" to expand more than is allowed by the wafer expansion (Fig. 4-54b). The result is a compressive stress, σ, which increases at the following rate as a function of temperature:[170]

$$d\sigma /dT = -2 \times 10^7 \text{ dynes cm}^{-2} \text{ °C}^{-1} \qquad (4-15)$$

As a result, when the Al film is heated to temperatures in excess of 300°C, the compressive stresses in the film become very high.

The second factor involves the low melting point of Al (660°C), and the consequent high rate of vacancy diffusion in Al films. Hillock growth takes place as a result of a vacancy-diffusion mechanism.[171] Vacancy migration occurs as a result of the vacancy-concentration gradient arising from the stresses; in addition, the rate of diffusion increases very rapidly with increasing temperature. Hillock growth can thus be visualized as a mechanism that relieves the compressive stress in the Al film through the process of vacancy migration away from the hillock site, both through the Al grains and along grain boundaries. Figure 4-53c shows the sizes and the densities of hillocks on an Al:Si:Cu thin film on which a dielectric film was deposited at different temperatures (between 300° to 400°C).[172]

The most significant hillock-related problem in IC manufacturing occurs in multimetal-interconnect structures. In such structures, hillocks cause interlevel (as well as intralevel) shorting when they penetrate the dielectric layer that separates neighboring metal lines.

Interlevel shorts can occur when a conformal-CVD dielectric is deposited over a hillocked metal film. When a photoresist is spun onto the dielectric layer, the resist will be thinner over the hillocks (Fig. 4-54a). During the via-etch process, the thin resist may be completely removed in these spots and the dielectric will then be exposed to the etching ambient. Under some circumstances the dielectric film may be completely etched away, allowing the exposed hillock to form a short to the next level of metal. Even if the dielectric is not completely removed, it may become so thin that low dielectric breakdown may occur between Metal 1 and Metal 2.

Intralevel shorts can occur if the hillocks protrude from the sides of the metal lines and cause overlying dielectric layers to delaminate (Fig. 4-54b). The hillocks can then continue to grow until they make contact with an adjacent-metal line. Such problems become more severe as the spacing between metal lines becomes narrower. The presence of hillocks also reduces film reflectivity, introducing alignment difficulties and making the microlithography process less controllable.

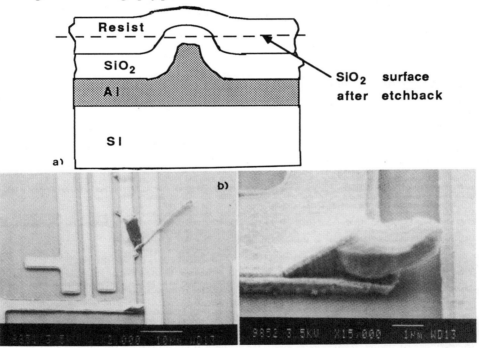

Fig. 4-54 (a) Schematic drawing of the thinning of photoresist over a hillock in the Al. During etching of the dielectric layer to open contacts the thinner resist may be completely eroded, causing the dielectric layer over the hillock to be etched. (b) Hillocks protruding from the sides of the metal lines, which may lead to intralevel shorts.[176] This paper was originally presented at the Fall 1988 Meeting of The Electrochemical Society, Inc. held in Chicago, IL.

The following are several approaches that have been developed to alleviate the problem of hillock formation:

• *Adding elements that have a limited solubility in Al as alloys to the film.* The alloying element is added in excess of the solid solubility limit, causing precipitation in the grain boundaries. This can "plug" the grain boundaries, inhibiting vacancy migration and thereby inhibiting the rate of hillock growth. Copper has been found effective in this application. (Note that since electromigration is also a grain-boundary diffusion process, alloys that exhibit improved electromigration resistance also tend to be less prone to hillock growth.) In general, this approach reduces, but does not eliminate hillock formation.

• *Depositing a capping layer of W or Ti on top of the Al film.*[173,174,175] Note, however, that the cap layer may increase the propensity of the films to form sidewall hillocks, which can lead to more intralevel shorting. A Ti cap layer exhibited the fewest lateral hillock failures, according to one report.[176]

• *Layering Al films with one or more Ti layers* (Fig. 4-55a).[177] Apparently, the Al film must include some Si in order for hillock-free films to be produced. The resistivity of layered Al:Si/Ti films, however, is ~50% higher than the resistivities of Al:Cu or Al:Si films. In addition, unless a diffusion barrier is present between the layered Al film and the Si substrate, Si will be taken up from the substrate until the Ti layers are completely transformed to silicides.[178]

• *A W layer can be selectively deposited onto the patterned Al lines.* This can be accomplished because W preferentially nucleates on Al surfaces, allowing the layer to cover the sidewalls of the lines, as well as the top surface (Fig. 4-55b).[179,180] A thin (80 nm) W layer deposited onto a 150-nm-thick Al layer resulted in a Metal 1 film in which hillocks on vertical and lateral sidewalls were effectively suppressed and the electromigration resistance of the lines was greatly improved. A hydrogen-reduction process (carried out at 300°C to prevent hillocks from growing on the Al surfaces) was used to deposit the W at a rate of 2.5 nm/min. Another report indicated that such selectively W-clad Al films demonstrated dramatically improved reliability, even with 0.6-μm-wide lines.[181]

• *A low compressively stressed intermetal-dielectric layer of PECVD SiO_2 is deposited so rapidly that hillock formation is not initiated until sufficient SiO_2 material has been deposited to suppress the formation.*[182] In this process, hillocks still form − but if the start of the deposition can occur within 10 seconds of the wafer's being placed on the hot susceptor, and if the deposition rate can exceed 300 nm/min, very few hillocks > 0.5 μm in height will occur (Fig. 4-55c).

• *A dielectric deposition process can be used whose temperature of deposition is lower than the temperature at which hillocks form.* Examples of such processes include:

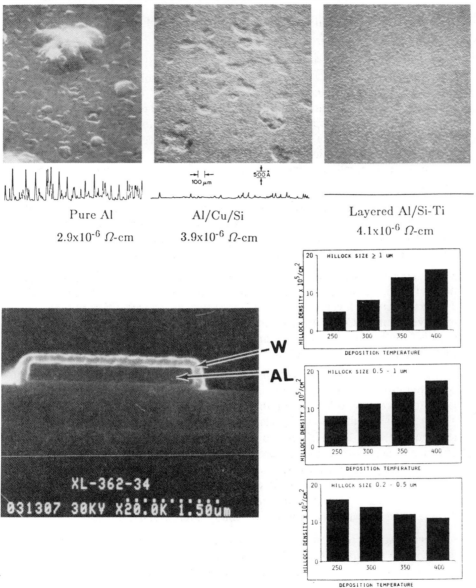

Fig. 4-55 (a) Comparison of hillocks formed in pure Al, Al/Cu/Si, and layered Al/Si:Ti films following an anneal at 450 °C for 30 min.[177] (© 1985 IEEE). (b) CVD W selectively deposited on an Al runner.[179] (© 1986 IEEE). (c) Hillock distribution on cold-sputtered Al after deposition of 0.9-μm SiO_2 at various temperatures. The time at temperature prior to deposition was held constant at one minute.[182] (© 1987 IEEE).

- PECVD TEOS film, deposited at 330°C. (Figure 4-54c shows that the resulting hillock size and density are both small.)[172]
- ECR deposition (see section 4.4.10).
- Photo-CVD SiO_2.[183]
- Afterglow-CVD of SiO_2.[24]
- Anodized Al deposition.[184]

Once such dielectric films are in place, they suppress hillock growth during later thermal cycling.

• *Use of refractory metal films such as W or Mo (which exhibit much less propensity to form hillocks at 400°-500°C) in place of Al.*

• *Use of an ionized cluster beam deposition process to deposit smooth $Al/CaF_2/Si$ films.* According to the report on this process, the films remained hillock free up to temperatures of 500°C.[185]

4.7.4.2 Dielectric Void Reliability Problems. If the voids are opened following an etchback step, they can trap moisture or photoresist residues that can cause long term reliability problems. In addition, metal may be deposited into the voids that can be very difficult to remove by etching, thus producing shorting between neighboring metal lines.

4.8 PASSIVATION LAYERS

Following patterning of the final metal layer, a *passivation layer* is deposited over the entire top surface of the wafers. This is an insulating, protective layer that prevents mechanical and chemical damage during assembly and packaging. The desired properties of the passivation materials are given in Table 4-5. In general, the thicker the passivation layer the better, since a thicker layer will provide better protection and improve the electromigration resistance of underlying Al lines. On the other hand, because thicker CVD films (especially silicon nitride films) have a higher tendency to crack, there is normally an upper limit to the thickness.

The final mask, called the *pad mask* or *bonding contact mask,* is used to define patterns corresponding to the regions in which electrical contact to the finished circuit will be made. These patterns in a resist layer allow openings in the passivation layer to be etched down to Al areas on the circuit called bonding pads (see chap. 5, Fig. 5-16). Either wet or dry etching can be used to etch the passivation layer. Since the dimensions of the pads are so large (i.e., normally 100 x 100 μm), wet etching is still frequently used to etch PSG films, while silicon nitride films are more easily etched by means of a dry etching process.

Phosphorus-doped, low-temperature CVD SiO_2 films were the first passivation layers to be used. The phosphorus is added to the SiO_2 to reduce the stresses in the film (and to thereby decrease the tendency of the film to crack), as well as to improve the gettering properties of the film (with respect to sodium ions and other fast-diffusing metallic contaminants). The higher the phosphorus concentration, the better these characteristics will be.

On the other hand, if more than 6 wt% phosphorus is added to the film, corrosion can become a serious problem, especially in the case of chips mounted in plastic, nonhermetic packages. Water vapor can rapidly penetrate the plastic packaging material, transporting with it contaminants from the surface of the package. If the PSG contains excess phosphorus, the moisture can react with it to form phosphoric acid (HPO_3), which will eventually penetrate the film. As noted earlier, electrochemical corrosion of the Al lines can lead to metallization failure. While the transport of contaminated moisture through the PSG is a relatively slow process, if cracks or defects (e.g., pinholes) exist in the film, the water vapor will be able to penetrate it much more rapidly.

Silicon nitride has also been used as a passivation-layer material because it provides an impermeable barrier to moisture and mobile impurities (e.g., sodium) and also forms a tough coat that protects the chips against scratching. Its high dielectric constant is not a disadvantage for this application, since the passivation layer is deposited on top of the last metal layer.

However, because the passivation layer must be deposited over Al films, only PECVD silicon nitride can be used for this application (since it is deposited at ~300°C). Unfortunately, PECVD nitride films normally exhibit a high mechanical stress (~6-8 x 10^8 Pa), which can cause cracks in the film during heating after deposition (especially at steps). The high compressive stresses in the films have also been shown to enhance void formation in Al interconnects (see section 4.7).

In addition, PECVD silicon nitride tends to be nonstoichiometric and contains substantial quantities of atomic hydrogen (10-30 at%). Large quantities of hydrogen have been found to accelerate hot-electron aging effects in MOS devices (see section 5.6.6). It has also been reported that the hydrogen from PECVD nitride is responsible for the formation of bubbles or cavities at the metal–plasma nitride interface when Al:Si alloys are used as the metallization. These appear after the 450°C anneal that is carried out following nitride deposition. It is supposed that the hydrogen reacts with the Si precipitates in the Al film to form gaseous compounds that produce the bubbles.[186,187] It has been recommended that a low-hydrogen-content passivation film be used, if nitride passivation is selected for MOS technologies with gate lengths of less than 1.5 μm (in which hot-electron degradation is significant).

Work has been done to develop processes for growing nitride films with low hydrogen concentration and stress, and more is now understood about the relationship of the film properties to such deposition conditions as rf frequency, power, and bias.[188] Lower H_2 concentrations in the nitride have been obtained when the film is formed with SiH_4-N_2 mixtures rather than with conventional SiH_4-NH_3 mixtures.[189] Fluorinated nitride (F-SiN) films have been developed that exhibit only 0.6% of hydrogen (in the form of Si-H).[190,191]

The use of PECVD silicon-oxynitride films (deposited with SiH_4, NH_3, N_2, and N_2O mixtures) as alternative passivation materials has also been investigated, since they exhibit nearly the same the moisture and sodium barrier characteristics of nitrides.

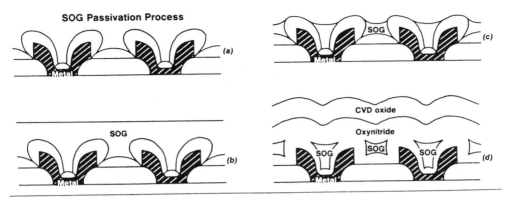

1. *(a) Deposition of the first oxynitride layer, (b) SOG applied, (c) after SOG etch back, and (d) finished passivation process after second oxynitride and CVD oxide depositions.*

Fig. 4-56 Using SOG film as a part of the passivation overcoat improves EPROM reliability.[256] Reprinted with permission of Semiconductor International.

While not quite as good as those of nitrides, the characteristics are better than those of oxides.[192] However, their stress is between that of APCVD oxide (tensile) and plasma nitride and oxide (compressive).[193] In addition, because the stress is a function of applied rf power, pressure, and bias, it is possible to optimize the stress by using bias. Ideally, a very low-stress dielectric oxynitride film can be formed that will still maintain good diffusion-barrier properties. (It is important to characterize the stress over the entire temperature range of operation to which the dielectric film will be subjected, since the stress can exhibit hysteresis effects. These may be due to structural changes in the film due to loss of material during heating.) Finally, it is reported that PECVD oxynitride films can be formed that contain considerably less H_2 (~one-half, in one report) than do PECVD nitrides.[194]

Another, more recently adopted approach to the formation of passivation layers involves multilayer passivation coating. An initial coating of PECVD oxide is deposited, followed by a PECVD nitride. The oxide layer reduces the mechanical stress (~40%) and the hydrogen content of the passivation layer, while the nitride protects the device against handling, humidity, and mobile ions. This inorganic bilayer may be followed by a polyimide layer that is several microns thick (especially useful in automated bump-bonding processes) and a thick layer of silicone gel, or similar material, for cushioning and for void elimination during die bonding.

In a second variation of this technique applied to EPROMs, a sandwich oxyntride-(etchedback) SOG-oxynitride layer is first formed. This film is then covered with a low phosphorus- content CVD-oxide layer to complete the composite passivation film (Fig. 4-56). Sandwiching the SOG film between the two oxynitride layers reduces the occurrence of voids and seams in the passivation layer. These voids and seams caused degraded passivation film coverage, which in turn correlated with increased EPROM array failures after steam stressing.[256]

Table 4.5   Desired Properties of a Passivation-Layer Material

1. Provides good scratch protection to underlying circuit structures. In general, the thicker the passivation layer the better, subject to cracking and patterning restrictions.
2. Impermeable to moisture, as moisture is one of the main catalysts for corrosion.
3. Exhibits low stress, preferably compressive ($\sim 5 \times 10^8$ dynes/cm^2).
4. Conformal step coverage.
5. High thickness uniformity.
6. Impermeable to sodium atoms and other highly mobile impurities.
7. Easily patterned.
8. Good adhesion to conductors, as well as to the interlevel dielectric beneath the last level of metal.

4.9 SURVEY OF MULTILEVEL METAL SYSTEMS

As noted earlier, with NMOS IC technology it was possible to exploit the polysilicon layer as an extra level of interconnect, while in bipolar technology it was necessary to develop a two-level-metal system in order to obtain comparable flexibility of interconnect routing. As a result, the problems of two-level-metal systems (primarily, the implementation of low-temperature planarization techniques) first had to be tackled by bipolar IC manufacturers. When CMOS replaced NMOS as the dominant MOS VLSI technology, CMOS ICs also required a two-level-metal system, since the polysilicon could not perform the function of a local interconnect level as effectively as it had in NMOS (see chaps. 2 and 6). However, the polysilicon gate structures and the nonrecessed LOCOS field-oxide steps in CMOS created an even more difficult topography for two-level-metal CMOS systems than for bipolar systems.

4.9.1 Bipolar Double-Level-Metal Systems

The first example we present is that of a structure described in 1984 by Ghate et al. of Texas Instruments.[195] The Metal-1 pitch is 4 μm, and the Metal-2 pitch is 6 μm. Metal 1 is a 575-nm-thick bilayer film of Ti:W covered with Al:Cu, and the contact to silicon is made by self-aligned PtSi formed in the contact holes. Metal 2 is also a bilayer film of Ti:W and Al:Cu, 775 nm thick. The intermetal-dielectric layer is a 600-nm-thick PECVD oxide layer in which 1.1-μm vias are opened to allow contact between Metal 2 and Metal 1. No planarization of the intermetal dielectric was reported for this DLM process.

A second example, detailed by Bergeron et al. of IBM, uses a bilayer PECVD silicon-nitride/polyimide film as the intermetal dielectric. Smoothing of the underlying metal topography is achieved through use of the polyimide.[196] The Metal-1 pitch is 5 μm and the Metal-2 pitch is 7.0 μm. Metal 1 and Metal 2 are both Al:4%Cu films defined by lift-off, and Metal 2 is 2 μm thick. The bilayer intermetal-dielectric film is etched by means of two masking and etching steps. In the first, an oversized via opening in the polyimide is defined and etched through an isotropic dry-etch step. In the second, a smaller via in the nitride is defined and dry etched.

4.9.2 CMOS Double-Level-Metal Systems

The following DLM processes are listed according to the generation of CMOS technology for which they were reported to be designed, as specified by the minimum MOS device gate length.

4.9.2.1 Nonplanarized DLM (2.0-μm CMOS). An example of a two-level-metal CMOS interconnect structure that does not rely on any planarization methods was presented by Smith et al. of Intel (Fig. 4-57a).[197] The Metal-1 pitch is 5.0 μm, and the Metal-2 pitch is 6.0 μm. Metal 1 and Metal 2 are Al:Si films, and the intermetal dielectric is an oxynitride composite film. Metal 2 is used to strap the polysilicon lines in order to reduce the word line delay in SRAMs built with this technology. Via-hole sloping is achieved by implanting the top surface of the oxynitride. Note that contact between poly and the Si substrate is made with *butted contacts* (see chap. 3).

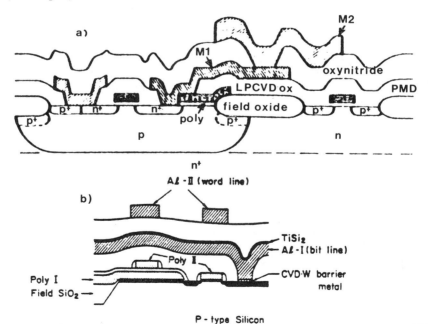

Fig. 4-57 (a) Example of a double-level-metal CMOS process with no planarization.[197] (© 1984 IEEE). (b) DLM process with a CVD W barrier metal and TiSi$_2$ hillock-suppression layer on top of the Metal-1 Al layer.[200] (© 1984 IEEE).

4.9.2.2. Nonplanarized DLM: CVD W Metal (2.0-μm NMOS).

A two-level-metal NMOS process that uses CVD W for both Metal 1 and Metal 2 was described by Mikkelson et al. of Hewlett-Packard.[198] The Metal-1 pitch is 3.0 μm; the Metal-2 pitch is 8 μm and the Metal-2 film thickness is 1.8 μm. The intermetal dielectric is CVD SiO$_2$. The tungsten is deposited by means of a blanket CVD process, and an underlying "glue layer" of CVD WSi$_2$ is used to provide good metal-layer adhesion to the SiO$_2$. No planarization was reported, possibly because of the good step coverage provided by the CVD W process.

4.9.2.3 Resist Etchback, Bias-Sputtered SiO$_2$, and SOG DLM for 1.5-μm CMOS.

Four examples of two-level-metal processes for 1.5-1.25-μm CMOS are the following:

1. Barton and Maze of Hewlett-Packard described a DLM process that uses Al metallization and a sacrificial etchback dielectric planarization process.[199] The Metal-1 pitch is 4.0 μm, and the Metal-2 pitch is 4.0 μm. A bilayer of silicon oxynitride and silicon nitride is partially planarized through a resist etchback process (see section 4.8). The vias are 1.6 x 1.6 μm and can be placed anywhere except over contact holes.

2. Another two-level-metal, double-polysilicon process (for 1-Mbit DRAMs) was presented by Shibata et al. of Toshiba.[200] A low-resistivity polycide structure is used for Poly 1 (for bit lines); polysilicon is used for Poly 2, but was strapped with Metal 2 to reduce the delay of the word lines. A 50-nm-thick selective W film is used to make contact to the n^+ source and drain regions (and to serve as a diffusion barrier layer). A resist-etchback technique provides planarization of the PECVD SiO$_2$ intermetal dielectric. A layer of TiSi$_2$ is deposited onto Metal 1 lines to prevent hillock formation during processing (Fig. 4-57b).

3. Monk et al. of Westinghouse outlined a DLM process that employs bias-sputtered SiO$_2$ as the intermetal dielectric.[201] The first-level metal is a trilayer film of Ti:W, Al, and Ti. The Ti top layer is used to prevent hillock formation. Contact to the silicon is made by self-aligned PtSi formed in the contact holes. The Metal-1 pitch is 3.0 μm, and the Metal-2 pitch is 4.0 μm. The intermetal sputter-deposited SiO$_2$ serves to smooth the steps of the dielectric layer by forming steps with 45° angles over the metal edges. A two-step etch process is used to form tapered vias. Metal 2 consists of a bilayer of Al covered with Ti. The Ti serves as an antireflection coating during lithography, as well as an anti-hillock-forming layer. An overetch of Metal 2 is used to clear stringers that remain in the most severe lithography.

4. The final CMOS DLM example uses a nonetchback sandwich SOG process as the planarization method. It also uses PtSi contacts and Ti:W/Al:Cu/Ti:W tri-layers for Metal 1 and Metal 2. The passivation film is a PECVD oxide/PECVD nitride bilayer. The intermetal dielectric consists of an underlayer of PECVD

SiO_2 (200-nm thick), a layer of carbon-containing SOG, and a 500-nm top layer of PECVD SiO_2.[202]

4.9.2.4 Nonsacrificial Layer Etchback DLM (1.0-μm CMOS).

A two-level-metal process that uses a nonsacrificial etchback step of a plasma-enhanced TEOS SiO_2 layer was outlined by Thoma et al.[172] and by Hills et al.,[203] all of of AT&T Bell Laboratories. It was designed for use in a twin-tub, 1.0-μm CMOS technology. The nonetchback planarization provides smoothing of the intermetal- dielectric layer, as was described in section 4.4.3. The Metal-1 pitch is 2.0 μm, and the Metal-2 pitch is 3.5 μm. Metal-1 thickness is 0.5 μm, and 1.5-μm-wide vias are used. (The dielectric etchback process was described in section 4.4.3.)

A more recent, and slightly more complex version of this process is presented in reference 235. Here, a resist-etchback step is used to planarize the PETEOS layer, and a proprietary technique is used to slope the sidewalls of the first Al layer. Vias are sloped with an isotropic/anisotropic etch, and the nominal via depth varies from 800 to 1400 nm. Adequate step coverage is achieved in even the deepest vias when a 400-nm-deep isotropic etch is used. This is an example of a relatively simple DLM process that can be extended to triple-level-metal structures.

4.9.2.5 Alternative CMOS DLM Process with Ti:W/Mo as Metal 1.

DLM processes for most nonsubmicron CMOS technologies use Al alloys for Metal-1 and Metal-2 layers. Although CVD W has been extensively investigated as an alternative metallization material for overcoming the problems of Al conductor layers (especially for Metal 1), another, less widely used approach employs a bilayer of Ti:W/Mo. A report of a complete DLM process that uses such a bilayer has recently been presented by Hawley et al. of Xicor.[204] The Mo offers excellent electromigration and hillock resistance and alleviates the problem of junction spiking.

In this process, Ti:W makes contact to the Si. In order to establish low contact resistance to PMOS source and drain regions, it's necessary to perform a 625°-700°C sinter step following Metal 1 deposition. An etchback sandwich SOG dielectric layer is used as the intermetal dielectric film, and Al:Cu is used as the Metal-2 film. The Ti:W/Mo layer is dry etched in an SF_6 + Ar mixture. An in situ sputter-etch is performed before the Metal-2 Al:Cu film is sputter deposited, which yields a low contact resistance (1.25 $\mu\Omega$-cm^2) between Metal 1 to Metal 2.

4.9.2.6 DLM Processes for Submicron CMOS.

The following is a survey of some DLM processes for submicron CMOS. These are more complex than the processes used for less advanced CMOS technologies.

- The first example of a DLM process that utilizes resist-etchback planarization for both the PMD and the intermetal-dielectric layers was described by Chapman et al. of Texas Instruments (Fig. 4-58).[205] This process was designed to be used in a 0.8-μm CMOS technology. The Metal-1 and Metal-2 pitches are both 2.6 μm, and the contact and via sizes are 0.8 μm. The PMD layer is bilayer of LPCVD TEOS and PECVD SiO_2.

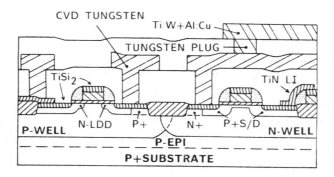

Fig. 4-58 Schematic cross section of a 0.8-μm CMOS DLM process.[205] (© 1987 IEEE).

The TEOS layer is deposited and then planarized by means of a resist-etchback technique prior to deposition of the phosphorus-doped PECVD SiO_2 layer. A blanket CVD W layer is deposited into the contact holes and also serves as the Metal-1 material. The CVD W makes contact to the self-aligned $TiSi_2$ contact structures. The intermetal dielectric is PECVD SiO_2, which is also planarized using a resist etchback technique. The vias are completely filled with selective CVD W.

The Metal-2 film consists of Ti:W covered with Al:Cu. The passivation layer is a plasma oxide/nitride bilayer film. Because no flow or reflow steps are used following $TiSi_2$ formation, low contact resistivity to the PMOS source/drain regions can be maintained by the $TiSi_2$.

• In the second process example, presented by Higelin et al. of Siemens,[206] planarization of both the PMD and the intermetal dielectric is performed. The PMD is LPCVD TEOS, which is planarized by using a resist-etchback process. A 200-nm-thick TEOS film is then deposited to ensure good isolation between poly and Metal 1. Blanket CVD W and etchback are used to fill the contact holes, with a Ti/TiN diffusion barrier under the W layer. Metal 1 consists of an Al:Si:Ti film. The intermetal-dielectric layer consists of a trilayer structure, with etched-back SOG or polyimide as the material in the middle. Blanket CVD W is used to fill the via, which is then etched back (using polyimide as the sacrificial layer). An Al:Cu:Si film is used as Metal 2.

• The third process was detailed by R. de Werdt et al.[207] and T. Doan et al,[208] both of Phillips Research Labs. It utilizes a single polysilicon layer with a $TiSi_2$ local interconnect structure to connect poly lines and diffused regions, selective CVD W as a contact-hole-filling material, and Al:2%Cu as both Metal 1 and Metal 2. Ti:W is used as a nucleation layer for the blanket CVD W and as an etchstop under the Al:Cu of Metals 1 and 2.

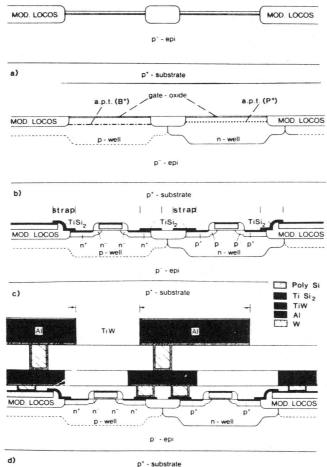

Fig. 4-59 Schematic cross section of submicron-CMOS DLM technology that uses: a fully planarized isolation structure; $TiSi_2$ over poly as a local interconnect; selective W as a contact hole plug; and Al:Cu for Metal 1 and Metal 2.[207]   (© 1987 IEEE).

In this process, the substrate is completely planarized at the conclusion of the isolation formation procedure (Fig. 4-59). The PMD layer is likewise completely planarized, by means of a conformally deposited LPCVD TEOS process and a resist-etchback technique. Following the PMD resist etchback, a thin layer of BPSG is deposited for gettering and dielectric-layer integrity. An etchback SOG sandwich dielectric approach is used to planarize the intermetal dielectric. The vias are etched to give sidewalls with slopes of 80°, which allows adequate filling by the sputter-deposited Al:Cu Metal-2 film. The Ti:W etchstop layer under Metal 2 allows a 70% overetch step to be used with only 0.1-μm CD line loss in

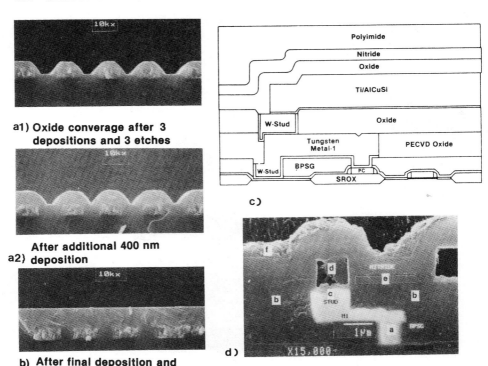

a1) **Oxide converage after 3 depositions and 3 etches**

a2) **After additional 400 nm deposition**

b) **After final deposition and planarization**

c)

d)

Fig. 4-60 Submicron-CMOS DLM process. (a) and (b) SEM photographs of CVD-oxide gap fill and planarization. (c) Cross section of the completed DLM structure. (d) SEM cross-sectional photograph of this structure.[128]   (© 1987 IEEE).

Metal 2. A bilayer passivation film is used that consists of a 0.4-μm-thick PECVD oxide and a 0.3-μm-thick PECVD nitride.

The polysilicon width and spacing of this process are 0.7 μm and 0.9 μm, respectively. The Metal-1 and Metal-2 pitches are both 2 μm, and the minimum size for contact holes and vias is 0.9 μm.

• The fourth example was presented by C. Kaanta et al. of IBM.[209] Blanket CVD W is used to form studs in the contact holes and vias and also to form the Metal-1 layer. Void formation is avoided through optimization of the deposition temperature, pressure, and relative gas flows (Fig. 4-60). TiN serves as the adhesion layer under the CVD W.[130]

A special dry-etch process of W was developed to prevent attack down the sidewalls of the contact holes and vias during W overetch. A Cl_2/O_2 gas mixture gives good selectivity to BPSG or other deposited oxides. A silicon nitride film is also deposited over the CVD W to protect the W during overetch, since the resist-masking layer is rapidly eroded by the dry-etch conditions needed to successfully etch the W.

A PECVD SiO_2 deposition/etchback process is used to fill the narrow spaces between adjacent metal lines (see section 4.4.7). As shown in Figs. 4-60a and b, a sequence of six thin CVD SiO_2 depositions – each followed by an etchback step – is needed to obtain filling of the narrow spaces. Another etchback technique (not specified in the paper) is used to completely planarize the surface. A final APCVD oxide is then deposited to provide a sufficiently thick dielectric film over the Metal-1 regions.

After the via holes have been opened and filled with CVD W (using an etchback step to planarize the W in the vias; Fig. 4-60c), the Ti/Al:Cu:Si Metal-2 film is deposited and patterned. A final passivation bilayer film of PECVD oxide and nitride is deposited and capped with a polyimide film.

The use of completely filled vertical contact holes and vias allows the use of unframed contact hole and via design rules, which significantly increases packing density.

• The fifth DLM process, proposed by H. Yamamoto et al. of Matsushita, is designed to be used in CMOS processes which require a 0.6-μm Metal-1 spacing and line width.[181] Al:Si Metal-1 and Metal-2 interconnects are clad by a 100-nm-thick layer of selective CVD W to improve electromigration and hillock resistance, as well as to increase the resistance of the lines to stress-induced voids. Selectively deposited W is also used to fill contact holes and vias (i.e., to completely fill the shallow ones and partially fill the deep ones). The intermetal dielectric is a photo-CVD and SOG multilayer film. The photo-CVD SiO_2 fills narrow spaces, and the SOG film fills the intermediate-dimension spaces.

4.9.3   Three-Level-Metal Systems

Most three-level-metal (TLM) systems have been designed for use in bipolar gate-array chips, probably for the same reasons that DLM processes were first developed in bipolar applications. Information on TLM processes for use with CMOS is also generally presented in less detail.

• An early TLM system for bipolar gate arrays was described in 1982 by Fried et al. from IBM (Fig. 4-61).[210] In this system, Metals 1, 2, and 3 were all Al:4%Cu films that were patterned by lift-off. The contact to silicon was made with PtSi and a Cr-Cr_xO_y diffusion barrier. The DM1 layer was bias-sputtered SiO_2, and DM2 and the passivation layers were PECVD SiO_2 that was etched back by bias sputtering.

• A second TLM system for a 10,000-gate ECL gate array was presented by Eggers et al. of Siemens in 1985.[211] This system utilizes a bilayer dielectric consisting of polyimide covered with PECVD nitride for DM1, DM2, and the passivation layer. The polyimide provides smoothing of the wafer topography. The nitride seals the polyimide, preventing it from absorbing moisture, and provides a surface to which Metals 2 and 3 can reliably adhere. The metallization for all three layers is an AlTiSi alloy, for good electromigration and hillock

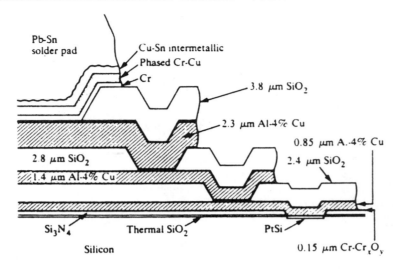

Fig. 4-61 Schematic of an early three-level-metal interconnect structure.[210] Copyright 1982 by International Business Machines Corporation. Reprinted with permission.

resistance. The pitches are 6 μm for Metal 1, 8 μm for Metal 2, and 10 μm for Metal 3. Reliability tests of circuits fabricated with this process and mounted in plastic packages showed excellent results.

• A third process was outlined in 1987 by Welch et al. of Texas Instruments.[212] Here, the premetallization dielectric topography is planarized by means of a resist-etchback technique, and a blanket CVD W process is then used to fill the contact holes. A planarizing resist is used to smooth the rough surface of the W film and the small dimples that remain over the contacts themselves. The etchback is stopped when 100-200 nm of W remain in the field regions. Ti:W is used as an adhesion layer for this W and for the W plugs in the contact holes. An Al alloy is used for the remainder of the Metal 1 layer. Blanket CVD W (including etchback) is also used to fill vias in DM1 and DM2.

• A fourth TLM process was presented by Moriya et al. of Toshiba in 1983.[213] In this structure, selective CVD W is used to fill the contact holes and make direct contact to Si, as well as to selectively fill vias in DM1 and DM2. A layer of $MoSi_2$ is deposited onto Metal 1 and Metal 2 to serve as a nucleating surface for the via-fill selective W process. Planarization of the CVD SiO_2 layers DM1 and DM2 is carried out by means of a resist-etchback technique.

• A fifth process uses CVD W for all metal layers, and polyimide capped with PECVD nitride as the intermetal dielectric layers.[214,216] Feasibility studies using this process produced working CMOS circuits.

• A sixth TLM process, described by Manos et al. of Motorola, again used a blanket-deposited and etched W layer as the first-level-metal layer.[237] A combination-sputter deposited/CVD TiN layer was used as a contact diffusion-barrier film. Since W is the first metal material, the vias between Metal 1 and Metal 2 can be reflowed. However, great care must be taken during this reflow step to prevent the formation of tungsten oxide. Good via filling and low-contact-resistance vias can be achieved using this process.

4.9.4   Four-Level Metal Systems

Two four-level-metal processes have been described as of this writing. The first, reported by Bartush of IBM,[216] was used to fabricate 10,000-gate bipolar gate-array chips. Instead of patterning and filling vias in the intermetal-dielectric layers, studs are formed using a *lift-off* technique. Studs may be stacked to achieve interconnection between conductors that are more than one level away from each other (Fig. 4-62). Sacrificial etchback of resist layers is reported to be the technique used to planarize the intermetal-dielectric films. The first film is apparently deposited by means of a process that can fill the narrow spaces between studs/Metal 1 lines and that can provide partial planarization (e.g., a bias-sputtered SiO_2 film, a PECVD TEOS film, or an ECR-deposited SiO_2 film). The partially planarized topography is then completely planarized through resist etchback. The intermetal-dielectric films are composite nitride/oxide layers. The nitride is used to ensure that all of the studs are exposed prior to deposition

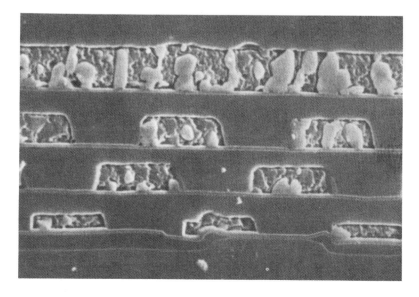

Fig. 4-62 SEM photograph of a cross section of a completed four-level-metal chip.[216]   (© 1987 IEEE).

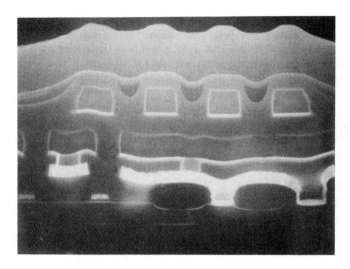

Fig. 4-63 SEM photograph of another completed four-level-metal interconnection structure.[236] (© 1988 IEEE).

of the next metal-level film, and it is left in place to become part of the bilayer film for subsequent intermetal dielectrics.

The second four-level-metal process was described by Nishida et al. of Hitachi (Fig. 4-63).[236] A minimum 0.6-μm feature size is employed. Sputtered W is used as the first level metal, and spin-on glass sandwich structures are used to planarize the intermetal-dielectric layers. Layered Al/TiN films were used for the other metal levels. A variety of circuits have been fabricated with this technology, including a 1-μm CMOS SRAM and a BiCMOS gate array with macro cells.

4.10 SUMMARY OF MULTILEVEL-INTERCONNECT-TECHNOLOGY REQUIREMENTS FOR VLSI

• *A planarized wafer topography at the end of the isolation-formation process* (see chap. 2 for discussion of various approaches that have been proposed to achieve this).

• *A planarized topography over the local-interconnect level* (e.g., over polysilicon or silicide). This can be accomplished through glass flow (probably by means of RTP, to maintain shallow junctions in devices) and etchback techniques.

• *Complete filling of contact holes and vias.* This will probably be achieved through the use of such processes as CVD of W or Al (blanket or selective), bias-sputtered planarized Al, laser-planarized Al, pillar formation, or selective electroless deposition.

• *Processes that result in complete global planarization of interlevel-dielectric surfaces.* Such processes are described in sections 4.4 and 4.6.

• *Unframed and vertically-etched vias.*

• *Interconnect materials with low resistivity, hillock resistance, high electromigration-resistance, and compatibility.*

• *An appropriate passivation layer.*

REFERENCES

1. R. W. Keyes, *Proceedings of the IEEE,* **69**, p. 267 (1981).

2. C. T. Sah, "Evolution of the MOS Transistor - From Conception to VLSI", *Proceedings of the IEEE,* October 1988, p. 1280.

3. W. Shockley, *Bell Systems Tech. J.,* **30**, NO. 10, p. 990, October 1951.

4. A. H. Dansky, *Electronics,* October 9, 1980, p. 46.

5. D. J. McGreivy in *"VLSI Technologies",* D. J. McGreivy and K. A. Pickar, eds., IEEE Computer Society Press, Los Angeles (1982), p. 185.

6. J. J. Lajza and J. L. Wendt, *Ext. Abs. Electrochem. Soc. Meeting,* Fall 1986, San Diego, CA, Abs. 356.

7. T. Saigo et al., *Abstracts of ISSCC,* 1983, p. 156.

8. K. Skidmore, *Semicond. Internatl.,* May 1988, p. 74.

9. K. Haberle et al., *Proceedings 5th Intl. IEEE VMIC Conf.,* Santa Clara, CA, p. 117, 1988.

10. T. Ohmi et al., *Proceedings 5th Intl. IEEE VMIC Conf.,* Santa Clara, CA, p. 135, 1988.

11. D. Pramanik, *Semiconductor Internatl.,* June 1988, p. 94.

12. R. A. Levy, P. K. Gallagher, and F. Schrey, *J. Electrochem. Soc.,* Feb. 1987, p. 430.

13. F.S. Becker and S. Rohl, *J. Electrochem. Soc.,* Nov. 1987, p. 2923.

14. P. B. Johnson and P. Sethna, *Semiconductor Internatl.,* October, 1987, p. 80.

15. D. Graveson and G. Grondin, "Quantitative Determination of Boron in BPSG", *Semiconductor Internatl.,* June, 1988, p. 140.

16. C. G. Magnella and T. Ingwersen, *Proceedings 5th Internatl., IEEE VMIC Conf.,* Santa Clara, CA, p. 366, 1988.

17. F. S. Becker et al., *J. Vac. Sci. and Technol.,* p. 732, Febuary 1987.

18. L. K. White et al., *Proceedings 5th Internat. IEEE VMIC Conf.,* Santa Clara, CA, p. 397, 1988.

19. B. Chin and E. P. van de Ven, *Solid State Technol.,* April, 1988, p. 119.

20. F. K. Moghadam and K. Suh, *Ext. Abs. of Electrochem. Soc. Meeting,* Fall 1988, Chicago, Il, Abs. No. 359, p. 369.

21. M. Kawai et al., *5th Internatl.VMIC Conf., Santa Clara, CA,* p. 419, 1987.

22. T. Fujita et al., *Proceedings 4th Internat. IEEE VMIC Conf.,* Santa Clara, CA, p. 285, 1987

23. C. Chiang et al., *Ext. Abs. of Electrochem. Soc. Meeting,* Fall 1988, Chicago, Il, Abs. No. 290, p. 418.

24. R. L. Jackson et al., *Solid-State Technol.,* April 1987, p. 107

25. J. Saraie et al., *J. Electrochem. Soc.,* Nov. 1987, p. 2805.

26. L. Rothman, *J. Electrochem. Soc.,* **127**, p. 2216, p. (1980).

27. G. W. Ray and P. W. Marcoux, *Proceedings 2nd Internatl IEEE VMIC Conf.,* Santa Clara, CA, p. 52, 1985.

28. P. B. Johnson and P. Sethna, *Semicond. Internatl.,* October, 1987, p. 80.

29. D. Bui et al., *Proceedings 4th Annual IEEE VMIC Conf.,* Santa Clara, CA, p. 385, 1987.

30. J. Mercier, *Tech. Proceedings Semicon/East* 1986, p. 22.

31. D. W. Freeman et al., *Ext. Abs. of Electrochem. Soc. Meeting,* Fall 1988, Chicago, Il, Abs. No. 240, p. 337.

32. R. J. Kopp, *Semicond. Internatl.,* January 1989, p. 54.

33. A. L. Wu, *Proceedings 2nd Intl IEEE VMIC Conf.,* Santa Clara, CA, p. 145, 1985.

34. T. Abraham, *Proceedings 4th Intl IEEE VMIC Conf.,* Santa Clara, CA, p. 115, 1987.

35. S. Mayumi et al., *Proceedings 4th Intl IEEE VMIC Conf.,* Santa Clara, CA, p. 79, 1987.

36. S. J. H. Brader and S. C. Quinlan, *Proceedings 3rd Intl IEEE VMIC Conf.,* Santa Clara, CA, p. 58, 1986.

37. M. J. Thoma et al., *Proceedings 4th Internatl. IEEE VMIC Conf.,* Santa Clara, CA, p. 20, 1987.

38. J. S. Mercier et al., *J. Electrochem. Soc.,* May, 1985, p. 1219.

39. M. Khan et al., *Proceedings 2nd Annual IEEE VMIC Conf.,* Santa Clara, CA, p. 32, 1985.

40. W. W. Yao et al., *Proceedings 2nd Annual IEEE VMIC Conf.,* Santa Clara, CA, p. 38, 1985.

41. S. J. H. Brader et al., *Proceedings 3nd Annual IEEE VMIC Conf.,* Santa Clara, CA, p. 93, 1986.

42. S. D. Senturia, "Fundamental Polyimide Property Measurements", presented at Polyimides for IC Fabrication, a seminar sponsored by the Semiconductor Materials Group of DuPont Co, Jan. 14, 1987.

43. J. J. Lajza and J. L. Wendt, *Ext. Abs. of Electrochem. Soc. Meeting,* Fall 1986, San Diego, CA, Abs. No. 356.

44. H. Eggers, H. Fritzsche, and A. Glasl, *Proceedings 2nd Intl IEEE VMIC Conf.,* Santa Clara, CA, p. 163, 1985

45. R. W. Wu, R. L. Alley and D. D. Kessler, *Proceedings 4th Intl IEEE VMIC Conf.,* Santa Clara, CA, p. 292, 1987.

46. C. Mitchell, *Proceedings of the 1st Intl IEEE VMIC Conf.,* Santa Clara, CA, p. 130, 1984.

47. Y. Misawa et al., *IEEE Trans. Electron Dev.,* March, 1987, p. 621.

48. F. Faupel et al., *Ext. Abs. of Electrochem. Soc. Meeting,* Fall 1988, Chicago, IL, Abs. No. 231, p. 323.

49. C. Y. Ting et al., *J. Vac. Sci and Technol.,* **15**, No. 3, p. 1105, 1978.

50. H. Okabayashi, *1984 Symposium on VLSI Technology,* p. 20, Japan Soc. App. Physics and IEEE Electron Devices.

51. L. J. Fried et al., *IBM J. Res. and Develop.* Vol. 26, No. 3, May 1982, p. 362.

52. Y. Hazuki and T. Moriya, *IEEE Trans. Electron Dev.,* March, 1987, p. 628.

53. S. Fujii et al., *IEEE Trans. Electron Dev.* Nov. 1988, p. 1829.

54. T. Abraham, *Proceedings 4th Intl. IEEE VMIC Conf.,* Santa Clara, CA, p. 115, 1987.

55. G. C. Smith and A. J. Purdes, *J. Electrochem. Soc.,* Nov. 1985, p. 2721.

56. B. Lee, A. Pierfederici, and E. C. Douglas, *Proceedings 4th Intl. IEEE VMIC Conf.,* Santa Clara, CA, p. 85, 1987.

57. Y. Homma and S. Tunekawa, *J. Electrochem. Soc.,* Oct. 1988, p. 2557.

58. E. J. McInerney and S. C. Avazino, *IEEE Trans. Electron Dev.,* March, 1987, p. 615.

59. L. Koyama and M. Thomas, *Proceedings 2nd Intl. IEEE VMIC Conf.,* Santa Clara, CA, p. 45, 1985.

60. C. Kaanta et al, *Proceedings 5th Intl IEEE VMIC Conf.,* Santa Clara, CA, p. 21, 1988.

61. W. Geiger and A. Sharma, *Proceedings 3rd Intl IEEE VMIC Conf.,* Santa Clara, CA, p. 128, 1986.

62. A. C. Adams and D. D. Capio, *J. Electrochem. Soc.,* **128,** 423, (1981).

63. A. N. Saxena and D. Pramanik, *Solid-State Technol.,* Oct. 1986, p. 95.

64. C. Jang et al., *Proceedings 4th Intl IEEE VMIC Conf.,* Santa Clara, CA, p. 357, 1987.

65. G. Ray and P. Marcoux, *Proceedings 2nd Intl IEEE VMIC Conf.,* Santa Clara, CA, p. 52, 1985.

66. G. E. Gimpelson and C. L. Russo, *Proceedings 1st Intl IEEE VMIC Conf.,* New Orleans, 1984.

67. A. Schlitz and M. Pons, *J. Electrochem. Soc.,* Jan 1986, p. 178

68. A. Shepela and B. Soller, *J. Electrochem. Soc.,* March, 1987, p. 714.

69. P. L. Pai, *Proceedings 5th Intl. IEEE VMIC Conf.,* Santa Clara, CA, p. 108, 1988.

70. S. Fujii et al., *IEEE Trans. Electron Dev.,* Nov. 1988, p. 1829.

71. M. D. Bui et al., *Proceedings 4th Intl IEEE VMIC Conf.,* Santa Clara, CA, p. 385, 1987.

72. B. Vasquez and R. Goodner, *Proceedings 4th Intl IEEE VMIC Conf.,* Santa Clara, CA, p. 394, 1987.

73. L. de Bruin and J. M. F. G. van Laarhovenin, *Proceedings 5th Intl IEEE VMIC Conf.,* Santa Clara, CA, p. 404, 1988.

74. D. Barton and C. Maize, *Semiconductor Internat.,* Jan. 1986, p. 98.

75. T. Sagio et al., *IEEE J. Solid-State Circuits,* SC-**18,** p. 578, (1983).

76. G. W. Hills and H. P. W. Hey, *Ext. Abs. of Electrochem. Soc. Meeting,* Spring 1988, Atlanta, GA, Abs. No. 124, p. 186.

77. D. L W. Yen and G. K. Gopal, *Proceedings 5th Intl. IEEE VMIC Conf.,* Santa Clara, CA, p. 85, 1988.

78. S. N. Chen et al, *Proceedings 5th Intl IEEE VMIC Conf.,* Santa Clara, CA, p. 306, 1988.

79. S. Morimoto and S. Q. Grant, *Proceedings 5th Intl IEEE VMIC Conf.,* Santa Clara, CA, p. 411, 1988.

80. J. Chu et al, *Proceedings 5th Intl. IEEE VMIC Conf.,* Santa Clara, CA, p. 474, 1988.

81. P. E. Riley and A. Shelley, *J. Electrochem. Soc.,* May 1988, p. 1207.

82. P.- L. Pai et al., *J. Electrochem. Soc.,* Nov. 1987, p. 2829.

83. C. H. Ting et al., *Proceedings 4th Intl. IEEE VMIC Conf.,* Santa Clara, CA, p. 61, 1987.

84. M. Nakamura and R. Kanzawa, *J. Electrochem. Soc.,* **133,** p. 1167, (1986).

85. H. M. Naguib et al., *Proceedings 4th Intl. IEEE VMIC Conf.,* Santa Clara, CA, p. 93, 1987.

86. J. K. Chu et al., *Proceedings 3rd Intl. IEEE VMIC Conf.,* Santa Clara, CA, p. 474, 1986.

87. K. L. White et al., *Proceedings 5th Intl. IEEE VMIC Conf.*, Santa Clara, CA, p. 397, 1988.

88. P. Elkins et al., *Proceedings 3rd Intl. IEEE VMIC Conf.*, Santa Clara, CA, p. 101, 1986.

89. L. B. Vines and S. K. Gupta, *Proceedings 3rd Intl. IEEE VMIC Conf.*, Santa Clara, CA, p. 507, 1986.

90. A. Rey et al., *Proceedings 3rd Intl. IEEE VMIC Conf.*, Santa Clara, CA, p. 491, 1986.

91. D. L. W. Yen and G. K. Rao, *Proceedings 5th Intl. IEEE VMIC Conf.*, Santa Clara, CA, p. 85, 1988.

92. S. N. Chen et al., *Proceedings 5th Intl. IEEE VMIC Conf.*, Santa Clara, CA, p. 306, 1988.

93. C. H. Ting et al., *Ext. Abs. of Electrochem. Soc. Meeting*, Fall 1988, Chicago, IL, Abs. No. 257, p. 366.

94. M. Kawai et al., *Proceedings 5th Intl. IEEE VMIC Conf.*, Santa Clara, CA, p. 419, 1988.

95. C. Chiang et al., *Proceedings 4th Intl. IEEE VMIC Conf.*, Santa Clara, CA, p. 404, 1987.

96. G. Hausamann and P. Mokrisch, *Proceedings 5th Intl. IEEE VMIC Conf.*, Santa Clara, CA, p. 293, 1988.

97. V. Grewal, A. Gschwandtner, and G. Higelin, *Proceedings 3rd Intl. IEEE VMIC Conf.*, Santa Clara, CA, p. 107, 1986.

98. R. M. Brewer and R. A. Gasser, *Proceedings 4th Intl. IEEE VMIC Conf.*, Santa Clara, CA, p. 404, 1987.

99. C.- K. Hu et al, *Proceedings 3rd Intl. IEEE VMIC Conf.*, Santa Clara, CA, p. 491, 1986.

100. S. Morimoto and S. Q. Grant, *Proceedings 5th Intl. IEEE VMIC Conf.*, Santa Clara, CA, p. 411, 1988.

101. H. Kojima et al., *Proceedings 5th Intl. IEEE VMIC Conf.*, Santa Clara, CA, p. 411, 1988.

102. K. Machida and H. Oikawa, *J. Vac. Sci. and Technol.*, B Vol. 4, p. 818 (1986).

103. S. V. Nguyen and K. Albaugh, *Ext. Abs. of Electrochem. Soc. Meeting*, Fall, 1988, Abs. No. 302, p. 437.

104. C. Chiang and D. B. Fraser, *Ext. Abs. of Electrochem. Soc. Meeting*, Spring 1989, Abs. No. 179.

105. D. R. Denison, C. Chiang, and D. B. Fraser, *Ext. Abs. of Electrochem. Soc. Meeting*, Spring 1989, Abs. No. 180.

106. S. Nakamura and S. Nakayama, *Ext. Abs. of Electrochem. Soc. Meeting*, Fall, 1988, Abs. No. 303, p. 439.

107. S. Gupta et al., *Semiconductor Internatl.*, Sept, 1987, p. 126.

108. D. R. Denison, and L. D. Hartsough, *Microelectronics Mfg. and Test*, Nov.1987, p. 6.

109. F. Moghadam, Private Communication.

110. S. Mehta et al., *Proceedings 3rd Intl. IEEE VMIC Conf.*, Santa Clara, CA, p. 418, 1986.

111. M. E. Burba et al, *J. Electrochem. Soc.*, October 1986, p. 2113.

112. T. H. Daubenspeck, P. Sukanek, and E. J. White, *Ext. Abs. of Electrochem. Soc. Meeting*, Spring 1988, p. 173.

113. B. Jucha and C. Davis, *Proceedings 5th Intl. IEEE VMIC Conf.*, Santa Clara, CA, p. 165, 1988.

114. M. J. Kim et al., *IEEE Trans. Electron Dev.*, July, 1985, p.1328.

115. J. Hems and A. McGeown, *Technical Proceedings, Semicon Europa*, 1987, *p. 158.*

116. L. T. Lamont, *Technical Proceedings, Semicon Europa*, 1987, p. 148.

117. R.C. Ross and J. L. Vossen, *Appl. Phys. Letts.*, A3, (1984), p. 239.

118. D. W. Skelly and A. Gruenke, *J. Vac. Sci. Technol.*, A4, (1986), p. 457.

119. R. J. Saia et al., *J. Electrochem. Soc.*, April, 1988, p. 936.

120. R. Blewer and V. A. Wells, *Tech. Dig. IEDM*, 1984, p. 852.

121. E. K. Broadbent and C. L. Ramiller, *J. Electrochem. Soc.*, 131, 1427, (1984).

122. Ph. Lami and Y. Pauleau, *J. Electrochem. Soc.*, April 1988 p. 980.

123. R. A. Levy and M. L. Green, *J. Electrochem. Soc.*, Feb. 1987, p. 37C.

124. T. Ohba, S.-I. Inoue, and M. Maeda, *Tech. Dig. IEDM*, 1987, p. 213.

125. H. Kotani et al, *Tech. Dig. IEDM*, 1987, p. 217.

126. G. C. Smith and B. Jucha, *Proceedings 3rd Intl. IEEE VMIC Conf.*, Santa Clara, CA, p. 403, 1986.

127. N. S. Tsai et al., *Tech. Dig. IEDM*, 1988, p. 462.

128. C. Kaanta et al., *Tech. Dig. IEDM*, 1987, p. 209.

129. G. Higelin et al., *Proceedings 3rd Intl. IEEE VMIC Conf.*, Santa Clara, CA, p. 443, 1986.

130. P.-I. Lee, J. Cronin, and C. Kaanta, *Ext. Abs. of Electrochem. Soc. Meeting*, Fall, 1988 p. 367, Abs. No. 258.

131. E. Bertagnolli et al., *Proceedings 5th Intl. IEEE VMIC Conf.*, Santa Clara, CA, p. 324, 1988.

132. F. Y. Robb and K. W. Ginn, *Ext. Abs. of Electrochem. Soc. Meeting*, Fall, 1988, p. 367.

133. J. J. Lee and D. C. Hartman, *Proceedings 4th Intl. IEEE VMIC Conf.*, Santa Clara, CA, p. 193, 1987.

134. H. Itoh, T. Moriya, and M. Kashiwagi, *Solid-State Technol.*, November, 1986, p. 83.

135. K. Y. Ahn, *Proceedings 5th Intl. IEEE VMIC Conf.*, Santa Clara, CA, p. 125, 1988.

136. J. R. Creighton, *J. Electrochem. Soc.*, Jan. 1989, p. 271.

137. D. M. Brown et al., *Tech. Dig. IEDM*, 1986, p. 66.

138. T. Moriya et al., *Tech. Dig. IEDM*, 1983, p. 550.

139. R. H. Wilson et al., *J. Electrochem. Soc.*, July, 1987, p. 1867.

140. F. K. Moghadam and K. Suh, *Proceedings 5th International VMIC Conference*, Santa Clara, CA, 1988, p. 345.

141. N. F. Raley and D. L. Losee, *J. Electrochem. Soc.*, Oct. 1988, p. 2640.

142. M. J. Cooke et al., *Solid State Technol.*, Dec. 1982, p. 62.

143. T. Amazawa and H. Nakamura, *Ext Abs. of 18th Conf. Solid State Devices and Materials*, p. 755, 1986.

144. T. Amazawa, H. Nakamura, and Y. Arita, *Tech. Dig. IEDM*, 1988, p. 442.

145. R. A. Levy, M. L. Green, and P. K. Gallagher, *J. Electrochem. Soc.*, 131, p. 2175 (1984).

146. J. Osborne and T. J. Magee, "Advances in Planarization of Aluminum Films Using Excimer Laser Technology", available from XMR Inc., Santa Clara, CA.

147. D. B. Tuckerman and A. H. Weisberg, *IEEE Electron Dev. Letts.*, January, 1986, p. 1.

148. R. Mukai et al., *Proceedings 5th International VMIC Conference,* Santa Clara, CA, 1988, p. 101.

149. C. H. Ting and M. Paunovic, *J. Electrochem. Soc.*, Feb 1989, p. 456.

150. P. -L. Pai et al., *Proceedings 5th International VMIC Conference,* Santa Clara, CA, 1988, p. 331.

151. C. S. Wei et al., *Ext. Abs. of Electrochem. Soc. Meeting,* Spring, 1988, Abs. No. 156, p. 239.

152. C. S. Wei et al., *Tech. Dig. IEDM,* 1988, p. 446.

153. T. A. Bartush, *Proceedings 4th International VMIC Conference,* Santa Clara, CA, 1987, p. 41.

154. U. S. Patent # 4, 004, 004; Franco et al.

155. M. T. Welch and C. Garcia, *Proceedings 3rd International VMIC Conference,* Santa Clara, CA, 1986, p. 450.

156. P. E. Riley and E. D. Castel, in "Multilevel Metallization Interconnect and Contact Technologies", p. 194, L. B. Rothman and T. Herndon, Eds. The Electrochemical Soc. Softbound Series, PV-87-4, Pennington, NJ, (1987).

157. E. K. Broadbent et al., *IEEE Trans. Electron Dev.,* July, 1988, p. 952.

158. D. C. Thomas et al., *Tech. Dig. IEDM,* 1988, p. 466.

159. P.- L. Pai, and W. G. Oldham, *IEEE Trans. Semiconductor Mfg.,* Feb. 1988, p. 3.

160. P.-L. Pai, C. H. Ting, and W. G. Oldham, *Proceedings 5th International VMIC Conference,* Santa Clara, CA, 1988, p. 108.

161. A. R. Nyaiesh and L. Holland, "Effects of Gas Composition on the Discharge & Deposition Characteristics when Magnetron Sputtering Aluminum", *Vacuum,* **31,** No. 8/9, 1981, p. 371.

162. J. Klema, R. Pyle, and E. Domangue, *Proceedings IEEE Internatl. Reliability Phys. Symp.,* 1984, p. 1.

163. G. C. Smith and R. B. Jucha, *Proceedings 3rd International VMIC Conference,* Santa Clara, CA, 1986, p. 403.

164. S. L. L. Ng and P. Merchant, *Proceedings 4th International VMIC Conference,* Santa Clara, CA, 1987, p. 186.

165. J. Yue, W. P. Funsten, and R. V. Taylor, *Proceedings IEEE Internatl. Reliability Phys. Symp.,* 1985, p. 126.

166. P. H. Singer, "Update on Thin Metal Line Cracking", *Semiconductor International,* Feb. 1989, p. 48.

167. K. Hinode, I. Asano, and Y. Homma, *Proceedings 5th International VMIC Conference,* Santa Clara, CA, 1988, p. 429.

168. H. Katto and S. Shimizu, *Ext. Abs. Electrochem. Soc. Meeting,* Fall, 1988, Abs. No. 310, p. 450.

169. S. K. Groothuis and W. H. Schroen, *Proceedings IEEE Internatl. Reliability Phys. Symp.,* 1987, p. 1.

170. A. K. Sinha and T. T. Sheng, *Thin Solid Films,* **48,** (1972), p. 117.

171. P. J. Chaudary, *Appl. Phys.*, **45**, 4339, (1974).

172. M. J. Thoma et al., *Proceedings 4th Internatl. IEEE VMIC Conf.*, Santa Clara, CA, p. 20, 1987.

173. S. Mak et al., *Proceedings 3rd International VMIC Conference*, Santa Clara, CA, 1986, p. 65.

174. B. W. Shen et al., *Proceedings 3rd International VMIC Conference*, Santa Clara, CA, 1986, p. 191.

175. A. K. Brown et al., *Proceedings 4th International VMIC Conference*, Santa Clara, CA, 1987, p. 426.

176. N. P. Armstrong, "Improved Electromigration Performance in Al:4%Cu Using a Range of Refractory Caps", *Ext. Abs. Electrochem. Soc. Conf.*, Fall, 1988, p. 448.

177. D. S. Gardener et al., *Proceedings 2nd International VMIC Conference*, Santa Clara, CA, 1985, p. 102.

178. K. Hinode et al., *IEEE Trans. Electron Dev.*, March, 1987, p. 700.

179. H. P. Hey et al., *Tech. Dig. IEDM*, 1986, p. 50.

180. J. L. Yeh et al., *Proceedings 4th International VMIC Conference*, Santa Clara, CA, 1987, p. 132.

181. H. Yamamoto et al., *Tech. Dig. IEDM*, 1987, p. 205.

182. E. P. van den Ven, R. S. Martin, and M. J. Berman, *Proceedings 4th International VMIC Conference*, Santa Clara, CA, 1987, p. 434.

183. S. Fujita et al., *Proceedings 4th International VMIC Conference*, Santa Clara, CA, 1987, p. 285.

184. H. Harada et al., *Tech. Dig. IEDM*, 1986, p.46.

185. I. Yamada and T. Takagi, *Proceedings 4th International VMIC Conference*, Santa Clara, CA, 1987, p. 415.

186. H. L. Peek and R. A. M. Wolters, *Proceedings 3rd Intl. IEEE VMIC Conf.*, Santa Clara, CA, p. 165, 1986.

187. T. Kikkawa et al., *Appl. Phys. Lett.*, **50**, 1987, p. 1527.

188. R. S. Martin and E. P. van den Ven, "RF Bias to Control Stress and Hydrogen in Plasma Nitride Deposition", *Proceedings 5th Intrnatl. IEEE VMIC Conf.*, Santa Clara, CA, p. 93, 1986.

189. Sz. Fujita et al., *IEDM Tech. Dig.*, 1984, p. 630.

190. D. L. Flamm et al., *Solid State Technol.*, p. 43, March, 1987.

191. Sz. Fujita and A. Sasaki, *J. Electrochem. Soc.*, Oct. 1988, p. 2566.

192. V. Dunton, *Proceedings 2nd Intrnatl. IEEE VMIC Conf.*, Santa Clara, CA, p. 259, 1985.

193. W. A. P. Claassen et al., *J. Electrochem. Soc.*, July, 1986, p. 1458.

194. W. R. Knolle et al., *J. Electrochem. Soc.*, May. 1988, p. 1211.

195. P. B. Ghate et al., *Tech. Dig. IEDM*, 1984, p. 126.

196. D. L. Bergeron et al., *IEEE Internatl. Reliability Physics Symp.*, 1984.

197. R. J. Smith et al., *Tech, Dig. IEDM*, 1984, p. 56.

198. J. Mikkelson et al., *IEEE Internatl. Solid-State Ckts. Conf.*, 1981, p. 106.

199. D. Barton and C. Maze, *Semicond. Internatl.*, January, 1985, p. 98.

200. T. Shibata et al., *Tech. Dig. IEDM*, 1984, p. 75.

201. J. L. Monk et al., *Proceedings 5th Internatl. IEEE VMIC Conf.*, Santa Clara, CA, p. 59, 1988.

202. J. Nulty, G. Spadini, and D. Pramanik, *Proceedings 5th Internatl. IEEE VMIC Conf.*, Santa Clara, CA, p. 453, 1988.

203. G.W. Hills et al., *Proceedings 5th Internatl. IEEE VMIC Conf.*, Santa Clara, CA, p. 35, 1988.

204. F. Hawley et al., *Proceedings 5th Internatl. IEEE VMIC Conf.*, Santa Clara, CA, p. 142, 1988.

205. R. A. Chapman et al., *Tech. Dig. IEDM*, 1987, p. 362.

206. G. Higelin et al., *Proceedings 5th Intrnatl. IEEE VMIC Conf.*, Santa Clara, CA, p. 29, 1988.

207. R. de Werdt et al., *Tech. Dig. IEDM*, 1987, p. 532

208. T. Doan et al., *Proceedings 5th Internatl. IEEE VMIC Conf.*, Santa Clara, CA, p. 13, 1988.

209. C. Kaanta et al., *Proceedings 5th Internatl. IEEE VMIC Conf.*, Santa Clara, CA, p. 20, 1988.

210. L. J. Fried et al., *J. IBM Research and Devel.* May, 1982, p. 362.

211. H. Eggers, H. Fritzsche, and A. Glasl, *Proceedings 2nd Internatl. IEEE VMIC Conf.*, Santa Clara, CA, p. 163, 1985.

212. M. T. Welch, R. E. McMann, and M. L. Torreno, *Proceedings 4th Internatl. IEEE VMIC Conf.*, Santa Clara, CA, p. 51, 1987.

213. T. Moriya et al., *Tech. Dig. IEDM*, 1983, p. 550.

214. D. Crook et al., *Proceedings 4th Internatl. IEEE VMIC Conf.*, Santa Clara, CA, p. 33, 1987.

215. C. C. Beatty and D. D. Kessler, *Proceedings 4th Internatl. IEEE VMIC Conf.*, Santa Clara, CA, p. 163, 1987.

216. T. Bartush, *Proceedings 4th Interntl. IEEE VMIC Conf.*, Santa Clara, CA, p. 41, 1987.

217. D. J. Desbiens, "Use of Bias Sputtered Quartz as a Planarizing Layer in Multilevel Metallization", *Ext. Abs. Electrochem. Soc. Conf.*, Fall, 1984, p. 605.

218. W. M. Siu, *Ext. Abs. Electrochem. Soc. Conf.*, Spring, 1989, p. 179.

219. S. Samukawa et al., *Ext. Abs. Electrochem. Soc. Conf.*, Spring, 1989, p. 244.

220. D. B. Tuckerman and A. H. Weisberg, *Solid State Technol.*, April, 1986, p. 129.

221. H. M. Liaw and C. Seelbach, *Ext. Abs. Electrochem. Soc. Conf.*, Spring, 1989, p. 341, Abs. No. 230.

222. C.-S. Wei, V. Murali, and D. B. Fraser, *Ext. Abs. Electrochem. Soc. Conf.*, Spring, 1989, p. 273, Abs. No. 188.

223. H. Kotani, *Extended Abs. of 20th Conf. on Solid State Devices and Materials*, p. 565, Tokyo, 1988.

224. P. E. Riley et al., *Ext. Abs. Electrochem. Soc. Conf.*, Spring, 1989, p. 272, Abs. No. 187.

225. R. J. Hopkins, T.A. Baldwin, and S. K. Gupta, *Ext. Abs. Electrochem. Soc. Conf.*, Spring, 1989, p. 257, Abs. No. 177.

226. S. Ito et al., *Ext. Abs. Electrochem. Soc. Conf.*, Spring, 1989, p. 258, Abs. No. 178.

227. D. A. Webb, A. P. Lane, and T. E. Tang, *Ext. Abs. Electrochem. Soc. Conf.,* Spring, 1989, p. 262, Abs. No. 181.

228. T. H. Daubenspeck et al., *Ext. Abs. Electrochem. Soc. Conf.,* Spring, 1989, p. 262, Abs. No. 181.

229. P. Y. Chang, *Semiconductor International,* Nov. 1984.

230. R. L. Kramer, *Ext. Abs. Electrochem. Soc. Conf.,* Spring 1989, p. 182, Abs. No. 131.

231. K. M. Kearney, "ECR Finds Applications in CVD", *Semiconductor Internatl.,* March 1989, p. 67.

232. J. M. F. G van Laarhoven et al., *Proceedings 6th Internatl. IEEE VMIC Conf.,* Santa Clara, CA, p. 129, 1989.

233. P. -L. Pai and C. H. Ting, *Proceedings 6th Internatl. IEEE VMIC Conf.,* Santa Clara, CA, p. 258, 1989.

234. H. W. Piekaar, L. F. Tz. Kwakman, and E. H. A. Granneman, *Proceedings 6th Internatl. IEEE VMIC Conf.,* Santa Clara, CA, p. 122, 1989.

235. C. A. Fieber et al, *Proceedings 6th Internatl. IEEE VMIC Conf.,* Santa Clara, CA, p. 122, 1989.

236. T. Nishida et al., *Proceedings 6th Internatl. IEEE VMIC Conf.,* Santa Clara, CA, p. 19, 1989.

237. P. Manos et al, *Proceedings 6th Internatl. IEEE VMIC Conf.,* Santa Clara, CA, p. 40, 1989.

238. Y. Koubuchi et al., *Proceedings 6th Internatl. IEEE VMIC Conf.,* Santa Clara, CA, p. 419, 1989.

239. J. J. Hsieh et al., *Proceedings 6th Internatl. IEEE VMIC Conf.,* Santa Clara, CA, p. 411, 1989.

240. B. Woratschek et al., *Proceedings 6th Internatl. IEEE VMIC Conf.,* Santa Clara, CA, p. 309, 1989.

241. R. Liu, K. P. Cheung, and W. Y. -C. Lai, *Proceedings 6th Internatl. IEEE VMIC Conf.,* Santa Clara, CA, p. 329, 1989.

242. E. K. Broadbent et al., *Proceedings 6th Internatl. IEEE VMIC Conf.,* Santa Clara, CA, p. 336, 1989.

243. D. Moy et al., *Proceedings 6th Internatl. IEEE VMIC Conf.,* Santa Clara, CA, p. 26, 1989.

244. S. Mehta and G. Sharma, *Proceedings 6th Internatl. IEEE VMIC Conf.,* Santa Clara, CA, p. 80, 1989.

245. L. Forester, A. L. Butler, and G. Schets, *Proceedings 6th Internatl. IEEE VMIC Conf.,* Santa Clara, CA, p. 72, 1989.

246. H. W. M. Chung, S. K. Gupta, and T. A. Baldwin, *Proceedings 6th Internatl. IEEE VMIC Conf.,* Santa Clara, CA, p. 373, 1989.

247. R. V. Joshi et al., *Proceedings 6th Internatl. IEEE VMIC Conf.,* Santa Clara, CA, p. 113, 1989.

248. C. H. Ting and P. -L. Pai, *Ext. Abs. Electrochem. Soc. Conf.,* Fall, 1989, p. 354, Abs. No. 249.

249. P. -L. Pai and C. H. Ting, *Ext. Abs. Electrochem. Soc. Conf.,* Fall, 1988, p. 364, Abs. No. 256.

250. B. Gorowitz, R. J. Saia, and E. W. Balch, "Methods of Metal Patterning and Etching", in *VLSI Electronics,* Vol. 15.

251. R. E. Oakley et al., *Proceedings 1st Internatl. IEEE VMIC Conf.,* New Orleans, LA, p. 23, 1984.

252. "Applied Enters Tungsten Deposition," *Electronic News,* September 11, 1989, p. 46.

253. H. G. Tompkins and C. Tracy, "Desorption from Spin-On Glass," *J. Electrochem. Society,* August, 1989, p. 2328.

254. T. B. Gorczyca et al., *J. Electrochemical Soc.,* September 1989, p. 2765.

255. L. D. Molnar, *Semiconductor International,* August 1989, p. 92.

256. "SOG Improves EPROM Reliability," *Semiconductor International,* July 1989, p. 16.

257. D. Alugbin et al., "Low Contact Reistance of Selective CVD W over Ti:W-capped Al," *Proceedings 6th Conf. on CVD W, Cu, and Other Advanced Metals for ULSI and VLSI,* Materials Research Society, 1990.

258. A. K. Sinha, *J. Vac. Sci. Technol.,* **19,** 778, (1981).

259. D. M. Brown, M. Ghezzo, and J. M. Pimbley, *Proceedings of the IEEE,* December 1986, p. 1678.

260. N. Owada et al., *Proceedings 2nd Internatl. IEEE VMIC Conf.,* santa Clara, CA, p. 173, 1985.

261. T. Fukuda et al., *Tech. Dig. IEDM,* !989, p. 665.

262. T. Hariu et al., *Proceedings IEEE IRPS,* 1989, p. 210.

263. D. Pramanik and S. Chen, *Tech. Dig. IEDM,* 1989, p. 673.

264. J. B. Price, Spectrum CVD, private communication.

265. B. Davari et al., *Tech. Dig. IEDM,* 1989, p. 61.

266. A. C. Bonora, "Silicon Wafer Process Technology: Slicing, Etching, Polishing," *Semiconductor Silicon 1977,* Electrochemical Society, Pennington, N. J., p. 154.

1

CHAPTER 5

MOS DEVICES AND
NMOS PROCESS INTEGRATION

This chapter deals with NMOS process integration. To provide background information for this topic, a review of the physics of long-channel MOS devices (i.e., whose channel lengths are greater than 2 μm) and the basics of MOS circuit design is presented. Also included is a discussion of the relationship between desired device performance and process technology. (More rigorous treatments of MOS device physics are found in references 1 - 5.)

The history of NMOS processing is also presented, emphasizing how the obstacles to fabricating reliable NMOS ICs were overcome. A detailed example of the fabrication sequence of a typical NMOS inverter circuit culminates this discussion. The chapter concludes with a description of short-channel and hot-carrier effects in MOS devices, together with processing techniques developed to combat the problems they cause.

5.1 MOS DEVICE PHYSICS

5.1.1 The Structure and Device Fundamentals of MOS Transistors

A perspective view of an *n-channel MOSFET* is shown in Fig. 5-1a, while additional details of its structure and its circuit symbol are given in Figs. 5-1b and 5-1c, respectively. The device has a *gate terminal* (to which the input signal is normally applied), as well as *source* and *drain* terminals (across which the output voltage is developed, and through which the output current flows, i.e., the drain-source current, I_D). The gate terminal is connected to the gate electrode (a conductor), while the remaining terminals are connected to the heavily doped source and drain regions in the semiconductor substrate.

A *channel* region in the semiconductor under the gate electrode separates the source and drain. The channel (of length L and width W) is lightly doped with a dopant type opposite to that of the source and drain. The semiconductor is also physically separated

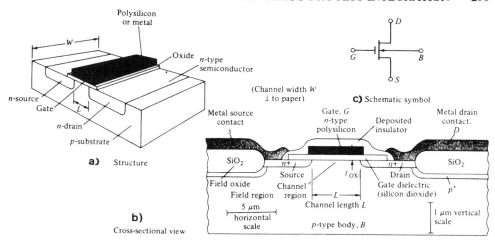

Fig. 5-1 (a) Structure of an MOS device. (b) Cross sectional view. (c) Schematic symbol.[7] From D. A. Hodges and H. G. Jackson, *Analysis and Design of Digital Integrated Circuits,* Copyright, 1983 McGraw-Hill Book Co. Reprinted with permission.

from the gate electrode by an insulating layer (typically, SiO_2), so that no current flows between the gate electrode and the semiconductor.

As shown in Fig. 5-1b, MOS transistors are symmetrical devices, which means that the source and drain are interchangable. In NMOS circuits, however, *the more positive of these two electrodes is normally defined to be the drain*, and thus the input signal is defined as the voltage between the gate and source terminals ($V_{in} = V_G$).

In simplest terms, the operation of an MOS transistor involves the application of an input voltage to the gate electrode, which sets up a transverse electric field in the channel region of the device (5-2a). By varying this transverse electric field, it is possible to modulate the longitudinal conductance of the channel region. Since an electric field controls current flow, such devices are termed *field-effect transistors (FETs)*. They are further described as *metal-oxide-semiconductor (MOS) FETs* because of the thin SiO_2 layer that separates the gate and substrate.

The substrate (or body) of the MOS transistor is a silicon wafer; this wafer also provides mechanical support for the finished circuit. In addition, an external electrical connection (or *terminal)* can be made to the body, making the MOS transistor a four-terminal device. (In later sections we will see how the transistor behavior is impacted if a bias is applied between the source and body terminals of the device.)

The top surface of the body consists of *active* or *transistor* regions as well as *passive* or *(field)* regions. The active regions are those in which transistor action occurs; i.e., the channel and the heavily doped source and drain regions. Conduction between separate active regions must be prevented (see chap. 2). A thick oxide layer (0.5-1.0 μm) is normally grown over the field regions as one of the measures to achieve this goal.

If no gate bias is applied, the electrical path between source and drain consists of two back-to-back two *pn* junctions in series. If a drain bias is applied such that the source-

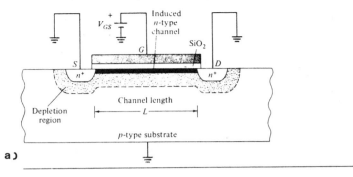

a)

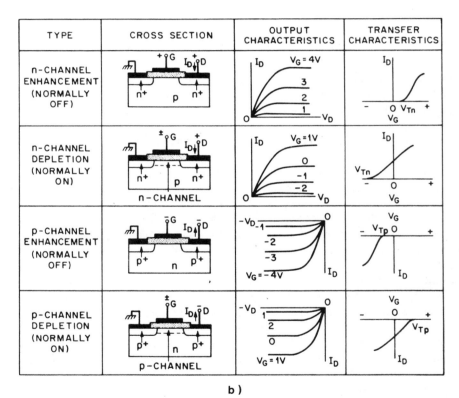

b)

Fig. 5-2 (a) Idealized NMOS cross section with positive V_{GS} applied showing depletion regions and the induced channel.[7] From D. A. Hodges and H. G. Jackson, *Analysis and Design of Digital Integrated Circuits*, Copyright, 1983 McGraw-Hill Book Co. Reprinted with permission. (b) Cross sections and output and transfer characteristics of four types of MOSFETs. [7] From S. M. Sze, *Semiconductor Devices - Physics and Technology*, Copyright, 1985, John Wiley & Sons. Reprinted with permission.

body and drain-body junctions remain reverse-biased, I_D will consist of only the reverse-bias diode leakage current and hence will be considered negligibly small.

When positive bias is applied to an NMOS transistor gate, electrons will be attracted to the channel region and holes will be repelled. (Holes are the majority carriers in the channel of the *p*-type body when no gate bias exists.) Once enough electrons have been drawn into the channel by the positive gate voltage to exceed the hole concentration, the region behaves like a *n*-type semiconductor. Under these circumstances, an *n*-type channel connects the source and drain regions (Fig. 5-2a). Current will flow if a voltage, V_{DS}, is applied between the source and the drain terminals. The voltage-induced *n*-type channel does not form unless the voltage applied to the gate exceeds the *threshold voltage*, V_T.

MOS devices such as those just described, in which no conducting channel exists when $V_G = 0$, are referred to as *enhancement-mode* (or *normally OFF*) transistors (see Fig. 5-2b). With NMOS enhancement-mode transistors, a positive gate voltage, V_G, greater than V_T must be applied to create the channel (or to turn them *ON*), while to turn *ON* PMOS enhancement-mode devices, a negative gate voltage (whose magnitude is $>V_T$) must be applied. Note that in NMOS transistors a positive voltage must also be applied to the drain to keep the drain-substrate reverse-biased, while in PMOS devices this voltage must be negative.

On the other hand, it is also possible to build MOS devices in which a conducting channel region exists when $V_G = 0$ V (see Fig. 5-2b), and such MOS devices are described as being *normally ON*. Since a bias voltage to the gate electrode is needed to deplete the channel region of majority carriers (that is, the channel is eradicated as long as the bias is applied), such devices are also commonly called *depletion-mode* devices. NMOS depletion-mode devices require a negative gate voltage to be turned *OFF*, while corresponding PMOS devices require positive gate voltages.

5.1.2   The Threshold Voltage of the MOS Transistor

If the source and body of an MOS transistor are both tied to ground ($V_{SB} = 0$), the threshold voltage, V_T, of the transistor can be found from the following equation (note that $V_T = V_{T0}$ when $V_{SB} = 0$):

$$V_{T0} = \varphi_{ms} - 2\varphi_f - Q_{tot}/C_{ox} - Q_{BO}/C_{ox} \qquad (5-1)$$

where φ_{ms} is the workfunction difference (in V) between the gate material and the bulk silicon in the channel, φ_f is the equilibrium electrostatic potential in a semiconductor (in V), Q_{BO} is the charge stored per unit area (C/cm^2) in the depletion region (when the voltage between source and body is zero), C_{ox} is the gate-oxide capacitance per unit area (F/cm^2), and Q_{tot} is the total positive oxide charge per unit area present at the interface between the oxide and the bulk silicon (see section 5.3.3).

Expressions have been established for these various quantities in terms of doping concentrations in the material, physical constants, device structure dimensions, and temperature. They are:

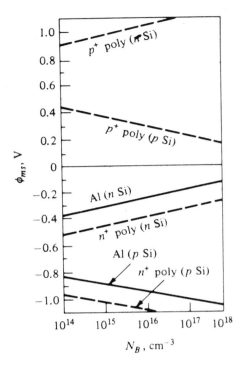

Fig. 5-3 Work-function difference ϕ_{ms} versus doping for degenerate polysilicon and Al electrodes.[80] Reprinted with permission of Solid-State Electronics.

$$\varphi_f = kT/q \; \ln (n_i /N_A) \quad \textit{(p-type semiconductor)} \qquad (5 - 2a)$$

$$\varphi_f = kT/q \; \ln (N_D /n_i) \quad \textit{(n-type semiconductor)} \qquad (5 - 2b)$$

where N_A is the acceptor concentration in a p-type semiconductor (cm^{-3}), N_D is the donor concentration in an n-type semiconductor, and n_i is the intrinsic carrier concentration in the semiconductor. Note that n_i is a strong function of temperature. For silicon at 300 K, however, $n_i = 1.45 \times 10^{10}$ cm^{-3}. Therefore, at 300 K:

$$\varphi_{ms} \text{ (metal gate, with Al as gate electrode)} = \varphi_{f(sub)} - \varphi_{f(Al)} = \varphi_{f(sub)} - (+0.6 \text{ V}) \text{ (5 - 3a)}$$

$$\varphi_{ms} \text{ (Si gate)} = \varphi_{f(sub)} - \varphi_{f(gate)} \qquad (5 - 3b)$$

Figure 5-3 shows the value of φ_{ms} for various substrate-doping values if an aluminum gate electrode is used. Note also that in silicon-gate NMOS, the doping of the polysilicon gate material is usually n-type, with $N_D \cong 10^{20}/\text{cm}^3$, so that $\varphi_{f(gate)}$ in this case is +0.59 V. Figure 5-4 shows the value of the threshold voltage of n^+ silicon gate MOS transistors (both PMOS and NMOS) versus substrate doping concentration, assuming either a 65-nm, a 25-nm, or a 15-nm thick gate oxide, and $Q_{tot} = 0$.

Q_{BO} is found from:

$$Q_{BO} = -\sqrt{2 q N_A \varepsilon_{si} \left| -2\varphi_f \right|} \qquad \text{NMOS} \qquad (5-4a)$$

$$Q_{BO} = +\sqrt{2 q N_D \varepsilon_{si} \left| -2\varphi_f \right|} \qquad \text{PMOS} \qquad (5-4b)$$

and C_{ox} is calculated from:

$$C_{ox} = \varepsilon_{ox} / t_{ox} = 3.9 \, \varepsilon_o / t_{ox} \qquad (5-5)$$

where $\varepsilon_{si} = 11.8 \, \varepsilon_o$, ε_o is the permittivity of vacuum, and ε_{ox} and t_{ox} are the permittivity and thickness of the gate oxide.

It is easy to become confused about the signs of the various terms in the threshold voltage equation. Equation 5-1 gives correct results for NMOS and PMOS if the signs shown in Table 5.1 are used for each of the parameters. When calculating the various terms, the table can be used to insure that the value of each parameter is entered into the equation with the correct sign (e.g., φ_{ms} will have a negative value for n^+ Si gate NMOS devices).

Table 5.1 Signs in the Threshold-Voltage Equation[7]

Parameter	NMOS	PMOS
Substrate	p-type	n-type
φ_{ms}		
Metal gate	-	-
n^+ Si gate	-	-
p^+ Si gate	+	+
φ_f	-	+
Q_{BO}	-	-
Q_{tot}	+	+
γ	+	+
C_{ox}	+	+
Source-body voltage, V_{SB}	+	+

EXAMPLE 5-1: Find the threshold voltage of an NMOS silicon-gate transistor that has substrate doping $N_A = 10^{15}/cm^2$, gate doping $N_D = 10^{20}/cm^3$, gate-oxide thickness $t_{ox} = 100$ nm, and $Q_{tot} = q \, (1 \times 10^{11}/cm^2)$. Note that the values for t_{ox} and Q_{tot} are typical of the values that existed in early NMOS transistors.

SOLUTION:

$\varphi_{f(sub)} = kT/q \ln [n_i/N_A] = - 0.26 \ln [10^{15}/1.4\times10^{10}] = - 0.29$ V

$\varphi_{ms} = \varphi_{f(sub)} - \varphi_{f(gate)} = - 0.29 - kT/q \ln [10^{20}/1.4\times10^{10}] = - 0.88$ V

$\varepsilon_{ox} = 3.9\varepsilon_0 = 3.5\times10^{-13}$ F/cm; $C_{ox} = \varepsilon_{ox}/1\times10^{-5} = 35\times10^{-9}$ F/cm^2

$Q_{BO} = - [2\times 1.6\times10^{-19} \times 10^{15} \times 1.04\times10^{-12} \times |- 0.58 \; |]^{1/2} = - 1.4\times10^{-8}$ C/cm^2

$Q_{BO}/C_{ox} = - 1.4\times 10^{-8}/35\times10^{-9} = - 0.4$ V

$Q_{tot}/C_{ox} = 1.6\times10^{-19} \times 1\times10^{11}/35\times10^{-9} = 0.46$ V

$V_{TO} = - 0.88 - (- 0.58) - (- 0.4) - (0.46) = - 0.36$ V

Note that a negative value of V_{TO} is yielded, implying that this transistor would be *ON* at $V_G = 0$ V (would therefore behave as a *depletion-mode* device). The parameters used in this calculation of V_T are typical of the NMOS device parameters encountered in the early days of MOS ICs. Hence, at that time it was not possible to easily or reliably fabricate enhancement-mode NMOS transistors. We shall later show how this problem was overcome so that enhancement-mode NMOS devices could be manufactured.

5.1.3   Impact of Source-Body Bias on V_T (Body Effect)

As mentioned earlier, the MOS transistor is a four-terminal device, insofar as a contact can also be made to the body region. A bias, V_{SB}, can also be applied between the source and body (e.g., with the source being tied to ground, as shown in Fig. 5-5) and such a bias will have an impact on V_T. If $V_{SB} = 0$, inversion will, of course, occur when the voltage drop across the semiconductor equals $2\varphi_f$. If $V_{BS} < 0$ V, the semiconductor still attempts to invert when the voltage drop across it reaches $2\varphi_f$. However, any inversion-layer carriers that do appear at the semiconductor surface migrate laterally into the source because this region is at a lower potential. Thus, the surface potential must be lowered to $2\varphi_f - V_{SB}$ in order for inversion to occur, implying that the threshold voltage required to achieve inversion must be increased.

Hence, for either enhancement-mode or depletion-mode MOSFETs, V_T becomes *more positive* for n-channel transistors and *more negative* for p-channel transistors as V_{SB} is increased. A simple, quantitative way to view this effect is to assume that biasing changes the inversion point in the semiconductor from $2\varphi_f$ to $2\varphi_f - V_{SB}$. The threshold-voltage equation (Eq. 5-1), which was applicable when $V_{SB} = 0$, is in turn modified to

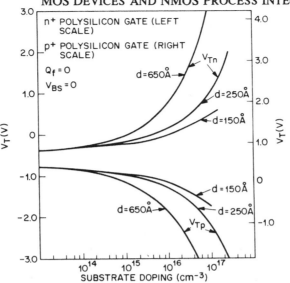

Fig. 5-4 Calculated threshold voltages of n-channel (V_{Tn}) and p-channel (V_{Tp}) transistors as a function of their substrate's doping, assuming n^+ polysilicon gate (left scale) and p^+ polysilicon (right scale). Curves for gate oxide thicknesses d of 150 Å, 250 Å, and 650 Å are shown. From S. M. Sze, Ed., *VLSI Technology*, 2nd. Ed., Chap. 11, "VLSI Process Integration." Copyright, 1988 Bell Telephone Laboratories. Reprinted with permission of McGraw-Hill.

$$V_T = V_{T0} + \gamma \left(\sqrt{\left| -2\,\varphi_f + V_{SB} \right|} - \sqrt{2 \left| \varphi_f \right|} \right) \qquad (5-6)$$

where the parameter γ (referred to as the *body-effect coefficient* or *body factor*, with units of $V^{1/2}$) is given by

$$\gamma = \frac{\sqrt{2\,q\,\varepsilon_{si}\,N_A}}{C_{ox}} \qquad \text{NMOS} \qquad (5-7a)$$

and

$$\gamma = \frac{\sqrt{2\,q\,\varepsilon_{si}\,N_D}}{C_{ox}} \qquad \text{PMOS} \qquad (5-7b)$$

5.1.4 Current-Voltage Characteristics of MOS Transistors

We will now present the equations that describe the large-signal current-voltage (I-V) characteristics of *long-channel* MOS transistors, assuming an NMOS device with its source grounded and with bias voltages V_{GS}, V_{DS}, and V_{BS} applied as shown in Fig.

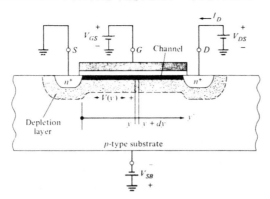

Fig. 5-5 NMOS device with bias voltages V_{GS}, V_{DS}, and V_{BS} applied.[7] From D. A. Hodges and H. G. Jackson, *Analysis and Design of Digital Integrated Circuits*, Copyright, 1983 McGraw-Hill Book Co. Reprinted with permission.

5-5. (Any modifications that occur in the I-V characteristics due to short-channel effects will be described section 5.5.)

In the simplest MOS model, if V_G is smaller than V_T, no channel exists, and no current is assumed to flow between the source and drain. If V_G is greater than V_T, a conducting channel is present, and V_{DS} causes a drift current (I_D) to flow from drain to source. For small values of V_{DS}, the drain current I_D is linearly related to V_{DS}. In this so-called *linear region* of operation, the equation that describes I_D is

$$I_D = k/2 \left[2(V_G - V_T) V_{DS} - V_{DS}^2 \right] \qquad \begin{array}{l} V_G \geq V_T \\ V_{DS} \leq (V_G - V_T) \end{array} \qquad (5\text{-}8)$$

where k (the so-called *device transconductance parameter*) is defined as

$$k = \mu_n \, C_{ox} \, (W/L) \qquad\qquad (5\text{-}9)$$

and μ_n is the surface mobility of electrons in the channel.

As the value of V_{DS} is increased, the induced conducting-channel charge decreases near the drain. When V_{DS} equals or exceeds V_G - V_T, the channel is said to be pinched off. Increases above this critical voltage produce little change in I_D, and Eq. 5-8 no longer applies. The value of I_D in this region is then given by the following

$$I_D = k/2 \, (V_G - V_T)^2 \qquad V_G \geq V_T; \;\; V_{DS} > (V_G - V_T) \qquad (5\text{-}10)$$

This is the so-called *saturation region* of operation.

A plot of I_D versus V_{DS} (with V_G as a parameter) for a long-channel NMOS transistor as described by Eqs. 5-8 and 5-10, is shown in Figure 5-6. If the value of V_G is smaller than V_T, the device is said to be in *cutoff*. In the model given here, I_D is assumed to be zero in cutoff (i.e., no *subthreshold current,* I_{Dst}, flows). We shall see in a later section that this assumption is not completely correct (even in long-channel devices), and that in short-channel devices such currents can cause severe problems.

5.1.5 The Capacitances of MOS Transistors

It can be shown that the switching speed of MOS digital circuits is not limited by the channel transit time (i.e., the time required for a charge to be transported charge across the channel), but by the time required to charge and discharge the capacitances that exist between device electrodes and between the interconnecting lines and the substrate. Figure 5-7 shows the significant capacitances between the electrodes of an MOS transistor. The capacitance from gate to other electrodes [C_G] is, to a first approximation, the sum of C_{GB}, C_{GS}, and G_{GD}. Furthermore, since C_{GS} and C_{GD} are small in silicon-gate technology (as the gate and source/drain regions are self-aligned), we can treat C_G as a constant, essentially determined by

$$C_G = W L C_{ox} = W L \varepsilon_{ox} / t_{ox} \qquad (5\text{-}11)$$

The capacitance per unit area of the source/body and drain/body junctions (C_{SB} and C_{DB}, respectively) are calculated using the parallel-plate capacitance formula with a plate spacing of W.[1] The larger the doping of the substrate, the larger the value of these capacitances. As an example, in the case of zero bias and a doping concentration of $10^{16}/cm^3$, the junction capacitance is approximately 10 nF/cm^2.

5.2 MAXIMIZING DEVICE PERFORMANCE THROUGH DEVICE DESIGN AND PROCESSING TECHNOLOGY

Having identified the parameters that determine V_T and the I-V characteristics of MOSFETs, we next discuss how to link device design and fabrication procedures in

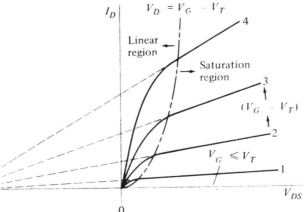

Fig. 5-6 NMOS device with I_D - V_{DS} characteristics.

order to achieve optimum device and circuit performance.

The desired device properties of MOSFETs include the following: high-output current drive (large value of I_D), high and stable transconductance (g_m), predictable and stable threshold voltage (V_T), fast switching speed, very small subthreshold current (I_{Dst}), high gate-oxide breakdown voltage, high drain/body breakdown voltage, low source/drain-to-body capacitances, high field-region threshold voltage and punchthrough voltage values, and high reliability of device operation. Note that the models used to select the process conditions that provide such optimum device behaviors are Eqs. 5-1, 5-8, and 5-10. In practice, these equations accurately describe the dc circuit behavior of *long-channel MOS devices*. (In section 5.5 we'll see how this behavior is modified in short-channel devices, as well as what device design and processing constraints result.)

5.2.1 Output Current (I_D) and Transconductance (g_m)

Equations 5-8 and 5-10 predict how I_D in MOSFETs can be impacted by various device parameters. Since increasing the *gate width* linearly increases I_D, one option when large drive-currents are needed, is to increase the dimension of the gate width. However, when minimum-sized devices must be used (e.g., for maximum packing density), this option cannot be implemented, and the other parameters that can influence I_D in Eqs. 5-8 and 5-10 must be considered.

From the dependence of I_D on $k' = \mu C_{ox}$, both the mobility of the carriers in the channel and the gate oxide capacitance should be as high as possible. Since electron mobility is greater than hole mobility, circuits using NMOS devices will exhibit higher performance than those built with PMOS devices. In fact, *NMOS transistors of the same width as PMOS transistors will indeed provide roughly 2.5x the current drive.* In addition, the mobility of carriers decreases as the doping concentration of the channel increases. Hence, lightly doped channel regions are also favored.

Because the value of C_{ox} is inversely proportional to t_{ox}, as thin a gate oxide as possible (commensurate with oxide breakdown and reliability considerations) is normally used. I_D is also inversely proportional to channel length, and minimum channel lengths are therefore desirable. On the other hand, if the channel lengths become too short, they adversely impact device operation in other ways, as described in section 5.5.

Although the equations indicate that ($V_G - V_T$) should be as large as possible, V_G is usually fixed by system specifications and material limitations, and cannot be changed by the device or circuit designer. Similarly, V_T is selected primarily by other circuit considerations (such as adequate noise margin in digital circuits).

The transconductance in saturation, $g_{msat} = dI_D/dV_G$, can be expressed simply by way of the following equation

$$g_{msat} = 2 \mu_n C_{ox} (W/L) [V_G - V_T] \qquad (5\text{-}12)$$

Hence, we can maximize g_{msat} by varying the process and device parameters in the same manner as discussed for I_D.

5.2.2 Controlling the Threshold Voltage through Process and Circuit Design Techniques

In many MOS IC applications, it is critical to be able to establish and maintain a uniform and stable value of V_T (i.e., the value of V_T should not vary with time or with device-operating conditions). An example of the importance of being able to control this parameter involves semiconductor memory devices. In these circuits, charge flows from the memory cells to the sense amplifier. The sense amplifier is a delicately balanced flip-flop whose voltage-sensing capability is directly related to the threshold-voltage variation between the transistors (see chap. 8). Hence, such circuits would not function reliably without the presence of a highly uniform and stable threshold voltage in the circuit devices.

The factors that impact threshold voltage (when $V_{SB} = 0$) are given in Eq. 5-1. Examining each term of this equation will enable us to determine which device parameters can be adjusted to provide practical control of V_T.

The φ_{ms} term depends on the work-function difference between the gate, $\varphi_{f(gate)}$, and the semiconductor, $\varphi_{f(sub)}$. For metal and heavily doped silicon gates, $\varphi_{f(gate)}$ is constant while the parameter $\varphi_{f(sub)}$ depends on the substrate doping, but only in a logarithmic manner. Hence, each factor-of-10 increase in substrate doping will change the φ_{ms} term by only 2.3 kT, or ~0.06 V (kT = 0.026 V at 300 K). Thus, changes in the substrate-doping concentration produce changes in V_T through the φ_{ms} term which are too small to provide the required degree of threshold-voltage control. The next term ($2\varphi_f$) also changes only slightly as a result of changes in the substrate doping concentration (for the same reason given for the φ_{ms} term); thus, the $2\varphi_f$ term is also not of much use in controlling V_T.

Since every attempt is made to keep Q_{tot} as low as possible through various processing procedures, the Q_{tot}/C_{ox} term is very small in modern MOS devices. Hence, it must also be ruled out as a candidate for controlling V_T.

While is true that C_{ox} can be varied (primarily by changing the gate oxide thickness), this parameter is not a practical vehicle for controlling V_T in active devices, since the gate oxide is normally made as thin as possible to achieve maximum I_D. In the field regions, however, large V_T values are needed to prevent inversion under the field oxide. A thick field oxide makes C_{ox} small, allowing V_T to be increased Thus, the C_{ox} is one of the parameters normally used to control V_T in the field regions of the circuits.

This leaves the Q_{BO} term as the remaining candidate for controlling V_T in active devices. Equation 5-4 indicates that the doping concentration can indeed provide a large change in V_T. The signs of Eq. 5-1 indicate that increasing the substrate doping (i.e., N_A in n-channel devices, and N_D in p-channel devices) will increase V_T. Thus, to increase V_T in Example 5-1 in this manner, it would be necessary to increase the substrate doping concentration above $N_A = 10^{15}$ cm^{-3}. Figure 5-7 shows how V_T can be controlled by changing the substrate doping concentration for various gate-oxide thicknesses (assuming that Q_{tot} is kept small enough that it is not a significant contribution to V_T).

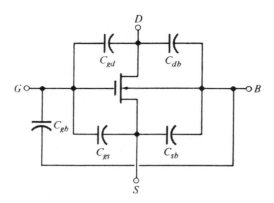

Fig. 5-7 MOS transistor capacitances. $C_G = C_{GS} + C_{GB} + C_{GD}$.[7] From D. A. Hodges and H. G. Jackson, *Analysis and Design of Digital Integrated Circuits*, Copyright, 1983 McGraw-Hill Book Co. Reprinted with permission.

Significant increases in substrate-doping concentration give rise to lower junction-breakdown voltages, larger junction capacitances, and lower carrier mobilities, making such substrate doping concentration increases undesirable. Yet, prior to the development of ion implantation in the early 1970s, adjustment of substrate doping was the only practical *processing approach* for significantly controlling V_T in active devices.

A *circuit approach* known based on applying a bias between the source and body (and hence known as *body biasing)* can also control V_T. When body biasing is used V_T is given by Eq. 5-6. Figure 5-8 gives an example of how the application of V_{SB} can change the V_{TO} value in an NMOS device.[14] The intercept of the channel conductance at the V_G axis corresponds to V_{TO}, which in this example is ~0.8 V As V_{SB} is

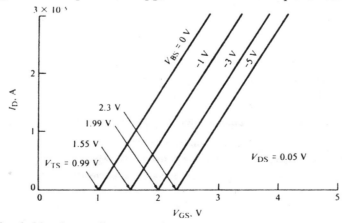

Fig. 5-8 Threshold voltage adjustment using substrate bias.

increased from -1 V to -16 V, V_T increases up to +3.25 V. However, body biasing is an added complication, and it is now avoided as a technique for adjusting V_T whenever possible in favor of the newer *ion implantation V_T-adjust method*.

We nevertheless include a discussion of body biasing because it pre-dated ion implantation as a method of controlling V_T. C. T. Sah points out that the availability of the technique allowed IBM to implement MOS memory devices instead of core memory for the first time in the IBM-370/158 mainframe computer in 1973.[60] The access time of NMOS RAMs at that time was competitive with magnetic core memory (~1 μs), whereas PMOS RAMs were slower. Body biasing allowed the higher-performance NMOS devices to function as enhancement-mode devices, even though their threshold voltages without body biasing would have caused them to behave like depletion-mode devices.

Another important reason for describing body biasing is that the body effect impacts devices in MOS IC logic circuits even when no deliberate attempt is made to apply an external bias. For example, when a logic gate containing a transistor as the load device has a logic 1 output (see for example Fig. 5-13), the voltage at the source of this device is different from that of the body. Hence, the device is subject to a non-zero V_{SB}, and V_T no longer equals V_{TO} value. The degree to which the change occurs depends on the *body factor*, γ. Since the smallest change in V_T is generally desired, small values of γ are preferred. Equation 5-7 indicates that γ depends on substrate doping, and that lower doping concentrations provide smaller values of γ, providing further impetus for lightly doped substrates in MOS ICs.

The value of γ also decreases with channel length below 1 μm. For example, γ in NMOS devices decreases by about 50% as L_{eff} is decreased from 1 μm to 0.6 μm, while for PMOS the decrease is about 30%.[8]

5.2.3 Subthreshold Currents (I_{Dst} when $V_G < |V_T|$)

The basic MOS device model neglects all free charges in the channel until the magnitude of the gate voltage exceeds V_T. This is a valid approximation for most subthreshold bias conditions because the free-charge densities in the channel change exponentially with the channel voltage. When this approximation is used, it implies that *no current flows between the drain and source* if $V_G < |V_T|$. As V_G approaches the value of V_T, (corresponding to a condition of *weak inversion*), however, the magnitude of I_D is not well defined by this simple approach. In fact, it is observed that $I_D \neq 0$ if V_G is close to (but still less than) the value of V_T. The current which is observed in such cases is therefore referred to as *subthreshold current,* I_{Dst}.

The values of I_{Dst} can be well predicted in long-channel MOSFETs by modifying the basic MOS model to take into account the fact that the minority carrier concentration at the Si surface is greater than the value at equilibrium, but still less than the bulk doping concentration.[19] The results of this modified model indicate that the magnitude of I_{Dst} is essentially independent of V_{DS} but is exponentially proportional to V_G.

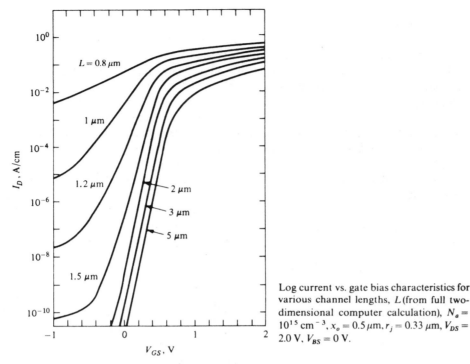

Log current vs. gate bias characteristics for various channel lengths, L (from full two-dimensional computer calculation), $N_a = 10^{15} \, cm^{-3}$, $x_o = 0.5 \, \mu m$, $r_j = 0.33 \, \mu m$, $V_{DS} = 2.0 \, V$, $V_{BS} = 0 \, V$.

Fig. 5-9 Example of subthreshold MOSFET I_D - V_{GS} curves for various channel lengths.[19] (© 1979 IEEE).

Since I_{Dst} can be accurately modeled in long-channel MOSFETs (Fig. 5-9), circuit designers can readily calculate the gate bias required to ensure a given allowable subthreshold leakage current. *Typically, to ensure that I_{Dst} will be negligibly small, the bias applied to the gate should be 0.5 V below V_T.*

Since I_{Dst} is exponentially proportional to V_G, if log I_{Dst} is plotted versus V_G the result on a semilogarithmic plot will be a straight line for values of V_G below V_T. The slope of the I_{Dst} versus V_G line, when plotted in this manner, is characterized by the *subthreshold swing*, S.S., where

$$S.S. \ = \ \Delta V_G \, / \, \Delta \log I_{Dst} \qquad\qquad (5\text{-}13)$$

Hence, S.S. is the change in V_G that produces a decade increase in I_{Dst}. A small value of S.S. is desirable, since it indicates maximum control of the gate over the channel current. Typical values of subthreshold swing in long-channel MOSFETs are around 90 mV/decade. Such values can be achieved by building devices that have a low value of substrate doping concentration (which gives a larger depletion width) and a thin gate oxide. (A more detailed treatment of the derivation of the I_{Dst} dependence on V_G can be found in references 1 and 3.)

Note that when the channel length gets small, the values of I_{Dst} are larger than than those predicted by the modified long-channel MOSFET model. This is due to so-called *short-channel effects* (discussed in section 5.5). However, measurements of the subthreshold swing can be used to detect the onset of these short-channel effects (*punchthrough* and *drain-induced barrier lowering*). Since the measurable I_{Dst} is the sum of both the normal subthreshold and the short-channel subthreshold current components, an increase in the value of S.S. will signal the onset of these effects.

5.2.4 Switching Speed

The switching speed of the logic gates in an MOS IC is limited only by the time required to charge and discharge the capacitances between device electrodes and between the interconnecting lines and ground (or other lines). At the circuit level the propagation delay is frequently limited by the interconnection-line capacitances. At the device level, however, the gate delay is determined primarily by the channel transconductance, the MOS gate capacitance (C_G) and the other two MOS parasitic capacitances, C_{DB} and C_{SB} (as defined in the previous section). If these capacitance values can be reduced, the device switching speed will be increased.

The gate capacitance is decreased by decreasing the gate area (although decreasing the gate oxide thickness increases its value). The dominant parasitic capacitance on the device level, however, is that due to C_{SB} and C_{DB} (i.e., junction capacitances). An analytical study has shown that these junction capacitances account for up to 50% of the total capacitance in logic gates.[9] Therefore, reductions in these capacitances should produce corresponding decreases in the gate delay. In general, *the lower the doping concentration in the body, the lower the junction capacitances.*

5.2.5 Junction Breakdown Voltage (Drain-to-Substrate)

The source and drain regions are very heavily doped to minimize their resistivities. Thus, the breakdown voltage of the drain-to-substrate junction will be determined by the lighter doping concentration of the body. As seen in Fig. 5-10a (which shows breakdown voltage of a one-sided *pn* junction as a function of the lighter doping concentration), the breakdown voltage decreases as the doping increases. Thus, *lightly doped substrates also yield high junction breakdown voltages.*

Junction curvature enhances the electric field in the curved part of the depletion region (Fig. 5-10b), and this effect reduces the breakdown voltage below that predicted by one-dimensional junction theory. A rectangular source or drain region (formed either by diffusion or ion implantation) has regions with both cylindrical and spherical curvature.[10,11] Figure 5-10a also shows the effect of junction curvature on the breakdown voltage of a one-sided step junction in silicon. It can be seen that as the junction depths get shallower (making the radius of curvature, r_j, of the spherical and cylindrical structures smaller), the breakdown voltage is significantly reduced, especially for low substrate impurity concentrations.

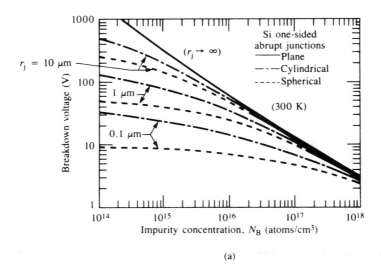

(a)

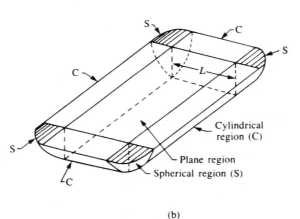

(b)

Fig. 5-10 (a) Abrupt *pn* junction breakdown voltage versus impurity concentration on the lightly doped side of the junction for both cylindrical and spherical structures. r_j is the radius of curvature.[10] (© 1957 IEEE). (b) Formation of cylindrical and spherical regions by diffusion through a rectangular window. From S. M. Sze, *Semiconductor Devices - Physics and Technology*, Copyright, 1985, John Wiley & Sons. Reprinted with permission.

5.2.6 Gate-Oxide Breakdown Voltage

High-quality SiO_2 films will typically break down at electric fields of 5-10 MV/cm (the exact value is a function of oxidation and anneal conditions, oxide charges, surface crystallographic orientation, surface preparation and a number of other factors). This corresponds to 50-100 V across a 100-nm-thick oxide. Present day 5 V processes use gate-oxide thicknesses of 15-100 nm. Below around 5 nm (at less than 3 V), there is a finite probability that electrons will pass through the gate by means of a quantum-mechanical tunneling effect. For proper device operation, the tunneling current must be small. This effect therefore sets a fundamental lower limit of about 5 nm for the thickness of the gate oxide. A search for alternative gate dielectric materials to mitigate this limitation is being conducted; this issue will be discussed in the CMOS chapter.

Oxide breakdown may also occur at electric-field values smaller than those given above, as a result of process-induced flaws in the gate oxide. Such defects include: metal precipitates on the silicon surface prior to oxide growth (see Vol. 1, chap. 2); high defect density in the silicon lattice at the substrate surface (e.g., stacking faults and dislocations, see Vol. 1, chap. 2); pinholes and weak spots created in the gate oxide by particulates; thinning of the oxide during growth caused by the Kooi effect (see chap. 2); and oxide wearout due to failure mechanisms related to hot-electron injection (this topic will be covered in Vol. 3).

5.2.7 High Field-Region Threshold-Voltage Value

A high value of threshold voltage in the field region is needed to keep the parasitic field channels between adjacent active devices from being turned on. This topic is described in great detail in chapter 2.

5.3 THE EVOLUTION OF MOS TECHNOLOGY (PMOS AND NMOS)

Following the introduction of MOS integrated circuits in the 1960s, MOS technology evolved through several stages. At first, PMOS was the dominant technology, but it was supplanted by NMOS in the early 1970s. NMOS in turn was largely replaced by CMOS in the mid-1980s. (The evolution of CMOS will be described in chap. 6.) In addition, the drive to shrink feature dimensions led to smaller device sizes in all IC technologies. In MOS technology, the main dimension that summarized the shrinkage in each new generation was the *minimum gate length* of the transistors. NMOS overtook PMOS roughly when gate lengths reached ~6 μm, while CMOS became the dominant technology at gate lengths of ~1.5-2 μm.

Although the *physical gate length*, L, of the MOS device is commonly used to identify each generation of technology, this practice can be misleading in that it may convey the impression that this dimension is also the minimum dimension of all other device features fabricated in the technology. In fact, the physical gate length does not necessarily reflect the dimensions of the other design rules (many, if not most, of which are larger than the physical gate length dimension).

In addition, the physical gate length does not accurately represent the electrical or *effective* channel length (L_{eff}), which in fact is given by

$$L_{eff} = L - 2x_{jl}$$

$$(5 - 14)$$

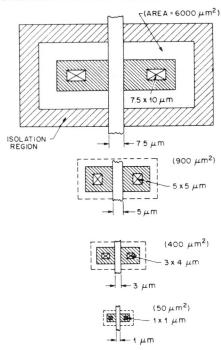

Fig. 5-11 Reduction in the area of the MOSFET as the gate length (minimum feature length) is reduced.[12] Reprinted with permission of Solid State Technology.

where x_{jl} is the lateral distance that the source or drain junction extends under the gate. For example, in a 1.2 μm MOS technology, the source and drain junctions both extend ~0.125 μm under the gate region, so L_{eff} in this technology is actually 0.95 μm. Furthermore, in CMOS technologies, the minimum physical length of PMOS devices is generally longer than that of NMOS devices, for reasons that will be discussed later. Figure 5-11 shows the reduction in MOS area as the gate length is reduced.[12] A longer and more detailed history of the evolution of the MOS transistor has been published by C. T. Sah, who describes both the evolution of the field-effect devices themselves and the technology that has been developed to fabricate these devices.[60]

5.3.1 Aluminum-Gate PMOS

The first MOS ICs built in the mid-1960s were implemented with *p*-channel enhancement-mode devices whose threshold voltages were approximately -4 V. (An example of an early PMOS IC was the Intel 256-bit DRAM, introduced in the late-60s with devices having gate lengths of 12 μm.) Such early MOS integrated circuits were built on silicon wafers of <111> orientation and used Al as the gate electrode material. These choices grew out of the experience gained in manufacturing bipolar ICs prior to the development of MOS ICs. When MOS IC technology was first being implemented, semiconductor manufacturers transferred their bipolar fabrication know-how to the newer technology.

Since precise control of dopant diffusion in <111> Si was a mature bipolar process, <111> wafers were also the logical choice for building MOS ICs. Similarly, since aluminum metallization was already being implemented, it was natural to adapt Al as the MOS gate electrode material. Unknowingly, these choices worsened the problems which prevented NMOS devices from being used to produce early MOS ICs, and instead, forced the use of PMOS devices in these circuits.

The earliest MOS circuits exhibited two serious limitations, partly as a result of being implemented with p-channel devices. First, the PMOS devices had a V_T of -4 V, which required a power-supply voltage of -12 V for the drain supply (a value that was incompatible with the +5 V power-supply voltage used in bipolar digital [TTL] ICs). Second, the circuits were very slow (e.g., a PMOS flip-flop could operate at 500 kHz to 1 MHz, while a bipolar flip-flop could operate at 5-10 MHz). It was known that the latter problem was due to the low surface mobility of holes in the channel and that electron mobility in silicon is nearly three times as large. Therefore, NMOS circuits would have been able to provide significantly improved performance. Nevertheless, the decision to manufacture ICs with PMOS devices was dictated primarily by the existence of large (and quite variable) oxide-charge densities in the early MOS technologies.

These oxide charges were generally positive, and positive voltages on the gate tend to accumulate n-type surfaces (but will deplete or invert p-type surfaces). The oxide charges present in early MOS devices were often large enough to cause inversion in NMOS devices fabricated with reasonably thin gate oxides, even when $V_G = 0$. A thicker gate oxide could have been used to increase V_T, but this would also have degraded g_m to such a degree that circuits built with such devices would have been even slower than those built with PMOS devices. Thus, only *depletion*-mode high performance NMOS devices could be reliably manufactured. Since *enhancement*-mode devices are needed for most applications, this presented a major difficulty.

In addition, because the oxide charges were also capable of inverting p-doped substrate regions under field oxides, it was difficult to reliably isolate n-channel devices. This problem was made worse by the depletion of boron during thermal oxidation, because it reduced the boron concentration at the p-type surface. In summary, at the outset of the MOS era, it was not possible to reliably manufacture integrated circuits with n-channel enhancement-mode MOSFETs.

On the other hand, since positive oxide charges tend to accumulate an n-type surface, they merely increase the negative voltage that is required to turn on a p-channel device. Hence, the manufacture of enhancement-mode PMOS devices was possible despite of the presence of the oxide charges. Nevertheless, it was clear that processing innovations were needed if the potential benefits of MOS (i.e., increased packing density, lower power consumption, TTL power-supply voltage compatibility, and process simplicity) were to be fully realized.

5.3.2 Silicon-Gate MOS Technology

One of the key process innovations for MOS ICs was the use of heavily doped polysilicon as the gate electrode in place of Al (see Vol. 1, chap. 6 for more information on polysilicon thin films). The development of this *silicon gate technology* improved the fabrication of MOS ICs in the following ways:

- Since aluminum must be deposited following completion of all high-temperature process steps (including drive-in of the source and drain regions), the gate electrode must be separately aligned to the source and drain. This alignment procedure adversely affects both packing density and parasitic overlap capacitances between the gate and source/drain regions. Since polysilicon has the same high melting point as the silicon substrate, it can be deposited prior to source and drain formation. Furthermore, the gate itself can serve as a mask during formation of the source and drain regions (by either diffusion or ion implantation). The gate thereby becomes nearly perfectly aligned over the channel, with the only overlap of the source and drain being that due to lateral diffusion of the dopant atoms. This *self-alignment* feature simplifies the fabrication sequence, increases packing density, and reduces the gate-source and gate-drain parasitic overlap capacitances.

- The threshold voltage of PMOS devices is reduced by the use of a polysilicon gate, since the φ_{ms} is less negative (see Eqs. 5-2). For PMOS devices on <111>- Si, the threshold voltage is reduced from roughly -4 V to -2 V. This smaller threshold voltage value enabled PMOS ICs to become compatible with TTL (bipolar) ICs, allowing MOS to be designed into many digital systems that operated at TTL-defined power supply voltage levels (i.e., 0 V to 5 V).

- The ability of polysilicon to withstand high temperatures also permits it to be completely encapsulated by an SiO_2 layer. This allows the polysilicon film to function as an interconnect path, in addition to serving as the gate electrode. By taking advantage of this new interconnection structure (without having to use a second layer of metal, as was necessary with bipolar ICs), it was possible to give MOS ICs an additional level of interconnection that could be crossed by the usual metal layer, or even by another polysilicon layer. This eased the problem of routing the electrical paths among the devices of an IC, thereby facilitating the layout of compact digital integrated circuits. (Techniques for establishing contact between the polysilicon layer and substrate are described in chap. 3, section 3.11.1). The ability of polysilicon to withstand high temperatures was also exploited to allow the dielectric (e.g. phosphorus-doped SiO_2) that covers it to be flowed, thereby making a significantly smoother surface topography for metallization layers.

The greatest disadvantage of polysilicon as a gate material compared to Al is its significantly higher resistivity. Even when doped at the highest practical concentrations, a 0.5-μm-thick polysilicon film has a sheet resistance of about 20 Ω/sq (compared to ~0.05 Ω/sq for a 0.5-μm-thick Al film). The resulting high values of interconnect line resistance can lead to relatively long RC time constants (i.e., long propagation delays) and severe dc voltage variations within a VLSI circuit. Consequently, the formation of refractory metal silicide layers on top of polysilicon layers (which results in so-called *polycide* films) was developed to reduce the severity of this drawback. Such polycide films can provide sheet resistances of 1 Ω/sq, at the expense of more complex processing (see Vol. 1, chap. 11 for more information on polycides). Despite of the above limitation, the development of silicon-gate technology proved to be the most important contribution to MOS technology during the reign of PMOS.

5.3.3 Reduction of Oxide-Charge Densities

Another set of important advances allowed the magnitude of the positive oxide-charge densities to be reduced. These oxide charge reduction techniques can be summarized as those involving *cleanliness, gettering, annealing,* and *replacing <111> wafers with <100> wafers.*

Cleanliness and gettering techniques reduced the densities of *mobile ionic charge,* which are due to the incorporation of ionized alkali metal atoms (Na^+, K^+) in the gate oxide. Na contamination can be controlled through clean gate oxidation processing. Gettering is used to prevent any Na^+ that enters the gate oxide from significantly degrading the V_T (see Vol. 1, chap. 7). Although tests for Na contamination of MOS devices must be routinely performed to ensure that accidental contamination does not occur, it is possible to establish fabrication procedures in which the instability of V_T due to mobile ionic charge is less than 0.1 V.

Another source of positive oxide charge is the *interface trap charge.* Again, as described in Volume 1, chapter 7, two annealing techniques were discovered that reduced this charge to acceptably low levels (i.e., to the low 10^{10} cm^{-2} eV^{-1} range). At these levels the contribution to the threshold voltage of the device is acceptably small for modern MOS devices.

The final source of positive oxide charge is the so-called *fixed oxide charge* which is located in the transition region between the silicon and the SiO_2 (see Vol. 1, chap. 7). It was learned that this charge can be reduced by the use of proper annealing techniques and that the lowest fixed oxide charge densities (~10^{10} cm^{-2}) are obtained on <100> wafers. As a result, the production of MOS ICs was shifted from <111> to <100> starting material.

In summary, the threshold voltage is negatively shifted by an amount proportional to the sum of these three oxide charge densities, Q_{tot}. When Q_{tot} was quite high this had an extremely important influence on MOS production. Through an enormous effort by the semiconductor community (industry and universities), each of the positive oxide charge density values was reduced. As a result, Q_{tot} can now be kept to less than 5 x 10^{10} charges /cm^2, and in currently used MOS devices, the oxide charge contribution to threshold voltage is minimal. As an example, the change in threshold voltage (ΔV_T) due to a Q_{tot} of q (5×10^{10} cm^{-2}) in an MOS device with a 20-nm-thick oxide is:

$$\Delta V_T = Q_{tot}/C_{ox} = Q_{tot}\, x_o\, / 3.9\, \varepsilon_o$$

$$= (1.6 \times 10^{-19}\, C)\, (5 \times 10^{10}\, cm^{-2})\, (2 \times 10^{-6}\, cm)\, /(3.5 \times 10^{-13}\, F/cm) \approx 0.05\ V.$$

The primary reason that thermal SiO_2 is used as the gate insulator in almost all MOSFETs is that it exhibits the best interface with silicon (where "best" means that the interface has a very low concentration of interface fixed charges and traps). In fact, if a phenomenon such as the tying up of the dangling silicon bonds at the silicon surface by SiO_2 did not exist, and if an annealing process was not discovered for reducing the remaining bonds and traps to an acceptable level, MOS devices would have remained merely a laboratory curiosity.[61] Experimentation, however, is still being conducted to determine the suitability of other materials as gate insulators; this is discussed in chapter 6 which deals with CMOS.

EXAMPLE 5-2: Recalculate the threshold voltage of the NMOS transistor considered in Example 5-1 of section 5.1.2 when the oxide thickness, t_{ox}, is reduced to 15 nm, and the total oxide-charge density is reduced to 5×10^{10} cm^{-2}.

SOLUTION: $\varphi_{f(sub)} = -0.29$ V; $\varphi_{ms} = -0.88$ V;

$$\varepsilon_{ox} = 3.9\varepsilon_o = 3.5 \times 10^{-13}\ F/cm; \quad C_{ox} = \varepsilon_{ox}/15 \times 10^{-6} = 2.3 \times 10^{-7}\ F/cm^2$$

$$Q_{BO} = -[2 \times 1.6 \times 10^{-19} \times 10^{15} \times 1.04 \times 10^{-12} \times |-0.58\ |]^{1/2} = -1.4 \times 10^{-8}\ C/cm^2$$

$$Q_{BO}/C_{ox} = -1.4 \times 10^{-8}/2.3 \times 10^{-7} = -0.06\ V$$

$$Q_{tot}/C_{ox} = 1.6 \times 10^{-19} \times 5 \times 10^{10}/2.3 \times 10^{-7} = 0.46\ V$$

$$V_{TO} = -0.88 - (-0.58) - (-0.06) - (0.03) = -0.27\ V$$

Note that the threshold voltage is still negative, and that this device would still be *ON* if the applied gate bias was $V_G = 0$ V.

EXAMPLE 5-3: Repeat the threshold-voltage calculation in Example 5-2 for an NMOS transistor whose oxide thickness (t_{ox}) is increased to 500 nm. This oxide thickness is typical of the field-oxide thickness between MOS devices on an IC. Hence, it allows us to calculate V_T for the parasitic NMOS field-region device.

SOLUTION: $\varphi_{f(sub)} = -0.29$ V; $\varphi_{ms} = -0.88$ V;

$$\varepsilon_{ox} = 3.9\varepsilon_o = 3.5 \times 10^{-13}\ F/cm; \quad C_{ox} = \varepsilon_{ox}/5 \times 10^{-5} = 7 \times 10^{-9}\ F/cm^2$$

$Q_{BO} = - [2 \times 1.6 \times 10^{-19} \times 10^{15} \times 1.04 \times 10^{-12} \times |-0.58 |]^{1/2} = -1.4 \times 10^{-8} \, C/cm^2$

$Q_{BO} / C_{ox} = -1.4 \times 10^{-8} / 7 \times 10^{-9} = -2 \, V$

$Q_{tot} / C_{ox} = 1.6 \times 10^{-19} \times 5 \times 10^{10} / 7 \times 10^{-9} = 1.14 \, V$

$V_{TO} = -0.88 - (-0.58) - (-2) - (1.14) = 0.66 \, V$

This shows that an increase in the oxide thickness in the field will increase the V_T of the NMOS device in Example 5-2 so that it is now an enhancement-mode device. Unfortunately, if $V_G = 5$ V, this device would still turn on, and thus the parasitic field device would conduct.

5.3.4 Ion Implantation for Adjusting Threshold Voltage

The development of ion implantation for V_T adjustment removed the last obstacle to reliable production of *n*-channel devices for MOS ICs, because this procedure made it possible to select the substrate-doping value without having to consider its impact on V_T. Substrate doping could now be selected strictly on the basis of optimum device performance, since V_T became separately adjustable by means of ion implantation. In addition, since dopants could be selectively implanted into the field regions, high-performance NMOS circuits could also be reliably fabricated on lightly doped substrates (i.e., without the possibility of inadvertent inversion of the surrounding field regions).

This technique of adjusting V_T involves implantation of boron, phosphorus, or arsenic ions into the regions under the oxide of a MOSFET. The implantation of boron causes a positive shift in the threshold voltage, while phosphorus or arsenic implantation causes a negative shift. For shallow implants, the procedure has essentially the same effect as placing an additional "fixed" charge at the oxide-semiconductor interface. To first order, the threshold-voltage change (ΔV_T) is thereby estimated from[1]

$$\Delta V_T = q \, N_I / C_{ox} \qquad (5 - 15)$$

where N_I is the dose of the implanted ions (atoms/cm^2) introduced into the silicon near its surface. For example, Eq. 5-15 predicts that when $N_I = 5 \times 10^{11}$ ions/cm^2 and $t_{ox} = 25$ nm, a shift in V_T of 0.58 V will be produced. Exact modeling can be performed to calculate the actual threshold voltage shift more accurately. Figure 5-12 graphically shows the results of such modeling calculations.[16,75]

The V_T-adjust implant is usually done through the gate oxide layer. When the correct implant energy for the gate-oxide thickness being used is selected, the peak of the implant will occur at the oxide-silicon interface. After the implant-activating anneal, the implanted distribution is broader than the as-implanted profile. Calculating the effect of the implant on V_T is greatly simplified by approximating the actual distribu-

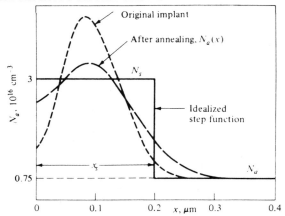

Fig. 5-12 Doping profile of implanted region beneath the gate oxide. The original implant is broadened by thermal annealing. The step doping is used to estimate the threshold voltage shift achieved using ion implantation.[75] Reprinted with permission of IBM J. of Research and Development.

tion via a "box" distribution (in which the implanted dopant is assumed to have a constant density from the surface to a depth of x_i.) Figure 5-12 shows that V_T does not change greatly as x_i is varied. Thus, the first-order approximation of the threshold-voltage change, ΔV_T (which ignored the depth of the implant as a parameter), will therefore give reasonably good estimates of ΔV_T. This calculation is often accepted in practice.

Ion implantation can also be used to fabricate depletion-mode MOSFETs. Depletion-mode NMOS devices (*i.e.* in which $V_T < 0$ V) are commonly used in NMOS logic circuits (see the following section). In order for the required negative threshold voltage for a depletion-mode NMOS device to be produced, n-type impurities are implanted to form a built-in channel between the source and drain. The dose required to shift the threshold voltage by the desired value may also be estimated using Eqs. 5-1 and 5-15.

EXAMPLE 5-4: If the NMOS transistor of Example 5-2 is to be used in an application that requires a $V_T = 1.0$ V, calculate the boron dose needed to adjust V_T to this value.

SOLUTION: The threshold voltage of the device in Example 5-2 is -0.27 V. We wish to have a V_T of 1.0 V. Thus V_T must be shifted (ΔV_T) by 1.27 V. Using Eq. 5-15, we see that the boron dose needed to cause this ΔV_T is

$$N_I = \Delta V_T\ C_{ox}/q\ =\ [\ 1.27\ \text{V} \times 2.3 \times 10^{-7}\ \text{F/cm}^2\ ]/1.6 \times 10^{-19}\ \text{C}$$

$$=\ 1.8 \times 10^{12}\ \text{boron atoms /cm}^2.$$

5.3.5 Isolation Technology for MOS

Although MOS transistors are inherently self-isolating devices it is still necessary to prevent the formation of spurious channels between MOS devices (see chap. 2). This can be accomplished with the combination of a thick field oxide between the devices and a high surface concentration under the field oxide.

Prior to about 1970 the process for obtaining thick oxide regions in the field involved growing an oxide to the desired thickness on the wafer surface and then etching windows into it. This approach caused severe steps in the wafer topography, which were difficult to cover with subsequent metal layers. The introduction of LOCOS isolation in 1970 substantially overcame this problem (see chap. 2). The smoothly tapered step from the edge of the active region to the top of the field oxide in LOCOS permits overlying conductors to be easily deposited on such steps without the occurrence of significant thinning. In addition, with the development of the threshold-voltage adjustment process via ion implantation, high surface concentrations of boron could also be selectively placed under the field oxide regions. These two advances made it feasible to reliably isolate devices in NMOS circuits.

EXAMPLE 5-5: If the parasitic field NMOS transistor of Example 5-3 is implanted with the boron implant dose calculated in Example 5-4 before the field oxide is grown, determine V_T of this device (assume that no boron is lost because of segregation effects during the oxide growth).

SOLUTION: The threshold voltage of the device in Example 5-3 is 0.66V. Using Eq. 5-15, we see that a boron dose of 1.8×10^{12} atoms /cm^2 would cause a ΔV_T of

$$\Delta V_T = q\,N_I/C_{ox} = 1.6 \times 10^{-19}\ C \times 1.8 \times 10^{12}\ cm^{-3}/7 \times 10^{-9}\ F/cm^2$$

$$= 41\ V$$

Thus, the V_T of the parasitic field device would be (0.66 V + 41 V), or almost 42 V. This example shows that a combination of a thicker field oxide and a channel-stop implant dose can increase the threshold voltage of the parasitic field device to sufficiently large values.

5.3.6 Short-Channel Devices

As MOS channel lengths got smaller than about 3 μm, so-called short-channel effects began to become increasingly significant. As a result, device design and, consequently, process technology had to be modified to take these effects into account so that optimum device performance could continue to be obtained. Short-channel effects and their impact on processing in will be discussed in section 5.5.

5.4 PROCESS SEQUENCE FOR FABRICATING NMOS INVERTERS WITH NMOS DEPLETION-MODE LOADS

This section will describe the process sequence used to fabricate silicon-gate NMOS digital integrated circuits. A simple inverter circuit with enhancement-mode pull-down and depletion-mode pull-up NMOS transistors is used in the example. Many other logic circuits can also be implemented with this *enhancement-depletion* (E-D) NMOS process sequence.

A brief outline of how the inverter functions from a circuit point of view is given at the outset to help define some of the characteristics of MOS IC circuits. The process flow described in this section represents a relatively simple NMOS technology. In fact, many techniques have been developed to improve the performance and packing density of MOS circuits beyond those that can be produced by this process sequence. These include alternatives to LOCOS isolation structures; thin gate oxides (and alternative gate oxide materials); shallow source/drain junctions; spacers (for forming lightly-doped drain structures and salicides); punchthrough prevention implants; double polysilicon; 2-level (or more) metallization; and self-aligned contact structures. While most of these techniques are discussed in this chapter, others are discussed elsewhere in the book (e.g., isolation technology in chap. 2, salicides and self-aligned contact structures in chap. 3, and multilevel metallization technology in chap. 4).

5.4.1 Operation of an NMOS Inverter with a Depletion-Mode Load

The most common applications of MOS transistors are in integrated circuit digital logic gates and memory arrays. Several types of circuits have been developed to implement logic gates in MOS ICs, with each circuit type characterized by the type of *load device* it utilizes. The class of logic-gate circuits that has become standard in most NMOS digital ICs is based on enhancement-depletion (E-D) NMOS technology, and such E-D logic gates are the basis for most NMOS microprocessors, microprocessor peripheral devices, and static NMOS memories.

The inverter is the fundamental logic gate. E-D NMOS inverters are composed of two transistors; an enhancement mode MOSFET called the *driver*, which is switched *ON* and *OFF* by the input signal; and a depletion-mode MOSFET, called the *load*. The circuit diagram of this inverter is shown in Fig. 5-13a, and an example of a layout (as it would appear on the completed wafer) is shown in Fig. 5-13b. The cross-sectional view of this structure is shown in Fig. 5-13c.

The load connects the power supply voltage V_{DD} and the output of the inverter. The gate of the load transistor is electrically connected to its source region, so that V_G in the load transistor always equals zero. Since depletion-mode devices are always *ON* when $V_G = 0$, tying together the gate and drain ensures that the device is always *ON*. The driver transistor has its source connected to ground, while its drain region is electrically connected to the source region of the load transistor and the output. The threshold voltage of the driver transistor, V_{TE}, is selected to be between 0 and V_{DD},

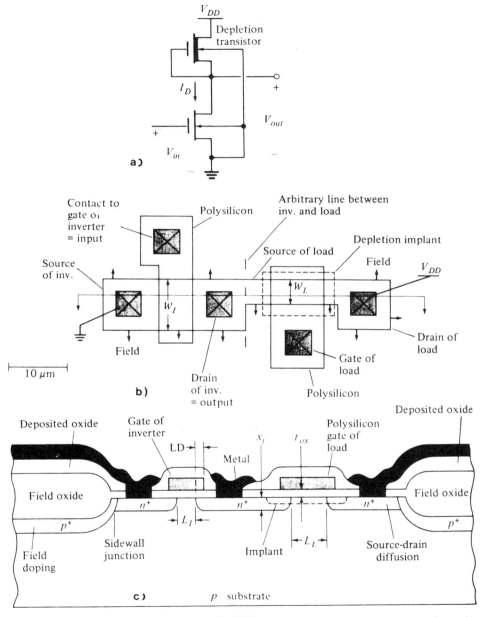

Fig. 5-13 NMOS inverter, depletion MOSFET load. (a) Schematic representation. (b) Layout. (c) Cross sectional view.[7] From D. A. Hodges and H. G. Jackson, *Analysis and Design of Digital Integrated Circuits*, Copyright, 1983 McGraw-Hill Book Co. Reprinted with permission.

while the threshold voltage of the load transistor (since this is a depletion-mode device), V_{TD}, is selected to have a negative voltage value.

The input signal to the inverter (a voltage, V_{in}) is fed to the gate of the driver transistor, and the output signal is the voltage level at the output node, V_o. When there is a logic 0 input signal (low voltage at V_{in}), V_G at the driver transistor is $<V_{TE}$. In this case, there is no conducting channel between source and drain, so the impedance between the output and ground is very high. Since the depletion transistor is *ON*, however, the output is electrically connected to V_{DD}, and V_o rises (or is *pulled up*) to logic 1 (close to V_{DD}).

When the input voltage to the inverter is logic 1 (close to V_{DD}), the gate voltage applied to the driver is greater than V_{TE}, thus turning the driver transistor *ON*. A low-impedance path then exists between the output node and ground. Hence, the driver transistor can conduct a large current with a small voltage drop across it, allowing the output to go to logic 0. This demonstrates how the logic level at the input is inverted at the output of the circuit.

The desired characteristics of digital logic gates as IC elements are the following: fast switching speed (i.e., small propagation-delay time); low power dissipation; small size (i.e., minimum silicon area); and high noise margin. We'll now see how the NMOS inverter just described is designed to meet these requirements. (Note: the circuit design considerations used to arrive at the configuration presented here can be found in texts dealing with IC design.[7,17,18] Since our interest here is in how these design choices impact the processing of the device, readers interested in their justification are advised to consult the references mentioned.)

In order to obtain maximum packing density (i.e., most logic gates per area of substrate) with minimum power consumption* an E-D NMOS inverter would be designed in the following manner:[17] The driver transistor would have a gate area whose dimensions would be the minimum that could be fabricated in that generation of technology. The depletion load transistor would have an effective *channel length* (L_{effD}) that would be four times the effective channel length of the driver device (L_{effE}) and a minimum gate width.

As a simple example, let us assume an inverter is to be fabricated with E-D NMOS technology, and that the minimum manufacturable feature dimension is 5 μm. Let us also assume that a 1-μm lateral diffusion occurs under each side of the gate. In this case, L_{effE} is 3 μm and thus L_{effD} should be 12 μm. This means that the drawn gate lengths of the driver and load devices would be 5 μm and 14 μm respectively. The drawn gate widths of both devices would be 5 μm (since these are the dimensions between the walls of the field oxide).

If a 5 V power supply is assumed, the threshold voltages of the driver and depletion load transistor (V_{TE}, and V_{TD}) would be selected to be approximately 1.0 V and -3.0 V, respectively. (These two V_T values are chosen in order to give a high noise margin without severely impacting switching speed.) To conserve space, *buried contacts* between the polysilicon gate and the silicon substrate would be employed. (The procedure used for fabricating such contacts is presented in chap. 3.)

* But as a result, with less than maximum switching speed,

While the E-D NMOS technology has some important advantages over other IC technologies (particularly high packing density), it does have some drawbacks that become extremely serious as the number of devices on the chip gets very large. The most important of these is the high total power that is consumed. The origin of this power consumption arises from the operation of the NMOS inverter. When the inverter (and similarly other logic gates) has a low output state, both driver and load are *ON*, allowing current to flow from V_{DD} to ground. The power consumed by each inverter in a low-output state is the product of this current and V_{DD}. Thus, if a series of inverters are connected together, 50% of them will be drawing power at all times. When the devices get small enough, the power density on the chip becomes so large that it becomes necessary to replace NMOS with CMOS, since this technology consumes much less power per logic gate.

5.4.2 Process Sequence of a Basic E-D NMOS IC Technology

Figure 5-14 is a flow-chart representation of the sequence of steps that were used to fabricate typical E-D NMOS digital integrated circuits for gate lengths down to about 3 μm.[76] Figures 5-15a through 5-15j show what occurs on the wafer as this sequence of process steps is followed. The process being illustrated is a seven mask process (including the passivation pad mask, even though this final pad mask is not shown). The E-D NMOS inverter described in the previous section is used here as a vehicle for showing how device features on the wafer surface are created during the course of the process flow.

5.4.1.1 Starting Material. The starting material is a lightly doped (~5×10^{14}-10^{15} atoms/cm^2) p-type <100> silicon wafer (substrate). As described earlier, the lightly doped substrate is chosen to provide low source/drain-to-substrate capacitance, high source/drain-to-substrate breakdown voltage, high carrier mobility, and low sensitivity to source-substrate bias effects. A backside gettering process, such as implanting with Ar, to create crystalline-damaged regions that will trap mobile impurities during subsequent heat steps during the process may be used prior to the next step (see Vol. 1, chap 2).

5.4.1.2 Active Region and Field Region Definitions. The first task in this processing sequence is to define the active device and field regions on the wafer surface. This is done by selectively oxidizing the field regions so that they are covered with a thick field oxide, using the LOCOS process. The steps involved in this task are those of boxes 3-9 in Fig. 5-14, and are illustrated in Figs. 5-15a and 5-15b.

A thin pad oxide (20-60 nm thick) is first thermally grown or CVD-deposited on the wafer surface as a stress-relief layer. This is followed by the deposition of a CVD nitride layer (100-200 nm thick). *Mask #1* is then used to expose a resist film that was spun on after the nitride deposition (Fig. 5-15a2). After exposure and development, the resist layer remains behind only in the regions that will be the active device regions

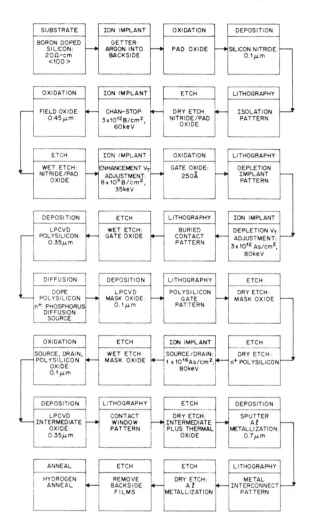

Fig. 5-14 Main steps in an *n*-channel, polysilicon-gate MOS IC process flow.[76] (© 1980 IEEE).

(Fig. 5-15b1). Next the nitride and pad oxide are anisotropically dry-etched away in the regions not covered by the resist (field regions). Thus, after the removal of the resist, the active areas are covered with the nitride/pad-oxide layer (Fig. 5-15b3).

In the next step, a boron implant (10^{12}-10^{13} atoms/cm^2, 40-80 keV) is performed to create *channel stops* in the field regions. The nitride/pad oxide layer now acts as a mask (Fig. 5-15c1) to prevent the boron from penetrating the silicon in the active areas. (Note that in some processes the resist is not removed until after the channel-stop

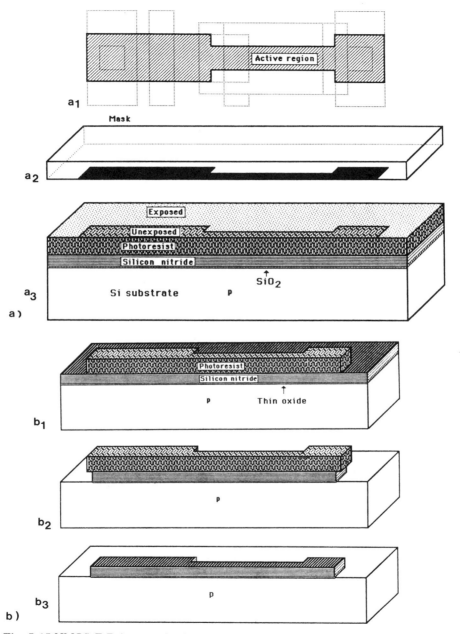

Fig. 5-15 NMOS E-D inverter fabrication sequence. (a) Patterning of the active region. (b) Patterning the silicon nitride-pad oxide layers. From W. Maly, *Atlas of IC Technologies*, Copyright 1987 by the Benjamin/Cummings Publishing Company. Reprinted with permission.

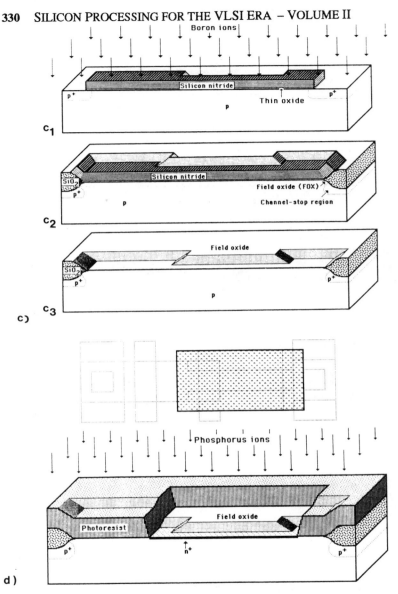

Fig. 5-15 (c) Active region formation and channel-stop implant. (d) Implantation of the channel of the depletion-mode transistor. From Maly, *Atlas of IC Technologies.*

implant, as the nitride/pad oxide layer may be too thin to act as an effective implant mask. In fact, with the patterned resist still in place, the channel stop may be implanted through the nitride/pad oxide layer, which in this case would not be etched until the implant is performed.)

A thermal-oxidation step is then performed to grow a thick (0.5-1.0 μm) field oxide over the regions where the nitride has been etched away. In this process, the field oxide is self-aligned to the channel stops. During the oxide growth, some lateral oxidation also occurs under the nitride edges, forming the *bird's beak* structure (see chap. 2 for more details concerning this effect, as well as other problems that arise in connection with the field-oxide growth step). After the field oxide has been grown, the remaining nitride and pad oxide are stripped, leaving the active areas with exposed silicon surfaces for further processing (Fig. 5-15c3).

5.4.1.3 Gate-Oxide Growth and Threshold-Voltage Adjust Implant

In the next major step the gate oxide is grown, and the threshold voltages of the enhancement-mode and depletion-mode transistors of the inverter are adjusted through ion implantation. The growth of the gate oxide is a critical step, as a defect-free, very thin (15-100-nm), high-quality oxide without contamination is essential for proper device operation. The gate oxide is grown only in the exposed active region (the field-oxide thickness is actually increased slightly as a result of this oxide-growth step). As noted earlier, the drain current in an MOS transistor is inversely proportional to the gate-oxide thickness (for a given set of terminal voltages). As a result, the gate oxide is normally made as thin as possible, commensurate with oxide breakdown and reliability considerations.

In order for a high-quality gate oxide to be obtained, the surface of the active area is wet-etched to remove any residual oxide. A sacrificial oxide is often deliberately grown on the exposed active areas after field oxidation to remove any dry-etch induced damage or unwanted nitride (due to the Kooi effect, see chap. 2).[65] After such oxides have been stripped, the gate oxide is grown slowly and carefully, usually through dry oxidation in a chlorine ambient (see Vol. 1, chap. 7).

The threshold-voltage adjust implant of the enhancement-mode devices is performed next. In this step, boron is implanted though the gate oxide (10^{12}-10^{13} atoms/cm^2, 50-100 keV), but the ions are not given enough energy to penetrate the field oxide. No mask is used in this step. (Note that in many processes, another pre-gate oxide oxide is grown, through which this implant is performed. It is again stripped off following the implant, and the gate oxide is then grown.)

Next, the depletion-mode devices of the circuit are given their threshold-voltage implant dose (Fig. 5-15d). The areas of the depletion-mode transistor channels are implanted with phosphorus or arsenic ions ($\sim 10^{12}$ atoms/cm^2, 100 keV) to give a threshold voltage of about -3.0 V. The implant dose is adjusted so that it over-compensates for the previous boron threshold-voltage-adjust implant, thus making the surfaces *n*-type. A negative threshold voltage is thus yielded, as required to establish a depletion-mode device. Photoresist (patterned by the use of *Mask #2*) is used to selectively allow the depletion-mode transistor channels to be implanted. The ions cannot penetrate the resist to reach active areas below. Likewise, the ions cannot penetrate any field oxide that is exposed by the resist opening. Hence, the location of the depletion transistor channel is defined by the intersection of the *Mask #2* window and the active region.

Buried contacts are then opened in the gate oxide using *Mask #3* (Fig. 5-15e). This opening in the gate oxide must be provided wherever it is desired to have polysilicon electrically contact the active silicon area (details of buried-contact formation are described in chap. 3). Since the polysilicon is deposited on the gate oxide, it will remain isolated from the substrate below unless a special opening is cut in the gate oxide. With *Mask #3*, resist covers the entire wafer except in those areas where the buried contact is desired. The gate oxide can then be etched from these regions, uncovering the silicon below.

5.4.1.4 Polysilicon Deposition and Patterning.

A layer of polysilicon (typically 0.4-0.5 μm thick) is next deposited by CVD over the whole wafer (see Vol. 1, chap. 6 for more information on the properties of polysilicon and its deposition process). Either ion implantation or diffusion with phosphorus is then used

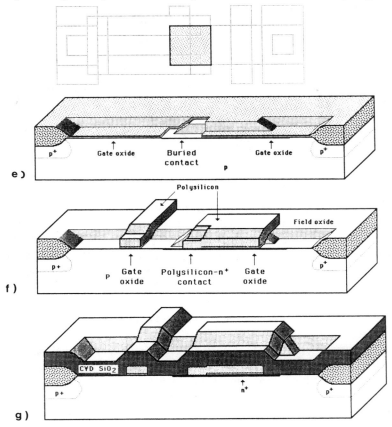

Fig. 5-15 (e) Buried contact etching. (f) Patterning of the polysilicon layer followed by gate oxide etching. (g) Deposition of the CVD SiO_2 layer followed by the diffusion of the drain and source regions. From Maly, *Atlas of IC Technologies.*

to dope the polysilicon to a sheet resistance of 20-30 Ω/sq. This resistance is adequate for MOS circuits with gate lengths ≥ 3 μm. For smaller devices, polycide layers (i.e., composite layers of refractory metal silicides and polysilicon) can be used to reduce the sheet resistance to ~ 1 Ω/sq (see Vol. 1, chap. 11). Using a polycide gives us the benefits of both silicon-gate and metal-gate technologies.

The gate structure and polysilicon interconnect structures are then patterned using *Mask #4* (Fig. 5-15f). Following exposure and development of the resist, the polysilicon film is etched (in current technology this is done by means of a dry-etch process). This is a critical etch step for several reasons. First, the channel length of the device depends on the gate length, because of the self-aligned nature of the silicon gate technology. Hence, the gate-length dimension must be precisely maintained across the entire wafer, and from wafer to wafer. Second, the profile of the etched poly gate structure should be vertical; this will prevent variation of channel lengths by the penetration of the ions of the thinner regions of the gate sidewalls during formation of the source/drain regions by ion implantation. Third, to achieve the above goals, an anisotropic polysilicon etch process must be employed. This type of process, however, requires overetching to remove the locally thicker regions of polysilicon that exist wherever it crosses steps on the wafer surface. During the overetch time, areas of the thin gate oxide are exposed to the etchants. Thus, it is necessary to use a polysilicon etch process that is highly selective with respect to SiO_2.

5.4.1.5 Formation of the Source and Drain Regions. Once the gate has been fabricated, the source and drain regions can be formed. This is normally done by ion implantation without the use of a lithography step (Fig. 5-15g). The gate and the field oxide act as masks to prevent the ion implantation from penetrating to the silicon substrate below. Therefore, only the active regions covered by the gate oxide (and no gate polysilicon), are implanted. An n^+ implant is used, with an energy that is insufficient to penetrate the gate-poly or field-oxide layers (arsenic is typically used, with a dose of $\sim 10^{16}$ atoms/cm^2 and an energy of 30-50 keV). As noted earlier, the source and drain are thereby *self-aligned* to the gate, and the dimension of the polysilicon gate thus plays a major role in the defining of the MOS gate length.

Following the source/drain implant, an anneal (or drive-in) step is performed to activate the implanted atoms and to position the source/drain junctions as desired. During this step, some of the phosphorus doping of the polysilicon outdiffuses into the silicon substrate wherever a buried contact opening the gate oxide has been cut. This diffusion (which occurs both vertically and laterally into the silicon below) forms a heavily doped n^+ region under the polysilicon in the buried-contact exposed region. The lateral diffusion of the implanted source/drain dopant thereby becomes electrically connected to the n^+ region under the polysilicon buried-layer region. In this manner, an electrical connection between the polysilicon and the silicon is established at the buried contact locations. In some processes, the junction formed by the buried-contact dopant outdiffusion from the polysilicon is deeper than the source/drain junctions, while in others it is not as deep.

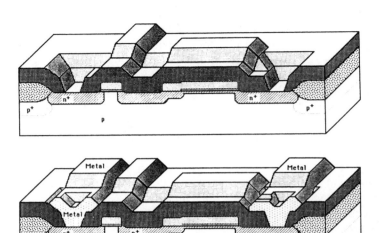

Fig. 5-15 (h) Contact cuts of the E-D inverter. Metallization of the E-D inverter. (i) Completed NMOS E-D inverter structure. Figures 15a through 15i from W. Maly, *Atlas of IC Technologies,* Copyright 1987 by the Benjamin/Cummings Publishing Company. Reprinted with permission.

The source/drain drive-in step also plays a part in determining the effective channel length (L_{eff}). That is, if the lateral junction depth is x_{jl} (which is primarily determined by the the lateral diffusion during the drive-in step, because the lateral straggle of arsenic at 30 keV is only ~5 nm, see Vol. 1, chap. 9), L_{eff} will be decreased by $2x_{jl}$ from the gate length at the mask level. Note that the channel *width* is also reduced by the bird's beak encroachment into the active area (see chap. 2). Thus, the actual width, W_w, of an MOS device is $W_w = W - \Delta W$, where W is the width at the mask level and ΔW is the channel-width shrinkage during processing.

The depth of the source and drain is thus a critical dimension, but the doping concentration is not as important. (A discussion of shallow source/drain junction formation techniques is presented in chap. 3, section 3.10.) To a first approximation, the device characteristics will not depend on the doping concentration value, provided it is sufficiently heavy.

A diffusion step may be used to dope the source/drain regions. In some of these cases the dopant source of the diffusion is the CVD oxide layer that is deposited after the gate has been defined (see next section).

5.4.1.6 Contact Formation. After the source/drain regions have been formed, a CVD process is used to deposit a layer of doped SiO_2 (glass), about 1 μm thick, onto the wafers (see Vol. 1, chap. 6). The dopant in the SiO_2 is either phosphorus (in which case the material is referred to as *phosphosilicate glass),* or both phosphorus and boron (making it a layer of *borophosphosilicate glass).* In some processes a thin thermal oxide is grown on the polysilicon prior to deposition of the glass layer. Nevertheless, a thick layer of SiO_2 cannot be thermally grown because of the excessive redistribution of the impurities that would take place during such growth. Hence, a lower-temperature CVD process must be used to get a sufficiently thick oxide.

The doped CVD glass layer plays several roles in the fabrication and operating aspects of the circuit. First, it acts as an insulating layer between the polysilicon and the metal to be deposited. Second, it reduces the parasitic capacitance of the interconnect metallization layer. Third, the addition of the phosphorus to the glass makes the layer an excellent getter of Na ions (recall that contamination by Na can cause instabilities in the V_T of the MOS devices). The phosphorus-doped glass immobilizes the otherwise mobile Na atoms within the CVD layer, preventing them from reaching the gate oxide and altering the threshold voltage. Finally, the dopants in the glass make it viscous at elevated temperatures (1000-1100°C for PSG, and 800-950°C for BPSG, see Vol. 1, chap. 6), allowing the layer of doped glass to be flowed after it is deposited. Through this procedure, a rounding of the contours of the glass and a smoothing out of any sharp steps is achieved. This produces better step coverage of the metal (which is deposited next) over the otherwise severe wafer topography. The high-temperature glass-flowing step also serves to activate the source/drain implanted junctions and drive them to their desired positions.

Contact openings are next created by a lithography-and-etch step (Fig. 5-15h). *Mask #5* is used to define contact opening patterns in a photoresist film, and a dry-etch process is then used to open the contact windows through the CVD SiO_2 to the underlying polysilicon and the n^+ regions in the silicon.

The contact-opening step can be critical, as the contact size and alignment limit the minimum size of the device. The source and drain regions must be large enough for the contact to fit, with allowance for alignment tolerance. If the contact opening exposes a part of the substrate, the drain or source will be shorted to the substrate. Likewise, any overlap of the source/drain contact opening and the gate will cause the gate to be shorted to the source or drain.

To keep the transistor as small as possible, the contact window is usually made at the minimum size achievable with the given process. In some processes, the exposed silicon in the contact is redoped to prevent shorting between the source/drain areas and the substrate (see chap. 2). Also note that the gate contact (i.e., Al to polysilicon) is often made outside the active device area to avoid possible damage to the thin gate oxide.

After the contact etch is completed and the resist is stripped, the doped CVD glass is again subjected to another flowing step. This procedure rounds the corners of the top of the oxide windows so that metal step coverage into the contact windows is improved. The process is called *reflow* of the contact windows (see chap. 3 and Vol. 1, chap. 6).

5.4.1.7 Metallization Deposition and Patterning. After the contacts have been opened, the metallization layer is deposited (~1 μm thick). Because the metal

layer is highly conductive, it is used whenever possible to interconnect circuit elements and to carry large amounts of supply current. The metal interconnect lines that are fabricated must have sufficient thickness, width, and step coverage to keep the current density in each line below the value that could produce electromigration failure (see chap. 4). In addition, the spacing between adjacent metal lines must be kept large enough that the lines will never touch, even under worst-case process variations.

Although evaporation was the method employed to deposit Al in the early days of MOS, it has generally been replaced by sputtering. To a great extent, the change was made because Al alloys with tightly controlled compositions became the materials of choice for the metal layer. Sputtering allows alloys to be deposited with much better compositional fidelity (see Vol. 1, chap. 10).

The metal alloy that was eventually chosen for NMOS is Al:1wt% Si. The silicon is added to the aluminum film to prevent spiking of the contacts during subsequent annealing steps (see chap. 3). In CMOS, such Al:Si alloys are being phased out as the metallization material for reasons that are discussed in the CMOS chapter and in chapters 3 and 4. Either wet-chemical etching or dry etching is used to pattern the Al film, using *Mask #6* (Fig. 5-15i and 5-15j).

Following the patterning of the metal, the Al-silicon contacts are alloyed. This step brings the Al and the n^+ silicon into intimate contact, since it allows the thin native SiO_2 layer that is likely to exist at the Al-Si interface to be reduced by the Al (see chap. 3). Such intimate contact between Al and n^+Si establishes a low-resistance ohmic contact. The anneal process exposes the wafer to a 375-500°C temperature in an H_2 or

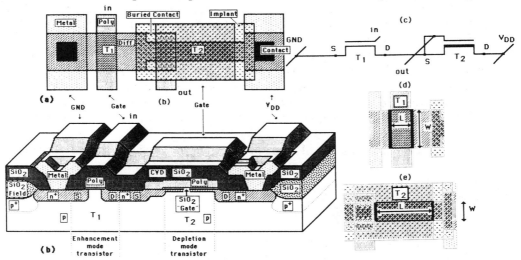

Fig. 5-15 (j) E-D inverter. 1) Composite drawing of the layout. 2) Cross section of complete structure. 3) Electrical diagram. 4) The enhancement transistor. 5) The depletion transistor. From Maly, *Atlas of IC Technologies.*

$N_2 + H_2$ (5%) ambient for about 30 minutes. As a result, this step may also be used as the annealing process for reducing the interface trap density in the gate oxide that was introduced by earlier processing steps (see Vol. 1, chap. 7).

5.4.1.8 Passivation Layer and Pad Mask. Finally, a *passivation* (or *overcoat*) layer, such as CVD PSG or plasma-enhanced CVD silicon nitride, is put down onto the wafer surface. This layer seals the device structures on the wafer from contaminants and moisture, and also serves as a scratch protection layer.

Openings are etched into this layer so that a set of special metallization patterns under the passivation layer is exposed. These metal patterns are normally located in the periphery of the circuit and are called *bonding pads* (Fig. 5-16). Bonding pads are typically about 100 x 100 μm in size and are separated by a space of 50 to 100 μm. Wires are connected (bonded) to the metal of the bonding pads and are then bonded to the chip package. In this way connections are established from the chip to the package leads.

The bonding-pad openings are created by patterning the passivation layer with *Mask #7*. If a PSG layer is used, the phosphorus (2-6 wt%) in the glass not only causes the PSG to act as a getter for Na but also prevents the glass film from cracking. Care must be taken to ensure that not more than 6% phosphorus is incorporated into the PSG, as this can cause corrosion of the underlying metal if moisture enters the circuit package (see Vol. 1, chap. 10). When silicon nitride is used, care must be taken to ensure that the deposited nitride film exhibits low stress (either tensile or compressive), so that it will not crack, since cracking would compromise the sealing capability of the film.

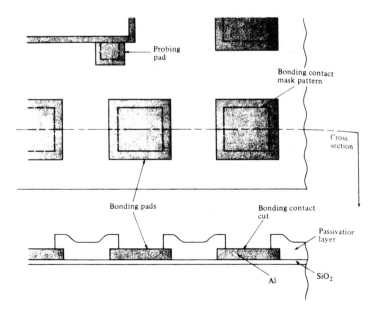

Fig. 5-16 Passivation layer and bonding pad openings. (Note, cross-section not to scale.)

5.5 SHORT-CHANNEL EFFECTS AND HOW THEY IMPACT MOS PROCESSING

The device characteristics of MOSFETs (such as threshold voltage, subthreshold currents, and I-V characteristics beyond threshold) are well predicted by Eqs. 5-1 through 5-10, if the channel lengths of the transistors are "long" (i.e., if they exceed 2 μm in length). A guide to the design of such long-channel MOSFETs is given in reference 70.

For shorter channel devices, however, a series of effects arise that result in significant deviations from the values predicted by the long-channel models. Such *short-channel effects* become a dominant part of MOS device behavior when channel lengths decrease below 2 μm. The effects are briefly described here to provide readers with a basis for understanding the processing steps that have been developed to mitigate their adverse impact on device performance. So-called *hot-carrier effects* will be described separately in the following section, even though these have only been observed in short-channel devices. More details on short-channel effects in *p*-channel CMOS devices are also given in chapter 6. A guide to the design of submicron channel MOSFETs (those with channel lengths ≥ 2 μm), is presented in reference 62.

Short-channel effects can be divided into the following categories: (a) those that impact V_T; (b) those that impact subthreshold currents; and (c) those that impact I-V behavior beyond threshold.

5.5.1 Effect of Gate Dimensions on Threshold Voltage

5.5.1.1 Short Channel Threshold Voltage Effect. V_T becomes less well predicted by Eq. 5-6 as the dimensions of the gate are reduced, and the error becomes significant when the dimensions are reduced to less than 2 μm. To get good agreement with measured data, a term $|\Delta V_T|$ must be subtracted from the V_T value obtained from Eq. 5-6. Thus, as the device length is reduced, the measured value of V_T of *n*-channel enhancement-mode devices becomes *less positive* than that given by Eq. 5-6, while for *n*-channel depletion-mode devices, V_T becomes *more negative*. For *p*-channel (enhancement-mode) devices, V_T becomes *less negative*.

The discrepancy arises because the equations for V_T given earlier in the chapter are based on one-dimensional theory.[20] It is assumed that the space charge *under* the gate is a function of only the vertical electric field, E_x (and is thereby influenced only by the charge *on the gate*). If the channel length is long, this is a reasonable assumption, as the influence of the drain and source junctions on the quantity of charge in the channel can be neglected. However, as the channel length approaches the dimensions of the widths of the depletion regions of the source and drain junctions, these depletion regions become a greater part of the *channel-depletion region* (Fig. 5-17a). Thus, some of the channel-depletion region charge is actually linked to the charge in the depletion region within the source and drain structures, rather than being linked to the gate charge. Hence, some of the channel region is partially depleted without any influence of the gate voltage. (In the extreme, if the two built-in depletion regions spanned the entire channel length when no voltage existed between source and drain, they could deplete all

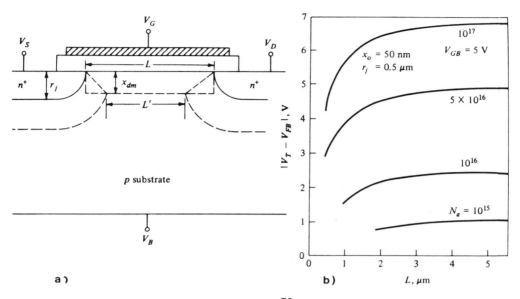

Fig. 5-17 (a) Yau's model of charge sharing.[73] (b) Theoretical threshold voltage as a function of channel length for various substrate doping concentrations.[73] Reprinted with permission of Solid-State Electronics.

of the channel region at the Si-SiO$_2$ interface.) Since some of the channel is depleted without the need to apply a gate bias, *less gate charge is required to invert the channel in short-channel devices than in a long-channel device with comparable substrate doping.*

A relatively simple equation that predicts the threshold lowering (ΔV_T) in terms of the device parameters is

$$\Delta V_T = \frac{q\, N_A\, W_{max}\, x_j \left(\sqrt{ \left\{ 1 + \frac{2\, W_{max}}{x_j} \right\} } - 1 \right)}{C_{ox}\, L} \qquad (5\text{-}16)$$

where W_{max} is the maximum depth of the depletion region in the channel, and x_j is the junction depth of the source drain regions.[74]

This effect is important because in order to be able to establish slightly positive V_T values (e.g., < +1 V) in long-channel NMOS transistors with lightly doped channels, it is necessary to increase the doping concentration at the surface of the channel. Consequently, in order to allow short-channel enhancement-mode NMOS devices to be fabricated with the same V_T value, the substrate doping concentration must be further increased. Since the magnitude of this effect increases as the device length is reduced (as is predicted by Eq. 5-16 and illustrated in Fig. 5-17b), it will be necessary to progressively increase the substrate doping concentrations as devices are made smaller, in order to maintain suitably positive values of V_T.

Application of a drain voltage causes the drain depletion region to extend into the channel region, where it acts as an additional substrate bias, and reduces V_T.[21]

5.5.1.2 Narrow Gate-Width Effect on Threshold Voltage.

In contrast to the short-channel-length effect, devices with narrow channel-widths require that such a positive-value correction term be *added* to Eq. 5-6 to give good agreement with the calculated values.[22] This effect is primarily due to the encroachment of the channel-stop dopants under the edges of the sides of the gate (see chap. 2). This has the effect of doping the channel at these edges more heavily than at the center (Fig. 5-18a). Thus, it requires more charge on the gate to invert the channel than if such encroachment did not occur, and this causes a shift in V_T from its predicted value (Fig. 5-18b).[23] On the other hand, if the voltage on the gate is held constant, the edges of the channel will have a higher V_T than the center of the channel. Because $V_G - V_T$ is smaller, narrow-channel devices will thus conduct less drain current. Several schemes are also described in chapter 2 for reducing the channel-stop encroachment and thus reducing the narrow-width effect.

Narrow-width effects are still observed even if there are no channel-stop implants. These arise because the relatively thinner depletion region under the field-oxide device distorts the depletion region under the gate oxide and prevents the formation of an

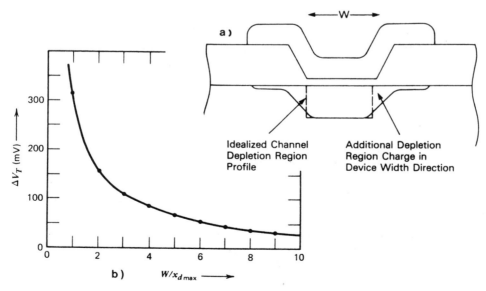

Fig. 5-18 (a) Schematic depiction of the narrow-channel effect. (b) Threshold voltage shift ΔV_T caused by the narrow-channel effect for a MOSFET with $N_A = 10^{15}$ cm^{-3}, and $t_{ox} = 50$ nm.[22]

inversion layer at the two edges. Although this leads to a slightly higher V_T, the effect is much less severe than that observed in devices with heavy channel-stop implants.

5.5.2 Short-Channel Effects on Subthreshold Currents (Punchthrough and Drain-Induced Barrier Lowering)

In section 5.2.3 we described the nature of subthreshold current flow (I_{Dst}) in MOSFETs, noting that a specific value of the *subthreshold-swing* parameter (S.S.) can be attributed to such "normal" I_{Dst} currents in long-channel devices. In short-channel MOSFETs, however, larger I_{Dst} values are observed at lower voltages than predicted by long-channel device models: one manifestation is an increase in the value of S.S. Note that even relatively small values of I_{Dst} can limit the transistor's ability to isolate nodes in a dynamic circuit or can allow excess current in static inverters. Hence, care must be taken to minimize I_{Dst}. Two of the primary causes of increased I_{Dst} are *punchthrough* and *drain-induced barrier lowering* (DIBL).

Punchthrough is normally observed when the gate voltage is well below V_T. It occurs as a result of the widening of the drain depletion region when the reverse-bias voltage on the drain is increased. The electric field of the drain may eventually penetrate into the source region and thereby reduce the potential energy barrier of the source-to-body junction (Fig. 5-19).[23] When this occurs, more majority carriers in the source region have enough energy to overcome the barrier, and an increased current then flows from source to body. Some of this current is collected by the drain, thereby increasing I_{Dst}. In general, punchthrough current begins to dominate I_{Dst} when the drain and

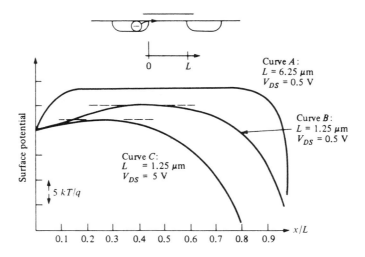

Fig. 5-19 Surface potential in the channel for devices with different channel lengths.[23] (© 1979 IEEE).

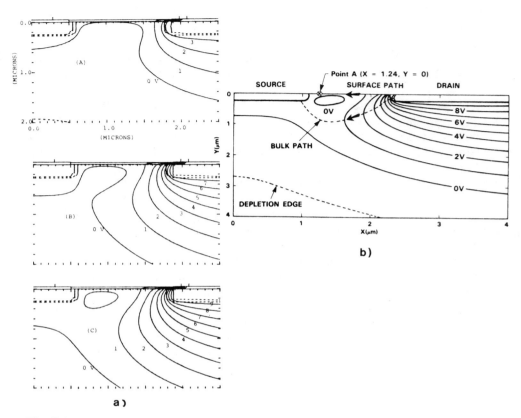

Fig. 5-20 (a) 2-D potential profile of an *n*-channel MOSFET with a drain bias of: 1) 3 V; 2) 7 V; 3) 9 V. Channel length = 1 μm. (b) Simulation of the potential profile of an *n*-channel MOSFET with a gate and drain bias of 0 and 9 V respectively. The surface DIBL and bulk punchthrough paths are indicated. From K. M. Cham, *et al., Computer Aided Design and VLSI Device Development.* Copyright 1986 Kluwer Academic Publishers. Reprinted with permission.

source depletion regions meet, and it can be suppressed by keeping the total width of the two depletion regions smaller than the channel length.[24]

Calculations of the potential in the bulk channel region in devices that use ion implantation to adjust V_T indicate that the barrier is lowest away from the Si-SiO$_2$ interface (usually at almost the same depth as the source/drain junction depths). That is, the V_T-adjust implant increases the doping concentration near the surface of the channel, causing the drain depletion region to be wider in the bulk than it is near the Si-SiO$_2$ interface. As a result, punchthrough current flows *below* the surface (Fig. 5-20). Consequently, the gate voltage has less control over the subthreshold current (i.e., even with sufficient gate voltage to turn off the channel, I_{Dst} can still flow in such devices).

An enhancement-mode device which is not turned off when $V_G = 0$ loses its ability to function as a switch.

Similarly, the application of a drain voltage in short-channel devices can also cause drain-induced barrier lowering (DIBL). That is, the drain voltage can cause the *surface potential* to be lowered (Fig. 5-21).[23, 71] As a result, the potential energy barrier at the *surface* will be lowered, and the subthreshold current in the channel region at the Si-SiO_2 interface can be increased (5-21b). This implies that I_{Dst} at the surface due to DIBL is expected to become larger as the gate voltage approaches V_T.

These two effects illustrate the complexity involved in modeling the overall subthreshold I-V behavior of short-channel MOSFETs. That is, both punchthrough current (in the bulk), as well as DIBL-induced current (at the surface), may simultaneously contribute to I_{Dst}.

To prevent punchthrough current in short-channel devices, the substrate doping can be increased to decrease the depletion-layer widths. These widths can be estimated using the formula for the width of a one-sided step junction:

$$W = \sqrt{\frac{2 K_s \varepsilon_o (|V_A| + V_{bi})}{q N_B}} \qquad (5-17)$$

where the built-in voltage, V_{bi}, given by

$$V_{bi} = 0.56 + (kT/q) \ln (N_B/n_i) \qquad (5-18)$$

and where V_A is the total applied voltage and N_B is the doping concentration of the body. Figure 5-22 gives the depletion-layer width of *pn* junctions as a function of doping and applied voltage.

However, increasing the substrate doping also increases the source-to-body and drain-

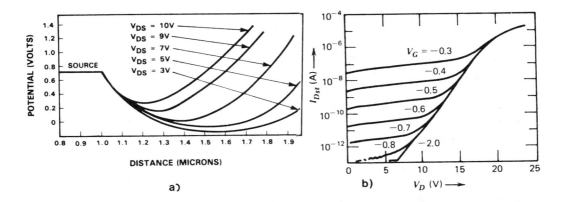

Fig. 5-21 (a) DIBL versus drain bias for short-channel MOSFET. (b) Experimental low-current characteristics for a MOSFET with L = 2.1 μm, $V_{SB} = 0$.[81] (© 1974 IEEE).

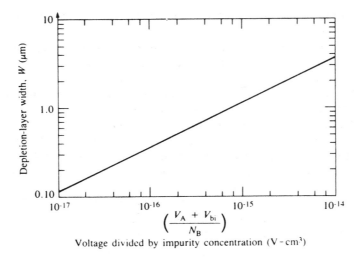

$$\left(\frac{V_A + V_{bi}}{N_B}\right)$$

Voltage divided by impurity concentration (V - cm³)

Fig. 5-22 Depletion-layer width of a one-sided step function as a function of doping and applied voltage calculated from Eqs. 5-17 and 5-18.

to-body junction capacitances, as well as the body factor. In addition, it reduces the breakdown voltages of the source/drain junctions. To avoid these drawbacks, an additional boron implant (whose peak concentration is located at a depth near the bottom of the source-drain regions) can be performed. This additional doping reduces the lateral widening of the drain-depletion region below the surface without increasing the doping under the junction regions. With such implants, the component of the punchthrough current can be suppressed to well below the normal I_{Dst} current of the device.

For example, in Fig. 5-23, a 1.2-μm device with a body doping of 1.9×10^{15} cm^{-3} without such a punchthrough-stopping implant shows a large value of I_{Dst} even when $V_G = 0$ V (curve A). This indicates that the device is already exhibiting punch-through.[26] Implants of boron with a dose of 8×10^{11} atoms/cm^2 and different energies are then performed in an attempt to reduce I_{Dst} to the values exhibited by a long-channel device (curve B). If the implant is too shallow, the extra implant has the effect of shifting the V_T of the device to well beyond the desired value. When the energy is increased so that the implant is sufficiently deep, the value of I_{Dst} drops to that exhibited by the long-channel device. At the same time, the surface concentration remains essentially unchanged, so that V_T is not appreciably shifted.

In another example, the S.S. of a device without a punchthrough-prevention implant is measured as its length is varied (Fig. 5-24a). At an L_{eff} of ~0.85 μm the S.S. starts to increase, indicating that punchthrough current begins to dominate I_{Dst}. By adding an implant step that places boron atoms in the dashed subsurface region shown in Fig. 5-24b, the punchthrough component of I_{Dst} is suppressed so that it is not observed until L_{eff} becomes nearly as small as 0.5 μm.

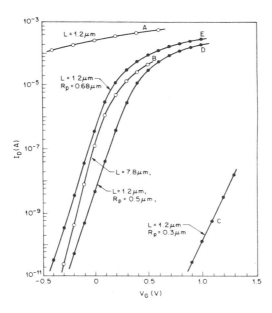

Fig. 5-23 Drain current versus gate voltage for n-channel devices with a substrate doping of 1.9×10^{15} atoms/cm^3, source/drain junctions 0.47-μm deep, 575 Å gate oxide, drain voltage of 5 V, and V_{BS} of 0 V. Devices A and B have no channel implant, and devices C and E have a boron channel implant of 8×10^{11} atoms/cm^2 at various energies.[26] (© 1978 IEEE).

5.5.3 Short-Channel Effects on I-V Characteristics

The I-V characteristics of short-channel devices are significantly altered in three ways. First, the combined effects of reduced gate length and gate width produce a change in V_T. Second, the channel length is modulated by the drain voltage when the device is in saturation (i.e., $V_{DS} > [V_G - V_T]$), causing an increase in device gain over that predicted in an ideal long-channel device (*channel-modulation effect*). Third, the mobility of the carriers in the channel is reduced by two effects, which also reduces I_D. (The two effects are the *mobility-degradation factor*, due to the gate field, and the *velocity-saturation factor*).

Figure 5-25 shows the I-V characteristics of an MOS device.[28] The curves in Fig. 5-25a are those of an ideal long-channel MOS device, while those in Fig. 5-25b show the effect of adding the channel modulation factor. Figure 5-25c shows the combined effect of adding the mobility degradation factor those of Fig. 5-25b. The velocity saturation factor also has the effect of making the both I_D and g_m independent of channel length in silicon MOS transistors for $L_{eff} \leq 1.25$ μm. More details on these effects are provided in suitable device physics texts.[1,3,28]

A general guide to the design of short-channel MOSFETs is given in references 62 and 69. In addition, a simple *engineering model* for short-channel devices has also been developed.[68] Its purpose is to provide a simple picture of the essential electrical behaviors of the short-channel MOSFETs from the perspective of a circuit designer That is, this engineering model relates the terminal voltages to the drain current, much as Eqs. 5-8 and 5-10 yield the I-V characteristics for long-channel MOS devices. Consequently, device designers who need to relate device and process parameters to circuit parameters should also find this model useful.

5.5.4 Summary of Short-Channel Effects on the Fabrication of MOS ICs

In the first section of this chapter, we showed that the use of lightly doped substrates generally produced optimum device behavior in long-channel MOS transistors. In this

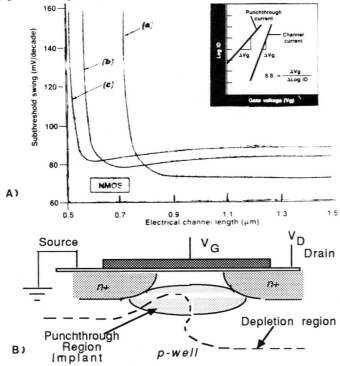

Fig. 5-24 (A) Subthreshold slope versus electrical channel length for NMOS devices (V_{Tn} = = 0.7 V), having a common threshold adjustment implant and punchthrough implant doses of: of: (a) zero; (b) 2×10^{11} cm^{-2}; and (c) 3×10^{11} cm^{-2}.[27] Reprinted with permission of Semiconductor International. (B) NMOS cross-section with implant placed into punchthrough region.

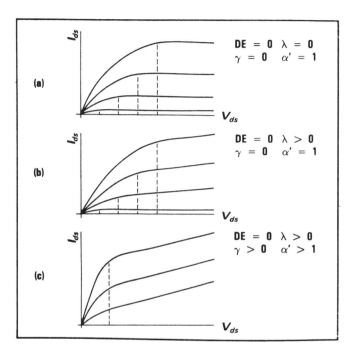

Fig. 5-25 The I-V curves of an MOS device showing the the effects of progressively increasing short-channel behavior. (a) Long-channel behavior. (b) With channel-length modulation. (c) Addition of velocity saturation.[28] (© 1986 IEEE).

section we noted that higher substrate doping is needed to overcome some of the detrimental impacts of short-channel effects. Thus, trade offs need to be made in selecting the proper substrate doping-concentration values to achieve optimum short-channel MOS device performance. Some of these trade offs are discussed by Kakumu,[29] who points out that a higher substrate doping concentration produces decreased ring-oscillator gate delay in submicron CMOS because of increased junction capacitances and decreased carrier mobility (due to increased impurity scattering).

5.6 HOT-CARRIER EFFECTS IN MOSFETs

If device dimensions are reduced and the supply voltage remains constant (e.g., 5 V), the lateral electric field generated in MOS devices increases. If the electric field becomes strong enough, it can give rise to so-called *hot-carrier* effects in MOS devices. This has indeed become a significant problem in NMOS devices with channel lengths smaller than 1.5 μm (and in PMOS devices with submicron channel lengths).[30] Hot-electron effects are more severe than hot-hole effects because of the higher electron mobility. Therefore, we begin our discussion by considering hot-electron effects in *n*-channel devices. At the end of this section we will also discuss the impact of hot-carrier effects on *p*-channel devices.

The maximum electric field, E_M, in a MOSFET occurs near the drain during saturated operation. A rigorous calculation of the field near the drain is a complex procedure, requiring a computer-aided solution of the two-dimensional Poisson equation, with the results of one such analysis being shown in Fig. 5-26. Nevertheless, the value of E_M can be estimated from[32]

$$E_M = (V_{DS} - V_{Dsat}) / m \qquad (5\text{-}19)$$

where

$$m = 0.22 \, t_{ox}^{1/3} / x_j'^{1/2} \qquad (5\text{-}19a)$$

and t_{ox} is the gate oxide thickness and x_j' approximately corresponds to the source/drain junction depth. Although V_{Dsat} depends on L_{eff}, the dependence is weaker than a linear

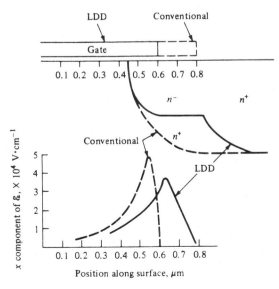

Fig. 5-26 Magnitude of the electric field at the Si-SiO$_2$ interface as a function of distance: L = 1.2 μm, V_{DS} = 8.5 V, V_{GS} = V_T.[38] (© 1980 IEEE).

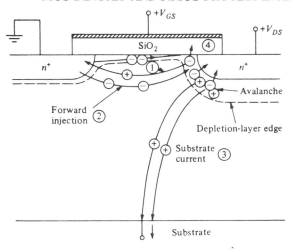

Fig. 5-27 Hot-carrier generation and current components. ①- Holes reaching the source. ②- Electron injection from the source. ③ - Substrate hole current. ④ - Electron injection into the oxide.

relationship (especially if L_{eff} >1 μm). Thus, we can see that E_M increases as device dimensions shrink, but this is due to thinner gate oxides and shallower junctions, as well as to the reduction in L_{eff}.

Regardless of the factors that increase their magnitude, such high electric fields cause the electrons in the channel to gain kinetic energy and become "hot" (i.e., their energy distribution is shifted to a much higher value than that of electrons which are in thermal equilibrium with the lattice). Such hot electrons (which become hot near the drain edge of the channel, since that is where E_M exists) can cause several effects in the device. First, those electrons that acquire ≥1.5 eV of energy can lose it via *impact ionization,* which generates electron-hole pairs. The total number of electron-hole pairs generated by impact ionization is *exponentially dependent on the reciprocal of the electric field, -* $1/E_M$. In the extreme, this electron-hole pair generation can lead to a form of *avalanche breakdown* (*Process 1,* shown in Fig. 5-27). Second, the hot holes and electrons can overcome the potential energy barrier between the silicon and the SiO_2 (~3.1 eV), thereby causing *hot carriers to become injected into the gate oxide.* Each of these events brings about its own set of repercussions.

5.6.1 Substrate Currents Due to Hot Carriers

When electron-hole pairs are created by impact ionization, the electrons are normally attracted to the drain, and they add to the drain current. The *holes,* on the other hand, enter the substrate and constitute a part of the parasitic substrate current (*Process 3* in Fig. 5-27). This substrate current, I_{sub}, can itself produce several problems:

• If a substrate bias-generator circuit is included on-chip, the output of the bias generator will be less negative with increasing substrate current.

• If some of these holes are collected by the source (instead of by the body contact), *and* this collected hole current causes a voltage drop in the substrate material on the order of 0.6 V, the substrate-source *pn* junction will conduct significantly. Electrons will then be injected from the source to the substrate, just like electrons injected from emitter to base of an *npn* transistor (the *forward injection* shown in Fig. 5-27). These electrons can, in turn, gain sufficient energy as they travel toward the drain to cause additional impact ionization and create new electron-hole pairs. A positive-feedback mechanism thus exists, one that can sustain itself if the drain voltage exceeds a certain value. This is observed externally as a form of breakdown, referred to as a *snapback breakdown*. A particularly clear explanation of this effect, including the reason for the negative-resistance, or "snapback," portion of the curves, is given in reference 3.

• As some of the holes are accelerated through the drain-substrate depletion region, they may acquire enough energy to cause secondary impact ionization. This will create electrons far from the drain region. Some these electrons, instead of being collected by the drain, may escape the drain field and instead travel (sometimes hundreds of microns) to other nodes on the chip. This may lead to a reduction of the storage time of dynamic circuit nodes in DRAMs (i.e., manifested as a degradation of the refresh time).[33] This excess electron current is reported to be around 10^{-4} times smaller than the substrate ionization impact current itself.

• I_{sub} may induce latchup in CMOS circuits.

The magnitude of the substrate current depends exponentially on the value of E_M. An example of how I_{sub} depends on decreasing channel length is seen in Fig. 5-28.[34] This shows the maximum I_{sub} generated at a voltage of 5 V versus the effective channel length for MOS transistors processed with the same technology. The magnitude of I_{sub} would increase even more rapidly with shrinking L_{eff} if the oxide thickness and junction depth were scaled together with the channel length.

5.6.2 Hot-Carrier Injection into the Gate Oxide

The hot holes and electrons that are injected into the gate oxide cause another set of problems. First, some of these carriers pass to the gate electrode (mostly hot holes) and thus constitute a gate current, I_G, typically in the fA (10^{-15} A) range. For higher, but still nondestructive, biases, the gate current can grow rapidly to become several pA (10^{-12} A).

However, some fraction of the hot carriers injected into the gate oxide do not reach the gate electrode. This is because the gate oxide contains empty electron states (also known as *traps*), which can be filled by the injected hot carriers. Such occupied traps generally become electron traps, even if they are initially filled with holes. As a result, there is a negative charge density caused by the trapping of the hot carriers in the oxide. This *trapped charge* acts like a contribution to the fixed oxide charge term in the expression for the device threshold voltage. Furthermore, the trapped charge accumulates with time. Due to the polarity of the trapped charge, the resulting shift in the *n*-channel device threshold voltage is *positive* (i.e., V_T increases in NMOS devices). If I_G becomes of the order of pA, the trapping of a fraction of the electrons injected into the oxide can become a significant effect (the fraction that gets trapped is ~1 in 10^6 injected electrons). As a result, this increase in *negative* stored charge can lead to a permanent change in V_T in the MOSFET. (The means by which charge in the oxide impacts V_T was described earlier.)

EXAMPLE 5-6: V_T Shift due to Hot-Electron Effects. Given a short-channel NMOS device with $t_{ox} = 20$ nm, a gate *width* of 5 μm, and a gate current of 1 nA that is momentarily caused by hot carrier injection. The current flows through a region of the gate oxide near the drain end which is 5 μm wide, and 0.2 μm long. Assume that 1 in 10^6 electrons of the gate current becomes trapped at an average distance of 0.1 t_{ox} from the Si-SiO$_2$ interface. How long does the gate current have to flow to change V_T by 0.1 V in the region where the injection is occurring?

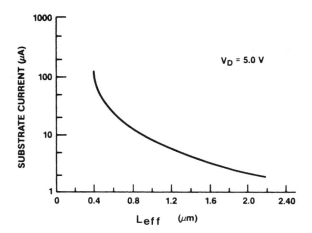

Fig. 5-28 The maximum substrate current due to impact ionization produced at a drain voltage of 5 V vs effective channel length for 250Å gate oxide transistors.[34] (© 1986 IEEE).

SOLUTION: The gate current of 1 nA produces a current density across the injection area of

$$10^{-9} \text{ nA } /1 \times 10^{-8} \text{ cm}^{-2} = 0.1 \text{ A cm}^{-2}$$

Since 1 in 10^6 electrons become trapped, the rate of charge trapping, J_{ot}, in the gate oxide is $0.1 \times 10^{-6} = 1 \times 10^{-7}$ C sec^{-1} cm^{-2}. The shift in V_T due to the trapped charge in the oxide can be found from

$$\Delta V_T = (1/C_{ox})(0.9 \ t_{ox}/t_{ox}) \Delta Q_{tot}$$

or, solving for ΔQ_{tot},

$$\Delta Q_{tot} = C_{ox} \Delta V_T /0.9 = [1.7 \times 10^{-7} \times 0.1]/0.9$$

$$= 1.89 \times 10^{-8} \text{ C/cm}^2$$

The time, t, necessary to trap this quantity of charge is

$$t = \Delta Q_{tot}/J_{ot} = 1.89 \times 10^{-8}/1 \times 10^{-7} = 0.19 \text{ sec}$$

5.6.3 Device-Performance Degradation Due to Hot-Carrier Effects

The increase that occurs in V_T leads to other changes in the MOS characteristics. First, the saturation current decreases because $(V_G - V_T)$ becomes smaller (Fig. 5-29a).[77] Second, as the substrate current increases, the transconductance decreases. Finally, since the trapped charge accumulates with time, the device performance will become unacceptable for a given application after a certain time of device operation.

A device lifetime, τ, can therefore be defined by selecting the maximum percentage of allowed degradation in the critical device parameter. It has been found, however, that this lifetime (regardless of whether V_T or g_m is monitored) can be related to I_{sub} by a power-law relationship (i.e., τ is observed to be inversely proportional to I_{sub} when plotted on a log-log plot). Figures 5-29b and c show τ as a function of I_{sub} for: (a) τ is defined as a 10-mV shift in V_T; and (b) τ is defined as a 10% degradation in g_m.

This has led to accelerated testing techniques that stress the devices with higher drain voltages than would be used in normal operating conditions.[35] A larger I_{sub} is thereby produced, leading to shorter times to device failure (i.e., possibly defined by a 10-mV shift in V_T or a 10% degradation in g_m). The time to failure under normal operating drain voltages (τ) is then extrapolated from these data using curves as shown in Figs. 5-29b and 5-29c.[36,37] Device design can then be modified to yield a desired lifetime (typically, 10 years) under normal operating conditions.

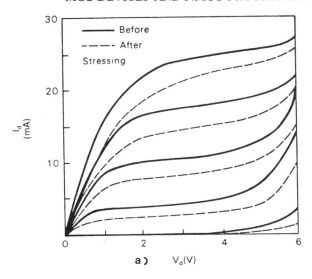

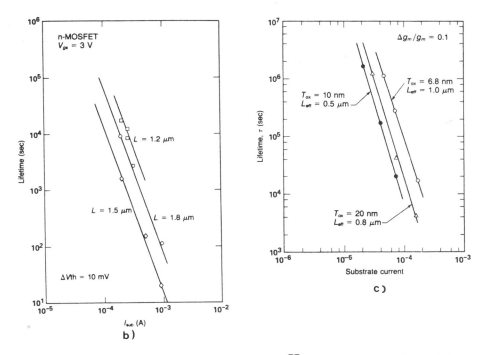

Fig. 5-29 (a) Degradation of I_{DSsat} after stressing.[77] (© 1984 IEEE). Degradation of other device-performance parameters as a function of substrate current (I_{sub}): (b) Lifetime defined as a 10-mV shift in V_T; (c) Lifetime defined as a 10-percent degradation in g_m.

5.6.4 Techniques for Reducing Hot-Carrier Degradation

Device performance degradation from hot electron effects can be reduced by a number of techniques. These include the following:

- The voltages applied to the device can be decreased (e.g., by lowering the power-supply voltage from 5 V to 3.3 V). The decision to implement this reduction, however, is not in the hands of the device designer or fabricator. The issue of reduced supply voltages for submicron MOS technologies is covered in chapter 6, section 6.7.2.

- The time the device is under the voltage stress can be shortened (e.g., by using a lower duty cycle and clocked logic).

- Appropriate drain engineering design techniques, which results in special drain structures that reduce hot electron effects in MOS devices (i.e., double-diffused drains and lightly-doped drains), can be implemented.

- The density of trapping sites in the gate oxide can be reduced through the use of special processing techniques.

- Thin, lightly doped epitaxial-layers on low-resistance substrates can be employed to shunt away substrate current and help eliminate the problem of impact-ionization-induced latchup.

We will next discuss the details of the latter three approaches since their implementation is the task of device and process engineers.

5.6.5 Lightly Doped Drains (LDD)

It has been determined that hot-carrier effects will cause unacceptable performance degradation in NMOS devices built with conventional drain structures if their channel lengths are less than 2 μm. To overcome this problem, such alternative drain structures as *double-diffused drains* and *lightly doped drains* (LDDs) must be used. The purpose of both types of structures is the same – namely, to absorb some of the potential into the drain and thus reduce E_M. Double-diffused drains, however, are less effective for short-channel devices (i.e., ≤1.25 μm) because they cause deeper junctions and more overlap capacitance. We will thus restrict our discussion to LDDs. Table 5-2 shows the evolution in AT&T's Twin-Tub CMOS technology, with respect to such drain structures.[30]

In the LDD structure, the drain is formed by two implants (Fig. 5-30). One of these is self-aligned to the gate electrode, and the other is self-aligned to the gate electrode on which two oxide *sidewall spacers* have been formed. (An extensive report on the details of sidewall spacer technology for both MOS and bipolar devices can be found in ref. 66.) The purpose of the lighter first dose is to form a lightly doped section of the

Table 5.2 Evolution of Device Structures in AT&T's Twin-Tub CMOS Technology Development[30] (Twin-Tub VI announced in 1989)[83]

	Twin-Tub I	Twin-Tub II	Twin-Tub III	Twin-Tub IV	Twin-Tub V	Twin-Tub VI
Design Rule	3.5 μm	2.5 μm	1.75 μm	1.25 μm	0.9 μm	0.6 μm
L_{eff}	2 μm	1.5 μm	1.3 μm	1.0 μm	0.75 μm	0.4 μm
t_{ox}	600 Å	350 Å	250 Å	200 Å	150 Å	125Å
Device Structure	Conventional	Conventional	DDD	LDD	N&P LDD	N&P LDD

drain at the edge near the channel. In NMOS devices, this dose is normally $1\text{-}2\times10^{13}$ atoms/cm^2 of phosphorus.

The value of E_M is reduced by this structure because the voltage drop is shared by the drain *and* the channel (in contrast to a conventional drain structure, in which almost

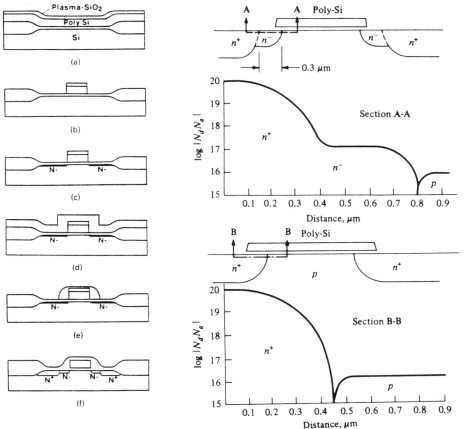

Fig. 5-30 (a) Process sequence used to form lightly doped drain (LDD) structures. (b) Doping profile in an LDD structure taken through section A − A. (c) Doping profile in a conventional drain structure taken through section B − B.

the entire voltage drop occurs across the lightly doped *channel* region). Figure 5-25 shows the electric-field profile at the drain of a MOSFET, both with and without an LDD structure.[38] We see that the electric field can be reduced by about 30-40%. Since the hot-electron-induced gate currents are exponentially dependent on E_M, this is sufficient to reduce currents by many orders of magnitude. As a result, the stability of the device is greatly increased. To ensure a high quality interface under the sidewall-spacer oxide, it is important that the thermally grown gate oxide remain in place after the polysilicon gate etch. This requires a polysilicon etch process with high selectivity to oxide, since the gate oxide is typically less than 20 nm.

The heavier, second dose forms a low resistivity region of the drain region, which is also merged with the lightly doped region. (In NMOS devices this implant is typically arsenic at a dose of about 1×10^{15} atoms/cm^2.) Since it is further away from the channel than would be the case in a conventional drain structure, the depth of the heavily-doped region of the drain can be made somewhat greater without adversely impacting the device operation (e.g., 0.3 μm deep versus 0.18 μm deep). The increased junction depth lowers both the sheet resistance and the contact resistance of the drain (see chap. 3). Deeper junctions also provide better protection against junction spiking.

The disadvantages of LDD structures are their increased fabrication complexity compared to conventional drain structures and the increased parasitic resistance of the source and drain regions caused by the lightly doped regions of the drain. This increase in parasitic channel resistance results in devices that dissipate more power for a constant applied voltage.

The effect on the I_D-V_{DS} curves of the MOSFET due to the additional voltage drop across the two lightly doped regions is that I_D does not saturate until a higher value of V_{DS} is applied. Due to the extra series resistance, the total channel conductance is appreciably lower in the linear region, but only slightly lower in the saturation region. This is because the channel region in saturation already has a high channel resistance, while in the linear region the resistance is much lower. The additional series resistance of the LDD therefore does not significantly increase the total resistance of the MOSFET in saturation, and I_D remains less affected. Consequently, the drain current is reduced more significantly in the linear region than in the saturation region.)

The I-V curves for LDD devices are consequently recognizable by their characteristically round shape, as shown in Fig. 5-31, curve 6. These curves also illustrate progressively the other short-channel effects described in section 5.5, beginning with the I-V curve of an ideal long-channel MOS device (Curve 1). Thus, curve 6 shows the I-V curve of a short-channel MOSFET with an LDD.

5.6.5.1 Drain Engineering for Optimum LDD Structures.

A proper LDD design should provide adequate hot-carrier protection while not introducing excessive source/drain resistance, which would degrade the device performance. The design and fabrication effort undertaken to achieve such an optimum LDD structure is referred to as *drain engineering*; this as we shall see, is a relatively complex task. A thorough approach to designing adequate hot-electron protection would entail addressing all of the following objectives:

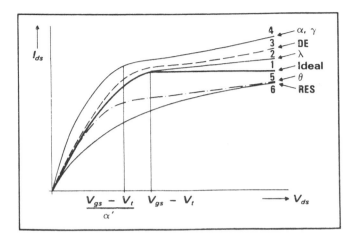

Fig. 5-31 Progressive influence of short-channel effects and LDD on the characteristics of an ideal, long channel MOSFET (Curve 1). Curve 6 shows the I-V curve of a short channel MOSFET that is also fabricated with an LDD, while Curve 5 shows the characteristic of the same device without an LDD.[28] (© 1986 IEEE).

1. Reducing the maximum electric field in the silicon as much as possible.
2. Ensuring that the injection position (i.e., the E_M point) is located under the gate edge.
3. Ensuring that the impact ionization region is pushed far below silicon surface to reduce the possibility of hot carriers reaching the $Si-SiO_2$ interface.
4. Separating the point where the electric field in the silicon is a maximum from the point of maximum current flow in the channel.
5. Minimizing the increase in the parasitic resistance due to the LDD structure.

The degree by which E_M is reduced (*Objective #1*) depends primarily on the doping value of the lightly doped extension of the drain, as well as on its length. If the doping level is too high, the value of E_M is not sufficiently reduced, and hot-electron protection is not provided. If the level is too low, the drive-current capability of the device is severely reduced, and the surface will easily be depleted by the hot carriers that do get trapped in the gate oxide above it (i.e., the device will again be vulnerable to hot-electron degradation). A model which calculates the electric field in LDD structures has been published.[72] It concludes that the lightly doped region should be long enough to attenuate the electric field to a value that is below the critical ionization field, but should still be short enough to keep the series resistance from becoming excessive. In early LDD structures in NMOS this meant that the primary parameters selected were the n^- length and its doping concentration.[39]

A process for forming LDD structures in *either* or *both* PMOS and NMOS structures in CMOS with only two photomasks was reported by Parrillo, et al.[40] An extension of this process also uses removable TiN gate sidewall spacers and incorporates self-alignment of the lightly doped regions to their respective gates without overcompensation (Fig. 5-32). Another example LDD structure in 0.8-μm CMOS is given in reference 41.

To keep the location of the E_M point under the gate (*Objective #2)* and yet maintain high drivability requires the proper combination of gate-drain (g/d) overlap length and spacer length. The g/d overlap length should be long enough to let the E_M point lie under the gate, and the spacer length should be short enough to let the n^+ region reach under the gate. In general, for process controllability, if a 0.3-μm-thick polysilicon film is used for the gate, a 0.2-μm-wide spacer is selected.

To attain *Objective #3,* a metal-coated LDD structure has been developed.[42] It uses a deeper n^- phosphorus profile than the n^+ source-drain (arsenic), in order to steer the maximum current path away from E_M (Fig. 5-33b).

A buried LDD structure has also been proposed to reach *Objectives #3 and #4*. This structure uses a retrograde LDD profile, with the peak concentration below the Si surface (Fig. 5-33c). Besides a reduction in substrate current similar to the metal-coated LDD, this profile also suppresses hot-carrier injection by driving the current away from the surface and shifting the avalanche region further into the bulk silicon.

The buried LDD approach together with an additional shallow arsenic implant (Fig. 5-34d), may allow NMOS transistors of 0.6 μm to be built with adequate hot-electron protection and with minimal the increase in parasitic resistance due to the LDD structure (*Objective #5).*[30,43,44] The idea is to form an abrupt profile at the silicon surface underneath the spacer to reduce the series resistance. The buried LDD profile is

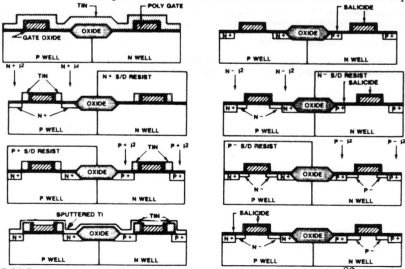

Fig. 5-32 Process sequence showing the TiN removable spacer process.[82] (© 1989 IEEE).

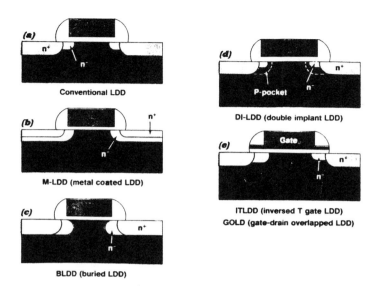

Fig. 5-33 Conventional LDD drain engineering and some of its variations – all designed to reduce hot-carrier effects – are shown.[30] (© 1986 IEEE).

used to force the main current path and the impact-ionization region deep into the silicon. There are two peaks in the lateral electric-field distribution, and the maximum current path is through the saddle point of these two high field regions. Thus, this device structure improves both the current drive and the hot-carrier resistance. A further modification of the buried LDD is the *sloped-junction LDD (SJLDD)*, shown in Fig. 5-33e.[45] In this structure a 165-keV phosphorus implant is used to form the lightly doped drain region; this provides improved device lifetime under high field stress. The main reason for the improvement is that hot-carrier generation is driven further away from the Si-SiO_2 interface by the junction of the SJLDD.

5.6.5.2 Asymmetrical Characteristics of LDD MOSFETs.

Formation of the lightly doped regions of LDDs is accomplished by means of ion implantation, with the polysilicon gate used as the implant mask. This can cause asymmetric doping of the source and drain regions,[46,64] with such asymmetry occurring as a result of implanting off-axis (typically, at an angle of 7°) to avoid channeling (see Vol. 1, chap. 9). This produces lateral shadowing (S in Fig. 5-34) of the substrate on one side of the poly gate, and penetration of dopants through the leading corner of the poly on the other side. The problem is compounded by etch processes that produce a re-entrant angle in the poly gate sidewall (Fig. 5-34 - poly etch process A.) The impact of this effect on the device characteristics can be examined by considering the implants to the lightly doped and the heavily doped regions separately.

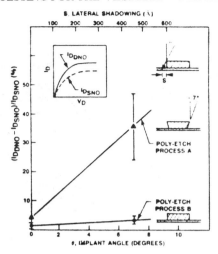

Fig. 5-34 Percent asymmetry in saturated drain currents ($V_{DS} = V_{GS} = 5$ V) versus source/drain implant angle for LDD NMOS devices having reentrant- and vertical-wall polysilicon gates. Insert schematically shows source-non-overlap (SNO) and drain-overlap (DNO) I-V behavior.[40] (© 1986 IEEE).

If the implant that forms the lightly doped region of the drain (usually phosphorus in NMOS devices) is done off-axis, the as-implanted region in the shadow of the gate will have its edge displaced from the gate edge. As a result, there will be *no overlap* of the implanted region and the gate edge. Even after a drive-in step, the overlap of this edge with the gate will be less than the overlap of the nonshadowed implant region edge and the gate. If the less-overlapped side is used as the source, the extra resistance introduced will reduce the drain current (I_{Dsno} in Fig. 5-34), thus degrading circuit speed. However, if the drain side has less overlap, the extra resistance will not affect I_D (curve I_{Ddno} in Fig. 5-34). The asymmetry in drive current when the same device is connected in these two ways can be as much as 40%; this can be catastrophic in circuits that require closely matched device characteristics.

The problems arising from such n^- implant-shadowing can be alleviated by one of the following measures:

• Insuring that the poly-etch process produces vertical sidewalls in the poly gates.

• Avoiding excessive reoxidation of the polysilicon after its definition (note that this also prevents a gap from being established between the drain region edge and the edge of the gap). A dry-oxidation step is often employed to reduce the accelerated oxidation rate on n^+ polysilicon.

• Implanting with a vertical implant through a screen oxide[27] (which reduces the channeling that would otherwise occur when a vertical implant is performed).

• Using a poly-etch process that produces a slight positive bevel to the poly sidewall profile, and combining this etch with a smaller angle (3° vs 7°) off-axis implant.[46]

Let us next examine how the implantation of the heavily doped region impacts the structure of the LDD. If an off-axis implant is again used with the heavily doped implant, asymmetric implant effects can again alter the LDD structure. That is, on the side of the gate that is *not* shadowed, the arsenic atoms penetrate the leading corner of the spacer and thereby enter the substrate under the spacer to some extent. If the spacer is too narrow, this can wipe out the lightly doped drain-extension region. When this happens, the device can lose the hot-carrier protection that was to be provided by the LDD structure, and it will become vulnerable to degradation by hot-carrier effects. To avoid this, the spacer must be sufficiently wide. Some guidelines for choosing the proper spacer width to deal with this problem are given in reference 47.

The general problem involving the gate-to-drain overlap with respect to its impact on the MOSFET characteristics was reviewed by Ko, et al.[48] It was observed that the critical dimensions in drain structures having weak overlap (WO) to the gate were only tens of nm in devices in which $L_{eff} = 1$ μm. In general, the drain current decreases as the overlap is weakened, a double hump appears in the substrate current of asymmetrical WO devices, and the reliability of the WO devices can be lower than devices having adequate overlap. To avoid random degradation of device performance by WO effects that arise as a result of process variations, stringent process-control measures must be exercised when submicron devices are fabricated.

5.6.6 The Impact of IC Processing on Hot-Carrier Device Degradation

The lifetime of a device in which the hot-electron degradation impacts device performance can be increased by keeping the number of trapping centers in the gate oxide to a minimum. In essence, this reduces the density of states that the hot electrons injected into the oxide can occupy. Maintaining a minimum number of trapping centers can be achieved in several ways during the device-fabrication process sequence. First, the gate oxide should be grown by a process that produces low interface-trap densities (such processes are described in Vol. 1, chap. 7). It has also recently been reported that the incorporation of optimized amounts of fluorine during the gate-oxide growth process suppresses interface-state generation during injection.[49] Second, the degradation of the oxide caused by damage during process steps carried out in a plasma environment should be minimized (damaging processes include plasma-enhanced CVD of oxides and nitrides, RIE, and sputter deposition of metal). Finally, the amount of hydrogen that is incorporated at the Si-SiO$_2$ interface should be reduced. The latter two topics will be considered in more detail here.

In typical CMOS processing, most of the damage created by RIE or ion implantation is annealed out at a high temperature (above 800°C) before metallization. Once the first layer of aluminum is deposited, however, the maximum annealing temperature is limited to ~450°C. It has therefore been reported that the RIE of the second layer of metal in multilevel metal processes deteriorates the device-aging characteristics. It has been found that while a subsequent anneal performed at 375°C is ineffective in annealing out this damage,[50] an anneal at 450°C improves device-aging characteristics.

Excess hydrogen at the Si-SiO$_2$ interface is also reported to be a culprit in increasing the density of the interface states. In most cases, hydrogen is used to fill the dangling bond, forming Si-H at the Si-SiO$_2$ interface. However, the Si-H bond can be easily broken by injected hot electrons. Furthermore, excess hydrogen introduced during processing can diffuse to the interface and lead to enhanced bond-breaking behavior. Such excess hydrogen can arrive during the final sinter step before passivation (i.e., in H$_2$ or N$_2$ + 5% H$_2$), or during the deposition of a conventional silicon nitride passivation layer. It has therefore been reported that sintering in pure N$_2$ produces devices with a lower device degradation rate.[51] Similarly, it has been reported that when a fluorinated nitride (F-SiN) is used as a passivation layer,[50,52] devices exhibited slower degradation rates than those seen in devices with conventional SiN passivation layers. F-SiN films can be produced by incorporating NF$_3$ in the deposition process to form an Si-F bonding structure instead of Si-H in this film.

The presence of fluorine in the gate oxide (possibly inadvertently originating from a BF$_2^+$ source/drain implant implant) has recently been reported to increase the hot electron resistance of devices fabricated with such oxides.[74] Other reports confirm that a deliberate incorporation of fluorine into the gate oxide (from implanting fluorine into a polysilicon film, and then diffusing it into the gate oxide) produces a more hot-electron resistant interface.[78,79]

5.6.7 Hot-Carrier Effects in PMOS Transistors

Hot-carrier effects are not significant in PMOS transistors at channel lengths greater than 1 μm because the impact-ionization rate of holes is 3 to 4 orders of magnitude lower than that of electrons at a given electric field.[53] At submicron channel lengths, however, hot-electron effects in PMOS start to become important. It has been reported that two hot-carrier effects predominate in such PMOS devices. Both are caused by hot electrons that are generated by impact ionization and are then injected into the oxide, becoming trapped. Hot holes do not appear to cause significant effects unless a device is stressed at large magnitudes of V$_G$. In PMOS devices, the gate-bias polarity favors electron injection (which is opposite to that in NMOS) but retards the injection of holes.

In the first degradation effect, these electrons are trapped near the drain and shorten the effective channel length. This reduction in L$_{eff}$ is even more severe in the buried-channel PMOS devices (used in CMOS technologies that have an n^+ polysilicon gate), because their buried-channel nature makes them more vulnerable to source-drain punchthrough,[54] and consequently increases subthreshold leakage. This effect is called *hot-electron-induced punchthrough* (HEIP).

The second effect was observed in PMOS devices with p^+, as well as n^+, polysilicon gates. In these devices, the injected and trapped hot electrons also produce V$_T$ shifts (which reduce the magnitude of V_T) and increases in the transconductance, g_m. Nevertheless, contrary to what occurs in NMOS, in PMOS these effects tend to saturate. This is explained by the fact that the electrons trapped near the drain reduce the electric field present there in PMOS devices.[55] As a result, this effect does not appear to limit PMOS devices fabricated with p^+ polysilicon gates as long as L_{eff} is ≥ 0.6 μm.[56] Still, LDD structures appear to be needed for p-channel devices fabricated with n^+ *polysilicon gates* when L_{eff} gets to be 0.8 μm or less.[54,57]

The LDD structure also helps to reduce subthreshold leakage in PMOS devices, where excessive junction depths caused by lateral diffusion have been a problem. The use of LDD basically increases L_{eff}, hence compensating for the L_{eff} reduction caused by hot-electron injection. A new LDD structure for PMOS devices, called a *halo LDD*, is described in reference 63. In this structure, a deeper phosphorus implant is placed below the lightly doped drain-extension p-type implant. The punchthrough resistance of the PMOS device is reported to be significantly improved by this LDD structure.

5.6.8 Gate-Induced Drain-Leakage Current

Another type of leakage current between drain and substrate in thin gate-oxide (12-28 nm) MOS devices has been observed at drain voltages much lower than the breakdown voltage.[58,59] This is not really a hot-electron-induced effect, but since it uses LDD structures to reduce its magnitude, we describe it in this section. The basis of this current is band-to-band tunneling that occurs through the gate oxide into the deep depletion layer in the gate-to-drain overlap region. It has been reported that in order for this leakage current to limited to less than 0.1 pA/μm^2, the oxide field in the gate-to-drain overlap region must be limited to 1.9 MV/cm. LDD structures are effective in reducing this leakage current, but only if the doping in the n^- extensions is less than 10^{18}/cm^3. Buried LDD structures with the peak of the n concentration at several hundred angstroms beneath the Si-SiO$_2$ interface appear to be even more effective than conventional LDD structures in reducing the effect. This effect is predicted to become more of a problem as devices continue to be scaled - that is, if a 10-nm-thick gate oxide is used, the E-field will be less than 1.9 MV/cm in the oxide only if a 3.1 V power supply is used.

REFERENCES

1. R. S. Muller and T. I. Kamins, *Device Electronics for Integrated Circuits,* 2nd Ed., New York, John Wiley & Sons, 1986.

2. R. F. Pierret, *Field-Effect Devices*, Vol. IV in the Modular Series on Solid-State Devices, Reading, MA, Addison-Wesley, 1983.

3. D. K. Schroder, *Advanced MOS Devices*, Vol. VII in the Modular Series on Solid-State Devices, Reading, MA, Addison-Wesley, 1987.

4. Y. P. Tsividis, *The Operation and Modeling of the MOS Transistor,* New York,   McGraw-Hill Book Co., 1987.

5. E. H. Nicollian and J.R. Brews, *MOS Physics and Technology,* Wiley-Interscience, New York, 1982.

6. S. M. Sze, *Physics of Semiconductor Devices,* 2nd Ed. Wiley, New York, 1981.

7. D. A. Hodges and H. G. Jackson, *Analysis and Design of Digital Integrated Circuits,* McGraw-Hill Book Co., New York, 1983.

8. R. A. Chapman et al., *Tech. Dig. IEDM ,* 1987, p. 362.

9. R. J. Bayruns et al*., IEEE J. Solid-State Circuits,* SC-19, 1984, p. 755.

10. H. L. Armstrong, "A Theory of Voltage Breakdown of Cylindrical P-N Junctions," *IRE Trans. Electron Dev.,* ED-4, 15 (1957).

11. A. S. Grove, *Physics and Technology of Semiconductor Devices,* Wiley, New York, 1967, Chap. 6

12. E. C. Douglas, "Advanced Process Technology for VLSI Circuits," *Solid State Tech.,* 24, May 1981, p. 65.

13. C. M. Osborn and E. Bassous, *J. Electrochem. Soc.,* 122, 89, (1975).

14. S. M. Sze, *Semiconductor Devices: Physics & Tech.,* Wiley, New York, 1985.

15. N. Kotani and S. Kawazu, *Solid-State Electronics,* 22, p. 63, (Jan. 1979).

16. T. Yamaguchi et al., *IEEE Trans. Electron Dev.,* ED-31, 1984, p. 205.

17. C. Mead and L. Conway (Eds.), *Introduction to VLSI Systems,* Reading, MA, Addison-Wesley, 1980.

18. A. Mukherjee, *NMOS and CMOS VLSI Systems Design,* Prentice-Hall, Englewood Cliffs, N .J., 1986.

19. N. Kotani and S. Kawazu, *Solid-State Electronics,* 22, 63 (January 1979).

20. L. D. Yau, *Solid-State Electronics,* 17, 1059, (1974).

21. L. M. Dang, *IEEE J. Solid-State Circuits,* SC-14, 358 (1975).

22. G. Merkel, *Process and Device Modelling for IC Design* (F. Van de Wiele, W. L.Engle, and P.G. Jespers, Eds.), Noordhoff, Leyden (1977), p. 705.

23. R. R. Troutman, "VLSI limitations from Drain-Induced Barrier Lowering," *IEEE Trans. Electron Dev.,* ED-26, p.461, 1979.

24. J. Zhu et al., "Punchthrough Current for Submicron CMOS VLSI," *IEEE Trans. Electron Dev.,* Feb. 1988, p. 145.

25. K. M. Cham, *et al., Computer Aided Design and VLSI Device Development,* Kluwer Academic Publishers, Boston, 1986, p. 194.

26. H. Nihara, et al., "Anomalous Drain Current in NMOS and its Suppression by Deep Ion Implantation," *IEDM Tech. Dig.,* p. 487 (1978).

27. L. C. Parrillo, "CMOS Active and Field Device Fabrication," *Semiconduct International,* April 1988, p. 64.

28. C. Duvvury, "A Guide to Short-Channel Effects in MOSFETs," *IEEE Circuits and Systems Magazine,* Nov. 1986, p. 6.

29. M. Kakamu et al., "Power Supply Voltage for Future CMOS VLSI in Half- and Sub Micrometer," *Tech. Dig. IEDM,* 1986, p. 399.

30. M. L. Chen, "CMOS Hot-Carrier Protection with LDD," *Semiconductor Internat.,* April 1988, p. 78

31. P. K. Ko, "Hot Electron Effects in MOSFETs," Doctoral Thesis, Dept. of EECS, University of California, Berkeley, June 1972.

32. T. Y. Chan, P.K. Ho, and C. Hu, *IEEE Electron Dev. Letts.*, **EDL-6**, p. 551, 1985.

33. P. K. Chatterjee, *Tech. Dig. IEDM,* 1979, p. 14.

34. M. L. Woods,"MOS VLSI Reliability and Yield Trends," *Proceedings IEEE*, Dec. 1986, p. 1715.

35. P. Yang and S. Aur, "Modelling of Device Lifetime due to Hot Carrier Effects," *Proc. of 1985 Internat. Symp. on VLSI Tech.*, p. 227.

36. C. Hu et al., *IEEE J. Solid-State Circuits,* **SC-20**, 1985, p. 295.

37. E. Takeda and N. Suzuki, "An Empirical Model for Device Degradation due to Hot Carrier Injection," *IEEE Trans. Electron Dev. Letts.*, **EDL-4**, p. 111, 1983.

38. S. Ogura et al., *IEEE Trans. Electron Dev.*, **ED-27**, 1980, p. 1359.

39. H. Mikoshiba, "Comparison of Drain Structures in n-Channel MOSFETs," *IEEE Trans. Electron Dev.*, Jan. 1986, p. 140.

40. L. C. Parrillo et al., *IEDM Tech. Dig.*, 1986, p. 244.

41. R.A. Chapman et al., *IEDM Tech. Digest*, 1987, p. 362.

42. Y. Tsunashima et al., *Proc. of VLSI Symposium*, 1985, p. 114.

43. T. Toyoshima et al., *Proc. of VLSI Symposium*, 1985, p. 362.

44. T. Noguchi, *Tech. Dig. IEDM*, 1986, p. 730.

45. S. Jain et al., *IEEE Electron Dev. Lett.*, Oct 1988, p. 539.

46. T. Y. Chan et al., *Electron Dev. Lett.*, Jan. 1986, p. 16.

47. J. R. Pfiester and F.K. Baker, "Asymmetrical High Field Effects in Submicron MOSFETs", *EDM Tech. Dig.*, 1987, p. 51.

48. P. K. Ko et al., *Tech Dig. IEDM*, 1986, p. 292.

49. E. F. Da Silva, *Tech. Dig. IEDM*, 1987, p. 848.

50. M. -L. Chen, "Hot Carrier Aging in Two-Level Metal Processing," *Tech. Dig. IEDM*, 1987, p. 55.

51. F. C. Hsu, *Proc. of VLSI Symp.*, 1985, p. 108.

52. S. Fujita et al., *Tech. Dig. IEDM,* 1985, p. 64.

53. J. Y. Chen, "CMOS-The Emerging VLSI Technology," *IEEE Circuits and Devices Magazine*, March, 1986, p. 16.

54. M. Koyanagi et al., *IEEE Trans. Electron Dev.*, **ED-34**, 1987, p. 871.

55. C. Hu et al., *IEEE J. Solid-State Circuits*, SC-20, 1985, p. 295.

56. Y. Hiruta et al., *Tech. Dig. IEDM*, 1986, p. 718.

57. T. Mizuno et al., *Tech. Dig. IEDM*, p.726.

58. T. Y. Chan, J. Chen, P.K. Ko. and C. Hu, *IEDM Tech. Dig.* 1987, p. 718.

59. C. Chang and J. Lien, *IEDM Tech. Dig.* 1987, p. 714.

60. C. T. Sah, "The Evolution of the MOS Transistor," *Proceedings of the IEEE,* Oct. 1988, p. 1280.

61. P. Balk, "Effects of Hydrogen Annealing on Silicon Surfaces," presented at the *Electrochem. Soc. Spring Meeting,* May, 1965. Ext Abs. No. 109, p. 237.

62. M.-C. Jeng et al., "Design Guidelines for Deep-Submicrometer MOSFETs," *Tech. Dig. IEDM,* 1988, p. 386.

63. M. L. Chen et al., *Tech. Dig. IEDM*, 1988, p. 390.

64. R. W. Gregor, "Consequences of Ion Beam Shadowing in CMOS Source/Drain Formation," *IEEE Electron Dev. Letts.*, Dec. 1986, p. 677.

65. I. Ahmed, H. Naguib, and C. Gomez, *Ext. Abs. ECS Meeting,* Fall 1988, Abs. No. 282, p. 405.

66. S. H. Dhong and E. J. Petrillo, *J. Electrochem. Soc.,* Feb. 1986, p. 389.

67. F. Hsu and K. -Y. Chiu, *IEEE Electron Dev. Letts.,* **EDL-5**, p. 162, 1984.

68. K. - Y. Toh, P.-K. Ko, and R. G. Meyer, *IEEE J. Solid-State Ckts,* Aug. 1988, p. 950.

69. J. W. Brews et al., "Generalized Guide for MOSFET Miniaturization," *IEEE Electron Dev. Lett.,* **EDL-1**, p. 2, 1980.

70. R. H. Dennard et al., "Design of Ion-Implanted MOSFETs with Very Small Physical Dimensions," *IEEE J. Solid-State Circuits,* **SC-9**, Oct. 1974, p. 256.

71. S. G. Chamberlain and S. Ramanan, *IEEE Trans. Electron Dev.,* Nov. 1986, p. 1745.

72. K. Marayam, J. C. Lee, and C. Hu, *IEEE Trans. Electron Dev.,* July 1987, p. 1509.

73. L. D. Yau, "A Simple Theory to Predict the Threshold Voltage in Short-Channel IGFETs," *Soild-State Electronics,* vol. 17, p. 1059, 1974.

74. P. Wright et al., *IEEE Electron Device Letts.,* August 1989, p. 347.

75. V. L. Rideout, F. H. Gaensslen, and A. LeBlanc, "Device Design Consideration for Ion-Implanted *n*-Channel MOSFETs," IBM J. Res. Dev., p. 50, (1975).

76. R. Siqusch et al., *Tech. Dig. IEDM,* 1980, p. 429.

77. F. H. Hsu and K. Y. Chiu, *Tech. Dig. IEDM,* p. 96, 1984.

78. P. Wright et al., "Hot-Electron Immunity of SiO_2 Dielectrics with Fluorine Incorporation," *IEEE Electron Device Letters,* August 1989, p. 347.

79. Y. Nishioka et al., "Hot-Electron Hardened Si-Gate MOSFET Utilizing F Implantation," *IEEE Electron Device Letters,* April 1989, p. 141.

80. W. M. Werner, *Solid-State Electronics,* **17**, 769, (1974).

81. R. R. Troutman, *IEEE J. Solid-State Circuits,* **SC-9**, 55 (April 1974).

82. J. R. Pfiester et al., *Tech. Dig. IEDM,* 1989, p. 781.

83. C.-Y. Lu et al., *IEEE Trans. Electorn Dev.,* November, 1989, p. 2530.

CHAPTER 6

CMOS PROCESS INTEGRATION

Complementary MOS (CMOS) is so-named because it uses both *p*- and *n*-type (complementary) MOS transistors in its circuits. (Figure 6-1 depicts a CMOS inverter.) Since CMOS technology is significantly more complex than NMOS with respect to the device physics and fabrication issues, the discussion here will include a thorough introduction to these subjects. (Note that an excellent comprehensive text on CMOS, dealing with both circuit and design issues, has recently been published.)[1]

6.1 INTRODUCTION TO CMOS TECHNOLOGY

6.1.1 The Power-Dissipation Crisis of VLSI, and How CMOS Came to the Rescue

NMOS remained the dominant MOS technology as long as the integration level of devices on a chip was sufficiently low. It was inexpensive to fabricate, very functionally dense, and faster than PMOS. The earliest NMOS technologies required only 5 masking steps (including the pad mask).* On the other hand, NMOS logic-gates (e.g., inverters) draw dc power during one of the inverter states. Therefore, an NMOS integrated circuit will draw a steady current even when being operated in the standby mode (i.e., even when no signal is being propagated through the circuit). Consequently, as the number of logic gates on the chip grows, the current being drawn (and hence the power being dissipated) also increases. Although this was always a limitation of NMOS for such applications as space-borne or portable electronic systems, it did not represent a drawback for most other applications when the number of devices on a chip was rela-

---

* Early NMOS logic gates used butted contacts as well as enhancement-mode transistors for both the driver and the load. Depletion-mode loads and buried contacts came later. Hence, of the seven masks of the E-D NMOS process described in chapter 5, only five were required in early NMOS technology. This resulted not only in lower manufacturing costs per wafer, but also in higher yields. Thus, the cost of manufacturing NMOS circuits was brought even lower.

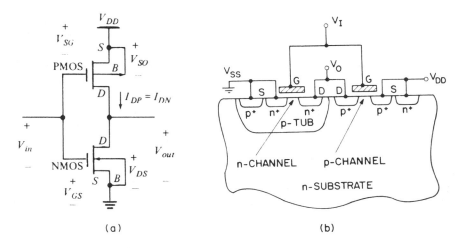

Fig. 6-1 CMOS inverter. (a) Circuit schematic. (b) Device cross section.

tively small. Such was the situation at the level of device integration that existed up to the mid-1970s.

With the dawning of the VLSI era, however, power consumption in NMOS circuits began to exceed tolerable limits. A lower-power technology was needed to exploit the VLSI fabrication techniques. CMOS represented just such a technology.

From a quantitative perspective, the ascendancy of CMOS was the inevitable result of the two-hundred-fold increase in functional density, and the twenty-fold increase in speed of integrated circuits between 1968 and 1987. For example, in 1969, 256-bit SRAM circuits (e.g., the Intel 1101) used 12-μm design rules to create the six-transistor SRAM cells; each cell occupied 20,600 μm^2. In 1987, however, SRAM cells using 1.2-μm design rules occupied only 150 μm^2. (The memory access-time in these respective memory circuits decreased from 1 μs to 46 ns.) By 1987 the decreased size and attendant increase in chip size allowed 256-kbit SRAMs to be built.

To take another example: The Intel 4004 4-bit microprocessor, introduced in 1971, had 2300 devices and was built in PMOS. The 8086 model, a 16-bit microprocessor introduced in 1978, had 29,000 devices and was built in NMOS, and the 80386, introduced in 1985, had 275,000 devices and was built in CMOS.

Chips can dissipate a maximum of about 5 W of power and still be used in conventional, but expensive, IC ceramic packages. In order for the much less expensive plastic packages to be used, however, the maximum power dissipation is limited to about 1 W. The Intel 8086 dissipated around 1.5 W of power when operated at 8 MHz. Thus, by the late 1970s it was already possible to manufacture NMOS chips whose power dissipation approached unacceptably high values. (Note that when the 8086 was later reissued in CMOS technology under the model number 80C86, its power consumption dropped to about 250 mW.)

In a CMOS inverter (unlike in an NMOS inverter) only one of the two transistors is driven at any one time. This means that when a CMOS inverter is not switching from one state to the other, a high-impedance path exists from the supply voltage to ground, regardless of the state the inverter is in. Hence, virtually no current flows, and almost no dc power is dissipated. CMOS thus allows the manufacture of circuits that need only several microwatts of standby power (Fig. 6-2).

The problem of power dissipation can also be considered from both a *chip* perspective and a *system* perspective. From the chip perspective, if microprocessors of the 32-bit generation were built in NMOS, they would dissipate 5 to 6 W of power. This would lead to severe heating and reliability concerns. In addition, expensive ceramic packages would be needed to house such chips. When such microprocessors are designed in CMOS, the power dissipation decreases to about 1 W.

From the system perspective, let's consider memory chips. Although a 1-Mbit DRAM may consume only 120 mW of power in NMOS, it consumes even less (~50 mW) in CMOS. Since there may be thousands of memory chips in a system (versus only a few microprocessors), the ramifications of lower power-dissipation are significant. Smaller power-supplies and smaller cooling fans are but two of these ramifications.

6.1.2 Historical Evolution of CMOS

Although CMOS is now the dominant integrated-circuit technology, for much of its 25-year history it was considered to be only a runner-up for the design of MOS ICs. The pairing of complementary *n*- and *p*-channel transistors to form low-power ICs was originally proposed by Sah and Wanlass in 1963.[2] The first CMOS ICs were fabricated

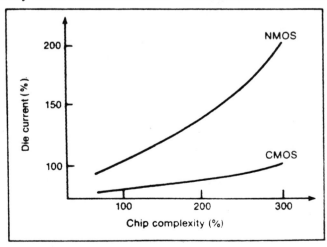

Fig. 6-2 As circuit complexity increases, NMOS power consumption rises to levels that eventually prevent further growth. In contrast, CMOS power consumption increases only slightly as the device count on a chip rises.

in 1966, and subsequent development of the technology was spearheaded by the RCA Corporation. The earliest volume commercial application was the use of the CMOS logic-inverter in the frequency divider circuits of digital watches.

CMOS technology at that time had many disadvantages compared to PMOS and, later, to NMOS. The drawbacks included significantly higher fabrication cost, slower speed, susceptibility to latchup, and much lower packing density. As a result, until the 1980s CMOS was limited to applications that could only be implemented with the technology's lower power dissipation (e.g., watches and calculators), or very-high noise margin (e.g., radiation-hard circuits). Furthermore, the advances made in NMOS fabrication were not rapidly transferred to CMOS, and for many years CMOS lagged behind the advanced Si-gate-NMOS and bipolar technologies. Except for the special applications mentioned above, it lay dormant for nearly a decade.

At the time CMOS circuits were first fabricated, the processes of ion implantation and local oxide isolation (LOCOS) had not yet been developed. In addition, metal gates were being used for MOS devices and control had not yet been established over the large and quite variable positive oxide charges in the gate oxide. As a result, p-well technology was the only means of fabricating CMOS.

While PMOS enhancement-mode transistors could be successfully fabricated in a lightly doped (e.g., 10^{15} cm^{-3}) n-substrate with a V_T of about -2 V, NMOS enhancement-mode transistors could not be fabricated on lightly doped p-substrates because the V_T of NMOS metal-gate transistors on such substrates is negative. In addition, the problems of oxide charge and the segregation of boron at the field-oxide/silicon-substrate interface, when combined with necessity of having to use relatively-thin field oxides (in order to be able to achieve adequate step coverage over nonrecessed field-oxide steps; see chap. 2), made it likely that parasitic channels would be established in the field regions between NMOS devices built on lightly doped substrates. Therefore, the only reliable way to manufacture enhancement-mode NMOS transistors for CMOS inverters was on regions with boron surface concentrations high enough to overcome these problems..

The p-well approach to building CMOS provided such regions, since the well had to be doped about ten times as heavily as the substrate for adequate control of doping in the well to be achieved. As a result, p-well technology became established in the companies that pioneered CMOS technology. Long after the problems of fabricating NMOS on lightly doped substrates had been solved (i.e., through control of the gate-oxide charge and of V_T by ion implantation) most of these companies continued to use p-well technology to design new circuits. While this was primarily due to historical inertia, p-well technology did have a few advantages over n-well technology (as will be discussed in section 6.2.2)

The packing-density limits of the early p-well CMOS technology, however, were responsible for its poor performance. The packing density was much worse than that of NMOS, primarily because the n- and p-devices had to be surrounded by guard rings (n^+ or p^+ diffusions that surround the device, as shown in Fig. 6-3)[114] to prevent the inversion of the field regions.

Before LOCOS isolation and ion implantation became available, guard rings were needed to provide adequately large values of V_T in the field regions. However, their use

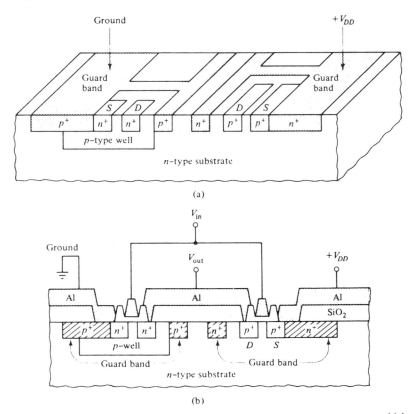

Fig. 6-3 CMOS structure with guard bands. A CMOS inverter circuit is depicted.[114] From W. C. Till and J. T. Luxton, *Integrated Circuits: Materials, Devices, and Fabrication.* Copyright 1982 Prentice-Hall. Reprinted with permission.

results in very large-area devices.* As a result, until oxide isolation and channel-stop implants were developed, interconnect capacitance and resistance severely degraded the speed performance.† Once guard rings were no longer needed, however, the devices could be brought closer together, and the speed of CMOS circuits improved dramatically.

When it became apparent in the late 1970s that the increases in power density and dissipation would make it impossible to design future generations of MOS circuits in

---

* The drain regions must also be isolated from the guard rings by a lightly doped region to prevent avalanche or Zener breakdown, as seen in Fig. 6-3. This requirement exacerbates the area penalty when guard rings are used.

† However, it should be noted that the fabrication of NMOS devices in heavily doped *p*-wells also increased the device junction-capacitances and reduced the magnitude of the drive current. This further decreased the performance of the circuits, but to a lesser degree than did the interconnect capacitance.

NMOS, the companies that had been stubbornly continuing to use NMOS for design and fabrication finally began to consider CMOS. It was natural for them to seek a technology that was compatible with the modern high-production-volume Si-gate-NMOS processes that they had successfully perfected. Since n-well CMOS offers near compatibility with such processes, and since it allows n-channel transistor performance to be optimized (through fabrication in the lightly doped p-substrate regions), it became the technology of choice for many companies that had formerly been manufacturing NMOS integrated circuits.[8]

It became evident, however, that neither p-well nor n-well would be the optimum choice for submicron CMOS. Instead, it appeared that *twin-well CMOS* would be more effective. As a result, many such processes were subsequently developed.

We will examine the various well technologies with regard to their advantages and disadvantages in modern CMOS processes. Circuit designers and process engineers should be aware of the trade-offs. The advances developed through the refinement of NMOS were incorporated into CMOS technology, and the performance was thus dramatically improved over that exhibited by primitive CMOS circuits.

6.1.3 Operation of CMOS Inverters

The CMOS inverter (for which the circuit schematic and a sample layout are shown in Fig. 6-1a and 6-4, respectively) uses enhancement-mode transistors for both the NMOS driver and the PMOS load transistors. The gates of the two transistors are connected and serve as the input to the inverter. The common drains of each device are connected to the output of the inverter, and we assume that the inverter is driving some load capacitance, C_L (e.g., the input to another CMOS logic gate). Note that both the source and the body of the NMOS transistor are connected to ground, while those of the PMOS transistor are connected to V_{DD} (e.g., 5 V).

The threshold voltage of the NMOS driver transistor, V_{Tn}, is positive (e.g., $V_{Tn} = 0.8$ V), while that of the PMOS load transistor, V_{Tp}, is negative (e.g., $V_{Tp} = -0.8$ V). Figure 6-5a shows the inverter's output voltage, V_o, as a function of the input voltage V_{in}. This curve is known as the *static voltage input/output characteristic*, or the *transfer characteristic* of the gate.

When $V_{in} = 0$, $V_{GSn} = 0$ and the NMOS transistor is *OFF* (since V_{GSn} is <0.8 V). The PMOS transistor, however, is *ON*, since $V_{GSp} = -5$ V, which is much more negative than -0.8 V. Thus, when $V_{in} = 0$, C_L is charged to V_{DD} through the turned-on PMOS load transistor, and $V_o = 5$ V.

When $V_{in} = 5$ V, the NMOS transistor is turned *ON*, since $V_{GSn} = 5$ V (which is >0.8 V). The PMOS transistor is turned *OFF*, since $V_{GSp} = 0$ V (which is more positive than -0.8 V). Consequently, when $V_{in} = 5$ V (V_{DD}), the output is connected to ground through the turned-on NMOS driver, allowing C_L to be discharged. Since the PMOS device is off, C_L will be completely discharged, and V_o will be 0 V.

The most important property of the CMOS inverter is that when the gate is sitting quiescently in either logic state ($V_o = V_{DD}$ or 0), one of the transistors is *OFF* and the

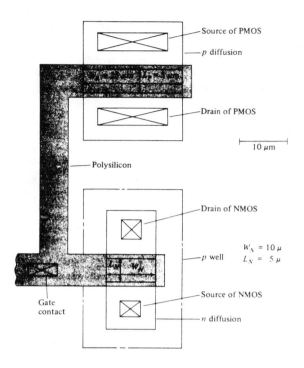

Source of PMOS

p diffusion

Drain of PMOS

10 μm

Polysilicon

Drain of NMOS

p well $W'_N = 10 \mu$
$L_N = 5 \mu$

Source of NMOS

Gate
contact

n diffusion

Fig. 6-4 CMOS inverter layout. From D. A. Hodges and H. G. Jackson, *Analysis and Design of Digital Integrated Circuits*, Copyright 1983, McGraw-Hill Book Co. Reprinted with permission.

current conducted between V_{DD} and ground is negligible (i.e., it is equal to the leakage current of the *OFF* device). This feature can be seen in Fig. 6-5b, which plots the current through the inverter, I_{DD}, as a function of V_{in} (solid curve). The power dissipated in the static (or standby) mode is then determined by the product of the leakage current and the supply voltage. Since the leakage current of an MOS transistor in cutoff (subthreshold current) is so small, very little power is consumed in the static mode. Another important feature is that the V_o swings all the way from V_{DD} to 0 as the inverter changes state; this characteristic of the output-voltage swing is referred to as swinging from *rail to rail*.

Two other operational features of the CMOS inverter must also be mentioned. First, as can be seen in Fig. 6-5b, significant current is conducted through the inverter only when both transistors are *ON* at the same time (i.e., during switching). Therefore, most of the power dissipation is due to the charging and discharging of C_L. In fact, it can be shown that the power dissipation is essentially given by $f C_L V_{DD}^2$, where f is the switching frequency.

Second, because the CMOS inverter output voltage can swing from rail to rail, it can inherently provide excellent noise margins. Noise margins are usually defined in terms of the *logic-gate threshold voltages,* V_{OH}, V_{OL}, V_{IH}, and V_{IL} (which are *not* the same parameters as *device threshold-voltages, see Fig. 6-5a*). The noise margins NM_L and NM_H (Fig. 6-5b) are defined according to the following equations:

$$NM_L = V_{IL} - V_{OL} \qquad\qquad (6\text{-}1a)$$
$$NM_H = V_{OH} - V_{OL} \qquad\qquad (6\text{-}1b)$$

Briefly, the argument for Eq. 6-1a is that the logic gate (e.g., *Inverter 1* in Fig. 6-6a) should provide a maximum "low output" that is less than the maximum "low input" that the subsequent logic gate (e.g., *Inverter 2*) can accept without causing the output voltage of *Inverter 2* to be driven into the ambiguous portion of the transfer characteristic. The difference between V_{IL} and V_{OL} (the noise margin) then specifies how much noise voltage can be tolerated at the input node of Inverter 2 before its output will be driven to a voltage value that is logically undefined.

In CMOS, V_{OL} approaches zero, and V_{OH} approaches V_{DD}. Because of the steep transition region in the transfer characteristic, V_{IL} and V_{IH} can be designed so that noise margins are on the order of $V_{DD}/4$ for expected process variations and operating temperatures. Thus, for a 5-V CMOS technology, V_{NML} can be 1.25 V, whereas in NMOS, V_{NML} is typically only 0.3 V.

For the performance of the logic gates to be maximized, the threshold voltages of the *p*- and *n*-channel transistors should be comparable (and, ideally, of equal magnitude). In addition, the threshold voltages should be as small as possible. The minimum values selected for V_{Tn} and V_{Tp} will be determined by the subthreshold leakage current, I_{Dst}.

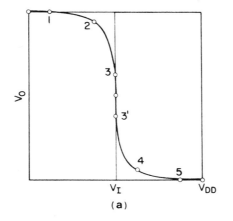
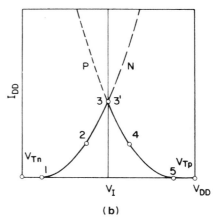

Fig. 6-5 (a) Output (V_O) versus input (V_I) voltage of CMOS inverter. (b) Current through inverter as a function of input voltage (solid curve); I-V characteristics of *n*- and *p*-channel transistors (dashed curves). The numbers correspond to different points on the inverter transfer characteristic.[115]

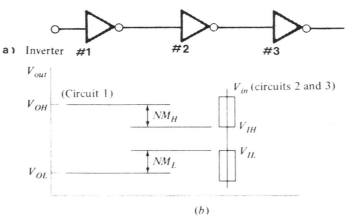

Fig. 6-6 (a) String of inverters connected in series. (b) Definition of noise margins.

For example, if the guideline described in the section on the subthreshold current in MOS devices is used (section 5.2.3 in chap. 5), a minimum value of V_{Tn} would be 0.5 V, to keep V_{OL} at least 0.5 V below V_{Tn} and to maintain sufficiently small I_{Dst}.

6.1.4 Advantages of CMOS

As noted earlier, the most important advantage of CMOS is its significantly reduced power density and dissipation.[3,4,5,6] There are other advantages as well, which fall into the following main categories:

- device/chip performance
- reliability
- circuit design
- cost issues

These advantages, as well as the disadvantages of CMOS, will be discussed in this section.

6.1.4.1 Device/Chip Performance Advantages.

• Although logic gates designed in CMOS are larger than those of NMOS (primarily because of the increased spacing needed to isolate n-channel from p-channel devices), this packing density penalty is becoming less important, for several reasons.

First, the CMOS gate uses a PMOS device rather than a depletion-mode NMOS device as the load. The PMOS device is usually made about twice as wide as the NMOS device in order for symmetrical driving capability to be achieved. The NMOS depletion-mode load, however, is four times as large as the NMOS driver. Hence, the area of the two CMOS devices is actually smaller than the area of the two NMOS devices in an inverter.

Second, since the devices are built with submicron dimensions, the difference in the output drive capabilities (I_D) of identically sized PMOS and NMOS devices is decreased due to velocity saturation effects (Fig. 6-7a). The area of the CMOS inverter devices will thus grow proportionately even smaller.

Third, as devices become smaller, the fraction of the chip area required for interconnections becomes larger. Hence, the fact that the gate density is lower in CMOS in NMOS becomes less important. This is especially significant in random logic designs.

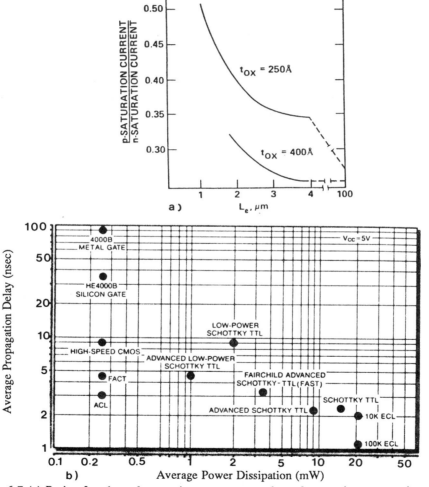

Fig. 6-7 (a) Ratio of *p*-channel saturation current to *n*-channel saturation current increases as the effective channel length of the devices shrinks, due to velocity saturation effects. (b) Comparative speed and power characteristics of various CMOS and bipolar logic families.

• Although the input capacitance of CMOS is larger than that of NMOS (since there are two MOS gates connected in parallel at the input of CMOS logic elements versus only one in NMOS), the interconnect capacitance is becoming more significant than the gate capacitance as gate sizes shrink and interconnect lengths grow with increasing chip size. NMOS gates are thus no longer significantly faster than CMOS gates. Furthermore, as the chip heats up from excessive power dissipation, the circuits slow down (due to degradation of carrier mobility). The lower power advantage might actually make a CMOS circuit operate faster than a comparable, but hotter-running, NMOS part.

• Since more devices can be placed onto a single chip in CMOS than in NMOS, less chip-to-chip driving is necessary, and the overall system speed is improved.

• The fan-out capacitance for MOS devices is much smaller than that of bipolar transistors. Therefore, while the transconductance is also much smaller, if the load capacitance (other than fan-out) is small, much less current is needed to charge the next MOS gate compared to a bipolar logic gate. As the channel length of MOS device shrinks, it may be possible to decrease the delay time required to charge other gates on the same chip. On the other hand, because the performance of bipolar devices is relatively insensitive to shrinking device dimensions, the delay times in CMOS are approaching those of bipolar ECL, but at much lower levels of power dissipation (Fig. 6-7b).

• CMOS can be operated over a wider range of V_{DD} values (e.g., 2-7 V) than NMOS.

• Threshold body-biasing sensitivity is less important in CMOS, and bootstrapping is not needed for transference of a signal through a string of inverter.

• The radiation hardness of CMOS circuits is much higher than that of NMOS.

6.1.4.2 Reliability Advantages.

• The heat generated during operation can raise the chip temperature to the point at which it becomes more prone to failure. Since CMOS circuits dissipate much less power, in most cases they should be inherently more reliable.

• Hot-carrier degradation of MOS devices should be decreased in CMOS for several reasons. First, since hot-electron effects are much less severe in PMOS devices, the load devices in CMOS will suffer less degradation than the depletion-mode load devices used in NMOS. Second, the CMOS gates do not draw static current, so long-term, cumulative hot–electron induced degradation will be smaller. Finally, unlike NMOS, CMOS generally does not use bootstrapping (which increases the electric field in the device and thus aggravates hot-electron degradation).

• Electromigration failures in the metal lines of the circuit are reduced, since no static current flows in the metal lines.

• The soft error rates (SERs) of DRAMs and SRAMs can be reduced by one to two orders of magnitude when the memory arrays are fabricated inside a well region where doping type is the opposite of the substrate's. The added SER protection arises because the reverse-biased well-substrate junction of CMOS creates a potential barrier against carriers generated in the substrate.

6.1.4.3 Circuit-Design Advantages.

• CMOS can achieve "static ratioless" logic design. Circuits that contain a p-channel transistor for every n-channel transistor are said to be "static," because such gates are triggered by the data path signal and do not require the use of an external clock. The design is said to be "ratioless" because the inverter voltage transfer characteristic does not depend strongly on the relative geometric sizes of the p and n transistors. By contrast, commonly used "ratioed" NMOS circuitry must have transistor widths and lengths that are chosen both to balance the currents between transistors and to ensure that this balance is maintained, given changes in operating temperatures, power-supply voltage variation, and day-to-day differences in the manufacturing process.

• As noted earlier, since the output of CMOS logic gates swings from rail to rail, excellent noise margins are inherently provided.

• The flexibility in selecting transistor geometry provided by the ratioless nature of CMOS makes it much easier for "uncommitted circuit" designs, such as gate arrays and standard cells, to be implemented. In gate arrays and standard cells, the number of input and output signals (fan-ins and fan-outs) are normally not known when the individual transistors are laid out. As a result, it is very difficult with PMOS or NMOS to create gate-array and standard-cell configurations that possess sufficient design flexibility. Instead, most standard cells and virtually all MOS gate arrays are implemented in CMOS.

• In NMOS, gating (or pass) transistors reduce the transmitted signal by the so-called *threshold loss*, whereas in CMOS such transmission gates leave the signal unchanged. As a result, signal regeneration can be less frequent in clocked CMOS circuits.

• CMOS allows both analog and digital functions with high circuit densities to be implemented on the same chip. This has stimulated the development of switched-capacitor techniques for analog-to-digital conversion and allowed their integration on a chip with a digital-signal algorithm.

• The circuit-design benefits of CMOS for analog applications are that the switches have no offset voltage, and that the area required for operational amplifiers is much smaller than that needed for NMOS op amps. That is, while an NMOS op amp might take 30 transistors, the same op amp in CMOS might take one-third the number of transistors, as well as one-third the area. Furthermore, CMOS op amps are three to five times smaller than bipolar op amps.

6.1.4.4 Cost Analysis.

• When CMOS and NMOS ICs were first being manufactured, CMOS required almost twice as many masking steps as NMOS. As NMOS processing grew more complex, however, the addition of depletion-mode loads, buried contacts, punchthrough-prevention implants, and lightly doped drains significantly increased processing costs. The costs of fabricating CMOS circuits, however, did not increase proportionately. Hence, the cost differential between the two technologies has steadily been reduced; at present the cost of manufacturing CMOS may be only 20 percent higher than that of manufacturing advanced NMOS circuits. The slightly increased cost of CMOS manufacturing is more than offset by the savings in design, packaging, system heat management, and reliability.

• The complexity of the design task is reduced in CMOS, lowering costs and allowing designs to be created more rapidly (which in turn, decreases costs further). Furthermore, since the time it takes to bring a product to market often has significant impact on market share and profitability, a shortened design time may represent a large increase in profit.

• Packaging costs can represent 25-75% of the total chip-manufacturing cost. Since the reduced power dissipation of CMOS allows the use of cheaper IC package technology, there is a significant cost savings with CMOS packaging compared to NMOS. In addition, the elimination of cooling measures and the reduced failure rates of CMOS result in lower costs. These savings translate into larger profit margins for both chip manufacturers and suppliers of electronic systems that use CMOS.

• Because grounding of the substrate can be done on the front side of the wafer, CMOS may not need back grinding or gold on the backside of the wafer, leading to further savings.

6.1.5 Disadvantages of CMOS

As is to be expected, CMOS also possesses disadvantages. Some of those listed here are inherent to all MOS circuits; these have been considered in in more detail in earlier sections on NMOS technology. Other disadvantages, however, are unique to CMOS and will be given more in-depth treatment in this chapter:

• Like NMOS, CMOS is susceptible to short-channel and hot-carrier effects when device channel lengths drop below about 2 μm (although as mentioned, the hot-carrier problem is reduced to some degree in CMOS). In addition, hot-electron effects in p-channel devices apparently do not become severe until channel lengths of below 1 μm are reached.

• CMOS has a somewhat lower packing density than NMOS.

• Static CMOS logic gates exhibit larger input capacitance than NMOS logic gates due to the additional input capacitance of the p-channel transistors, which are in parallel with the n-channel transistors.

• The need to simultaneously manufacture high-quality PMOS and NMOS devices on the same chip can give rise to processing difficulties.

• There are constraints on the scaling of PMOS devices manufactured with n^+ polysilicon gates.

• Well contacts must be provided, which takes up more chip area than required in NMOS.

• The well drive-in step requires a long process time (e.g., four hours or more at 1100°C).

• CMOS is susceptible to latchup, and hence needs guard bands or epi (see section 6.4.8). In addition, it is often very difficult to identify the exact location of the latchup site (i.e., special liquid-crystal or infra-red techniques must be used).

• Most current CMOS technologies use n^+-doped polysilicon as the gate material. An interconnect routing problem arises because the metal layer must be used when contact is made between this n^+ polysilicon and the p^+ source/drain region of PMOS devices.

• Like all other MOS technologies, CMOS is vulnerable to electrostatic-discharge damage.

6.2 THE WELL CONTROVERSY IN CMOS

There are many trade-offs involved in the optimization of a CMOS process. The choices revolve around the highly interrelated parameters of circuit performance, layout density, fabrication cost, and tolerance to latchup. As described in chapter 5, obtaining the best circuit performance from an MOS device involves maximizing the drive current and minimizing junction capacitances and body effect – all of which favor lower doping concentrations in the device body. Optimizing density, however, favors raising these same doping concentrations (to avoid punchthrough and to achieve high field thresholds). Higher density is thus achieved by allowing closer packing of adjacent n- and p-channel transistors. (Issues relating to CMOS isolation will be described in more detail in a later section.) Latchup tolerance can also be improved by spacing n- and p-channel transistors farther apart (see section 6.4), but this in turn lowers circuit density.

These complex interacting tradeoffs converge on several processing configurations that are determined at the outset of the processing sequence through the selection of the type of *well doping*. The ramifications of this choice must therefore be considered in more detail.[1,4,6]

6.2.1 The Need for Wells in CMOS

Both n- and p-channel transistors must be fabricated on the same wafer in CMOS technologies. Obviously, only one type of device can be fabricated on a given starting

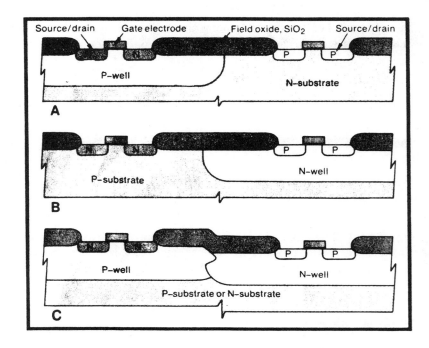

Fig. 6-8 CMOS circuits can be designed using (a) *p*-well, (b) *n*-well, or (c) twin-well technology.[3] (© 1983 IEEE).

substrate. To accommodate the device type that cannot be built on this substrate, regions of a doping type opposite that present in the starting material must be formed, as shown in Fig. 6-8. These regions of opposite doping, called *wells* (or sometimes *tubs*) are the first features to be defined on a starting wafer. This is done by implanting and diffusing an appropriate dopant to attain the proper well depth and doping profile. The well's doping type becomes the identifying characteristic of the CMOS technology.

Normally, a *p*-type substrate is connected to the most *negative* circuit voltage, and an *n*-type substrate to the most *positive*, to ensure that *pn* junctions are not forward-biased during circuit operation. Similarly, the respective well regions must also be connected to the appropriate circuit voltages to prevent forward-biasing of the junctions within the well (and between the well and substrate). Because the wells are totally junction isolated from the rest of the wafer, it is especially important that such *well contacts* be made. That is, it may still be possible to contact the substrate from the backside of the wafer even if no provision is made for a *substrate* contact from the top surface of the wafer. Such a backside connection, however, cannot be established to the *well* regions.

Because one of the device types must be located in the well, there has been some controversy as to which type of well should be used in CMOS-circuit fabrication. The performance of devices in the well will suffer as a consequence of the higher doping, exhibiting higher junction capacitance, increased sensitivity to body effect, and decreased transconductance (due to reduced carrier mobility). Furthermore, substrate currents (e.g., due to hot-carrier effects) will be harder to collect from the well regions. The issue thus becomes that of deciding which device type should be subjected to such performance degradation.

Some argue that the NMOS devices should be built in the substrate, where their better performance can be optimized (n-well CMOS). This argument is persuasive for companies that have had long experience in producing high performance NMOS, since they can merely transfer this technology to the building of NMOS devices in the p-substrate starting material. Furthermore, if a circuit-design technique rich in NMOS devices is used (e.g., Domino logic),[97] most of the devices on the chip will be NMOS, which will again favor the building of NMOS devices in the p-substrate (and PMOS devices in the n-well).

On the other hand, the hot-electron-induced substrate current is much higher in NMOS than in PMOS and, as noted, is harder to collect from the well regions than from the substrate. In addition, device technologists might argue that the better-performing NMOS devices can afford to have their circuit behavior somewhat degraded, as this will balance the performance between these and PMOS devices.

The next sections will outline the pros and cons of both p-well and n-well configurations, as well as those of the more complex well configurations that have been developed (e.g., twin-well and retrograde-well CMOS).

6.2.2 *p*-Well CMOS

The *p-well process,* illustrated in Fig. 6-8a, involves the creation of p-regions in an n-type substrate for the fabrication of NMOS devices. The p-wells are formed by implanting a p-type dopant into an n-substrate, at a high enough concentration to over-compensate for the substrate doping and to give adequate control over the p-type doping in the well. The starting n-type substrate, however, must also have sufficient doping to ensure that the characteristics of devices fabricated in the substrate regions are adequate (a minimum doping concentration of of 3×10^{14}-$1 \times 10^{15}/cm^3$ is required). The p-well doping must therefore be about five to ten times higher than the doping in the n-substrate. If the p-well is doped *too* highly, however, the performance of the n-channel devices will be degraded through lower carrier mobility, increased source/drain to p-well capacitance, and increased sensitivity to body-biasing effects.

As noted in section 6.1.2, p-well CMOS was the first type of CMOS that could be practically manufactured. The first companies to commercially offer CMOS components produced many designs in p-well technology. As this experience was spread throughout the industry, p-well CMOS became widely established.

There are several advantages of p-well over n-well CMOS. First, p-well technology may be the better choice for pure-static logic, in which a good balance between the performance of both MOS device types is beneficial. Second, it is attractive for applications that require an isolated p-region (such as those using an *npn* bipolar transistor as an on-chip driver or n-channel FETs for an analog input). Third, it is less susceptible to field-inversion problems than n-well CMOS, and can thus be slightly easier to fabricate. (We will describe later how p-well CMOS can use the well itself as a channel stop, whereas n-well CMOS must use a separate channel-stop process.) Fourth, if the so-called *retrograde-well* process (rather than a diffusion of a shallow implant) is used to form the wells, p-well technology is more feasible. It is easier to form a p-retrograde than an n-retrograde well, since boron ions penetrate deeper than arsenic or phosphorus ions at a given implant energy.

Finally, p-well CMOS may be better for fabricating SRAMs. Since the alpha-particle–induced soft-error rate (SER) becomes significant even in SRAMs if feature sizes are scaled to submicron dimensions, the cells should be made inside a well. The sensing of the state of an SRAM cell depends on the current provided by the cell. As a result, high-gain NMOS devices are more desirable for the pass gates and drivers in the cell, and must be built in a p-well.

The organization that provides university communities with IC fabrication services, the MOS Implementation Service (MOSIS), offers a standard p-well CMOS process (as well as standard NMOS and advanced CMOS [twin-well] processes). MOSIS cooperates with various IC-manufacturing vendors that fabricate the designs submitted to it. Circuits designed to the specifications of the standard p-well process can be executed by vendors, which serve as *silicon foundries* for MOSIS.

6.2.3 *n*-Well CMOS

In the n-well process, shown in Fig. 6-8b, the p-channel devices are formed in the more heavily doped n-well. As noted earlier, n-well technology became the choice of companies with extensive experience in building NMOS ICs.[8] Because the NMOS device could be fabricated in a lightly doped substrate, virtually all of the experience that had been amassed in fabricating high-performance NMOS could be transferred to an n-well CMOS process. As a result, virtually all EPROMs, microprocessors, and dynamic RAM designs in the generations of technology built with 1.25-2.0 μm dimensions were implemented with n-well CMOS.

This technology does have some disadvantages. First, as mentioned earlier, it is more sensitive to field-inversion problems than p-well CMOS. Second, it may be more difficult to build pure-static, high-performance logic circuits with n-well CMOS.

> **EXAMPLE 6-1:** An n-well CMOS process is to be developed for operation with a power-supply voltage of $V_{DD} = 5$ V. The substrate doping of the p-type wafers is 1×10^{15}/cm^3. The n-wells are to have an average dopant concentration of 1×10^{16}/cm^3. The p-channel MOSFET sources and drains are to have junction depths of 0.4 μm and an average dopant density of 10^{18}/cm^3. What is the minimum n-well depth needed to avoid vertical punchthrough to the substrate?
>
> **SOLUTION:** Vertical punchthrough will occur if the depletion region of the source/drain-to-well junction were to contact the depletion region of the well-substrate junction when $V_{DD} = 5$ V (see section 5.5.2).

The source-to-well junction is essentially a one-sided *pn* junction with a built-in voltage of $V_{bi} = 0.82$ V. From Fig. 5-20, we estimate that the depletion-region width extends into the *n*-well ~0.35 μm, since there is no applied voltage across the junction (i.e., both the source and the well are connected to V_{DD}). The *np*-junction to the substrate has a built-in voltage of 0.63 V, and the total depletion width of this junction at a 5 V bias is ~1.9 μm. We calculate that about 0.19 μm of this depletion region is in the *n*-well. The *n*-well must therefore be deep enough to accommodate the depth of the source junction (0.4 μm) as well as the sum of the depletion region widths in the well in order for vertical punchthrough to be avoided. While the total of these dimensions is 0.94 μm, it is good engineering practice to increase the depth of the well by about 50% to account for process variations. A reasonable well depth might therefore be 1.5 μm.

6.2.4 CMOS on Epitaxial Substrates

As will be described in the section dealing with latchup prevention (section 6.4.8.2), heavily doped substrates with a more lightly doped surface epitaxial layer have been utilized to suppress latchup in CMOS.[11] When such starting material is used with single-well CMOS, the epitaxial layer is doped to a concentration equal to that of the substrate in a nonepitaxial wafer used for that process. If a twin-well CMOS process is used, the epitaxial layer is doped to a level significantly lower than that required for building either the *p*- or *n*-channel MOSFETs (see section 6.2.5).

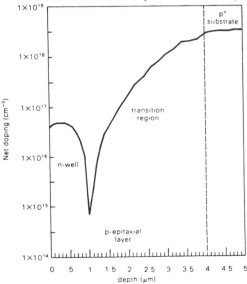

Fig. 6-9 A *p*-on-*p*-epi doping profile and diffused *n*-well profile measured after the entire CMOS process has been completed. The as grown epi thickness is 4 μm and the final lightly doped epi thickness is about 2 μm.[116] (© 1987 IEEE).

The epitaxial layer is made thicker than the well depth, since the dopants in the heavily doped substrate under the epi layer diffuse toward the surface as the well dopants are diffused toward the bulk (Fig. 6-9). Thus, some of the epitaxial layer becomes more heavily doped during the CMOS process flow. The process is designed so that the bottom of the well is eventually adjacent to the heavily doped substrate region.

Either n-epi on n^+ substrates or p-epi on p^+ substrates can be used, with each method having advantages and drawbacks. Because the problems with n-epi on n^+ are more serious, p-epi on p^+ is more widely used. The major limitation of the latter approach is that the outdiffusion of boron from the p^+ is much more severe than it is in n-epi on n^+. (The reason for this is that boron diffuses much more rapidly than antimony, which is the most widely-used n^+ dopant material.) Thus, a thicker p^-- epitaxial layer must be used.

In addition, the transition region between the p^+ substrate and the p^-- epi layer is thicker, producing a larger series resistance (R_{sub}), which in turn reduces latchup immunity. On the other hand, the p-on-p^+ material is less sensitive to process-induced defects, and the p-type substrate provides higher conductivity under NMOS devices. Such additional conductivity is desirable, since it reduces the voltage drops caused by the substrate currents (generated as a result of the hot-carrier effects in short-channel devices). It is especially important in the regions containing NMOS devices, since the hot-electron substrate current is much higher in such devices than in PMOS devices.

The n-on-n^+ epitaxial substrates also offer some advantages. First, SRAMs are often built on n-type substrates, because the p-well to n-substrate junction provides protection from radiation-induced discharging of the memory's n-type storage nodes in the p-well.[12] Second, retrograde p-wells are easier to implement because their implantation energy requirements are much lower. Third, the n-to-n^+ transition region is smaller than that in p-on-p^+ substrates, and the smaller value of R_{sub} provides improved latchup protection.

The limitations of n-epi on n^+ involve the process by which the heavily doped substrate is grown. Antimony (Sb) is the n-type dopant used, both because its diffusion coefficient is so low and because it exhibits much less lateral autodoping than arsenic (the other slow-diffusing n-type dopant). The problem with Sb is that its segregation coefficient, k_o, is very small (i.e., $k_{o,Sb} = 0.023$; see, Vol. 1, chap. 1), and thus a large quantity of Sb must therefore be put into the Si melt to ensure that a sufficiently heavily doped ingot will be produced. In even the most highly refined Sb there are high concentrations of unwanted heavy metals, which become incorporated into the growing silicon crystal. In addition, the oxygen content of the Sb-doped crystal is relatively small, due in part to the special growing conditions used when the Sb-doped ingot is pulled.[7,9,10] As a result, intrinsic gettering techniques that would getter the metals in the substrate are not as effective as they are in p-on-p^+ epi. For these reasons, n-on-n^+ substrates are less frequently chosen when epi-CMOS is implemented.[12]

A problem that exists with both types of epitaxial substrates for CMOS is that the wells cannot be made too deep, since the lateral diffusion would then take up too much area. On the other hand, if the wells are too shallow, vertical punchthrough will ensue.[13] A second problem is that of leakage current. Appreciable leakage current can

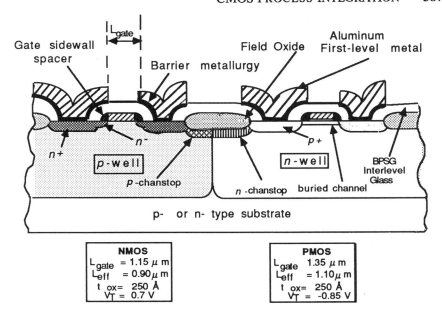

Fig. 6-10 Cross section of a twin-well 1.2-μm CMOS structure.[12] Reprinted with permission of Semiconductor International.

flow in the vertical path in two different situations. First, punchthrough can occur if the depletion regions of the p^+/n-well junction and the n-well/p^+-substrate junction touch each other. The problem is even more severe in the case of the heavily doped p^+ substrate, due to the high degree of boron outdiffusion. A major limitation is thus imposed on the minimum epitaxial-layer thickness. Second, it may be possible for the source regions to become biased to a potential below V_{SS}.

Another problem of epi-CMOS is back surface autodoping. For example p-type 20 W-cm (7×10^{14} boron/cm^3) epi on a 0.005 Ω-cm (2×10^{19} boron/cm^3) substrate is representative of epi for CMOS devices. If the substrate back is not sealed (e.g., using a sealing layer such as undoped silicon dioxide or silicon nitride), boron evaporation can contribute to autodoping on the front surface during the entire epi deposition cycle. This widens the epi/substrate interface and may even prevent the epi from reaching the 20 Ω-cm specified resistivity. Note that if silicon nitride is used as a sealing layer it should only be used in thin layers (e.g., less than 100 nm thick) since its high intrinsic stress causes it to bow the silicon.[113]

6.2.5 Twin-Well CMOS

With the twin-tub approach, two separate wells are formed for n- and p-channel transistors in a lightly doped substrate (Figs. 6-8c and 6-10). The substrate may be either a lightly doped wafer of n or p material, or a thin, lightly doped epitaxial layer on a heavily doped substrate. In either case, the level of surface doping is significantly lower than that required for building either the *p*- or *n*-channel MOSFETs. Each of the well dopants is implanted separately into the lightly doped surface region and is then driven in to the desired depth.

The doping profiles of each of the device types can be set independently, since the constraint of single-well CMOS does not exist (i.e., that the well doping must always be higher than the doping of the substrate in which one type of device is made). This was originally cited as an advantage of twin-well CMOS over single-well CMOS, with the argument made that both device types could thus be optimized.[14] This claim for 1-2 μm CMOS has been questioned by Chen,[4] who points out that in modern single-well CMOS processes, an additional implant is used to prevent punchthrough without the need to raise the entire substrate-doping concentration. By incorporating this additional implant, it is possible to build higher-performance devices than can be achieved with the twin-well approach, in terms of junction capacitance and sensitivity to body effect.

Twin-well CMOS does offer some significant benefits for devices with submicron-channel lengths (although these advantages are gained at the cost of greater process complexity).

The first, and most important advantage arises when devices with submicron channel lengths are fabricated. Since it is recognized that the two device types perform similarly as channel lengths approach 0.5 μm, it is useful to provide symmetrical *n*- and *p*-channel devices. Furthermore, at submicron dimensions the body doping of both transistor types must be raised significantly to prevent punchthrough and to maintain adequate threshold-voltage levels. Thus, the advantage of having one type of MOS transistor in a lightly doped region (to optimize its performance at the expense of the other) disappears. It is instead more beneficial to produce two types of active device wells, each with its own optimized doping profile (i.e., formed by separate implants into a lightly doped substrate).*

The second advantage of the twin-well process is that it is compatible with the technologies of either isolation by selective epitaxial growth (SEG) or trench isolation.[4] Both approaches restrict the lateral diffusion of the dopants in each of the wells. In addition, sidewall inversion along the trench is less likely when both device sidewalls are butted against a trench that has been formed in a more highly doped well.[15] Finally, when deep trenches are used with a thin epitaxial layer on a heavily doped substrate, latchup can be eliminated. The combined use of the twin-well process and advanced isolation techniques allows n^+ to p^+ spacing to be dramatically reduced in comparison to single-well technologies.

---

* It is not useful to start with a substrate doped to the optimum level needed for just *one* of the submicron devices. If this were done, a single well of much higher doping would have to be established for the other type of submicron device, which would unnecessarily degrade the device performance. It is possible, with the twin-well approach, to have both types placed in a well of optimum doping profile.

A third benefit is that the combination of epitaxial substrates and the twin-well process allows either substrate type to be chosen with no effects on transistor performance and essentially no change in process flow. Such flexibility is useful since some applications are best met with n^+ substrates and others with p^+ substrates. This advantage may be important if a single process is needed for implementing designs with different applications.

Finally, self-aligned channel stops can be easily implemented in the twin-well approach, allowing the spacing between n- and p-channel devices to be reduced. Although this spacing reduction is not as great as it would be if trenches or SEG were used, the process is much less complex with the twin-well method.

6.2.6 Retrograde-Well CMOS

Conventional wells are formed in single- and twin-well CMOS technology by implanting dopants and then diffusing them to the desired depth. However, the diffusion occurs laterally as well as vertically, which has the effect of reducing packing density. If a high-energy implant is used to place the dopants at the desired depth without further diffusion, much less lateral spread occurs (see Vol. 1, chap. 9). Such high-energy implants also cause the peak of the implant to be buried at a certain depth within the silicon substrate (depending on the implantation energy), and the impurity concentration decreases as it approaches the wafer surface. Since the well profile in this case is different from that of conventionally formed wells (in which the doping concentration is highest at the surface, and decreases monotonically with depth), such deeply implanted wells are known as *retrograde wells*. The retrograde-doping profile is retained by minimizing the temperature cycles of later process steps. Retrograde wells can be implemented on both bulk wafers and on epitaxial wafers.[5]

Besides the potential benefit of increased packing density, such wells offer the following advantages:[4]

- A retarded electric field is created in the parasitic vertical bipolar transistor (thereby providing some protection against latchup; see section 6.4).

- Susceptibility to vertical punchthrough is reduced.

- The conductivity in the bottom of the well is increased, which also provides some further latchup protection (as will be explained in section 6.4.5).

- A higher threshold voltage can be achieved in the field regions of the p-wells since the boron implant is done following field oxidation (i.e., the boron does not segregate out into the field oxide as it is grown).[16]

- Lateral diffusion of the boron is also eliminated, thereby reducing encroachment of the boron into the active regions.

A disadvantage of the retrograde-well approach is that both the junction capacitance and body factor are significantly increased. For example, in a simulation of a 32-bit CMOS arithmetic logic unit (ALU), it was found that the circuit delay increases by

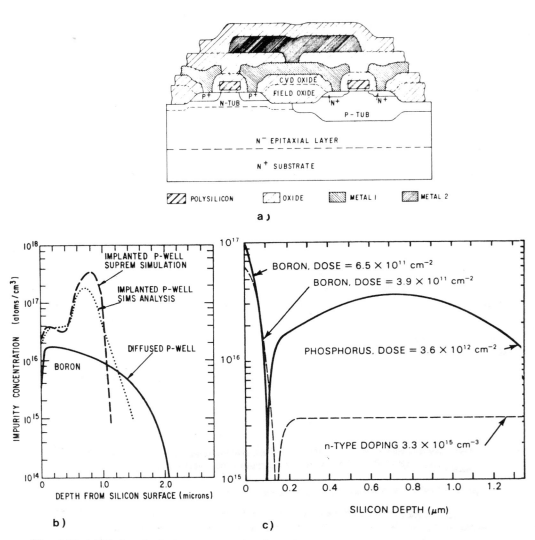

Fig. 6-11 (a) Twin-tub device cross section in which a retrograde *p*-well is used. General Electric AVLSI Process. (b) Retrograded *p*-well implanted impurity concentration profile. General Electric AVLSI process. Also shown is a conventionally thermally diffused well.[5] (© 1986 IEEE). (c) Simulated doping profiles in PMOS channel regions: dashed lines, lightly doped substrate; solid lines, retrograde *n*-well.[23] (© 1982 IEEE).

about 7% as the *p*-channel junction capacitance increases by 30 percent.[4] When the retrograde well is formed by means of a very high-energy implant, however, the doping concentration under the bottom of the source and drain regions is reduced, which reduces the junction capacitance. This implies that if a very-high-energy ion implanter (~1 MeV) were available as a manufacturing tool, the disadvantage of retrograde wells could be overcome.[17] In addition, the higher doping that will be required in the wells for fabrication of submicron CMOS devices may mean a less severe junction capacitance penalty.

Although both p-type[18] and n-type[19] retrograde wells have been demonstrated (see Figs. 6-11a and b,[5] and Fig. 6-11c,[23] respectively), the p-type technology has been more widely implemented. The reason for this is that a 700-keV (or greater) ion implanter is required for the formation of n-type retrograde wells. As of 1989 such machines were commercially available (e.g. NV 1003), but not yet found in ordinary production environments because of their relatively high cost. The p-type retrograde well, on the other hand, can be formed either by implanting singly ionized boron at 400 keV or doubly ionized boron at 200 keV (because of the larger projected range of boron compared with arsenic or phosphorus; see Vol. 1, chap. 9). Although the doubly ionized boron approach is achievable with conventional production implanters, it is not a trivial process to implement. In addition, a quadruple-well technology has been developed. This approach uses deep retrograde p- and n-wells, as well as shallow p- and n-wells (Fig. 6-12).[112]

A twin-retrograde-well 0.7-μm-CMOS process for fabricating 1-Mbit SRAMs has been reported.[74] The high energy implants also allow a restricted thermal budget to be used, thus reducing the up-diffusion from the p^+ substrate. This permits the use of a thinner epi layer (which also helps prevent latch-up), and also allows implantation of

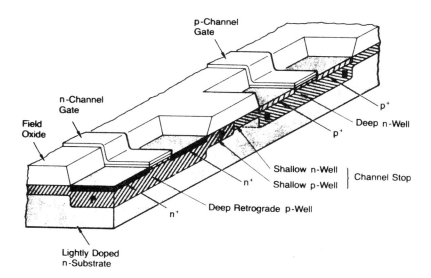

Fig. 6-12 Cross section of a quad-well CMOS device.[117] Reprinted with permission of VLSI Design.

the channel-stop dopants after the field oxide has been grown and planarized (thus minimizing lateral diffusion of the boron channel-stop dopants).

6.2.7 Summary of CMOS Well-Technology Issues

The selection of a particular CMOS process depends largely upon circuit applications, and to a lesser degree on technology evolution. For VLSI circuits with 1-2 μm design rules, n-well CMOS has been the more widely used single-well technology, both because n-channel devices can be optimized in the p-substrate and because most circuits use more NMOS than PMOS devices. In circuits that use transistor channel lengths smaller than 1 μm, twin-well and retrograde-well technology will become more attractive. As noted, the price will be increased process complexity.

The optimum process-integration design strategy is to first select the well technology (and to decide whether or not to use epitaxial substrates), and to then select the isolation method. Once these decisions have been made, the well depth and doping profile can be determined. The well depth will impact the lateral-diffusion distance of the well and the vertical-punchthrough voltage. The doping profile will affect transconductance, threshold voltage, source/drain punchthrough, junction capacitance, carrier mobility, source/drain-to-substrate breakdown, sensitivity to body effect, and hot-electron effects.

6.3 p-CHANNEL DEVICES IN CMOS

The fabrication of p-channel devices in CMOS presents some unique problems which arise from the need to build both NMOS and PMOS devices on the same chip. The problems revolve around the choice of a doping type for the polysilicon gate electrode and the impact of this choice on the threshold voltage and transistor action of PMOS devices.

6.3.1 PMOS Devices with n^+-Polysilicon Gates

As mentioned earlier, the threshold voltages of the n- and p-channel devices in a CMOS circuit should have comparable magnitudes for optimal logic-gate performance. To allow for maximum current-driving capability, the threshold voltages should also be as small as possible, with the minimum value dictated by the need to prevent excessive subthreshold currents under normal circuit operating conditions. For 5-V CMOS technologies, typical threshold voltages are 0.8 V for V_{Tn} and -0.8 V for V_{Tp}. Furthermore, the most common choice for the gate material is heavily doped n-type polysilicon. The work function of n^+ polysilicon is ideal for n-channel devices since these will yield threshold voltages of less than 0.7 V for reasonable values of channel doping and oxide thicknesses.*

---

* Note that the polysilicon layer may be combined with a layer of silicide for sheet resistance reduction; since the polysilicon remains as the underlayer of the polycide, the work function of the gate electrode will not be changed.

Figure 5-4, chapter 5 shows the value of V_T in devices manufactured with n^+ polysilicon gates (left scale)[18] as a function of doping and various gate-oxide thicknesses (Q_{tot} is assumed to be small enough that its effect on V_T can be ignored). The value of V_{Tn} is less than 1.0 V for gate-oxide thicknesses of 25 nm or less and a substrate doping of less than 10^{17} cm^{-3}. Thus, the threshold voltage of NMOS devices can easily be adjusted to the desired value of 0.7 V by means of ion implantation.

When n^+ polysilicon is used for the gate electrode of PMOS devices, however, it is not as easy to adjust V_{Tp} to -0.7 V. Figure 5-4 shows V_{Tp} (on the left scale for n^+ poly) as a function of substrate doping and gate-oxide thickness. In the doping range of 10^{15}-10^{17} cm^{-3}, V_{Tp} is already more negative than -0.7 V. Thus, implanting the n-doped body with more n-type dopant would only raise the *magnitude* of V_{Tp}, rather than bringing its value closer to the desired -0.7 V. To reduce the magnitude of V_T in PMOS devices using an n^+ polysilicon gate, it is necessary to implant the channel with a shallow layer of boron. The dose must be heavy enough to overcompensate the n-surface so that a p-region depleted of holes is formed. This shifts V_{Tp} toward more positive values by forming a compensating layer.

The fact that boron is implanted to adjust both V_{Tn} and V_{Tp} in CMOS circuits with n^+ polysilicon gates suggests that a single implant could be used instead of two separate implants. Figure 6-13 shows that this can be accomplished if the appropriate background dopings are chosen for the substrate and the well.[21] On the other hand, it

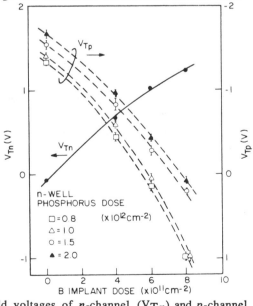

Fig. 6-13 Threshold voltages of n-channel (V_{Tn}) and p-channel (V_{Tp}) transistors as a function of boron threshold-adjustment dose. The CMOS structure uses an n-well implanted into a p-substrate (whose doping level is 6×10^{14} atoms/cm^3). V_{Tp} results are shown for various implant doses of the n-well.[21] (© 1980 IEEE).

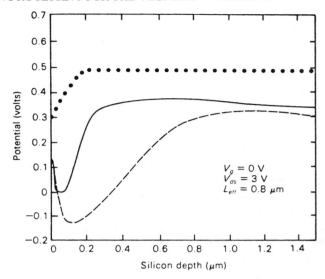

Fig. 6-14 Channel potential profiles in PMOS device in CMOS. *Dashed lines*: buried channel PMOS in a lightly doped substrate. *Solid lines*: buried-channel PMOS in a retrograde *n*-well. Dotted lines: surface channel NMOS drawn for comparison.[23] (© 1982 IEEE).

may be decided to use two separate implants in order to achieve better short channel behavior through individual optimization of the *n*- and *p*-channel devices.

6.3.1.1 Punchthrough Susceptibility.

PMOS devices in which boron is used to adjust V_T exhibit a high susceptibility to punchthrough effects, since the boron implant produces a *p*-layer with a finite thickness. The potential minimum in the channel is thus moved away from the Si-SiO$_2$ interface (Fig. 6-14), causing more current to flow below the surface of the device.[23] Such PMOS devices are referred to as *buried-channel transistors.* As seen in Fig. 6-14 (which gives the *calculated* variation of the potential as a function of distance below the surface), the potential minimum moves further into the substrate as the thickness of the implanted *p*-layer is increased.

As the potential minimum moves deeper below the surface, the susceptibility also becomes more pronounced (see section 5.5.2). This is illustrated in Fig. 6-15, which shows the results of a simulation[22] in which the lines of equipotential in the channel region were plotted for various depths of the channel junction Y_j (for a constant source/drain junction depth of 0.15 μm). As Y_j is increased from 0.1 to 0.2 μm, the drain electric field extends closer to the source for a constant gate and drain bias. Hence, this simulation predicts that more barrier lowering will occur as Yj is increased, leading to an increase in the punchthrough current.

The calculated predictions are supported by experimental data, as shown in Fig. 6-16, which plots the subthreshold I$_D$-V$_{GS}$ characteristics of the structures described in Fig. 6-15. The subthreshold swing (S.S.) has the smallest slope when Y_j is 0.2 μm (indicating that the largest punchthrough current flows in this case). In fact, when

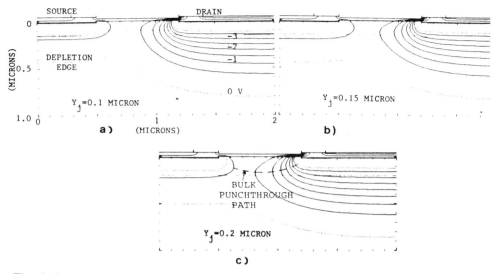

Fig. 6-15 Simulated 2-D potential profile for *p*-channel MOSFETs with $Y_j = 0.1$, 0.15, and 0.2 μm. The drain and gate bias are -3 V and 0 V, respectively.[22] From K. M. Cham, S.-Y. Oh, D. Chin, and J. L. Moll, *Computer-Aided Design and VLSI Device Development*, 2nd Ed., Copyright Klewer Academic, 1989. Reprinted with permission.

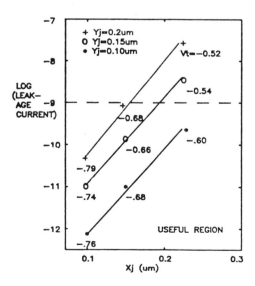

Fig. 6-16 Simulated subthreshold leakage current versus Y_j and X_j. $L_{eff} = 0.5$ μm, W = 50 μm, $t_{ox} = 25$ nm, $V_{GS} = 0$ V, and $V_{DS} = -3$ V.[22] From K. M. Cham, S.-Y. Oh, D. Chin, and J. L. Moll, *Computer-Aided Design and VLSI Device Development*, 2nd Ed., Copyright Klewer Academic, 1989. Reprinted with permission.

$V_{GS} = 0$, I_{Dst} is increased by two orders of magnitude as Y_j is increased from 0.1 to 0.2 μm.

Leakage currents due to punchthrough in PMOS devices can be a significant problem in some CMOS IC applications. For ULSI devices, even a small value of leakage current per device may not be tolerable. For example, it has been shown that in submicron PMOS transistors ($L_{eff} = 0.8$ μm), the punchthrough current will increase by two orders of magnitude if V_{DS} is increased from -1 V to -10 V.[24] Thus, if the punchthrough current were 1 nA at -1 V, it would be increased to 0.1 μA if V_{DS} were increased to -10 V. Such leakage would cause the dissipation of a few-tenths of a watt of power in a chip containing 1 million PMOS transistors.

The most obvious solution is to increase the PMOS device channel length. This is done in many CMOS technologies, and it is the reason that the minimum channel lengths of PMOS devices are often larger than those of the NMOS devices on the same chip. Another obvious technique is to make the *p*-buried layer as thin as possible. One of the reported approaches for achieving this involves implantation with BF_2^+ (which produces shallower boron layers than implantation with boron, see Vol. 1, chap. 9). Another approach is to use a high-energy *n*-implant (e.g., As at 400 keV) in order to place more *n*-type dopant atoms below the *pn* junctions (the more heavily doped regions absorb the drain voltage in a shorter distance, while simultaneously squeezing the channel *pn* junction toward the surface).[25]

To prevent a shallow implanted-boron layer from growing thicker, it is necessary to use a reduced thermal budget in order to restrict the process sequence following the implant in order to restrict boron diffusion. Specific steps for restricting boron redistribution include the following:[12]

1. Implant the boron through the gate oxide. This avoids the oxidation-enhanced diffusion of boron that would occur during the growth of the gate oxide if the implant were performed first. In this case, it is necessary to prevent the gate oxide from becoming contaminated during implant, either by material sputtered by the beam line or by vaporized resist material used as a mask against the implant. To prevent such contamination, a thin layer of polysilicon may be deposited on the gate oxide prior to the implant (in fact, immediately after the oxide is grown).[26]

2. After the implant has been performed, the remainder of the polysilicon film is deposited. This polysilicon is doped during the deposition step (at ~600°C) in situ with phosphorus, in order to avoid the 900°C phosphorus doping thermal cycle that would have to be used if the poly were doped following deposition.

3. A BPSG glass layer is used as the dielectric between the gate and the first level of metal. A significantly lower temperature can be used to flow BPSG than PSG (e.g., 850°C versus 1000°C). A lower temperature cycle can thus be used to smooth the surface topography (flow step) and gently taper the contact holes after etch (reflow step).

The use of a boron implant to adjust V_{Tp} becomes less feasible as devices use even thinner gate oxides, since larger doses of boron are needed. Y_j thus becomes deeper, and the punchthrough problem worsens. Solutions involving the use of gate electrodes other than n^+ polysilicon must therefore be explored. One alternative is to use p^+ polysilicon for PMOS devices; another is to use a material whose work function is very close to the mid-gap of silicon, thereby allowing symmetrical threshold voltages for n- and p-channel devices.[5]

6.3.2 PMOS Devices with p^+-Polysilicon Gates

When p^+ polysilicon is used for the gate material, V_{Tp} is shifted from the values that occur when n^+ polysilicon is used. Figure 5-4, chap. 5 plots V_{Tp} as a function of substrate doping when p^+ polysilicon is employed (using the scale on the right side of the figure). V_{Tp} can be seen to be less negative than -0.7 V over the substrate doping range of 10^{15}-10^{17} cm^{-3}. Thus, it can easily be made more negative through the implantation of phosphorus or arsenic into the channel. If p^+ polysilicon is used throughout the circuit, however, the NMOS device then has to be overcompensated in order for V_{Tn} to be reduced to sufficiently small values. (Note that some reports have been published on the use of p^+ polysilicon alone.[27,28]) This implies that it would be ideal to use both n^+ and p^+ poly gates on the same chip (with n^+ poly for NMOS devices, and p^+ poly for PMOS devices).[100]

Such a dual-doped poly approach, however, would add process complexity, and would also introduce other problems arising from the need to connect the two types of poly (e.g., at the input of an inverter). Such problems occur when a silicide overlayer (or *strap*) is used to make the connection between n^+ and p^+ poly (as a method to avoid a separate, space-consuming metal contact). Because the silicide strap provides a very rapid diffusion path for boron and arsenic, one type of poly can be counterdoped by the other when the device is subjected to high-temperature excursions. This counterdoping can occur to the degree that a region of poly can change doping types (i.e., from n^+ to p^+). In such cases, the threshold voltage of devices with counterdoped poly will be shifted from their designed value.

If the processing temperature is limited to 800°C after the two types of poly have been connected by the silicide, such lateral diffusion does not produce significant shifts in V_T.[29] On the other hand, temperatures of 900°C will produce sufficient counterdoping to significantly shift V_T. Since one of the last high temperature steps in CMOS is activation of the source/drain implants, this would mean either using a lower-temperature activation step, or performing the step prior to formation of the silicide layer. Process considerations for 0.5-μm CMOS technologies using both n^+ and p^+-poly gates are described in reference 85.

Another approach is to form a polysilicon electrode with an overlying conductor layer that suppresses such counterdoping. One such structure is a W-TiN-poly electrode structure.[105] The thin (30-nm) TiN film acts as a diffusion barrier to the dopants in the poly and also prevents reaction between the poly and the W. Very little of the dopant present in the poly is found to diffuse into the W, even following an anneal at 900°C for 1 hour.

Another problem encountered with p^+ poly gates when a thin gate oxide is used is poor V_T process control, due to penetration of the boron into the oxide (or further, into the silicon).[78] It is reported that boron will penetrate gate oxides that are ≤12.5 nm thick during a 900°C 30-minute post-implant anneal in N_2.[81] This implies that if p^+ poly gates are used, a lower processing temperature may be needed.[85]

If too low a temperature is used, however, the boron implanted into the polysilicon will not be sufficiently redistributed. The polysilicon dopant concentration at the polysilicon–gate oxide interface could thus end up being less than the mid-10^{19}/cm^3. This would make the work function of the polysilicon different from the desired degenerately doped value, creating V_T control problems in the MOS devices.

It has also been found that the presence of fluorine in the gate oxide worsens the boron penetration problem (Fig. 6-17a).[123, 124] Such fluorine can be introduced into the gate oxide if the PMOS source drain regions are implanted using BF_2. Elemental boron is therefore considered inherently superior to BF_2 as the implant species for surface-channel PMOS devices in CMOS technologies that use p-doped polysilicon. A study of enhanced boron diffusion through thin SiO_2 layers in a wet oxygen atmosphere is also reported in reference 107.

6.3.4 Gate Materials with Symmetrical Work Functions (for Both NMOS and PMOS Devices)

Because the larger work functions of molybdenum (4.7 V), tungsten, or refractory silicides produce low and nearly symmetrical threshold voltages for both PMOS and NMOS devices on moderately doped Si substrates, work has been conducted to investigate their suitability as gate-electrode materials.[30] For example, $TaSi_2$ gates have been successfully implemented.[31] Some of the benefits that such gate materials provide (besides symmetrical threshold-voltage values) include a reduction in subthreshold leakage currents and a decreased sensitivity to body bias (Fig. 6-17b).

Molybdenum and tungsten films deposited by means of magnetron sputtering have also been evaluated in terms of adhesion to SiO_2, mechanical stress, and compatibility with silicon processing techniques.[32,33] It was found that both could be deposited with low compressive stress by adjusting the deposition conditions of the sputter process. Both films can also exhibit good adhesion to SiO_2. (An additional advantage of Mo and W is that their resistivities are about 100 times lower than that of doped polysilicon and about 10 times lower than that of polycide gates.)

Some novel techniques had to be developed in order for compatibility to be established with the conventional Si-gate MOS process sequences. For example, it is necessary to use a wet H_2 ambient to oxidize the silicon without oxidizing the Mo[33] or W gates.[30] (This procedure is useful when a screen oxide is to be formed prior to the source/drain implant, but after the formation of the gate sidewall spacers used in LDD structure.) In addition, Mo and W form a layer of columnar grains, which makes such films susceptible to ion implant channeling along the grain boundaries. When the

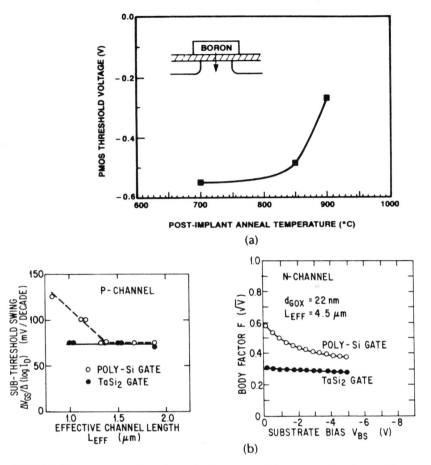

Fig. 6-17 (a) Boron penetration through the gate oxide causes the threshold voltage in BF$_2$- implanted PMOS devices to shift positive at anneal temperatures above ~800°C.[123] (© 1989 IEEE). (b) Sub-threshold characteristics versus effective channel length for poly-gate and TaSi$_2$-gate PMOS devices.[118] (© 1984 IEEE).

source/drain implantation is performed, some of the dopants may thus penetrate the gate film. This problem can be overcome by limiting the ion energy[33] or by depositing a thin layer of PSG (~60 nm thick) on the gate electrodes prior to implantation.[30]

Finally, V$_T$ control with refractory metal gates was once a problem because of the relatively high concentrations of radioactive impurities (e.g., U and Th) in Mo and W sputtering targets. Such impurities can also produce soft errors in large memories. Recently developed chemical-purification procedures have significantly reduced the concentrations of U and Th in these sputtering target materials.[34] Methods for depositing Mo[35] or W[98] by CVD offer another option for forming of high-purity films. A deep-submicron CMOS process that uses a W gate has been reported.[77]

6.4 LATCHUP IN CMOS

A major problem in CMOS circuits is the inherent, self-destructive phenomenon known as *latchup*.[36] Latchup is a phenomenon that establishes a very low-resistance path between the V_{DD} and V_{SS} power lines, allowing large currents to flow through the circuit. This can cause the circuit to cease functioning or even to destroy itself (due to heat damage caused by high power dissipation).

The susceptibility to latchup arises from the presence of complementary parasitic bipolar transistors structures, which result from the fabrication of the complementary MOS devices in CMOS structures. Since they are in close proximity to one another, the complementary bipolar structures can interact electrically to form device structures which behave like *pnpn* diodes. In the absence of triggering currents, such diodes act as reverse-biased junctions and do not conduct. It is possible, however, for triggering currents to be established in a variety of ways during abnormal (but nevertheless frequently occurring) circuit-operation conditions (*e.g.*, terminal overvoltage stress, transient displacement currents, ionizing radiation, or impact ionization by hot electrons). Since there are many such parasitic *pnpn* structures on a VLSI CMOS chip, it is possible to trigger any one of them into latchup.

The phenomenon of latchup is well understood,[37] and many approaches have been implemented to control or even eliminate it. However, since the problem is increasing in severity as device dimensions continue to shrink, new latchup suppression techniques will be needed.

Latchup is fairly complex and can be a difficult concept to grasp for readers not well-versed in device physics. We will therefore briefly review the device concepts relevant to the problem before discussing the processing, layout, and circuit design techniques that have been developed to solve it.

6.4.1 Parasitic *pnpn* Structures in CMOS Circuits

Our example device configuration will be a *n*-well CMOS technology. As shown in Fig. 6-18a, lateral *npn* and vertical *pnp* transistor structures are inevitably created in *n*-well CMOS as a result of the multiple diffusions needed to fabricate *p*- and *n*-channel devices. The emitter of the lateral *npn* transistor is the n^+ source and/or drain, while its base is the *p*-substrate and its collector the *n*-well. The p^+ source and/or drain comprises the emitter of the vertical *pnp* device, while the *n*-well forms its base and the *p*-substrate its collector. The equivalent resistances of the substrate (R_{sub}) and the well (R_w) are also important elements of the latchup structure.

Several major aspects of these parasitic devices can be seen by studying the CMOS device cross-sections shown in Figs. 6-18a and 6-18b. First, the base of the *npn* transistor is connected to the collector of the *pnp* transistor (i.e., they are part of the same *n* region in the CMOS structure), and the base of the *pnp* is connected to the collector of the *npn*. We can therefore draw a simplified equivalent circuit diagram of this connection (Fig. 6-18c). It can be seen that the base of each transistor is driven by the collector current of the other, forming a positive feedback loop. Next, we note that

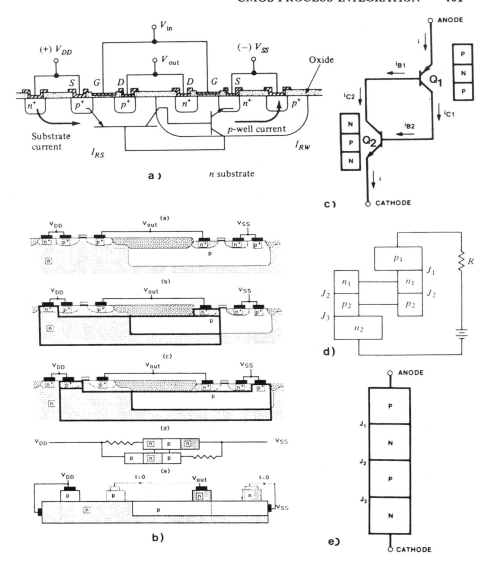

Fig. 6-18 (a) Cross-sectional view of a *p*-well CMOS inverter with parasitic bipolar transistors and lateral currents schematically shown.[4] (© 1986 IEEE). (b) Schematic showing the parasitic *npn* and *pnp* transistors in a *p*-well CMOS inverter. From W. Maly, *Atlas of IC Technologies*, Copyright 1987 by the Benjamin/Cummings Publishing Company. Reprinted with permission. (c) Simplified equivalent circuit diagram showing how these parasitic transistors are connected in the CMOS structure. (d) Schematic version of the circuit diagram of part (c). (e) *pnpn* diode obtained by merging the connected n_1 and p_2 regions shown in part (d).

the emitter of the *npn* is connected to V_{SS} (e.g., 0 V), and the emitter of the *pnp* is connected to V_{DD}. These external power-supply connections can also be added to the equivalent circuit diagram.

It should be noted that this simplified diagram ignores the parasitic resistances of the electrical paths through the bulk regions of the silicon, R_{sub} and R_w. These resistances, however, *are* very important in the understanding and control of latchup, and they will therefore be incorporated as soon as the simplified description of the device behavior is presented (Fig. 6-18c). The simple *pnpn* diode circuit to be used in the initial discussion turns out to be a "worst-case" possibility. That is, the addition of R_{sub} and R_w to the circuit model actually reduces the latchup susceptibility of CMOS. Since the two transistors are connected via their base and collector regions, by merging both the n_1 and the p_2 regions in Fig. 6-18d, we come up with a parasitic device structure (Fig. 6-18e).

6.4.2 Circuit Behavior of *pnpn* Diodes

Devices with structures like those shown in Fig. 6-18e and with external connections to the two end regions only, are known as *pnpn diodes*.* The terminal connected to the p_1 region is called the *anode*, and the terminal connected to the n_2 region is called the *cathode*. When an external voltage is applied with the anode voltage positive with respect to the cathode, and the resulting current I is measured, the I-V characteristic of this device is observed to have four distinct regions (Fig. 6-19); as follows:

Region 1. For voltages with values from *a* to *b*, very little current is observed to flow from anode to cathode, and the device is said to be in an *OFF, forward-blocking,* or *high-impedance* state. In this state, junctions J_1 and J_3 are forward-biased, and J_2 is reverse-biased. The externally applied voltage appears primarily across the reverse-bias junction.

Region 2. For voltages with values from *b* to *c*, the voltage across the reverse-bias junction, J_2, approaches the breakdown voltage of that junction. The current during this voltage excursion increases slowly up to the breakover voltage, V_{BO} (point *c*), at which point it suddenly increases abruptly.

Region 3. For voltages from *c* to *d*, the device exhibits a differential negative resistance (i.e., the current increases as the voltage sharply decreases). This is a transient state, which occurs as the device switches from the *OFF* state to *Region 4* operation (*ON* state).

Region 4. For voltages beyond *c*, the I-V characteristic exhibits *low-imped-*

---

* Note that the generic term *diode* simply means a device with two electrodes. Therefore, we should not expect the electrical behavior of all two terminal electronic devices to be alike. More specifically, we should not confuse *pnpn diodes* with *pn diodes*, with respect to electrical behavior. In fact, as we shall see, the I-V characteristic of the three junction *pnpn* diode is considerably different, and more complex, than that of the single-junction *pn* diode.

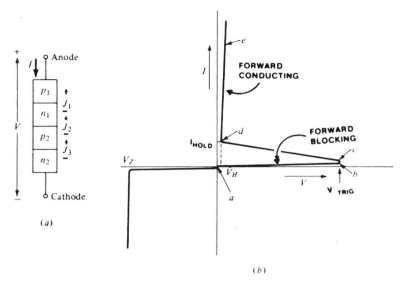

Fig. 6-19 (a) *pnpn* diode showing current reference direction and voltage polarities used in part b. (b) I-V characteristic of a *pnpn* diode.

ance behavior. The I-V curve in this region behaves very much like that of a single forward-biased *pn* junction and is known as the *ON*, (or *forward conducting*) state. In this state the junction J_2 is forward-biased, and the voltage across the entire device is on the order of 1.0 V. If the current through the device is reduced while it is operated in this state, the device will remain *ON* only as long as I exceeds I_H (the so-called *holding current*). If the current is reduced below I_H, the diode switches back to the high-impedance state. The voltage across the device when I_H is flowing is called the *holding voltage,* V_H.

Thus, a *pnpn* diode operated with a positive voltage between the anode and cathode is a bistable device that can switch from an *OFF* state to an *ON* state, or vice versa. If the external circuit can supply more current than I_H, the device will remain *latched* in the *ON* state as long as power is applied.

The parasitic *pnpn* structure in the CMOS circuits exhibits essentially the same I-V characteristic as the device just described. If the parasitic *pnpn* diode is triggered into operating in Region 4, and if the external circuit can provide the necessary holding current, the CMOS circuit will remain *latched up* in the *ON* state, even if the source of trigger current has been removed.

6.4.3 Device Physics Behavior of *pnpn* Diodes

Assume that a voltage source through a resistor once again applies a positive voltage to the *pnpn* diode, as shown in Fig. 6-19a, and that this produces a current I through the device. In *Region 1* the applied voltage forward-biases J_1 and J_3 and reverse-biases J_2.

If the emitter-base junction of a bipolar transistor is forward biased and the collector-base junction is reverse-biased, the transistor is in the *active* mode of operation. Thus, in the equivalent circuit of Fig. 6-18c, transistors Q_1 and Q_2 are both biased in the active mode when the diode is operated in *Region 1*. If a bipolar transistor is operated in the active region, the collector current of a transistor is given by $I_C = \alpha I_E + I_{CO}$. When this equation is applied to Q_1 and Q_2, then

$$I_{C1} = -\alpha_1 I + I_{CO1} \qquad\qquad (6-2a)$$

and

$$I_{C2} = -\alpha_2 I + I_{CO2} \qquad\qquad (6-2b)$$

where α_1 and α_2 are the common-base current gains of Q_1 and Q_2, respectively.

According to Kirchhoffs current law, the sum of the currents entering Q_2 must be zero. Using this law and the current components as shown in Fig. 6-18c, we get

$$I + I_{C1} - I_{C2} = 0 \qquad\qquad (6-3)$$

Combining Eqs. 6-2 and 6-3, we obtain

$$I = \frac{\left(I_{CO_2} - I_{CO_1}\right)}{1 - (\alpha_1 + \alpha_2)} \qquad\qquad (6-4)$$

It can be seen that as the sum of $\alpha_1 + \alpha_2$ approaches unity, the current I increases without limit. At this point the device is said to *break over*.

In bipolar transistors, the magnitude of α increases with collector current at low current levels (as shown in Fig. 6-20a). Since the value of I_C is increased as the avalanche breakdown voltage of the collector junction is approached, an increase in collector voltage can therefore lead to a significantly increase α.

When the collector currents are small, both α_1 and α_2 are much less than 1, and the current flowing through the diode is essentially the sum of the leakage currents, $I_{CO} = I_{CO2} - I_{CO1}$. As a result of the effects that lead to the increase of α, however, if the voltage across the *pnpn* diode is increased to a point near the collector-base junction breakdown voltage, the magnitude of the two alphas also increases. If the sum of the two alphas approaches unity, the current I begins to rise rapidly, which further increases the magnitudes of the alphas (this is an example of a *positive-feedback,* or *regenerative,* mechanism). When $\alpha_1 + \alpha_2 = 1$, breakover occurs, and *Region 3* operation sets in.

Beyond this point, device stability is provided by forward-biasing of the junction J2. Since both junctions in each of the two transistors are now forward-biased, both Q_1 and Q_2 are in the *saturation* region of operation. In saturation, the current gain of a bipolar transistor α again becomes smaller. As a result, once the diode enters the *ON* region, the transistors enter saturation to the degree necessary to maintain the condition of $\alpha_1 + \alpha_2 = 1$. The current I then increases to a value that is essentially limited by the external circuitry.

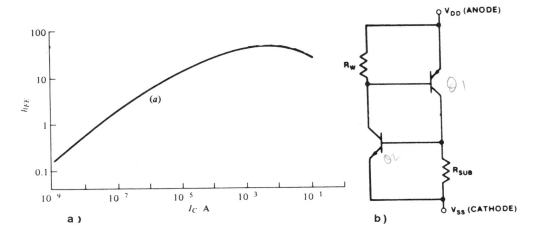

b)

Fig. 6-20 (a) Dependence of current gain on the collector current. (b) Equivalent circuit model of the latchup structure in CMOS, including the parasitic resistances, R_w and R_{sub}.

In the *ON* region, the voltage across the device is the algebraic sum of the voltages across the three forward-biased junctions. Since the voltage drop across the center junction J_2 is in the opposite direction of the voltage drops across J_1 and J_3, the total drop across the *pnpn* diode in the *ON* region is about 1.0 V.

To maintain the diode in the *ON* region, the condition of $\alpha_1 + \alpha_2 = 1$ must continue to be satisfied. The holding current is the minimum current at which this condition is still met. If the current through the device is reduced to less than I_H (or if the applied voltage drops below V_H), the diode switches back to the *OFF* region of operation.

We can also express the condition $\alpha_1 + \alpha_2 = 1$ in terms of the β of the device by adding $(\alpha_1 \alpha_2)$ to both sides of the equation and rearranging terms, to get

$$\alpha_1 \alpha_2 = \frac{(\alpha_1 \alpha_2)}{1 - \alpha_1 - \alpha_2 + \alpha_1 \alpha_2} \qquad (6\text{-}5)$$

This can be further rearranged to give us

$$\frac{\alpha_1}{(1 - \alpha_1)} \frac{\alpha_2}{(1 - \alpha_2)} = 1 \qquad (6\text{-}6)$$

Since $(\alpha_1 / 1 - \alpha_1) = \beta_1$, and $(\alpha_2 / 1 - \alpha_2) = \beta_2$, we finally get

$$\alpha_1 + \alpha_1 = \beta_1 \beta_2 = 1 \qquad (6\text{-}7)$$

We will refer to the expression $\beta_1 \beta_2 = 1$ as the *current-gain-product latchup condition for an ideal pnpn diode.*

6.4.4 Summary of Necessary Conditions for Latchup

Based on the principles of operation of *pnpn* diodes, the following four conditions must exist in a CMOS circuit in order for latchup to occur:

> *Latchup Condition #1.* The emitter-base junctions of both parasitic bipolar transistors must be forward-biased. (It commonly happens, however, that initially only one transistor has its emitter-base junction forward-biased. This condition may then supply the current necessary to forward-bias the emitter-base junction of the second bipolar transistor.)

> *Latchup Condition #2.* The product of the transistor gains must be sufficiently large to allow regenerative feedback. (We swill see that the minimum product of β_{npn} and β_{pnp} needed to induce latchup in real CMOS structures is usually much greater than 1. That is, the $\beta_{npn}\beta_{pnp}$ product needed to cause latchup = 1 in ideal *pnpn* devices, but >1 in actual CMOS structures.)

> *Latchup Condition #3.* The external circuit must be able to supply a voltage equal to or greater than the holding voltage, V_H, of the *pnpn* structure.

> *Latchup Condition #4.* There is a minimum latchup trigger time for which the stimulus must be present in order for the circuit to be latched.

6.4.5 Circuit Behavior of *pnpn* Structures in CMOS Circuits

In actual CMOS circuits, the parasitic *pnpn* device structures consist of two complementary parasitic bipolar transistors adjacent each other. The parasitic series resistances of the electrical path from the *n*-well contact to the *npn* collector, R_w, and of the path from the substrate contact to the *pnp* collector, R_{sub}, are also important circuit elements of the structure. Hence, the equivalent circuit that most accurately models the latchup structure in CMOS must include R_{sub} and R_w as well as the two parasitic bipolar transistors (Fig. 6-20b).

One of the effects caused by each of these two resistors is that a portion of the collector current of each transistor is siphoned away, reducing the base current. As a result of such current shunting, the effective current gains of the bipolar transistors are reduced. In fact, when finite R_{sub} and R_w are included in the circuit, Eq. 6-7 – in terms of the current gains β_{npn} and β_{pnp} – must be modified to more accurately express *Latchup Condition #2* for *pnpn* structures in CMOS circuits. For the latchup structures in actual CMOS, therefore, the inequality is expressed as

$$\beta_{npn}\beta_{pnp} > 1 + \frac{(\beta_{pnp} + 1)\left(I_{R_{sub}} + \dfrac{I_{R_w}}{\beta_{pnp}}\right)}{I - I_{R_{sub}} - I_{R_w}\left(1 + \dfrac{1}{\beta_{pnp}}\right)}. \qquad (6-8)$$

where I_{Rsub} and I_{Rw} are the currents that flow in R_{sub} and R_w, respectively. The minimum gain-product needed to induce latchup in CMOS circuits can thus be much greater than 1, depending on the values of the series resistances.

In general, the approaches to eliminating latchup can be divided into two categories: (1) those that reduce the bipolar transistor current gains, and (2) those that lower the value of the series resistances R_{sub} and R_w. If either of these approaches is successful, the latchup condition given by Eq. 6-8 will be harder to satisfy. In the extreme, for example, if R_{sub} and R_w could be made equal to zero, the two emitters would be short-circuited and would thus be prevented from ever turning on. Even if R_{sub} and R_w are only made smaller, the values of I_{Rsub} and I_{Rw} will increase, making the right side of the inequality larger. Therefore, various techniques have been developed to decrease the values of R_{sub} and R_w.

6.4.5.1 Value of β in CMOS Vertical Parasitic Bipolar Transistors.

The current gain of the vertical parasitic bipolar transistor in an actual CMOS structure depends on the well depth, the well doping concentration, and the built-in field in the well (see chap. 7). In typical 1-μm CMOS structures, the well is 1-2 μm deep and is doped to ~1×10^{16} cm^{-3}. In n-well technology, the current gain of the vertical pnp transistor, β_{pnp}, will be ~100. In p-well or twin-well technology, the current gain of the vertical npn transistor, β_{npn}, will be two to three times larger than that of the pnp device in n-well technology (due to the higher minority-carrier mobility in the base region).

The transient response of the bipolar transistors is also important because of the minimum latchup trigger time (*Latchup Condition #4*). In addition, latchup in a real circuit is normally induced by transient triggering. The transit time for minority carriers across the base region is a measure of the transient response of a bipolar transistor. The minimum latchup trigger time may be approximated by the sum of the vertical and lateral bipolar transit times.[40] A typical value of the transit time of a vertical transistor in a 1-μm n-well CMOS technology is several nanoseconds.

6.4.5.2 Value of β in CMOS Lateral Parasitic Bipolar Transistors.

The current gain of a lateral bipolar transistor is determined primarily by the layout spacing between the diffusion outside the well to the well edge, since this is the dimension of the transistor's base width (although the gain is also impacted by the field doping outside of the well and the well depth). Since the base width of the lateral transistor is usually much larger than that of the vertical transistor, the β value is generally an order of magnitude lower (e.g., β ranges from ~2 to 4 in CMOS structures whose n^+ to n-well spacings vary from 5 μm down to 2 μm).

The larger base width also means that the base transit time is also much greater. The poor current gain of the lateral bipolar transistor reduces the tendency of the CMOS structure to undergo latchup when dc signals are applied, while the longer base-transit time increases the minimum triggering time for transient-induced latchup. As the layout spacing shrinks in high-density ULSI, however, the β is increased and the base transit time is reduced.

6.4.6 Circuit and Device Effects that Induce Latchup

For latchup to occur, the emitter-base junctions of both of the parasitic transistors must be forward-biased. This *circuit condition* can be produced in the latchup structure of CMOS circuits in three ways, each of which can be triggered by various physical stimuli. In addition, virtually all latchup failures occur as a result of transient stimuli. To fully describe the various latchup causing mechanisms, we will refer to Fig. 6-18d (the equivalent circuit model), and to Fig. 6-18a (the CMOS inverter cross-section). The three scenarios[36,38] that lead to latchup are described in the next paragraphs.

6.4.6.1 An external stimulus forward-biases the emitter-base of one transistor, and its collector current then turns on the second transistor.
To explain latchup in such cases, let us assume that the externally applied stimulus is a *voltage overshoot at the output node of an* p-well *CMOS inverter driver circuit*. This is, in fact, the most common cause of latchup.[54]

The sequence of events is as follows: The *source* region of the PMOS device and the substrate contact are both connected to V_{DD} to ensure that the source-substrate junction is never forward-biased (Fig. 6-18a). The *drain* of the PMOS device, however, is connected to the output of the inverter. If this output is *low* (e.g., 0 V), the drain-substrate junction is reverse-biased, and no latchup can occur – that is, both of the structures that could be emitters for the parasitic *pnp* device (i.e., the source and the drain of the PMOS device) are at potentials lower than or equal to that of the base region. If the output state of the inverter is *high*, the drain-substrate bias should then equal 0 V; in this case there should still be no reason for latchup to occur (i.e., we assume that the output of the inverter is designed to reach V_{DD} in the high-output state).

If a voltage overshoot occurs at the output terminal, however, the output node experiences a condition in which V_{DD} is exceeded. (Overshoots and undershoots are both common, especially at input/output [I/O] device nodes, where signals tend to be noisy). If the overshoot causes the voltage at the drain region to exceed V_{DD} by more than about 0.6 V, the p^+-n drain-substrate junction becomes forward-biased. Holes are injected into the n-substrate (the base of the *pnp*; see Fig. 6-18a), from the p-type drain region of the output node. (This region behaves like a second emitter to transistor Q_1.) In essence, a triggering current flows through this emitter.

Some of the holes of this triggering current recombine in the base, while the remainder reach the p-well (the collector of the *pnp*). The latter represent the *pnp* transistor collector current, which now flows both into the base of Q_2 (I_{B2}) and through R_w to V_{SS}. The fraction of the current that flows through R_w causes a voltage drop across it; this voltage is also impressed across the emitter-base junction of the *npn* transistor (Fig. 6-18a).

If the voltage drop across R_w reaches 0.6 V, the *npn* device will be turned on. That is, the n^+ source (the emitter of the *npn*) will emit electrons into the *p*-well (the base of the *npn*). Some of these electrons will reach the *n*-substrate and will drift out of the V_{DD} terminal. If enough electron current exists in the *n*-substrate and if sufficient resistance, R_{sub}, exists between the V_{DD} contact and the p^+ source, an IR drop will develop, causing the p^+ source to inject holes into the *n*-substrate. This hole current adds to the initial hole current injected from the positively biased drain region. Thus, the positive-feedback scenario between the *pnp* and the *npn* transistors is triggered, and latchup is rapidly induced. A voltage undershoot when the output node of the inverter is in the low state will have the same net effect, except that the *npn* will be the first transistor to turn on.

Transient overshoot or undershoot voltages are a particular problem at the outputs of CMOS driver circuits. since impedance mismatches at the far ends of transmission lines or printed-circuit-board wiring traces result in reflections that return to the driver output node. It is therefore especially important to utilize stringent latchup-prevention measures at the input and output circuitry of CMOS chips (some combination of guard structures and multiple-well contacts should be used; see section 6.4.8.3).

6.4.6.2 An external stimulus causes current to flow through both bypass resistors, forward-biasing one or both bipolar transistors.
Triggering mechanisms that can cause currents to flow in both bypass resistors (R_{sub} and R_w) include avalanche breakdown of the well-substrate junction (J_2, in Fig. 6-18e), photocurrents due to ionizing radiation, and *n*-well displacement currents (due to the charging or discharging of the large well-to-substrate junction capacitance). As shown in Fig. 6-21, the effects of these mechanisms can be modeled by adding a current source, I_o, and a capacitance, C_{ws}, to the equivalent circuit model of Fig. 6-20b. It is evident that leakage or displacement currents can still flow in both resistors, even if the bipolar parasitic transistors are in cutoff.

Examples of the external stimuli that produce these currents are: (a) voltages across the power-supply terminals that exceed the breakdown voltage of J_2 (and hence cause avalanche breakdown current, I_o); (b) ionizing radiation that causes photogeneration leakage current, I_o; and (c) external voltage transients (e.g., the voltage step function that occurs when a chip is powered-up) that produce displacement currents in the course of charging and discharging C_{ws}.

If the current in the bypass resistors is large enough, it can turn on both transistors. Typically, however, one of the resistances is larger than the other, and the voltage drop across the larger resistance causes the transistor to which it is connected to be turned on first. The positive-feedback mechanism that induces latchup is then set in motion, and when *Latchup Condition #2* is satisfied (as given by Eq. 6-8), the circuit latches up.

6.4.6.3 Current is shunted through one of the parasitic transistors by some degradation mechanism, and the resulting collector current flows through the bypass resistor of the second transistor, turning it on.
The triggering mechanisms that effectively shunt

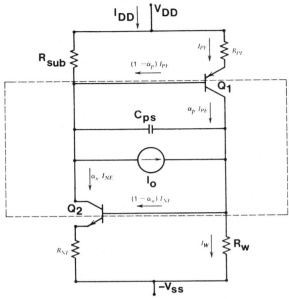

Fig. 6-21 Latch-up equivalent circuit including well-to-substrate capacitor, C_{ps}, and parasitic current source, I_o.

current through a low-impedance path from the emitter region to the collector region of one of the parasitic transistors include the following:

- Inversion of the field region between the source/drain regions of the MOS device in the substrate (emitter of lateral bipolar device) and the edge of the substrate (collector region).

- Punchthrough between the substrate region and the source/drain regions of the MOS device in the well.

- Avalanche ionization near the drain due to hot-electron effects. Note that in some cases this type of triggering mechanism can induce latchup even at voltages lower than the supply voltage, V_{DD}.

When one of these effects causes a large enough current to be shunted to the collector of one of the parasitic transistors, this current will flow through the bypass resistance of the second transistor, causing it to become turned on. Latchup will occur if the positive-feedback mechanism causes *Latchup Condition #2* to be satisfied.

6.4.7   Test Methods for Characterizing Latchup

The most common parameters used for characterizing latchup are trigger current (I_{trig}), holding current (I_H), and holding voltage (V_H). The trigger current is the current that must pass through the emitter-base junction of the *pnpn* device in order for latchup to

Table 6.1 Latchup Stimuli

I/O node voltage overshoot and undershoot	Type 1
Avalanche of well-substrate junction	
(extra-high voltage on some node)	Type 2
Photogeneration	
(ionizing radiation)	Type 2
Displacement current transient	
(voltage transient during power-up of chip)	Type 2
Inversion under field oxide	Type 3
Punchthrough between n^+ and n-well	Type 3
Hot-electron-induced current	Type 3

be induced (i.e., it is the current drawn from the power supply just before the test structure enters the latched state). Large values of I_{trig}, I_H, and V_H are desirable for reduced latch susceptibility. The efficacy of the processing and circuit-layout schemes designed to reduce the values of R_{sub} and R_w are therefore evaluated by measuring the values of I_{trig} and V_H. *If it is found that V_H is greater than the power-supply voltage, the circuit is said to be latchup-immune.* This assertion is based on *Latchup Condition #3*. Even if latchup is momentarily induced, it will not be sustained, because the power-supply voltage will be less than V_H.

The susceptibility of CMOS circuits to latchup is often determined experimentally by measuring the total current through the *pnpn* path while overstressing the anode voltage. With the source and *n*-well contact (Fig. 6-22) maintained at V_{DD}, the isolated voltage on the p^+ region in the *n*-well is raised above V_{DD}.[39] The value of I_{trig} is then measured for the circuit structure being evaluated. (Once a latchup has been triggered, the stressing voltage on the isolated region is returned to V_{DD}.) The value of V_H is measured by lowering V_{DD} after a latchup.

Techniques for studying the transient behavior of latchup have also been described. Such transient testing may provide better characterization of the latchup that occurs in

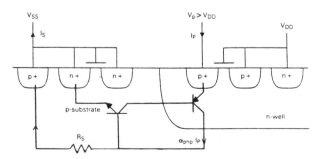

Fig. 6-22 Measurement technique for determining latchup triggering by p^+ overvoltage.

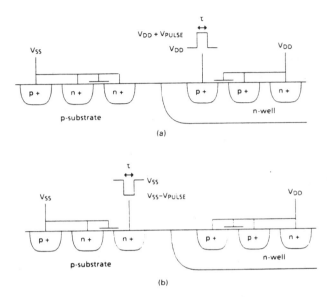

Fig. 6-23 Transient latchup measurement setups: (a) p^+ overvoltage triggering; and (b) n^+ overvoltage triggering.[116] (© 1987 IEEE).

actual circuit environments (*Latchup Condition #4*). One test involves the measurement of the *pnpn* diode current when different voltage ramp rates are impressed on the isolated p^+ region of Fig. 6-18a. This allows characterization of the displacement-current-triggering event, described as being one of the causes of Type 2 mechanisms. In another test setup, an overvoltage *pulse* is applied to the drain of the device described in Fig. 6-23. For a given pulse height, the pulse width is increased until latchup is triggered. As the pulse length approaches the bipolar transient response time, the pulse height needed to induce latchup increases. When the length is further reduced, latchup cannot be triggered, regardless of how large a pulse amplitude is applied.

A third transient test for latchup involves pulsing the base of one of the parasitic bipolar transistors.[40] Transient excitation with a pulse width shorter than the minimum regeneration time causes no latchup.

Another technique for evaluating the latchup hardness of a CMOS circuit design is given by Troutman,[36] who defines a differential latchup criterion. The following data need be gathered when this technique is used: the bypass resistance values; small-signal alphas of both parasitic transistors; base-emitter saturation currents for both transistors; and temperature.

6.4.7.1 Modeling Latchup in CMOS Technology.
Modeling of latchup has been attempted by a number of different groups. It has turned out to be a difficult undertaking, since the phenomenon involves bipolar transistors with strong injection

effects, with the structures also inherently distributed in nature. As a result, conventional one-dimensional device analysis is very difficult to apply.

6.4.8 Techniques for Reducing or Eliminating Latchup Susceptibility

Circuits are described as being either latchup-free or latchup-immune; the distinction between the two terms is important. *Latchup-free* refers to the ideal situation in which latchup will never occur under any circumstances. *Latchup-immune* refers to circuits that will not exhibit latchup under normal operating conditions, but that could be forced to do so through the application of sufficiently high voltages or the injection of high currents. While as of this writing there is no industrial standard for latchup hardness, a circuit is generally recognized as being latchup-immune if it can withstand an I_{trig} of more than 500 mA at the I/O pins.

Latchup immunity exists if V_H is less than the power-supply voltage per *Latchup Condition #3*. Many reports in the literature describe such a circuit as being *latchup-free*. However, Troutman, points out that even transient switching (unsustained latchup) can cause circuit problems and undesired power dissipation.[38] *Thus, he argues that the most effective way to prevent latchup problems is to ensure that the* pnpn *structures remain in the OFF state at all times.*

The approaches to reducing or eliminating susceptibility to latchup can be divided into three categories, as follows:

1. Processing techniques that reduce the current gains of the parasitic transistors to the degree that *Latchup Condition #2* cannot be satisfied.

2. Processing schemes that either reduce the values of R_{sub} and R_w or eliminate the parasitic *pnpn* diode structure.

3. Circuit-layout procedures that decouple the parasitic bipolar transistors.

6.4.8.1 Processing Techniques That Reduce Current Gains.
Several techniques have been utilized in an attempt to reduce β of the parasitic bipolar transistors (bipolar spoiling). The first group involves methods that physically separate the emitter and collector regions of the lateral transistor (in which the n^+ regions are kept far from the n-well border). Obviously, for high-density circuitry, merely increasing the spacing of these regions on the wafer surface is not a good solution. Another approach for keeping these regions apart is to use recessed oxides and/or trench structures to increase the distance carriers must travel from the n^+ region to the well. While these techniques do help, by themselves they provide inadequate latchup protection. (These techniques will be described in more detail in the sections dealing with device isolation).

The techniques in the second group attempt to reduce the minority-carrier lifetimes in the base. These include gold doping to produce trapping sites,[41] neutron irradiation to cause structural damage and resultant recombination centers,[42] and harnessing of the oxygen precipitates formed in the substrate bulk by internal gettering processes for use as efficient recombination centers for minority carriers.[43] The first two of these lead to increased leakage currents, while the third has little effect on lateral transistors with small base widths, since the precipitates do not form in the lightly doped epi. As a result, the techniques in this group have not been widely implemented.* The third type of technique involves the use of a retrograde well, which reduces the vertical bipolar gain by providing a high Gummel number and a retarded E-field in the base.[4] However, care must be taken to remove injected carriers so that lateral transistor action to the well edge is not increased.[39] This approach also requires the availability of a high-energy implanter.

The fourth type involves the use of silicided source/drains[45,46] to reduce the emitter efficiency of the parasitic bipolar transistors. These techniques have the disadvantage of degrading the gain of the MOS devices.[47]

In summary, the methods described above can help to reduce latchup susceptibility, but provide an insufficient degree of latchup protection. The remaining two categories of latchup-suppression techniques, have been found to be much more effective, and as a consequence, they have been much more widely implemented.

6.4.8.2 Processing Techniques That Reduce R_{sub} and R_W or Eliminate the *pnpn* Structure.

Techniques that reduce the series resistance values include the use of *epitaxial layers on heavily doped substrates* and the use of *retrograde implants*. Techniques that eliminate the *pnpn* path use either: (1) a combination of *deep trench isolation structures* and *epitaxial layers on heavily doped substrates*; or (2) *silicon-on-insulator* (SOI) substrates.

The use of lightly doped epitaxial layers on heavily doped substrates is effective in reducing latchup susceptibility for two reasons. First, the highly doped layer substantially reduces the value of R_{sub} by placing a low-resistance path for *majority carriers* in the substrate in parallel with the more lightly doped epi region in which the devices are formed. Hence, it very effectively shunts the lateral parasitic bipolar transistor. Second, any *minority carriers* injected into the epi layer that then diffuse into the highly doped substrate are more rapidly recombined there, so that fewer reach the collector of the lateral bipolar device.

Figure 6-24 shows the trigger current and the holding voltage as functions of n^+ to p^+ spacing, d, for various epi thicknesses.[48] Thin epi can be seen to dramatically improve latchup immunity from either standpoint. For a 12-μm epitaxial layer, the triggering current required for latchup is less than 1 mA when $d = 10$ μm. When the epitaxial thickness is reduced to 3 μm, d can be reduced to 5 μm, and I_{trig} is increased to 80 mA. The advantage diminishes as d approaches the epi thickness. However, the minimum epi thickness is limited by outdiffusion of impurity atoms from the heavily

---

* An approach to creating oxygen precipitates in a thin layer 2.5 μm below the surface by means of ion implantation of oxygen and epi growth has recently been reported.[44] Since these precipitates are positioned so that they reside in the base region of the lateral *pnp* device, they are effective in reducing its alpha by decreasing the minority-carrier lifetime in the base.

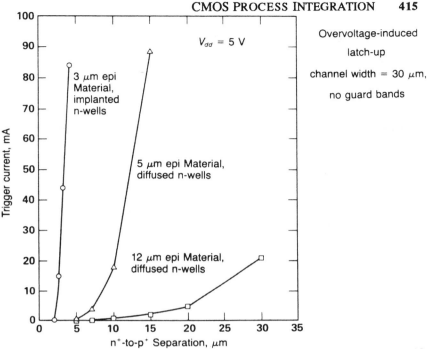

Fig. 6-24 Triggering current versus n^+-to-p^+ separation in an n-well CMOS structure.[48] (© 1985 IEEE).

doped substrate during epi deposition and subsequent wafer processing (see section on autodoping in Vol. 1, chap. 5). That is, due to this effect the *effective* epi thickness is thinner than the *initial* epi thickness. (The effective epi thickness is defined as the thickness of the epi layer when the doping concentration is less than some specific number – e.g., 1×10^{18} cm^{-3}.) The fact that n-epi on n^+ produces a sharper epi interface than p-epi on p^+ implies that n-epi should provide better latchup protection.[4] But as noted in section 6.2, on well technology in CMOS, this benefit is obtained only at the price of accepting the other drawbacks inherent to n-epi on n^+ wafers.

The disadvantages of epi CMOS include increased wafer cost, lower breakdown voltages, and increased leakage currents.[49] In addition, epi-layer growth may generate defects that will lead to reduced chip yields.[50]

With respect to *retrograde implants*, it has been found that retrograde wells can be effective against latchup for several reasons.[4] In p-well technology, the retrograde implant is used to form the p-well, with the following advantages being gained:

• The retrograde profile gives a high doping concentration deep in the well, which reduces the gain of the vertical bipolar device.

• The heavier doping near the bottom of the well reduces R_w, thereby helping to decouple the bipolar transistors.

- The retrograde well can minimize thermal processing following epitaxial-layer deposition. Epi layers with thinner *effective* thicknesses may therefore be possible if retrograde wells are combined with epitaxial layers.

- If the combination of the two techniques is used, a process that also reduces *both* R_{sub} and R_w is obtained.

In *n*-well technology, the retrograde implant is a *p*-type implant that is placed beneath the NMOS devices in the substrate.[51] This creates a p^+ layer in the substrate that has a much smaller transition region than a *p*-epi-on-p^+ substrate, thus providing even better latchup protection.

Trench isolation, as described in chapter 2, allows n^+ and p^+ regions to be placed close to one another in CMOS. If the trenches are deeper than the well regions, they provide physical separation among device types. However, when trenches are used without epitaxial layers on heavily doped substrates, their main contribution to latchup immunity is that they force carriers in the lateral transistor base region to travel greater distances to reach the collector. As a result, trench isolation alone does not significantly increase I_{trig} or V_H. The major advantage of using trenches on their own is the increase in latchup response time, which results in a substantial improvement in *transient upset* prevention. When deep trenches are combined with a lightly doped epitaxial layer on a heavily doped substrate, however, latchup-free structures are produced.[11,96] As minority carriers injected into the substrate attempt to diffuse toward the collector, the deep trench forces them into the heavily doped substrate. There they rapidly recombine, substrate and hence never arrive at the collector region.

Silicon-on-insulator technology is also receiving wide attention, since in addition to providing electrical isolation between the MOS devices, it results in latchup-free CMOS structures. Each MOS device is isolated from neighboring devices by the insulating layer; as a result, the *pnpn* path is no longer present.

6.4.8.3 Circuit-Layout Techniques for Decoupling Parasitic Bipolar Transistors.
Two types of structures can be incorporated into the circuit layout that will provide latchup protection: *guard structures,* and *substrate and well contacts.*

Guard structures are heavily doped diffused regions that encircle the well. Troutman states that they are the most effective layout procedure available for providing latchup protection,[39,54] and this claim is supported by the simulated and experimental evidence of Menozzi et al.[52] Guard-ring structures can be fabricated either outside the well (in which case the guard surrounds the outer edge of the well region), or inside it (in which case the well is placed between the active device regions and the well border). The guard can be either a *minority-carrier* or a *majority-carrier* structure.

Minority-carrier guard rings have a doping type opposite to that of the region in which they are formed (Fig. 6-25). Normally, these guard rings are placed in the substrate outside of the well edge (e.g., an *n*-type diffusion in the *p*-substrate of an *n*-well CMOS technology). Since the guard is connected to the power supply in such a way that the *pn* junction it forms with the substrate is reverse-biased, this type of guard

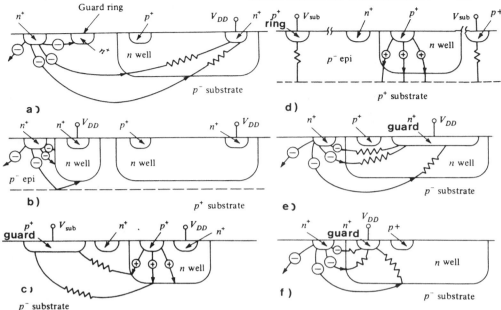

Fig. 6-25 [(a) and (b)] Minority-carrier guard in substrate: (a) n^+-diffusion guard; (b) Deeply diffused guard in epitaxial CMOS. [(c) and (d)] Majority-carrier guard in substrate: (c) p^+-diffusion guard for reducing substrate sheet resistance; (d) Contact ring is preferable to p^+-diffusion guard in epi-CMOS. (e) Majority-carrier guard in well. (e) n^+-diffusion guard to reduce n-well sheet resistance. (f) n^+-diffusion guard to steer current away from vertical *pnp* emitter. [36] From R. R. Troutman, *Latchup in CMOS Technology*, Copyright Klewer Academic, 1989. Reprinted with permission.

ring will collect any minority carriers in the substrate that diffuse into its depletion region. The carriers are thus prevented from reaching the well, and as a result, the collector current of the lateral bipolar transistor is reduced. The main role of minority-carrier guards in preventing latchup is to provide this reduction in effective current gain.

If a thin epitaxial layer is used together with a deeply diffused guard, the effectiveness of the guard is increased (Fig. 6-25b).[36] That is, the path of minority-carrier diffusion is narrowed (by the reflecting boundary where the high and low doped regions in the substrate meet); this forces the carriers closer to the guard ring, where they can be intercepted. In addition, any minority carriers that enter the heavily doped substrate recombine more rapidly there.

Majority-carrier guard rings, on the other hand, are doped with the same type of dopant as the regions in which they are formed. While they can be implemented outside the well, they are usually more effective when placed inside. The function of this type of guard is to provide a shorter (and, hence, a lower-resistance) path for the carriers that constitute the collector current in the well (or substrate). The majority carriers in the collector must drift to the power-supply node through R_w (or R_{sub}) once they enter the collector region, and the guard rings effectively reduce the value of R_w and R_{sub} by placing smaller resistances in parallel with these series resistances.

The reason that the majority guard is usually placed within the well is that the value of R_w is generally higher than that of R_{sub}. The effective lateral resistance can be quite high in shallow wells, since the source/drain diffusions in the well pinch the lateral current paths. Thus, a guard ring in the well that reduces the distance between the well edge and the well contact can be very effective in reducing the large value of R_w. A way to implement three guard structures with only two diffusions is to place one guard so that it overlaps the well and substrate boundary (Fig. 6-26).

Voltage drops along the power supply bus to which the guard ring is connected should never be allowed to become large enough to initiate a latchup condition. For example, if the bus to which the guard is attached has a sufficiently large resistance that the voltage of the n^+ guard in the n-well drops to 4 V, when the output rises to 5 V, the emitter tied to this higher voltage will turn *ON*, initiating a potential latchup event.

Troutman also emphasizes that a substrate-contact majority-carrier ring should be mandatory for all chips.[36,39] This would minimize lateral bypass resistance by distributing substrate majority carriers. When used with epitaxial CMOS, such contact rings can reduce lateral bypass resistance to below 1 Ω; in addition, they are as effective as backside substrate contacts in eliminating latchup. In fact, the need for multiple majority guard rings (which would surround each of the well regions) can be eliminated through the use of a single substrate contact ring (as long as it is used together with wafers that have a thin epi layer on a highly doped substrate).

The major limitation of guard rings is that they decrease overall circuit density. In the input/output circuits of CMOS, however, the MOS transistors must be made quite large in order for sufficient off-chip drive-current capability to be provided. Thus, the area penalty imposed by guard-ring use around these circuits is usually quite small. Because of the noisy signals encountered at the I/O nodes of a chip, the use of guard rings to suppress latchup in such circuitry is widespread.[54] Each of the output devices in the substrate is typically surrounded by a majority-carrier guard ring (tied to V_{DD} in p-well CMOS, and to V_{SS} in n-well).

Note that n-well CMOS generally provides better latchup immunity than p-well, for two reasons.[8] First, since the mobility of majority carriers is higher in an n-well than

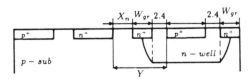

Fig. 6-26 Implementation of two guard structures with only one diffusion by placing one guard so that it overlaps the well–substrate boundary.[52] (© 1987 IEEE).

in a p-well, R_w in the n-well is lower than in a comparable p-well. Furthermore, R_w is usually much higher than R_{sub}, especially in thin wells that are pinched by source and drain diffusions. Thus, the lower value of R_w in the n-well can be of significant help. Second, because hot-electron-induced substrate currents are much higher in NMOS than PMOS devices, it is better to have the NMOS devices in the substrate. Such currents are more easily collected from here than from the well region, especially if an epitaxial layer on a heavily doped substrate is used.

The second type of circuit-layout technique to be discussed is that of *substrate and well contacts*. The use of guard rings in the well may represent too large an area penalty because in high density circuits multiple well contacts can serve as an alternative approach to improving latchup immunity. However, this approach is not as effective as the use of guard rings.[52]

The number and placement of power-supply contacts to both the well and the substrate also impact the effectiveness of this approach. Increasing the number of well contacts to V_{DD} reduces latchup susceptibility, since the resistor length of R_w for each FET in the well is reduced when a well contact is in its proximity. The contact spacings should therefore be no more than two squares apart, to ensure that no local high-resistance regions exist. The well contacts should be hard-wired to the power-supply connection (ground or V_{DD}) with a metal stripe. In this way, any injected charge will be shunted to the supply rail through a low-impedance path that will not contribute to the well's already relatively high lateral resistance.

The closer the contacts are to the source regions of the MOS devices, the smaller the potential drop during current flow, leading to increased latchup immunity. The use of *butted source-substrate* and *butted source-well* contacts, in which the n^+ and p^+ regions are contiguous and connected to the same potential has thus become popular.[101] While such butted contacts also conserve area, they are limited to FETs that operate with grounded sources.

6.5   CMOS ISOLATION TECHNOLOGY

In CMOS ICs, *like* kinds of devices within a given well must be isolated in the same manner as the devices in either NMOS or PMOS circuits (i.e., through a combination of a thick field oxide and channel-stop doping). However, the isolation requirements of CMOS technology extend beyond those of either PMOS or NMOS alone, in that in CMOS it is also necessary to isolate the p- and n-channel devices from one another. The isolation of p-channel from n-channel devices must satisfy two requirements: (1) any possible leakage currents that could flow between adjacent PMOS and NMOS devices must be suppressed; and (2) the susceptibility of CMOS to latchup must be minimized.

In CMOS structures, the isolation spacing between the n- and p-channel devices is defined as the total of the distances between the edge of the n^+ region of the n-channel device and the edge of the well, and the edge of the p^+ region of the p-channel device and the edge of the well (or, in other words, the n^+ to p^+ spacing – Fig. 6-27).

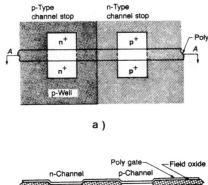

a)

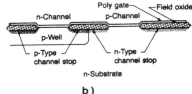

b)

Fig. 6-27 (a) Layout top view of the isolation region between n-type and p-type transistors in CMOS. (b) Cross sectional view of a CMOS inverter showing the channel stops that are designed to prevent the surface under the field oxide from inverting.[53] (© 1986 IEEE).

Basic isolation between p- and n-channel devices in CMOS is established by both a reverse-biased well-substrate junction and by the regions under the field oxide (as long as they are not inverted). This isolation, however, may not be perfect. Unwanted current flow can arise due to junction breakdown, the formation of leakage paths between devices, or latchup. The key *leakage paths* involve the *parasitic MOS field devices* created in the field regions between the devices.* Leakage currents can arise if these well-border parasitic field devices conduct prematurely, as a result of surface inversion, or punchthrough below the surface.** As noted, the *latchup* effect in CMOS arises as a

* The parasitic NMOS field transistor at a p-well border consists of the n^+ source, the p-well body, the n-substrate drain, and the polysilicon gate runner over the field oxide, as shown in Fig. 6-27b. Similarly, the parasitic PMOS field transistor at this well border consists of the p^+ source, the n-substrate body, the p-well drain, and the poly runner over the field oxide.

** In Fig. 2-4 of chapter 2 we described how punchthrough can occur between the source and drain of the same device, or between the source/drain regions of neighboring devices of the same channel type. In this case, we refer to the punchthrough between the border of the substrate and source/drain regions in the well.

result of the *parasitic bipolar devices*. Various isolation techniques have therefore been developed to prevent both leakage and latchup. Device and process simulators used to model such CMOS isolation structures are described in chapter 9.

Interest in *n*- and *p*-channel isolation techniques is very keen, because such isolation requires much more area than between like types of devices. For example, in early single-well CMOS technologies, the isolation spacing required was typically about three times the diffusion depth of the well.[8] Thus, for a 4-μm well depth, a minimum n^+ to p^+ spacing of ~12 μm was needed. In early twin-tub CMOS technologies, which used dual-channel stops, a minimum of 3-9 μm of lateral space was necessary for effective isolation.* In chapter 2, however, we showed that isolation between *like* kinds of MOS devices can be accomplished with isolation spacings of only 1.0-1.5 μm. The large area penalty of *p*- to *n*-channel device isolation is one reason why CMOS technologies using conventional isolation methods cannot achieve as high a packing density as NMOS.

This isolation-spacing requirement is due in part to the processes used to fabricate CMOS devices. Wells are typically driven in quite deeply to ensure that enough charge exists below the transistor to prevent vertical punchthrough to the substrate. This results in both a lateral diffusion of the well dopant and a reduction in the surface concentration near the border of the well (Fig. 6-28).   The channel stop doping in the

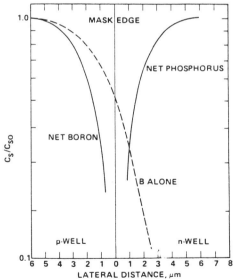

Fig. 6-28 Impurity surface concentration near the border of the two tubs in twin-tub CMOS.[14] (© 1980 IEEE).

* A 3-μm minimum space between n^+ and p^+ regions was reported for a non–trench isolated, twin-well CMOS process. To achieve such tight spacing, high-pressure oxidation was used to grow the field oxide. This was more effective in preventing interdiffusion of the impurities from the two wells than an atmospheric-pressure field-oxidation process would have been.[29]

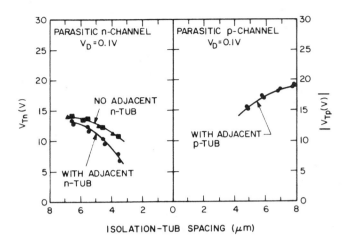

Fig. 6-29 Threshold voltage of n- and p-channel parasitic device as a function of the separation between the transistor edge and the tub border.[14] The upper curve in the left-hand graph shows the parasitic n-channel threshold-voltage reduction near the tub border when no adjacent n-tub is present. When an n-well is present, the threshold voltage is reduced (lower curve), since the interdiffusion of the two types of impurities reduces the net surface concentration of each dopant near the well border.[14] (© 1980 IEEE).

substrate at the edge of the well is also substantially reduced, due to the compensation between the acceptor and donor atoms; this causes a reduction in the threshold voltage of the parasitic MOS field transistors. We will next describe this phenomenon in more detail for the case of twin-tub CMOS.

Figure 6-29 shows the threshold voltage of each type of parasitic device as a function of the separation between the transistor edge and the tub border.[14] The upper curve in the left-hand graph shows the parasitic n-channel threshold-voltage reduction near the tub border when no adjacent n-tub is present. When an n-well is present, the threshold voltage is reduced (lower curve), since the interdiffusion of the two types of impurities (Fig. 6-28) reduces the net surface concentration of each dopant near the well border. Unless the net dopant-reduction effects are somehow counteracted, the spacing between adjacent n^+ and p^+ devices must be kept quite large (e.g., >10 μm) to prevent inversion beneath the field oxide.

If smaller spacings between n- and p-channel devices are to be possible, the channel-stop doping concentration must somehow be increased, particularly in the substrate regions in n-well CMOS. During field oxidation, boron segregates into the oxide, while phosphorus piles up at the silicon surface. As a result, in an n-well process a separate p-type channel-stop implant must be added to increase the surface concentration of the lightly doped p-substrate. Without such an implant, inversion between the n-channel devices is likely to occur. In p-well technologies the well itself is an adequate n-type channel stop, because heavier boron doping exists in the p-well, and the concentration of phosphorus is increased at the surface of the n-substrate during field oxide growth.

In twin-tub processes, two channel stops are needed to reduce the isolation distance as much as possible. Figure 6-30, which is based on a 2-D device model,[53] shows the impurity contours in the isolation region of a CMOS twin-tub process (with retrograde wells).

The AT&T Twin-Tub V technology,[55] is an example of an advanced CMOS process uses a LOCOS-based approach to isolate like devices. The starting material is a p-epi layer on a p^+ substrate. The technology is implemented with a single-mask, self-aligned twin-well process that uses two channel-stop implants (as well as two separate well implants).

First, phosphorus and arsenic are sequentially implanted into the n-well areas, forming a high/low doping profile (channel-stop and well; Fig. 6-31). Since arsenic diffuses much more slowly than phosphorus, it will remain near the surface of the well during the drive-in step. This high arsenic concentration provides an effective channel stop for p-channel devices and also protects against punchthrough.

Next, a relatively thick masking oxide is selectively grown over the n-well region, and the p-well implant (boron) is then performed. Because the implant is blocked by the oxide that covers the n-wells, it enters the silicon only in the p-well regions (a self-aligned implant step). Both well regions are then driven in (with the thick oxide over the n-wells being retained), and a second boron implant (which will serve as the p-well channel stop) is carried out. This implant is kept shallow because a high pressure LOCOS process is subsequently used to grow the field oxide, minimizing the lateral and vertical impurity-profile spreading. Once again, the implant is self-aligned to the p-well regions by the presence of the oxide on the n-wells (Fig. 6-31). The masking oxide is then removed, and a nitride masking layer is deposited and patterned to cover the active areas. Finally, the field oxide is formed.

This process makes it possible to achieve adequate isolation and a reduced tendency

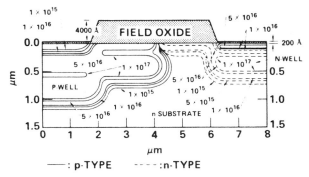

Fig. 6-30 2-D net-impurity contours in the isolation region of a retrograde-well CMOS structure.[53] (© 1986 IEEE).

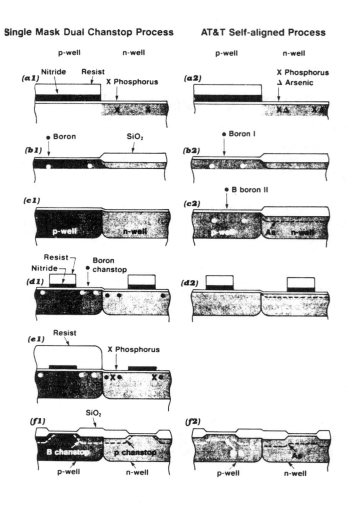

Fig. 6-31 (a1) - (f1) Twin-tub dual-field implant process.[119] (a2) - (f2) Self-aligned twin-tub and field-implant process using high-pressure field oxidations.[62]

toward latchup with smaller device separations than are possible with an atmospheric-pressure oxide-growth process. Nevertheless, a 7-μm spacing must be used between the n^+ and p^+ regions in the Twin-Tub V structure to provide a V_H greater than the power-supply voltage (5 V). Since this spacing still represents a significant layout-density limitation, alternatives to the LOCOS-only process are being vigorously pursued.

In another, more recent study using n-well technology, high-field channel-stop implant doses for the substrate (yielding a peak concentration of 5×10^{16} boron

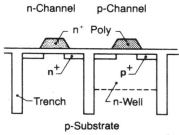

n-Channel   p-Channel

Fig. 6-32 Schematic of an *n*-well CMOS with trench isolation.[4] (© 1986 IEEE).

atoms/cm^3) allowed n^+ to p^+ spacings of 2.4 μm with conventional wells, and 1.8 μm spacings with retrograde wells.[106]

6.5.1   Trench Isolation for CMOS

The two major alternative CMOS-isolation approaches are *trench isolation*[58] and *selective epitaxial growth* (SEG) isolation. Trench-formation technology, discussed in chapter 2, is applied to CMOS to attack the problems of latchup and punchthrough. Its main advantage is that latchup can be completely eliminated if a process employing a thin epitaxial layer on a heavily doped substrate is used, and if the trench is allowed to penetrate to the heavily doped region.[11] Trenches of micron and submicron widths have been used to reduce p^+ to n^+ spacings to 2-2.5 μm.[57] In addition, when trenches deeper than the well depth are used, they replace the reverse-biased *pn* junction as the isolation structure at the well sidewalls (Fig. 6-32).

As of 1988, active devices can not yet be set against the trench sidewall, due to channel-inversion problems associated with the vertical trench sidewall (Fig. 6-33).[56] The sidewall inversion is caused by the horizontal parasitic MOS device, with the well acting as the gate electrode and the trench dielectric acting as the MOS gate oxide (with a thickness equal to the trench width). The voltage across this parasitic device is 5 V

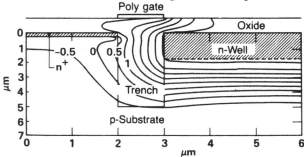

Fig. 6-33 The corresponding two-dimensional potential contours showing sidewall inversion; the simulation is done at Q_b =5x10^{10} cm^{-2}, N_A = 6x10^{14} cm^{-3}, V_{GD} = 3 V, V_{GS} = 0 V, V_{SB} = -1 V.[120] (© 1983 IEEE).

under normal CMOS operating conditions. This gate voltage and the narrow (e.g., 1-μm) trench width can easily cause inversion along the sidewall, outside the well but facing it.

There are two reasons why *p*-wells exhibit more severe trench-sidewall inversion problems. First, some of the needed boron doping segregates into the trench oxide during the thermal oxidation step. The second reason is the presence of the fixed charge, which is normally positive.

Once sidewall inversion occurs, *n*-channel devices with regions butted to the same sidewall become electrically connected by a path along the sidewall. Two obvious solutions are to leave a separation between the n^+ and the sidewall or to increase the trench width, with the end result of either approach being an increase in the isolation spacing. Because of this limitation, the minimum n^+ to p^+ isolation distance in trench isolation is 2-2.5 μm (1988).

Trench isolation has several other disadvantages, as follows:

• Fabrication is much more complicated than in LOCOS, making this an expensive alternative approach.

• For adequate filling and planarization, only one trench width can be utilized (unless SEG refill of the trenches is used).

• Another type of isolation is also needed (usually LOCOS) for most inactive parts of a chip. If a trench deeper than the well depth were used to isolate each device within the well region, each device would be isolated from the substrate. To keep the device bodies from floating, it would be necessary to provide each one with its own well contact, rather than using a single contact for the entire well. Chip area would thus be wasted, defeating the purpose for which the technology was adopted.

6.5.2 Isolation by Selective Epitaxial Growth for CMOS

Various schemes using SEG have been explored as CMOS isolation alternatives (see chap. 2 for more details on SEG technology). One of the first, *selective-etch-and-refill-with epi* (SEREPI), uses an epi refill of recesses in silicon, with the sidewalls passivated prior to refilling. In one approach,[59] the well regions were anisotropically etched to a 5-μm depth, and a 200-nm thermal oxide was grown on the recess surfaces. This oxide was removed from the bottom of the recess with an anisotropic dry etch, leaving it on the sidewalls. Arsenic buried layers of arsenic were formed by implantation into the bottoms of the wells, and epitaxial silicon was selectively grown in the wells. By controlling the growth rate, it was possible to refill the wells so that the surfaces were level with the outside substrate.

The SEREPI structure offers the advantages of minimum isolation area and also the flexibility of dopant-profile control in the epitaxial silicon in the wells. A similar process, reported by Kasai et al.,[60] uses a sidewall layer consisting of a 250-nm-thick composite film of oxide and nitride (see Fig. 2-34b, chap. 2). After these layers have been removed anisotropically through dry etching, a sacrificial oxide is grown on the bottom of the trenches and is then wet-etched away to remove any dry etching damage. Finally, an SEG step is performed.

While it is possible to make the n^+ to p^+ isolation distance as small as 0.25 μm, with either approach, some problems exist. First, LOCOS must still be used to isolate devices from one another within the same well. Second, unless a deep boron implant is added to the n-channel side of the isolation, sidewall leakage or inversion can occur. Finally, the 0.25-μm distance between p^+ and n^+ regions can become a limitation for masking of opposite-type implants if n^+ and p^+ junctions are to be placed on opposite sides of the 0.25-μm-thick dielectric.

Another SEG-based approach for isolating CMOS has been described by Manoliu and Borland,[15] and by Stivers.[61] In this approach, windows are anisotropically etched into a 1-2-μm-thick surface-layer oxide (see Fig. 2-43, chap. 2). Separate implants are used to form p^+ layers in those recesses designated to be p-wells, and n^+ layers in those that are to be n-wells. The oxide recesses are then refilled, using an SEG step in which sidewall inversion in the p-well is suppressed by tailoring the doping profile of the boron doping (i.e., this region is doped with $1 \times 10^{17}/cm^3$ in the active area of the n-channel devices). As a result, no subthreshold "kink" is observed in the turn-off current (such kinks in the subthreshold current curve are due to leakage from source to drain along the sidewall of the isolation structure).

The two buried layers increase the latch-up immunity of the CMOS devices considerably. Furthermore, this approach eliminates the need for an additional LOCOS isolation. Although this technique was only tested for minimum spacings of 2 μm, it is estimated that it will provide adequate isolation for spacings down to 1 μm. Finally, good planarity and gate oxide integrity were observed.

The technique of *retrograde wells* has also been investigated for implementing reduced isolation spacing and increased latchup immunity in CMOS (as well as for enhancing other device characteristics).[16,19] In this approach, the wells are formed by means of a high-energy implant, which is performed following formation of the field oxide. Because the peak of the implant is well below the silicon surface, the impurity concentration in the well decreases as it approaches the surface (hence the name "retrograde well").

It is easier to implement p-well CMOS with retrograde wells, since boron has a higher implant range in silicon than does phosphorus, and boron implants thus require a lower ion-implantation energy in order to achieve a similar depth. Because the well is formed after the field oxidation, retrograded p-wells can have higher NMOS field-threshold voltages compared to those exhibited in conventionally-formed wells. (In the latter, some of the boron segregates into the oxide during field oxidation, leaving less boron present at the silicon surface. This reduces the field-region threshold voltage.) In addition, the lateral diffusion of boron that occurs when the channel-stop implant is done prior to field oxidation is eliminated.

There are several drawbacks to the retrograde well approach. First, it produces devices with a high junction capacitance (C_j) and a high body-effect coefficient (γ) due to the relatively high doping concentration below the surface in the wells. These effects decrease the speed of circuits manufactured with such devices. Second, this approach may require the use of implanters with accelerating voltages greater than 400 keV. While such machines are available, their low beam current makes them incompatible with high-volume manufacturing needs. Note that doubly ionized boron can be implanted with an accelerating voltage of 200 keV to obtain the same result, and that such a voltage is within the range of ordinary implanters. However, the beam current in this case remains low (see Vol. 1, chap. 9.)

6.6 CMOS PROCESS SEQUENCES

Since there are so many options available for designing a CMOS process flow, no "standard" approach has been adopted. The process flows presented here are merely illustrations of the general sequence of process steps and of the types (and number) of masking layers used. We will first present a simple single-well (in this case, *n*-well), single-level-metal CMOS process. This type of process flow might be used in the fabrication of CMOS circuits with a minimum feature size of 1.25-2.0 μm. We will then describe a twin-well, double-level-metal process that uses LDD structures. With additional enhancements such as trench or SEG isolation, the filling of contact holes and vias with CVD metal, and novel interlevel dielectric planarization methods, such a process would likely be used in the manufacture of CMOS circuits with submicron dimensions.

6.6.1 Basic *p*- and *n*-Well CMOS Process Sequences

A basic single-well CMOS process can be implemented in either *p*-well or *n*-well technology using eight masking levels. Figure 6-34 shows seven of the eight masking levels of a basic *p*-well process. Figure 6-35 illustrates the *front end* masking levels of an *n*-well process. (The *back end* steps of a process sequence are those that begin with the contact masking step. Thus, the term *front end processing*, refers to all of the steps up to that point.) We will describe an example *n*-well CMOS process in more detail.

The NMOS devices in the *n*-well technology are formed in the lightly doped *p*-substrate ($\leq 1 \times 10^{15}$/cm^3), while the PMOS devices are formed in the more heavily doped *n*-well ($\sim 1 \times 10^{16}$/cm^3). The starting material is either a lightly doped <100> *p*-type wafer or a heavily doped <100> p^+ wafer with a thin (5-10-μm thick), lightly doped *p*-type epitaxial layer at the surface. A process for enhancing the gettering capabilities of the wafer may be employed before feature formation on the wafer surface is begun (see Vol. 1, chap. 2).

The *n*-well regions are the initial features formed on the starting material. First, a thermal oxide is grown and a CVD nitride film deposited. Then *Mask #1* is used to pattern windows in these layers, through which phosphorus for the *n*-well is implanted. Since the implantation process is unable to place the phosphorus ions deeply enough into the silicon, these impurities must be driven in to the appropriate depth during

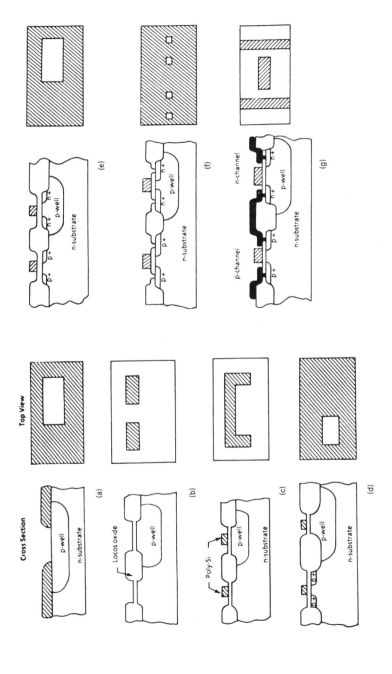

Fig. 6-34 Cross sections and top views of a typical *p*-well CMOS process at all mask levels: (a) *p*-well mask; (b) active-area mask; (c) poly-gate mask; (d) p^+ mask; (e) n^+ mask; (f) contact mask; (g) metal mask.[1] From J. Y. Chen, *CMOS Devices and Technology for VLSI.* Copyright Prentice-Hall, 1989. Reprinted with permission.

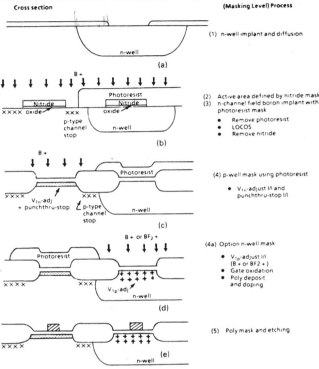

Fig. 6-35 Cross sections and masking levels and associated processes of an *n*-well CMOS front-end process.[1] From J. Y. Chen, *CMOS Devices and Technology for VLSI.* Copyright Prentice- Hall, 1989. Reprinted with permission.

subsequent high temperature cycles. An oxide is also grown on the *n*-well regions during the drive-in step. At the conclusion of the drive-in processes the surface concentration in the well is ~1×10^{16}/cm^3, and the impurity concentration gradient within the well is also rather small. Note that the redistribution of the well dopant occurs laterally as well as vertically.

Next, a boron threshold-adjust implant is carried out. There is no resist mask for this step, as the thin oxide or oxide/nitride layer covering the silicon wafer surface protects it from contamination. As was described in section 6.3.1, this single implant can provide a correct V_T adjustment for both the NMOS and the PMOS devices.

The surface is then stripped of its oxide and nitride/oxide layers, and a new pad-oxide/nitride layer (needed for LOCOS) is formed. *Mask #2* is then used to pattern this layer to define the active device and field regions. A boron channel-stop implant is performed for the *p*-substrate field regions. Although no separate mask is used, the boron implanted into the *well field regions* is not of sufficient concentration to significantly alter the *n*-concentration there. In addition, when the field oxide is grown the phosphorus piles up beneath it in the well regions (see Vol. 1, chap. 7), while some of the boron implanted during the channel-stop implant segregates out into the field oxide. Hence, the surface concentration of phosphorus in the well remains high enough that a separate channel-stop implant is usually not needed for the well region.

The field oxide is then grown, after which the nitride/oxide layer is removed from the active device regions (see chaps. 2 and 5). Next, a gate oxide is grown, in the same manner as that described in chapter 5 for NMOS devices. (Note that a *sacrificial pre-gate oxide* is frequently grown and stripped prior to the growth of the actual gate oxide.)

The deposition of polysilicon by CVD for the gate layer is then carried out. This layer is subsequently doped with phosphorus to form an n^+ polysilicon gate material. The resistivity of the polysilicon should be as small as possible, since this layer also serves as an interconnect structure. (In more advanced processes, even lower resistivities are achieved by forming a silicide layer on top of the polysilicon layer.) *Mask #3* is used to pattern the polysilicon.

Masks #4 and *#5* are then used to selectively implant the source/drain regions of the PMOS and NMOS devices, respectively. The polysilicon protects the channel region under the gate from being implanted. Arsenic is preferable for the n^+ regions so that shallow junctions and minimum lateral diffusion under the gate can be obtained. (It is typically implanted to a dose of $3\text{-}6\text{x}10^{15}$ cm^{-2}, and with an energy of 40-60 keV.) Boron is often implanted as BF_2^+ (for shallow junction formation) at doses of $1\text{-}5\text{x}10^{15}$ cm^{-2} and energies of 30-50 keV. Typical values of source/drain sheet resistance should be below 30 Ω/sq. Ohmic contacts to the n-well and p-substrate are formed simultaneously with the implants that create the NMOS and PMOS source/drain regions, respectively. These implants are then annealed with a short thermal process at a moderate temperature (e.g., 900-1000°C). The gate oxide that covers the source and drain regions during the implant is usually later stripped and regrown, since it has been damaged by the heavy implants and may have been contaminated by the RIE step used to pattern the poly.

A CVD doped oxide (with the dopants being either phosphorus, or boron + phosphorus) is deposited to a thickness of 50-100 nm, to serve both as a dielectric between the polysilicon and metal layers, and as a gettering layer. This layer is flowed (see Vol. 1, chap. 6) to improve the wafer-surface topography with respect to metal step coverage.

Mask #6 is used to open contact windows in the CVD oxide so that connections can be made between the metal layer and the silicon and polysilicon. Following a *reflow step*, an aluminum (or aluminum-silicon) layer is deposited and patterned (using *Mask #7*). The wafers are subjected to a final anneal step, which is followed by the deposition of the passivation layer. *Mask #8* is then used to open windows in this layer.

An important aspect of this simple CMOS process is that it uses only one more mask than the E-D NMOS process described in chapter 5. The three extra masks in the CMOS sequence (two masks for the source/drain implants, and one for the well mask) replace the depletion-mode implant mask and the buried-contact mask of the E-D NMOS process. Thus, the process complexity of a modern simple CMOS process is not much greater than that of advanced NMOS.

6.6.2 Twin-Well CMOS Process Sequence

This section describes an advanced ten-mask, twin-well CMOS process (Fig. 6-36). Again the process sequence options are many, so there is no one standard process. What is described here is an academic, hypothetical baseline process sequence that *almost certainly does not exist exactly as presented*, and that is useful primarily as a vehicle for illustrating a general sequence of steps in advanced CMOS-circuit fabrication. Where the process steps are not significantly different from those of the single-well process, details will not be repeated; instead readers will be referred to the appropriate sections of the book.

6.6.2.1 Starting Material.

The starting material in an advanced twin-well process is typically a <100>-orientation, heavily doped substrate on which a thin (5-10-μm thick), lightly doped epitaxial layer has been grown. Although n-epi-on-n^+ substrates have some advantages for a few special applications (e.g., for building CMOS SRAMs; see chap. 8), p-epi-on-p^+ substrates are the more common choice, because they are less sensitive to process-induced defects (see section 6.2).

6.6.2.2 Formation of Wells and Channel Stops.

The twin wells are the first features to be formed, assuming that neither trench isolation nor SEG isolation is being used. (Note that, as described earlier, trench isolation would require an additional masking step.) The well-formation procedure can be carried out in a number of ways. The most obvious method is to use two masking steps, each of which blocks one of the well implants. A single masking-step procedure, however, has been developed,[63] and that is probably the most commonly used approach (see Fig. 6-31).

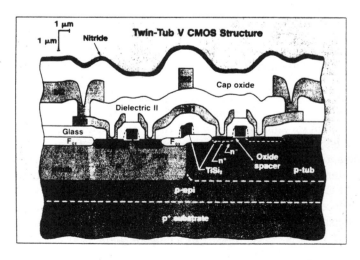

Fig. 6-36 Cross section of AT&T's Twin-Tub V CMOS structure.[121] Reprinted with permission of Semiconductor International.

In this method, a single mask is used to pattern a nitride/oxide film that has been formed on the bare silicon surface (*Mask #1*). The openings in the film become the *n*-well regions, and phosphorus is then implanted into them (e.g., at 80 keV). Next, a thermal oxide is grown on these regions, to a thickness that is sufficient to block the boron implant used to form the *p*-wells. After stripping of the oxide/nitride layer, the silicon in the *p*-well regions is exposed, while the *n*-well regions are covered with the implant-blocking oxide. Thus, during the *p*-well implant (e.g., at 50 keV), the boron penetrates the silicon only in the desired areas.

Next the wells are driven in (e.g., at 1100°C for 500 min). At the conclusion of the drive-in steps, the concentration in the wells for a 0.8-μm-CMOS process could be ~1×10^{16}/cm^3 for the *p*-well, and ~3×10^{16}/cm^3 for the *n*-well. The *n*-well might have a higher doping concentration, to improve the punchthrough performance of the PMOS devices and to eliminate the need for a separate channel-stop step for the *n*-well. A higher concentration in both wells would still produce devices with relatively low capacitances at the bottoms of the source/drain-to-well junctions.

The channel-stop implant procedure is also usually included in the part of the processing sequence in which the wells are formed. In one reported twin-well process,[26] only a *p*-well channel-stop implant (boron) is used, because the doping in the *n*-well is high enough (3×10^{16} cm^{-3}) that a separate channel stop implant is not required. In this case, an unmasked boron implant is done following both well implants, but prior to field-oxide growth.

In a second approach,[55] an additional mask can be used to provide both *n*-well and *p*-well channel stops (Fig. 6-31a). Note that in this procedure, the boron channel stop is implanted into both the *n*-well and *p*-well field regions; the phosphorus channel stop must be increased to compensate for the presence of the boron in the *n*-well field regions. The disadvantages of this method include the additional alignment step between the channel stops and the well masks, the interdiffusion that occurs during the oxidation step, and the asymmetry of the doping profiles. (The latter is due to the increased phosphorus concentration that must overcompensate the nonselective boron channel stop).

A maskless variation of this method has been developed to overcome the above drawbacks (Fig. 6-31b).[55,64] In this sequence, both arsenic and phosphorus are implanted into the *n*-well regions. This places both the dopants that form the well and the dopants that form the channel stop into the *n*-well regions prior to implanting of the *p*-wells. An oxide is then selectively grown on the *n*-well regions, and the boron dopant for the *p*-wells is implanted. After the *n*-wells have been driven in, the oxide over the *n*-wells is retained, and a second boron implant is carried out. This implant serves as both a channel stop in the *p*-well *field* regions and a punchthrough-prevention implant in the *active* regions of the *p*-well.

Various approaches have been proposed to prevent excessive redistribution of the boron channel stop implant during field-oxide growth. These include:

1. Co-implanting of Ge with the boron, which has been found to reduce the boron diffusion constant.[65]

2. Implanting of the field regions with Cl to increase the field-oxide growth rate (thus reducing the amount of time the wafers must be exposed to the high temperature).[66]

3. The use of high-pressure oxidation to reduce the thermal cycle that must be employed to grow the field oxide.[12,26,55] This also reduces the boron channel-stop dopant loss to the growing field oxide, as well as the interdiffusion of the wells (As a result, smaller isolation spacings between *n*- and *p*-channel devices can be used).

6.6.2.3 Definition of Active and Field Regions.
In a conventional advanced CMOS process, a standard LOCOS process (described in detail in chaps. 2 and 5) is used to define the active and field regions (*Mask #2*) and to grow a thick thermal oxide over the field regions. Enhanced LOCOS approaches, which overcome some of the limitations of conventional LOCOS, can be used as alternative processes to form isolation structures in the field regions. Some that have been reported as having been integrated into submicron CMOS processes include the SILO process[67] and a poly-buffered LOCOS process[26] (see chap. 2 for details on both of these). A high-pressure oxidation process for growing the field oxide, was reported to have been used in both of these processes.

6.6.2.4 Gate-Oxide Growth and Threshold-Voltage Adjustment.
The gate-oxide growth process is essentially the same as that used in the NMOS process described in chapter 5. The threshold-voltage implant process can be accomplished with either a single implant step, using boron (see section 6.3.1), or two separate implants (in which case additional masking steps are required). A separate selective punchthrough-prevention implant for the NMOS devices is usually needed, especially for submicron devices; this step also requires the use of an extra mask. The V_T-adjust and punchthrough-prevention implants can also be done either before or after the gate-oxidation step. To keep the boron V_T implant shallow (i.e., to improve the PMOS punchthrough characteristics, as described in section 6.3.2), implanting is sometimes done through the gate oxide. To protect the gate oxide from contamination (as well as to keep the implant shallow), part of the polysilicon can be deposited prior to the implant step,[26,67] with the remainder deposited afterward.

Alternative gate-oxide materials are being studied as replacements for SiO_2. Gate oxides of less than 10 nm will not be practical unless the gate voltage is also reduced, for two reasons. First, in order to prevent tunneling, the minimum gate oxide thickness must be $2V_{DD}/E_{ox}(max)$, where $E_{ox}(max)$ is 6 MV/cm. For V_{DD} = 3 V, the minimum thickness is thus 10 nm. The second reason has to do with the reliability and burn-in techniques used in accelerated lifetime testing. It becomes very difficult to screen out weak devices from good ones when the breakdown-voltage criterion defining a weak oxide falls within the range of the distribution exhibited by the good devices. This situation is encountered at oxide thicknesses below 10 nm.[68]

If a material with a larger dielectric constant than that of SiO_2 (3.9) could be used, the gate dielectric could be thicker, while the same capacitance per unit area could be maintained. Extensive work has been conducted on such materials, including nitrided oxides, thermal nitrides (dielectric constant = 8), and tantalum pentoxide, Ta_2O_5 (22). Rapid thermal processing has been explored as a means for forming the nitrided oxides, since the high temperatures needed for their formation can be produced by rapidly heating and cooling the wafers.

The Ta_2O_5 films have been prepared through thermal oxidation of a tantalum layer,[69] by reactive sputtering,[70] and CVD deposition.[71] Several reports indicate that these films exhibit excellent characteristics, and thus show promise for potential applications in advanced integrated circuits.[72,75]

In addition, a novel technique for forming ultrathin (15 nm), low-defect SiO_2 gate layers has been been reported.[76] In this approach, a 5-nm thermal oxide is first grown. A 5-nm CVD oxide is then deposited, and finally another 5-nm oxide is thermally grown, *under* the CVD layer (Fig. 6-37). The layer is able to grow beneath the CVD oxide because the oxidizing species diffuse through the first two layers to the Si-SiO_2 interface (see Vol. 1, chap. 7). During the second thermal oxidation, the top CVD oxide is also densified. This combination layer exhibits a low defect density because the growth-induced micropores (which are the major factor contributing to the defect density in gate dielectrics), are misaligned. If they were to extend completely through the oxide film, these micropores would be potential paths for rapid diffusional mass transport, as well as for current leakage.

6.6.2.5 Polysilicon Deposition and Patterning.

The process of forming the n^+ polysilicon gate structure involves the deposition of a polysilicon layer and patterning with *Mask #3*. The steps followed are essentially the same as those in the basic CMOS process. In most advanced CMOS processes the polysilicon film is overlaid with a refractory metal silicide to form a *polycide* structure (see Vol. 1, chap. 11). In some cases, a *salicide* (self-aligned silicided gate and source/drain regions – see Vol. 1, chap. 11) process is employed to reduce the parasitic resistance of the source and drain regions, as well as that of the gate material. In most such cases, the salicide

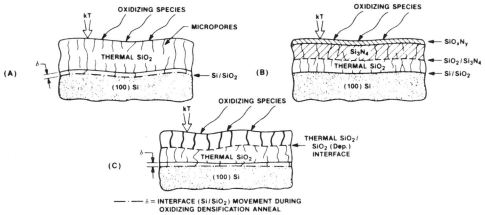

Fig. 6-37 Effects of oxidizing densification anneal: (a) SiO_2 films; (b) SiO_2- Si_3N_4 dual dielectric; (c) thermal SiO_2 - deposited SiO_2 multilayered stacked films.[76] (© 1988 IEEE).

formation is carried out after the source and drain have been implanted.

The polysilicon can be doped either after deposition by diffusion or ion implantation, or *in situ* during deposition. The former approaches were used in most MOS processing prior to submicron device generations. For submicron devices, in which a reduced thermal budget becomes more imperative, *in situ* doping becomes more attractive, (since the long, high-temperature drive-in step can be eliminated).[103] The redistribution of channel and source/drain implants is thus minimized, allowing improved control of threshold voltage, punchthrough, and L_{eff}.

As was described in sections 6.3.3 and 6.3.4, alternative materials to n^+ polysilicon (or to polycides with n^+ polysilicon as the underlayer) have been investigated as gate materials (e.g., *p^+ and n^+ polysilicon*, or refractory metal gates). Each alternative process introduces some changes into the process flow; nevertheless it is predicted that as device dimensions shrink, such measures will need to be used to maintain adequate device performance.

Etching of the gate structure can be a difficult step, especially when gate lengths are in the submicron range. First, the dimensions must be accurately and uniformly produced across the wafer, as I_D is strongly dependent on the gate dimensions. Second, the sidewalls must be vertical in order to reduce asymmetric I_D characteristics in devices built with LDD structures (see section 5.6.5.2). Third, the formation of vertical sidewall gate structures implies the need for a completely anisotropic etch step (see Vol. 1, chaps. 15 and 16). In turn, such a step requires that the etch process be highly selective against etching the thin underlying gate oxide layer.

6.6.2.6 Formation of Source/Drain Regions.

In advanced-CMOS processes, the gate lengths are short enough that LDD structures must be used to minimize hot-electron effects, especially in NMOS devices. Therefore, the procedures outlined in section 5.6.5 are used to fabricate such LDD structures. If these structures are used only for the NMOS transistors, *Masks #4* and *#5* are used to allow the sources and drains of the two transistor types to be selectively implanted.* Note that if LDD structures are also used for the PMOS devices, two additional masking layers may be needed (6-38a).

A removable-spacer LDD process for both NMOS and PMOS devices has been reported (Fig. 6-38b) that does not require the use of any other masks than the two needed to selectively form the sources and drains of the two transistor types.[67,73] In the removable spacer process, the heavily doped source/drain implant is performed first, with the spacers in place. After the spacers have been removed, the implant that forms the lightly doped drain regions is carried out (Fig. 6-38b). This process can be used to provide LDD structures for one or both transistor types, as desired. Polysilicon is used as the material of the removable spacers in one approach,[67,99] while a low-temperature

---

* A procedure that requires only one masking step for creating both types of source and drain regions has also been reported. In this approach, the heavy boron implant is carried out nonselectively. The mask is then used to cover the PMOS device active regions. An n-type implant heavy enough to overcompensate for the boron implant in the active regions of the n-channel devices is used to form the sources and drains of the NMOS transistors.

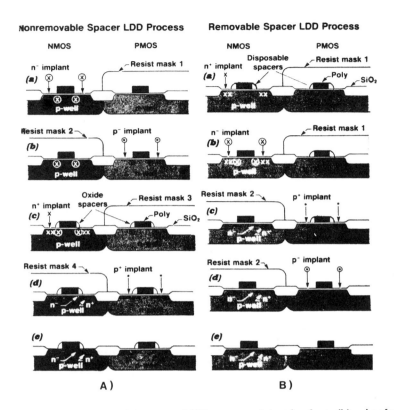

Fig. 6-38 (A) Non-removable oxide spacer LDD process: (a) n^- implant, (b) p implant, (c) after LDD spacer etch and n^+ implant, (d) after n^+ and p^+ implant, (e) after p^+ anneal. **(B)** Removable spacer LDD process: (a) after spacer etch and n^+ implant, (b) after NMOS spacer removal and n^- implant, (c) p^+ implant, (d) after PMOS spacer removal and p^- implant, (e) after p^+ anneal.[67] (© 1986 IEEE).

CVD oxide (over a thin polysilicon film) is used in the other.[73]

Various techniques have been developed for forming the shallow source/drain junctions needed for submicron CMOS devices. First, As is implanted for the n-channel devices and BF_2^+ for the p-channel devices, since both species have very shallow projected ranges at implant energies of 30-50 keV (see Vol. 1, chap. 9). These implants are usually performed through a screen oxide to protect the source/drain regions from contamination during the implant procedure. Second, preamorphization of the silicon by implantation with Si or Ge reduces channeling and helps produce shallow junctions. It is necessary, however, to diffuse the implanted species past the layer of implant damage that cannot be annealed out, in order to prevent junction leakage (see Vol. 1, chap. 9). RTP techniques have been explored as a means of carrying out these anneal and diffusion thermal cycles. Shallow $p^+ n$ junction formation through the use of diffusion (i.e., by utilizing RTP and either a solid diffusion source[82] or a spin-on diffusion source[83]) has also been reported. The use of an antimony implant to obtain shallow n^+p junctions with n^+ layers of lower resistivity than is possible with arsenic has also been described.[81]

Note that the screen oxide is usually stripped off following the source/drain implant, since it is damaged (and may be contaminated) by the heavy implants. A new oxide is grown over the source/drain regions and on the sidewall of the etched polysilicon electrode (and this step is referred to as *poly reoxidation*). During this thermal cycle, some (but not all) of the implantation damage is annealed out. The oxide formed on the poly sidewall shifts the remaining damage away from the gate edge to maintain the integrity of the thin gate oxide (Fig. 6-39).[104,110] At the same time, the sidewall oxidation produces a *gate bird's beak*, or GBB, under the polysilicon edge. This reduces the the gate-to-drain overlap capacitance and relieves the electric-field intensity at the corner of the gate structure. However, because the GBB encroachment can degrade transconductance of submicron MOSFETs and impact their subthreshold swing and threshold voltage,[109] this poly reoxidation process must be optimized for fabrication of submicron MOSFETs.[111]

An approach to forming shallow junctions that uses $CoSi_2$ source/drain junctions has been reported.[29] In this approach, the $CoSi_2$ is formed before the source and drain junctions are created by means of a heavy ion implantation step. The implant is then performed so that the damaged layer, which would have occurred in the silicon crystal, is kept within the $CoSi_2$ layer. As a result, it is necessary to drive the implanted impurities just far enough so that they enter the silicon region to form the required *pn* junction. If a salicide process is used, the device has a cross-section, as shown in Fig. 6-40.

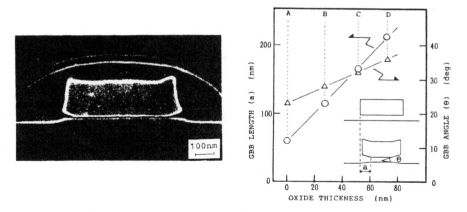

Fig. 6-39 (a) SEM micrograph of a gate structure with a GBB. (b) Profiles of GBB versus oxide thickness grown during re-oxidation.[104] This paper was originally presented at the Spring 1989 Meeting of The Electrochemical Society, Inc. held in Los Angeles, CA.

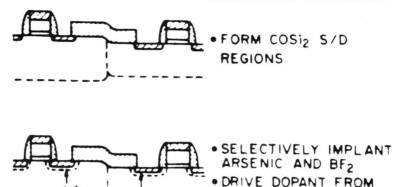

- FORM CoSi$_2$ S/D REGIONS

- SELECTIVELY IMPLANT ARSENIC AND BF$_2$
- DRIVE DOPANT FROM CoSi$_2$ TO FORM N$^+$ AND P$^+$ JUNCTIONS

Fig. 6-40 Cross section of CMOS structure in which CoSi$_2$ is first formed on the source and drain regions, and then As and BF$_2^+$ is selectively implanted into the CoSi$_2$. These dopants are then driven out of the CoSi$_2$ and into the Si substrate to form shallow p^+ and n^+ source/drain regions.[29] (© 1986 IEEE).

Another approach is to create so-called *elevated source-drains*. A thin (e.g., 200-nm) epitaxial layer of silicon can be selectively deposited onto the exposed source-drain areas of the MOS transistor, following the implantation of the lightly doped region of the LDD structure and formation of the spacers (Fig. 6-41a). This method has been used for the NMOS devices of a 1 Mbit SRAM.[79] In this case, a heavy BF$_2^+$ implant is done such that the SEG film becomes heavily doped; the boron penetrates to the substrate during an RTP anneal that follows the implant. In this way, elevated, heavily doped, shallow source/drain regions are formed. The source-drain junction depths in the substrate are less than 0.2 μm deep (Fig. 6-41b). As noted earlier, the gate oxide that covers the source and drain regions is usually etched away and regrown following the implant step.

6.6.2.7 CVD Oxide Deposition and Contact Formation. Following the formation of the source and drain regions (and the salicide layers), a doped oxide is deposited by CVD, with procedures very much like those of the basic NMOS and CMOS processes (see section 5.4.1.6).

Contact windows are opened in this CVD layer to allow electrical connections to be made between the Metal 1 layer and the source/drain, gate, and substrate and well contact regions (*Mask #6*). Again, the details are similar to those of the basic NMOS and CMOS processes.

Advances in contact technology for CMOS include the use of barrier layers to prevent spiking through the shallow junctions, and the filling of contact holes by such materials as CVD W, polysilicon, or even SEG. In addition, as device dimensions decrease, the parasitic resistances of the source and drain regions become more significant. More details on all of these topics are provided in chapter 3.

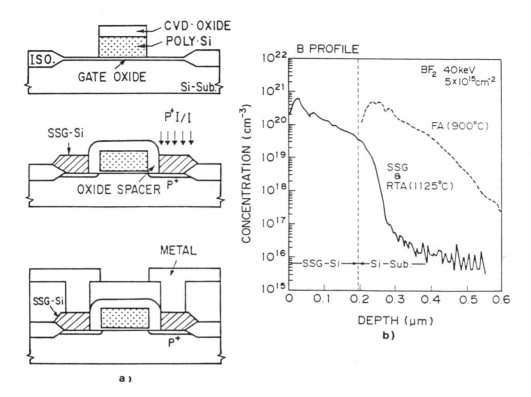

Fig. 6-41 Elevated source/drains. (a) Key process steps in forming elevated source/drains using SEG. (b) Comparison of the doping profile in source/drain regions of a PMOS device after a BF_2^+ implant directly into the regions (and a conventional furnace anneal) versus a BF_2^+ implant into the SEG regions and an RTP anneal.[79]

6.6.2.8 Metal 1 Deposition and Patterning.

The issues involved in the deposition and patterning of the Metal 1 layer are not significantly different from those described in chapter 5 for the NMOS process. *Mask #7* is used to pattern the Metal 1 layer. (For more information on the metal-layer deposition and patterning processes, see chapter 4 of this volume, and Vol. 1, chap. 9.)

Despite its higher resistivity compared to Al, tungsten's superior electromigration properties make it more appropriate for certain applications. As a result, CVD W has been selected as the Metal 1 material for a variety of circuit designs.

6.6.2.9 Intermetal Dielectric Deposition/Planarization and Via Patterning.

Following the Metal 1 patterning, an intermetal dielectric must be deposited to electrically isolate Metal 1 from the Metal 2 layer. A variety of materials and deposition processes, have been utilized (see chap. 4).

However, deposition of this layer may make the wafer topography too severe to allow the Metal 2 film to be deposited with adequate step coverage. A number of *planarization* processes have been developed to overcome this problem (see chap. 4). If such techniques are successfully implemented, the wafer topography will be relatively planar, and any steps on its surface will be gently sloped rather than severely vertical.

It is necessary to open vias in the intermetal dielectric layer so that an electrical connection can be established between Metal 2 and Metal 1 at desired locations (*Mask #8*). Metal 2 must be deposited with adequate step coverage into the vias.

A number of techniques have been studied to improve coverage of Metal 2 in the vias, including the following (see chap. 4):

- Sloping the via walls by means of the via-etch process.

- Filling the vias with a blanket W or polysilicon deposition, and then etching back to provide a planar surface.

- Filling the vias with a selective deposition of W or Al.

- Increasing the step coverage into the vias through bias sputtering and heating of the substrate during deposition.

- Laser melting the Metal 2 film to increase step coverage.

6.6.2.10 Metal 2 Deposition and Patterning. The processing issues of depositing and patterning (*Mask #9*) the Metal 2 layer are discussed in chapter 4.

6.6.2.11 Passivation Layer Deposition and Patterning. The passivation-layer deposition and patterning (*Mask #10*) issues are the same those of the basic CMOS or NMOS processes.

6.7   MISCELLANEOUS CMOS TOPICS

6.7.1 Electrostatic-Discharge Protection

The input signals to an MOS IC are fed to the gates of MOS transistors. If the voltage applied to the gate insulator becomes excessive, the gate oxide can break down. The dielectric breakdown strength of SiO_2 is approximately 8×10^6 V/cm; thus, a 15-nm gate oxide will not not tolerate voltages greater than 12 V without breaking down. Although this is well in excess of the normal operating voltages of 5-V integrated circuits, voltages higher than this may be impressed upon the inputs to the circuits during either human-operator or mechanical handling operations.

The main source of such voltages is triboelectricity (electricity caused when two materials are rubbed together). A person can develop very high static voltage (i.e., a few hundred to a few thousand volts) simply by walking across a room or by removing an integrated circuit from its plastic package, even when careful handling procedures are followed. If such a high voltage is accidentally applied to the pins of an IC package, its discharge (referred to as *electrostatic discharge*, or ESD) can cause breakdown of the gate oxide of the devices to which it is applied. The breakdown event may cause sufficient damage to produce immediate destruction of the device, or it may weaken the oxide enough that it will fail early in the operating life of the device (and thereby cause device failure). A more detailed description of ESD failures in semiconductor devices can be found in reference 84.

All pins of MOS ICs must be provided with protective circuits to prevent such voltages from damaging the MOS gates. The need for such circuits is also mandated by the increasing use of VLSI devices in such high-noise environments as personal computers, automobiles, and manufacturing control systems. These protective circuits, normally placed between the input and output pads on a chip and the transistor gates to which the pads are connected, are designed to begin conducting or to undergo breakdown, thereby providing an electrical path to ground (or to the power-supply rail). Since the breakdown mechanism is designed to be nondestructive, the circuits provide a normally open path that closes only when a high voltage appears at the input or output terminals, harmlessly discharging the node to which it is connected.

Four types of circuits are used to provide protection against ESD damage, as follows:

1. diode breakdown

2. node-to-node punchthrough

3. gate-field-induced breakdown

4. parasitic *pnpn* diode latchup.

Often, a combination of protection methods is used, with a breakdown diode and one of the other protection devices connected in parallel with the gate being protected.

6.7.1.1 Diode Protection. Protection is obtained by using the diode-breakdown phenomenon to provide an electrical path in the silicon substrate that consists of a diffused resistor region (of a doping type opposite to that of the substrate). This diffused region is connected between the input pad and the gate (Fig. 6-42a). If a reverse-bias voltage greater than the breakdown voltage of the resultant *pn* junction is applied, the diffusion region (which otherwise works as a resistor), operates as a diode and undergoes breakdown. Furthermore, this diffused region will also clamp a negative-going transition at the chip input to one diode drop below the substrate voltage. In CMOS technologies, an additional protection diode can be added by utilizing the *pn* junction that exists between a p^+ node and the body region of the PMOS device (an *n*-type region, that is connected to V_{DD}). This diode is utilized as a protection device when a connection is made between the pad and a p^+ region. (Note that this second diode will clamp positive-going transitions to one diode drop *above* V_{DD}.)

6.7.1.2 Node-to-Node Punchthrough. As defined in section 5.5.2, *punchthrough* is the phenomenon by which the depletion region surrounding the drain of an MOS device extends along the channel of the device and contacts the depletion region of the source of the device. While the current that results from punchthrough is ordinar-

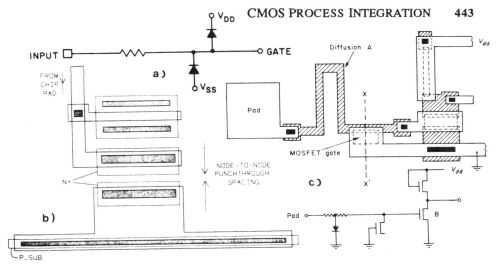

Fig. 6-42 MOS protective circuits: (a) Diode; (b) Punchthrough device; (c) Layout of an input protection structure containing both devices.[122] From T. E. Dillinger, *VLSI Engineering*. Copyright Prentice-Hall, 1988. Reprinted with permission.

arily an unwanted effect, it is harnessed in ESD protection circuits for beneficial purposes. The source and drain regions of an ESD punchthrough protection structure are placed close enough together that punchthrough occurs at a voltage lower than that of the gate-oxide breakdown (Fig. 6-42b). Such a structure is layed out and tested to meet a specific ESD transient reliability measure. Once it has been verified as functioning according to the design specifications, it can be connected to each chip signal I/O pad.

6.7.1.3 Gate-Controlled Breakdown Structure.
The electric field near the corner of an MOS device's drain node is strongest at the surface, since the depletion region is narrowest at this point[85] and the entire voltage across the depletion region must be dropped over a very short distance. The fact that the gate voltage can increase the strength of the electric field at the corner of the drain is the principle used in designing the gate-controlled breakdown structure (Fig. 6-43).

As the reverse-bias voltage applied to the drain is increased, the gate-to-drain voltage also increases (note that the gate is tied to ground). Because the presence of the gate reduces the breakdown voltage of the junction near the corner of the drain, a relatively small voltage can cause the junction to break down at this point. At breakdown, the junction conducts a large current; in this device, the current flows from the drain to the substrate. The actual input voltage at which breakdown occurs can be controlled by altering the oxide thickness. This produces a large voltage drop across the relatively high resistance of the diffusion area, reducing the voltage applied to the gate of the MOSFET.

Such structures were the workhorses of ESD protection devices. It has been found, however, that processing enhancements that increase the device performance of small-dimension devices sometimes have a negative impact on the failure resistance of the

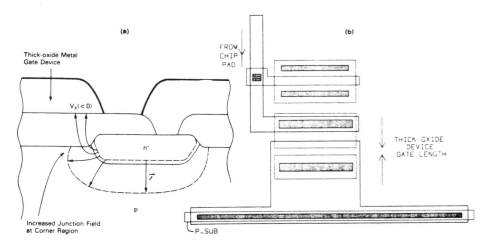

Fig. 6-43 (a) Cross-section, and (b) layout of a thick-oxide metal-gate device for node-to-substrate breakdown and input-pad protection.[122] From T. E. Dillinger, *VLSI Engineering*. Copyright Prentice-Hall, 1988. Reprinted with permission.

ESD protective devices. Specifically, silicided source/drain regions and LDD structures can degrade the ESD performance of the gate-controlled breakdown structures so that the voltage at which *these devices themselves fail* is reduced from 4 kV to 1 kV.[86,87] As a result, ESD protection structures can be rendered largely ineffective by the silicided regions, or else the circuits can be caused to fail by ESD in new failure modes as a result of the addition of performance-enhancement features.[86,88]

6.7.1.4 *pnpn*-Diode ESD Protection for Advanced CMOS Circuits.
Even as devices are made smaller, a fixed minimum volume of silicon must always be used to dissipate the power associated with the current flowing through the protective devices to ground. If the power is dissipated in too small a volume of silicon, the silicon can be heated beyond its melting point; this can destroy the device, even if the circuit-protection device is operating properly. As a result, a minimum unscalable area (Fig. 6-44) will be required in order for the same degree of protection to be maintained through each generation of scaled technology.[68,89]

Effective protection of VLSI must be provided for both automated and human handling. ESD from a machine is typically modeled as a high-voltage (4-kV), short-duration pulse (i.e., the *EIAJ machine model [MM] test method*), while ESD from a human operator is usually modeled as a longer, lower-voltage (300-V) but higher-current pulse (e.g., the *Mil. Std. human body model [HBM] test method*). Thus, if the gate-controlled breakdown structure itself fails at 1 kV, it ceases to provide protection against HBM ESD.[87]

To solve this problem on the inputs of a CMOS chip, it has been proposed that the parasitic lateral *pnpn* structure inherent in all CMOS circuits be exploited (Fig. 6-45a)

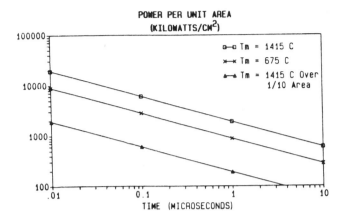

Fig. 6-44 Wunsch and Bell power curves for silicon melt, aluminum melt, and silicon melt for nonuniform current density.[89] (© 1968 IEEE).

as an ESD structure (see section 6.4). Such a structure – which is deliberately configured to latch up at a voltage lower than that required to damage the input MOS gate oxide – is built into each input circuit on the chip (D1 in Fig. 6-45b). Thus, this parasitic *pnpn* structure serves as the primary ESD structure of the input circuit by

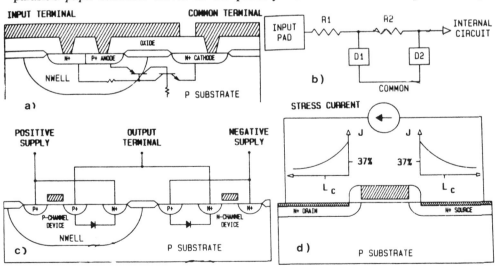

Fig. 6-45 (a) Lateral *pnpn* input protection structure showing parasitic and increased area at *n*-well and *p*-well junction resulting in lower power density. (b) Input-protection circuit showing primary protection elements R1 and D1 and secondary protection elements R2 and D2. (c) Representation of dual-diode output circuit cross section. Effectiveness depends on parasitic diode impedance as well as capacitance between supplies. (d) Cross section of MOS thin-oxide device showing current density at silicide and silicon interface.[87] (© 1979 IEEE).

providing a low-impedance path when it is turned on. It also offers increased area for power dissipation. Secondary ESD structures, however, must also be used on each input circuit (e.g., D2 and R2 in Fig. 6-45b) to provide effective protection until D1 achieves its low-impedance conducting state.

On the outputs of the chip, the parasitic diodes that exist in the CMOS output circuit are used to provide ESD protection, much as described in section 6.7.1.1 (Fig. 6-45c).[87] Such dual-diode CMOS outputs have proved to be effective when the resistance of the parasitic diode in the *p*-channel device is minimized and the failure threshold of the *n*-channel device is maximized. As shown in Fig. 6-45d, the stress current density in the diode of the silicided *n*-channel device is very nonuniform due to the silicided source/drain regions. As a result, the diode can fail if too high a current density passes through only a small fraction of the area of the silicided contact.

6.7.2   Power-Supply Voltage Levels for Future CMOS

As MOS has evolved, device scaling has been practiced to increase the speed and packing density of MOS ICs. Although scaling of the electric field to maintain a constant electric field (CE) would have promoted both high performance and high reliability, in practice a constant voltage (CV) has been maintained in order to maintain TTL compatibility and produce higher circuit performance. Such nonscaling of the voltage leads to reliability problems, caused by hot-electron effects and oxide breakdown. Scaling of the power supply voltage is thus desirable. M. Kakamu has calculated the optimum power-supply voltage as a function of shrinking line width, and has shown that this voltage can be scaled down in proportion to the square root of the design-rule shrinkage without sacrificing circuit performance.[90] The relationships given between the supply voltage and the design rule are

$$V_{DD} = 6.1 \times (L/2)^{1/2} \text{ (for conventional drain structures)} \qquad (6-9)$$

and

$$V_{DD} = 8.2 \times (L/2)^{1/2} \text{ (for LDD structures)} \qquad (6-10)$$

where L (μm) is the design rule and V_{DD} is the power-supply voltage (volts). These equations predict that at dimensions of 0.6 μm, the power supply voltage will have to be reduced to 3.3 V in order for reliable device operation and circuit performance to be maintained if conventional drain structures are used. On the other hand, if LDD structures are used, it is predicted that V_{DD} = 5 V will still be feasible for 0.7-μm dimensions. A recent report indicates that these equations may be slightly pessimistic, in that MOSFETs with gate lengths of 0.6 μm that operate from a 5-V supply have been successfully fabricated.[102] Through optimization of the LDD structures of these devices, a lifetime of more than 10 years in the face of hot-carrier degradation effects, is predicted.

6.7.3   Low-Temperature CMOS

Since it is possible to build MOS devices with minimum dimensions of less than 0.5 μm, low-temperature operation of CMOS (e.g., at 77°K) has been actively investigated. At low temperatures, the MOS devices exhibit lower subthreshold leakage, higher carrier mobility (which yields improved speed performance), and a steeper logarithmic current-voltage slope. It has been found that the normalized propagation delay in CMOS logic gates is reduced by a factor of nearly 2 or 3 when the devices are operated at 77°K.[91,92,93,94] The major reason for the slow commercial introduction of such systems has been the difficulty associated with liquid-nitrogen refrigeration.

6.7.4   Three-Dimensional CMOS

Since the limits of two-dimensional CMOS may soon be approached, a logical direction for continued advances in device scaling is three-dimensional devices. CMOS is well-suited for 3-D integration, since low-power circuits will be mandatory in order for heat dissipation problems to be avoided. The 3-D approach will offer the benefits of increased packing density and higher speeds (due to shorter interconnects).

The inability to fabricate single-crystal silicon over insulating layers, however, has historically been an obstacle to the implementation of 3-D ICs. Recent advances in SOI technologies show promise of being able to surmount this obstacle and much work is currently being done to develop 3-D CMOS.[95]

REFERENCES

1. J. Y. Chen, *CMOS Devices and Technology for VLSI*, Englewood Cliffs N.J., Prentice-Hall, 1989.
2. F. M. Wanlass and C. T. Sah, *IEEE Int. Solid-State Circ. Conf.*, February 1963.
3. R. D. Davies, "The Case for CMOS", *IEEE Spectrum*, October 1983, p. 26.
4. J. Y. Chen, "CMOS-The Emerging Technology," *IEEE Circuits and Systems Magazine*, March 1986, p. 16.
5. D. M. Brown, M. Ghezzo, and J. M. Pimbley, "Trends in Advanced CMOS Process Technology," *Proceedings of the IEEE*, December 1986, p. 1678.
6. W. C. Holton and R. K. Calvin, "A Perspective on CMOS Technology Trends," *Proceedings of the IEEE*, December 1986, p. 1646.
7. S. Prussin, private communication.
8. R. Chwang and K. Yu, "CHMOS-An *n*-Well Bulk CMOS Technology for VLSI," *VLSI Design*, Fourth Quarter, 1981, p. 42.
9. W. Wijaranakula et al., *J. Electrocchem. Soc.*, December, 1988, p. 3113.
10. H. Tsuya et al., *Japanese J. Appl. Phys.*, **22**, L16, (1983).
11. T. Yamaguchi et al., *IEEE Trans. Electron Dev.*, **ED-31**, 205 (1984).
12. L. C. Parrillo, "CMOS Active and Field Device Fabrication," *Semicond. Internat.*, April 1988, p. 64.

13. A. G. Lewis et al., "Vertical Isolation in Shallow *n*-Well CMOS Circuits," *IEEE Electron Dev. Letts.*, March 1987, p. 197.

14. L. C. Parrillo et al., "Twin-Tub CMOS," *IEDM Tech. Dig.*, 1980, p. 752.

15. J. Manoliu and J. O. Borland, "A Submicron Buried-Layer Twin-Well CMOS SEG Process," *IEDM Tech. Dig.*, 1987, p. 20.

16. R. D. Rung, C. J. Dell' Oca, and L. G. Walker, "A Retrograde *p*-Well for High-Density CMOS," *IEEE Trans. Electron Dev.*, **ED-28**, p. 1115, 1981.

17. R. A. Martin and J. Y. Chen, "Optimized Retrograde *n*-Well for 1 μm CMOS," *Proc. Custom Integ. Circ. Conf.*, 1985, p. 199.

18. S. R. Combs "Scalable Retrograde *p*-Well CMOS Technology," *IEEE Trans. Electron Dev.*, **ED-28**, p. 346, 1981.

19. Y. Taur et al., *J. Solid-State Circuits*, **SC-20**, p. 123, 1985.

20. S. M. Sze, Ed., *VLSI Technology*, 2nd. Ed., New York, McGraw-Hill, 1988, p. 491.

21. T. Ohzone et al., *IEEE Trans. Electron Dev.*, **ED-32**, 1789, (1980).

22. K. M. Cham, S.-Y. Oh, D. Chin, and J. L. Moll, *Computer-Aided Design and VLSI Device Development*, Boston, Klewer Academic Publishers, 1986 p. 182.

23. G. J. Hu et al., *IEDM Tech. Dig.*, 1982, p. 710.

24. J. Zhu et al., *IEEE Trans. Electron Dev.*, February 1988, p. 145.

25. A. Schmitz and J. Y. Chen, "Design, Modelling, and Fabrication of Submicron CMOS Transistors," *IEEE Trans. Electron Dev.*, **ED-33**, p. 148, (1986).

26. R. A. Chapman et al., "A 0.8 μm CMOS Technology," *Tech. Dig. IEDM*, 1987, p. 362.

27. L. C. Parrillo et al., *Tech Dig. IEDM*, 1984, p. 418.

28. K. M. Cham et al., *IEEE Electron Dev. Lett.*, January 1986, p. 49.

29. S. J. Hillenius et al., "A Symmetric Submicron CMOS Technology," *Tech. Dig. IEDM*, 1986, p. 252.

30. Y. Pauleau, "Interconnect Materials for VLSI Circuits, Part-1," *Solid State Technology*, February 1987, p. 61.

31. V. Schwabe, F. Neppl, and E. P. Jacobs, *IEEE Trans. Electron Dev.*, **ED-31**, 1984, p. 988.

32. S. Iwata et al., *IEEE Trans Electron Dev.*, **ED-31**, 1984, p. 1174.

33. R. F. Kwasnick et al., *IEEE Trans. Electron. Dev.*, September 1988, p. 1432.

34. H. Oikawa and T. Amazawa, *Proc. 3rd. Intl. VLSI Symp., ECS Meeting*, May, 1985, p. 131.

35. D. M. Brown et al., *IEEE Trans. Electron Dev.*, **ED-18**, 1971, p. 931.

36. R. R. Troutman, *Latchup in CMOS Technology*, Klewer, Boston, MA., 1986.

37. G. J. Hu, "A Better Understanding of CMOS Latchup," *IEEE Trans. Electron Dev.*, **ED-31**, p. 62, 1984.

38. R. R. Troutman, "Latchup in CMOS Technologies," *IEEE Circuits & Sys. Mag.*, May 1987, p. 15.

39. A. G. Lewis et al., *Tech. Dig. IEDM*, 1986, p. 248.

40. R. D. Rung and H. Momose, *IEEE Trans. Electron Dev.*, **ED-30**, 1983, p. 1647.

41. W. R. Dawes and G.F. Derbenwick, "Prevention of CMOS Latchup by Gold-Doping," *IEEE Trans. Nucl. Sci.*, **NS-23**, 1976, p. 2027.

42. J. R. Adams and R. J. Sokel, Presented at the1979 Nuclear and Space Radiation Effects Conf., Santa Cruz, CA., July, 19, 1979.
43. J. O Borland and T. Deacon, "Advanced CMOS Epitaxial Processing for Latchup Hardening," *Solid-State Technology*, p. 123, August 1984.
44. S. Ratanphanyarat et al., *Tech. Dig. IEDM*, 1987, p. 744.
45. F. S. Lai et al., *IEDM Tech. Dig. 1985*, p. 513.
46. M. L. Chen et al., *Tech. Dig. IEDM*, 1986, p. 256.
47. S. Swirhun et al., "Latchup-Free CMOS Using Guarded Schottky Barrier PMOS," *Tech. Dig. IEDM*, 1984, p. 402.
48. R. A. Martin et al., *IEDM Tech. Dig.*, 1985, p. 403.
49. A. G. Lewis, *IEEE Trans. Electron Dev.*, Oct. 1984, p. 1472.
50. D. Takacs et al., *Tech. Dig. IEDM*, 1983, p. 159.
51. K.W. Terrill et al., *Tech. Dig. IEDM*, 1984, p. 406.
52. R. Menozzi et al., "Layout Dependence of CMOS Latchup," *IEEE Trans. Electron Dev.*, Nov. 1988, p. 1892.
53. J. Y. Chen and D. E. Snyder, "Modeling Device Isolation in High-Density CMOS," *IEEE Electron Dev. Letts.*, February 1986, p. 64.
54. L. Herman, "Controlling CMOS Latchup," *VLSI Design*, April, 1985, p. 100.
55. M. -L. Chen et al., *Tech. Dig. IEDM*, 1986, p. 256.
56. K. M. Cham and S.-Y. Chiang, "A Study of Trench Surface Inversion Problem for the Trench CMOS Technology," *IEEE Electron Dev. Lett.*, Sept., 1983, p. 303.
57. Y. Nitsu et al., "Latchup Free CMOS Structure Using Shallow Trench Isolation," *Tech. Dig. IEDM*, 1985, p. 509.
58. R. D. Rung, H. Momose, and Y. Nagakubo, "Deep Trench Isolated CMOS Devices," *IEDM Tech. Dig.*, 1982, p. 6.
59. T. I. Kamins and S. Shiang, *Electron Dev. Letts.*, December 1985, p. 617.
60. N. Kasai et al., *IEEE Trans. Electron Dev.*, June 1985, p. 1331.
61. A. Stivers et al., *10th Internat. Conf. on CVD, ECS*, **87-8**, p. 389.
62. R. D. Rung et al., *IEEE Trans. Electron Dev.*, **ED-28**, 1981, p. 1115.
63. L. C. Parrillo et al., "Twin-Tub CMOS II," *IEDM Tech. Dig.*, 1982, p. 706.
64. S. J. Hillenius and L.C. Parrillo, U.S. Patent No. 4,554,726, Nov. 25, 1985.
65. J. Pfiester and J. Alvis, "Improved CMOS Field Isolation Using Ge/B Implantation," *IEEE Elect. Dev. Lett.*, August 1988, p. 391.
66. J. Pfiester, *Electron Dev. Letts.*, November 1988, p. 561.
67. L. C. Parrillo et al., *Tech. Dig. IEDM*, 1986, p. 244.
68. M. H. Woods, "MOS VLSI Reliability and Yield Trends," *Proceedings of IEEE*, Dec. 1986, p. 1715.
69. G. S. Ohrlein, *J. Appl. Physics*, **59**, 1587 (1986).
70. B. W. Shen et al., *Tech. Dig. IEDM*, 1987, p. 582.
71. M. Saitoh, T. Mori, and H. Tamura, *Tech. Dig. IEDM*, 1986, p. 680.
72. K. Ogiue et al., "Technology Improvement for High Speed ECL RAMs," *Tech. Dig. IEDM*, 1986, p. 468.
73. J. Pfiester, *IEEE Electron Dev. Letters*, April, 1988, p. 189.
74. R. de Werdt et al., *Tech. Dig. IEDM*, 1987, p. 532.

75. S. G. Byeon and Y. Tzeng, *Tech. Dig. IEDM*, 1988. p. 722.

76. P. K. Roy et al., *Tech. Dig. IEDM*, 1988, p. 714.

77. N. Kasai, N. Endo, and A. Ishitani, *Tech. Dig. IEDM*, 1988, p. 242.

78. L. Manchanda, *Intl. Reliability Symp.*, 1986, p. 183.

79. K. Hashimoto, Presented at the Third Annual Applied Materials Inc.,*"Innovations in Epitaxial Technology for Advanced Device Structures"* Seminar, December 15, 1988, Palo Alto, CA.

80. G. Sai-Halasz and H. B. Harrison, *Electron Dev. Letts.*, September 1986, p. 534.

81. M. L. Chen et al., "Constraints in P-Channel Device Engineering for Submicron CMOS Technologies," *Tech. Dig. IEDM*, 1988, p. 390.

82. K. -T. Kim and C. -K. Kim, *IEEE Electron Dev. Letts.*, December 1987, p. 569.

83. E. Ling et al., *IEEE Electron Dev. Letts.*, March, 1987, p. 96.

84. B. A. Unger, *Intl. Reliability Physics Symp.*, 1981, p. 193.

85. A. S. Grove, *Physics and Technology of Semiconductor Devices*, Wiley, New York, 1967, Section 10.5.

86. K.- L. Chen et al., *Tech. Dig. IEDM*, 1986, p. 484.

87. R. N. Rountree, *Tech. Dig. IEDM*, 1988, p. 580.

88. C. Duvvury et al., "ESD Protection in 1 μm CMOS," *Intl. Reliabilty Physics Symp.* 1986.

89. D. C. Wunsch and R. R. Bell, "Determination of Threshold Failure Levels of Semi-conductor Diodes and Transistors due to Pulse Voltages," *IEEE Trans. Nucl. Sci.*, NS-15, p. 244, Dec. 1968.

90. M. Kakumu et al., *Tech. Dig. IEDM*, 1986, p. 399.

91. J. D. Plummer, "Low Temperature CMOS Devices and Technology," *Tech. Dig. IEDM*, 1986, p. 378.

92. J. Watt and J. D. Plummer, *Tech. Dig. IEDM*, 1987, p. 393.

93. J. Y-C. Sun et al., *Tech. Dig. IEDM*, 1986, p. 236.

94. J. S. T. Huang and J. W. Schrankler, *IEEE Electron Dev.*, January 1987, p. 101.

95. Y. Akasaka, "Three-Dimensional IC Trends," *Proc. IEEE*, December 1986, p. 1703.

96. M. H. El-Diwany et al., *Tech. Dig. IEDM*, 1987, p. 917.

97. R. H. Krambeck, C. M. Lee, H. F. S. Law, *IEEE J. Solid-State Circuits*, SC-17, (1982), p. 614.

98. M. Wong and K. Saraswat, *IEEE Electron Dev. Letts.*, November 1988, p. 579.

99. C. -Y. Yang et al., *Ext. Abs. of the Electrochemical Soc. Meeting*, Spring, 1988, Abs. No. 138, p. 207.

100. C. Y. Wong et al., *Tech. Dig. IEDM*, 1988, p. 238.

101. G. J. Hu and R. H. Bruce, *IEEE Electron Dev. Letts.*, EDL-5, p. 211, 1984.

102. H.-M. Mulhoff, K. H. Kusters, and H. Melzner, *Ext. Abs. of the Electrochemical Soc. Meeting*, Spring, 1989, Abs. No. 133, p. 186.

103. S. M. Kugelmass and J. P. Krusius, *Ext. Abs. of the Electrochemical Soc. Meeting*, Spring, 1989, Abs. No. 135, p. 188.

104. M. Kishimoto et al., *Ext. Abs. of the Electrochemical Soc. Meeting*, Spring, 1989, Abs. No. 137, p. 191.

105. P.-H. Pan, J. G. Ryan, and M. A. Lavoie, *Ext. Abs. of the Electrochemical Soc. Meeting*, Spring, 1989, Abs. No. 138, p. 193.

106. A. G. Lewis et al., *IEEE Trans. Electron Dev.*, June 1987, p. 1337.

107. Y. Sato, K. Ehara, and K. Saito, *J. Electrochem. Soc.*, June, 1989, p. 1777.

108. C. Mead and L. Conway, *Introduction to VLSI Systems,* Addison-Wesley, Reading, MA, 1980.

109. J. R. Pfeister et al., *IEEE Electron Dev. Letts.,* August 1989, p. 367.

110. C. Y. Wong et al., "Process Induced Degradation of Thin Oxides," in *Proc. 1stInt. Symp.Ultra Large Scale Integration Sci. Technol.,* S. Broydo and C. M Osburn, Eds., Electrochem. Soc., 1987, p. 155

111. C. Y. Wong et al., *IEEE Electron Dev. Letts.,* September, 1989, p. 420.

112. J. Y. Chen, *IEEE Trans. Electons Devices,* ED-31, p. 910 (1984).

113. K. Tanno, F. Shimura, and T. Kawamura, *J. Electrochem. Soc.,* 128, 395, (1981).

114. W. C. Till and J. T. Luxton, *Integrated Circuits: materials, devices, and fabrication,* Prentice-Hall, Englewood Cliffs, N. J., 1982.

115. B. Hoefflinger and G. Zimmer, in "J. Carrol, ed., *Solid-State Devices 1980, Institute of Physics Conf. Ser. 57,* from the 10th European Solid-State Device Research Conf., Sept. 1980.

116. A. G. Lewis et al., *IEEE Trans. Electron Dev.,* **ED-34,** 2156, (1987).

117. J. Y. Chen, *VLSI Design,* July 1984, p. 78.

118. H. Momose et al., *Tech. Dig. IEDM,* 1984, p. 706.

119. L. C. Parrillo, G. W. Reutlinger, and L. K. Wang, U.S. Patent No. 4,435,895, March 31, 1984.

120. K. M. Cham et al., *Tech. Dig. IEDM,* 1986, p. 296.

121. M. L. Chen, *Semiconductor International,* April 1988, p. 78.

122. T. E. Dillinger, *VLSI Engineering.* Englewood Cliffs, N.J., Prentice-Hall, 1988.

123. F. K. Baker et al., *Tech. Dig. IEDM,* 1989, p. 443.

124. J. M. Sung et al., *Tech. Dig. IEDM,* 1989, p. 447.

1

CHAPTER 7

BIPOLAR AND BICMOS
PROCESS INTEGRATION

This chapter deals with bipolar and BiCMOS process integration. A review of the physics of bipolar transistors is provided first.[1-3] Next is a discussion of the relationship between device performance and process technology, followed by a discussion of the evolution of bipolar IC process technology. (A book dealing exclusively with the device physics of bipolar transistors has recently been published.)[151]

In the second half of the chapter we consider BiCMOS technology, emphasizing process integration. The details of two process sequences are discussed – one for the fabrication of high-performance, digital-BiCMOS ICs, and the other for the fabrication of analog/digital BiCMOS ICs. (Reference 4 provides information on an excellent, recently published book dealing with the device physics, process technology, and circuit design of BiCMOS technology.)

7.1 BIPOLAR-TRANSISTOR STRUCTURES FOR INTEGRATED CIRCUITS

The bipolar transistor is an electronic device with two *pn* junctions in very close proximity. Idealized one-dimensional representations of bipolar transistors are shown in Figs. 7-1a and 7-1b. It can be seen that there are three device regions: an *emitter, a base* (the middle region), and a *collector*. In the ideal structure, the two *pn* junctions (called the *emitter-base* [E-B] and *collector-base* [C-B] junctions) are in a single bar of semiconductor material, separated by a distance designated as W_B. Modulation of the current flow in one *pn* junction by means of a change in the bias of the nearby junction is called *bipolar-transistor action*.

External leads can be attached to each of the three regions, as depicted in Fig. 7-1, and external voltages and currents can be applied to the device from these leads. Figure 7-1 also shows the circuit symbols of the devices with each lead labeled. If the emitter and collector are doped *n*-type and the base is doped *p*-type, the device is called an *npn* transistor (Fig. 7-1a). If the opposite doping configuration is used, the device is referred to as a *pnp* transistor (Fig. 7-1b).

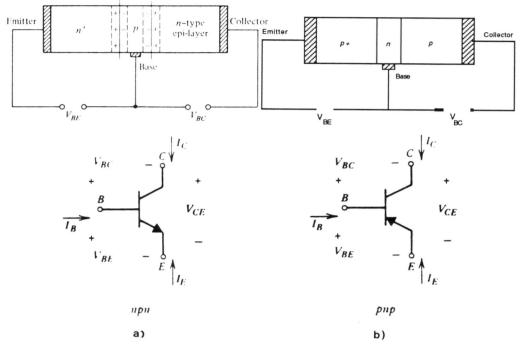

Fig. 7-1 Idealized one-dimensional model and circuit symbol of: (a) an *npn* bipolar transistor, and (b) a *pnp* bipolar transistor.

7.1.1 The Transistor Action

Transistors are used as either amplifying or switching devices. In the first application, the transistor's function is to faithfully amplify small ac signals; in the second, a small current is used to switch the transistor from an *ON* to an *OFF* state and back. These capabilities exist because of *transistor action,* which will be described in detail in this section. (Note that since the mobility of minority carriers – i.e., electrons – in the base region of *npn* transistors, is higher than that of holes in the base of *pnp* transistors, higher-frequency operation and higher-speed performances can be obtained with *npn* devices. As a result, the latter make up the vast majority of bipolar transistors used to build ICs.)

7.1.1.1 Basic Bipolar-Transistor Physics.
The energy bands pertaining to the idealized *pnp* transistor at equilibrium are shown in Fig. 7-2a. Under equilibrium, no net current flows across either the E-B or C-B junction, and hence no currents flow in the external leads (e.g., the emitter, base, and collector currents – I_E, I_B, and I_C, respectively – are all equal to zero).

Since external voltages may be applied to the transistor, either junction may be forward- or reverse-biased. If the E-B junction is forward-biased and the C-B junction is

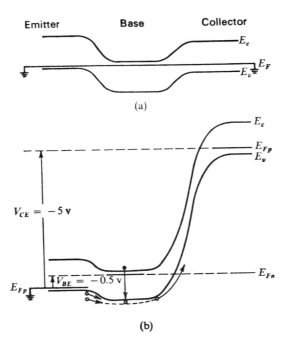

Fig. 7-2 Energy band diagram for a *pnp* transistor: (a) In equilibrium; (b) With the E-B junction forward biased and the C-B junction reverse biased (forward-active mode).

reverse-biased, the energy-band diagram of the device is as shown as in Fig. 7-2b. This mode of operation is called the *forward-active mode*. Forward-biasing of the E-B junction lowers its potential-energy barrier by qV_{BE}, whereas reverse-biasing of the C-B junction increases the potential energy barrier of this junction by qV_{CB} (where q is the charge of one electron – 1.6×10^{-19} C – and V_{EB} and V_{CB} are the bias voltages applied to the E-B and C-B junctions, respectively).

In an *npn* transistor operating in the forward-active mode, a large number of electrons are injected into the *p*-type base as a result of the forward bias applied to the E-B junction. If the distance between the two junctions is small enough, most of these electrons will reach the reverse-biased C-B junction. The electric field there will sweep the electrons across the junction, where they will be *collected* in the *n*-type collector, giving rise to an I_C current that is almost as large as the forward-bias current of the E-B junction, I_E. Hence, a large current can flow in a reverse-biased junction due to the existence of a nearby forward-biased junction, but only if the two junctions are close enough physically to interact in the manner described. *This effect represents bipolar-transistor action.**

---

* The conceptualization of transistor action and the invention of a device in which such action could be implemented (the *bipolar transistor*) together represent one of the most

7.1.1.2 Bipolar-Transistor Current Gain.

Not all of the electrons injected into the base will arrive at the collector. Some will recombine in the base region as they move through it. In addition, the forward bias applied to the E-B junction will cause some holes to be injected into the emitter region from the base. These two phenomena give rise to a current that flows in the base lead of the transistor.

The total emitter current (I_E) consists of those electrons that reach the collector (I_C) plus those that flow out of the transistor through the base lead (I_B):

$$I_E = I_C + I_B. \tag{7-1}$$

Two important parameters that characterize the transistor action described above are the *common-base current gain*, α, defined as

$$\alpha \equiv I_C/I_E \tag{7-2}$$

and the *common-emitter current gain*, β, defined as

$$\beta \equiv I_C/I_B. \tag{7-3}$$

From Eqs. 7-1, 7-2, and 7-3 it can be shown that α and β are related to each other by

$$\beta = \frac{\alpha}{1-\alpha} \tag{7-4}$$

The value of α must be less than unity, but in well-designed transistors it closely approaches this value (e.g., $\alpha > 0.9$). Accordingly, β will be much greater than unity in such transistors (typically, between 30 and 200).

Figure 7-3b shows the current-voltage output characteristics of an *npn* transistor for the common-emitter circuit configuration depicted in Fig. 7-3a. These characteristics imply that the transistor can be operated in modes other than the forward-active mode. When it is operated in that mode, however, a small change in I_B will bring about a β-times-larger change in I_C. Thus, when the transistor is used as an amplifying device (with I_B acting as the input signal to be amplified and I_C as the output signal), its operation is normally restricted to the forward-active mode.

As just noted, however, the E-B and C-B junctions may not always be biased so that the transistor is operating in this mode. Two other operating modes are also normally possible. In the first (the so-called *cutoff mode* of operation), the voltage across each junction is reverse-biased. In cutoff, the amount of leakage current flowing between the collector and emitter is so small that in many applications it can be neglected. The bottom portion of the I-V curves shown in Fig. 7-3b represents the cuttoff region.

significant feats in the history of device electronics. As a result of the investigations that led to these accomplishments the Nobel Prize in physics was awarded to the bipolar transistor inventors, William Shockley, John Bardeen, and Walter Brattain.

In the second operating mode, both junctions are forward-biased (the so-called *saturation* mode). As a result, the emitter-collector voltage (V_{CE}) approaches zero (note that if $V_{CE} > 0$, the E-B junction of *npn devices* is more forward-biased than the C-B junction). The left-hand portion of the characteristic curves of Fig. 7-3a represents this region of operation.

In saturation, however, the forward-biased C-B junction injects charge carriers into the base (as does the emitter), and the collector current thus becomes the net of two opposing flows of charge carriers. Therefore, once the device has entered the saturation mode, I_C decreases as V_{CE} decreases. Furthermore, in saturation the base current no longer controls the collector current, and current varies with power-supply voltage according to Ohm's law.

The transistor is operated in the cutoff and saturation modes when it is being used as a switching device. By controlling a small current (I_B), it becomes possible to turn a larger one (I_C) *ON* and *OFF*. If the base current is made zero, the transistor is in cutoff, $I_C \sim 0$, and V_{CE} is equal to the power-supply voltage. The electrical path between the emitter and collector leads of the transistor thus appears as an *OFF* switch. When the base current is large enough, I_C is increased to the degree that the transistor goes into

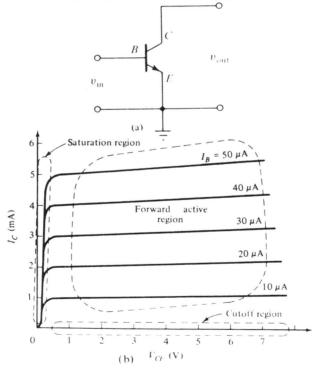

Fig. 7-3 (a) Common-emitter circuit configuration. (b) Current-voltage (I-V) output characteristics of *npn* transistor in common-emitter (C-E) configuration. Copyright, 1967, John Wiley and Sons, Inc. Reprinted with permission from ref. [11].

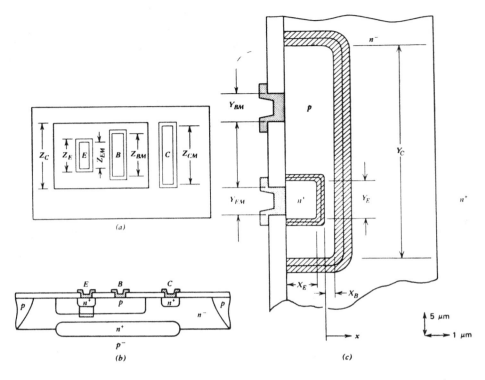

Fig. 7-4 Top view (a) and cross sections [(b) and (c)] of a representative *npn* transistor. The region dominating the transistor action is shaded in (b) and the area bounded by the base diffusion is rotated 90° and expanded in (c); *x* and *y* scales are indicated for (c) only. Copyright, 1986, John Wiley and Sons, Inc. Reprinted with permission from ref. [1].

saturation. Under this mode of operation, the ratio of the output voltage to the output current is small (i.e., V_{CE} is small and I_C is large). Hence, the current path between the emitter and collector leads of the transistor appears like the current path that exists in an *ON* switch.

7.1.2 Integrated-Circuit Transistor Structures

The physical structure of bipolar transistors used in ICs is quite different from the idealized 1-D structures shown in Fig. 7-1, primarily because the universal method used to fabricate bipolar transistors is the *planar process*. A top view of such a structure with junction isolation is shown in Fig. 7-4a; cross-sectional views are shown in Figs. 7-4b and 7-4c. The specific device dimensions, doping concentrations, and process sequences used to produce such devices will be described later, but it is useful to emphasize here that one of the most critical parameters in transistor fabrication is the width of the base region (W_B in Fig. 7-4c). Transistor action is critically dependent on the proximity of the two interactive junctions, and in bipolar transistors this dimension is less than 10 μm (in fact, in high-performance bipolar transistors, W_B is less than 0.5 μm).

Because W_B is so much smaller than the other dimensions in the device, transistor action is essentially confined to the shaded region shown in the cross-sectional view in Fig. 7-4b (as shown on a larger scale in Fig. 7-4c, this is the region bounded by the width of the emitter, Y_E). This implies that for many purposes the transistor can still be well approximated by the one-dimensional device structure in Fig. 7-1a. It should also be noted that the collector region in most bipolar *npn* IC transistors consists of a heavily *n*-doped layer buried beneath a lightly doped epitaxial layer, shown in Fig. 7-4b.

7.2 DIGITAL CIRCUITS USING BIPOLAR TRANSISTORS

A common application of bipolar transistors is in digital logic gates and memory arrays. Several types of circuits have been developed to implement logic gates in bipolar ICs; the most popular are transistor-transistor logic (TTL) and emitter-coupled logic (ECL). Although the detailed circuit behavior of these logic families lies beyond the scope of this text, it is useful to present the information that is relevant to our discussion of processing. For an excellent source of information concerning the analysis and design of digital logic circuits (both bipolar and MOS), see reference 5.

7.2.1 Basic Bipolar-Transistor Inverter Circuits

To define the language of digital bipolar ICs we present a description of a basic bipolar inverter circuit. This inverter consists of a bipolar transistor – connected in a common-emitter configuration – and two external resistors. The transistor is the driver element and is switched *ON* and *OFF* by the input signal; the collector resistor, R_C, is the *load*. A circuit diagram of the inverter is shown in Fig. 7-5a, and a three-dimensional cross-sectional view of the structure is given in Fig. 7-5b.

The load resistor connects the power-supply voltage, V_{CC}, and the output of the inverter. The input signal (V_{in}) is fed to the base of the transistor; the output signal is the voltage level at the output node, V_o. Figure 7-5c depicts the inverter's voltage-transfer characteristic. When there is a logic 0 input signal, V_{in} is less than the turn-on voltage, $V_{BE(on)}$, and the transistor is in the cutoff mode (i.e., the voltage across E-B is less than V_{BE}, and the voltage across C-B is reverse-biased). In this case, I_C is essentially zero; however, since the output is electrically connected to V_{CC} through R_C, V_o rises (or is *pulled up*) to logic 1 (close to V_{CC}).

When the input voltage increases above the value of $V_{BE(on)}$, the driver transistor is turned *ON*. The transistor first enters the forward-active mode, because V_{CB} is reverse-biased. In this mode, the collector current is related to the base current by $I_C \cong \beta I_B$. As the input voltage increases, I_B also increases. This produces a decrease in V_o, given by ($V_o = V_{CC} - \beta I_B R_C$). If the input voltage becomes high enough, V_o will become so small that the transistor will enter the saturation mode (both junctions would have to be forward-biased to allow V_o to drop below V_{BE}). In the saturation, region V_o will remain at a constant low value (i.e., V_{CEsat}, which is typically between 0.1 and 0.2 V) as the voltage is increased further. This demonstrates how the logic level at the input is inverted at the output of the circuit.

As noted in chapter 5, the desired characteristics of digital logic gates as IC elements are the following: fast switching speed (i.e., short propagation-delay time); low power dissipation; small size (i.e., minimum silicon area); and high noise margin. The basic bipolar inverter circuit just described does not fulfill such requirements as well as do more complex digital bipolar circuits, and hence it is not used in VLSI circuits. Instead, inverters (and other digital logic gates) are implemented with other digital circuits, which will be discussed next. Note, however, that the basic bipolar IC inverter circuit requires the availability of resistors as part of the integrated circuit. Such IC resistors are also required in other bipolar-logic gates. A discussion of the fabrication of IC resistors is given in Appendix A.

7.2.2 Bipolar Digital-Logic-Circuit Families

Many types of bipolar digital logic families have been developed for LSI and VLSI

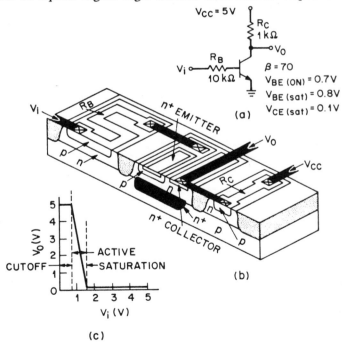

Fig. 7-5 Bipolar transistor inverter. (a) Circuit diagram. (b) Perspective view of the inverter structure. (c) Voltage transfer characteristics. From D. A. Hodges and H. G. Jackson, *Analysis and Design of Digital Integrated Circuits,* Copyright 1983, McGraw-Hill Book Co. Reprinted with permission from ref. [5].

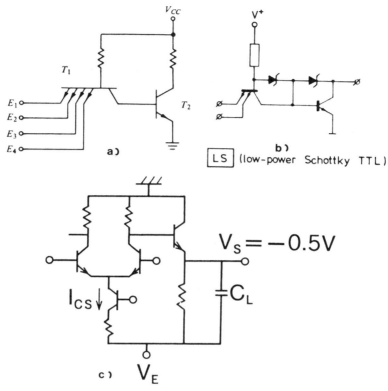

Fig. 7-6 Equivalent circuits of basic bipolar gates in the following logic families: (a) Transistor-transistor logic (TTL), (b) Low-power Schottky TTL, (c) Emitter-coupled logic (ECL).

circuits. A detailed analysis of their circuit characteristics can be found in reference 6 (which also provides a survey of the most important of these families), and in reference 5 (which, as noted earlier, contains a thorough treatment of transistor-transistor-logic [TTL] and emitter-coupled logic [ECL] circuit behavior).

In this section we present the electrical characteristics of TTL and ECL circuits. This background information provides a rationale for some of the device-design and process approaches used to fabricate bipolar and BiCMOS ICs.

Figure 7-6a shows the circuit diagram of a standard TTL gate, while Figs. 7-6b and 7-6c give the circuit diagrams of a low-power Schottky TTL gate and a basic emitter-coupled logic gate (ECL), respectively. The two TTL circuits are operated from a power-supply voltage of 5 V, but the ECL gate uses a -5.2 V supply.

Table 1-1 lists some of the other important properties of these logic families (including the 10K and 100K ECL families, of which the latter is the more advanced). The logic levels of TTL and ECL are quite different (e.g., 2.4 to 0.4 V in standard-TTL, but -0.9 V to -1.7 V in 10K ECL). Thus, if a digital system is built with both

Table 7-1 Typical Electrical Characteristics of Bipolar Logic Families, at T_A = 25 °C.

	V_{OL}/V_{OH}	V_{IH}/V_{IL}	Power Dissipation per Gate	Propagation Delay Time
TTL Families (Supply volts = 5.0 V, Logic swing = 2.2 V)				
Series 74 (Standard)	2.4 V/0.4 V	2.0 V/0.8 V	10 mW	10 ns
Series 74S (Schottky)	"	"	20 mW	3 ns
Series 74LS (Low-power Schottky)	"	"	2 mW	10 ns
Advanced Schottky TTL (ALS)				
Series 54/74F (Fast)	"	"	4 mW	2.5 ns
Series 54/74 AS (Adv. Schottky)	"	"	20 mW	1.5 ns
Series 54/74 ALS (Low-power)	"	"	1 mW	4 ns
ECL Families (Supply volts = - 5.2 V, Logic swing = 0.8 V)				
10K	-0.9 V/-1.7 V	-1.2 V/-1.4 V	24 mW	2 ns
100K	"	"	40 mW	0.75 ns

TTL and ECL logic gates, the circuits must be interfaced correctly (i.e., it is necessary to voltage-level-shift the output from one of the families to be compatible with the input of the other). This level-shifting function is performed by circuits known as ECL-to-TTL and TTL-to-ECL *translators.*

Some other interesting comparisons between the logic families can be made. First, ECL gates are faster than TTL gates, but they consume more power. (The speed advantage of ECL is primarily due to two factors: the transistors are prevented from entering the saturation mode, and the output logic swings are smaller.) ECL gates are also faster than CMOS gates; thus, for applications where ultimate speed is the most important requirement, ECL has been, and remains, the technology of choice (for example, ECL circuits with gate delays as small as 35 ps have been demonstrated).[7]

However, as shown in Fig. 7-7 (which gives the projected delay as a function of the switch current of the scaled ECL circuit), in order for circuit delay to be in proportion to the lithographic dimension, the circuit's current – and hence its power dissipation – must remain essentially constant in scaling. With the power dissipation per gate in the milliwatt range, the integration of ECL logic chips will remain severely limited. For instance, a 10,000-gate ECL IC, with each gate dissipating 5 mW, would dissipate a total of 50 W!

Low-power Schottky TTL technology was found to provide slower performance than ECL, but to consume much less power. For MSI and LSI chips, Schottky TTL

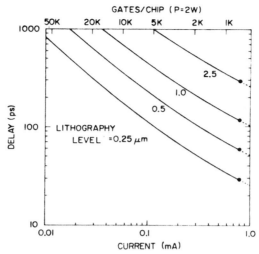

Fig. 7-7 ECL gate delay as a function of switch current and emitter-stripe width, predicted by bipolar scaling theory.[14] (© 1979 IEEE).

offered the compromise of faster speed than CMOS, with lower power dissipation than ECL. In the VLSI era, however, CMOS was eventually able to provide speeds comparable to Schottky TTL, but with much smaller power-dissipation levels (see chap. 6, Fig. 6-7b). CMOS has thus tended to displace Schottky TTL technology. Today, ECL remains an important digital logic family, and BiCMOS-based circuits also incorporate bipolar devices in logic circuits. A relatively new family, *non-threshold logic* (NTL; see Fig. 7-8) has recently been drawing attention as a possible replacement for ECL, since NTL-with-emitter-follower circuits exhibit speeds comparable to those of ECL with lower power dissipation. NTL circuits as fast as 30 ps at 1.48 mW have been demonstrated.[8]

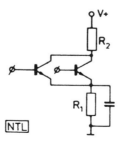

Fig. 7-8 Circuit diagram of non-threshold logic gate (NTL).

7.3 MAXIMIZING BIPOLAR-TRANSISTOR PERFORMANCE THROUGH DEVICE DESIGN AND PROCESSING TECHNOLOGY

The desired device characteristics of bipolar transistors include the following: high current gain (large value of β), high-frequency ac operation, fast switching speed, high device-breakdown voltages, minimum device size (to achieve high functional density), and high reliability of device operation.

In order for high-frequency ac performance and fast switching speed to be achieved, the parasitic resistances of the transistor (i.e., R_E, R_B, and R_C), and the parasitic junction capacitances (i.e., C_{EB}, C_{CB}, and C_{CS}) must be minimized. In addition, high-level injection effects (e.g., the Kirk effect) should be avoided. Finally, for faithful amplification of ac signals, the Early voltage must be high.

This section describes how these characteristics can be obtained through device design and processing technology.

7.3.1 Current Gain

As mentioned earlier, two of the figures of merit used to describe the performance of bipolar transistors are the current gains α and β. If the leakage current across the reverse-biased C-B junction is a negligibly small fraction of I_C, α can be expressed as the product of the *emitter-injection efficiency*, γ, and the *base-transport factor*, α_T:

$$\alpha = \gamma \, \alpha_T \qquad (7-5)$$

Here γ is defined to be the ratio of the current in the base due to injection of electrons from the emitter, I_{En}, and the total emitter current (which is the sum of I_{En} and the current due to injection of holes into the emitter from the base, I_{Ep}:

$$\gamma \equiv \frac{I_{En}}{(I_{En} + I_{Ep})} \qquad (7-6)$$

and α_T is defined as the fraction of I_{En} that reaches the C-B junction, as follows:

$$\alpha_T \equiv \frac{\text{electron current reaching the collector}}{\text{electron current injected into the base}}$$

Accordingly, β will be given by

$$\beta = \frac{\gamma \, \alpha_T}{(1 - \gamma \, \alpha_T)} \qquad (7-7)$$

For a transistor with a large β (i.e., $\alpha \rightarrow 1$), we can write

$$\frac{1}{\beta} \cong 1 - \gamma\,\alpha_T \cong \frac{1}{2}\left(\frac{W_B}{L_B}\right)^2 + \left(\frac{D_E GN_B}{D_B\,GN_E}\right) \qquad (7\text{-}8)$$

where the first term is due to recombination within the base and the second to the injection into the emitter. In these terms, W_B is the base width (defined as the distance between the edges of the two depletion regions in the base); L_B is the minority-carrier diffusion length in the base; D_B and D_E are the diffusion coefficients of carriers in the base and emitter, respectively; and GN_B and GN_E are the base and emitter Gummel numbers (so named because they were first described by H. K. Gummel). These two parameters are defined as

$$GN_B \equiv \int_{base} N(x)\,dx \quad \text{and} \quad GN_E \equiv \int_{emitter} N(x)\,dx \qquad (7\text{-}9)$$

and represent the total number of active impurity atoms in the base and emitter regions (per cm^2). The minimum value of GN_B is $\sim 10^{12}$ cm^{-2}, because lower values would produce transistors with unacceptably small punchthrough-voltage values. (Typical values of N_B are in the $10^{12} - 10^{13}$ atoms/cm^2 range.) The typical range of GN_E is 10^{15} cm^{-2} to 10^{16} cm^{-2}. Thus, with Eq. 7-8 used to find β, the maximum theoretical current gain in conventionally designed bipolar transistors is seen to be $\sim 10^4$.

For large current gain, Eq. 7-8 should be as small as possible. This in turn implies that the GN_E/GN_B ratio should be high, W_B should be small, and L_B should be large. A large GN_E/GN_B ratio also implies that the emitter must be much more heavily doped than the base. In modern, high-performance transistors, $W_B \ll L_B$ (i.e., L_B is greater than 10 μm for typical base doping concentrations, and W_B is less than 1 μm in low-voltage logic devices). Hence, the second term in Eq. 7-8 determines the current gain.

On the other hand, when the emitter is as heavily doped as it is in modern transistors, the silicon band gap in the emitter region is reduced and the recombination rate of minority carriers is increased. Both effects lower the emitter-injection efficiency. As a result, observed current gains in transistors with conventionally formed emitters (e.g., emitters formed by direct diffusion or ion implantation of the emitter dopant into the substrate) are not as high as those predicted by Eq. 7-8. If even higher doping concentrations were used in the emitter, the result would be decreased emitter-base junction breakdown voltages and increased emitter-base junction capacitances. Advanced emitter structures have been developed to overcome this problem and these will be discussed in section 7.7.

Note that in analog applications a high value of β (≥ 100) is generally desired, but in digital applications a lower value is often acceptable (in some cases as low as 20).[9] On the other hand, in fast-switching transistors (which will be discussed next), small base-storage times and small intrinsic base resistances are needed. To reduce the base-storage times, the base width must be minimized, subject to base punchthrough. The minimum base doping to avoid punchthrough is proportional to $(1/W_B)^2$; therefore, the minimum GN_B is proportional to $1/W_B$ and needs to be large if W_B is small. Since smaller

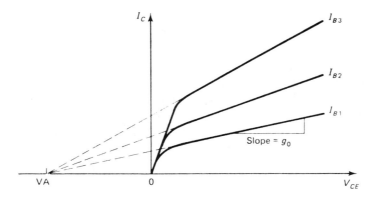

Fig. 7-9 Effect of base-width modulation on the C-E I-V output characteristics. The intercept on the voltage axis of the extrapolated curves is the Early voltage (V_A).

current gains can be tolerated in digital applications, some gain can be traded off by increasing the doping in the base to achieve a narrower W_B for a high switching speed.

7.3.2 The Early Voltage

The base width in bipolar transistors decreases as the C-B junction voltage increases because the depletion region increases in width. This effect is called *base-width modulation* (or the *Early effect*, after the scientist who first correctly explained its basis). The degree to which this effect impacts the operating characteristics of a bipolar transistor is represented by the Early voltage, V_A (volts). In the collector-emitter configuration, V_A is given by

$$V_A = I_C / (\partial I_C / \partial V_{CE}). \tag{7-10}$$

The dependence of $\partial I_C / \partial V_{CE}$ on I_C results in typical transistor output characteristics as shown in Fig. 7-9. Extrapolation of the characteristics of Fig. 7-9 back to the V_{CE} axis gives an intercept that represents V_A.

A large value of V_A (> 100 V) is desired in analog bipolar circuits for two reasons:

1. The *open-circuit voltage gain*, a_o (defined as the small-signal low-frequency voltage gain in the common-emitter configuration), is approximately found from[10]

$$a_o \cong q V_A /kT. \tag{7-10a}$$

Since a_o is the maximum voltage gain that can be obtained from a bipolar transistor, it is a significant parameter in analog circuits. Because Eq. 7-10a indicates that a_o depends only on V_A and T, a large value of V_A will permit larger voltage gains to be achieved. (For a value of V_A = 50 V and room-temperature operation, $a_o \cong 2000$.)

2. Since $\Delta V_{CE}/\Delta I_C = V_A/I_C = r_o$ (where r_o is the small-signal output resistance of the transistor in the common-emitter configuration), small values of V_A imply a smaller output resistance, which is generally undesirable. Analog applications therefore require a minimum V_A of 30 volts, while a lower voltage is normally acceptable for digital applications (15-20 V).[22]

The value of V_A is increased by raising the doping level in the base. Higher doping decreases the penetration of the depletion region as a function of increasing C-B reverse-bias voltage. In advanced bipolar transistors the doping levels are typically high enough (i.e., so as to prevent punchthrough, as will be described in section 7.3.4.2) that V_A is not a critical parameter in digital designs.

7.3.3 High-Level Injection Effects (Kirk Effect)

As the collector-current density (J_C) in an *npn* transistor increases, the density of electrons being transported across the C-B space-charge region also increases (i.e., this density is calculated from J_C/qv_l, where v_l is the electron scattering-limited velocity, $\sim 10^7$ cm/s). When the density of electrons crossing the C-B junction becomes comparable to the doping on the collector side of the space-charge region (a condition corresponding to *high-level injection*), the total charge in this region becomes significantly reduced from that under low-level injection conditions, leading to a lower electric-field gradient in the C-B junction. Since the collector voltage is constant, the integration of the electric field will remain the same, with or without current. However, with a lower maximum electric-field strength, the space-charge region edge in the base moves toward the collector, effectively increasing the base width. This phenomenon is known as the *base push-out effect** (or *Kirk effect*, after the author of the first paper analyzing it).[23]

At first the base-width increase is small, and the current density at the onset of the Kirk effect, J_1, is given by

$$J_1 = q\, v_l\, N_{epi} \qquad\qquad (7\text{-}11)$$

where N_{epi} is the doping concentration in the epi layer region of the collector. In practical device operation J_1 is easily exceeded in many applications. For example, in a transistor that has a 2 x 5-μm emitter and an epi doping concentration of 5×10^{15} cm^{-3}, Eq. 7-11 indicates that J_1 will be reached when $I_C = 0.8$ mA.

As the current is increased beyond J_1, the maximum electric-field strength continues to decrease. It eventually reaches zero, which represents a collapse of the original collector-base region. The base then spreads over the collector epitaxial layer and becomes very large. The current density at which this takes place is given by

$$J_{Ck} = q\, v_l\, (N_{epi} + \{2K_s\varepsilon_o V_{CB}/qx_C\}) \qquad\qquad (7\text{-}12)$$

*A more rigorous analysis of this effect is presented in chapter 7, reference 1.

where x_C is the thickness of the epi layer from the metallurgical C-B junction to the edge of the epi/buried-layer boundary.

The increase in base width leads to an increase in GN_B, and hence to a decreased current gain at high collector currents. In addition, the larger base width reduces f_T and the transistor speed. The transistor doping profiles must thus be optimized with this effect in mind. Once a transistor structure has been designed, operation at current levels above J_{Ck} for the device should be avoided.

From the equation used to calculate J_{Ck}, it is seen that this value can be raised by increasing the doping in the collector. Unfortunately, this also produces a decrease in the BV_{CBO} and BV_{CEO} of the device, as well as an increase in the collector-base junction capacitance, C_{CB}. Since transistors cannot be fabricated with smaller breakdown voltages than those required by the system power-supply voltage levels, the breakdown voltage constraints will set the maximum value of N_{epi}.

On the other hand a tradeoff between J_{Ck} and C_{CB} can be made, based on the operating point of the transistor. That is, if the transistor is to be operated at low current densities, the speed will be limited by C_{CB}. Thus, C_{CB} should be minimized (by reducing N_{epi}), since the Kirk effect will not impact the speed performance. However, for high-current applications, the speed will be limited by high-level injection effects, and C_{CB} can thus be allowed to rise when N_{epi} is increased in order to minimize speed degradation due to the Kirk effect.

Finally, it should be noted that in digital ICs, where BV_{CEO} is smaller than in analog ICs, a larger value of collector doping can be tolerated. This also implies that higher collector currents can be sustained in such devices without the onset of the Kirk effect.

7.3.4 Operating-Voltage Limits in Bipolar Transistors

The mechanism of avalanche breakdown in a *pn* junction limits the maximum reverse-bias voltages that can be applied to a junction and is also responsible for some of the maximum operating voltage values in bipolar transistors. The breakdown voltage of the C-B junction with $I_E = 0$ (BV_{CBO}) is due to this mechanism. The doping on the lightly doped side of the collector typically determines BV_{CBO} (except in cases of *reachthrough,* as will be discussed in section 7.2.4.1).

In C-B junctions in planar bipolar transistors, however, the *pn* junction is formed by diffusion into the silicon at openings in the surface SiO_2 layer. Thus, impurities will diffuse downward and sideways under the edges of the SiO_2 layer. Hence, the junction has a plane region with nearly cylindrical edges (as can be seen in Fig. 7-4c). Such junction curvature enhances the electric field in the curved part of the depletion region (see chap. 5, Fig. 5-10a), which reduces the breakdown voltage below that predicted by one-dimensional junction theory. Figure 5-10a also shows that as the junction depths get shallower (making the radius of curvature, r_j, of the spherical and cylindrical structures smaller), the breakdown voltage is significantly reduced, especially for small substrate-impurity concentrations. Thus, the measured value of BV_{CBO} is less than the value of the breakdown voltage that would be predicted if the junction were only one-dimensional (as would be the case if the transistor had a structure like that shown in Fig. 7-1a).

The breakdown voltage of the E-B junction with $I_C = 0$ (BV_{EBO}) is also limited by avalanche breakdown. Since the emitter is so heavily doped, it is the base region that is the lightly doped side of this junction. Since the base is more heavily doped than the collector, BV_{EBO} is much smaller than BV_{CBO} and is typically $6 - 8$ V. Since the base is formed by diffusion from the surface, the doping in the base region is highest there so the E-B junction will undergo breakdown first in this area. One of the reasons for performing a drive-in step following predeposition when a diffusion step is used to form the base is to reduce the surface concentration of boron, which thus increases BV_{EBO} to a sufficiently high value.

While high doping in the base region adjacent to the emitter sidewalls is advantageous for reducing the base series resistance, R_B (see section 7.5.2.6), if the doping is too high, BV_{EBO} becomes too small. Tradeoffs must therefore be made in the selection of the doping value in this region. The curvature of the junction enhances the electric field and reduces BV_{EBO} in the same manner as described for BV_{CBO}.

For a bipolar transistor with a voltage applied between the emitter and collector and in which $I_B = 0$, breakdown occurs at BV_{CEO}. In this situation, the breakdown may be due either to avalanche breakdown multiplied by the effect of transistor action, or to punchthrough.

In the first instance, the E-B junction will be very slightly forward-biased as a result of the base region's acquiring a potential that is intermediate between the potentials of the collector and emitter. Thus, the collector current will also be greater than I_{CBO} (the leakage current of the C-B junction when $I_E = 0$) because most of the carriers injected from the slightly forward-biased E-B junction will be collected by the collector. This increased current will precipitate avalanche breakdown at a smaller applied C-B voltage, which implies that BV_{CEO} must be smaller than BV_{CBO}. An empirical expression has been derived that relates BV_{CEO} and BV_{CBO} (plane):[11]

$$BV_{CEO} = \frac{BV_{CBO}(plane)}{(\beta)^{\frac{1}{n}}} \qquad (7 - 13)$$

where BV_{CBO} (plane) is the collector-base breakdown voltage along the planar (flat) portion of the base region. The empirical parameter n has been found to have values of between 3 and 6.

7.3.4.1 Reachthrough Breakdown. Since most bipolar transistors used in ICs are fabricated with collector regions that consist of a lightly doped epitaxial layer over a heavily doped buried layer, BV_{CBO} may in some cases be limited by *reachthrough breakdown*. If the epitaxial-layer thickness, t_{epi}, is too small, under increasing reverse-bias voltage the depletion region of the C-B junction will entirely penetrate (or "reach through") the epitaxial layer before the predicted BV_{CBO} voltage has been reached. If the voltage is increased beyond this point, the depletion region must enter the heavily doped buried layer, resulting in BV_{CBO} values smaller than those

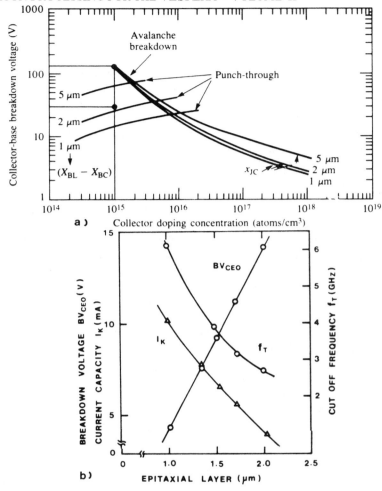

Fig. 7-10 (a) Collector-base-junction breakdown voltage as a function of collector-doping concentration with collector-base junction depth and punchthrough limits as parameters.[125] Reprinted by permission of the publisher, The Electrochemical Society, Inc. (b) Epitaxial layer thickness dependence of cutoff frequency, f_T, breakdown voltage BV_{CEO}, and knee current, I_K.[12] (© 1987 IEEE).

predicted merely by the doping in the epi layer. Figure 7-10a gives the collector-base junction-breakdown voltage as a function of the collector-doping concentration, with the collector-base junction depth and the thickness of the lightly doped region of the epi layer ($x_{BL} - x_{BC}$) as parameters.[125] Figure 7-10b shows how a decrease in t_{epi} causes BV_{CEO} to decrease in a bipolar transistor designed for use in 5-V digital ICs.[12] Section

7.12.1.2 gives another example of a BV_{CEO} calculation in which this effect is considered.

7.3.4.2 Punchthrough Breakdown.

In some cases the value of BV_{CEO} can be limited by *punchthrough* rather than by avalanche breakdown. If the base width is too narrow and/or the doping level in the base is too small, the depletion region of the C-B junction will penetrate the base region and join the E-B junction depletion region before the avalanche-limited value of BV_{CEO} has been reached. (Section 7.2.2 describes such base-width modulation.) If the collector voltage is increased beyond this value, the potential-energy barrier at the E-B junction will be lowered by ΔV, as shown in Fig. 7-11. As a result, a large emitter current (and correspondingly, a large collector current) can flow in the device, even with no avalanche breakdown. (Note that a large current flow across the C-B junction when both the E-B and C-B junctions are reverse biased is a manifestation of breakdown.)

Only a slight increase in the collector voltage beyond that needed to merge the two depletion regions is required for a large increase in I_C. Thus, punchthrough appears as a form of device breakdown. To prevent punchthrough, appropriate combinations of base doping and base width for the circuit operating voltages must be used. Most bipolar transistors are designed so that avalanche breakdown of the C-B junction will occur before punchthrough.

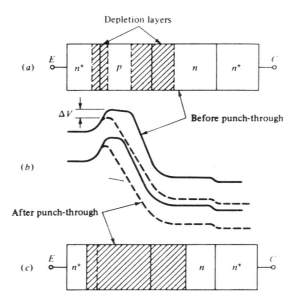

Fig. 7-11 Punchthrough in an *npn* transistor: (a) space-charge regions before punchthrough, (b) energy-band diagram (and space-charge regions) after punchthrough. Copyright, 1988, McGraw-Hill Book Company. Reprinted with permission from ref. [2].

7.3.4.3 Breakdown Voltage and High-Level Injection Limits in Advanced Bipolar Transistors.

To maintain the sufficiently high current levels needed to achieve minimum gate delays (see section 7.3.8), the current density is increased as the emitter area is scaled down. Higher collector doping levels must be used to avoid the Kirk effect. Furthermore, higher doping concentrations in the base are needed to avoid punchthrough of the narrow base regions. Thus, as of 1990, the base and collector doping concentrations in advanced bipolar devices are typically on the order of 1×10^{18} cm^{-3} and 1×10^{17} cm^{-3}, respectively. The resulting peak electric field in the C-B junction is high enough to produce intolerably large avalanche-multiplication currents at low C-B reverse bias voltages (e.g., less than 3 V). Lu and Chen describe this problem in detail in reference 13.

7.3.5 Parasitic Series Resistances in Bipolar Transistors

Due to the planar structure of IC transistors, the current that flows from the emitter to the collector must travel parallel to the surface of the wafer after it has been transported vertically through the base. It must then flow upward to a contact at the wafer surface. Because the current-flow paths within the transistor have significant resistivity, parasitic series resistances exist in each of the three bulk regions of the device: a collector series resistance, R_C; a base series resistance, R_B; and an emitter series resistance, R_E. Each must be as small as possible.

7.3.5.1 Collector Series Resistance, R_C.

A low value of R_C is important in both high-frequency ac circuits and high-speed digital circuits. In the former, a large R_C value would decrease the cutoff frequency and introduce a voltage drop that would increase V_{CEsat}. In the latter, a large R_C value would reduce both f_T (see section 7.3.7) and the maximum current-drive capability. Figure 7-12 shows the components of R_C.

Several approaches are used to minimize R_C. First, the lightly doped epitaxial layer used as the collector is doped as much as can be tolerated with regard to breakdown-voltage and junction-capacitance considerations. This means that in low-voltage digital devices (in which a low value of V_{CEsat} is critical but lower operating voltages are

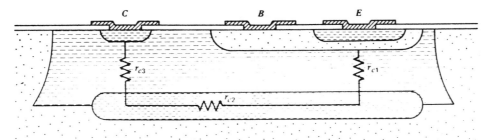

Fig. 7-12 Components of collector resistance, R_C. Copyright, 1984, John Wiley and Sons, Inc. Reprinted with permission from ref. [10].

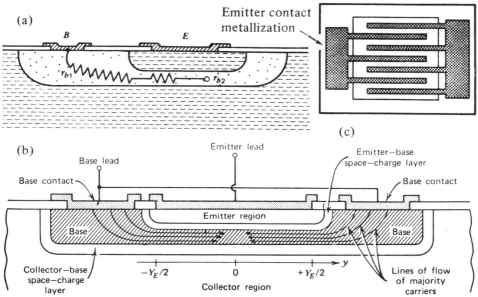

Fig. 7-13 Base-resistance components for the *npn* bipolar transistor. Copyright, 1984, John Wiley and Sons, Inc. Reprinted with permission from ref. [10]. (b) Cross section of a transistor under forward active bias. The base current is supplied from two side contacts and flows toward the center of the emitter causing the base-emitter voltage to vary with position. Copyright, 1986, John Wiley and Sons, Inc. Reprinted with permission from ref. [1].

used), the epi layer is doped somewhat more heavily than it is in analog transistors, which are operated at higher supply voltages. Second, heavily doped buried layers (subcollectors) are used beneath the more lightly doped collector regions to provide a parallel high-conductivity path while maintaining the high C-B breakdown-voltage capability provided by the lower-doped epi.

Inclusion of the buried layer reduces R_C from the kilohm range to a few hundred ohms in typical devices. For some applications, however, even this value of R_C is too high, so an additional, heavily doped, deep n^+ region is diffused beneath the collector contact until it reaches the buried layer. Such deep-collector contacts (also called *plugs* or *sinkers)* can reduce R_C to the order of 10 Ω. Details of the process steps used to form buried layers and deep-collector contacts will be discussed later in this chapter in the sections describing process sequences for bipolar and BiCMOS technologies. Chapter 2 of reference 10 gives a first-order-calculation technique for R_C when the sheet-resistance values of the buried and epi layers and their device-structure dimensions are known.

7.3.5.2 Base Series Resistance, R_B. Low values of R_B are also needed to achieve adequate gain at high frequencies and to obtain high speeds in digital bipolar ICs. As illustrated in Fig. 7-13a this resistance consists of two parts. The first part is the resistance of the so-called *extrinsic* (or *inactive) base region,* R_{B1} (which makes up the path between the base contact and the edge of the emitter region). The second part, R_{B2}, is the resistance between the edge of the emitter and the site within the *intrinsic* (or *active)* base region at which the current is actually flowing. R_{B2} is also known as the *base-spreading resistance.*

One technique for reducing R_{B1} is to use a separate process step to selectively add dopants to the extrinsic base region in order to increase its conductivity. If the extrinsic base is doped too heavily, however, excessive emitter-base capacitance will result along the emitter sidewall, and a low emitter-base breakdown voltage may also be produced. Self-aligned techniques have been exploited to reduce R_{B1} by decreasing the distance between the base contact and the intrinsic base region. This approach is very effective and will be discussed in more detail later.

7.3.5.3 Base-Spreading Resistance, R_{B2} (and Emitter Current Crowding). R_{B2} is not well modeled by a single resistor, for two reasons. First, this effect is actually distributed (or spread) throughout the intrinsic base region, and two-dimensional effects are important.* Second, the effect of *current crowding* sets in, even at moderate current levels, causing most of the carriers to be injected into the base at the periphery of the emitter diffusion (since the base current in the narrow base flows from the edges of the emitter to the center [parallel to the wafer surface; see Fig. 7-13b]). An ohmic drop thus occurs between the emitter's edge and the center. Because the base is so narrow, the distributed resistance along it is quite high, and the value of the ohmic drop can cause the base-emitter voltage drop to vary with position along the E-B junction. The forward bias of the emitter junction will be largest at the edge of the emitter diffusion. Therefore, most of the emitter current will be injected into the base near the edge, and the remainder will decrease to a minimum at the center.

Although current crowding can have the beneficial effect of reducing R_B (in the extreme, the effective value of R_B approaches R_{B1}), it can also give rise to localized heating at current levels that might be tolerable only if the current were uniformly distributed. Furthermore, the Kirk effect can occur at lower currents because of the uneven current density across the active region of the device.

The most effective solution is to distribute the current along a relatively long emitter edge, thereby reducing the current density at any one point. In bipolar transistors designed for low-noise and/or high-frequency applications, the emitter-edge length is maximized by designing the emitter layout in the form of many long, narrow stripes with base contacts between them (this layout approach is sometimes called an *interdigitated geometry,* see Fig. 7-13c).

Since R_{B1} and R_{B2} are both inversely proportional to the *emitter* length (i.e., a longer emitter implies a wider horizontal base region) and R_{B2} is proportional to the emitter width, it is preferable for emitters to be long and narrow (i.e., with width-to-length ratios of one-third, or less).

---

* An analysis that rigorously formulates two expressions for the effective value of R_{B2} is given in reference 47.

7.3.5.4 Emitter Series Resistance, R_E. Until recently, bipolar device performance has not been significantly impacted by R_E. Since the emitter is a shallow, heavily doped region, the series resistance of its bulk region is quite small. In addition, the contact resistance of ohmic metal-semiconductor contacts has also been small enough that it has not increased R_E to values that would significantly impact the performance of analog or digital ICs. However, the introduction of polysilicon-emitter structures for advanced bipolar ICs has brought with it the penalty of an increased emitter resistance, and its value in such structures can significantly degrade their performance. This topic will be discussed further in section 7.7.

7.3.6 Parasitic Junction Capacitances in Bipolar Transistors

The parasitic capacitances associated with the E-B, C-B, and collector-substrate (C-S) junctions of the planar bipolar transistor can limit the high-frequency and switching characteristics of these devices. We will describe in section 7.3.9 how these capacitances impact the propagation delays of bipolar digital ICs.

The C-B and E-B capacitances in junction-isolated bipolar transistors are composed of the capacitances of the C-B and E-B junctions along the flat bottom portions and along the sidewalls. The C-S capacitance in junction-isolated bipolar transistors with buried-layer collectors consists of three parts: (1) that of the junction between the buried layer and the substrate; (2) that of the sidewall of the isolation diffusion; and (3) that between the epitaxial material and the substrate. In general, the value of the capacitance is proportional to the area of the junction and the doping concentration on the lightly doped side. (Techniques for first-order calculations of the values of these capacitances in junction-isolated transistors are given in chapter 2 of reference 10.) Since junction-isolated bipolar transistors for digital ICs are operated at smaller power-supply voltages than those used in analog ICs, thinner epi layers are used. Hence, the areas of the transistors are smaller, resulting in lower values of parasitic capacitance.

The benefit of reduced parasitic capacitance is extended even further if oxide or trench isolation is used instead of junction isolation. In such structures the sidewall-capacitance components are reduced (or eliminated), and the ability to use smaller feature sizes reduces the area of the devices (and hence the planar-capacitance components). Examples of parasitic capacitance values (as well as of the device layouts) for each of these four types of IC bipolar transistors are given in Fig. 7-14.

7.3.6.1 Storage Capacitances in Bipolar Transistors. The capacitances associated with the storage of minority carriers in the bulk regions under conditions of forward bias can also significantly impact the switching speed of digital bipolar circuits. This topic is discussed further in section 7.3.9.3.

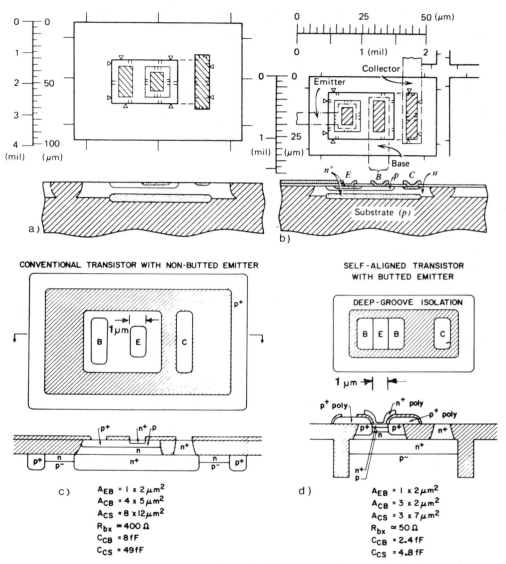

Fig. 7-14 Top views (showing dimensions), cross-sections and device structure characteristics of an: (a) junction-isolated amplifying (35 V) bipolar transistor, 10-μm minimum dimensions;[126] (b) junction-isolated digital (5 V) bipolar transistor, 5-μm minimum dimensions;[126] (c) oxide-isolated, digital (5 V) bipolar transistor, 1-μm minimum dimensions;[26] and (d) trench-isolated, digital (5 V) bipolar transistor, 1-μm minimum dimension.[26] {(a) and (b) Copyright, 1972 by Litton Educational Publishing, Inc., Reprinted with permission of Van Nostrand Reinhold Company from ref. [127].} {(c) and (d) (© 1986 IEEE).}

7.3.7 Bipolar Transistor Unity-Gain Frequency, f_T

Another important characteristic of bipolar transistors is their ability to provide current gain at high frequencies. The figure of merit that is typically used to describe the high-frequency behavior of a bipolar transistor is the *unity-gain frequency*, f_T, which is the frequency at which the current gain of the transistor decreases to unity. An approximate expression for f_T is given by

$$f_T = \frac{1}{2\pi\tau} \qquad (7\text{-}14)$$

where τ is the *transit time* of the transistor. An approximate expression for τ is

$$\tau = \frac{W_B{}^2}{\eta\,D_B} + (C_{CB} + C_{CS})\,R_C \qquad (7\text{-}15)$$

The first term $(W_B{}^2/\eta D_B)$, called the *base transit time* (τ_B), represents the average time per carrier spent in diffusing across the neutral base region of width W_B. D_B is the diffusion coefficient of minority carriers in the base, and the parameter η is associated with the doping profile in the base. For a uniformly doped base region, $\eta = 2$, whereas in a graded base, η can be larger by a factor of 10. The second term is the delay associated with the charging of the capacitances connected to the collector node (C_{CB} and C_{CS}) through R_C.

For a minimum value of τ to be obtained, several features of the bipolar transistor structure must be optimized. The base width is made as narrow as possible, and buried layers and deep-collector contacts are used to minimize R_C. In addition, light doping, oxide isolation, and reduced device area are all employed to decrease the values of C_{CB} and C_{CS}.

Note that τ in fact also depends on I_C. For medium to high values of I_C, the time required to charge C_{CB} and C_{CS} becomes small, and thus approaches a constant value (τ_B). At low values of I_C, the term involving C_{CB} and C_{CS} dominates. For very high I_C values, however, τ again increases as a result of high-level injection effects (Kirk effect).

7.3.8 First-Order *npn* Device Design

A first-order method for designing bipolar IC transistor structures, proposed by Solomon and Tang,[14] is graphically depicted in Fig. 7-15 (which also summarizes the discussion of bipolar-device parameters presented in this section).[15]

The circuit applications specify several characteristics that set the device design boundaries, including collector-current density, current gain, BV_{CEO}, R_B and τ_B. The epi layer thickness and doping concentration are then selected based on these requirements and on Kirk effect considerations (with the emitter area also used as a parameter to establish the collector current density). The base profile is then optimized

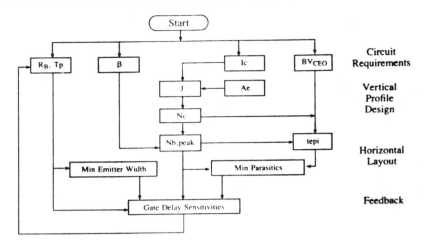

Fig. 7-15 First order *npn* device design. The vertical profile is set to optimize device delay, while optimal horizontal design minimizes the parasitics. From [14] © 1979 IEEE.

based on the design decisions and on the constraints of meeting the target β, R_B, and τ_B specifications.

7.3.9 Switching-Speed Behavior in Bipolar ICs

MOS and bipolar transistors have separate advantages with respect to their use in digital circuits. Circuits built with CMOS consume less power than bipolar circuits and can be fabricated with higher packing densities, while those built with bipolar transistors can operate at higher speeds with minimum-size transistors. In addition, small voltage swings are possible with bipolar gates, resulting in a relatively small speed degradation due to on-chip wiring capacitance. Finally, the power-supply voltage and the gate output-voltage swings of bipolar digital circuits can be minimized independent of transistor size. Bipolar circuits are thus preferred for applications in which ultimate speed performance is desired.

7.3.9.1 Propagation-Delay-Time Calculation in Bipolar Transistors.
Historically, the unity-gain cutoff frequency, f_T, has also been widely used as the figure of merit for describing the switching speed of bipolar transistors. Unfortunately, as shown in Fig. 7-16a, this parameter often does not correlate well with the gate delays measured in actual ECL circuits. An alternative delay expression has been derived by Stork[16] that agrees very well with reported data of ECL ring-oscillator delays (Fig. 7-16b). The objective in deriving this figure of merit was to obtain a simple delay expression that is most accurate when the minimum delay is approached.

For the simplified case in which the emitter-follower resistance equals the load resistance, Stork's expression for the minimum delay, τ_s, of the switching action of a bipolar transistor, is given by

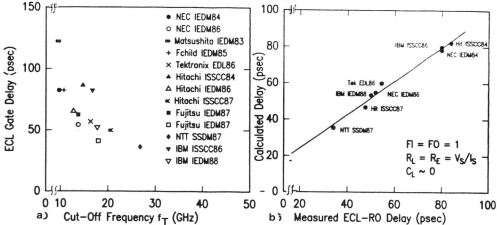

Fig. 7-16 (a) Scatter plot of ECL ring-oscillator speed: the correlation of published delay data with f_T is poor. (b) Calculated versus measured/published ECL ring-oscillator delay: the correlation is generally within 10%.[16] (© 1988 IEEE).

$$\tau_s = \sqrt{\left\{1 + \left(\frac{2\alpha R_B I_{EE}}{\Delta V}\right)\right\} \left(\frac{1}{2\pi f_T}\right)(3C_{CB} + C_{CS})\left(\frac{\Delta V}{I_{EE}}\right)} \qquad (7-16)$$

where ΔV is the total output-voltage swing of the gate, and I_{EE} is the current of the of the ECL gate's current source (see Fig. 7-6c).

7.3.9.2 Propagation Delay in Digital MOS versus Digital Bipolar Circuits.

Figure 7-17 shows two stages of a ring oscillator built with arbitrary devices. The logic swing is given by $\Delta V = V_H - V_L$. The load capacitance (C_L) is the sum of the junction capacitances (C_j), the on-chip wiring capacitance (C_w), and the input capacitance (C_{in}) of the device in the second stage being driven. In the case of a bipolar ring oscillator, C_{in} is a nonlinear capacitance with charge $Q = I_C \tau_B$ (where τ_B is the transit time of the bipolar transistor). With an MOS ring oscillator, the capacitance is the (linear) gate capacitance, C_g. As a first order approach, neglecting device parasitic series resistances, the propagation-delay times through the stages of these two circuits are given by

$$\tau_{dbipolar} \sim \{(C_j + C_w)\Delta V / I_C\} + \tau_B \qquad (7-17)$$

and

$$\tau_{dMOS} \sim \{(C_j + C_w)\Delta V / I_{DS}\} + \{C_g \Delta V / I_{DS}\} \qquad (7-18)$$

Since the collector current of the bipolar transistor is exponentially dependent on V_{BE} (Fig. 7-17b), the current in the bipolar ring oscillator can be increased by several orders of magnitude by a small change in V_{BE} (e.g., ~60 mV at room temperature) without the need to increase the transistor size. According to Eq. 7-17, this means that the ideal propagation-delay time can be decreased to τ_B, by increasing the value of I_C and/or reducing ΔV.

It is much more difficult to increase I_{DS} in the case of MOS circuits, due to the quadratic current/voltage characteristic (see chap. 5, Eq. 5-10, in section 5.1.4). While I_{DS} could be increased by making all of the MOS devices in the circuit larger, this would also increase C_g. In practice, the performance of digital circuits with minimum-size transistors is found to be strongly affected by the on-chip wiring capacitances. These capacitances will in fact prevent the minimum ideal propagation delay of the circuit from ever being reached.[6]

In summary, it can be said that the two main advantages of digital bipolar circuits over their digital MOS counterparts are that (1) optimum speed can be obtained with minimum size transistors, and (2) it is feasible to use small gate-output-voltage-swings as a result of the exponential dependence of output current on input voltage. This implies that digital bipolar circuits will continue to have an edge in the IC market whenever ultimate speed performance is required.

7.3.9.3 General Switching-Speed Behavior of Digital Bipolar Circuits.
In general, the propagation-delay time $\tau_{dbipolar}$ decreases when the current

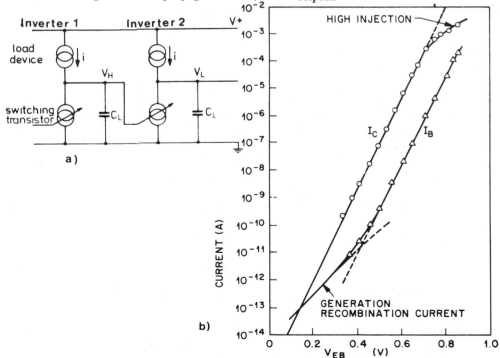

Fig. 7-17 (a) Two basic inverter stages of a ring oscillator built with arbitrary devices. C_L represents the total capacitance. (b) Collector and base currents as a function of V_{BE}.

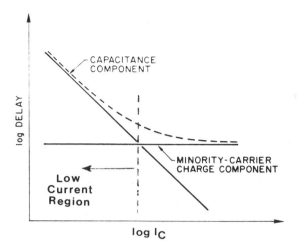

Fig. 7-18 Schematic of delay-versus-I_C characteristics of a typical bipolar logic circuit. It is assumed that the Kirk effect has been avoided by the proper tailoring of the collector doping profile.[26] (© 1986 IEEE).

per logic gate increases. At low current levels the storage of minority carriers can be ignored, and the switching speed is determined only by C_L (the sum of the junction and on-chip wiring capacitances). In such cases, $\tau_{dbipolar}$ is approximately equal to $C_L \Delta V/I_C$. If the series resistances are negligible, $\tau_{dbipolar}$ decreases linearly with increasing current, as shown in the low-current part of Fig. 7-18.

If the current is sufficiently increased, the storage of minority carriers begins to play a dominant role. Eventually, the minimum propagation-delay time, $\tau_{dbipolar}(min) = \tau_B$, is reached. The Kirk effect can cause $\tau_{dbipolar}$ to increase with increasing current, but the effect can be minimized by adjusting the collector doping profile (see section 7.2.3).

In practice, the parasitic series resistances of the device, R_B and R_C, can also significantly increase $\tau_{dbipolar}(min)$. At high currents, high values of R_B or R_C produce a large increase in $\tau_{dbipolar}(min)$ because as the minority-carrier storage increases, the available current to discharge this charge is limited by the series resistances. If the wiring capacitance makes C_L relatively large, $\tau_{dbipolar}(min)$ can increase even more as it begins to be limited by RC-time effects.

It can be concluded that the performance of digital bipolar circuits is optimized by doing the following:

- Minimizing C_L and ΔV.

- Minimizing the minority-carrier storage (by realizing a high f_T for the switching transistors and eliminating saturation effects – e.g., by using Schottky clamps or non-saturating logic circuits, such as ECL).

• Reducing the parasitic series-resistance values to acceptably low values.

The parasitic capacitances in junction-isolated (and even oxide-isolated) non–self-aligned bipolar transistors are so large that it is necessary to use rather large gate currents in order to reach propagation-delay times that are limited by the minority carrier storage component. For this reason, bipolar circuits are customarily associated with high power dissipation. The scaling of the device area with self-aligned structures, together with reductions in the parasitic-junction-capacitance values through the use of trench isolation, offers the promise of dramatically reduced power per gate in bipolar circuits. These two advances will be described in sections 7.8 and 7.9.

7.4 NON–OXIDE-ISOLATED BIPOLAR *npn* TRANSISTOR STRUCTURES

Three types of planar transistor structures were developed in early bipolar integrated circuits, all of which used a reverse-biased *pn* junction to provide isolation between collectors on the same chip. These device structures, named after the processes used to fabricate them are as follows: (1) *standard buried collector* (SBC) transistors (Fig. 7-19a), (2) *collector-diffused isolation* (CDI) transistors (Fig. 7-19b), and (3) *triple-*

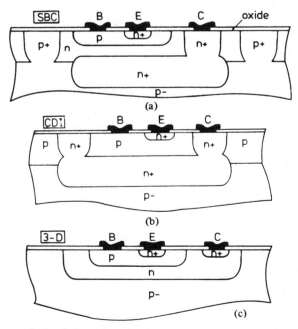

Fig. 7-19 Cross sections of three junction-isolated processes: (a) Standard buried-collector (SBC) process, (b) Collector-diffused isolation (CDI) process. (c) Triple-diffused (3D) process.[12] (© 1981 IEEE).

diffused (3D) transistors (Fig. 7-19c). The SBC transistor has been most widely used and a detailed description of the process flow used in its fabrication will be presented in section 7.5. The CDI process is described in chapter 2, section 2.1.1.2, and the 3D process is also discussed here.

7.4.1 Triple-Diffused (3D) Process

A cross-section of an *npn* bipolar transistor fabricated with the triple-diffused process is shown in Fig. 7-19c.[17] No buried layers or epitaxial layers are used. Instead, the collector is formed by means of ion implantation (and diffusion) directly into the substrate, as are the base and emitter regions. The chief advantages of this process are its simplicity (which allows for high yields and low manufacturing costs), and the fact that smaller-area transistors can be fabricated than is possible with the SBC process, when the same design rules are used (since no *p*-type isolation diffusions are needed to surround each collector). The benefit of process simplicity also makes the 3D transistor the bipolar structure of choice in low-cost digital BiCMOS technology (see section 7.12.1.1). An oxide-isolated, triple-diffused process has also been developed.[18] The incorporation of high-pressure oxidation to improve the properties of devices fabricated using this process has also been described.[19]

Higher values of BV_{CEO} can also be obtained with smaller 3D devices than with either CDI or thin-epi-layer SBC transistors (the latter of which are used for 5-V digital ICs). Thus, 3D transistors have been used in such applications as analog-to-digital converters with greater than 8-bit resolutions (such circuits require the use of operating voltages greater than 5 V). The main disadvantage of 3D transistors is their high value of R_C, which results from the absence a buried-layer subcollector.

7.5 STANDARD-BURIED-COLLECTOR (SBC) PROCESS

7.5.1 Characteristics of *npn* Transistors Fabricated with the SBC Process

A transistor fabricated with the SBC process has a collector region that consists of a heavily doped n^+ buried layer and a lightly *n*-doped epitaxial layer (Fig. 7-19a). The collectors are isolated from one another by *p* diffusions that extend all the way through the epi layer to contact the *p*-substrate. The base and emitter regions are formed by diffusions after the isolation structures have been created.

Until the mid-1970s the planar, standard-buried-collector process with junction isolation was the workhorse for the fabrication of both analog and digital bipolar ICs, since devices fabricated with this process had the best mix of characteristics for the largest number of applications.

Because it is easier to control the base width in the SBC than in the CDI process, higher current gains can be obtained (the graded doping profile in the base also helps to

Table 7-2   Typical Design Parameters for Junction-Isolated IC BJTs.

	Amplifying (Junction-Isolated)	Switching (Junction-Isolated)	Switching (Oxide-Isolated)
Epitaxial Film			
Thickness	10 μm	3.0 μm	1.2 μm
Resistivity	1 Ω-cm	0.3 – 0.8 Ω-cm	0.3 – 0.8 Ω-cm
Buried Layer			
Sheet resistance	$\sim$ 20 $\Omega/\square$		$\sim$ 30 $\Omega/\square$
Up diffusion	2.5 μm	1.4 μm	0.3 μm
Emitter			
Diffusion depth in base	2.5 μm	0.8 μm	0.25 μm
Sheet resistance	5 $\Omega/\square$	12 $\Omega/\square$	30 $\Omega/\square$
Base			
Diffusion depth	3.25 μm	1.3 μm	0.5 μm
Sheet resistance	100 $\Omega/\square$	200 $\Omega/\square$	600 $\Omega/\square$
Substrate			
Resistivity	$\sim$ 10 Ω-cm		$\sim$ 5 Ω-cm
Orientation	(111)		(111)

produce such high gains). In addition, the use of an *n*-type epitaxial layer together with buried layers allows BV_{CBO} to be made higher than is possible in the CDI process, since the collector doping at the edge of the junction can be made much lower. Furthermore, R_C is much lower than it is in transistors fabricated with the 3D process (which does not use buried layers).

As will be explained more fully later, junction-isolated SBC transistors are unsuitable for digital VLSI circuits and have been replaced for this application by oxide-isolated SBC devices. Nevertheless, the junction-isolated SBC process continues to be the principal bipolar process for the fabrication of analog ICs operated with power-supply voltages in excess of 10 V.

As shown in Table 7-2, SBC transistors used in analog ICs (primarily for amplification functions) are fabricated with different device structures than those used in digital ICs (in which the devices are used for switching). The differences result mainly from the higher operating-voltage levels of analog ICs (i.e., in digital ICs the typical power-supply voltage is only 5 V). As a result, analog-IC transistors must have thicker epitaxial layers and lighter collector-doping levels (at the edge of the base-collector junction).

Since digital ICs must often operate at higher current levels, the collector doping level must be increased in order avoid device performance degradation caused by the Kirk effect. For example, epitaxial layers in analog junction-isolated bipolar-SBC ICs have thicknesses of between 7 and 15 μm and resistivities of 1 Ω-cm, while their digital counterparts have epi thicknesses of between 2.0 and 5.0 μm and resistivities of 0.3-0.8 μm. The doping profile of a digital SBC transistor with an arsenic-doped emitter along a coordinate perpendicular to the surface and passing through the emitter, base, and collector is shown in Fig. 7-20. In this example, a single implant is used to form both the extrinsic and intrinsic base regions. Because the diffusion of the base region is significantly altered by the diffusion of the arsenic as it forms the heavily doped emitter, the intrinsic-base profile does not decrease smoothly into the silicon.

As an example of how the thickness and doping concentration in the epi layer impacts the breakdown voltage of the device, circuits operated at power-supply voltages of 36 V must use devices in which BV_{CEO} exceeds this value. Using Eq. 7-13 and assuming that n = 4, β = 100, and the base-junction depth is 3 μm, for a BV_{CEO} of 36 V, we calculate that BV_{CBO}(plane) must correspondingly be greater than 120 V. However, BV_{CBO} is usually determined by avalanche in the curved region of the collector, which means that a lower value is more correct for this equation. From Fig. 5-10b, for the case of a base junction depth of 3 μm (which also gives rise to a junction whose periphery has a radius of curvature of 3 μm), a BV_{CBO} (plane) of 120 V is reduced to ~90 V. An impurity concentration in the collector of approximately 3×10^{15} atoms/cm^3 (equivalent to a resistivity of ~1 Ω-cm) is needed to obtain a BV_{CBO} of 90 V.

The thickness of the depletion region must also be large enough to accommodate the depletion layer associated with the collector-base junction. At 36 V, it is calculated (using Eq. 5-17, chap. 5) that the depletion thickness is approximately 6 μm. Since Table 7-2 indicates that the buried layer diffuses upward approximately 2.5 μm during subsequent processing, and the base diffusion is approximately 3 μm deep, a total epitaxial-layer thickness of 12 μm is required for a 36-V circuit. For circuits with lower operating voltages, thinner and more heavily doped epitaxial layers can be used, because this reduces R_C.

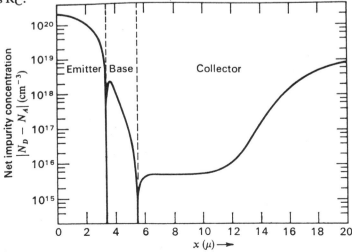

Fig. 7-20 The impurity distribution in an SBC bipolar transistor.

If thinner layers are used, BV_{CEO} will be limited by reachthrough of the base region instead of by avalanche multiplication. That is, the depletion layer in the lightly doped collector will hit the n^+ buried layer at voltages less than that of BV_{CEO}. At this point, further depletion-layer expansion will occur in the base, and any increase in the collector-base voltage will quickly cause breakdown to occur (see the second set of curves in Fig. 7-10b).[24] For example, if the epi layer is 8.5 μm thick in the device just described, the value of $x_{BL} - x_{BC}$ will be 3 μm, and BV_{CBO} thus will be limited to a value of 30 V.

7.5.1.1 Limitations of Junction-Isolated SBC Transistors for VLSI Circuits.
These transistors have several limitations. First, their packing density is relatively low since so much silicon area is taken up by inactive isolation regions. (Although the window defining the the isolation diffusion may be a minimum size at the surface, the total width of the isolation region there is determined by lateral diffusion. For example, if the epitaxial layer is 12 μm thick and the minimum feature size is 3 μm, the isolation region will approach 24 μm in width (assuming that lateral diffusion is approximately 80% of the vertical diffusion). Thus, the width of the isolation region ends up about twice the thickness of the epitaxial layer in order for the acceptor dopants to penetrate through this layer. As a result, transistors fabricated with the SBC process are not small enough to meet the packing density requirements of VLSI circuits.

Second, these large isolation-region areas give rise to large parasitic capacitances, which degrade the speed performance.

Third, lateral *pnp* transistor action with minority-carrier collection in the isolation-diffusion region can occur, possibly leading to latchup. This parasitic *pnp* transistor is normally cut off, because the collector-substrate junction and the collector-base junctions are usually reverse-biased (i.e., the collector-substrate junction is kept at reverse bias by having the *p*-doped substrate connected to the most negative dc voltage in the electronic circuit). If the *npn* transistor saturates, however, the collector-base junction becomes forward-biased, and the parasitic lateral *pnp* enters its active region. The resultant conducting transistor diverts current from the collector circuit to the substrate; under some circumstances latchup can even occur. All three of these limitations can be reduced by using oxide isolation instead of junction isolation, as is being done in high-performance, 5-V digital bipolar ICs.

Fourth, since the base region is formed by a single diffusion, the parasitic series base resistance will often be high enough to limit high-performance operation. A separate extrinsic base-formation process is used in advanced bipolar structures to overcome this problem.

7.5.2 Standard-Buried-Collector (SBC) Process Flow

7.5.2.1 Starting Material.
The starting material is a *p*-type, lightly doped wafer ($\sim 10^{15}$ cm^{-3}). The substrate doping is light enough to minimize the parasitic collector-to-substrate depletion-layer capacitance, but heavy enough to prevent it from being

changed to n-type during subsequent processing. In the early days of IC processing (111) wafers were used, but (100) is now the standard.

7.5.2.2 Buried-Layer Formation.

Heavily doped n^+ regions called *buried layers* or *subcollectors* are the first features to be formed. These regions are used to reduce R_C by providing a low-resistance path from the collector contact to the active portion of the transistor.

The process is begun by growing an oxide (0.5-1.0 μm thick) on the wafer. *Mask #1* and an etch step are used to open windows in the oxide wherever the buried layer is to be formed (Fig. 7-21a). Since this is the first pattern on the wafer, no alignment is necessary. Instead of phosphorus, either arsenic or antimony is selectively introduced into the silicon (these impurities are used because each has high solid solubility and a small diffusion coefficient). The predeposition step can be performed using either chemical diffusion (in which case the surface oxide acts as a mask to prevent the n-dopant from entering the substrate, except where windows exist), or ion implantation ($\sim$30 keV, $\sim 10^{15}$ atoms /cm^2; in this case the resist may remain on the wafer as a mask layer during the implant). Even if implantation is used, a thermal step is also employed to activate the implanted impurities and anneal out the implant damage.

In either case, the oxide that was initially grown on the wafer is stripped following the introduction of the dopants into the substrate and prior to the drive-in step. The anneal/drive-in step is then conducted in an oxidizing ambient, which simultaneously

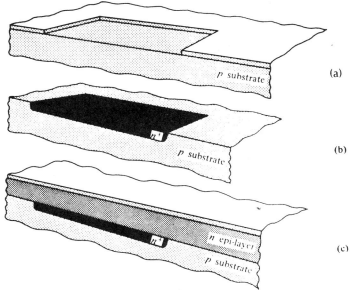

Fig. 7-21 (a) Oxide windows are opened for n^+ buried layer deposition. (b) Oxide layer is removed in preparation for epitaxial growth. (c) n-type epitaxial layer and oxide layer are grown. Copyright, 1982, Prentice-Hall. Reprinted with permission from reference [127].

causes a new oxide to be grown on the wafer surface. The oxide grows faster over the heavily doped n-regions than over the more lightly doped p-regions. Since the oxide consumes silicon to a depth of $0.44t_{ox}$ (where t_{ox} is the grown oxide thickness), when this new oxide is etched away, 100-200-nm-high steps exist on the bare silicon wafer at the edges of the n^+ regions (Fig. 7-21b). These silicon steps propagate through the epitaxial layer and thus become the alignment marks that allow subsequent mask levels to be aligned with the buried layer.

Although antimony has lower solid-solubility and requires higher anneal/drive temperatures (1250°C as compared with ~1000°C for arsenic), it has historically been more popular as a buried-layer dopant because it exhibits less autodoping during epitaxy and subsequent thermal cycles. On the other hand, since As atoms are closer to the size of Si atoms, As provides a better fit to the silicon lattice. This results in less strain and, hence, in fewer defects in the subsequently grown epi film. In addition, it has has been found that arsenic autodoping can be decreased if the epi layer is deposited at reduced pressures.[25] The recent commercial introduction of reduced-pressure epitaxial reactors has renewed interest in the use of As as an emitter dopant.

The sheet resistance of the buried layer should be as low as possible in order to reduce R_C. After the dopant is introduced into the substrate, it is driven in; this spreads out the doping profile so lowers the impurity concentration. One benefit of a lower doping concentration is higher carrier mobility, which results in lower sheet-resistance values. Furthermore, if the surface-doping concentration is too high, excessive outdiffusion into the more lightly doped collector will occur during epi growth. This can also cause defects in the epitaxial layer. Following the drive-in step, the wafer oxide is stripped.

To ensure a clean, defect-free surface, some silicon is removed from the surface during the HCl etch step in the subsequent epitaxial growth step. The loss of this highly doped layer increases the buried-layer sheet resistance and so must be taken into account. (Typical buried-layer sheet-resistance values are between 15 and 50 Ω/sq.)

7.5.2.3 Epitaxial Growth. Following the stripping of the oxide, a lightly doped n-type epitaxial layer (10^{15}-10^{16} atoms/cm^{-3}) is deposited. This layer serves to "bury" the heavily doped subcollector regions Fig. 7-21c). Because of its small diffusivity, arsenic is most often used as the epi-layer dopant. The minimum thickness and maximum doping concentration of this layer are determined by the avalanche breakdown and reachthrough limitations on the value of BV_{CEO} (as was discussed in section 7.2.4).

The lighter the epitaxy doping, the smaller the collector-base capacitance of the device (which is an important parameter limiting the high-frequency performance). The minimum doping concentration, however, is generally limited by one of two factors:

1. If the transistor is to be operated at high collector currents, the conductivity of lightly doped n-epitaxy layer becomes modulated by the collector current. This causes the base-collector junction to be pushed out into the lightly doped epitaxial layer (Kirk effect; see section 7.2.3), resulting in gain degradation as well as in a reduction in the high-frequency performance. To avoid device performance degradation due to the Kirk effect, the doping in the collector should be greater

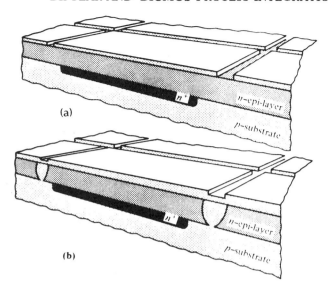

Fig. 7-22 (a) Oxide windows are opened for p-type isolation diffusion. (b) Oxide layer is regrown over p-type isolation regions. Copyright, 1982, Prentice-Hall. Reprinted with permission from reference [127].

than J/qv_1, where J is the maximum expected collector-current density in the device and v_1 is the saturation velocity ($\sim 10^7$ cm/s).

2. Even if the Kirk effect is not important, too light a doping concentration can be difficult to control in the epitaxy growth process, due to autodoping by the buried layer.

The epi layer in early SBC transistors was deposited by means of the hydrogen reduction of $SiCl_4$. However, because this process must be carried out at $\sim 1200°C$ in order for sufficiently rapid growth rates to be achieved, it has been replaced in most production processes by the thermal decomposition of dichlorosilane, SiH_2Cl_2 (which yields faster growth rates at $\sim 1050°C$; see Vol. 1, chap. 5). This process has the added benefit of having chlorine present during the epitaxial film growth.

The thermal decomposition of silane (900-1000°C) is often used in high-speed bipolar devices. In such devices, the epi layer thickness is very small (1.0-1.5 μm) and the processing temperature is kept low so that the epi/substrate junction will be as sharp as possible. However, there are disadvantages to using silane. First, the presence of impurities in the gas phase (even at relatively small concentrations) can cause epi defects. Second, the low stability of silane makes it susceptible to gas-phase nucleation of particles (which can then "rain down" and become incorporated in the growing film).

7.5.2.4 Formation of Isolation Regions. Following epitaxial growth, a new masking oxide for subsequent diffusions is grown on the top surface of the

epitaxial layer. *Mask #2* is used to form the patterns of the isolation regions (Fig. 7-22a). After windows have been etched in the oxide, a boron predeposition and drive-in step is performed. (The boron concentration at the surface is the solid solubility of boron in silicon at the predeposition temperature; e.g., 4×10^{20} cm^{-3} at 1200°C.) The purpose of this step is to isolate the collectors of the transistors from one other with reverse-biased *pn* junctions. Thus, the acceptor concentration in the newly diffused regions must be higher than the donor concentration in the epi layer, and the junction depth must at least equal the epi thickness in order for complete isolation to be achieved. In fact, it is customary to overdrive the isolation diffusion to prevent the depletion region from extending underneath the isolation diffusion.

The result is the formation of *n*-type islands that are completely surrounded by *p*-type material. Because thick epi layers (e.g., >10 μm) are used in analog SBC transistors, the diffusion requires several hours at temperatures of 1200°C. The isolation-diffused layer has a sheet resistance of from 20 to 40 Ω/sq. (Additional details regarding junction- isolation-structure fabrication are described in chap. 2, section 2.1.1.1).

Following diffusion, a low-temperature-oxidation and etch step is used to remove the boron-rich layer from the surface of the isolation regions, and another oxidation step is used to grow a new oxide layer over the isolation-diffusion openings (Fig. 7-22b).

As noted in chapter 2, the n^+ buried layer regions are usually not allowed to intersect the *p*-isolation diffusions, as this would decrease the breakdown voltage of the collector-substrate junction and increase its junction capacitance. Thus, a layout design rule specifies the minimum spacing between the n^+ region and the isolation diffusion. (In fact, the minimum buried-layer-to-buried-layer separation is determined primarily by the buried-layer-to-isolation spacing.)

In addition, successful fabrication of the isolation structures can be confirmed by stripping the oxide over the isolation diffusion and probing two adjacent structures. If complete isolation has been achieved, no current will flow unless the applied voltage exceeds the breakdown voltage of the isolation junction. If even a small gap exists at the bottom of the isolation, leakage current will occur. Poor isolation results in catastrophic device failure, since all of the transistor collector regions are shorted together.

As a final note, the windows opened by the isolation mask are generally reopened during the base-diffusion step in order to increase the *p*-impurity concentration at the surface. This is done because the surface can become depleted of boron as a result of impurity redistribution and dopant segregation during the intervening oxidations.

7.5.2.5 Deep-Collector Contact Formation (Optional).

Even with the use of buried layers, in some applications the value of R_C is still too high. To reduce R_C even further, a deep n^+ contact diffusion (also referred to as a *plug,* or *sinker*) is used. The diffusion once again penetrates the epitaxial layer to contact the underlying buried layers (Fig. 7-23). Such deep-collector-contact diffusions are particularly useful for reducing the *ON*-state output voltage of saturating-type digital logic circuits.

In junction-isolated SBC devices, this contact is formed by means of a diffusion process, whereas in oxide-isolated bipolar devices – which use thinner epitaxial layers –

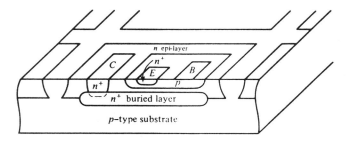

Fig. 7-23 Deep n^+ collector diffusion reduces R_C. Copyright, 1982, Prentice-Hall. Reprinted with permission from reference [127].

ion implantation (followed by diffusion) is used. Phosphorus is chosen for this application since it is a faster-diffusing impurity than arsenic. In any case, an additional mask must be employed to allow the dopant to be selectively introduced only into the collector regions.

Besides the additional masking step, a second penalty for using this approach is that a relatively large collector-to-base spacing must be maintained to avoid degradation of BV_{CBO}, since significant lateral diffusion of the deep n^+ will occur. In addition, in analog bipolar SBC transistors this process is usually performed before the base-formation step, since the diffusion must penetrate a thick epi layer. In digital bipolar SBC devices, however, the step may be carried out after the base-predeposition step. The deep-collector-contact and base-drive-in steps may then be carried out simultaneously.

7.5.2.6 Base-Region Formation.

Formation of the base region, a critical process in bipolar transistor fabrication, begins with the use of the base mask (*Mask #3*) to pattern windows in the oxide where such regions are to be created (Fig. 7-24a). The base must be aligned so that the collector-base and collector-substrate depletion regions do not merge, following diffusion at the surface. The minimum allowable spacing between the isolation regions and the base region can be determined based on a knowledge of the applied voltages and the epitaxial-layer doping concentration. Additional space must be allowed to accommodate the alignment tolerances. In some cases, the base diffusion is also used to create diffused resistors for the circuit; the values of these resistors depend not only on the diffusion conditions, but also on the dimension of the opening made during etching.

The β of the transistor is heavily dependent on the base width and the doping profile. Thus, the base-junction depth and base-doping profile must be not only appropriately selected but also tightly controlled during production. As described earlier, the base region is made thinner and less heavily doped than the emitter. Since β (which is typically set to around 100), is inversely proportional to GN_B, the specified value of β sets a constraint on the *maximum* base doping allowed.

The *minimum* doping level in the base is set by a separate set of constraints. If the level is too low, the transistor will not be able to withstand the reverse-bias voltage

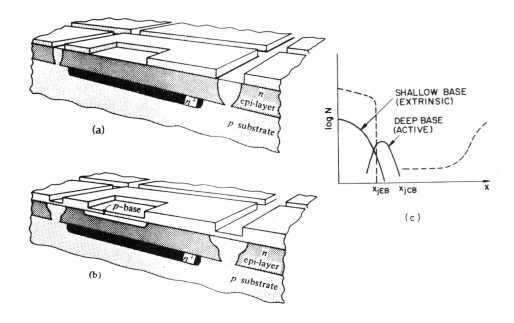

Fig. 7-24 (a) Oxide windows are opened for *p*-type base formation. (b) Base diffusion is completed and oxide is regrown. Copyright, 1982, Prentice-Hall. Reprinted with permission from ref. [127]. (c) Doping profile showing results of double-base implant.

applied across the C-B junction, and punchthrough will ensue. Also, as the base charge is reduced the transistor exhibits smaller values of V_A, which may be undesirable in some circuit applications. Furthermore, the parasitic series resistance in the base region becomes a problem at low base-doping concentrations. Finally, if the concentration at the surface is less than ~5×10^{16} cm^{-3}, poor metal-semiconductor contacts to the base region can occur. Typical values used for GN_B are between 10^{12} and 10^{13} atoms/cm^2.

The choice of base width, W_B, is made based on similar considerations. That is, small W_B values also produce shorter minority-carrier diffusion times across the base, improving performance. Given the constraints on GN_B, the value of W_B is minimized to improve the base-transit time. To allow for such narrower base widths, however, the doping level must be increased to prevent punchthrough. In high-performance bipolar transistors, the base-doping profile is graded so that a higher doping concentration will exist at the emitter-base junction, resulting in a built-in electric field that aids the transit of minority carriers.

The base may be formed through diffusion (i.e., a boron predeposition and drive-in step) or ion implantation (~10^{12} boron atoms/cm^2) and diffusion. At the end of the process, the depth of the base is ~3 μm in analog SBC transistors and ~1.0 μm in digital SBC devices. The corresponding sheet-resistance values are 100 and 200 Ω/sq.

The drive-in step in either case is done in an oxidizing ambient so that an oxide is grown over the base region for subsequent masking (Fig. 7-24b).

One reason for performing the drive-in step is to lower the impurity concentration at the wafer surface. Avalanche breakdown of the emitter-base junction will tend to occur at the device surface, because the impurity concentration is highest there. Since such breakdown is dependent on the doping concentration of the more lightly doped side of a *pn* junction, the drive-in step reduces the value of the surface-doping-concentration that exists after the predeposition step. Higher emitter-base breakdown voltages can thus be achieved.

In most advanced SBC processes, a separate procedure (involving an additional masking step and a selective-doping step) is used to create an *extrinsic* base region with higher doping than that in the intrinsic base region. This provides a lower value of R_{B1} (with typical sheet resistances reduced to 100-200 Ω/sq). However, if the extrinsic base is too heavily doped, a low value of BV_{EBO} will result, and an excessively high C_{EB} value will be produced.

A double base-implant procedure is typically used to create the intrinsic and extrinsic base regions separately.[20] A higher-energy, lower-dose (deep-base) implant is used to establish the intrinsic base properties (Fig. 7-24c),[21] while a lower-energy, higher-dose (shallow-base) implant is used to produce the extrinsic-base characteristics. An additional mask is used to selectively implant the additional dopants in the extrinsic base regions.

7.5.2.7 Emitter-Region Formation.

The formation of the emitter region is the last step in the fabrication of the transistor. The emitter is heavily doped to obtain high current gain and to minimize R_E (e.g., for an emitter with a Gummel number of ~10^{16} atoms/cm^2 and an E-B junction depth of 0.5 μm, the concentration is ~2×10^{20}/cm^3). In addition, as the width of the base is decreased to improve performance, the junction depth of the emitter must be decreased to maintain base-width control and reproducibility. For example, in analog-SBC transistors in which the base width is on the order of 0.5 μm, this depth is around 2.5 μm, while in junction-isolated digital SBC transistors (which have a base width of ~0.3 μm) it is less than 1 μm. (In advanced digital oxide-isolated, self-aligned SBC transistors, in which the base width is less than 0.1 μm, the emitter depths are smaller than 0.1 μm.) Control of the E-B junction depth is obviously a key step in determining bipolar-transistor behavior. Minor process variations will produce significant changes in β.

The emitter mask (*Mask #4)* is used to open the emitter-region windows in the wafer-surface oxide. The opening must lie wholly within the base region (7-25a). A shallow, high-concentration diffusion (or shallow, high-dose implant) of phosphorus or arsenic is used form the emitter structure. The drive-in that follows the predeposition (or implant) step is done in an oxidizing ambient so that oxide will cover the entire wafer. Since the thermal-oxidation rates are enhanced over heavily doped silicon regions, this oxidation cycle can be relatively short. (Note that the collector contact regions are also opened and simultaneously doped during the emitter formation step, allowing low-resistance, ohmic contacts to be made to the collector.)

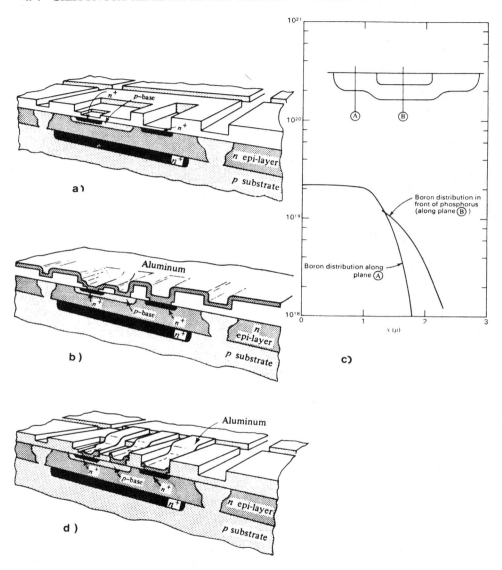

Fig. 7-25 (a) Oxide windows are opened for n^+ emitter formation. (b) n^+ emitter diffusion is completed and oxide layer is regrown. (c) The *emitter push effect*; anomalous diffusion due to a second high concentration phosphorus diffusion.[128] (d) Completed *npn* transistor; aluminum is removed as needed to complete the circuit. Copyright, 1982, Prentice-Hall. Reprinted with permission from ref. [127].

Phosphorus was used as the dopant in early bipolar transistor emitters, since the diffusion coefficients of boron and phosphorus have about the same value. Therefore, as the phosphorus is diffused, the base-collector junction should move downward at about the same rate. However, control of β is complicated by the *emitter push effect* which occurs when phosphorus is diffused at high concentrations. Because this effect enhances the diffusion of boron in the intrinsic region beneath the emitter, the C-B junction is deeper in the intrinsic base region than in the extrinsic base regions (Fig. 7-25b; see also Vol. 1, chap. 8). As a result, the base width becomes wider than it would be if the emitter push effect played no role.

Since they allow shallow E-B junctions to be fabricated more easily, arsenic emitters are therefore now more commonly used. In addition, the concentration-dependent behavior of arsenic diffusion in silicon allows for more-abrupt doping profiles than can be obtained with phosphorus. If ion implantation is used, a low-energy, high arsenic dose ($\sim 10^{16}$ cm^{-2}) is utilized.

The active area of the transistor (i.e., the region directly beneath the emitter) is a very small fraction of the total device area in SBC transistors (roughly 5%, as can be estimated from Fig. 7-26a). The rest of the area is needed to make contacts, support depletion layers, and provide isolation. In fact, the isolation region takes up the majority of the device area. Thus, its minimization represents an important issue in developing high-performance devices, not only for density improvement but also for junction-capacitance reduction as well. This limitation of junction-isolated transistors has led to the development of oxide-isolation and trench-isolation structures, as will be described in later sections.

7.5.2.8 Contact and Interconnect-Layer Formation.
Part of the emitter drive-in step is done in an oxidizing ambient, so that an oxide will be grown over the exposed emitter and collector contact regions. *Mask #5* (the *contact* mask) is then used to pattern the areas of the wafer surface where holes will be cut in the oxide to allow for electrical connection of the device. (Contact openings are also made to the base, the collector, and the isolation diffusion with the contact mask.)

Since the emitter contact window must lie wholly within the emitter region (Fig. 7-25c), the emitter area must extend at least one alignment tolerance beyond the contact edge in all directions. The size of the contact thus controls the minimum emitter size. For this reason, contacts are usually fabricated at the smallest dimension that the masking process can reproducibly create.

The actual contact structure in SBC transistors has most commonly been an Al-Ti:W-PtSi-Si contact. (The details of the fabrication such contact structures are presented in chap. 3, section 3.6.1.)

A bilayer film about 1 μm thick, consisting of Ti:W and Al (or Al alloy), has been the most commonly used interconnect layer. This film is generally sputter deposited, then patterned using *Mask #6*. The metal mask is designed with sufficient overlap that the metal covers the entire contact window (7-25d). Following etching of the interconnect film, a passivation layer is deposited (see chap. 4, section 4.8), and an anneal step is performed (see chap. 3, section 3.4.2.7). Opening of the pads is accomplished using *Mask #7* and an etch step.

The basic SBC process is thus seen to be more lengthy and complex than the basic five-mask NMOS process described in chapter 5 (which also implies higher production costs per wafer). Furthermore, the use of a deep collector contact and a separate process for forming a more heavily doped extrinsic region would involve the use of two additional masks; if a double-level-metal process were implemented, yet another two masks would be required.

7.5.2.9 Washed Emitters. The *washed-emitter contact-formation process* allows the device area to be scaled down without a scaling of the design rules. While conventional contact masking and etching steps are used to open the base and collector contacts in this approach, the emitter areas remain covered by resist during the contact-opening step.

The resist is then stripped and the wafer is dipped in a dilute HF solution for a short time in a process known as *washing*. This washing step removes the oxide that has formed over the emitter during the emitter drive-in step, without significantly thinning the other oxides on the wafer surface. (Since the oxide grown over the heavily doped emitter regions contains a high concentration of phosphorus, it etches much more rapidly than normal in an HF-containing etchant.)

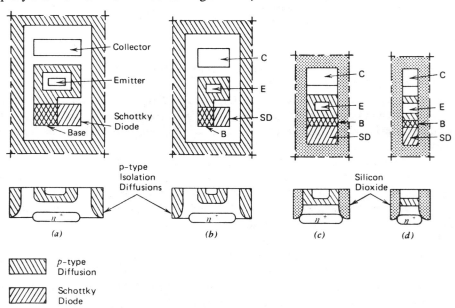

Fig. 7-26 Composite layout and cross-sections of Schottky-clamped bipolar transistors. (a) Conventional junction-isolated transistor. (b) Washed-emitter junction-isolated transistor. (c) Oxide-isolated washed-emitter transistor. (d) Walled-emitter oxide-isolated transistor.[129]

The entire emitter-diffusion area is thus available as a contact window for the subsequently deposited metal (compare Figs. 7-26a and 26b). Since no extra area is needed for possible misalignment of the contact hole to the emitter-diffusion pattern, the emitter dimensions can be made as small as the minimum lithographic line width.

The risk of using washed emitters is that the metallization may short out the emitter-base junction at the periphery of the window. Since the emitter dopant diffuses laterally under the thicker oxide that exists over the base during emitter formation, the distance of this diffusion must be great enough that the E-B junction at the surface is not exposed by the washing step.

7.5.2.10 Schottky Contacts. A metallization technique involving the use of Schottky-barrier-diode clamps to the collector region can be used to keep bipolar transistors out of saturation. If the output voltage of a bipolar transistor without a Schottky clamp becomes small enough, the C-B junction becomes strongly forward-biased (i.e., the transistor enters the deep-saturation region of operation). As a result, a large number of minority carriers are injected into the base from the collector. Removal of these carriers when the device must be brought out of deep saturation is time consuming, and thus slows down the switching speed of the transistor.

To reduce the degree of collector-base-junction forward-biasing, a Schottky-barrier diode to the collector can be formed (as shown in Fig. 7-27a). Schottky diodes turn on at a smaller forward-bias voltage (e.g., ~0.3 V for a PtSi Schottky diode to n-type Si) than do pn junction diodes (~0.7 V). Since the Schottky and C-B pn diodes are in

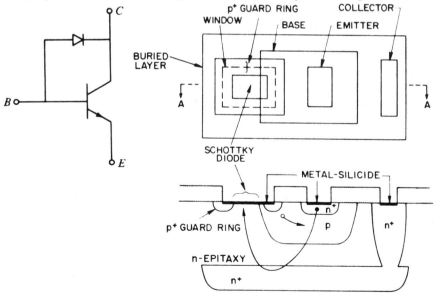

Fig. 7-27 Schematic top and side view of Schottky-clamped transistor (with p^+ guard ring) to prevent saturation. Copyright, 1983, Bell Telephone Laboratories, Inc. Reprinted with permission.

parallel, the voltage across the C-B junction is held (*clamped*) at the turn-on voltage of the Schottky diode, and hence cannot become forward-biased enough to conduct significant current.

Figure 7-27b shows the layout of a junction-isolated SBC bipolar transistor that includes a Schottky diode. The PtSi layer simultaneously contacts the *n*-epitaxy and the *p*-base regions. Hence it forms an ohmic contact to the base and a Schottky contact to the collector. A p^+ guard ring can also be used around the periphery of the Schottky diode to raise its reverse-breakdown voltage, as described in chapter 3. $MoSi_2$ Schottky diodes for bipolar LSIs have also been described.[46,48]

7.6   OXIDE-ISOLATED BIPOLAR TRANSISTORS

As discussed in section 7.5.1.1, the large size of junction-isolated SBC transistors makes them unable to meet the functional density requirements of VLSI applications, and also causes them to possess high parasitic-capacitance values. While digital transistors exhibit these limitations to a lesser degree than do their analog counterparts, the drawbacks of junction isolation still prevent digital SBC transistors from being suitable for VLSI ICs. By replacing junction isolation with oxide isolation, however, it becomes possible to shrink the the size and propagation-delay times of bipolar transistors so that they can be used to implement VLSI ICs.

The evolution of oxide-isolated bipolar transistors has occurred in several stages, each of which resulted in smaller device sizes and lower parasitic capacitances. In addition, the use of oxide isolation has reduced the current gain of the parasitic *pnp* transistor, decreasing the likelihood that the parasitic *pnpn* structure will latch up (a potential problem inherent in bipolar ICs).

Details of the fabrication of LOCOS-based, fully recessed field-oxide-isolation structures are presented in chapter 2, sections 2.3 and 2.5. A variety of similar processes for bipolar ICs, each with its own acronym, have been reported (e.g., Isoplanar, OXIS, Planox, OXIM, and ISAC). In early oxide-isolated bipolar structures, the field oxides exhibited relatively large bird's beak and bird's head features. More advanced processes have substantially reduced these undesired effects (see chap. 2). Figure 7-28 illustrates a process sequence for fabrication of high-performance, oxide-isolated bipolar transistors, and Fig. 7-29 shows a three-dimensional view of a completed device. With the exception of the oxide-isolation structure fabrication, most of the details of this process are identical to those of the junction-isolated SBC process described in section 7.4.2.

To ensure complete isolation between adjacent collectors, it is necessary that the field oxide be thick enough to penetrate the entire depth of the epitaxial layer. The oxide must also make contact to the n^+ buried layer. Because the collector is surrounded by an oxide collar, the sidewall contribution to the collector-to-substrate capacitance is essentially eliminated. (Note that while the growth of field oxides greater than ~1.5 μm is impractical, if thicker epitaxial layers need to be employed, a combination of oxide and junction isolation may be a viable option. One name given to such a combination process is *recessed oxide isolation* [ROI].) The isolation oxide is also formed over a

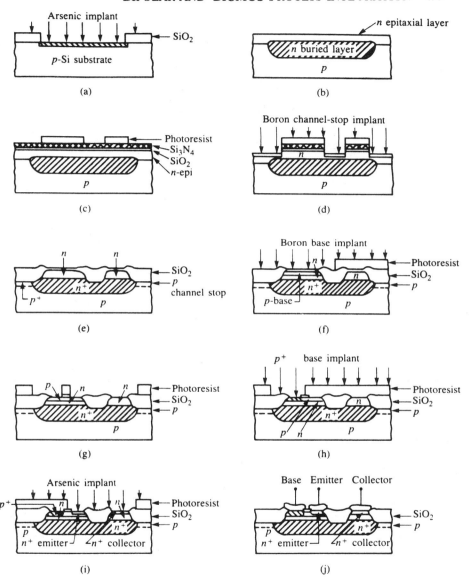

Fig. 7-28 Process sequence for a high-performance oxide-isolated bipolar transistor. (a) Buried-layer formation; (b) Epitaxial layer growth; (c) Mask for selective oxidation; (d) Boron implant prior to recessed oxide growth; (e) Selective oxidation; (f) Base mask and boron implantation; (g) Emitter, base contact, and collector contact mask; (h) p^+ base contact implantation; (i) Arsenic implantation for emitter and collector contact; (j) Structure completed with multilayer metallization. Copyright, 1985, John Wiley and Sons, Inc. Reprinted with permission from ref. [130].

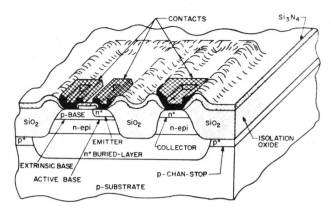

Fig. 7-29 Three-dimensional view of oxide-isolated bipolar transistor. Copyright, 1980, John Wiley and Sons, Inc. Reprinted with permission from ref. [131].

fraction of the active device area (i.e., between the collector and base contacts, to eliminate unnecessary area and capacitance between them; see Fig. 7-29).

A boron channel-stop implant is used beneath the field oxide to prevent the lightly doped substrate *p*-regions beneath the field oxide from inverting, as this would destroy the isolation between adjacent collectors.

In early oxide-isolated bipolar device structures, oxide surrounded the periphery of the collector as noted above. In addition, the base region butted up against the oxide on all sides, and C_{CB} was thus reduced. The emitter, however, was still wholly enclosed within the base region (a so-called *nonwalled emitter*). Nevertheless, since the collector-metal and the base-metal stripes could overlap the sidewall oxide, and the collector n^+ and the *p* base contacts could extend over the isolation region, it was possible to reduce the total device area to somewhat less than 50% of that needed by a junction isolated device (see Fig. 7-26c).

In the second generation (Fig. 7-26d), the size was further reduced by allowing two sides of the emitter (and in some device structures, even three sides) to abut the oxide-isolation region. Devices fabricated with this approach are referred to as *walled-emitter transistors*. The value of also C_{EB} in walled-emitter transistors is also reduced. Nonwalled emitter transistors will have smaller R_{B2} values than walled-emitter transistors using the same minimum dimension, but they will also be larger.

7.7 ADVANCED BIPOLAR-TRANSISTOR STRUCTURES FOR VLSI AND ULSI

Although CMOS has been the major driving force in the rapid progress in silicon technology in recent years, it is less well known that there has also been rapid progress in bipolar technology.[26] The advances here have been due to a number of process breakthroughs, including polysilicon emitters, self-aligned bipolar structures, deep-trench isolation, and heterojunction-bipolar transistors.

7.8 ADVANCED EMITTER STRUCTURES

As pointed out in section 7.3.1, the maximum theoretical current gain in conventionally designed bipolar transistors is $\sim 10^4$. However, such high gains are not obtained in actual devices, since the heavy doping in the emitter reduces the band gap and increases the recombination rate of minority carriers. Both effects lower the emitter-injection efficiency, which also implies that attempts to increase the current gain by raising the emitter-doping levels to even higher levels would be to no avail. In addition, higher doping concentrations would lead to smaller values of BV_{EBO} and larger E-B junction capacitances. Finally, when the emitter junction depths are scaled below 200 nm (to allow control to be maintained over the base width as it is reduced), the minority-carrier diffusion length becomes larger than the emitter depth, resulting in further reduced current gain.[27]

Two approaches to circumventing these problems have been pursued, each based on the use of alternative emitter structures: polysilicon emitters, and heterojunction bipolar transistors (HBTs).

7.8.1 Polysilicon Emitters

Besides being formed through ion implantation of dopants into the substrate, emitters can be formed through outdiffusion of dopant from a polysilicon layer in direct contact with the monocrystalline substrate. In 1979, it was reported that the current gain of transistors fabricated with such polysilicon emitters is three to seven times greater than that of transistors with conventional emitters.[28] Figure 7-30 gives an example of such enhanced device performance, comparing the current gains of polysilicon-emitter *npn*

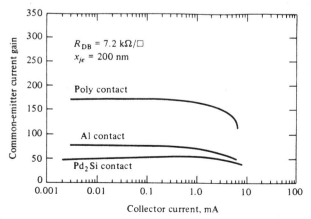

Fig. 7-30 Current gain as a function of collector current for an emitter 200-nm deep contacted either by Pd$_2$Si, Al, or 100 nm polysilicon.[28]   (© 1979 IEEE).

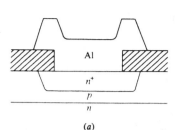 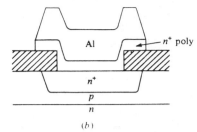

Fig. 7-31 Emitter contacts: (a) Al, (b) n^+ polysilicon + Al.

transistors with those of two types of transistors fabricated with conventional emitters. Each transistor is fabricated with a shallow (200-nm-deep) emitter and a base width of 100 nm.[29] (The polysilicon layer in such structures typically has a thickness of between 150 and 250 nm and is covered with an Al layer; see Fig. 7-31.) The transistors with polysilicon emitters exhibit higher gains than those made with direct metal contacts to the emitter.

In applications where higher transistor current gains do not significantly increase performance (e.g., digital circuits), the enhanced current gain can be traded off for an increase in the base doping level. This allows the base width to be decreased without the occurrence of premature punchthrough (a significant benefit, because as noted earlier, narrower base widths yield transistors with higher maximum switching speeds.) Polysilicon-emitter *npn* transistors with f_T's of 17-30 GHz and sub-50-ps ECL gate delays have been produced.[30]

The development of polysilicon emitters has also been pursued for other important reasons:

• The lateral and vertical dimensions of the emitter can be scaled in a coordinated manner, allowing the peripheral component of the E-B junction capacitance to be maintained at a reasonable value.

• Defects in the vicinity of the emitter-base junctions of conventional ion-implanted emitters give rise to severe yield problems.[31] These defects are associated with the residual damage caused by the emitter-formation implant (i.e., damage that remains even after post-implant annealing; see Vol. 1, chap. 9). In polysilicon emitters, this problem is avoided, since the dopants are implanted into the poly-Si and the emitter-base junction is formed by diffusion into undamaged silicon.

• The incidence of Al spiking of the emitter-base junctions is significantly decreased.

7.8.1.1 Models That Describe Polysilicon-Emitter Behavior. Three
models have been proposed to explain why polysilicon-emitter contacts increase the current gain:

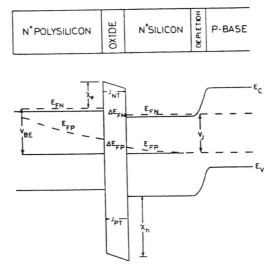

Fig. 7-32 Band diagram of a polysilicon emitter transistor with an interfacial oxide.[33] (© 1989 IEEE).

1. *Reduced-Mobility Model.* This model postulates that the current gain increases because the polysilicon film increases the length of the emitter, while the low minority-carrier mobility in the polysilicon retards the transport of the injected minority carriers. However, this early model is no longer considered correct, as it fails to explain many of the other characteristics observed in poly-Si emitters.[32]

2. *Interfacial-Oxide Tunneling Model.* A layer of oxide can be formed on the single-crystal-silicon substrate during the processing that precedes polysilicon deposition.   The tunneling model assumes that this native oxide acts as a tunneling barrier to hole injection into the polysilicon (Fig. 7-32).[33]

This model predicts current gain improvements of ten or more, depending on the thickness and height of the tunneling barrier. It also predicts an increase in emitter resistance due to the transport of majority carriers through the interfacial oxide. This is consistent with the high values of emitter resistance observed in polysilicon emitters, especially in devices with deliberately grown interfacial oxides. It appears that tunneling through an interfacial oxide plays a significant role in poly-Si emitters if the oxide is more than about 10 Å thick.

3. *Interface Dopant-Segregation Model.* This model describes why increased current gains are observed for poly-Si emitters with oxide-free interfaces (and is also apparently applicable to poly-Si-emitters with interfacial oxides thinner than ~10 Å – e.g., the native oxides that unavoidably form even at "clean" poly-Si interfaces). The model is based on the premise that dopants in polycrystalline silicon segregate to the gain boundaries (see Vol. 1, chap. 6). Since the poly/mono interface is essentially a large continuous grain boundary, dopant from

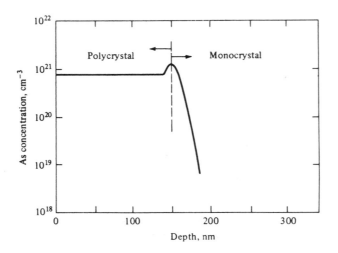

Fig. 7-33 Measured doping of arsenic concentration in a polysilicon emitter contact. (© 1979 IEEE).

the polysilicon should segregate to it. SIMS-analysis results confirm that such dopant pileup does occur (Fig. 7-33).[34] This gives rise to a potential-energy barrier that hinders the injection of holes into the emitter during forward bias. The blocking action of the barrier also decreases the slope of the hole concentration in the emitter, leading to a smaller hole current and an increased current gain.[35, 36]

7.8.1.2 Process Technology for Polysilicon-Emitter Fabrication.

Either phosphorus or arsenic can be used to dope the polysilicon. Arsenic is selected for transistors in which an f_T of more than 10 GHz is required. The smaller diffusion coefficient of As in Si permits the formation of shallower emitter regions and narrower bases, while the emitter push effect that occurs when phosphorus is diffused makes control of narrow base widths more difficult in phosphorus-doped poly emitters (see Vol. 1, chap. 8). However, the latter are less sensitive to the presence of interfacial oxide layers than are those doped with arsenic. Thus, phosphorus-doped polysilicon emitters yield transistors with lower emitter resistances but wider base widths, making them suitable for bipolar processes that can tolerate *npn* transistors with less-than-optimum performance characteristics. (Polysilicon emitters doped with phosphorus typically exhibit minimum emitter interfacial contact resistances of 20 Ω-μm^2 compared to ~60 Ω-μm^2 for arsenic-doped emitters.)[37]

The presence of an intentionally grown oxide layer at the interface (typically ~14 Å thick)[38] leads to higher current gain than if oxide-free interfaces are contacted (actually, "oxide-free" implies that only native oxides thinner than 10 Å are present). Since the intentional oxide layers are normally formed through a chemical oxidation of the silicon in a solution of ammonium hydroxide (which is one of steps of the RCA-prefurnace wafer cleaning procedure; see Vol. 1, chap. 15), they are sometimes referred to as *RCA-clean interfacial layers.*

Unfortunately, as noted above, interface oxides also represent a significant electrical barrier for *majority carriers;* if the layers are continuous they increase the minimum emitter resistance by more than an order of magnitude.[39] The negative impact on device performance caused by such high polysilicon-emitter resistance is significant.[40] For small-area emitters this resistance severely limits the speed and transconductance of the device; consequently, emitter contact resistances of less than 50 Ω-μm^2 are required for high-performance bipolar transistors.

Annealing the emitter at temperatures of 1000°C can reduce the emitter resistance by breaking up the interfacial oxide (i.e., this causes the oxide layer to become discontinuous) and diffusing the emitter deeper into the substrate. However, these improvements are obtained at the costs of lower current gains and larger base width. Thus, attempts to integrate such chemically formed oxide layers into polysilicon-emitter fabrication processes have been abandoned.[41]

Nevertheless, it has been reported that high current gains can still be obtained if no intentional oxide is grown, but if instead the silicon substrate surface is cleaned using an HF dip and DI rinse just prior to polysilicon deposition.[39] A native-oxide layer still forms at the interface (with an average thickness of slightly less than 8 Å (see Fig. 7-34a).[42] This layer is also effective in blocking minority-carrier injection into the emitter (which further reduces the base current). Annealing at temperatures of 900°C or more causes some of this oxide layer to be broken up (Fig. 7-34b), and also causes some areas of the polysilicon layer to epitaxially realign with the substrate Si.

Nevertheless, the nonuniform and variable nature of even this non–intentionally grown interfacial region has represented a source of poor reproducibility,* especially as device regions have been scaled down.[43] However, several processing techniques, have been studied for reliable production of poly-Si emitters that exhibit low interfacial contact resistances and that also effectively block minority-carrier transport. These techniques include the following:

- The emitter drive-in step is carried out by means of RTP at 1050°C[36] (following a careful cleaning procedure and HF dip just prior to poly deposition). This allows a shallow (~50-nm deep) As-doped emitter region to be formed without disruption of the continuous grain boundary at the poly-mono interface (i.e., during *furnace* drive-in steps, some of the poly is epitaxially realigned with the mono-Si, reducing the effectiveness of the barrier). The emitter resistance following this procedure is reduced to 30 Ω-μm^2. In a second approach, a 1050-1150°C, 10-sec RTP step was carried out following a 900°C furnace anneal. This also produced enhanced current gain and reduced emitter resistance.[44]

---

*   In analog bipolar ICs, where low and reproducible values of R_E are required, the formation of emitters by means of direct implantation and metal contact to the substrate is still more widely practiced.

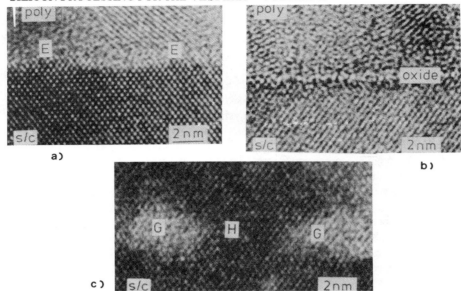

Fig. 7-34 (a) High-resolution-TEM micrograph of a Si surface given an HF etch prior to polysilicon deposition, an As implant and a drive-in of 30 mins. Interfacial oxide is not continuous. (b) Si surface given HF etch prior to polysilicon deposition, but with no heat treatment yet. A thin (~4 Å thick), continuous oxide is present. (c) Same surface, but after an interface anneal at 1000°C and full emitter processing. Interface oxide is broken up, and consists of balls to 50 Å in diameter. (after Wolstenholme et al.,[42] copyright 1987 AIOP).

• The polysilicon deposition is performed at 275°C using a PECVD process.[45] This approach produces highly reproducible enhanced current gains at relatively low post-deposition anneal temperatures (900°C).

7.8.2   Heterojunction Bipolar Transistors (HBTs)

As the vertical dimensions of the bipolar transistor are scaled even further, it is predicted that some serious device-operational limits will be encountered, even if polysilicon emitters are used. The following scenario will illustrate this point: When the base thickness of the bipolar transistor is decreased, the doping level must be increased in order to control punchthrough and to maintain a sufficiently low value of R_{B2}. Since the doping level in the emitter must remain high in order for a sufficient emitter-injection efficiency to be maintained, C_{EB} will increase. The current density will thus need to be increased so that the propagation delay associated with this extra capacitance can be overcome. To handle the increased current density without suffering the Kirk effect, it will be necessary to raise the doping in the collector. Consequently, even though such a device will have a higher f_T, it will have a narrower operating-current range. Furthermore, the increased doping levels will reduce the values of BV_{EBO} or BV_{CEO} to such a degree that it will not be possible to use the device in many practical applications.

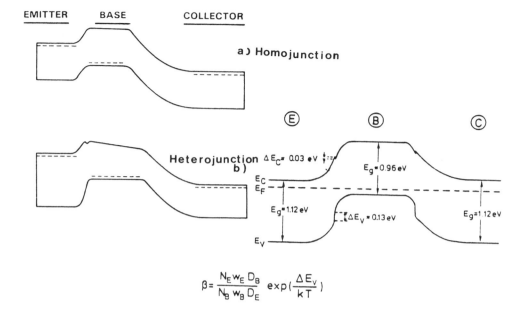

Fig. 7-35 Band diagrams of representative *npn* (a) homojunction and (b) heterojunction bipolar transistors.[49] (© 1989 IEEE).

One actively studied approach to overcoming the above problems is to build transistors with emitter materials whose band gaps are larger than that of the material used in the base (i.e., *heterojunction transistors*). This technique appears promising because the emitter-injection efficiency, γ, depends on the band-gap difference according to the equation[49]

$$\gamma = (n_e v_e / p_b v_h) \exp(\Delta E_g / kT) \qquad (7 - 19)$$

where n_e is the emitter electron concentration, v_e is the effective velocity of electrons injected into the base, p_b is the base hole concentration, v_h is the corresponding hole velocity injected into the emitter, and ΔE_g is the difference in the band gaps of the emitter and base. (From Fig. 7-35, it can be seen that the larger band gap in the emitter gives rise to a larger potential-energy barrier for holes going from the base to the emitter than for electrons going from the emitter to the base.) Thus, if the band gap of the emitter in silicon devices is larger than that of the base, when ΔE_g is 0.2 eV or more, $\exp(\Delta E_g/kT)$ will exceed 2000. As a result, a sufficiently high emitter-injection efficiency will be assured, no matter what doping levels exist in the base and emitter.

For example, doping levels in the base can be increased to above 10^{19} cm^{-3}, and in the emitter they can can be dropped to 5×10^{17} cm^{-3} (Fig. 7-36a). Under such conditions, many other device characteristics are also improved (among the improvements are a lower base resistance, a higher punchthrough voltage, a higher Early voltage, and a decreased E-B junction capacitance). However, for such devices to be successful in silicon technology, the technology must be compatible with advanced silicon bipolar processing.

Most approaches to heterojunction fabrication are based on growing an epitaxial layer on the Si substrate, such that the epi material has a different band gap than the Si, but also has a lattice constant that is at least approximately matched to the lattice constants of the Si substrate. If such deposited layers are sufficiently thin, the requirement for exact lattice-match is relaxed. That is, thin layers with an almost-matched lattice constant will deform so that they can accommodate to the lattice constant of the substrate. In this deformed condtion they are either "coherently strained" or "psuedomorphic" layers. The latter are thermodynamically stable for thicknesses below a strain-dependent "critical" value. However, it is important that subsequent processing temperatures be restricted to prevent the stress of the layers from being relaxed through the formation of an array of misfit dislocations.

From the point of view of process integration, the most attractive approach to building devices with hetero-E-B junctions is to increase the band gap of the emitter (since this represents the only change from a conventional bipolar process). Several methods for implementing this approach have been attempted, including the use of the following materials as the wider band gap emitters: GaP;[50] semi-insulating polycrystalline-silicon (SIPOS, $E_g \cong 1.7$ eV);[51] oxygen-doped Si-epitaxial films (OXSEF, $E_g \cong 1.7$ eV);[52] β-SiC ($E_g \cong 2.2$ eV);[53] and phosphorus-doped hydrogenated microcrystalline silicon (μc-Si, $E_g \cong 1.7$ eV).[54]

Despite an intense effort to develop such processes, no entirely successful fabrication procedure has yet been demonstrated. Although numerous high-gain transistors have been reported, these devices exhibited other problems, including high-temperature requirements, high bulk and contact resistivities, high emitter resistances, or poor stability during the multilevel-interconnect fabrication steps. The most promising candidate of this group is the μc-Si emitter. Fujioka et al. reported achieving reduced emitter resistances with such structures and built ECL gate arrays with them using a non-self-aligned emitter structure.[54] Basic gate delays of 295 ps at 6.5 mW/gate were obtained.

An alternative approach is to use a material in the base of the transistor that has a smaller band gap than that of Si. Layers of Si to which Ge has been added exhibit such reduced band gaps (Fig. 7-36b). For example, psuedomorphic epitaxial Si_xGe_{1-x} alloy layers of up to 100 nm in thickness can be formed with a Ge content of 20%, and the ΔE_g between the emitter and base in Si transistors fabricated with such base layers (i.e., Si_xGe_{1-x}-base HBTs) can be as large as 0.1 eV. An excellent, highly detailed review of the science and technology of such devices is presented in reference 55.

Psuedomorphic epitaxial Si_xGe_{1-x} layers can be deposited by means of molecular beam epitaxy (MBE) or chemical vapor deposition (CVD). The most successful CVD methods have involved *ultrahigh-vacuum CVD*[56] or *limited-reaction-processing (LRP) CVD*.[57] Note that appropriate levels of boron doping must also be incorporated into these films during the epi growth process of the base region.

It has been calculated that HBTs with narrow band-gap Si_xGe_{1-x} bases will be able to achieve very high performances (e.g., f_T's as high as 75 GHz have been predicted for a 20% Ge concentration in a 100-nm base, a 0.2-μm-wide Si emitter, and a collector current density of 1×10^5 A/cm^2).[58] Conversely, Si$_x$Ge$_{1-x}$-base HBTs offer the potential benefits of the same performance characteristics as those of conventional homojunction transistors, but with a relaxed profile (i.e., the base can be doped more heavily for a lower base-sheet resistance, or it can be made thicker to achieve lower doping levels with higher breakdown voltages).

The chief drawback of the reduced–band-gap/base approach is that it is more difficult to integrate into existing advanced bipolar processes. The metastable nature of the Si$_x$Ge$_{1-x}$ films places constraints on the Ge content, layer thickness, subsequent annealing cycles, and thermal oxidation. On the other hand, this method does not suffer from the other problems exhibited when wide band-gap emitter materials are used. In addition, to provide the ultrathin base profiles for future bipolar devices, the conventional ion-implantation approaches are becoming increasingly difficult to implement successfully, while the low-temperature epitaxial processes show great promise.

While Si$_x$Ge$_{1-x}$-base HBT devices have demonstrated excellent performance potential

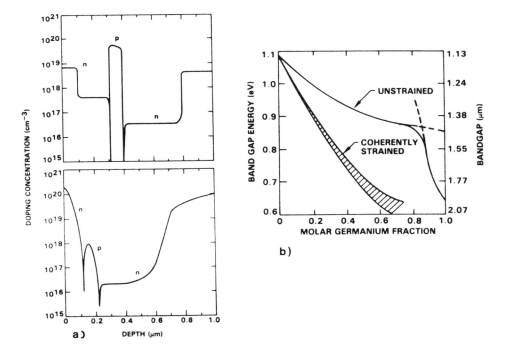

Fig. 7-36 (a) Representative doping profiles for *npn* heterojunction bipolar transistors.[49] (© 1989 IEEE). (b) Band gap energy versus alloy composition for Si-Ge alloys.[132] (after People and Bean,[42] copyright 1986 AIOP).

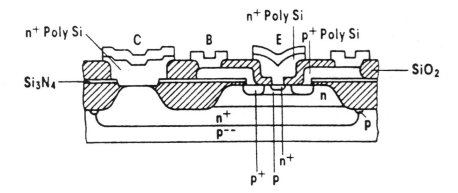

Fig. 7-37 Cross-section of a double-polysilicon self-aligned bipolar transistor.[133] (© 1986 IEEE).

to date, they must still pass several tests with respect to process integration, manufacturability, and long-term reliability before they can be incorporated into mainstream IC production.[55]

7.9 SELF-ALIGNED BIPOLAR STRUCTURES

Although oxide isolation reduces the bipolar device size by requiring less isolation area, it does not reduce the distances between the contacts within the device. That is, relatively large inactive device areas are needed within the isolation boundaries to accommodate mask misalignment errors. (For example, the optical-stepper alignment accuracy is about ±0.4 μm). However, these same inactive regions not only waste valuable silicon area but also cause parasitic capacitances and resistances that degrade transistor performance. The most effective techniques developed to reduce the areas of these regions involve the use of *self-aligned* (SA) structures. The chief disadvantage of SA structures is that they increase process complexity; in addition, their smaller tolerances make them difficult to manufacture with high yield. Process-development efforts in this area will thus focus on these two issues.

7.9.1 Double-Polysilicon Self-Aligned Structures

The most widely used SA approach employs a double-polysilicon process to form an emitter that is self-aligned to the base, as shown in Fig. 7-37.[59,60] The first layer of polysilicon (*poly 1*) is doped p^+ and serves as the base electrode, while the second (*poly 2*) is doped n^+ and makes contact to the transistor's emitter region. The emitter and the extrinsic base regions are formed by means of dopant outdiffusion from poly 1 and poly 2, respectively.

Figure 7-38 shows the sequence of steps that form this self-aligned structure.[59] Following the growth of the field oxide and the removal of the oxide from the surface of the active regions, poly 1 is deposited and heavily *p*-doped with boron. This layer is then covered with a CVD oxide (Fig. 7-38a). The emitter mask is used to pattern the emitter-area regions, and a dry-etch process is used to produce openings in the CVD oxide and poly 1 (Fig. 7-38b). A thermal oxide is then grown over the etched structure, and a relatively thick oxide (on the order of 0.1-0.4 μm) is grown on the vertical sidewalls of the heavily doped poly. The thickness of this oxide determines the spacing between the edges of the base and emitter contacts. The extrinsic base regions are also formed during the thermal oxide growth step, as a result of the outdiffusion of boron from the poly 1 into the substrate. Because the boron diffuses laterally as well as vertically, the extrinsic base region will be able to make contact with the intrinsic base region that is formed next (and that will be located under the emitter contact).

The parasitic series resistance of the extrinsic base region, R_{B1}, will now consist of the sum of the resistances of the poly 1 electrode, R_{B1poly}, and of the path between the edges of the base and emitter contact, R_{B1sub}. Since the latter distance is so small, R_{B1sub} is negligibly small. The doping of poly 1 can be much heavier than that of the extrinsic base region in the substrate, and the value of R_{B1poly} can therefore be made much smaller than R_{B1} in a non-SA transistor.

Following the oxide-growth step, the intrinsic base region is formed through the ion implantation of boron (Fig. 7-38c). This serves to self-align the intrinsic and extrinsic base regions. Following a cleaning of the contact to remove any oxide layer present, poly 2 is then deposited and implanted with As or P. A shallow emitter region is then formed through dopant outdiffusion from poly 2 (this critical step will be discussed in greater detail in section 7.8.1.2). The use of RTP for the base and emitter out-diffusion steps has also been reported to facilitate the formation of shallow E-B and C-B junctions.[60]

This SA structure allows the fabrication of emitter regions smaller than the minimum lithographic dimension. When the sidewall-spacer oxide is grown, it fills the contact hole to some degree (since the thermal oxide occupies a larger volume than the original volume of polysilicon; see Vol. 1, chap. 7). Thus, a 0.8-μm-wide opening will shrink to about a 0.4-μm width if a 0.4-μm-thick sidewall oxide is grown on each side. This is another reason why SA structures can be made smaller than conventionally fabricated devices.

7.9.1.1 Limitations of Double-Polysilicon SA Structures. A major obstacle to producing small W_B values in advanced bipolar devices is the channeling tail of the intrinsic-base boron implant (see Vol. 1, chap. 9). Furthermore, the use of the normal 7° off-axis implant to reduce such channeling tails results in shadowing effects that adversely impact the device's perimeter-punchthrough and tunneling-leakage-current characteristics (Fig. 7-39).[62] Several techniques have been reported for overcoming this drawback.

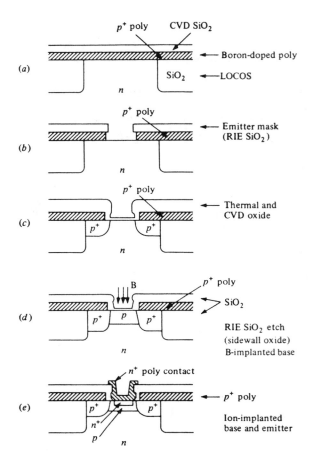

Fig. 7-38 Process sequence for fabricating double-polysilicon self-aligned *npn* transistors.[59] (© 1980 IEEE).

In one approach, the direct implantation of boron into the substrate is eliminated.[63] Instead, after the sidewall spacer oxide has been grown, poly 2 is deposited and implanted with boron. Following an outdiffusion step to form the intrinsic base region, an As implant into the poly 2 is performed. A second diffusion then forms the emitter region. This approach has produced *npn* transistors with base widths of 100 nm, emitter junction depths of 50 nm, and f_T's of 16 GHz.

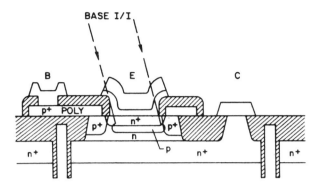

Fig. 7-39 Device cross-section of an advanced polysilicon self-aligned bipolar transistor. The base profile is offset with respect to the emitter profile due to the sidewall shadowing effect.[62] (© 1987 IEEE).

It should be noted that both diffusions cannot be done simultaneously, since boron is nearly immobile in the presence of high arsenic concentrations.[64] Successive diffusion of both dopants, as shown in Fig. 7-40a, solves this problem and yields very narrow base widths. On the other hand, since the poly is partially recrystallized during the first boron drive-in step, the implanted As no longer diffuses as rapidly, resulting in a shallower As profile with a steeper slope.

In another approach, the thickness of the poly 1 layer is reduced so that the intrinsic base implant can penetrate it.[65] Another method also eliminates the base-implant step, forming the emitter by means of a selective epitaxial process (Fig. 7-40b).[66] This approach was extended by using a low-temperature-epi (LTE) process for forming the base, and an epitaxial-lateral-overgrowth (ELO) process for forming the extrinsic base (i.e., instead of using *p*-doped polysilicon for the latter).[136] *NPN* transistors with 60-nm-wide epitaxial-base regions and 7.5 kΩ/sq intrinsic bases sheet resistances were successfully fabricated with this method. Yet another method involves the deposition of borophosphosilicate glass (BSG) into the etched emitter window and the outdiffusing of the base dopant from the BSG to form the intrinsic base (Figs. 7-41a).[67] The BSG film is then stripped before the polysilicon-emitter film is deposited. An advanced BSG-based process was subsequently described in which the epitaxial collector region under the intrinsic base was implanted with phosphorus to suppress the Kirk effect and W was selectively deposited over the emitter polysilicon to reduce the emitter resistance (Fig. 7-41b).[137] Transistors with a 40 GHZ f_T were fabricated with this approach. Other approaches are also described in section 7.8.2.

As noted earlier, in the SA structures that use polysilicon-base electrodes, the extrinsic-base-region series resistance is the sum of R_{B1sub} and R_{B1poly}. The latter is a function of the polysilicon sheet resistance. With 300-nm-thick polysilicon used, an acceptably low value of 50 Ω for R_{B1poly} has been measured.[68] However, as the polysilicon is reduced in thickness (to avoid topography problems in submicron processes), R_{B1poly} will increase. A potential solution is to replace the polysilicon with a polycide. One example of such an approach has been reported by Tashiro et al., who formed a PtSi on the polysilicon toward the end of the process.[69] Silicided base regions in single-poly SA structures will be described in section 7.8.2.

There are three other problems associated with the double-poly SA process. First, the silicon surface can be damaged during the poly 1 etch.[70] Second, the topography of the double-poly structure has severe steps, which give rise to increased complexity in the metallization and lithography processes. Third, perimeter effects play an important role in the device characteristics (analyses of such effects are presented in references 71 and 72). Increasingly complex structures have been devised to solve these problems.

7.9.1.2 Current-Gain Degradation Due to Sidewall Injection in SA Bipolar Structures.
Injection of minority carriers into the base occurs at the sidewalls (or perimeter) of the emitter, as well as at the emitter bottom. If the injections from these two regions are considered separately, it can be seen that the current gain due to transistor action from the sidewall injection is much smaller than that due to the bottom injection. This is because both the base width and the doping

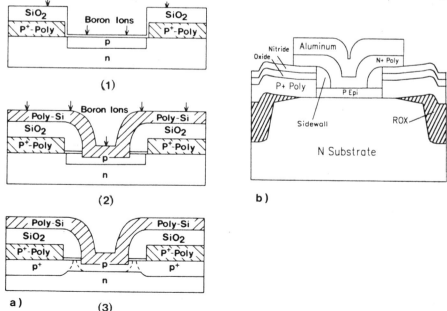

Fig. 7-40 (a) Double diffusion of the polysilicon to form the intrinsic base and the emitter regions without ion implantation into the Si substrate.[63] (b) Cross-section of selective-epitaxy base transistor.[66] (© 1988 IEEE).

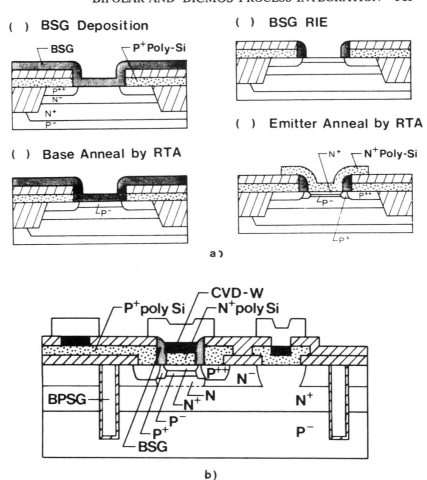

Fig. 7-41 (a) Key processing steps for borosilicate-glass, self-aligned transistor formation.[67] (© 1987 IEEE). (b) Cross section of the advanced BSG transistor.[137] (© 1989 IEEE).

concentration are smaller in the base region beneath the emitter than they are along its sidewalls. In bipolar transistors in which the minimum emitter dimension is greater than about 1.5 μm, the fraction of the current injected at the emitter sidewalls is so small that the overall current gain, β, is not significantly reduced by the perimeter injection. However, in bipolar devices in which the emitter widths are 1 μm or less, the perimeter-to-area ratio of the emitter increases to such a degree that β is significantly reduced by the perimeter-injection effects.

For example, a 65% reduction in current gain of has been reported in devices with an emitter width of 0.5 μm, compared to those having 4.25-μm-wide emitters (and identical emitter lengths).[73] This effect can become even more severe in SA structures in which the spacing is very small between the heavily doped extrinsic base and the emitter edge. An obvious remedy is to continue to scale the depth of the emitter so that the perimeter remains a small fraction of the total emitter area. Careful design of the base profile at the edges of the emitter is also mandatory.

7.9.1.3 Linkup-Region Formation.

As noted, the connection between the extrinsic and intrinsic base regions in the double-polysilicon SA bipolar structure (i.e., the *linkup,* or *graft-base,* region) is critical. The space between the edge of the base and the emitter contacts is determined by the thickness of the sidewall oxide spacer. The amount of overlap between the two base regions depends on this thickness and on the amount of lateral diffusion from the p^+ poly. Control of these two parameters is vital in order to fabricate these SA structures with high yield.

If a heavy overlap is produced between these regions in the linkup region (Fig. 7-42a), low values of BV_{EBO}, β, and f_T are exhibited. On the other hand, if the space between the edge of the intrinsic base implant and the extrinsic base diffused regions is too great (Fig. 7-42b), the result will be anomalously high values of R_{B1} and low values of *perimeter punchthrough voltage* (i.e., between the emitter and collector through the lightly doped link region of the base). If the structure is formed according to the process indicated in Fig. 7-38, a proper choice of sidewall-spacer thickness (Fig. 7-42c) thus involves a trade- off between these device characteristics.[74] A study of such trade-offs by Sawada[75] and Chuang et al.[65] measured the effects of several different poly 1 outdiffusion cycles on the device characteristics.

An alternative approach is to use a separate procedure to form the linkup region.[76] In one such process, the contact hole is etched, and a thin oxide is grown by RTP on the poly sidewalls and the epi surface. A shallow BF_2^+ implant is used to form the lightly doped linkup region, and only then are oxide sidewall spacers produced (Fig. 7-40). Following the deposition of poly 2, a boron implant and diffusion are used to form the intrinsic base, and an arsenic implant and an anneal step are employed to create the emitter region. (A "local" collector implant to reduce the channeling tail of the boron linkup implant, as described in section 7.9.2, may also be performed in the same process step in which the boron-linkup implant is carried out.) Another approach for separately forming the link-up region is given in reference 77.

7.9.2 Single-Polysilicon Self-Aligned Bipolar Structures

Several other SA techniques that use only a single polysilicon layer (and thus offer the advantage of somewhat reduced process complexity) have also been developed. The first takes advantage of the fact that regions of highly doped silicon oxidize much more rapidly than lightly doped ones. A heavily doped, patterned polysilicon-emitter structure (beneath which an intrinsic base region has already been formed, shown in Fig. 7-43a) is oxidized at the same time as the bare, lightly doped substrate.[78] A much thicker oxide is grown on the poly, and a low-energy boron implant is used to form the extrinsic base region. The boron is prevented from penetrating into the polysilicon by the thick oxide,

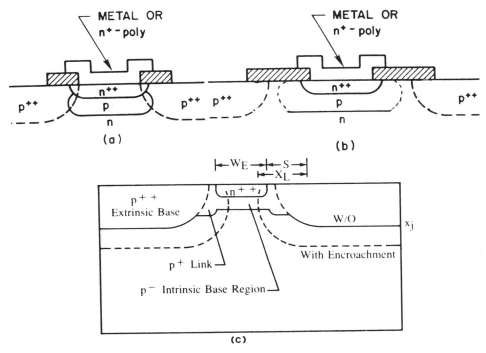

Fig. 7-42 (a) Heavy overlap between the intrinsic and extrinsic base regions. (b) Wide space between intrinsic and extrinsic base regions.[71] (c) Selecting the appropriate overlap, based on various tradeoffs − s is the spacing from extrinsic base implant mask to emitter: x_L is the lateral diffusion from the edge of the implant mask; x_j is the junction depth of the extrinsic base.[134] (© 1987 IEEE).

but it can enter the silicon substrate through the thinner oxide grown there. The arsenic outdiffuses from the poly during the oxide-growth step to form the emitter, allowing the base to be self-aligned to the intrinsic base and the emitter diffusion.

In a slightly more complex version of this approach, a linkup boron implant is performed following etching of the 250-nm thick poly layer (Fig. 7-44).[79] An oxide spacer is then formed around the poly emitter, and the p^+ extrinsic-base implant is performed. The emitter is diffused during the anneal step following the extrinsic base implant. The base and emitter resistances are further reduced through the use of a salicide technique by which a silicide is formed on all of the exposed silicon.

An enhancement of the above process involves the deposition of an amorphous silicon layer in place of poly.[80] Following the formation of the linkup and extrinsic base regions, the amorphous silicon is implanted with boron. (Since the silicon is amorphous, the formation of a channeling tail is avoided.) The intrinsic base is formed through out-diffusion from the amorphous silicon. During this step, the silicon crystallizes with a larger grain size and different grain orientations than those of

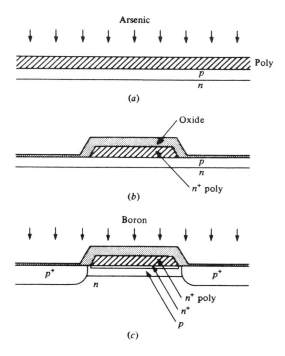

Fig. 7-43 Forming a single-polysilicon self-aligned transistor.[78] (© 1985 IEEE).

polycrystalline silicon. The altered poly grain structure results in a shallower E-B junction and a narrower base width than were the case when polysilicon was used as the diffusion source. This approach produced devices with good high-frequency characteristics.

Another single-poly process that uses removable sidewall spacers to self-align the linkup region to the intrinsic and extrinsic base regions has been described by Chen et al.[77,87] As shown in Fig. 7-45,[87] the resulting topography is much less severe than that in double-poly SA structures. This approach was used to build sub-50-ps ECL circuits.

Another single-poly SA process, called SPOT (Single **PO**ly **T**ransistor) has been used to fabricate structures with minimum feature sizes of 1.25 μm.[81] The primary goal in designing this structure was to create a simple, high-yielding process without seriously degrading device performance. A cross-section of a SPOT device is shown in Fig. 7-46. Following the conventional formation of buried collectors, deep collector contacts, and intrinsic base regions, a 300-nm-thick layer of polysilicon is deposited. Poly resistors and p^+ electrode regions are implanted with boron, and a nitride layer is deposited. The poly electrode patterns are defined through PR patterning and RIE of the nitride and poly. Following stripping of the resist, the exposed substrate silicon and poly sidewalls are thermally oxidized to a thickness of 150 nm. A self-aligned extrinsic base region is formed by means of boron implantation though this oxide. The nitride is then removed, and arsenic is implanted into the emitter and collector regions of the poly. Following a final emitter drive-in step, PtSi is formed over the exposed regions of the poly electrodes. A two-level-metal process completes the process sequence.

In this simpler approach, however, the etching and oxidation may form a nonvertical boundary between the polysilicon and the adjacent oxide (Fig. 7-47). For example, since selective oxidation between the emitter and base is used to provide electrical isolation, the shape of the polysilicon over the emitter may be trapezoidal rather than rectangular following oxidation.

Because dopant is usually implanted near the surface of the poly and diffused through it into the underlying silicon substrate to form the emitter region, in trapezoidal poly structures dopant atoms must diffuse laterally as well as vertically to act as an emitter-forming dopant source. The resulting junction may be shallower near the emitter's corner than near the center, which can lead to degradation of the transistor gain, especially in devices with narrow emitters.[82]

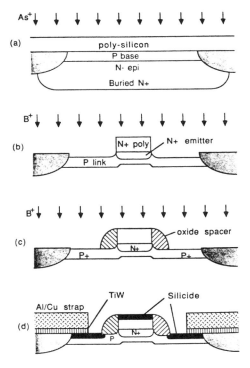

Fig. 7-44 Single-polysilicon self-aligned transistor with extra link-up implant.[79] (© 1988 IEEE).

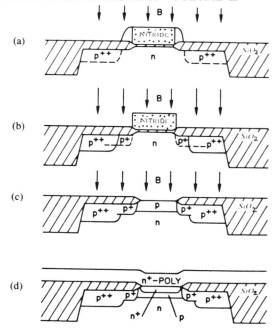

Fig. 7-45 Alternative approach for fabricating single-polysilicon self-aligned transistor.[87] (© 1988 IEEE).

7.9.3 SIdewall-Base-COntact Structures (SICOS)

Another self-aligned structure that can potentially result in even smaller nonactive device regions than the double- or single-poly SA structures is the sidewall-base-contact structure (SICOS; Fig. 7-48).[83,84] In the structure shown, only the active transistor region is formed in the single-crystal-silicon epitaxial layer. The linkup base regions are formed through the diffusion of boron from heavily doped polysilicon layers. A buried oxide beneath these polysilicon base electrodes significantly reduces C_{CB}. Figure 7-49 shows the steps used to fabricate the SICOS structure shown in Fig. 7-48a. ECL gates with a minimum propagation-delay time of 63 ps have been fabricated using this structure.

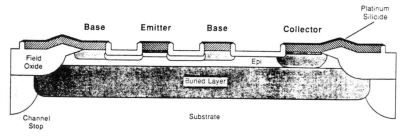

Fig. 7-46 Cross-section of the SPOT transistor.[81] (© 1988 IEEE).

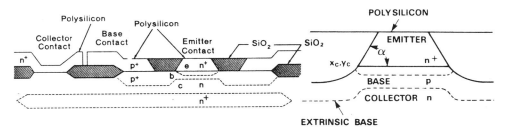

Fig. 7-47 Cross-section of the emitter region in SPOT transistor showing trapezoidal shape, which leads to nonuniform distribution of dopant across emitter area.[82] (© 1989 IEEE).

A SICOS structure in which a submicron linkup region is provided (to permit control of such problems as E-B forward tunneling and low E-B-junction breakdown voltages) has recently been described.[138] A cross section of this structure, called the *Best Alignment with Sidewall Contact (BASIC)* device is shown in Fig. 7-48b.

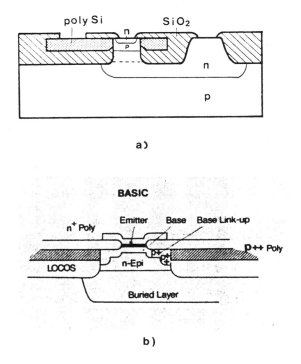

Fig. 7-48 (a) An advanced SICOS transistor.[84] (© 1979 IEEE). (b) Cross section of the BASIC SICOS device.[138] (© 1989 IEEE).

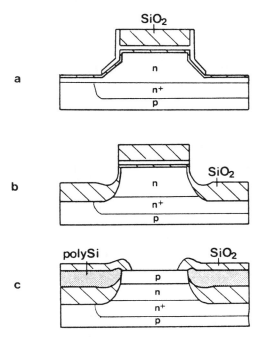

Fig. 7-49 Key process steps for fabricating a SICOS transistor.[84] (© 1979 IEEE).

7.10 TRENCH-ISOLATED BIPOLAR TRANSISTORS

Because trench-isolated bipolar transistors greatly reduce the device area and the associated parasitic collector-substrate capacitance, they allow a significant increase in the density of bipolar circuits and contribute to a reduction in the power-delay product. In addition, the deep trenches produce larger collector-to-collector breakdown voltages. Such trenches also decrease the latchup susceptibility of the device structures. Figure 7-50 shows a cross-section of a trench-isolated self-aligned bipolar structure.[87] Deep trenches are used to isolate the collector regions, and LOCOS is used to form the field-oxide regions that abut the edges of the base and emitter regions (as is done in oxide-isolated bipolar transistors with walled base and emitter regions).

Several reports of such structures have been published.[76,85,86] Still another report describes a bipolar isolation approach that uses 5-μm-deep trenches to isolate the collector regions and shallower BOX isolation structures as the field oxides.[77] This technology exhibits a 33-V breakdown between isolated collector regions. (The details of fabricating deep trench structures for isolation purposes are presented in chapter 2, section 2.6.3.)

Another paper dealing with the scaling property of trench isolation capacitance and its impact on high-performance ECL bipolar circuits has been published.[139] It shows that the trench isolation capacitance has a distinct dependence on the trench width, and that

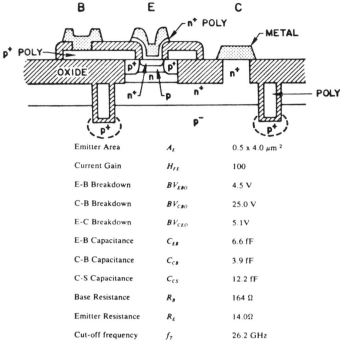

Emitter Area	A_E	$0.5 \times 4.0 \ \mu m^2$
Current Gain	H_{FE}	100
E-B Breakdown	BV_{EBO}	4.5 V
C-B Breakdown	BV_{CBO}	25.0 V
E-C Breakdown	BV_{CEO}	5.1V
E-B Capacitance	C_{EB}	6.6 fF
C-B Capacitance	C_{CB}	3.9 fF
C-S Capacitance	C_{CS}	12.2 fF
Base Resistance	R_B	164 Ω
Emitter Resistance	R_E	14.0Ω
Cut-off frequency	f_T	26.2 GHz

Fig. 7-50 Cross-section of a self-aligned, trench-isolated transistor.[87] (© 1987 IEEE).

the type of trench used is also significant. Four types of trenches (in order of decreasing capacitance: (1) Open-bottom, p^+-poly fill; (2) Close-bottom, p^+-poly fill; (3) Close-bottom, undoped poly fill; and (4) Close-bottom, oxide fill), were studied.

7.11 BiCMOS TECHNOLOGY

BiCMOS is a technology that integrates both CMOS and bipolar device structures on the same chip. This capability can be exploited in a number of ways to produce circuits with capabilities that exceed those that are possible when only one of the device types is used.

For example, while CMOS can implement low-power, high-density digital ICs with TTL outputs, such circuits are slower than ECL-based ICs implemented with bipolar devices (especially when driving large capacitive loads). Such loads are encountered on-chip when long interconnects are driven or high fan out occurs, as well as when signals must be sent off-chip.

Bipolar transistors, on the other hand, can deliver large drive currents, allowing heavy loads to be rapidly charged. Bipolar digital circuits implemented with ECL gates can also operate with small logic swings and high noise immunity. Conversely, ECL gates exhibit high power consumption, poor density, and limited circuit options, and they do provide TTL outputs (i.e., an ECL output is produced).

BiCMOS offers the benefits of both bipolar and CMOS circuits. By appropriately trading off the characteristics of each technology, the speed-versus-power middle ground (not achievable with either bipolar or CMOS ICs alone) can be reached. For example, high-density, low-power logic arrays can be integrated with high-speed bipolar drivers to provide gate arrays that are faster than comparable CMOS versions, but that consume far less power than ECL arrays having approximately the same density. BiCMOS can also offer analog and digital functions on the same chip. Furthermore, either a TTL or ECL interface can be produced.

These benefits, however, are attained at the expense of longer chip-fabrication times and higher costs (due to the more complex technology-development and chip-manufacturing tasks). Thus, BiCMOS is not being pursued as a replacement for CMOS in digital systems with low and medium performance. However, the benefits of BiCMOS are often perceived as being significant enough to justify the extra cost for many high-performance digital applications, as well as in mixed analog/digital systems (although it should be noted that this opinion is not universally held).

7.11.1 Device and Circuit Advantages of BiCMOS

Integrating CMOS with bipolar devices can provide superior performance for a variety of circuit applications (particularly those that require circuits capable of operating between the very-high-speed but power-hungry ECL, and the very-high-density, medium-speed CMOS regimes). Advantages of BiCMOS include the following:

• Digital BiCMOS logic gates can provide more drive current than can comparably sized pure CMOS gates, and as a result, they can operate at higher speeds when large capacitive loads must be driven.

• The dc power consumed by BiCMOS gates can be as small as that consumed by pure CMOS gates, and the ac power be can even smaller.

• Digital BiCMOS circuits can be fabricated with either a TTL or ECL output.

• In the move from a pure CMOS to a BiCMOS process, the performance of the purely CMOS circuits can be increased without the need to reduce the minimum feature sizes of current-generation technology. Thus, the existing fabrication tools can be used for at least one more generation. For example, while 1-Mbit CMOS SRAMs have been built with access times of 9 ns, these have required the use of 0.5-μm design rules. 1-Mbit BiCMOS SRAMs with the same access times have been built using 0.8 μm design rules. As of this writing, the task of fabricating parts with 0.5-μm dimensions is quite risky, in part because it entails the use of new, expensive production equipment that is also difficult to work with.

7.11.1.1 Comparison of BiCMOs and CMOS Propagation-Delay Times. BiCMOS gates exhibit smaller propagation-delay times when driving large

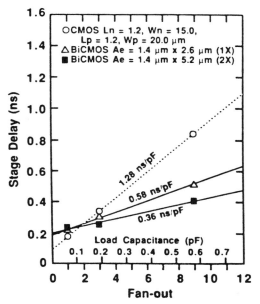

Fig. 7-51 CMOS and BiCMOS propagation delay times at 5 V versus load capacitance and fanout.[89] (© 1987 IEEE).

capacitance loads than do pure CMOS gates of equal area. This is demonstrated in Fig. 7-51, where it is seen that the slope of the τ_{pd}-versus-C_L curve is smaller for BiCMOS gates than for CMOS gates.[89] It should be noted that under no-load (and very light loading) conditions, τ_{pd} is indeed smaller if CMOS gates are used. However, as the value of C_L grows larger, the reverse becomes true. In the example shown, when the loads exceed ~0.3 pF, BiCMOS is faster.

While heavy loads are normally encountered when gates must drive large wiring-capacitance or off-chip loads, they also exist in logic circuits when fan-out exceeds 2. Figure 7-51 also indicates that the delay times of BiCMOS gates do not increase as rapidly as those of CMOS gates with increasing C_L. This means that approximately constant values of τ_{pd} can be maintained, independent of loading conditions.

The basis of this advantage of BiCMOS can be explained with the aid of Figs. 7-52a and 7-52b, which show CMOS and BiCMOS inverters driving the same load capacitance, C_L. To a first order, τ_{pd} of an inverter can be expressed as:

$$\tau_{pd} = \tau_o + \Delta V_o \, C_L / \Delta I \qquad (7 - 20)$$

where τ_o is an intrinsic delay time of the gate itself. For small values of C_L, CMOS gates are faster than BiCMOS gates because the bipolar stage of the BiCMOS gate represents a load to the CMOS portion of the gate that drives it. Roughly speaking, if the parasitic junction capacitances of the bipolar transistor are larger than C_L, more time may be required for the CMOS portion of the gate to charge these capacitances than if only the C_L of the following gate had to be charged.* The value of C_L at which the BiCMOS-gate delay becomes equal to that of a comparably sized CMOS gate is important. This point changes as a function of many device parameters (e.g., L_{eff}, W_B, f_T, g_m, and C_j), as well as of external operating conditions (e.g., temperature). The degree to which the C_L-crossover value can be reduced without the need for complex processing measures is one of the factors determining the cost-effectiveness of BiCMOS in digital systems. (Note that for small values of C_L, this fact also implies that the use of CMOS inverters may yield higher performance.)

At sufficiently large values of C_L, however, the second term of Eq. 7-20 dominates. In such cases, τ_{pd} becomes inversely proportional to the transconductance ($g_m = \Delta I/\Delta V_{in}$) of the driver device in the gate, as shown by

$$\tau_{pd} \sim \Delta V_o\, C_L\, /g_m\, \Delta V_{in} \qquad (7\text{-}21)$$

and when $\Delta V_o = \Delta V_{in}$,

$$\tau_{pd} \sim C_L/g_m. \qquad (7\text{-}22)$$

Since it is well known that for equal-area devices bipolar transistors exhibit g_m values that are much larger than those of MOS transistors,[10] the values of τ_{pd} in BiCMOS gates should be smaller than those in comparably sized CMOS gates.

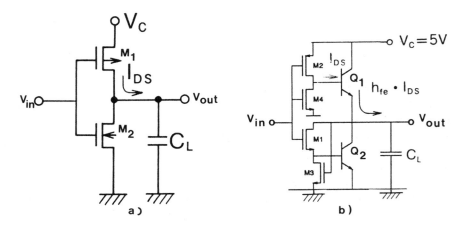

Fig. 7-52 (a) CMOS logic gate. (b) BiCMOS logic gate.

* When the basic digital inverter shown in Fig. 7-17a switches by having the input go from high to low, the current supplied by the load element charges C_L. In this case, $\Delta V_o/\Delta t = (\Delta Q_o/\Delta t)C_L = IC_L$. In the case of the CMOS inverter, the load element is an MOS device, and I $= I_{DS}$. In the case of a BiCMOS gate with a totem-pole output, as shown in Fig. 7-52c, the load element is a bipolar transistor, and $I = I_C$. More rigorous analyses of BiCMOS-gate propagation delays can be found in references 61 and 90.

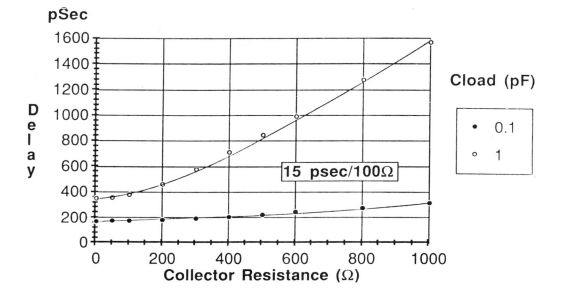

Fig. 7-53 BiCMOS gate delay versus R_C.[91] (© 1987 IEEE).

Another way to explain this result is to note that the CMOS inverter can only provide a drive current I_{DS} for charging C_L, whereas the BiCMOS inverter can supply a current βI_{DS} where β is the current gain of the bipolar device (I_{DS} represents the base current to the bipolar load element in the BiCMOS inverter.) Note that the effective β can be much less than the small-signal β exhibited by bipolar transistors (e.g., the effective β can be as small as 5-7, compared to a typical small-signal β of 100 or more). The lower β value applicable to this case results because during switching the collector resistance, R_C, of the bipolar load element causes the transistor to go into saturation. Less drive current is then available to charge C_L. Figure 7-53 shows how an increase in R_C results in greater τ_{pd} values.[91] Reference 90 gives more details about the changes in the BiCMOS gate-drive current during different stages of the switching cycle.

The difference between the g_m of bipolar and MOS transistors also gives BiCMOS some other advantages. The g_m of bipolar transistors is given by

$$g_{mbipolar} = qI_C/kT \qquad (7\text{-}23)$$

whereas the g_m of MOS transistors is given by

$$g_{mMOS} = (2\,\mu\,C_{ox}\,I_{DS}\,Z/L)^{1/2} \qquad (7\text{-}24)$$

Since $g_{mbipolar}$ depends on I_C, and I_C is directly proportional to the area of the emitter (A_E), an increase in A_E will lead to a linear increase in g_m. On the other hand, g_{mMOS}

only increases as the square root of Z/L. Thus, for an MOS device with a minimum gate width, L, increasing the width Z (which directly increases the device area) will produce a smaller increase in g_m than will an equal increase in the size of the bipolar driver. Thus, bipolar I/O buffer circuits can be much smaller than CMOS drivers while still providing the same drive current.

In addition, the g_{mMOS} is sensitive to variations in the process conditions (i.e., to variations in t_{ox} and L), whereas $g_{mbipolar}$ is essentially independent of such process variations (i.e., variations in A_E impact the value of I_C only in a logarithmic manner). In this sense, the BiCMOS technology exhibits better process stability.

Finally, because the mobility of carriers decreases as $T^{-1.5}$, the value of I_{DS} is reduced by 40% at 125°C from its 25°C magnitude. This effect does not impact BiCMOS gates as severely, again making the technology more stable than CMOS.

7.11.1.2 Power Consumption of BiCMOS versus CMOS Gates.
BiCMOS gates such as those shown in Fig. 7-52 behave like CMOS gates when they are not being switched, insofar as no significant dc current is drawn. When switching occurs, however, the power-dissipation behavior of the two gates differs. In both cases, all transistors are *ON* during the transition, implying that significant power is being dissipated due to conduction between V_{CC} and V_{SS}. In CMOS gates, this path is through the NMOS and PMOS transistors. In BiCMOS gates, the bipolar transistors are also *ON*, providing another low-impedance path between the supply rails, and thereby dissipating additional power. When driving another gate, however, the larger drive current of the BiCMOS gate causes the driven gate to spend less time in the transition region, which saves power. In addition, since BiCMOS gates generally do not swing from rail to rail, the smaller logic-signal swing results in a lower ac-power-dissipation level.

7.11.1.3 Capability of Providing TTL or ECL Outputs From a BiCMOS Chip.
If a digital system uses both ECL and TTL logic levels, the transfer of a signal from an ECL to a CMOS chip (or the other way around) requires a separate IC to translate the different voltage levels (see section 7.2.2). Since a BiCMOS chip can include the translation function, the result is a system with reduced chip count. Where low manufacturing costs are a major factor, BiCMOS may thus be chosen over ECL (albeit at the cost of some performance loss).

7.11.1.4 Increases in Process Complexity Associated with BiCMOS.
Additional processing steps must be added to either a baseline CMOS or a baseline bipolar process to create BiCMOS. In most cases, CMOS processes are modified, and 3 or 4 masks must be added to implement a high-performance digital or analog/digital BiCMOS process. These additional masking levels not only increase the cycle times of the process runs, but also decrease the manufacturing yield. Both factors contribute to an increase in relative processing costs (for a 12-mask CMOS process, the increase in die cost has been estimated to be 25-35%).[92] On the other hand, the process complexity of CMOS is also projected to continue to increase, while the number of

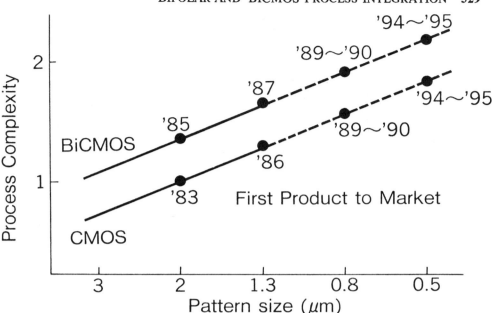

Fig. 7-54 Process complexity versus pattern size for CMOS and BiCMOS technologies.[93] (© 1987 IEEE).

masks needed to integrate bipolar transistors into a CMOS process is predicted to remain at 3 to 4. Thus, the *relative* added complexity of BiCMOS fabrication is projected to decline (Fig. 7-54).[93]

7.11.1.5 Extending Process-Equipment Life by Fabricating BiCMOS. The capital-equipment costs required for manufacturing CMOS with submicron device dimensions is increasing exponentially with decreasing dimensions (Fig. 7-55).[94] This implies that each incremental decrease in the circuit feature size requires a new generation of of ever-more-expensive production equipment.

Since BiCMOS does not have to be scaled as aggressively as CMOS to achieve the same speed performance, the existing generation of processing equipment can be used to fabricate circuits with performance capabilities of the next-generation CMOS technology. In addition, because the requirement of smaller power-supply voltages for 0.5-μm CMOS degrades performance,[95] switching to BiCMOS and staying with 0.8-μm design rules means that the power-supply voltage can be maintained at 5 V for a longer time period.

7.12 CLASSIFICATION OF BICMOS TECHNOLOGIES

Early (mid-1970s) BiCMOS chips were primarily "niche" products (such as BiCMOS metal-gate op amps and power ICs). The application of BiCMOS to mainstream IC products did not occur until the mid-1980s, when 5-V digital BiCMOS technologies were developed for logic (e.g., gate arrays) and memory (SRAM) applications as fabrication of device features smaller than 2 μm became possible. Such small feature sizes permitted the implementation of functional densities that would have consumed too much power if they had been fabricated using bipolar technologies alone. (At larger feature sizes, pure bipolar digital ICs dissipated high – although still tolerable – power levels, and BiCMOS thus did not yet represent a significant advantage.) However, when it became feasible to fabricate devices with feature sizes smaller than ≤1.5 μm, BiCMOs made it possible to implement circuits with these densities which consumed less power than bipolar ICs, but offered speeds faster than than those possible with CMOS alone.

Three categories of BiCMOS ICs have emerged: (1) low-cost, medium-speed 5-V digital; (2) high-performance, higher-cost 5-V digital; and (3) analog/digital. The distinctions among these are based on differences in the process flows used to produce them.

Low-cost BiCMOS parts are fabricated with slightly modified, single-well CMOS processes (i.e., only one or two mask levels are added to a baseline CMOS process). *High- performance digital BiCMOS ICs* are fabricated with twin-well, CMOS-based process flows that are significantly modified from the baseline CMOS process (e.g., three or four mask levels are typically added to the baseline flow). *Analog/digital BiCMOS ICs* are fabricated with processes designed to accommodate the larger voltage levels of analog applications (e.g., 10-30 V for low-voltage analog circuits, and more than 30 V for power applications). Analog/digital processes must also permit the production of the resistors, capacitors, and lateral *pnp* transistors needed in analog circuits, in addition to allowing the fabrication of *npn* bipolar and CMOS structures. Analog-circuit requirements are such that they can still be met by devices built with less aggressive design rules than those needed in digital circuits (i.e., 1.5-2.0 μm are sufficient for most applications).

Table 1-3 compares these three categories in terms of process complexity, with respect both to one an another and to a CMOS and bipolar technology of the same minimum-feature-size generation.[96] (It is assumed that all the processes compared use double-level metal, and that the MOS devices are fabricated with LDDs.)

Table 7-3 Comparison of BiCMOS Technology Process Complexity.

| Step | CMOS | BiCMOS | | | Bipolar |
		H. Perf.	Low cost	Analog	
Masks	12	15	13	16	13
Etches (RIE)	11	12	11	12	11
Epi	Optional	Required	Optional	Required	Required
Furnace	16	19	16	19	16
Implant	8	12	9	13	7
Metal	2	2	2	2	2
Total	49	61	51	63	50

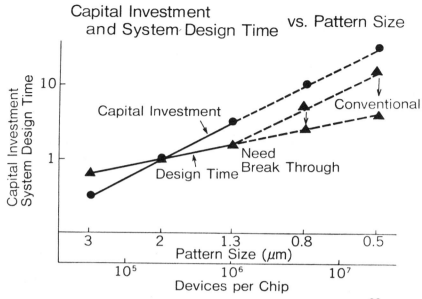

Fig. 7-55 Capital investment and system design time versus pattern size.[93] (© 1987 IEEE).

Since this book is essentially restricted to mainstream process-integration issues, the bulk of the BiCMOS process-integration discussion will be devoted to 5-V digital technology. However, issues involved in the fabrication of 10 to 30-V analog-digital BiCMOS technology will also be covered (see section 7.12.2).

7.12.1 Digital BiCMOS Technology

Digital ICs require high device density, low power dissipation, small propagation-delay times, and high noise immunity. Since the power-supply voltages of digital systems are normally 5 V or less, processes that produce devices with relatively small breakdown voltages can be used.

7.12.1.1 Low-Cost Digital BiCMOS Technology. As noted earlier, the first digital BiCMOS chips were designed with arrays of CMOS logic gates and bipolar I/O circuits. Since the I/O drivers could be designed to provide adequate drive current by increasing the size of the *npn* transistors, it was acceptable to use bipolar transistors whose performance characteristics were inferior to those used in high-speed ECL circuits (in fact, in applications where the on-chip bipolar transistors are restricted to totem-pole-buffer applications, no significant speed performance is gained by using the bipolar transistors fabricated with a high-speed digital process).[97] Such early digital BiCMOS ICs still represented an incremental performance improvement over their CMOS counterparts. However, as long as the bipolar devices were restricted to the I/O circuitry, the maximum benefits of a high-performance BiCMOS technology were not achievable.

Since even these early circuits contained many more CMOS than bipolar devices (due to the power-dissipation limitations of the latter), BiCMOS technologies have tended to evolve from CMOS processes. The initial approach used was to graft added steps to well-established CMOS processes to produce bipolar devices without degrading the characteristics of the CMOS transistors. Since less-than-optimum bipolar-transistor characteristics were acceptable in early BiCMOS designs, it was realized that such transistors could be produced by means of a baseline CMOS process with the addition of only one or two mask levels. As a result, only a small increase in fabrication costs was incurred.

The simplest low-cost BiCMOS technologies add only one mask to an existing n-well CMOS process (Fig. 7-56).[93] The added mask level is used to selectively produce lightly doped p regions, which serve as the p-bases of the bipolar transistors (~1 μm deep, with a doping level of about 1×10^{17} atoms/cm^3). The n-well fabrication step also forms the collector regions of the npn transistors. The heavy implant used to form the NMOS source/drain regions is employed to produce the emitter regions and collector contacts of the bipolar transistors. The p^+ source/drain implant is also used to create the base contact, since this serves to lower the series base resistance, R_B. Semirecessed LOCOS structures isolate both the CMOS and bipolar devices. High-performance CMOS gates are achieved with this approach. The starting substrate is either a p-wafer or a p-epi layer on a p^+ wafer.

Such simple processes are also called *triple-diffused (3D) BiCMOS processes*, because they produce triple-diffused bipolar transistor structures. As noted earlier, the main drawback of 3-D bipolar transistors is the large collector resistance. (When the n-well is used as the collector, the collector sheet resistance is typically around 2-kΩ/sq.) The large R_C values that result cause internal debiasing of the base-collector junction at high currents (*quasi-saturation*), and this reduces the drive current (which in turn increases the gate delay). R_C in such transistors can be reduced by increasing the transistor size. Doing so, however, increases the collector and emitter junction capacit-

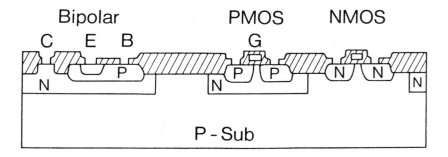

Fig. 7-56 3-D BiCMOS structure cross section.[93] (© 1987 IEEE).

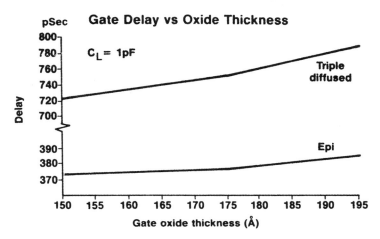

Fig. 7-57 BiCMOS inverter gate delay as a function of gate oxide thickness for 3-D and twin-well BiCMOS gates, $V_{CC} = 5$ V, and $L = 1$ μm.[135] Reprinted with permission of Semiconductor International.

ances, so that the bipolar drivers must then be driven with larger values of I_{DS} from the CMOS gates. To produce sufficiently large I_{DS} values, the performance of the CMOS gates must be optimized; this implies the use of thin gate oxides. Figure 7-57 shows the gate delay of a 3D BiCMOS inverter as a function of gate-oxide thickness.

A second, somewhat more complex BiCMOS fabrication process adds two masking layers to a baseline n-well CMOS process. We refer to this method as the *collector-diffusion isolation (CDI) BiCMOS process*, since the bipolar transistors have the same structures as those resulting from the CDI process described in chap. 3, section 2.1.1.2.

In this process, n^+ buried layers are formed in a p-substrate beneath the n-wells. (The use of heavily doped n^+ layers beneath the n-well not only reduces collector resistance but also reduces the susceptibility to latchup; see chap. 6.) A thin p-epi layer, which will serve as the base region, is then grown. The n-wells are formed next and are driven to a sufficient depth so that they merge with the buried n^+ layers. The merged n-regions also form the low-sheet-resistance collector regions and the isolation structures of the npn transistors. To provide collector regions with even smaller R_C values, a second additional mask is used to create a deep n^+ connection between the surface and the buried n^+ layers. Note that the bipolar collector regions are isolated from one another by the p-substrate and p-epi regions.

The CDI BiCMOS approach improves the performance of the npn transistors by decreasing R_C. However, the performance is still not as high as it is in ECL gates for several reasons:

- The base is formed from the relatively lightly doped epi layer, resulting in a large R_B value.

- The emitter region (created during the ion-implantation step that forms the NMOS source/drain regions) does not yield transistors with as high a performance as is the case when polysilicon emitters are used.

- The minimum spacing of n^+ to n^+ regions is limited by the lightly doped p-regions that separate them. Thus, the maximum packing density is smaller with the CDI process than with submicron-CMOS processes (which do not employ local n^+-buried regions).

7.12.1.2 High-Performance Digital BiCMOS Technology. This
technology strives to produce CMOS and bipolar devices with performances equal to those of devices found in pure high-performance CMOS and bipolar ICs (so-called *uncompromised* device structures). Two basic approaches have been taken: (1) modification of a p-well CMOS process through the addition of three masking steps, and (2) modification of a twin-well CMOS process through the addition of three or four masking steps (Fig. 7-58).

In the first process, a lightly doped n-epi layer is deposited on a p-substrate in which n^+ regions have been selectively produced (Fig. 7-59a). This forms the collector regions of the bipolar transistors (as is done in the *standard-buried collector [SBC]* bipolar process described in section 7.4). Thus, such processes are sometimes referred to as *SBC BiCMOS processes.*

The p-well step is also used to form the junction-isolation structures of the bipolar transistor collectors (also akin to the procedure used in conventional junction-isolated

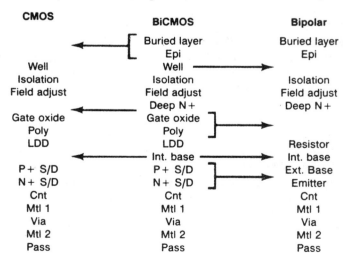

Fig. 7-58 Process flow for high-performance BiCMOS technology.

*Alternatively, trench isolation could be used, and such a process has in fact been reported.[88] Of course, process complexity is increased (the process just mentioned requires 18 masking steps).

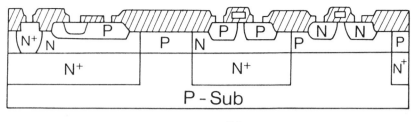

a)

Bipolar   Isolation   PMOS   NMOS

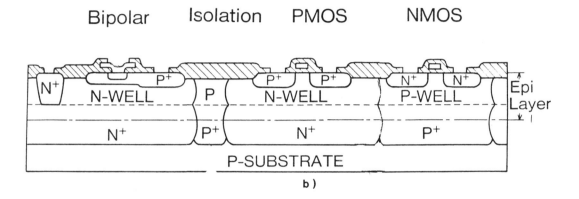

b)

Fig. 7-59 (a) Cross-section of SBC BiCMOS structure. (b) Cross-section of twin-well BiCMOS structure.[93] (© 1987 IEEE).

SBC processes). The three masking steps added to the p-well CMOS process are the n^+ buried layer mask, the collector deep-n^+ mask, and the base p-mask.[89,99]

While the SBC process is simpler than the modified twin-well process, the device performance is lower, for two reasons. First, the packing density is limited by the low p-substrate doping level (that is, the collector-to-collector spacing must be wide enough to prevent punchthrough from one bipolar collector to another). While the substrate doping could be raised to allow closer device spacing, this would increase the collector-to-substrate capacitance.* Second, the n-epi layer must be counterdoped to isolate the n-well regions and to form p-wells for the NMOS devices. However, adding sufficient p-dopant to the 0.3–1.0 Ω-cm, 1.5–2.0 μm-thick n-epi layer can result in not only processing problems but also in a reduction of the NMOS performance through mobility degradation.

Therefore, modified-twin-well BiCMOS processes (Fig. 7-59b) have been the most widely used for achieving high-performance, high-density digital BiCMOS ICs. Many such processes have been reported.[100,101,102]

7.12.1.3 Device-Design Issues of the Modified Twin-Well BiCMOS Process.
The design of the bipolar and CMOS device structures in BiCMOS circuits follows many of the guidelines presented earlier (both in this chapter and in chapters 5 and 6) describing the relationships between device characteristics and processing in NMOS and CMOS ICs. This section focuses on the impact of BiCMOS process integration on bipolar and CMOS device design.

The three circuit-performance parameters that must be optimized in a digital BiCMOS technology are speed, power dissipation, and noise margins. These parameters are dependent not only on device characteristics but also on the circuit architecture (which also plays a role in determining the ratio of FETs to *npn* transistors on the chip). Thus, device optimization is normally carried out after the circuit-architecture and application operating conditions have been specified. The optimum set of device characteristics is then evolved by making the appropriate trade-offs in the device-structure design. A fabrication process sequence can then be developed.

Many tradeoffs must be made to obtain the smallest possible values of τ_{pd}. For example, if the device dimensions and doping concentrations were selected merely to yield minimum values of τ_{pd}, many other device parameters would likely be degraded to an intolerable degree. Some speed-performance benefits must therefore be sacrificed. As noted, such trade-offs will also be a function of the circuit architecture.

For example, if it is decided to restrict the bipolar *npn* transistors to the I/O circuits and voltage regulators, relatively low-performance *npn* transistors may be acceptable (i.e., transistors with large parasitic junction capacitances and R_C values could be used). However, the MOS devices would then have to be able to supply a large I_{DS} so that the high capacitive loads of such *npn* transistors could be rapidly charged. This implies that short-channel FETs with thin gate oxides would be desirable.

On the other hand, if advanced, self-aligned *npn* transistors were used, the circuit speed could be limited by on-chip wiring instead of by the parasitic capacitances of the bipolar transistors. In such cases, ease of manufacturing considerations would make it advantageous to use a process that produced FETs with thicker gate oxides (and consequently, small I_{DS} values).

To make this kind of design decision it is necessary to analyze the gate delay as a function of length and oxide thickness should be analyzed (using device and circuit simulators). As was shown in Fig. 7-57, the use of thinner gate oxides in a BiCMOS process that uses low-performance bipolar transistors (3D) will reduce the gate delay to some degree, whereas the gate delay is almost independent of gate-oxide thickness in a process that uses higher-performance *npn* transistors (designated as *epi* transistors in Fig. 7-57).

The use of high-speed bipolar transistors in digital BiCMOS ICs will certainly improve the propagation delay of BiCMOS logic gates. However, the incremental improvement in τ_{pd} versus f_T of the *npn* transistors in the gate must also be studied. For example, Fig. 7-60 shows that as f_T is increased, τ_{pd} decreases.[103] However, for f_T's beyond about 8 GHz, the incremental improvement in τ_{pd} is small. Since significantly greater process complexity is required to fabricate *npn* transistors with f_T's of 14 GHz than 8 GHz, such an extra processing effort would most likely not yield

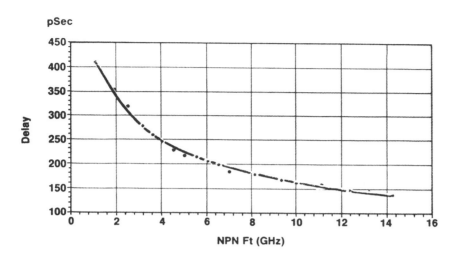

Fig. 7-60 Example of BiCMOS gate delay time versus the f_T of the *npn* transistors.[135] Reprinted with permission of Semiconductor International.

much improvement in speed performance.

The major design trade-offs in high-performance digital BiCMOS involve the characteristics of the epitaxial layer and the well profiles. From the perspective of bipolar device needs, the minimum epi thickness, $t_{epi(min)}$, is determined by the breakdown voltage of the bipolar transistors (BV_{CEO}), the collector capacitance, and the manufacturing controllability. As shown in Fig. 7-10b, for a BV_{CEO} of 7 V, a minimum epi thickness of ~1.25 μm is needed.[104] On the other hand, the epi layers should not be too thick. That is, thicker epi layers will reduce f_T and increase R_C, leading to decreases in drive-current capability. In addition, thicker epi layers require longer *n*-collector-diffusion heat cycles, and *p*-isolation diffusion times. The latter considerations generally set an upper bound on the epi-layer thickness.

From the perspective of CMOS device requirements, the use of *p*-buried layers beneath the NMOS devices for adjacent *n*-well isolation dictates that $t_{epi(min)}$ be large enough to prevent the NMOS body-effect coefficient and junction capacitance from being excessively increased (due to the updiffusion of the *p*-buried layers). However, by keeping the *p*-buried layer dose low and the epi layer sufficiently thick, it is possible to decouple these NMOS parameters from the *p*-buried-layer parameters.

The *n*-well in twin-well BiCMOS not only impacts the PMOS devices but also serves as part of the collector structure of the *npn* bipolar devices. Thus, its profile is derived through the following tradeoffs: From the bipolar-device perspective, the well should be doped heavily enough to prevent the Kirk effect, yet lightly enough to provide an adequately high value of BV_{CBO}. (A more heavily doped well will also cut off the ion-implantation-channeling tail of the base, thus producing a smaller base width and consequent higher f_T.) The specifications of minimum drive-current capability, f_T, and

BV_{CBO} dictated by the circuit application will drive the optimization of the well-profile design.

Both the n-well and p-well profiles must be selected so that the following device characteristics are optimized: (1) threshold voltage, (2) punchthrough voltage, (3) source/drain junction capacitance, and (4) body effect. A blanket-boron implant is typically used to provide the desired V_T for both the NMOS and PMOS devices. While the doping profile in the well beneath the surface must be high enough to prevent punchthrough, such higher concentrations will lead to larger source/drain capacitance and body effect values (see chap. 5). Compromises are therefore necessary.

An example of the details involved in the design of an n-well profile for a 0.8-μm digital BiCMOS technology is given in references 105 and 106.

7.12.1.4 A Process Sequence for Fabricating High-Performance 5-V Digital BiCMOS ICs.

The following process sequence was developed at Texas Instruments to implement a 0.8-μm high-performance digital BiCMOS technology and has been described in several reports.[105,107] Figure 7-58 shows the twin-well CMOS baseline flow and the additional process steps that have been added to integrate bipolar devices. Four masking steps are added to an advanced 0.8-μm, 12-mask CMOS process[108] to integrate high-performance bipolar transistors. Since most of the steps are essentially identical to those used in a twin-well CMOS process (described in detail in chap. 6), we will focus here on the steps that have been added to incorporate the bipolar devices. Note that of the 16 masking steps used in this BiCMOS process, the extra ones are #1, #4, #5, and #6.

The starting material is a lightly p-doped (~10 Ω-cm) silicon wafer with a (100) orientation. Buried n^+ layers (with doping levels exceeding 1×10^{19} atoms/cm^3) are selectively formed as described in the discussion of the SBC process in section 7.5.2 (Fig. 7-61a). *Mask #1* is used to pattern the buried-layer regions. Note that the lateral diffusion that occurs during the buried-layer anneal step (following the As or Sb implant) limits the spacing between adjacent subcollector regions and thereby also effectively sets the minimum n-well-to-n-well spacing.

Next, a selective boron implant ($1-2 \times 10^{13}$ cm^{-2}) is performed (Fig. 7-61b) which places additional p-dopant between the n^+ buried layers.* A self-aligned approach requiring only one mask to create both the n^+- and p-buried layers (as also described in ref. 106) is widely used, but two-mask processes can also be employed.

Following formation of the two types of buried layers, all oxide is removed from the wafer. A thin (1.0-1.5-μm), near-intrinsic epi layer (i.e., containing fewer than 1×10^{15} dopant atoms/cm^3) is then deposited. The use of such low background-doping levels in the epi layer makes it unnecessary to perform excessive counterdoping during n- or p-

---

* These p-buried-layer regions serve to maintain adequate punchthrough breakdown-voltage values between adjacent n^+ buried layers. The p-concentration must thus be high enough to provide isolation between the buried n^+ layers, but not so great as to produce high sidewall capacitance for the n^+ buried layer.[106]

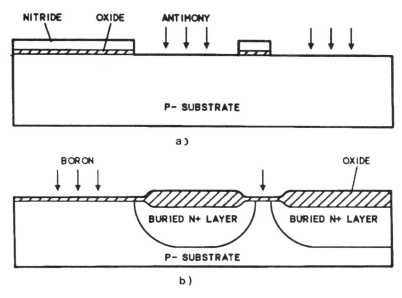

Fig. 7-61 (a) Formation of the n^+-buried layers. (b) p-implant self-aligned to the n^+ buried layers to form p-buried layers.[107] Reprinted by permission of Semiconductor International.

well formation (which would degrade mobility). As described in the previous section, the choice of epi thickness is dependent on the particular circuit-performance requirements.

Nevertheless, the growth of a lightly doped, thin epi layer over two types of buried layers presents some difficult challenges for the epitaxial deposition process. Autodoping for both types of dopant must be minimized to avoid the need for excessive counterdoping in the wells. It has been shown that arsenic autodoping can be suppressed through the use of reduced-pressure epitaxy, but that, conversely, boron autodoping increases at lower pressures (Fig. 7-62).[109,110,123] Antimony autodoping is only slightly reduced by a reduction in pressure. Compromises must thus be made in the selection of optimal epitaxial-growth conditions.

A novel two-step epi process has been reported[111] in which a high-temperature (1150°C), low-pressure (10 torr) SiH_2Cl_2-epi cap is deposited first, followed by a low-temperature (900°C), low-pressure (10 torr) SiH_4-epi-layer deposition. This process produces an abrupt epi substrate transition and minimal arsenic and boron autodoping.

The steps for forming the twin wells are conducted next (Fig. 7-63a). The doping profiles in the wells are tailored to optimize both the CMOS and bipolar device characteristics, as was described in the previous section. A single-mask process can be used to form both well types, as described in chapter 6 (*Mask #2*). In the TI process, a dual-phosphorus implant (deep and shallow) is used to form the n-well. Following the implant steps that introduce the well dopants, a well drive-in step is performed.

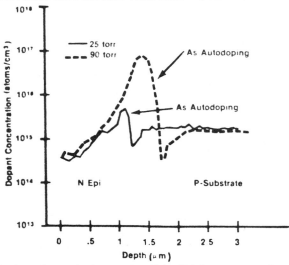

Fig. 7-62 Arsenic lateral autodoping peak off buried layer comparing 90 torr to 26 torr, 1080 °C (high-temperature/low-pressure) epi process.[110] Reprinted with permission of Solid State Technology.

The active regions are then defined using *Mask #3*. A semirecessed LOCOS-isolation process is used to form the field oxide between the active regions (as was also described in chap. 6, section. 6.6). A boron channel stop is implanted prior to the field-oxide growth (Fig. 7-63b). In some processes, advanced semirecessed isolation structures are used (e.g., poly-buffered LOCOS or SILO), with a high-pressure field-oxidation growth step included to minimize dopant redistribution.

Following the stripping of the nitride, *Mask #4* is used to pattern the deep n^+ collector contact. A deep phosphorus implant and subsequent drive-in/anneal step are used to form the deep-collector-contact structure. A long thermal drive would have to be used to diffuse the implanted dopants all the way through the epi layer. However, because the lateral diffusion of the dopants from such a thermal cycle would be excessive for submicron BiCMOS processes, only a reduced thermal cycle can be tolerated. Unfortunately, this results in increased collector resistance.

To overcome this problem, a polysilicon-plug deep collector contact has been developed. A trench is etched in the silicon down to the n^+ buried layer. This trench is refilled with heavily in-situ-doped polysilicon to form a low-resistance plug to the buried collector. The sidewalls are covered with a dielectric to prevent lateral diffusion (Fig. 7-64). The use of this structure also reduces the area of the bipolar transistor by ~15% compared to devices that use conventional deep n^+ collector contacts. Following completion of this step, *Mask #5* is used for patterning the *p*-base and diffused resistor regions. A boron implant and drive-in is used to form these regions.

In some processes, an oxide is selectively grown over the base and resistor regions prior to the base implant. The boron is then implanted through the oxide (Fig. 7-65a).

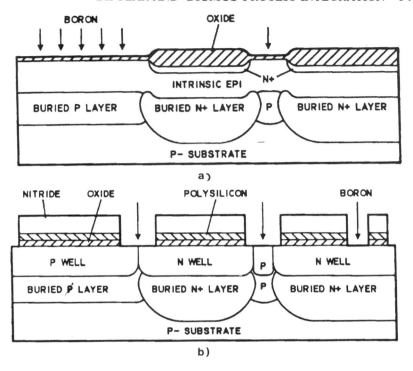

Fig. 7-63 (a) *p*-well implant self-aligned to oxide-masked *n*-well. (b) Channel-stop implant after etch of field isolation regions.[107] Reprinted with permission of Semiconductor International.

In this approach, the oxide-growth thermal cycle does not impact the base-doping profile. In other processes the oxide is deposited by means of CVD and is then selectively removed from the non–base/resistor regions. The purpose of the oxide (regardless of how it is formed) is to screen the base regions from the V_T-adjust boron implant that follows the base implant. That is, no mask is used during the V_T-adjust implant, but the oxide over the base regions prevents it from affecting the base. Following the base implant, the most recently formed oxide is removed.

In processes that use a polysilicon emitter, the emitter may be formed next. In the example TI process, this is accomplished through a *split-poly* process. A gate oxide is grown and is immediately capped by an initial poly layer. *Mask #6* and an etch step are then used to create openings down to the silicon substrate (i.e., by etching through the polysilicon and gate oxide). A second layer of poly is then deposited to make contact with the substrate. A blanket phosphorus implant is used to dope the poly so that a later thermal step will cause the phosphorus to diffuse and form the emitter region (see section 7.8.1). With the emitter formed in this fashion, the gate oxide of the MOS devices is protected by the first layer of poly from damage and/or contamination during the dry-etch, resist-strip step following the opening of the emitter contact. Other high-performance BiCMOS processes that have used directly implanted emitters have also been reported.[112]

The polysilicon film is next patterned and etched using *Mask #7* and an anisotropic dry-etch process. *Mask #8* is used to selectively implant the phosphorus dose for the *n*-type LDD region. A CVD-oxide layer is then deposited and anisotropically etched back to form sidewall spacers on the patterned poly lines. *Mask #9* is used to selectively implant the NMOS source and drain regions, the collector and *n*-well contacts, and the NMOS gate-polysilicon lines. *Mask #10* is employed to allow selective implantation of the PMOS source and drain regions, the substrate and *p*-well contacts, the extrinsic base region of the bipolar transistor, and the PMOS gate-polysilicon lines (Fig.7-65b).* Following these implants, a 900°C drive-in/anneal step is used to simultaneously form the diffused emitter and extrinsic base and to activate and drive-in the dopants of the NMOS and PMOS source/drain regions.

A silicidation of the polysilicon layers and the source and drain regions with a $TiSi_2$ is then performed, and a TiN local-interconnect layer is formed at the same time. As described in chapter 3, section 3.11.2.4, patterned TiN straps (~15 Ω/sq) are used as extensions of the source/drain regions and as local interconnects. *Mask #11* is used to pattern the TiN.

A double-level metal interconnect is then implemented. *Mask #12* is used to pattern contact holes in the CVD oxide deposited following the TiN processing. After Metal 1 is deposited, it is patterned with *Mask #13* and dry etched. *Masks #14* and *#15* are used to pattern the vias in the intermetal dielectric and the Metal 2 layer, respectively (Fig. 7-65c). *Mask #16* is the pad mask.

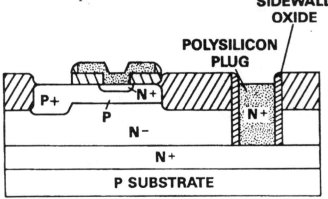

Fig. 7-64 Cross section of a bipolar transistor with a n^+ polysilicon plug used to minimize the collector series resistance. From A. R. Alvarez, Ed., *BiCMOS Technology and Applications*. Copyright, 1989, Kluwer Academic Publishers. Reprinted with permission.

---

* Note that in this example process dual-doped gate-polysilicon lines are used. A dual-doped poly process can be successfully implemented because a TiN local interconnect layer is used to strap *p*-doped and *n*-doped poly lines together. The diffusion-barrier properties of TiN prevent counterdoping of one poly type by the other; see chap. 3, section 3.11.2.3.

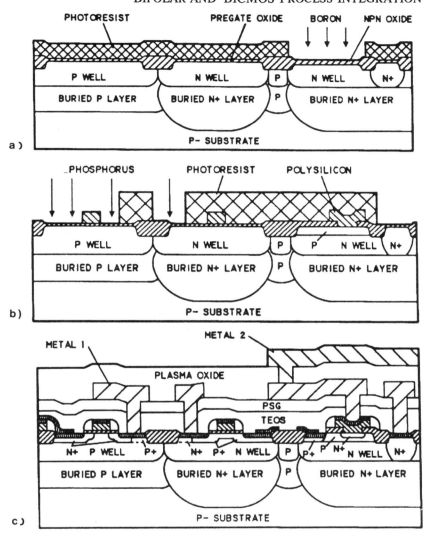

Fig. 7-65 (a) Patterned p^--type implant for forming base and resistor. (b) Phosphorus is implanted to form a shallow, n-type LDD region that is self-aligned to the edge of the NMOS gate. (c) Device cross-section following completion of double-level metal interconnect.[107] Reprinted with permission of Semiconductor International.

7.12.2 Process Integration of Analog/Digital BiCMOS

Many electronic systems employ both analog and digital circuits, and thus having the option of producing both types of circuit functions on the same chip can provide significant benefits. For example, CMOS can be used to minimize dc power dissipation and provide high-impedance FET inputs for sample-hold operations. Bipolar devices can not only provide high current gain and extended gain-bandwidth capabilities, but they can also be used to minimize noise factors and provide good on-chip voltage references. This section describes the issues involved in designing BiCMOS processes that permit simultaneous fabrication of high-performance analog and digital functions. The applications of such chips will be described in section 7.12.3.4.

The differences between analog/digital BiCMOS and 5-V digital-BiCMOS processes stem primarily from the fact that analog functions generally operate over a much wider range of power-supply voltages (e.g., higher than 10 V) and power-dissipation levels. Since these higher voltages exceed some of the maximum operating limits of devices fabricated with 5-V-compatible BiCMOS processes (e.g., hot-carrier-degradation effects and gate-oxide breakdown and punchthrough voltages in MOS devices, and BV_{CEO} and BV_{EBO} values in bipolar devices), the device structures and process sequences must be significantly modified to make them suitable for analog/digital applications.

Two subcategories of analog-compatible BiCMOS processes have emerged to address these applications: (1) medium-voltage (10-30 V) processes that modify a baseline CMOS flow, and (2) high-power (>30 V and >1 A) processes based on power-bipolar process flows. Since the topic of power-IC technology is outside the scope of this text, our discussion will be limited to the process-integration issues of medium-voltage analog/digital BiCMOS ICs.

Medium-voltage analog/digital processes (10-30 V) seek to optimize the performance of both CMOS and bipolar devices in the face of voltages that are higher than those encountered in strictly digital applications.* Some CMOS speed performance must nevertheless be traded off to gain reliable operation at the increased voltage levels (e.g., thicker gate oxides are needed to withstand the higher gate voltages, and I_{DS} will thus be reduced.) A number of other necessary compromises are discussed in the following section, followed by details of a low-to-medium-voltage analog/digital BiCMOS process flow.

Analog functions also require passive circuit components (which preferably also exhibit small temperature and voltage coefficients), and the process sequence must be able to accommodate their fabrication. Resistors fabricated in polysilicon are preferred for precision applications because they exhibit smaller temperature coefficients than do resistors made in diffused regions of the substrate (e.g., 500-1500 ppm/°C versus 2500-3000 ppm/°C, respectively). Appendix A deals with the general aspects of IC resistor fabrication, and section 8.1.2.3 in chapter 8 gives additional details concerning the fabrication of high-valued polysilicon resistors.

High-performance capacitors are also key elements for analog CMOS and BiCMOS technology, especially in the areas of A/D converters and switched-capacitor filters. Polysilicon-polysilicon MOS capacitors find the widest use because they exhibit smaller

---

* Most high-performance, medium-voltage analog/digital BiCMOS circuits require 10-15-V operating voltages in order to maintain high signal-to-noise ratios and to permit the use of cascading in analog designs, whereas lower-performance circuits used in automotive applications are typically operated between 15 and 30 V.

parasitic effects than do polysilicon-silicon capacitors. The polysilicon films must be heavily doped to minimize depletion effects in the capacitor electrodes.

A recent report, however, has described polysilicon-silicon capacitor structures for single-poly BiCMOS processes.[113] These capacitors exhibit high capacitance per unit area, low voltage coefficients (e.g., lower than 50 ppm/V), and good dielectric quality. The capacitor interplate dielectric can be an oxide that is grown simultaneously with a gate oxide, or it can consist of oxide-nitride-oxide (ONO) layers that need one more mask than that of the oxide structure. These capacitor structures can also be integrated into a BiCMOS process in such a way that the capacitor implant (which is a high-dose, high-energy n^+ implant that forms a junction with a depth of about 1.2-1.6 μm) can also form the collector plug without the need for extra masks.

7.12.2.1 Process-Integration Issues of Medium-Voltage Analog BiCMOS.

For analog-function circuit design, the most important device characteristics are those of the *npn* transistors. High gain is required to reduce input bias current into the transistor and to prevent the loading of a previous circuit stage. A value of 100 or more is usually satisfactory, and f_T values of at least 5 GHz for 10-V ICs and 2 GHz for 20-V ICs are desirable for high-frequency applications. (Such β and f_T values can be achieved with the example 20-V process sequence presented in section 7.12.2.2.) The transistors must also have low R_C values (e.g., less than 100 Ω) to allow high-current operation, as well as high Early-voltage (V_A) values (e.g., greater than 50 in a 10-V process and 25 in a 20-V process), since, as noted in section 7.3.2, the intrinsic small-signal voltage gain of an amplifier is proportional to V_A. Although low R_C values (e.g., less than 100 Ω) are achieved with the use of buried n^+ layers, thin epitaxial layers, and deep collector plugs, analog voltage requirements force the use of thicker epitaxial layers (which also reduce the f_T of the transistor). Similarly, large V_A values are obtained with the use of higher base Gummel numbers, but this also degrades the value of f_T.[114] Tradeoffs must thus be made to arrive at optimum *npn* device designs.

Isolated *pnp* transistors are also needed in some analog circuits. Although lateral *pnp* transistors are usually "free" devices, their speed performance is consequently poor. That is, while they can exhibit values of β as high as 100 and large Early-voltage values, these transistors are slow (f_T = 5-10 MHz) because of the need for wide base regions to prevent emitter/collector punchthrough. As result, their use must be avoided in high-frequency signal paths. (A process that allows vertical *pnp* transistors to be integrated into a 5-V analog/digital BiCMOS process is described in ref. 115.)

The higher operating voltages also limit the minimum gate lengths of MOS devices to 2-3 μm in analog/digital BiCMOS to prevent drain-induced barrier lowering (see chap. 5, section 5.5.2) and excessive channel-length modulation. Reduced junction capacitance, which implies the use of lightly doped wells, is thus desirable. Furthermore, since analog designs often use both positive and negative supply voltages, isolated CMOS structures are desirable. Finally, thicker gate oxides must be used (e.g., a minimum gate-oxide thickness of 30 nm must be used to allow reliable operation at 10 V).

Thicker field oxides are also usually needed to provide field-region threshold-voltage values that exceed the maximum operating voltage. Likewise, the CMOS devices must be protected against hot-carrier effects that arise as a result of exposure to higher supply voltages. Latchup is another concern due to the high substrate currents common in many analog BiCMOS designs.

7.12.2.2 An Example of an Analog/Digital BiCMOS Process. The

following process sequence for fabricating medium-voltage analog/digital BiCMOS ICs, developed by Texas Instruments,[116,117] it is used to manufacture ICs capable of operating from 20-V supplies. The first-generation process uses 3-μm design rules (although these are predicted to be scaled to less than 2 μm by the third generation). Typical parameters for the devices are the following: β of 80, f_T of 2.5 GHz, and breakdown voltages of 25, 40, and 8 V, respectively, for BV_{CEO}, BV_{CBO}, and BV_{EBO}.

The process is a modified n-well CMOS process with four additional masking steps to permit the integration of high-performance bipolar transistors. The bipolar devices are of the collector-diffused isolation (CDI) type, which means that n^+-buried layers and a p-epi layer are used. (Note that the buried layers are also placed beneath the n-wells to increase latchup immunity.) Bipolar transistors fabricated with CDI need no deep p^+-isolation diffusions (unlike npn transistors fabricated with the SBC process), yet they exhibit much lower lower R_C values than do 3-D bipolar transistors. They can therefore be made smaller than transistors produced with the SBC process. In CDI, the n-well region serves as the npn collector region, and the p-epi layer provides the sidewall junction-isolation area for the collectors of the bipolar transistors and the n-wells.

The starting material is a p-type wafer with (100) orientation. The wafer surface is first oxidized. *Mask #1* is then used to open windows in the oxide wherever n^+ buried layers are to exist. These buried layers are formed by means of an antimony implant and diffusion, and the oxide is then stripped. (Since the formation of the n^+ buried layers takes place first, it does not impact the CMOS device profiles.)

The p-epi layer is deposited next. The resistivity and thickness (t_{epi}) of this layer impact both the MOS and bipolar device properties, so some trade-offs must be made to achieve optimum characteristics in both device types. Typically, however, the resistivity of the epi layer is selected to be identical to that of the substrate doping employed in the n-well CMOS process, so that the CMOS device characteristics will not be impacted by the epi resistivity value.

The value of t_{epi} is more difficult to specify, since it depends on such device dimensions and processing factors as the n-well junction depth, the n-well doping level, the n^+ buried-layer updiffusion, the maximum operating voltage, the maximum β of the npn transistors, and the thickness of the epi layer that is removed during subsequent oxidations.

The selection of the epi thickness is begun by satisfying the requirement that the n-wells must extend down to contact the n^+ buried layers (to establish low-resistance collector regions). This dictates the *maximum* epitaxial-layer thickness ($t_{epi\,(max)}$) for a given n-well depth (as illustrated in Fig. 7-66). The fact that the n^+ buried layers typic-

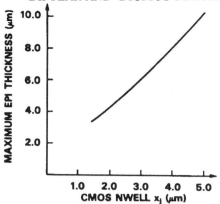

Fig. 7-66 The maximum p-epitaxial layer thickness is a function of the CMOS n-well diffusion depth.   Approximately 1.0 μm of the p-epitaxial layer is removed through oxidations in this example.[107] Reprinted with permission of Semiconductor International.

ally up-diffuse a distance of 80-100% of the well depth, also plays a role in determining $t_{epi\,(max)}$.

Once $t_{epi(max)}$ has been determined from the well-formation design considerations, the *minimum* epi-layer thickness ($t_{epi(min)}$) must be determined. This value depends on three interrelated parameters: the maximum β that is to be exhibited by the *npn* transistors, the maximum operating voltage, and the n^+ buried-layer updiffusion. We next illustrate how to determine $t_{epi\,(min)}$ based on these factors.

The maximum operating voltage of a bipolar circuit is usually limited by the BV_{CEO} breakdown mechanism, which depends on BV_{CBO} (plane) and β of the transistor, according to Eq. 7-13.

The value of BV_{CBO} (plane) is dependent on the epitaxial thickness. As the thickness is reduced, the vertical distance between the base and the edge of the up-diffused n^+ region is decreased, which reduces BV_{CBO} (plane). Figure 7-67a gives an example of the doping profile in a typical *npn* transistor fabricated with a base-collector junction depth of 1.0 μm, a 4.5-μm-deep *n*-well, and an 8.0-μm-thick *p*-epi layer. Figure 7-67b then illustrates how BV_{CBO} in such a device varies as the epi thickness is decreased. With Eq. 7-13 and the information from Fig. 7-67b used, $t_{epi(min)}$ can be selected once the maximum operating voltage and maximum *npn* β have been specified.

Example: Assume that an analog/digital BiCMOS process is used to produce bipolar *npn* transistors with doping profiles as shown in Fig. 7-67a. The ICs must be able to withstand a maximum operating voltage of 15 V. The maximum β of the transistors is designed to be 150. Select an appropriate value for the epi-layer thickness, assuming that the parameter n in Eq. 7-13 has a value of 4.

Solution: Using Eq. 7-13, we calculate that in order for BV_{CEO} to exceed the maximum operating voltage of 15 V, the minimum value of BV_{CBO} (plane) would have to be 53 V. From Fig. 7-67b, we see that the $t_{epi(min)}$ needed to achieve this BV_{CBO} value would be 7.0 μm.

Following the epitaxial deposition, the wafer is oxidized again, and *Mask #2* is used to open windows in the oxide wherever *n*-well regions are to exist. The *n*-wells are formed by means of ion implantation and diffusion (Fig. 7-68a). The wells are partially diffused with one thermal cycle, and the subsequent deep n^+ collector cycle completes the *n*-well drive-in. The *n*-well doping levels are set by parameters that produce high-performance CMOS devices.

Mask #3 is used to produce the patterns of the deep n^+ collector regions, and these regions are then formed (e.g., through the diffusion of phosphorus alone, or through a high-energy, high-dose implant of phosphorus, followed by diffusion).[113] Since a high temperature is needed for the diffusion, this step is also performed before the channel-stop steps to prevent adverse impact on CMOS device characteristics.*

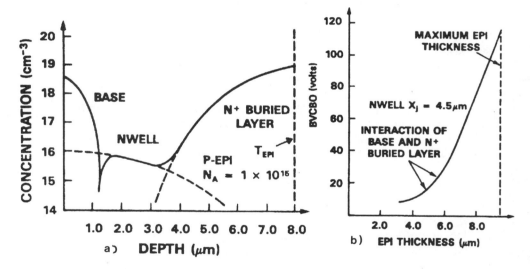

Fig. 7-67 (a) BiCMOS *npn* doping profiles for BV_{CBO} (plane) calculations. The solid line shows the net concentration while the dotted lines show the original profiles. (b) BV_{CBO} (plane) as a function of *p*-epitaxial layer thickness for a 4.5 μm *n*-well process. From A. R. Alvarez, Ed., *BiCMOS Technology and Applications*. Copyright, 1989, Kluwer Academic Publishers. Reprinted with permission.

* Note that the use of deep diffusions to form the buried layers, wells and, n^+ collector contact wastes silicon area due to lateral diffusion. Trench isolation could be utilized to reduce this loss of area and thereby increase packing density. However, the cost of the added complexity of would have to be weighed against the benefit gained.

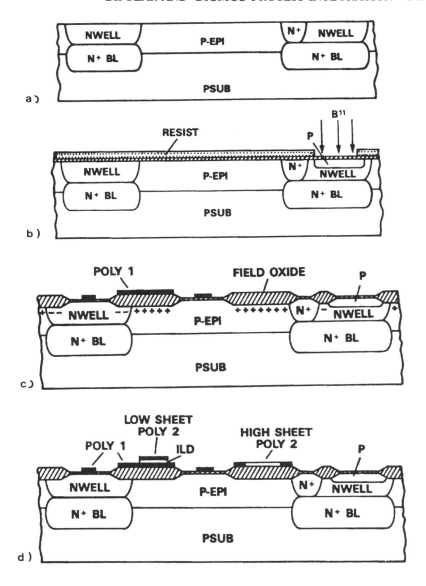

Fig. 7-68 (a) Formation of *n*-well and *n*⁺ collector regions following the *p*-epitaxial layer deposition. (b) Definition of the base regions. (c) Deposition and definition of the 1st polysilicon layer for CMOS gates and MOS capacitors. (d) Formation of the 2nd polysilicon layer for resistors and capacitor plates.

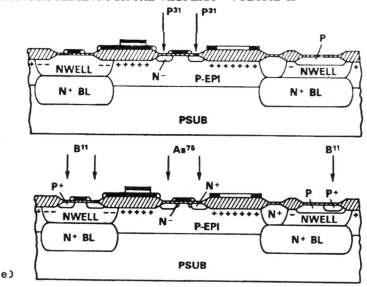

Fig. 7-68 (e) Formation of the n^+ emitter and collector regions. From A. R. Alvarez, Ed., *BiCMOS Technology and Applications.* Copyright, 1989, Kluwer Academic Publishers. Reprinted with permission.

After the oxide has been stripped, a thin pad oxide is grown. The base regions of the bipolar *npn* transistors are patterned with *Mask #4* and are then formed by means of ion implantation of boron (Fig. 7-68b). This is followed by an anneal in an inert atmosphere to activate the implant and reduce the implant damage (this step serves to reduce R_B).

A nitride layer is deposited over the pad oxide and *Mask #5* is used to define the active areas. After the nitride has been etched, a blanket phosphorus implant is carried out to produce a channel-stop dopant layer that will increase the threshold voltage of the parasitic field-oxide PMOS devices. Subsequently, another masking step is used (with *Mask #6*) to pattern the regions where a boron channel-stop implant is used to raise the threshold voltage of the parasitic NMOS field-oxide devices.

The resist is stripped, and a 1.0-μm-thick field oxide is grown. The nitride is then removed. A gate oxide of 35-50 nm is grown and a threshold-adjust implant carried out. The first level of polysilicon (*poly 1*) is deposited, doped, patterned (with *Mask #7*), and etched to produce the gate electrodes of the MOS devices and the bottom plates of the poly-to-poly capacitors (Fig. 7-68c).

Next, a 30-100-nm-thick capacitor interlevel-dielectric is formed over poly 1 (by means of either an oxide or combination oxide-nitride-oxide [ONO] film). A second layer of polysilicon (*poly 2*) is then deposited and is appropriately doped to produce a high sheet-resistance film suitable for fabricating high-valued resistors. *Mask #8* is then used to protect these high-resistivity regions against the implant that follows. The unprotected regions are implanted with a heavy dose of arsenic so that they can be used as the capacitor top plates and resistor contacts. *Mask #9* is then used to pattern poly 2. The subsequent etching of this layer completes the fabrication of the circuit resistor and capacitor structures (Fig. 7-68d).

The rest of the process follows conventional CMOS and bipolar process steps. *Masks #10* and *#11* are used to independently pattern the CMOS source and drain regions (with the p^+-base-contact region of the *npn* transistors also being formed with one of these masking steps). NMOS LDD structures are formed, with a phosphorus implant used to form the lightly doped extensions of the drain and a heavy arsenic implant used to produce the source/drain regions of the NMOS devices. A boron implant is utilized to create the source/drain regions of the PMOS devices (and the p^+ base-contact regions).

Mask #12 is used to pattern the emitter and collector contact regions (Fig. 7-68e). A separate phosphorus implant is employed to produce these regions, since attempts to use the source/drain implants to also form the emitter regions have resulted in lower MOS g_m values.[118] Furthermore, the deeper E-B junctions typically needed in analog devices are inconsistent with the requirements for high-density CMOS logic devices. This separate emitter-forming implant step allows both problems to be avoided, although at the cost of an additional masking step. It should also be noted that polysilicon emitters are also generally avoided in analog/digital processes because they yield higher (and relatively nonreproducible) emitter-contact resistance values than do directly-ion-implanted emitters. *Masks #13* and *#14* are used to pattern the contact holes and the metal interconnect layer, respectively. *Mask #15* is used to pattern an opening in the passivation layer. Note that if a double-level-metal process is used, the mask count would increase to 17.

7.12.3 BiCMOS Applications

7.12.3.1 Digital Logic Circuits and Gate Arrays. As noted earlier, one obvious way to integrate CMOS and bipolar devices on a single chip is to relegate the logic-function duties to high-density, low-power CMOS devices and to restrict the use of bipolar devices to the I/O circuits (Fig. 7-69).[119] This approach was used in early digital BiCMOS ICs.

More advanced, higher-performance digital BiCMOS integrated the bipolar transistors into the logic gate (Figs. 7-70). Like conventional CMOS gates, these *BiCMOS logic gates* consume no dc power (except for leakage current), but they offer the possibility of smaller values of τ_{pd} when gates of comparable size are fabricated. Thus, although an area penalty is incurred when bipolar drivers are added to a CMOS gate (~20% increase in area for a CMOS inverter), the gate is actually denser in terms of its load-driving capability. Furthermore, the percentage increase in gate area declines as the fraction of the gate composed of CMOS components increases. Consequently, in some large, high-performance BiCMOS gate arrays (e.g., those containing 30 – 100K gates), bipolar devices are for the most part incorporated into such larger cells as adders, shift registers,

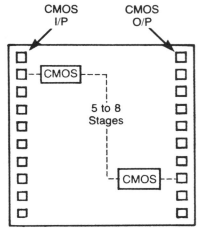

Fig. 7-69 BiCMOS gate array with CMOS gates and bipolar I/O.[119] Reprinted with permission of VLSI Design.

and ST flip-flops. In these cells, the overhead penalty of using BiCMOS is decreased to as little as 5%.

BiCMOS logic gates are also attractive for gate-array applications because the fan-out in logic circuits implemented with such gate arrays is frequently 3-5 (and, in some instances, even larger; see Fig. 7-71, which illustrates the typical fan-out distribution in a modern VLSI gate array).[120,121] Such large fan-out represents a heavy load capacitance, and BiCMOS gates can drive such loads faster than CMOS gates. Furthermore, the device size in gate arrays is typically uniform (for ease of physical design). Therefore, the delay degradation per unit load for CMOS circuits generally differs with the circuit function. In BiCMOS, the load degradation is practically the

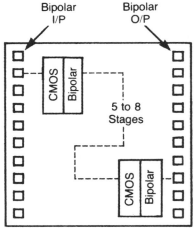

Fig. 7-70 BiCMOS gate array with BiCMOS gates and bipolar I/O.[119] Reprinted with permission of VLSI Design.

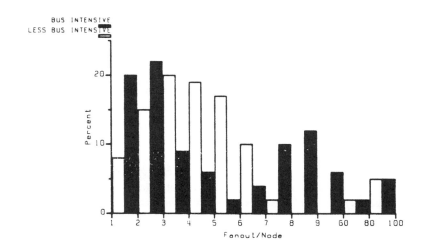

Fig. 7-71 Typical gate array fanout per internal output for options of low and high internal bus utilization.[120] (© 1988 IEEE).

same for all circuit functions, since the bipolar push-pull devices isolate the CMOS circuits from the loading. This makes the semicustom design task easier.

In some arrays, macrocells with uncommitted CMOS and bipolar drivers are offered (Fig. 7-72).[122] The user can connect the transistors in each cell to provide an appropriate MOS and bipolar device mix for each logic gate. Many gate functions can be implemented by connecting MOS devices within a single macrocell through the use

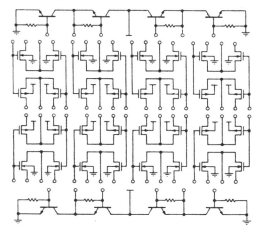

Fig. 7-72 Example of BiCMOS gate array macrocell.[122] Reprinted with permission of Electronics.

of local interconnects. Since bipolar drivers are available for heavily loaded on-chip nodes, the MOS devices do not have to be designed to drive global interconnects (as must be done in pure CMOS arrays). Hence, gate widths in MOS gate arrays can be made much smaller than those in CMOS.

The large signal swing in CMOS often results in noise spikes. BiCMOS implemented with ECL outputs offers much smaller signal swings, so switching noise can also be better controlled.

BiCMOS gate arrays with gate delays of 0.8-0.9 ns have been fabricated using 1.5-2.0-μm technology. If 0.8-μm design rules are used, gate delays as small as 0.3 ns are expected.

7.12.3.2 Interface Driver Circuits. The latest generations of 32-bit microprocessors are being fabricated in CMOS. The off-chip buses that must be driven by these microprocessors represent large capacitive loads. *Interface driver circuits*, which have traditionally been implemented in bipolar technologies, are used to increase bus-driving speeds. A typical 32-bit microprocessor system, however, may contain 10 or more interface devices. Just one bus driver at a time is activated, causing a typical driver to be in the disabled state as much as 90% of the time. When the interface units are purely bipolar, the disabled drivers are all burning up power, since they remain in the high-impedance state awaiting their turns to be activated.

However, when BiCMOS bus drivers using cells similar to those described in the previous section are utilized, the current drawn by the disabled drivers is much lower. In some cases, power savings of nearly 100% can be achieved. Since studies indicate that traditional bus interfaces consume as much as 30% of a system's total current, such a savings can be significant. Propagation-delay times for the BiCMOS drivers are also comparable to those in advanced bipolar technology (e.g., 3-5 ns).

7.12.3.3 BiCMOS SRAMs. BiCMOS SRAMs perform better than CMOS or bipolar SRAMs in certain applications (for example, large, high-speed ECL SRAMs [up to 64 kbits] have power-dissipation and yield problems [see chap. 8, section 8.2.2.1]). On the other hand, CMOS alone cannot be used to build high-performance, higher-density SRAMs, since the driving capability of CMOS is inferior to bipolar and since it is practically impossible to design input and output buffer circuits in CMOS that have an ECL I/O capability.

BiCMOS SRAMs, however, offer densities and power-dissipation levels close to those of CMOS SRAMs, but with higher operating speeds. Figure 8-8 in chapter 8 shows the speed-versus-density characteristics of bipolar, CMOS, and BiCMOS SRAMs.[34] ECL I/O BiCMOS can be seen to offer higher speeds than bipolar ECL at middle and higher densities (because of the severe speed, power, and density trade-off in bipolar ECL). In addition, BiCMOS SRAMs are somewhat faster than CMOS SRAMs at equivalent power and density for the same geometry. BiCMOS thus offers a range of speed and power trade-offs (from very fast, low-density BiCMOS at high power, to slightly slower, mid-density memory circuits at moderate power).

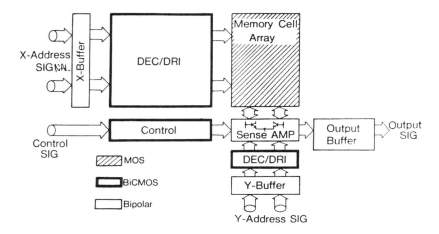

Fig. 7-73 Block diagram of a BiCMOS SRAM.[93] (© 1987 IEEE).

The memory array in a BiCMOS SRAM typically consists of NMOS cells in a p-well, so the die areas of CMOS and BiCMOS SRAMs are approximately the same. The decoders, word-line drivers, and write drivers are implemented with BiCMOS logic gates. (Control-logic functions with small fan-out, however, use CMOS.) The sense amplifiers, which require high input sensitivity, use pure bipolar circuits (Fig. 7-73).[123] In TTL BiCMOS SRAMs, the I/O buffers are typically BiCMOS gates, while in ECL BiCMOS SRAMs, bipolar ECL gates are used for I/O buffers.

7.12.3.4 Analog/Digital Applications.

BiCMOS technology can also be exploited to build ICs that can perform both analog and digital functions on the same chip (e.g., ASIC chips with arrays of analog and digital cells). A large fraction of the chip area is typically set aside for the CMOS cells (mostly logic gates) that are used to perform digital-signal-processing functions. The remainder of the chip (~15-20%) is dedicated to the analog cells that permit the chip to interface with the analog world. These analog cells include the following: I/O blocks (containing resistors and *npn* transistors); cells containing *npn* and *pnp* transistors for building op amps, reference voltage sources, and comparators; and cells with large, high-breakdown-voltage (e.g., 60-V) *npn* transistors. The latter are used to directly drive solenoids, lamps and LEDs and to rapidly discharge external capacitors. These ASIC chips can be used to implement such circuits as printer or camera interfaces, floppy-disk data separators, video-recorder controllers, and telecommunications circuits (e.g., PCM codec filters, ISDN transceivers, and modems).

Note that since the turn-on voltage of the bipolar transistor (V_{BE}) is much more precisely controlled than the threshold voltages of MOS devices (which are sensitive to manufacturing processes and device dimensions), well-matched pairs of bipolar transistors are much more easily obtained. Such well-matched pairs allow bipolar-IC operational amplifiers to exhibit offset voltages that are more than an order of magnitude lower than those found in MOS op amps. Thus, BiCMOS analog/digital circuits can offer the low-input offsets and high gains of bipolar-circuit components, together with high packing density and low power of CMOS. Combining such attributes permits high-speed, low-power A/D and D/A converters to be implemented.

7.12.4 Trends in BiCMOS Technology

It is generally agree that BiCMOS gates provide a factor of 2x improvement in speed performance (at nearly equal power dissipation) over CMOS gates of the same area when a 5.0 V power supply is used. This also implies that such circuits as SRAMs, gate arrays, and microprocessors will exhibit better performance when fabricated in 0.8-μm BiCMOS technology than with 0.5-μm CMOS. As BiCMOS is scaled to the 0.5 μm regime, however, several issues must be addressed. First, the smaller MOS devices in such BiCMOS gates will require lower power-supply voltages. At voltages smaller than 5.0 V, however, the performance of the BiCMOS gate is severely degraded.[143] Second, the scaling of the bipolar devices relative to the MOS devices in the BiCMOS gates must be optimized. Finally, new BiCMOS device structures will need to be developed. Progress in all of these areas has been reported.

7.12.4.1 BiCMOS Performance at 3.3 V. The problem of reduced BiCMOS gate performance at lower voltages has been attacked from two directions. The first uses on-chip voltage regulation to operate the on-chip 0.5-μm CMOS gates at lower voltages, while allowing the BiCMOS gates and the I/O interface to remain at the standard 5.0 V levels.[140] The other is to replace the conventional BiCMOS and ECL gates with new logic gates capable of operating at 3.3 V or less without suffering speed performance loss.[141,142,144] Figure 7-74a shows examples of such novel gates, i.e., a complementary-BiCMOS inverter and a BiNMOS inverter as described in reference 144. The dependence of the delay times of these gates on supply voltage is compared to those of conventional BiCMOS and CMOS inverters in Fig. 7-74b.

7.12.4.2 Scaling BiCMOS to Retain the Advantage of Speed-Performance. The optimal scaling of bipolar devices in BiCMOS gates to allow such gates to retain their relative speed-performance advantage over scaled CMOS has been studied.[145] It is found that bipolar transistors in BiCMOS gates can be scaled according to the same rules formulated for bipolar devices in ECL gates,[14,146] with the exception that bipolar transistors in BiCMOS require a higher collector doping (typically greater than 5×10^{16} cm^{-3}) then those used in ECL gates (typically lower than 2×10^{16} cm^{-3}).

7.12.4.3 New Process Advances in BiCMOS. With the scaling of devices into the submicron regimes, MOSFETs need increasing subsurface doping to control short-channel effects, while bipolar devices require greater collector doping to prevent the Kirk effect. Thus, the doping profile requirements of the well and collector regions will converge as BiCMOS circuits are scaled. This will relax some of the tradeoffs that have to made when designing submicron-BiCMOS processes.

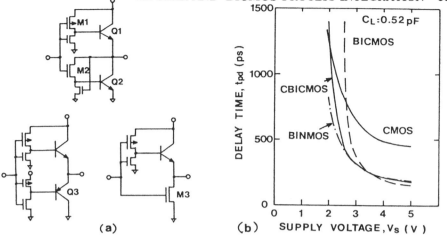

Fig. 7-74 (a) Conventional BiCMOS, CBiCMOS, and BiNMOS inverter circuits. (b) Dependence of delay time on supply voltage for various types of inverter circuits.[144] (Copyright 1989 IEEE).

Trench isolation will also be incorporated into more BiCMOS structures. Such trench-isolated BiCMOS processes have already been reported,[102] and a cross section of one structure is shown in Fig. 7-75a.

One of the ways of making BiCMOS gates that can operate at lower supply voltages is to decrease the E-B and C-B junction capacitances of the bipolar devices. This will allow the smaller drain currents provided by the MOS devices driven with lower gate voltages to still charge these capacitances in a short time. The E-B and C-B junction capacitances can be decreased by using polysilicon-self aligned bipolar devices, as described in section 7.9. Several BiCMOS processes incorporating PSA bipolar devices have been reported (Fig. 7-75b).[144, 147]

7.12.4.4 Analog-Digital BiCMOS Trends. In analog-digital BiCMOS circuits, the push to smaller feature sizes in the digital sections of the chip will call for shallow heavily-doped twin wells and very thin gate oxides. These conflict with the demands of higher operating voltages of the analog sections. Thus, separate wells and dual gate oxides may be needed. This may lead to very high mask count processes.

7.13 COMPLEMENTARY BIPOLAR (CB) TECHNOLOGY

Up to this point we have primarily considered bipolar (and BiCMOS) ICs fabricated with only one type of transistor (*npn* type). However, it is possible to design circuits which contain both *npn* and *pnp* transistors, and ICs having both such types are called *complementary bipolar (CB)* circuits. CB circuits have advantages for both digital and analog applications.

Digital CB circuits offer the potential of providing the high-speed performance of bipolar logic gates, but which are analogous to CMOS in that they dissipate very little

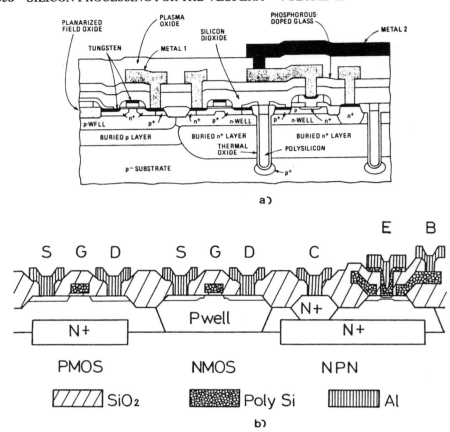

Fig. 7-75 (a) Cross section of a 0.8-μm BiCMOS process using trench-isolated bipolar devices.[102] (Copyright 1987 IEEE). (b) Cross section of a BiCMOS structure incorporating a PSA bipolar device.[147] (Copyright 1989, IEEE).

standby power. (An earlier bipolar logic family, *integrated injection logic* [I^2L], merged both the *pnp* and the *npn* transistors into a single structure, but used the *pnp* transistor only as a load device). Since both the *pnp* and the *npn* are involved in pull-up or pull-down during switching, the circuit performance is very much limited by the slower of the transistors.

Analog CB circuits would be able to improve upon the following limitations of an *npn*-only, high performance digital bipolar technology: the lack of high-speed, substrate-isolated *pnp* transistors, the lack of capacitors and low and high sheet resistance diffused resistors, and the inability to provide precision matching of components.

Because of these deficiencies, if such analog circuits as operational amplifiers (Op Amp) are fabricated with an *npn*-only bipolar technology, they exhibit severely limited frequency response behavior. If a high-performance CB process is used, the level shift and gain of the output stage of the Op Amp can be achieved with a *pnp* transistor (the bandwidth of the Op Amp is typically limited by the output stage), and significantly improves the amplifier performance. A host of other linear circuits, such as voltage regulators, phase-locked loop circuits and D/A converters also benefit from being fabricated with a CB process.

Until recently, the major drawback of implementing CB technology has been that they have been "lopsided" with slow *pnp* devices. The conventional lateral *pnp* devices that can be integrated as "free" devices in an otherwise *npn* technology (Fig. 7-76a) have very much lower f_T's than the *npn* transistors (e.g., 5-10 MHz), due to the large spacing that must be used to prevent emitter-collector punchthrough. Several other *pnp* structures have therefore been explored.

The first group of these are also lateral devices. Of them, the *epi-base lateral pnp* device (Fig. 7-76b) is available "free" in the self-aligned bipolar process.[148] At 2.2 μm, typical values of f_T are about 120 MHz. The *double-diffused lateral pnp* device (Fig. 7-76c) has a base whose width is given by the difference of the diffused emitter and base lateral-junction "depths." The double-diffused lateral *pnp* exhibits better performance than the epi-base lateral *pnp*, but is more complex to fabricate. Lateral *pnp* devices based in the SICOS structure described in section 7.9.3 have the smallest parasitic capacitance

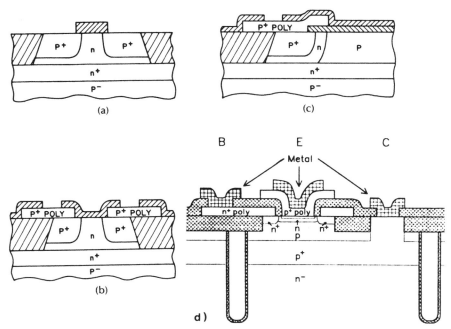

Fig. 7-76 (a) Conventional lateral *pnp* transistor. (b) Epi-base lateral *pnp* with a polysilicon emitter contact. (c) Double-diffused lateral *pnp* transistor. (d) Cross section of the 27-GHz vertical *pnp* transistor.[149] (Copyright 1989, IEEE).

and parasitic current component, and hence, the best current gain and performance among the lateral structures.

The second group of *pnps* are vertical devices. Although some have been built as non-self-aligned devices, the self-aligned vertical *pnp* transistors have the potential of providing the highest performance, but at the price of requiring the most complex process.

The real challenge has been to integrate vertical *pnp* devices into the high performance *npn*-only processes without significantly degrading the performance of the *npn* devices. Recently, a 27-GHz 20 ps double-polysilicon self-aligned vertical *pnp* transistor has been announced. It has features that allow it to be integrated into a more general CB process (Fig. 7-76d).[149] That is, these devices have 80-nm-wide implanted bases, 0.5 μm-wide emitters, optimized emitter and collector dopant profiles, and are fabricated on a thin *p*-type epi layer over a buried p^+ layer to provide high current-driving capability. A deep polysilicon-filled trench and field oxide provide device isolation.

AT&T has developed several CB processes for 5-V analog and mixed analog-digital applications. The most recent of these are the CBIC-V and CBIC-U (complementary bipolar IC) processes. In them, *pnp* transistors are isolated from the common *p*-substrate by means of junction isolation. The more-advanced CBIC-V process features vertical *pnps* with f_Ts of 5 GHz and *npns* with f_Ts of 13 GHz.[150] The process uses two layers of metal and selective epitaxial growth. Its predecessor, CBIC-U has *npns* operating at 4.5 GHz and *pnps* with an f_T of 3.75 GHz.

Another analog CB process designed for higher operating voltages (i.e., $BV_{CBO} = 36$ V) has been developed by Analog Devices. Its *npn* and *pnp* transistors have f_Ts of 600 MHz. For applications which require ultra-precise Op Amps that operate faster and produce wider bandwidth, dielectrically isolated CB processes have been developed.

REFERENCES

1. R. S. Muller and T. I. Kamins, *Device Electronics for Integrated Circuits*, 2nd Ed., New York: John Wiley & Sons, 1986.

2. E. S. Yang, *Microelectronic Devices,* New York: McGraw-Hill, 1988.

3. D. L. Pulfrey and N. G. Tarr, *Introduction to Microelectronic Devices,* Englewood Cliffs, N.J., Prentice-Hall, 1989.

4. A. R. Alvarez, Ed., *BiCMOS Technology and Applications*, Norwell, MA, Klewer Academic Publishers, 1989.

5. D. A. Hodges and H. G. Jackson, *Analysis and Design of Digital Integrated Circuits*, New York, McGraw-Hill, 1983.

6. J. Lohstroh, "Devices and Circuits for Bipolar (VLSI)," *Proc. of the IEEE,* vol. 69, no. 7, July, 1981, p. 812.

7. T. -Z. Chen et al., *IEEE Electron Dev. Lett.,* August 1989, p. 364.

8. S. Konaka et al., *Ext. Abs. 16th Int. Conf. on Solid-State Devices and Materials,* p. 2019, 1984.

9. J. Woo and J. Plummer, *IEEE Trans. Electron Dev.,* August 1988, p. 1311.

10. P. R. Gray and R. G. Meyer, *Analysis and Design of Analog Integrated Circuits*, New York, John Wiley and Sons, 1984, Eqs. 1-91, 1.113, 1.114, and 3.9.

11. A. Grove, *Physics and Technology of Semiconductor Devices*, New York: John Wiley & Sons, 1967, pp. 230-234.

12. T. Ikeda et al., *IEEE Trans. Electron Dev.*, June 1987, p. 1304.

13. P. -F. Lu and T. -C. Chen, *IEEE Trans Electron Dev.*, June 1989, p. 1182.

14. P. M. Solomon and D. D. Tang, "Bipolar Circuit Scaling," *Abs. IEEE Intl. Solid-State Ckts. Conf.*, 1979, p. 86-87.

15. A. R. Alvarez et al., *IEEE Trans. Computer-Aided Design,* vol. CAD-7, No. 2, p. 272, February 1988.

16. J. M. Stork, *Tech. Dig. IEDM*, 1988, p. 550.

17. J. Buie, "Improved Triple Diffusion means Densest ICs Yet,"*Electronics*, August 1975, p. 101.

18. D. Fuoss and T. Yuzuriha, *Dig. of IEEE Custom IC Conf.*, Portland, OR, 1987, p. 348.

19. D. Fuoss, *Ext. Abs. of Electrochem. Soc. Meeting*, Spring, 1989, p. 291.

20. L. C. Parrillo et al., *Tech. Dig. IEDM*, 1977, p. 265A.

21. R. S. Payne et al., *IEEE Trans. Electron Devices*, **ED-21**, 273, (1974).

22. *ibid.,* Reference 1, chapter 2.

23. C. T. Kirk, *IRE Trans. Electron Devices*, **ED-9**, 164 (1962).

24. C. C. Alen, L. H. Clevenger, and D. C. Gupta, *Journal Electrochem. Soc.,* May 1966, p. 508.

25. J. Borland and J. Hann, *Ext. Abs. of Electrochem. Soc. Meeting*, Spring, 1989, p. 218.

26. T. H. Ning and D. D. Tang, "Bipolar Trends," *Proceedings of the IEEE*, December 1986, p. 1669.

27. T. H. Ning and R. D. Isaac, *IEEE Trans. Electron Devices,* **ED-27**, 1980, p. 2051.

28. T. H. Ning and R. D. Isaac, *Tech. Dig. IEDM*, 1979, p. 473.

29. T. H. Ning, D. D. Tang, and P. M. Soloman, *Tech. Dig. IEDM*, 1980, p. 61.

30. A. Tahara et al., *IEEE Bipolar Circuits and Technology Meeting*, 1989, p. 169.

31. L. Parrillo et al., *IEEE Trans. Electron Dev.,* **ED-28**, p. 1508, (1981).

32. H. Schaber and T. F. Meister, "Technology and Physics of Polysilicon Emitters," *IEEE Bipolar Circuits and Technology Meeting*, 1989, p. 75.

33. P. Ashburn, "Polysilicon Emitter Technology," *IEEE Bipolar Circuits and Technology Meeting*, 1989, p. 90.

34. P. Ashburn and B. Soerowirjo, *IEEE Trans. Electron Dev.*, **ED-31**, 1984, p. 853.

35. C. C. Ng and E. S. Yang, *Tech. Dig. IEDM,* 1986, p. 32.

36. T. F. Meister et al., *IEEE Bipolar Circuits and Technology Meeting*, 1989, p. 86

37. B. Landau et al., *IEEE Bipolar Circuits and Technology Meeting*, 1988, p. 117.

38. P. Ashburn, D. Roulston and C. R. Sevakumar, "Comparison of Experimental and Computed Results on Arsenic- and Phosphorus-Doped Polysilicon Emitter Bipolar Transistors," *IEEE Trans. Electron Dev.*, June 1987, p. 1346.

39. G. L. Patton, J. C. Bravman, and J. D. Plummer, "Physics, Technology, and Modeling of Polysilicon Emitter Contacts for VLSI Devices'", *IEEE Trans. Electron Dev.,* November, 1986, p. 1754.

40. J. Stork and J. Cressier, " The Impact of Non-Ohmic Polysilicon Emitter Resistance on Bipolar Transistor Performance," *Symp.VLSI Technol., Dig. Tech. Papers,* p. 47, 1986.

41. S. -Y. Yung and D. E. Burk, *IEEE Trans. Electron Dev.,* September 1988, p. 1494.

42. G. R. Wolstenholme et al., "An investigation of the thermal stability of the interfacial oxide in polysilicon emitter bipolar transistors by comparing device results with high resolution electron microscopy observations," *Jnl. Appl. Phys.,* vol. 61, p. 225,(1987).

43. J. E. Brighton et al., *Proc. IEEE Bipolar Circuits and Technology Mtg.,* 1989, p. 121.

44. Y. Niitsu et al., *IEEE Bipolar Circuits and Technology Meeting,* 1989, p. 98.

45. R. Bagri et al., *IEEE Bipolar Circuits and Technology Meeting,* 1989, p. 63.

46. Y. Yamamoto et al., *IEEE Trans. Electron Dev.,* July 1985, p. 1231.

47. J. E. Lary and R. L. Anderson,"Effective Base Resistance in Bipolar Transistors," *IEEE Trans. Electron. Dev.,* November 1985, p. 2503.

48. A. K. Kapoor, M. E. Thomas, and M. B. Vora, *IEEE Trans. Electron Dev.,* June 1986, p. 772.

49. P. Asbeck, "Heterojunction Bipolar Transistors: Status and Directions," *IEEE Bipolar Circuits and Technology Meeting,* 1989, p. 65.

50. S. L Wright, H. Kroemer, and M. Inada, "Molecular beam epitaxy growth of GaP on Si," *J. Appl. Phys.,* vol. 55, p. 2916, April 1984.

51. Y. H. Kwark and R. M. Swanson, "N-type SIPOS and poly-silicon emitters," *Solid State Electronics,* vol. 30, p. 1121, November 1987.

52. M. Takahashi, M. Tabe, and Y. Sakakibara, *IEEE Electron Dev. Lett.,* October 1987, p. 475.

53. T. Sugii et al., *IEEE Electron Dev. Lett.,* February 1988, p. 87.

54. H. Fujioka et al., *Tech. Dig. IEDM,* 1988, p. 574.

55. S. S. Iyer et al., *IEEE Trans. Electron Dev.,* October 1989, p. 2043.

56. B. S. Meyerson, "Low-temperature silicon epitaxy by ultrahigh vacuum chemical vapor deposition," *Appl. Phys. Lett.,* vol. 48, no. 12, p. 797 (1986).

57. J. F. Gibbons, C. M. Gronet, and K. E. Williams, "Limited Reaction Processing: Silicon Epitaxy," *Appl. Phys. Lett.,* vol. 47, p. 721 (1985).

58. T. Won and H. Morkoc, *IEEE Electron. Dev. Lett.,* vol. 10, no. 1, p. 33, 1989.

59. D. D. Tang et al., *IEEE J. Solid-State Circuits,* **SC-15,** p. 444, (1980).

60. T. -C. Chen et al., "A Submicron High-Performance Bipolar Technology," *IEEE Electron Dev. Letts.,* August, 1989, p. 364.

61. G. P. Rosseel et al., "Delay Analysis for BiCMOS Drivers," *Proc. IEEE Bipolar Circuits and Technology Mtg.,* 1988, p. 220.

62. C. -T. Chuang, G. P. Li, and T. H. Ning, *IEEE Electron. Dev. Letts.,* July 1987, p. 321.

63. T. Yuzuhira, T. Yamaguchi, and J. Lee., *Tech. Dig. IEDM,* 1988, p. 748.

64. H. Schaber et al., *Tech. Dig. IEDM,* 1987, p. 170.

65. G. P. Li et al., *IEEE Trans. Electron. Dev.,* November 1988, p. 1942.

66. J. N. Burghartz et al., *IEEE Electron Dev. Lett.,* May 1988, p. 259.

67. H. Takemura et al., *Tech. Dig. IEDM,* 1987, p. 375.

68. T. H. Ning and D. D. Tang, *IEEE Trans. Electron Dev.,* **ED-31,** p. 409, 1984.

69. T. Tahiro et al., *Tech. Dig. IEDM,* 1984, p. 686.

70. H. K. Park et al., *IEEE Electron Dev. Letts.,* December 1986, p. 658.

71. G. P. Li et al., *Tech. Dig. IEDM*, 1987, p. 174.

72. C. -T. Chuang, *Tech. Dig. IEDM*, 1987, p. 178.

73. D. P. Verret and J. E. Brighton, "Two-Dimensional Effects in the Bipolar Transistor," *IEEE Trans. Electron Dev.*, November 1987, p. 2297.

74. C. T. Chuang, D. D. Tang, and E. Hackbarth, *IEEE Trans. Electron. Dev.*, July 1987, p. 1519.

75. S. Sawada, *Proc. IEEE Bipolar Circuits and Device Mtg.*, 1988, p. 206.

76. T. Gomi et al., *Tech. Dig. IEDM*, 1988, p. 744.

77. T. -C. Chen et al., *Tech. Dig. IEDM,* 1988, p. 740.

78. A. Cuthbertson and P. Ashburn, *IEEE Trans. Electron Dev.*, **ED-32**, p. 242 (1985).

79. J. L. de Jong et al., *Proc. IEEE Bipolar Circuits and Device Mtg.*, 1988, p. 202.

80. B. van Schravendijk et al., *Proc. IEEE Bipolar Circuits and Device Mtg.*, 1988, p. 202.

81. B. Y. Hwang et al., *Proc. IEEE Bipolar Circuits and Device Mtg.*, 1988, p. 28.

82. T. I. Kamins, *IEEE Electron Dev. Letts.*, September 1989, p. 401.

83. T. Nakamura et al., *IEEE Trans. Electron Dev.*, vol. ED-29, p. 596 (1982).

84. T. Nakamura et al., *Tech. Dig. IEDM*, 1986, p. 472.

85. T. Yamaguchi and T. H. Yuzuriha, *IEEE Trans. Electron Dev.*, May 1989, p. 890.

86. A. Tahara et al., *Proc. Bipolar Ckts. and Technol. Conf.*, 1989, p. 169.

87. T. -C. Chen et al., *IEEE Trans. Electron Dev.*, **ED-35**, August 1988, p. 1322.

88. T. Yamaguchi and T. H. Yuzuriha, *IEEE Trans. Electron Dev.*, May 1989, p. 890.

89. M. P. Brassington et al., *IEEE Trans. Electron Dev.*, **ED-36**, April 1989, p. 712.

90. G. P. Rosseel and R. W. Dutton, *IEEE J. Solid-State Circuits,* February, 1989, p. 90.

91. A. R. Alvarez, "BiCMOS Technology," 1987 IEDM Short Course on BiCMOS Technol., December 1987.

92. A. R. Alvarez, "Introduction to BiCMOS," Chap. 1, p. 1, in *BiCMOS Technology and Applications,* A. R. Alvarez, Ed., Klewer Academic Publishers, Norwell, MA., 1989.

93. K. Miyata, "BiCMOS Technology Overview," 1987 IEDM Short Course on BiCMOS Technology, December 1987.

94. K. Shibayama and Y. Akasaka, "Laboratory and Factory Automation for ULSI Developments and Mass Production", *Tech. Dig. IEDM*, 1988, p. 736.

95. R. A. Chapman et al., "0.5-μm CMOS for High Performance at 3.3 V," *Tech. Dig. IEDM*, 1988, p. 52.

96. A. Alvarez and D. W. Schucker, "BiCMOS Technology for Semi-Custom Integrated Circuits," *Proceedings Custom Integ. Cir. Conf.,* 1988, p. 22.1.1.

97. H. Klose et al., *IEEE Bipolar Circuits and Technology Meeting*, 1989, p. 98.

98. A. R. Alavarez et al., *Semiconductor International,* April 1989, p. 226.

99. R. Richards, "Fujitsu's BiCMOS Process," *Semiconductor Internat.*, June 1989, p. 104.

100. T. Ikeda et al., "High Speed BiCMOS VLSI Technology," *Tech Dig. IEDM*, 1986, p. 408.

101. S. Cosentino, "Motorola's BiCMOS Process," *Semiconductor Internat.*, June 1989, p. 102.

102. R. H. Havemann et al., "An 0.8-μm BiCMOS SRAM Technology," *Tech. Dig. IEDM,* 1987, p. 841.

103. T. Ikeda et al., *Proc. 1st. Int. Symp. on BiCMOS Technologies and Circuits,* Spring Meeting, Electrochem. Soc., May 1987.

104. T. Ikeda et al., *IEEE Trans. Electron Dev.,* June 1987, p. 1304.

105. R.A. Haken et al., "BiCMOS Process Technology," in *BiCMOS Technology and Apllications,* A. R. Alvarez, Ed., Norwell, MA, Klewer Academic, 1989.

106. R. A. Chapman et al., "Submicron BiCMOS Well Design for Optimum Circuit Performance," *Tech. Dig. IEDM,* 1988. p. 756.

107. R. A. Haken et al., *Semiconductor Internatl.,* June 1989, p. 96.

108. R. A. Chapman et al., *Tech. Dig. IEDM,* 1987, p. 362.

109. S. B. Kulkarni and A. A. Kozul, *Extended Abstr. Electrochem. Soc.,* **80-2,** 1351 (1980).

110. J. Borland et al., "Silicon epitaxial growth for advanced device structures," *Solid-State Technology,* January 1988, p. 111.

111. J. Borland and J. Hann, *Extended Abstr. Electrochem. Soc. Meeting,* Spring 1989, p. 21.

112. H. Iwai et al., *Tech. Dig. IEDM,* 1987, p. 28.

113. T. -I. Liou and C.-S. Teng., *IEEE Trans. Electron Dev.,* September 1989, p. 1620.

114. M. Nanba et al., *IEEE Trans. Electron Dev.,* July 1988, p. 1021.

115. D. de Lang et al., *Proc. IEEE Bipolar Circuits and Tech. Conf.,* 1989, p. 190.

116. S. Weber, "TI Soups up LinCMOS Process with 20-V Bipolar Transistors," *Electronics,* February 4, 1988, p. 59.

117. R. A. Haken et al., "BiCMOS Process Technology," in *BiCMOS Technology and Applications,* A. R. Alvarez, Ed., Norwell, MA., Klewer Academic, Chap. 3, p. 63.

118. C. Anagnostopoulos et al., *Tech. Dig. IEDM,* 1984, p. 588.

119. S. -C. Lee, D. W. Schucker, and P. T. Hickman, *VLSI Design ,* August 1984, p. 98.

120. J. McDonald et al., *Proc. IEEE Custom Integ. Circ. Conf.,* 1988, p. 13.4.1.

121. K. Deierling, "Digital Design," Chap. 5, in *BiCMOS Technology and Applications,* A. R. Alvarez, Ed., Norwell, MA., Klewer Academic, Chap. 3, p. 63.

122. B. C. Cole, "AMCC's BiCMOS Array Hits Record Gate Utilization," *Electronics,* February 4, 1988, p. 65.

123. M. Kubo et al., "Perspective on BiCMOS VLSI's," *J. Solid-State Ckts.,* February 1988, p. 8.

124. M. R. Goulding et al., "Low vs High Temperature Epitaxial Growth," *Semiconductor International,* May 1988, p. 90.

125. C. C. Allen, L. H. Clevenger, and D. C. Gupta, *J. Electrochem. Soc.,* 113, 508 (May 1966).

126. H. Camenzind, *Electronic Integrated Systems Design.* Copyright 1972 by Litton Educational Publishing, Inc., New York.

127. W. C. Till and J. T. Luxton, *Integrated Circuits: materials, devices, and fabrication,* Prentice-Hall, Englewood Cliffs, N. J., 1982.

128. K. H. Nicholas, *Solid-State Electron.,* **9,** 35, (1966).

129. D. Rice, "Isoplanar-S Scales Down for New Heights in Performance," *Electronics,* **52,** 137 (1979).

130. S. M. Sze, *Semiconductor Devices - Physics and Technology*, John Wiley & Sons, New York, 1985.

131. E. F. Labuda and J. T. Clemens, "Integrated Circuit Technology," in R. E. Kirk and D. F. Othmer, Eds., *Encyclopedia of Chemical Technology*, John Wiley & Sons, New York, 1980.

132. R. People and J. C. Bean, *Appl. Phys. Lett.* **48**, 538. (1986).

133. A. Wieder, *Tech. Dig. IEDM*, 1986, p. 8.

134. D. D. Tang et al., *IEEE Electron Dev. Lett.*, Vol.EDL-8, April 1987, p. 174.

135. A. R. Alvarez et al., *Semiconductor International*, May 1989, p. 226.

136. J. N. Burghartz et al., *Tech. Dig. IEDM*, 1989, p. 229.

137. M. Sugiyama et al., *Tech. Dig. IEDM*, 1989, p. 221.

138. J. van der Veldedn et al., *Tech. Dig. IEDM*, 1989, p. 233.

139. C. T. Chuang and P. F. Lu, *Tech. Dig. IEDM*, 1989, p. 799.

140. H. Fukuda et al., *Dig. Tech. Papers IEEE Int. Solid-State Circuits Conf.*, 1989, p. 176.

141. W. Heimsch et al., *Dig. Tech. Papers IEEE Int. Solid-State Circuits Conf.*, 1989, p. 112.

142. Y. Nishio et al., *Dig. Tech. Papers IEEE Int. Solid-State Circuits Conf.*, 1989, p. 117.

143. H. Momose et al., *Symp. VLSI Tech.*, 1989, p. 55.

144. A. Watanabe et al., *Tech. Dig. IEDM*, 1989, p. 429.

145. G. P. Rosseel and R. W. Dutton, *Tech. Dig. IEDM*, 1989, p. 795.

146. T. Ning et al., "Scaling Properties of Bipolar Transistors," *Tech. Dig. IEDM*, 1980, p. 550.

147. T. Yoshimura et al., *Tech. Dig. IEDM*, 1989, p. 241.

148. D. Tang et al., "73 ps bipolar ECL circuits," *Tech. Dig. ISSCC*, 1986.

149. J. Warnock et al., *Tech. Dig. IEDM*, 1989, p. 903.

150. A. Feygonson et al., *Proc. IEEE Bipolar Circuits and Tech. Conf.*, 1989, p. 173.

151. D. J. Roulston, *Bipolar Semiconductor Devices*, McGraw-Hill, New York, 1990.

1

CHAPTER 8

SEMICONDUCTOR MEMORY

PROCESS INTEGRATION

Memories store digital information (or data) in terms of *bits*, or binary digits (ones or zeros). Modern digital systems use *memory devices* to store and retrieve large quantities of digital data at electronic speeds. Early digital computers used magnetic-cores as the devices in fast-access memories. With the introduction of semiconductor memory chips in late 1960s, however, magnetic cores began to be replaced by integrated circuits (which implemented a much higher-density digital-memory function). This not only increased the performance capabilities of the memory, but also drastically decreased its cost. By the end of the 1970s, magnetic-core memories had been completely displaced as high-speed memory devices.

8.1 TERMINOLOGY OF SEMICONDUCTOR MEMORIES

Memory capacities in digital systems are usually expressed in terms of bits, since a separate storage device or circuit is used to store each bit of data. Each storage element is referred to as a *cell*. Memory capacities are also sometimes stated in terms of *bytes* (8 or 9 bits) or *words* (32 – 80 bits). Each byte typically represents an alphanumeric character. Every bit, byte or word is stored in a particular location, identified by a unique numeric address, and only a single bit, byte, or word is stored or retrieved during each cycle of memory operation.

Memory-storage capability is expressed in units of kilobits and megabits (or kilobytes and megabytes). Since memory addressing is based on binary codes, capacities that are integral powers of 2 are typically used. As a result, a memory device with a 1-kbit capacity can actually store 1,024 bits, and a 64-kbit device can store 65,536 bits.

In digital computers, the number of memory bits is usually 100 to 1000 times greater than the number of logic gates, which implies that the memory cost per bit must be kept very low. In addition, it is desirable for the memory devices to be as small as possible (since this will allow the highest density of cells on a chip), to operate at a high speed, to have a small power consumption, and to operate reliably.

Memory cells could be designed to possess a set of characteristics close to those of an ideal digital-logic-element. Such an ideal cell would be able to

1. perform the desired logic function;

2. robustly quantize the signal levels of the stored data;

3. exhibit a high degree of input-output isolation and fan-out; and

4. regenerate the stored logic-levels.

However, to enable each memory cell to possess all of these attributes would require the use of a complex circuit to implement each cell. Memory-cell design therefore involves trading off most of the desired properties of digital-logic devices in order to achieve a cell that is as simple and compact as possible. Consequently, the cell itself is not capable of outputting digital data in an electrical form compatible with the requirements of the remainder of the system. To restore the electrical characteristics of the cell's outputted data to adequate values, properly designed peripheral circuits (e.g., sense amplifiers, memory registers, and output drivers) are necessary. These circuits are designed to be shared by many memory cells. The trade-off thus made is that of a less-robust output signal from the cell, in exchange for a simple, compact memory cell design (consisting of only 1 to 6 transistors).

8.1.1 Random-Access and Read-Only Memories (RAMs and ROMs)

The most flexible digital memories are those that allow for data storage (or *writing*) as well as data retrieval (or *reading*). Memories in which both of these functions can be rapidly and easily performed, and whose cells can be accessed in random order (independent of their physical locations), are referred to as *random-access memories* (RAMs). *Read-only memories* (ROMs) are those in which only the read operation can be performed rapidly (although ROMs are generally configured so that their cells are also randomly accessible, and data can be entered into them). Entering data into a ROM, however, is referred to as *programming* the ROM, to emphasize that this operation is much slower than the writing operation used in RAMs.

8.1.2 Semiconductor-Memory Architecture

The organization of large semiconductor memories is shown in simplified form in Fig. 8-1.[1] The storage cells of the memory are arranged in an array consisting of horizontal rows and vertical columns. Each cell shares electrical connections with all the other cells in its row, and column. The horizontal lines connected to all the cells in the row are called *word lines,* and the vertical lines (along which data flows into and out of the cells) are called *bit* lines. Each cell therefore has a unique memory location, or address, which can be accessed at random through selection of the appropriate word and bit line. (Some memories are designed so that four or eight cells are accessed simultaneously.) Thus, in semiconductor memories such as that shown in Fig. 8-1, any cell can be accessed in random order, at a fixed rate, for the purpose of either reading

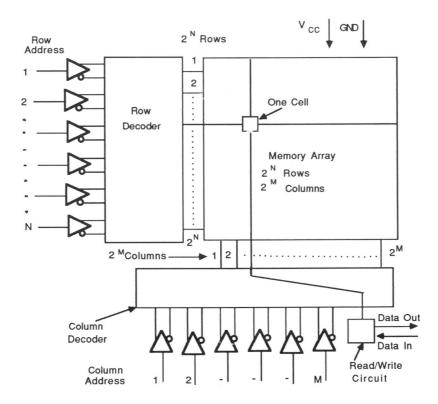

Fig. 8-1 Organization of a random access memory (RAM).

or writing data.   The array configuration of semiconductor memories lends itself well to the regular, structured designs favored in VLSI.

There are also a number of important circuits at the periphery of the array. The first such peripheral circuit is the *address decoder*. Two of these are used on each chip – one for the word lines, the other for the bit lines. These circuits allow a large number of word and bit lines to be accessed with the fewest number of address lines. Address decoders for this purpose have 2^n output lines, with a different one selected for each different n-bit input code. In later generations of memory circuits, address multiplexing was integrated on some memory chips to reduce the number of address pins by half. (Note that the address decoder was the first peripheral logic-circuit to be built on the memory chip).

The read/write control circuitry shown in Fig. 8-1 determines whether data is to be written into or read from the memory. Because such circuits also amplify and buffer the data signals retrieved from the cells, one of the important circuits in this subsystem is the *sense amplifier*. In dynamic memories that need periodic data refreshing, refresh circuitry may also be provided.

Note that most RAMs have only one *input-data lead* and one *output-data lead* (or only one combined *input/output lead*). Writing into and reading from such RAMs is done one bit at a time. Other RAMs have a number of input- and output-data leads, with the number determined by the word length of the system's data bus. ROMs, on the other hand, are typically organized so that the number of output-data leads (usually eight) is the same as the number of lines on the data bus. ROMs are programmed word by word (i.e., eight bits, or one byte, at a time) and are read from in the same manner.

8.1.3   Semiconductor-Memory Types

In semiconductor RAMs, information is stored on each cell either through the charging of a capacitor or the setting of the state of a bistable flip-flop circuit. With either method, the information on the cell is destroyed if the power is interrupted. Such memories are therefore referred to as *volatile memories*. When charge on a capacitor is used to store data in a semiconductor-RAM cell, the charge needs to be periodically refreshed, since leakage currents will remove it in a few milliseconds. Hence, volatile memories based on this storage mechanism are known as *dynamic RAMs*, or *DRAMs*.

If the data is stored (i.e., written into the memory) by setting the state of a flip-flop, it will be retained as long as power is connected to the cell (and no other write signals are received). RAMs fabricated with such cells are known as *static RAMs*, or *SRAMs*. Volatile RAMs can be treated as nonvolatile if they are provided with battery backup. Some DRAM and SRAM chips are even packaged together with a battery to facilitate implementation of this approach.

It is often desirable to use memory devices that will retain information even when the power is temporarily interrupted (or when the device is left without applied power for indefinite periods). Magnetic media offer such nonvolatile-memory storage. In addition, a variety of semiconductor memories have been developed with this characteristic. At present, virtually all such nonvolatile memories are ROMs. While data *can* be entered into these memories, the programming procedure varies from one type of ROM to the other (and none of them can be considered to be RAMs).

The first group of nonvolatile memories consists of those ROMs in which data is entered during manufacturing, and cannot subsequently be altered by the user. These devices are known as *masked ROMs* (or simply *ROMs)*. The next category consists of memories whose data can be entered by the user (*user-programmable* ROMs). In the first example of this type, known as a *programmable ROM, or PROM,* data can be entered into the device only *once*.

In the remaining ROM types, data can be erased as well as entered. In one class of erasable ROMs, the cells must be exposed to a strong ultraviolet light in order for stored data to be erased. These ROMs are called *erasable-programmable ROMs*, or *EPROMs*. In the final type, data can be *electrically erased* as well as entered into the device; these are referred to as *EEPROMs*. The time needed to enter data into both EPROMs and EEPROMs is much longer than the time required for the *write* operation in a RAM. As a result, none of the ROM types can at present be classified as fully functional RAM devices.

A few nonvolatile RAMs have been developed, but these have not yet reached the state of development at which they can be considered an important class of semiconductor memory.

8.1.4 Read-Access and Cycle Times in Memories

The two principal time-dependent performance characteristics of a memory are the *read-access time* and the *cycle time*. The first is the propagation delay from the time when the address is presented to the memory chip until data stored at that address is available at the memory output. The cycle time is the minimum time that must be allowed after the initiation of a read operation (or a write operation, in a RAM) before another read operation can be initiated. The minimum cycle times for reading and writing in a RAM are not necessarily equal, but for simplicity of design most systems employ a single minimum cycle time. For semiconductor RAMs, the read-access time is typically 50 – 90% of the read-cycle time.

8.1.5   Recently Introduced On-Chip Peripheral Circuits

Additional peripheral circuits have recently been added to the basic memory-organization structure shown in Fig. 8-1.   These circuits serve mainly to improve the manufacturability and testability of the chips. Those designed to increase manufacturability include redundancy circuits and error-correction circuits. Redundancy circuits allow some defective chips to be salvaged, while self-testing circuits reduce testing time.

Redundancy allows a defective row or column of cells to be replaced with a spare. Replacement techniques include the use of electrically or laser-blown fuses, or of one-time-programmable memory cells (which control on-chip multiplexers that switch in spare rows or columns). Redundancy measures typically improve manufacturing yields by factors of between 1.5 and 5.

Error-detection and correction techniques involve the addition of parity bits to allow the system to detect bad data, as well as circuitry to accomplish parity checking and error correction. This imposes an area penalty: for example, if one parity bit is added to each byte, the size of a 32k x 8k chip would be increased to 32k x 9k. If a failure is detected during programming, the parity bits can be sacrificed to act as a spare bit field for byte-wide applications. In DRAM designs, the use of error-correction coding (ECC) requires another 27% of memory cells. Because the ECC approach corrects soft errors as well as hard errors, the problem of soft errors can be reduced for the life of the product.

8.1.6   Logic-Memory Circuits

Special-purpose circuits that combine logic and memory on the same chip have recently been introduced. These fall into two categories: (1) memories that have some additional logic capabilities (*logic-in-memory circuits,* or *special-application RAMs*); and (2) logic circuits that contain some memory capability (*memory-in-logic circuits*).

A number of different special-application RAMs have been developed, including *video RAMs* (VRAMs) and *multiport SRAMs*.[45] Video RAMs are DRAMs designed to support the high-capacity requirements of frame buffers and display memories found in graphics terminals and systems. They have two input/output ports – one for random access (as in conventional RAMs), and one for serial-access. The serial port accesses the memory sequentially and performs the various serialization tasks necessary to drive cathode-ray tubes or other serial-data devices. The fast serial-readout-rate also enables quick refreshing of the graphics screen. The random-access port is driven by a graphics processor and is used to build the screens of displayed data (i.e., the random-access capability allows real-time updating of pictures).

Multiport SRAMs are being offered for use in multiprocessor systems. For example, dual-port SRAMs are becoming widely used to allow two independent logic circuits to simultaneously access one memory in a read and write mode. The two circuits can thus communicate with each other by passing data through the common memory. (Two processor-containing components of a digital system might be a CPU and a disk controller, or two processors working on two related but different tasks.) The use of the dual-port memory would eliminate the need for any special data-communication hardware.

Many applications for memory-in-logic circuits have also been envisioned, but these have only begun to be implemented. Some manufacturers of *application-specific circuits* (ASICs) do offer large memory blocks as part of prefabricated gate arrays, as well as increasingly larger memory blocks in their standard-cell libraries. Other, more advanced circuits are still being developed. For instance, more extensive use of memory will be needed in *image processing* and *coder-decoder* (CODEC) ULSI circuits. Image processing systems will be greatly enhanced when a processor is available with a memory of at least 2 Mbits that can store a frame of a picture. Similarly, CODECs used to move pictures could use such an image processor with an on-chip memory for data compaction. Finally, microprocessor chips are being built with on-board memories. (For example, the Intel 486 μP contains *cache memory* on the chip, which consumes ~40% of the chip area.) Even more memory integration might allow for an on-chip memory hierarchy. Products that could use such circuits include *point-of-sale terminals*, *smart cards*, and *telecommunications circuits*.

8.2   STATIC RANDOM-ACCESS MEMORIES (SRAMS)

SRAMs, the first type of semiconductor memory to be implemented, are referred to as *static memories* because they do not require periodic refresh signals in order to retain their stored data. The bit state in an SRAM is stored in a pair of cross-coupled inverters, which form a circuit known as a flip-flop. The voltage on each of the two outputs of a flip-flop circuit is stable at only one of two possible voltage levels, because the operation of the circuit forces one output to a high potential, and the other to a low potential. The memory logic state of the cell is determined by whichever of the two inverter outputs is high. Flip-flops maintain a given state for as long as the circuit receives power, but they can be made to undergo a change in state (i.e., to flip), through the application of a trigger voltage of sufficient magnitude and duration to the appropriate input. Once the circuit has settled into its new stable state, the trigger voltage can be removed. SRAM cells can be implemented in NMOS (Fig. 8-2a), CMOS (Fig. 8-2b), bipolar (Fig. 8-2c), or BiCMOS technologies.

The chief disadvantage of an SRAM cell is that it consists of at least six devices, as compared to only two for the dynamic-memory cell. Thus, even when the same set of design rules is used, an SRAM chip cannot be built with as many cells as a DRAM chip (as is illustrated in Figs. 8-3a and 8-3b).

On the other hand, SRAMs are the fastest semiconductor memories. Their speed is derived from the self-restoring nature of the flip-flop and the static peripheral circuits of the memory chip. Bipolar SRAMs are the fastest of all, and MOS SRAMs are the fastest among MOS memories. Bipolar SRAMs, however, dissipate much more power than CMOS SRAMs (e.g., 0.1-to-1.0 mW/bit versus ~25 μW/bit).

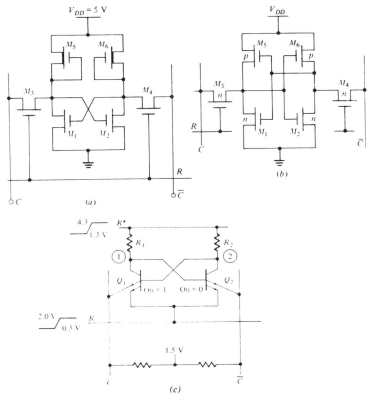

Fig. 8-2 Circuit schematic of (a) NMOS SRAM cell, (b) CMOS SRAM cell, (c) Bipolar SRAM cell.[1] From D. A. Hodges and H. G. Jackson, *Analysis and Design of Digital Integrated Circuits*, Copyright, 1983 McGraw-Hill Book Co. Reprinted with permission.

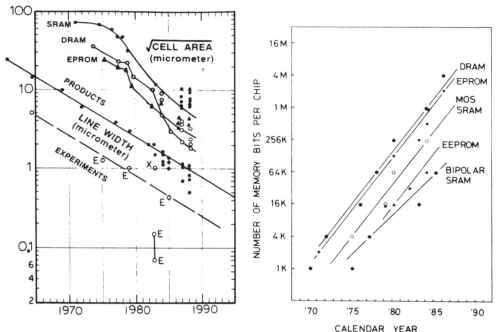

Fig. 8-3 (a) Trends of the laboratory and the production lithographic line width (two lower straight lines) and of the cell size (the square root of cell area) of production SRAM, DRAM, and EPROM cells.[3] (© 1988 IEEE). (b) Level of integration of semiconductor memories which have been presented at the *IEEE International Solid-State Circuits Conference* versus calendar year.[2] (© 1986 IEEE).

SRAMs are also characterized by their input/output (I/O) capability: one group has a TTL I/O capability, while the other has an ECL I/O capability (see chap. 7, section 7.2.2 for a comparison of TTL and ECL I/O signals). For the past several generations, high-speed CMOS technologies have been used to build TTL I/O SRAMs, while ECL I/O SRAMs have been implemented using bipolar technology. This is changing, since BiCMOS technology has begun to be used to fabricate both TTL and ECL I/O SRAMs. For example, 256-kbit ECL I/O SRAMs with 15-ns access times were commercially introduced in 1989, and 1-Mbit devices are expected shortly. TTL-I/O BiCMOS SRAMs are challenging incumbent CMOS and bipolar devices and already offer higher performance than all but the best of the CMOS SRAMs.

As SRAMs have evolved, they have undergone an increase in density. Most of this has been due to the use of smaller line widths (e.g., the 4-kbit MOS SRAM used 5-μm lines, while the 16-kbit, 64-kbit, 256-kbit, and 1-Mbit SRAMs were built with 3.0-μm, 2.0-μm, 1.2-μm, and 0.8-μm lines, respectively). The remainder of the density increase has been due to improvements in process technology, novel cell designs, and circuit innovations.[2]

Figure 8-2 indicates that one word line and two bit lines are connected to each SRAM cell: consequently two access transistors are also provided in each cell. In principle, it should be possible to achieve all memory functions using only one column line, one bit line, and one access transistor. In practice, however, normal variations in device parameters and operating conditions, make it difficult (or impossible) to obtain reliable operation at maximum speed using a single access-line to flip-flop cells. Therefore, the symmetrical bit lines (bit 0 and bit 1), are necessary. In a matrix of memory cells, each pair of bit-0 and bit-1 lines is shared by all memory cells in each column, and each word line is shared by all memory cells in each row.

8.2.1 MOS SRAMs

MOS SRAMs can be fabricated in either NMOS and CMOS, with early MOS SRAMs implemented in the former (Fig. 8-2a depicts a schematic diagram of an NMOS-SRAM cell). The load devices, M_5 and M_6, are depletion-mode NMOS transistors; the driver transistors, M_1 and M_2, and the access transistors, M_3 and M_4, are enhancement-mode NMOS transistors. The fully static CMOS SRAM (full-CMOS) cell shown in Fig. 8-2b was developed next. In this cell, the load devices are PMOS enhancement-mode transistors, while the other four transistors are enhancement-mode NMOS devices. Finally, a four-transistor SRAM cell was developed, with high-valued polysilicon resistors used as the load devices (poly-load cell). Since all the transistors in this cell are NMOS devices, the cells can be built in CMOS p-wells. When an array of such poly-load cells was fabricated in a p-well and combined with full-CMOS peripheral circuits, the resulting SRAMs demonstrated a significant decrease in power consumption compared to NMOS SRAMs (since most of the power in an SRAM is dissipated in the peripheral circuits, such as the off-chip drivers). These SRAMs also displayed a higher packing density than did SRAMs built with fully static CMOS cells.[4]

The full-CMOS cells dissipate less power than do the other types of MOS SRAM cells when in the standby mode of operation (i.e., when the cell is not being written into or read from). In the other types of cells, one inverter is always ON, and hence significant current is drawn from V_{DD} (much more, of course, in the six-transistor NMOS cell). In the full-CMOS cell, however, one transistor in each of the coupled inverters is OFF; thus, only junction-leakage current is drawn from V_{DD}. This current is approximately three orders of magnitude less than that drawn by the poly-load cell,[5] demonstrating that very little power is dissipated in the full-CMOS cell. In addition, the stability of the full-CMOS cell is high, since higher alpha-particle immunity and smaller junction leakage sensitivity is exhibited. The insensitivity to leakage allows operation at higher temperatures. Fully static CMOS SRAMs have been exploited as low-power, battery-backed memory devices in battery-operated consumer goods and portable office equipment.

The isolation of n-channel from p-channel devices means that the full-CMOS cell generally requires a larger area than does the poly-load cell. Furthermore, in order to establish contacts from the drain regions of the p-channel devices to those of the

n-channel devices (as well as to the n^+ polysilicon-gate materials), the metal layer must be used. In four-transistor/poly cells, there are no p-channel devices; so no n-channel-to-p-channel isolation is needed in the memory array of the chip. Buried contacts which take up less space can also be used to connect the drains of the access transistors and the gate of driver-transistors. Since it is also more costly to manufacture SRAMs with full-CMOS cells, four-transistor/poly-load cells have been used in the design of most high-density CMOS SRAMs. By 1989, 1-Mbit CMOS SRAMs were being offered commercially.

Alternative cell structures have also been investigated as a way to increase the density of SRAMs. One such approach is to stack the transistors on top of one another. For example, the full-CMOS cell can be built with the active p-channel transistor load stacked above the n-channel devices. The second layer of transistors can be fabricated on recrystallized silicon[12] or built using a hydrogen-passivated polysilicon transistor.[13] Although the processing techniques needed to fabricate such three-dimensional stacked structures are complex and difficult to control, these devices will become more attractive as the need to form higher-density structures becomes greater.

A novel SRAM cell based on the reverse base current of a bipolar transistor and consisting of only one bipolar transistor and one MOS transistor was described by Sakui et al.[14] Such a compact cell would allow SRAMs to be built with the same densities as DRAMs.

8.2.1.1   Circuit Operation of MOS SRAM Cells.

NMOS and CMOS SRAM cells all exhibit the same basic circuit behavior (Figs. 8-2a and b). When writing or reading data in such a cell is desired, the word line of the cell (which is held low in the standby state) is raised to V_{DD} (e.g., +5 V). This causes the enhancement-mode NMOS access transistors M_3 and M_4 to be turned *ON*. *Writing* is performed by forcing one of the bit lines low (e.g., close to 0 V), while maintaining the other at its standby value (about 3 V). For example, to write a *1*, the bit-*0* line must be forced low.*

When this occurs, M_1 turns *OFF* and its drain voltage rises due to the currents flowing through M_5 and M_3. When M_2 has been turned *ON*, the bit line can be returned to its standby level, leaving the cell in the state of storing a *1*. (The operation of writing a *0* is complementary to that just described).

For *reading* a *1*, the bit lines must both be biased at about 3 V. When the cell is selected, current flows through M_4 and M_2 to ground and through M_5 and M_3 to the bit *1* line. The gate voltage of M_2 does not fall below 3 V, so M_2 remains *ON*. The voltage of the bit-*0* line is thus reduced to less than 3 V, while the voltage of the bit-*1* line is pulled up above 3 V, since M_1 is *OFF* but M_5 is *ON*. As a result, a differential output signal exists between the bit-*0* and bit-*1* lines. This signal is fed to the sense amplifier, which in SRAMs is a differential amplifier capable of providing rapid sen-

---

* The cell must be designed so that the conductance of the access transistor is several times larger than that of the load transistor (i.e., comparing M_4 to M_6), in order for the drain of M_2 and the gate of M_1 to be brought below V_T.)

sing. Consequently, one of the bit lines needs to be only slightly discharged in order to generate a differential input signal large enough to drive the sense amplifier. To avoid a change in the state of the cell during reading, however, it is necessary for the conductance of M_2 to be around three times as large as that of M_4 so that the drain voltage does not rise above V_T. (The operation of reading a *0* is complementary to the one just described).

Among the most important factors limiting the maximum speed of MOS SRAMs are the delay associated with signal propagation through address buffers and decoders (which gets longer as the number of inputs and outputs increases), and the delay associated with the charging and discharging of the word and bit lines (which increases as the RC product of the word- and bit-line structures increases).

Bit-lines are typically formed in metal (Al), and hence their resistance is not a significant limitation. Word-lines, however, are normally implemented with polysilicon or polycide, so their higher resistance is considerably larger than that of the bit-lines. This then becomes one factor that limits SRAM speed. The parasitic capacitance of the word- and bit-line structures themselves, combined with the many paralleled access transistors (which are connected to each word and bit line), results in a large equivalent lumped capacitance on each of these lines.

Finally, there is a delay associated with the signal propagation through the sense-amplifier and data-output circuits. Considerable effort has been expended to develop high-speed sense amplifiers for SRAMs. In addition, a circuit technique known as *address-transition detection* (ATD) has also been used to speed up sensing in MOS SRAMs. In this technique, the bit lines are equilibrated upon detection of a change in the address input. (Sense-amplifier design, and the details of ATD are discussed in texts on VLSI circuit design, and so will not be further described here.

8.2.1.2 SRAM Cell Layout and Processing Issues. For maximum density to be achieved in a memory device, the cells must be be laid out in as small a size as possible. The size is determined by the cell's topology and by the design rules of the IC fabrication technology. A completed layout design represents the outcome of years of development, and a great deal of design experience. Figures 8-4a, 4b, and 4c are examples of SRAM-cell designs for a six-transistor NMOS SRAM cell,[6] a poly-load cell,[8] and an advanced full-CMOS cell,[7] respectively. These cell layouts also reflect some of the process enhancements that have made possible the improvements in SRAM performance, speed, and density.

Table 8-1 shows the evolution of the MOS SRAM; shrinking line widths and a variety of process enhancements can be seen to have been primarily responsible for the density and performance improvements. Figure 8-5 presents the same information graphically. The process enhancements are summarized in the following paragraphs describing the cells used in 1-Mbit SRAMs.

Several 1-Mbit CMOS SRAMs based on the poly-load cell were described in detail in the *IEEE Journal of Solid-State Circuits* (October, 1988), and a 4-Mbit SRAM (October, 1989).[109] The access time of these devices ranges from 7.5 to 18 ns (although the access times of commercially available 1-Mbit SRAMs in 1989 were

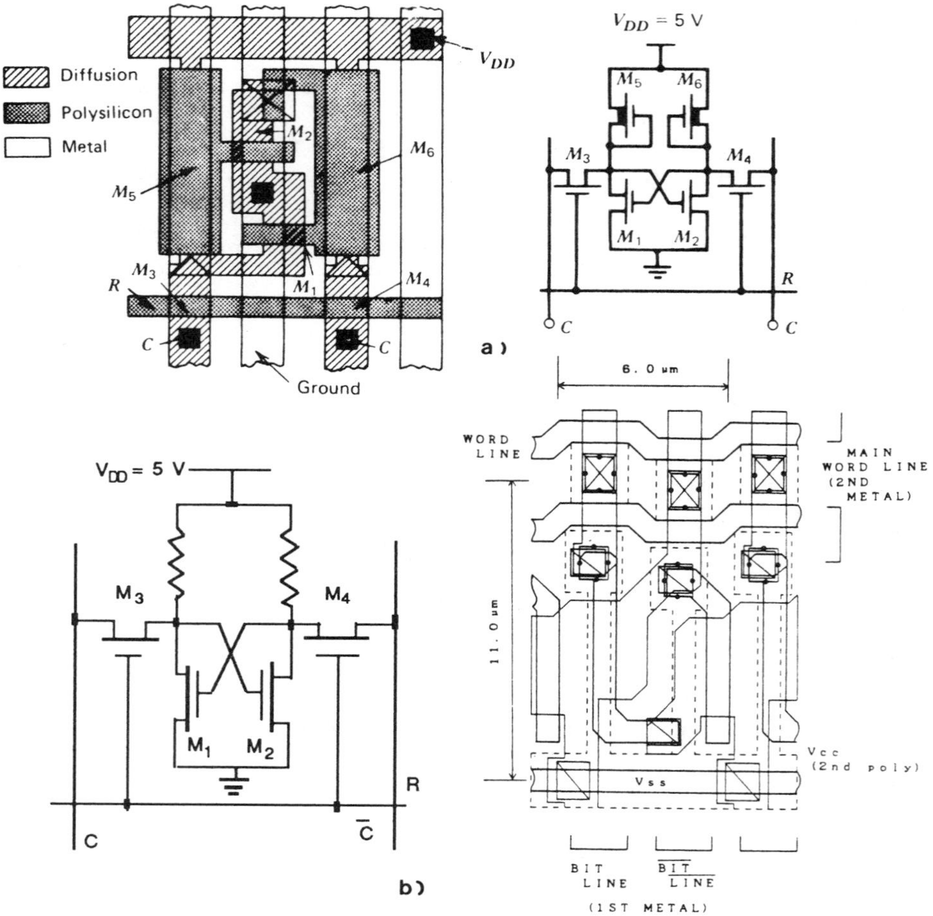

Fig. 8-4 (a) Schematic and layout of NMOS SRAM cell.[6] After R. Hunt, "Memory Design and Technology," in M. J. Howes and D. V. Morgan, eds., *Large Scale Integration*. Copyright 1981, John Wiley & Sons. Reprinted with permission. (b) Schematic and layout of poly load SRAM cell.[8] (© 1988 IEEE).

being given as 25-120 ns).[32] From the layout of the 7.5-ns CMOS SRAM cell (Fig. 8-4b), it can be seen that a double-polysilicon, double-level metal process is used, and that the cell area is 66 μm^2 (6.0 x 11 μm). The first level of polysilicon is a polycide structure, which is used for the V_{SS} power line in the memory array as well as for the gates of the MOS transistors. The second poly layer is used to form both the high-valued load resistors and the low-resistance V_{DD} lines. The bit lines are formed in Metal 1, and the word lines in Metal 2. A 0.8-μm twin well-CMOS process is used in which the channel lengths of the *n*- and *p*-channel transistors are 0.8 and 1.0 μm, respectively.

Other advanced poly-load–based CMOS SRAM designs include such process enhancements as trench-isolation structures, triple-level polysilicon, self-aligned con-

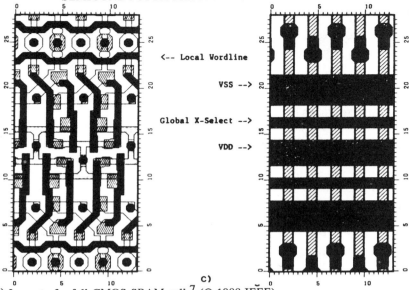

<-- Local Wordline

VSS -->

Global X-Select -->

VDD -->

Fig. 8-4 (c) Layout of a full-CMOS SRAM cell.[7] (© 1988 IEEE).

tacts, buried contacts, and spare rows and columns for redundancy.

A 1-Mbit CMOS SRAM based on a full-CMOS cell was reported in the same issue of the *IEEE Journal* mentioned above (Fig. 8-4b).[7]   It has a longer access time (25 ns) but consumes only 1 μW of power in the standby mode. Its memory-cell size is 5x12 μm, which implies that clever cell design and advanced processing techniques can produce fully static cells with sizes comparable to those of poly-load cells. (Note that the major reduction of cell area in such full-CMOS cells has been attributed to the use of a local interconnect layer[108] – chap. 3, section 3.11.2. The local interconnect layer allows the two inverters of the cell to be cross coupled with only two contacts per cell

Table   8-1.   Evolution of MOS SRAM Technology

Introduction Date		Size (bits)	Access Time	Minimum Feature Size	Process Enhancements
1969	PMOS	256 bit			Silicon Gate, CVD Oxide
1972	NMOS	1k		8 μm	Depletion-Mode Load
1975	NMOS	4k	4 ns (1988)	5 μm	Ion-Implant V_T Adjust
1978	NMOS	16k		3 μm	Plasma Etching /Wafer Stepper
1982	CMOS/NMOS	64k	15 ns	2 μm	Double-Poly
1985	CMOS/NMOS	256k	25 ns (1988)	1.2 μm	Polycide/Poly, LDD Structures
1988	CMOS/NMOS	1M	25 ns (1988)	0.8 μm	(Polycide/Poly, Double-Metal,
	Full CMOS	1M	25 ns (1988)		Twin-Well, LDD Structures)
1989	CMOS/NMOS	1M	10 ns		
	CMOS/NMOS	4M	25 ns	0.5 μm	3.3 V, Retrograde p-Well,
	BiCMOS	1 M	8 ns	0.8 μm	25 Mask Levels, Twin-Well

– compared to nine contacts in a cell implemented with double-level metal, but with no local interconnect level – and thus the cell size can be made signigicantly smaller.)

This SRAM is fabricated using a 14-mask process, and it employs a single level of polysilicon and two layers of metal. The poly layer is selectively doped p-type when it acts as the gate for PMOS devices, and n-type when it is a gate for NMOS devices. A silicide strap is also used to connect poly lines to diffused regions. Spare rows and columns are included for redundancy.

Another full-CMOS SRAM (256 kbits in size) has also been described.[9] This 35-ns access-time part uses TiN as a local-interconnect structure between gates and diffused regions, and 0.8-μm MOS devices. The 100-nm-thick TiN layer has a sheet resistance slightly lower than that of a 500-nm doped-poly layer (14 Ω/sq vs. 20 Ω/sq). The TiN makes contact to the TiSi$_2$ layer that is formed on the surfaces of both the diffusion and gate regions. Since TiN is also an effective diffusion barrier, it prevents the phosphorus dopant in the n^+ polycide structure from diffusing and counterdoping the diffused drain regions of the PMOS devices when a connection is formed between them. (The formation and properties of TiN as an interconnect and barrier material is discussed in greater detail in chap. 3).

Several 4-Mbit CMOS SRAMs are described in the October, 1989 issue of the *IEEE Journal of Solid-State Circuits*. The decrease in SRAM cell size as a function of minimum feature size is shown in Fig. 8-5.

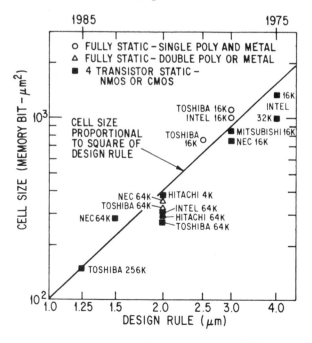

Fig. 8-5 MOS SRAM cell size versus design rule history.[116] (© 1986 IEEE).

8.2.1.3. High-Valued Polysilicon Load Resistors for MOS SRAMs

High-valued resistors are used as the load devices in the poly-load SRAM cell (Fig. 8-2b). In order to minimize power consumption and yet maintain an optimum soft-error rate, the load current of the cell is set to about 31 pA.[15] Very high-valued load resistors must be used to obtain such small load currents. For example, it has been calculated that 164 $G\Omega$ resistors must be used for 64-kbit and 256-kbit SRAMs, and 97 $G\Omega$ resistors are needed for 1-Mbit and 4-Mbit SRAMs (Fig. 8-6a).[15] Films made of materials with very high sheet resistances must be used to fabricate these load resistors to avoid the consumption of excessive area.

Undoped polysilicon films exhibit high sheet-resistivity values, making them good candidates for fabricating such structures (Fig. 8-6b). When undoped polysilicon films are implanted with arsenic in doses from $\sim 1\times10^{13}/cm^2$ to $1\times10^{15}/cm^2$, the sheet resistivity can be controlled from 10^4 Ω/sq up to about 10^{12} Ω/sq (Fig. 8-6c). Hence, to fabricate a high valued resistor (for example, a 97-$G\Omega$ resistor for a 1-Mbit SRAM cell), a polysilicon film with a sheet resistance of 26 $G\Omega$/sq can be used. This sheet resistance can be obtained with an As implant dose of $\sim 3\times10^{13}/cm^2$. The length of a 97-$G\Omega$ resistor fabricated in such a 50-nm-thick, 1.2-μm-wide line of polysilicon would be 4.0 μm.

Undoped polysilicon exhibits such high resistivity because some of the impurities in the films segregate to the grain boundaries and do not effectively produce free carriers. In addition, the grain-boundary regions trap some of the free carriers that are produced (see Vol. 1, chap. 6).

The I-V characteristics of the high-valued polysilicon resistors predicted by the trapping model of Lu et al. fit the experimental data fairly well if the resistor length is not too short.[16] On the other hand, lateral diffusion from adjacent higher-doped regions in the poly can significantly alter the resistance value if such diffusion takes place over a large enough fraction of the resistor length. Because the potential energy barrier to diffusion along the grain boundaries is lower than that in the bulk (see Vol. 1, chap. 8), the rapid diffusion of impurities along grain boundaries can bring impurities to the lightly doped poly regions, even at relatively low temperatures.

The effect just described can be important in the design of polysilicon-load SRAM cells. The resistors are normally formed in the second polysilicon layer, and the remainder of this layer is implanted with a much higher dose (so that it can serve as a low-resistance interconnect path). During this implant, the high-resistivity poly regions are covered with a mask to avoid impurity doping. The minimum size of the mask is limited by the effect of the lateral diffusion of impurities from the highly doped regions during the activation anneal of the polysilicon following ion implantation (e.g., 950°C for 30 min). Hence, a lower limit of about 3 μm was initially predicted for the length of such resistors.

A technique for reducing the extent of the lateral diffusion by implanting the polysilicon with a very heavy dose of oxygen ($\sim 1\times10^{22}/cm^3$) has been reported.[17] High-valued resistors can be fabricated with lengths as small as 0.8 μm (Fig. 8-6d). The oxygen apparently segregates to the grain boundaries, retarding the diffusion of the

	Memory size (bits)	Feature Size (μm)	Power Supply Voltage (V)	Load Current per Bit (pA)	Typical Memory Standby Current (μA)	Load Resistance (GΩ)	Memory Cell Size (μm²)	L/W of Load Resistor (μm)	Sheet Resistance (GΩ/□)	Thickness of Poly-silicon Resistor (nm)	Chip Size (mm²)
	[14] 64K	2.0	5.0	31	2	164	16×19 (304)	7.0/2.0	47	100	5.44×5.80
Estimated Value	256K	1.2	5.0	31	8	164	10×11 (110)	4.0/1.2	47	70	~6.5×7.5
Estimated Value	1M	0.8	3.0	31	33	97	7.0×7.5 (52.5)	3.0/0.8	26	50	~8.5×9.5
Estimated Value	4M	0.5	3.0	31	130	97	3.4×4.2 (14.3)	2.0/0.5	24	30	~8.5×10
Estimated Value	16M	0.25	1.5	31	520	48	1.7×2.1 (3.6)	1.0/0.25	12	30	~8.5×10

a)

b)

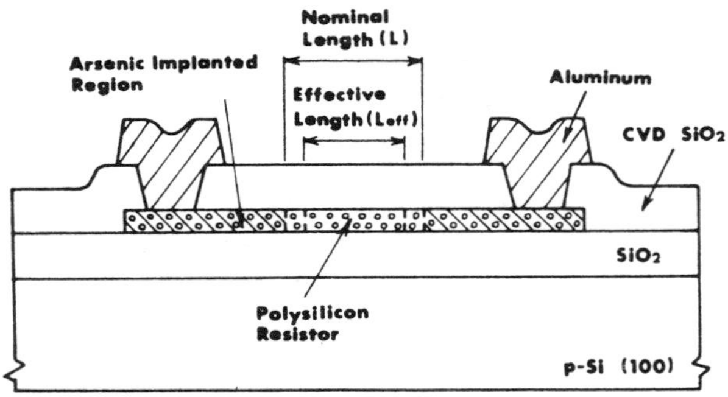

Fig. 8-6 (a) Comparison of parameters about the load resistors in SRAM cells. (b) Schematic cross section of a polysilicon resistor.

arsenic (and perhaps also increasing the potential barrier height by forming silicon-oxygen bonds).

The high sheet resistance of polysilicon-load resistors is also reduced by hydrogen diffusion into the polysilicon from plasma-deposited nitride passivation films. This was found to be controllable by sandwiching the polysilicon film with an LPCVD silicon-nitride film (which contains much less hydrogen than does plasma-deposited nitride; see Vol. 1, chap. 6).[19]

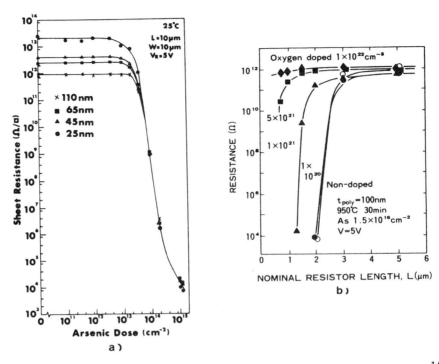

Fig. 8-6 (c) Sheet resistance versus arsenic dose for ion-implanted polysilicon resistors.[15] (© 1985 IEEE). (d) Resistance versus nominal resistor length for oxygen-doped and undoped polysilicon resistors. Resistance is normalized to 1-μm width.[17] (© 1987 IEEE).

8.2.1.4 Soft Errors in SRAMs.

SRAMs offer better resistance than DRAMs to both transient and total-dose radiation, making them better suited for some military and space applications. Until recently, the soft-error rates of SRAMs (see section 8.3.4) were negligible compared to those of DRAMs. However, as geometries have been scaled down to produce circuits of greater density, alpha-particle–induced soft-error rates have also become a concern in SRAMs.[33] Although p-well CMOS in itself raises the threshold against soft-error failures, the use of extra buried p-layers has also been explored as a way to reduce such errors by an additional three orders of magnitude.[10] In addition, the full-CMOS cells exhibit less susceptibility than poly-load cells to single-event upsets and soft errors.

CMOS/SOS technology provides inherently harder parts than does bulk CMOS, and SRAMs have thus been built in CMOS/SOS for such applications. A report detailing the causes of the increase in soft-error rates in densely packed SRAM cells is given in reference 11. Another report on the modeling of alpha-particle sensitivities of SRAMs indicates that state-of-the-art CMOS SRAMs from 64 kbits to 1 Mbit can be made to exhibit sufficient insensitivity to alpha particles.[18]

8.2.2 Bipolar and BiCMOS SRAMS

Although MOS SRAMs have achieved much higher device densities on a chip, as well as lower cost and lower power per bit, bipolar SRAMs using emitter-coupled logic (ECL) technology are still faster. Hence, they are chiefly used in applications where highest-speed operation is required (e.g., in the cache memory of high-speed computers). ECL SRAMs are classified into two groups: *high speed* (7-15 ns access times), and *ultra-high speed* (<7 ns access time).[24] The fastest 16-kbit bipolar SRAMs have access times of <4 ns,[20] and a subnanosecond 5-kbit bipolar SRAM has been reported.[21,31] As noted earlier, such bipolar SRAMs exhibit an ECL I/O capability.

The emitter-coupled cell shown in Fig. 8-2c is the most widely used bipolar SRAM cell. The load devices affect both the current in the standby mode and the saturation conditions of the driver transistors. In addition, since these devices also determine the read/write current, they have a significant impact on the access time. In early bipolar SRAMs the load for the flip-flop was simply a resistor, which caused the driver transistors to saturate when accessed. In more recent high-speed ECL SRAMs, another cell (as shown in Fig. 8-7a), has been used that utilizes a *pnp* transistor as the load. The advantage of this is that it allows for a smaller cell size, since the transistors are fabricated in parasitic elements. However, because the loads still allow the driver transistors to saturate, SRAMs which use such cells cannot achieve ultra-high speeds. Hence, such cells are used in medium-speed, high-density bipolar SRAMs.[23]

In ultra-high-speed ECL cells, saturation is avoided through the use of a Schottky diode in combination with resistors (the so-called *Schottky-barrier-diode, [SBD] switched-load-resistor cell;* Fig. 8-7b.[22] The load-resistance value in this cell is high during standby and low when active, making the current during standby about one-thousandth of that drawn by the cell during reading and writing.

Although this cell was invented quite some time ago, it did not gain wide acceptance because of its low density. Recent cell designs that incorporate trench isolation (to allow closer spacing of the transistors) and two levels of polysilicon (one for the interconnections and the other for the resistors) have allowed cells about half the size of conventional SBD cells to be realized.

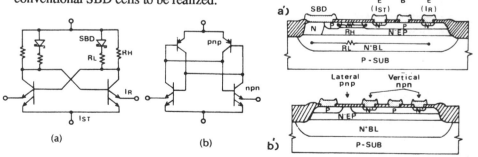

Fig. 8-7 Equivalent-circuit diagrams and cross-sectional views of (a) switched-load memory cell, and (b) cross-coupled *pnp* load cell.[2] (© 1986 IEEE).

In ultra-high-speed SRAMs, the Schottky diode speeds the switching response and increases the soft-error immunity by sustaining most of the stored charge on the SBD capacitance. A large SBD capacitance is therefore needed in order for high reliability to be maintained. The SBD capacitance, however, is quite small per unit area (2.8 $fF/\mu m^2$). An alternative cell that incorporates a separate Ta_2O_5 capacitor has been reported.[24,27] The use of this capacitor, which has a capacitance of 8.5 $fF/\mu m^2$, makes it possible to reduce the cell size by 30% compared to a conventional SBD cell.

Another ECL-SRAM cell using polysilicon diodes as the load elements has been developed.[113] The advantages of this approach include: compact cell size, very low standby current, very low parasitics, and small active junction area. Access times of 1.5 ns for a 1-kbit SRAM have been demonstrated.

Process innovations that have been incorporated to allow faster, higher-density SRAMs include the use of electromigration-resistant aluminum alloys (to permit higher current densities in metal interconnect stripes) and U-groove isolation (see chap. 2) which reduces the isolation width by a factor of three compared to fully recessed LOCOS isolation (see Fig. 2-35a).

Circuit techniques have also been used to enhance the performance of bipolar SRAMS. For example, read/write current has been concentrated in the active region of device operation, and the word delay has been reduced through the use of Darlington drivers. Table 8-2 summarizes the evolution of bipolar SRAM technology.

Table 8.2   Bipolar SRAM Evolution

Introduction Date	Size	Access Time	Load Device	Process Enhancements	Circuit Enhancements
1975	1 k	1.5 ns	Resistor	Al-Cu	Non-Saturated Read/Write Current
1978	4 k	2.2 ns	Schottky Diode		Darlington Drivers
1982	16 k	3.0 ns	*pnp* Transistor	U-Groove Isolation	
1986	64 k	5.0 ns			

8.2.2.1 BiCMOS SRAMs.

Although high-speed ECL SRAMs up to 64 kbits in size have been fabricated, such large bipolar SRAMs have power dissipation and yield problems. Power dissipation increases because each cell draws a minimum standby current of about 2 μA to maintain sufficient noise margins and immunity from alpha particle soft errors. Defects in the narrow base region make it difficult for high-yielding circuits containing 262,144 narrow bases to be produced (i.e., each cell of a 64-kbit ECL SRAM designed with *pnp* loads contains four transistors).

On the other hand, CMOS alone cannot be used to build such high-performance, higher-density SRAMs because the driving capability of CMOS is inferior to bipolar, and it is practically impossible to design an input and output buffer circuit in CMOS that has an ECL I/O capability.

SRAMs have been developed which combine both bipolar and CMOS devices on the same chip. A comparison of 64-kbit SRAMs built using ECL and BiCMOS techno-

Table 8-3   Comparison of 64-kbit Bipolar and BiCMOS SRAMs[24]

	1.3-μm BiCMOS	2.0-μm BiCMOS	1.2-μm Bipolar	Unit
Organization	64 k x 1	16 k x 1	64 k x 1	word x bit
Address Access Time	7	13	10	ns
Write Pulse Width	4	7	11	ns
Operating Power	350	500	1300	mW
Memory-Cell Size	97	230	524	μm^2
Die Size	20	30	55.4	mm^2

logies is given in Table 8-3.[24] More recently, 256-kbit BiCMOS SRAMs have been announced, with access times of 8-12 ns.[25,26] As noted earlier, BiCMOS SRAMs can be designed with TTL as well as ECL I/O capabilities.

In early BiCMOS SRAMs, moderate-speed bipolar transistors were integrated into what was essentially a CMOS technology in order to provide faster output buffers and sense amplifiers. In more recent designs, the CMOS devices are being fabricated on a basically high-speed bipolar chip. In an early 16-kbit BiCMOS SRAM, the cells of the memory array were poly-load cells, the peripheral circuits were CMOS, and the I/O buffers and sense amplifiers were bipolar circuits. In a more recent 256-kbit BiCMOS SRAM design, the memory array was implemented with full-CMOS cells, and bipolar sense amplifiers and ECL output buffers were used. This memory had a reported access time of 8 ns and could be operated with battery backup (i.e., since it draws only 1 μA during standby, a battery with minimal power can provide long-term backup).

Figure 8-8 shows the speed-versus-density characteristics of bipolar, CMOS, and BiCMOS SRAMs.[34] These curves indicate that bipolar technologies no longer offer the fastest performance at such densities, and that ECL I/O BiCMOS offers higher

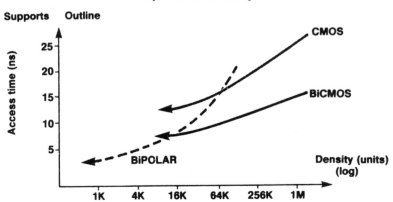

Fig. 8-8 Comparison of speed and density of CMOS, Bipolar and BiCMOS SRAMs.[34] Reprinted with permission of Semiconductor International.

speeds than bipolar ECL at middle and higher densities (because of the severe speed, power, and density trade-off in bipolar ECL). In addition, BiCMOS SRAMs are somewhat faster than CMOS SRAMs at equivalent power and density for the same geometry, and TTL I/O BiCMOS SRAMs are a little slower than ECL I/O SRAMs because of their larger output voltage swing. Finally, BiCMOS offers a range of speed and power trade-offs (from very fast, low-density BiCMOS at high power, to slightly slower, mid-density memory circuits at moderate power).

Recently, an 8 ns 1-Mbit BiCMOS SRAM[114] and a 3.5 ns 16-kbit ECL BiCMOS SRAM[115] have been reported.

8.3 DYNAMIC RANDOM ACCESS MEMORIES (DRAMS)

As noted earlier, *dynamic random access memories* (DRAMs) are so named because their cells can retain information only temporarily (on the order of milliseconds); even with power continuously applied. The cells must therefore be read and refreshed at periodic intervals. While the storage time may at first appear to be very short, it is actually long enough to allow for many memory operations between refresh cycles.

Despite of this apparently complex operating mode, the advantages of cost per bit, device density, and flexibility of use (i.e., both read and write operations are possible) have made DRAMs the most widely used form of semiconductor memory to date.

The earliest DRAMs used three-transistor cells and were fabricated using PMOS technology. Nevertheless, their introduction represented an immediate, dramatic decrease in the minimum semiconductor memory-cell size, since they could replace SRAMs based on a six-transistor cell. As a result, more cells per chip could be implemented. However, DRAM cells consisting of only one transistor and one capacitor were quickly implemented,[28] and such cells have been used in DRAMs ever since.

8.3.1 Evolution of DRAM Technology

The earliest MOS combinational logic networks, referred to as *static logic circuits,* operated without any need for periodic clock signals. However, it was recognized that clock signals could be used to advantage in combinatorial and sequential logic circuits. By introducing clock signals at arbitrary circuit nodes, it was possible to achieve faster operation, greater circuit density, and reduced power dissipation. Such logic circuits became known as *dynamic logic circuits.*

Data in these circuits was temporarily stored in dynamic registers (in the form of charge on the gate of an MOS transistor), rather than in static registers (which store data as the state of a flip-flop circuit). Thus, dynamic shift registers could be built with fewer transistors and, consequently, on much smaller areas of silicon than were needed for static shift registers. This allowed a dramatic increase in logic-circuit density.

A question arose, however, as to how long the gate of a MOS transistor could store charge before that charge would be lost through leakage currents. It turned out that at near room temperature, the charge could stored for more than 10 milliseconds. If a clock signal were to arrive at intervals significantly shorter than this, a large fraction of the initially stored charge would still remain on the MOS transistor gates. Therefore, if a clock signal were to be applied at least this frequently, dynamic registers could serve as efficient charge-storage nodes.

It was also quickly realized that if the dynamic stored-charge approach was practical in dynamic logic circuitry, it might also work for semiconductor-memory designs. In the case of a dynamic memory, however, a *refresh signal* had to be applied to each charged cell node at sufficiently frequent intervals (typically, every 4-8 ms), to allow the temporarily stored data on each cell to be retained indefinitely.

The concept of the DRAM was patented by Dennard of IBM in 1968, and the first commercial DRAM was introduced by Intel in 1970. The latter was built using a three-transistor cell (Fig. 8-9) in PMOS silicon-gate technology, while Dennard's patent used a one-transistor cell. In the three transistor cell, the charge is stored on the parasitic capacitance of the gate of transistor M_1. (Although the capacitance in this cell is a parasitic effect, it is drawn explicitly in the circuit schematic of Fig. 8-9 because it is essential for normal memory-cell operation.) The leakage current of the reverse-bias junction of the drain region of transistor M_3 discharges this capacitance over a period of several milliseconds or more. Hence, a periodic signal must arrive at the node in order for the charge stored on the capacitor to be maintained. Since one-transistor DRAM cells quickly replaced the three-transistor cells, the rest of our discussion will be restricted to them.

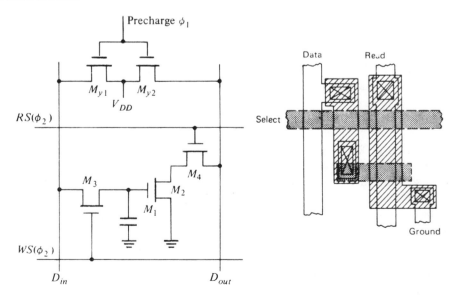

Fig. 8-9 Layout and circuit for a 3-transistor DRAM cell.

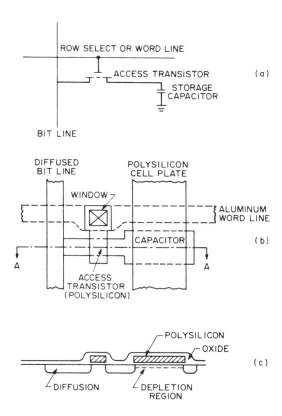

Fig. 8-10 Single-transistor DRAM cell and storage capacitor. (a) Circuit schematic. (b) Cell layout. (c) Cross section through A - A.[6] After R. Hunt, "Memory Design and Technology," in M.J. Howes and D.V. Morgan, eds., *Large Scale Integration.* Copyright 1981, John Wiley & Sons. Reprinted with permission.

8.3.1.1 One-Transistor DRAM Cell Design.   The influence of Dennard's one-transistor cell[28] (which actually consists of one transistor and one capacitor) is considered to be comparable to that of the invention of the transistor itself.[3] The design of this cell has been rendered in many versions since its invention, and we will begin the description of its evolution (which, by the way, is far from over), with a description of a simple cell that utilizes a polysilicon layer as one plate of the cell capacitor (Fig. 8-10). The first such cell was fabricated with 8-μm features, used 1280 μm^2 of silicon area, and was employed in the design of the 4-kbit NMOS DRAM. There have since been many variations of even this simple cell, with single or double layers of polysilicon, different methods of capacitor formation, and different materials used for the word and bit lines. These are reviewed in detail in reference 30.

In the cell shown in Fig. 8-11a, the capacitor stores the charge on the cell (storage capacitor), and the NMOS transistor allows the bit line to access the charge-storage

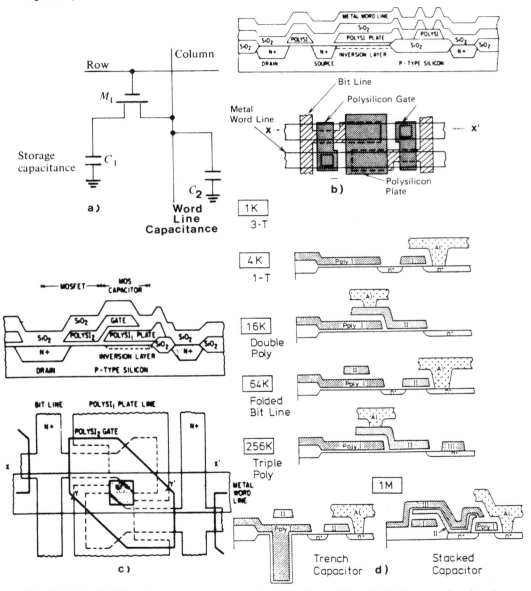

Fig. 8-11 (a) DRAM cell connections to word and bit lines. (b) and (c) Cross-sectional and layout views of (b) the single-poly-word-line/diffused-bit-line 4-kbit DRAM cell, and (c) the double-poly diffused bit line merged DRAM cell.[3] (After C. N. Berglund) (d) Structural innovations of DRAMs.[41] (© 1985 IEEE).

region of the capacitor. The storage capacitor consists of a polysilicon plate over a thin oxide film (which is the capacitor dielectric), with the semiconductor region under the oxide serving as the other capacitor plate. An n^+ diffused region in the semiconductor substrate serves as the bit line and an aluminum stripe as the word line. As can be seen in the cross-section of the cell (Fig. 8-11b), the bit-line diffused region makes contact with the n^+-diffused source region of the access transistor. A contact between the word line and the polysilicon gate of the access transistor is made, as also shown in this figure.

One widely implemented enhancement of this basic cell design is shown in Fig. 8-11c. In this modified cell, the floating drain region of the access transistor is eliminated and a second layer of polysilicon transfers the charge from the bit line to the storage capacitor. This not only allows the cell to be reduced in size, but also increases its storage capacity.[35] The disadvantage is that a double-polysilicon process must be used. As has often been the case in the evolution of VLSI, increased packing density and better performance are achieved at the price of somewhat greater process complexity. Figure 8-11d summarizes the structural innovations used as DRAMs have evolved.

8.3.1.2 Operation of the One-Transistor DRAM Cell. To study the operation of the cell in Fig. 8-11c, assume that the substrate is grounded and that 5 V are applied to the polysilicon top plate of the storage capacitor (which we'll refer to as the *plate electrode* of the capacitor). The semiconductor region under the polysilicon plate serves as the other capacitor electrode, and in an NMOS cell this *p*-type region is normally inverted by the 5-V bias. As a result, a layer of electrons is formed at the surface of the semiconductor, and a depleted region is created below the surface. (The electrode on which the charge is stored will be referred to as the *storage electrode*.)

To write a *one* into the cell, 5 V are applied to the bit line, and a 5-V pulse is simultaneously applied to the word line. The access transistor is turned *ON* by this pulse, since its V_T is about 1 V. The source of the access transistor is biased to 5 V, since it is connected to the bit line. However, the electrostatic potential of the channel beneath both the access-transistor gate and the polysilicon plate of the storage capacitor is less than 5 V, because some of the applied voltage is dropped across the gate oxide. As a result, any electrons present in the inversion layer of the storage capacitor will flow to the lower potential region of the source, causing the storage electrode to become a depletion region that is emptied of any inversion-layer charge. When the word-line pulse returns to 0 V, an *empty potential well* remains under the storage gate. This empty well represents a *binary one*, and it is shown as the deep-depletion space-charge region in Fig. 8-12.

For writing a *zero*, the bit-line voltage is returned to 0 V, and the word line is again pulsed to 5 V. With the access transistor turned *ON*, electrons from the n^+ source region (whose potential has been returned to 0 V) have access to the empty potential well (whose potential is now lower than that of the source region). Hence, the electrons from the source move to fill it, thus restoring the inversion layer beneath the poly plate. When the word-line voltage is returned to zero, the inversion-layer charge present

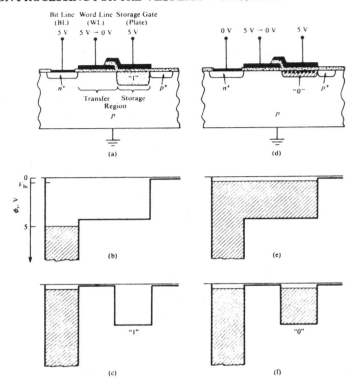

Fig. 8-12 A basic dynamic RAM cell, showing (a) a stored **one** and (d) a stored **zero**. The writing of a **one** is shown in (b) and (c), and the writing of a **zero** is shown in (e) and (f). From D. K. Schroder, *Advanced MOS Devices*. Copyright 1988, Addison-Wesley. Reprinted with permission.

on the storage capacitor is isolated beneath the storage gate. This condition represents a stored *binary zero*.

Note that when a one is stored, an empty well exists. This is not an equilibrium condition, since electrons that should be present in the inversion layer have been removed. As electrons are thermally generated within and nearby the depletion region surrounding the well, they will move to re-create the inversion layer. A stored one thus gradually becomes a stored zero as electrons refill the empty well. The nature of the planar one-transistor DRAM cells is that ones become zeros, and zeros remain zeros.

To prevent the ones from being lost, each cell must be periodically refreshed by the memory so that the correct data remains stored at each bit location. The time interval between refresh cycles is called the *refresh time*. The total leakage current of the cell must be low enough that the cell does not discharge and lose its memory state between refreshes. A typical guideline has allowed a 20% degradation of the charged state of a cell during the refresh time.[105] For example, if this interval is 8 ms and the charge stored on a cell is 10^6 electrons, the maximum allowed leakage current at maximum operating chip temperatures and worst-case voltages and timing conditions is ~4 pA. This specification can be achieved fairly easily at the average cell level. However, since the leakage currents in a large number of cells exhibit a distribution, the high side of the distribution must be monitored in the course of technology development – and manufacturing – to ensure that the maximum allowable current is not exceeded in any of the memory chip cells.

It is predicted that as memory sizes continue to increase, the maximum allowable leakage current will be reduced. Since alternative capacitor dielectric materials may be used in advanced DRAMs, the leakage currents across such dielectrics will be an important material characteristic. In addition, the access transistor of the cells will have to be turned off very "hard" when the memory cell is not being accessed, since any source-drain leakage can also discharge the storage capacitor.

8.3.1.3 Writing, Reading, and Refreshing DRAM Cells.
It's useful to consider the reading and writing of DRAM cells from a circuit perspective as well as from the device physics perspective just discussed. Figure 8-13 provides a simplified schematic of this operation.

Switches S1 and S2 are used to select the appropriate bit and word lines that are connected to the cell in question. Such switch closure is the response to an address

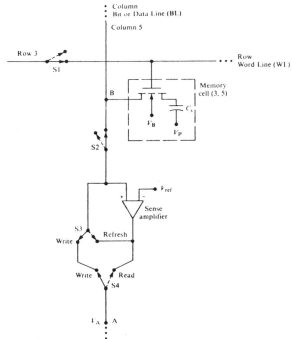

Fig. 8-13 Schematic diagram showing the writing, reading, and refreshing operations of a DRAM. After ref. [117], reprinted by permission of Scientific American, Inc.

signal fed to the bit (column) and word (row) decoders. The closure of S1 causes a "high" voltage to be applied to that particular word (row) line (in turn causing the access transistors controlled by this line to be turned *ON*).

If it is desired to write data into the cell at that point, an appropriate voltage must also exist on the correct bit line. In addition, switches S3 and S4 must be set to the write positions. This allows the voltage at point A on the bit line, V_A, to be applied to the source of the cell being addressed (point B). As long as Switch S1 is closed and the access transistor is turned on, the capacitor of the cell can be charged to $V_A - V_p$ (where V_p is the *storage-electrode potential*). When the write operation is completed, switches S1 and S2 are opened, and another cell can be written.

If it is desired to read data from a particular cell, the appropriate switches S1 and S2 are once again closed (it is assumed that all of the cells already contain the stored information). Switch S4, however, is set to the read position. The storage capacitor of the cell being read is now connected to one input of the sense amplifier. The sense amplifier is a *comparator circuit,* with its other input being connected to a reference voltage, V_{ref}. Therefore, if the cell-capacitor voltage is larger than V_{ref}, a logic *1* is read; if it is smaller, a logic *0* is read.

A logic *1* corresponds to the condition in which the storage electrode is depleted of its inversion layer charge. If a logic *1* is read immediately after the cell has been written, the signal will be strong. As time passes, thermal electron-hole pair generation will cause refilling of the empty potential well, thereby degrading the amplitude of the logic *1* signal. If too long a time elapses between the writing and reading, the inversion charge will be reestablished, and a logic *0* will be produced when the cell is read. Once a logic *0* is stored on a cell, however, it will continue to be read for as long as power remains applied to the capacitor plate electrode.

This indicates that the logic *1* must be periodically refreshed to allow it to be retained on the cell for indefinite time periods. Because it is not known what logic level is stored on each cell at any instant of time (especially since the cells are randomly read and written), it is mandatory that the entire memory be refreshed at periodic intervals (usually every few milliseconds). Furthermore, since reading a cell changes the charge on its capacitor, the cell must also be refreshed immediately following each read operation.

The refresh procedure is accomplished by switching S3 to the refresh position after the sense amplifier output has been set by the read operation. The output voltage of the sense amplifier will then write the appropriate information back onto the cell capacitor. If it is desired to refresh the entire memory, each cell can be read and refreshed. It is apparent, however, that data cannot be written while reading or refreshing is in progress.

One sense amplifier must be available for each bit line. Note that the sense amplifier is an extremely sensitive comparator (basically of the cross-coupled flip-flop type), and its design is critically important to the success of DRAM manufacture. Although the details of the design task are not our subject here, some aspects of sense-amplifier performance should be mentioned. When a cell is read, the charge stored on the cell capacitor is shared with the 10 to 20 times larger capacitance of the bit line (which, as we saw earlier, is a long conductor line connected to the sources of all of the cells in the column). After the time interval between refresh pulses has elapsed (e.g., 8 ms) the difference in stored voltage between a *1* and a *0* may be as small as 2 V. As a result, there may only be a 100-200 mV difference between the *1* and *0* signals applied to the sense-amplifier input.

8.3.1.4 DRAM-Cell Charge Storage and Capacitance.

A *one* must be clearly distinguishable from a *zero* when the read operation is performed. The zero is represented by the inversion charge present when the potential well is full. This quantity in an MOS capacitor, Q_s, is given by

$$Q_s = (V_{G'} - 2\varphi_f) C_{ox} \approx V_{G'} C_{ox} \qquad (8\text{-}1)$$

where $V_{G'}$ is the voltage applied to the gate, φ_f is the difference in potential between the intrinsic Fermi level (E_i) and the Fermi level (E_F), and C_{ox} is the capacitance of the capacitor oxide.* In order to pack a great many cells onto a DRAM chip, the cell size is made as small as possible. This implies that it is also desirable to make the area of the storage capacitor as small as possible. On the other hand, Q_s of the storage capacitor must be large enough to send a sufficiently strong signal to the sense circuitry and to provide sufficient immunity from soft errors (see section 8.3.5). Novel cell designs have been developed in an attempt to satisfy these apparently contradictory requirements (such designs will be discussed later).

EXAMPLE 8-1: Calculate the charge stored in the inversion layer when a *zero* is stored (both in units of *coulombs*, and in terms of the number of electrons present) when 5 V are applied to an MOS capacitor whose dimensions are 4 x 4 μm, and which has an SiO_2 dielectric that is 15 nm thick. Also, find its capacitance, C_s.

SOLUTION: $Q_s(0) \approx V_{G'} C_s = (\varepsilon \times A \times V_{G'}) / t_{ox}$

$$= (3.9 \times 8.85 \times 10^{-14} \text{ F/cm} \times 16 \times 10^{-8} \text{ cm}^2 \times 5 \text{ V}) / 1.5 \times 10^{-6} \text{ cm}$$

$$= 18.6 \times 10^{-14} \text{ C} = 186 \text{ fC, or since } q = 1.6 \times 10^{-19} \text{ C/electron}$$

$$Q_s = 1.15 \times 10^6 \text{ electrons; and}$$

$$C_s = Q_s / V_A = 37 \text{ fF.}$$

The above shows that the most important parameters involved in increasing the charge stored on the capacitor are the dielectric constant and thickness of the insulator, and the area of the capacitor.

The capacitance of the DRAM cell is also important. As described earlier, when the

---

* The approximation used in Eq. 8-1 is valid when $V_{G'} > 2\varphi_f$, which is the case for $V_{G'} = 5$V and $\varphi_f \approx 0.4$ V.

contents of the cell are sensed, the charge stored on it is "dumped" into the bit line connected to the sense amplifier. Because the bit-line capacitance (C_B) is typically 7 to 15 times larger than the cell capacitance, the capacitively-divided voltage difference applied to the sense amplifier (ΔV_{sa}) is substantially smaller than that existing in the cell alone. ΔV_{sa} is given approximately by

$$\Delta V_{sa} = (1/2) V_A (C_{ox}) / (C_{ox} + C_B) \qquad (8-2)$$

The minimum detectable voltage difference that the sense amplifiers of 1-Mbit DRAMs can detect (i.e., their *sensitivity*) is in the neighborhood of 150-200 mV. (Note that this sensitivity must be maintained under worst-case operating conditions of voltage, temperature, and noise, as well as worst-case variations of processing conditions.) It is predicted that the sensitivity of sense amplifiers will have to be significantly increased in order for higher-density DRAMs to be fabricated.

8.3.1.5 High-Capacity (Hi-C) DRAM Cells. Another technique for increasing the cell's charge-storage capacity without increasing its size was suggested independently by Sodini and Kamins,[37] and Tasch, et al.[106] This novel technique involves multiple ion implantations to increase the substrate doping in the local vicinity of the storage node.

A deep implantation of *p*-type impurities (boron) is first performed under the storage-plate area. This increases the substrate doping, which in turn increases the depletion-region capacitance of the storage capacitor. However, this single implant alone does not increase the charge-storage capacity of the cell, since the extra *p*-doping also reduces the difference in the surface potential between an empty and a full potential well.

To restore surface potential to its previous value it (without compromising the increased capacitance), a very shallow layer of *n*-type dopant (arsenic) is implanted under the storage plate area (Fig. 8-14). The implanted *donor* atoms, which are very close to the Si/SiO$_2$ interface, behave like a fixed positive oxide charge (and the presence of such oxide charge acts to increase the surface potential in NMOS structures). By increasing the depletion-region capacitance without simultaneously increasing the surface potential, the charge-storage capacity of these so-called *high capacity cells* (or *Hi-C cells*) is enhanced by about 50 percent compared to conventional cells.[35,104]   Although, this technique increases the storage capacities of planar capacitor structures in these types of one-transistor cells, the cell size that must be used in order for adequate charge-storage capacity to be obtained eventually becomes too large for this type of cell to be used in advanced DRAMs (i.e., >1 Mbit). New cell structures have thus been developed for larger DRAMs.

8.3.1.6 CMOS DRAMs. With the introduction of the 256-kbit DRAM, the design of DRAM circuits began to change from NMOS to a mixed NMOS/CMOS technology. The cells of the mixed-technology memory array are all built in a common well of a CMOS wafer. The access transistor and storage capacitor of each cell are usually still fabricated using NMOS technology, while the peripheral circuits are

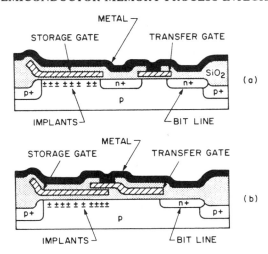

Fig. 8-14 High-capacity (Hi-C) dynamic RAM structure with arsenic (+) and deeper boron (-) implants. (a) One-transistor cell with single-level polysilicon. (b) Double-level polysilicon cell.[35] (© 1978 IEEE).

designed and fabricated in CMOS technology. The majority of 1-Mbit DRAM designs were executed in NMOS/CMOS technology, and this trend is expected to continue.

The advantages of NMOS/CMOS over NMOS DRAMs include lower power dissipation (i.e., by a factor of around 3) and smaller soft-error rates (see section 8.3.5). In addition, as power-supply voltages are reduced to allow smaller MOS transistors to be built, NMOS device design must become much more complex to allow the cells to function properly. CMOS on the other hand, can easily operate at such lower voltages. Finally, a circuit technique known as *static column decoding*, which significantly reduces the memory access time, can be successfully implemented in CMOS but not in NMOS (i.e., since to do so in NMOS would require a circuit dissipating excessively large standby power).[38]

In late 1989 1-Mbit BiCMOS DRAMs were introduced. Their access times of 40 ns placed them between the high-speed 1-Mbit SRAMs (access times of 20 ns), and the slower 1-Mbit CMOS DRAMs (access times of 60-80 ns).

8.3.2 Design and Economic Constraints on Advanced DRAM Cells

As the DRAM cell has been scaled down in size, the minimum amount of stored charge needed to maintain reliable memory operation has remained the same. This constant charge-storage value has had to be maintained within fabrication-cost constraints. From the system point of view, a new generation of DRAM will be embraced if it allows a density increase of about fourfold at the circuit board level, provided that this is accompanied by a cost reduction. To allow such an increase to be realized, the new

DRAM generation must be able to use the same size package as that used by the previous generation. This explains why the 300-mil package housed DRAMs for five generations. To squeeze enough cells onto a chip to allow this package to be retained implies that the cell size of the 1-Mbit DRAM cannot exceed 20 μm^2.

For future DRAM generations, however, increased memory size will be achieved by both increasing the chip size and shrinking the cell size. Since the chip size is predicted to increase by a factor of 1.5 from generation to generation, the cell area will therefore need to shrink to 40% of the size of the previous generation (which can be done by shrinking the minimum line width to 70% of that used previously). For 4-Mbit DRAMs, the cell size must therefore be no larger than 9 μm^2; for 16-Mbit DRAMs, no larger than 4 μm^2, and so on. To meet the signal-to-noise ratio constraints and the soft-error rate requirements, a minimum of ~200 fC of charge (~10^6 electrons) will have to be stored. For the fabrication process to be economically feasible, as few steps as possible should beyond those required to fabricate the transistors and interconnects should be added.

Recovery of the equipment and development costs of the 256-kbit, 1-Mbit, and 4-Mbit DRAMs has become the major factor dictating the three-year product-introduction and delivery cycle so that a profit can be generated from each DRAM generation. In a typical three-year cycle, fewer than 1 million chips would be shipped for sampling during the introduction year. In the first production year, 5 million chips would be shipped; production would increase to 50 million chips in the second year, and 500 million in the third. (Note that in the second year the market would be mainframe computers, and in the third year it would be personal computers.)[3] The three-year model predicts that 16-Mbit DRAMs will be introduced in 1992, and 64-Mbit DRAMs in 1995.

It was noted in section 8.3.1.3 that the cell's storage capacity could be increased by making the capacitor dielectric thinner, by using an insulator with a larger dielectric constant, or by increasing the area of the capacitor. The first two options are not currently viable, since capacitor dielectrics thinner than those now being used in DRAM cells (10 nm) will suffer leakage due to Fowler-Nordheim tunneling, and dielectrics with significantly larger dielectric constants than of SiO_2 have not yet been accepted for DRAM-cell application (although research work is under way to develop such higher-dielectric-constant materials).[39] One recent report described a plasma-CVD process for depositing high-quality Ta_2O_5 films.[119] (Ta_2O_5 exhibits a much higher dielectric constant than SiO_2 [22 vs. 3.9], but normally also suffers from a much higher leakage current.) However, by reacting $TaCl_5$ with N_2O under optimized plasma-CVD conditions, Ta_2O_5 films with a thickness that yielded capacitances equivalent to those of a 30-Å SiO_2 film, demonstrated very low leakage currents for up to 10-year operation at 3.3 V.

It should also be noted that since the 256-kbit DRAM generation bilayer films (consisting of both silicon nitride and SiO_2), have been used as the capacitor dielectric to increase cell capacitance (Fig. 8-15a). The higher dielectric constant of Si_3N_4 (twice as large as that of SiO_2) was responsible for this increase.

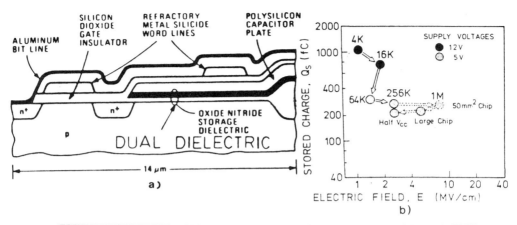

Fig. 8-15 Cross-sectional view of a planar-capacitor DRAM cell with a two-layer capacitor dielectric.[3] (© 1988 IEEE). (b) Trends in storage charge and electric field across capacitor insulator.[41] (© 1985 IEEE). (c) Comparison of DRAM production factors.[44]

One technique that did allow thinner dielectric films to be used was the *half-V_{CC}* approach.[40] That is, the *plate electrode* is biased to $V_{CC}/2$ (i.e., to 2.5 V when V_{CC} = 5 V), and the *storage electrode* is allowed to swing between 0 V and 5 V. As a result, the same quantity of charge can be stored on the capacitor, but the value of the electric field acting on the dielectric is only half the value that exists when the voltage between the two plates equals V_{CC}. This technique has been implemented in the fabrication of 1-Mbit DRAMs (Fig. 8-15b).

The third option (increasing the capacitor area) can be effective if the area is increased by forming the storage capacitor in a trench etched in the substrate or by using a stacked capacitor structure. Both approaches have been implemented, and many variations of such three-dimensional capacitors have been reported.

The planar capacitor structure used in the one-transistor DRAM cell described in section 8.3.1.1 was predicted to be usable up to the 256-kbit DRAM generation. In this generation, the capacitor consumes 30 to 40% of the cell area. It was generally agreed that beyond this, a three-dimensional capacitor structure would be needed in order for sufficient charge storage to be obtained. It turned out, however, that virtually all of the DRAM manufacturers elected to squeeze everything they could from the planar capacitor, and continued to use it to manufacture 1-Mbit DRAMs. This decision was due largely to the difficulty in achieving a reliable capacitor dielectric in a trench cell at the time 1-Mbit DRAMs were introduced. The use of both larger chip sizes and the half-V_{CC} plate-electrode voltage technique permitted the planar capacitor to perform adequately for 1-Mbit DRAMs. Reference 42 presents the details of a 1-Mbit DRAM technology using a 38-fF planar-capacitor structure in which the cell size is 37 μm^2.

As DRAM size increases, process complexity is expected to increase markedly as well. For example, a 1-Mbit DRAMs is reported to require ~18 masks and 350 processing steps, all of which could be successfully carried out in a Class 10 cleanroom (Fig. 8-15c). In comparison, the 4-Mbit DRAM is expected to need 20-25 masks and in excess of 450 processing steps, and will thus require a Class 1 cleanroom processing facility.[43,44] A detailed report on the technology issues that will need to be addressed in the design and fabrication of 64- and 256-Mbit DRAMs has recently been published.[105]

In 1989 1-Mbit CMOS DRAMs with access times ranging from 6-100 ns were being commercially offered. (The fabrication of a high speed 22-ns CMOS DRAM was announced in late 1989, but it was not being offered for sale.)[110] At that time, 4-Mbit DRAMs with access times of 80-120 ns were also being offered, and 16-Mbit CMOS DRAMs with access times as small as 45 ns were being reported.[111] Finally, 1-Mbit BiCMOS DRAMs with access times of 30 ns were being introduced.[112]

8.3.3 Trench-Capacitor DRAM Cells

8.3.3.1 Trench Capacitor Processing for DRAMs.
Trench-capacitor structures have been developed as a way to to achieve DRAM cells with larger capacitance values without increasing the area these cells occupy on the chip surface. (For example, the silicon-area reduction of a trench capacitor compared to a planar capacitor for the same specific capacitance is a factor of 18 or more. Specifically, a 4.0-μm-deep trench capacitor with surface dimensions of 0.87 x 2.4 μm will occupy less than 3 μm^2 of chip area but will have a capacitance of 40 fF.)[66] Many of the processing details involved in trench- capacitor fabrication are the same as those described in chapter 2, section 2.6.3, which deals with the process technology of trench-isolation structures. In this section we discuss those issues that are unique to the fabrication of trench capacitors used in DRAM cells.

There are several differences between the trench structures used for isolation and those used as DRAM capacitors. In the former, the dielectric film on the trench walls can be relatively thick, and the trench can be refilled with polysilicon or CVD SiO_2. In the latter, the insulator formed on the trench walls serves as the capacitor dielectric, and it must therefore be as thin as possible. Since the material that refills the trench serves as one plate of the capacitor, it must consist of highly doped polysilicon. Furthermore, in order for increased capacitance to be obtained through increases in trench depth (while all other parameters remain constant), the trench walls must be highly vertical. To allow for reliable refilling of the trenches, however, some trench sidewall slope must be allowed, and a compromise process that produces a nominal sidewall slope of 87° has been suggested.[46] Finally, to obtain such structures as Hi-C capacitors, the trench walls may need to be selectively doped.

Several techniques have been developed for achieving a dielectric capacitor film that is thin enough to provide both high capacitance and high reliability (that is, the dielectric must be able to provide the same equivalent breakdown voltage as the planar capacitor used in previous DRAM generations). First, composite dielectric films (e.g., thermally grown oxide and CVD nitride) are frequently used.[47,62,68] Since the nitride has a higher dielectric constant than SiO_2, a thicker composite film will yield the same capacitance as a thinner single SiO_2 layer. This thicker film prevents capacitor leakage due to dielectric breakdown or Fowler-Nordheim tunneling.

The growth of the thermal oxide film is also a key step. Unless preventative measures are taken, a thinner oxide will grow in the bottom corners (concave) and top corners (convex) of the trench. A higher electric field will exist across these regions, causing trench capacitors to exhibit higher leakage currents than planar capacitors.

This problem is avoided for the bottom corners by ensuring that the etch process produces a trench with rounded bottom corners (see chap. 2). In addition, an oxidation step for edge rounding and stripping is performed prior to the growing of the actual capacitor SiO_2 film. One report indicates that a 50-nm SiO_2 film is grown in this process and is then stripped in dilute HF (Fig. 8-16a).[48] In addition to smoothing out any sharp bottom corners, this step also removes any plasma damage from the trench walls.

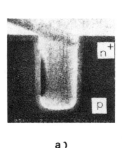

a)

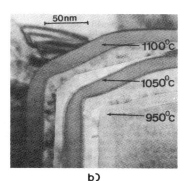

b)

Fig. 8-16 (a) Rounding-off oxidation can produce trenches with smooth bottom corners.[48] (© 1985 IEEE). (b) Rounding-off oxidation can also reduce the severity of the sharp upper corner of the trench.[50] This paper was originally presented at the Spring 1989 Meeting of The Electrochemical Society, Inc. held in Los Angeles, CA.

The electric field is intensified at the top corners of the trench because they are normally quite sharp after etch, and this magnifies the effect of any oxide thinning that may occur. (The extent of the electric-field intensification is modeled in reference 49.) The edge-rounding oxidation step mentioned above also increases the curvature radius of the top corner of the trenched Si surface (Fig. 8-16b), thus helping to produce a trench capacitor with low leakage currents under high electric fields. However, because it is necessary to use a process that allows viscoelastic flow during oxide growth (to relieve the stresses that inhibit oxide growth at the convex corners), a higher temperature oxidation process (*e.g.* 1100°C) is usually involved.[49,50] The use of rapid thermal processing (RTP) to grow the trench oxide has also been reported.[51] Good leakage-current behavior is exhibited when RTP cycles of 1150°C for 25 sec in O_2 were used to grow the trench oxide.

The polysilicon that fills the trench must also be highly doped to prevent depletion effects. In situ doping of the poly is thus necessary. The conventional process for in situ doping of polysilicon employs gaseous phosphine as the dopant source. Unfortunately, this reduces the polysilicon deposition rate by a factor of about 25 (see Vol. 1, chap. 6). Specially designed LPCVD furnaces with caged boats are needed to improve the process.[117] However, these furnaces have particulate problems and cannot be automated, making them incompatible with a high-volume fabrication environment. A recent report described the use of t-butylphosphine as an alternative doping source.[118] It can be used in standard, automated 100-wafer LPCVD furnaces to produce in-situ doped polysilicon films. A higher deposition rate can be achieved   (~20 Å/min), with adequate thickness uniformity. This material is also much less toxic than phosphine.

8.3.3.2 First-Generation Trench-Capacitor–Based DRAM Cells.
Trench structures for storage capacitor application in DRAMs were first reported in 1982-83 (Fig. 8-17a).[52] The processing technology that made these structures possible was anisotropic etching of Si by RIE. Earlier V-groove structures etched in Si by means of wet etching resulted in crystallographically produced sharp edges, which in turn degraded the gate-oxide integrity to the point where devices could not be reliably manufactured. One of the first tests that had to be met by RIE-etched trench capacitors was that of exhibiting breakdown characteristics equal to those of planar-type capacitors. As described in the previous section (and summarized in Fig. 8-17b), several reports showed that this could be achieved through the implementation of trench etching control measures, the use of edge-rounding procedures, or the use of combination films for the trench dielectric (e.g., thermal SiO_2 and CVD nitride).

In the first generation of trench-capacitor–based cells *the plate electrode of the storage capacitor is inside the trench, and the storage electrode is in the substrate.* The access transistor is a planar MOS transistor fabricated beside the trench capacitor, and the trenches are 3-4 μm deep. The cell size of the basic cells of this generation requires about 20 μm^2 of surface area, making the cells suitable for 1-Mbit DRAM designs. It was thought that with appropriate design-rule shrinkage, these cells would be appropriate for early 4-Mbit DRAM designs.[53]

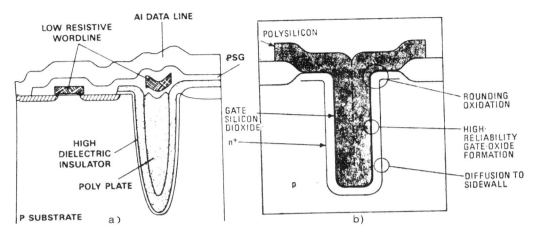

Fig. 8-17 (a) Basic DRAM trench capacitor structure.[52] (© 1982 IEEE). (b) Processing techniques used to insure fabrication of high-quality trench structures.

In one variation of this cell type, the plate electrode is grounded, and the substrate is biased between 0 and 5 V (which improves device isolation between adjacent cells).[54] The walls of the storage electrode (i.e., those in the *p*-type substrate) are doped *n*-type, creating a Hi-C type cell.

These first generation cells exhibit some disadvantages for smaller-sized DRAM cells. Since the charge is stored in a potential well in the substrate, if the cells are too close together, high leakage currents arise between adjacent cells (due to punchthrough or surface conduction). This problem can be alleviated through increased doping of the region between the cells or through the use of deeper, narrower trenches, but at the cost of creating other problems. First, the required doping in the substrate will lead to avalanche breakdown of the reverse-biased junction of the access transistors at spacings ≤0.8 μm. Second, deeper, narrower trenches are significantly more difficult to fabricate reliably and for practical trench dimensions the spacing limit is nearly reached for the cell sizes needed in 4-Mbit DRAMs. Further, since the storage node is in the substrate, there is no immunity to charge collection from alpha particles. Consequently, this type of trench capacitor is as vulnerable to alpha-particle–induced soft errors as cells made with planar storage capacitors. Several design modifications have been developed to increase capacitance without either making the trenches deeper or increasing cell size.

In the first modification, the plate electrode is folded around the sides of the storage electrode, creating a structure called the *folded-capacitor cell,* FCC (Figs. 8-18a and b).[55] A shallow trench is etched around most of the perimeter of the storage electrode. The plate electrode is deposited over this trench, much as a tablecloth is laid over a table top.[56] When both the sides and the planar area (tabletop) are covered, a capacitor with a larger area is obtained, and the capacitance is thereby increased (Fig. 8-18c).

Interestingly, the capacitor of this cell apparently utilizes *both* the planar- and trench-capacitor concepts. In addition, the cell's storage plate edges are electrically isolated

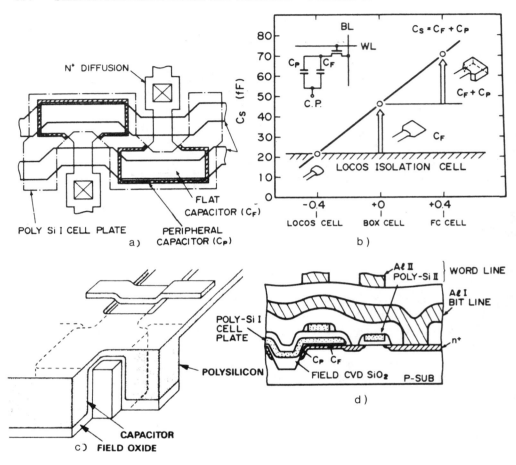

Fig. 8-18 (a) and (b) Top and perspective views of the folded capacitor cell (FCC).[55] (© 1984 IEEE). (c) Increase in capacitance by FCC. (d) Cross section of FCC showing CVD SiO₂ BOX-isolation structure.[56] (© 1986 IEEE).

from those of adjacent cells by means of a *BOX-type* isolation structure, rather than a LOCOS isolation structure (Fig. 8-18d). This increases the memory-array packing density (and in effect decreases the cell size), while also increasing the capacitance. An FCC cell size of 32 μm^2 with a 70-fF capacitor was reportedly used to fabricate 1-Mbit DRAMS. This cell appears to be scalable to 4-Mbit and 16-Mbit DRAM requirements.

In a second novel approach, the walls of the storage electrode were made to follow the outside edges of the cell perimeter, and the access transistor was placed inside (*Isolation VErtical Capacitor cell,* or IVEC, Fig. 8-19a).[57] A third invention folded the plate electrode around the storage electrode but used selective doping of certain trench walls to achieve isolation (i.e., the substrate trench walls that act as isolation structures were

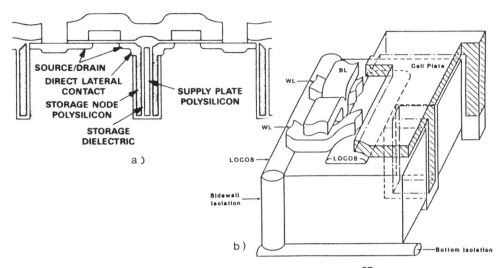

Fig. 8-19 (a) Isolation-merged VErtical Capacitor (IVEC) cell.[57] (© 1984 IEEE). (b) Perspective view of the FASIC cell.[58] (© 1987 IEEE).

selectively boron doped by means of oblique ion implantation) and storage (i.e., those walls used as the storage plate surfaces were arsenic doped by means of oblique ion implantation, creating a Hi-C storage capacitor). This latter cell was named the *folded bit-line adaptive sidewall isolated capacitor* (FASIC) cell (Fig. 8-19b).[58] FASIC cells can be made as small 10 μm^2 and with capacitances as large as 50 fF, making them suitable for use in 4-Mbit DRAMs. They require trenches of only 2 μm in depth.

8.3.3.3 Trench-Capacitor Structures with the Storage Electrode inside the Trench (Inverted Trench Cell).
One set of trench-capacitor designs sought to reduce punchthrough and soft-error problems by placing the plate electrode on the *outside* of the trench, and the storage electrode *inside* (Fig. 8-20a). Since the charge is stored inside the trench (which is therefore completely oxide isolated except in the region of lateral contact to the access transistor), it can leak only through the capacitor oxide or the lateral diffused contact.

Four examples of early approaches using such cell designs are the *buried-storage-electrode cell* (BSE) (Fig. 8-20b),[60] the *substrate-plate-trench cell*, (SPT) (Fig. 8-20c),[61] and the *stacked-transistor-capacitor cell*, (STT) (Fig. 8-20d).[62] In the first two, the plate electrode is heavily *p*-doped and is connected to the power supply, while the inside storage plate is heavily *n*-doped. Since the substrate is maintained at essentially an equipotential, the punchthrough problem exists only around the region through which the charge is introduced into the trench. Note that for heavily-doped storage electrodes (e.g., $>2 \times 10^{19}$ cm^{-3}), inversion will not occur at 5 V or less. Instead, the bias applied to the capacitor causes both plates of the capacitor to deplete; together with the oxide capacitor, these two depletion regions make this type of trench

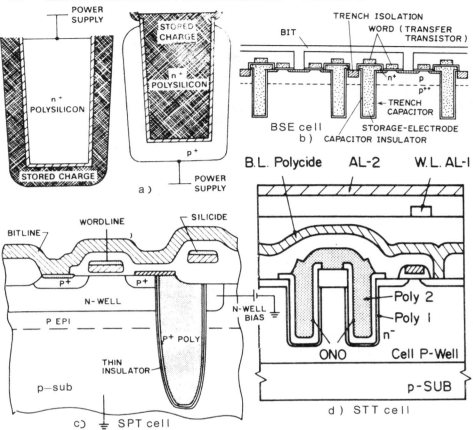

Fig. 8-20 (a) Mechanism of charge storage on outer and inner plates of the DRAM trench storage capacitor. (b) Cross section and process sequence of BSE cell.[60] (© 1985 IEEE). (c) Cross section of SPT cell.[61] (© 1985 IEEE). (d) Cross section of STT cell.[62] (© 1987 IEEE).

capacitor equivalent to three capacitor elements in series. Since the depletion regions grow with increasing voltage, the total trench capacitance decreases monotonically. The heavy doping of the plates therefore helps to maximize the cell capacitance. Finally, in such cells a *0* logic level is stored as 0 V and a *1* level as 5 V.

The problem with this type of cell is that the *gated-diode structure* shown in Fig. 8-21a can cause a significant leakage current to flow into the storage node, adversely affecting the cell's retention time. (The physics of the gated diode structure is treated in detail in reference 63.) An alternative cell (the IBM SPT cell) overcomes this problem by using PMOS access transistors and *p*-type doped inner-storage electrodes, and then creating the SPT cells in an *n*-well on a *p*-substrate (Fig. 8-21b).[64] As a result, the storage electrode gates the *n*-well-to-substrate junction, and the leakage current (as well as the *band-to-band tunneling-induced leakage current* generated in the bulk silicon)[65] is collected at the *n*-well contact instead of at the storage electrode. If such a cell is not built in a well (e.g., the BSE cell), the storage electrode will gate the junction formed by the storage electrode and the substrate, and the resulting leakage current will be collected by the storage electrode.

In the most advanced type of cell that does not use the substrate as the storage electrode, *both* the plate and storage electrodes are fabricated inside the trench opening, allowing both electrodes to be completely oxide-isolated. Lightly doped epitaxial layers on heavily doped substrates are not needed, and the cells will be free from punchthrough at arbitrarily small cell spacings. In addition, the soft-error rate will be reduced further than it is in the other inverted trench cells. However, these improvements are achieved through a substantial increase in process complexity.

Several such cells have been reported, including the *dielectrically encapsulated trench* (DIET) capacitor (Fig. 8-22a),[66] the *half-V_{CC} sheath-plate capacitor* (HPSC) (Fig. 8-22b),[67] and the *double-stacked capacitor* (DSP) (Fig. 8-22c).[68] The last has two polysilicon plates, one biased to V_{BB} and the other to $V_{CC}/2$. The capacitors formed by the lower poly plate and substrate (separated by the outer dielectric layer), and by the two poly layers (separated by the interpoly dielectric) act in parallel, almost doubling the cell's storage capacitance. A DSP cell of 6 μm^2 in size with trench depths of 4 μm is reported to exhibit a capacitance of 50 fF.

8.3.3.4 Trench-Capacitor Cells with the Access Transistor Stacked above the Trench Capacitor.
The access transistor occupies a significant fraction of the cell area in trench-transistor cell designs. When this transistor is a planar transistor and is placed alongside the trench capacitor, surface area must be devoted to both structures. Attempts to use short-channel lengths for the access

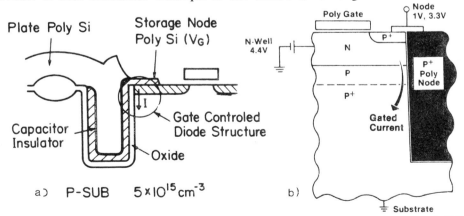

Fig. 8-21 (a) Cell structure with gate controlled diode. (b) Schematic representation of SPT cell bias conditions – *p*-substrate-to-*n*-well junction is gated by the polysilicon node.[64] (© 1987 IEEE).

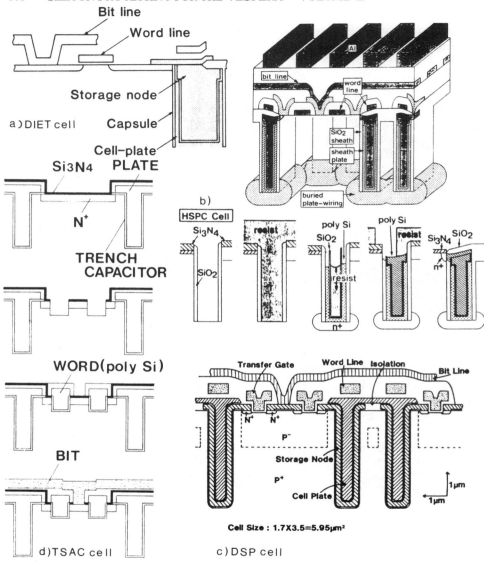

Fig. 8-22 (a) Cross section of DIET cell.[66] (© 1986 IEEE). (b) Perspective view and process sequence of the HPSCC cell.[67] (© 1987 IEEE). (c) Cross section of the DSP cell.[68] (© 1987 IEEE). (d) Fabrication process of the TSAC cell.[69] (© 1986 IEEE).

transistor have run up against the effects of drain-induced barrier lowering (see section 5.5.2).

One technique for overcoming this problem extends the gate length of the access transistor by forming a trench in the transistor channel (Fig. 8-22d). This reduces the area of the planar access transistor without decreasing its channel length.[69] Using this technique with a self-aligned contact structure, a cell size of 9 μm^2 was realized; such a cell is making it suitable for a 4-Mbit DRAM.*

A more efficient use of space would be to stack the transistor above the trench capacitor (and, if possible, to form a vertical-access transistor). Two examples of such cells are the *trench-transistor cell* (Fig. 8-23) and the *self-aligned epitaxy over trench cell* (SEOT) (Fig. 8-24).[71]

In the trench-transistor cell, the vertical-access (or trench) transistor is built in the top 2 μm of the trench. Its source is connected to the n^+ polysilicon storage electrode of the capacitor by a lateral contact, made by means of an oxide undercut etch and polysilicon refill. The drain, gate, and source of the trench transistor are formed by a diffused buried n^+ bit line, an n^+ polysilicon word line, and a lateral contact, respectively. The gate-oxide thickness is ~25 nm and the channel length is 1.5 μm. The transistor width is determined by the perimeter of the trench. The electrical behaviors of this trench transistor have been modeled, and the results are presented in reference 72. This cell has reportedly been used to build 4-Mbit DRAMs.

A *surrounding gate transistor* (SGT) cell that extends the trench-transistor cell approach has recently been reported (Fig. 7-23d).[123] This cell can be made smaller than the trench-transistor cell because it uses trench isolation for the bit-line isolation, rather than the LOCOS isolation used in the latter cell. The transistor and capacitor of this cell surround a silicon pillar. allowing the cell size to be shrunk to 1.2 μm^2 while still providing 30 fF storage capacitance. The SGT cell is being studied as a candidate for 64/256-Mbit DRAMs.

In the SEOT cell the storage electrode is first completely isolated from the substrate (Fig. 8-25a), and selective epitaxy is then grown. With the exposed Si area surrounding the trench acting as a seed, a single-crystal-silicon layer grows over the top of the trench (Fig. 8-25b). When the epitaxy growth is stopped before the lateral epitaxial film has grown completely over the trench, a self-aligned window is formed on top of the trench. The capping oxide on the top of trench surface is then etched, and a second epitaxial film is grown. A pyramidal window of polysilicon is formed on top of the exposed polysilicon in the trench; the material surrounding this pyramid is single-crystal silicon formed by means of lateral epitaxy. A planar surface is achieved after a specific minimum of epitaxial growth, and the isolation structure and MOS transistors are then fabricated. An 8-μm^2 cell size has been achieved using 0.85-μm design rules, making this cell suitable for 4-Mbit DRAMs. With some process improvements and design modifications, the cell appears to be scalable to 64-Mbit DRAM dimensions.

8.3.4 Stacked Capacitor DRAM Cells

Another approach that allows the cell to shrink in size without a loss of its storage capacity is that of stacking the storage capacitor on top of the access transistor, as

* A report that studied the design methodology and size limitations of submicron access transistors for DRAM applications is published in reference 70.

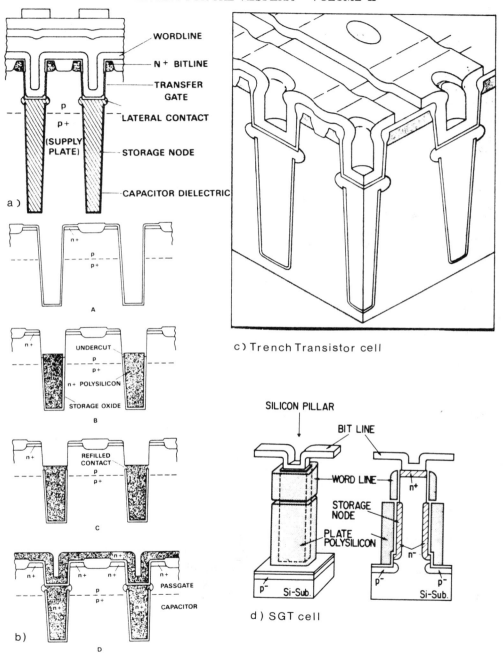

a)

WORDLINE

N⁺ BITLINE

TRANSFER GATE

LATERAL CONTACT

STORAGE NODE

CAPACITOR DIELECTRIC

p

p⁺ (SUPPLY PLATE)

b)

n+

p

p+

A

n+

UNDERCUT

p

p+

n+ POLYSILICON

STORAGE OXIDE

B

n+

REFILLED CONTACT

p

p+

C

n+   n+   n+   n+

n+   p   p+   n+

PASSGATE

CAPACITOR

D

c) Trench Transistor cell

d) SGT cell

SILICON PILLAR

BIT LINE

WORD LINE

STORAGE NODE

PLATE POLYSILICON

n+

n⁻

p⁻   Si-Sub.   p⁻   p⁻   Si-Sub.

Fig. 8-23 (a) Cross section; (b) perspective view; and (c) fabrication sequence of the trench transistor cell;[59] (© 1985 IEEE). (d) Schematic view of SGT cell.[123] (© 1989 IEEE).

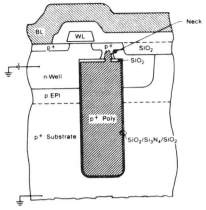

Fig. 8-24 Cross section of the SEOT cell.[71] (© 1988 IEEE).

shown in Fig. 8-26.[73] The lower electrode of the stacked capacitor is in contact with the drain of the access transistor, and the bit line runs over the top of the stacked capacitor. Although some stacked capacitor (STC) type cells have been used to fabricate

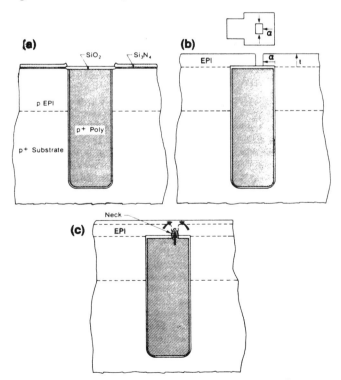

Fig. 8-25 Key processing steps of the SEOT technology.[71] (© 1988 IEEE).

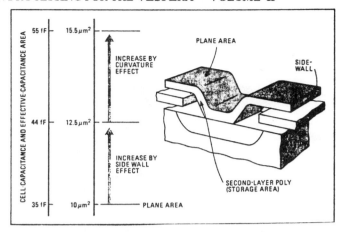

Fig. 8-26 Stacked capacitor (STC) cell structure.

256-kbit DRAMs, the minimum cell size required by conventional stacked-capacitor cells for adequate charge-storage capacity is too large for these cells to be used in larger DRAMs.

For STC cells to be made feasible for 1-Mbit DRAMs and beyond, an insulator with a larger dielectric constant than that of SiO_2 must be used, or novel cell structures must be developed. Although research is continuing into the use of such higher dielectric constant materials as tantalum pentoxide, currently acceptable insulators do not make *conventional* STC cells viable for 1-Mbit and larger DRAMs. However, several novel STC cell designs have been reported.

In the first of these, the contact hole used to connect the lower capacitor electrode to the drain of the access transistor is not filled up with polysilicon, as is the case in conventional STC cells (Figs. 8-27a and b).[74] Such filling up of the contact hole reduces the effective area of the capacitor, especially for holes with small dimensions. Instead, the contact hole is opened *after* the lower capacitor-electrode polysilicon film is deposited, and only a thin second polysilicon film is subsequently deposited into the hole (Fig. 8-27c). Good contact between the second film and the substrate is established by means of an ion-beam mixing implantation step following the thin-poly deposition (see section 3.4.2.5). This process produces a cell capacitance that is ~1.3 times as large as that obtained with a conventional STC. (Figure 8-27d shows an SEM photograph of the new cell.) A cell capacitance of ~35 fF with a cell size of 8.8 μm^2 is achieved with this design.

As shown in Fig. 8-27d, by trenching into the Si substrate it is possible to produce even more capacitance for a cell size of the same area (or a comparable capacitance for a smaller cell size). The trenched STC cell of Fig. 8-27d is predicted to be able to provide 30 fF of capacitance in a cell area of 1.3 μm^2, which would make it suitable for 64-Mbit DRAMs. A trench depth of ~1 μm is needed to achieve this capacitance value.

In the second novel STC cell, the bit lines are formed before the stacked capacitor is fabricated. In addition, the capacitor is laid out on a diagonal with respect to the bit and

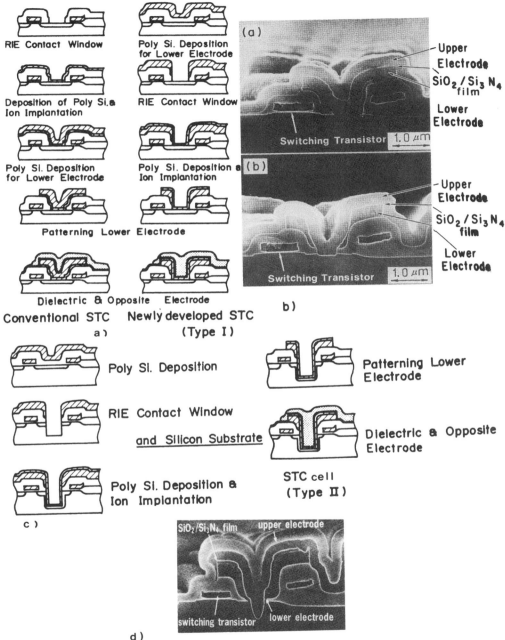

Fig. 8-27 (a) Process sequence of conventional and newly-developed type-I STC cells. (b) Cross-sectional SEM pictures of these two cells. (c) Process sequence of an advanced STC cell (type II). (d) SEM picture of this cell.[74] (© 1988 IEEE).

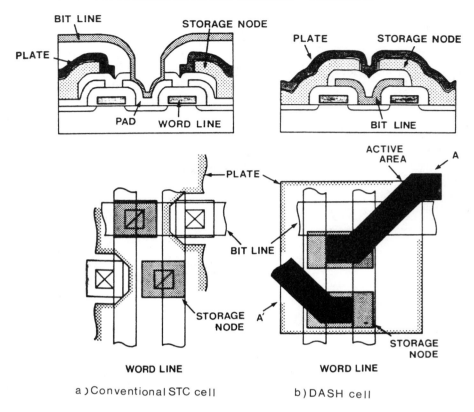

Fig. 8-28 Cross section and layout of: (a) conventional STC cell. (b) DASH cell.[75] (© 1988 IEEE).

word lines, which increases the cell area without increasing its size. This cell is thus called a *diagonal active-stacked-capacitor cell with a highly packed storage node (DASH)*.[75] Figure 8-28a shows the cross section of the DASH cell, and Fig. 8-28b compares the layouts of this type of cell with that of a conventional STC cell. The DASH cell yields a 35-fF capacitance for a cell size of 3.4 μm^2, and hence could be used in 16-Mbit DRAMs.

In the third novel STC cell, a unique fin structure is used to fabricate a capacitor with a high capacitance in a small area. (Figure 8-29a shows the fabrication process sequence.)[76] This structure would allow cells with two fins to be used in a 16-Mbit DRAM. When the bit lines are formed prior to formation of the fin structure (Fig. 8-29b), a cell structure can be obtained that has sufficiently high capacitance in a small enough area to allow fabrication of a 64-Mbit DRAM.

A fourth novel STC cell, called a *spread stacked capacitor* (SSC) cell,[122] the storage electrode is expanded into the neighboring 2nd memory cell area (Fig. 8-29c and d), and the storage electrode of the 2nd memory cell is expanded to the 1st memory-cell area.

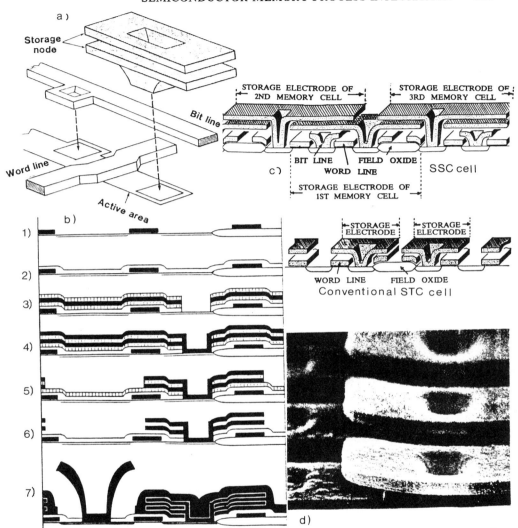

Fig. 8-29 (a) Schematic view of fin-STC structure. (b) Schematic view of fin structure and fabrication sequence.[76] (© 1988 IEEE). (c) Schematic cross sections of the SSC cell and a conventional-STC cell. (d) SEM photo of a spread-stacked capacitor.[122] (© 1989 IEEE).

This allows the storage capacitance of the SSC cell to be ~1.8x as great as that of the conventional STC cell, making a possible candidate for 64-Mbit DRAMs.

8.3.5 Soft-Error Failures in DRAMs

When a DRAM is tested, each cell is operated to verify that it functions correctly. A *hard fail*, or *hard error*, indicates that a particular array location repeatedly fails to output correct data values previously written into the location. Although there are a number of potential sources of hard fails, by and large they are caused by random physical defects. As noted earlier, various layout design approaches and on-chip error-detection and correction circuits (that replace failed bits with spares), have been implemented to eliminate faulty cells from a memory, allowing some defectively manufactured chips to be salvaged.

Soft errors, which are single-nonrecurring read errors on single bits of a memory array, are also a significant problem in DRAMs. A soft error is not a permanent error, in the sense that the cause is not a process defect. A write (then read) cycle in an array location that was previously in error carries no greater or lesser probability of error again than does a cycle in any other array location.

Although soft errors can be caused by such circuit related problems as supply-voltage noise, inadequate noise margins, and sense-amplifier imbalance, there is one specific *physical* failure mode that will cause soft errors, even when all circuit-related failure modes are eliminated. The cause of this failure mode of was identified in 1979 by May and Woods[77] as the alpha particles originating from the decay of uranium and thorium atoms. These radioactive atoms are naturally occurring trace impurities in the materials used to make IC packages, and the alpha particles they emit have energies in the 8-9 MeV range.

Since a bit of information is stored on a cell by the presence or absence of charge in the potential well of the storage capacitor, the number of electrons that distinguishes an empty well from a full one (and hence which differentiates between a logical *one* and a *zero*) is known as the *critical-upset charge.* The generally quoted value for this charge is 45-50 fC (2.5×10^5 electrons),[78] which is about 25% of Q_s (see section 8.3.1.8).

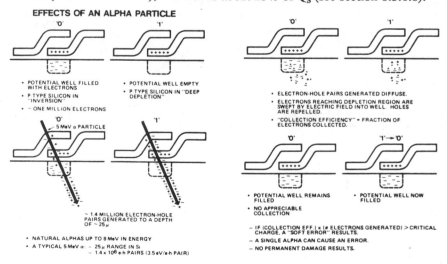

Fig. 8-30 Collection process of free carriers generated by an incident alpha particle. This shows that the cell is sensitive to the alpha particle hit when the storage node is depleted, which is when the cell is storing a **one** for the double-poly cell shown.[77](© 1979 IEEE).

If the temperature is raised, or if light is incident on the memory, the generation rate of electron-hole pairs in the substrate can be significantly increased. If enough electrons are produced that an empty well fills up in the interval between refresh pulses, a soft error will occur. Electron-hole pairs are also produced by ionizing radiation incident on a memory chip – specifically, when energetic alpha particles strike a semiconductor substrate (Fig. 8-30).

The electron-hole pairs produced as an alpha particle first passes through the chip's passivation layers are not collected by the empty well, implying that some fraction of the initial alpha-particle energy is lost before it reaches the substrate. Even if only half of the initial energy is retained, there will still be enough energy (4 MeV) to produce roughly 10^6 electron-hole pairs along a trajectory ~25 μm in length.[79] Any of the electrons generated within the potential well will be swept into the storage node by the depletion region electric field. Those electrons generated in the bulk that diffuse to the edge of the well will likewise be collected by the storage node. The remainder will recombine in the bulk. If a large enough fraction of the electrons generated by an alpha-particle strike is collected by the empty potential well, a soft error can result. The entire collection process occurs in a matter of microseconds.

8.3.5.1 Techniques Used to Reduce Soft-Error Rates In DRAMs.
The *soft-error rate* (SER), expressed as the number of errors per hour, describes the degree of susceptibility to soft-error phenomena. When the storage nodes on silicon devices were on the order of 25 μm in length, it was possible (but not likely) for all of the electrons from one strike to be collected on one node. However, since Q_S for these structures was also larger (e.g., 2×10^6 electrons), the SER was so low that the problem was not noticed. As devices decreased in size, SERs grew significantly larger.

Soft errors in DRAMs were first reported in 16-kbit DRAMs (1 soft-failure per 1000 hours of operation was typical), and they became an appreciable concern in the design and packaging of 64-kbit DRAMs and larger. When new DRAMs are designed, an attempt is made to ensure that the SER is comparable to the hard-error rate.

It was found that as long as a DRAM cell can store more than 6×10^5 electrons, the SER can be made comparable to the hard-error rate. For storage devices in DRAMs of up to 16 kbits, this was accomplished by scaling the oxide thickness to maintain adequate capacitance in the cell. The reliability trade-off in this approach was an increased hard failure rate due to increased defects in the thinner gate oxide. The Hi-C cell was invented as a means of increasing the storage capacity without increasing the capacitor size. An added benefit of this cell is that the doping gradient of the deep *p*-implant produces a potential energy barrier to the diffusing electrons, preventing some of them from reaching the depleted well of the storage capacitor. Hence, the SER of memories built with such cells is decreased. An additional advance was the implementation of the insulator of the storage capacitor with a dual dielectric film (SiO_2 and Si_3N_4), which increased the cell's capacitance (see section 8.3.1.5).

The same SER-reducing phenomenon is exploited when DRAMs are built in CMOS technology. That is, the entire array of NMOS DRAM cells is built in a *p*-well. The well-substrate junction acts as a reflecting barrier for diffusing minority carriers created outside of the well, preventing most of them from reaching the unfilled storage nodes. Putting the memory on a heavily doped substrate with a lightly doped epitaxial layer on the surface is also beneficial. Since minority carriers generated in the heavily doped substrate have much shorter lifetimes than those in the lightly doped regions, they are prevented from reaching the epitaxial region in which the cells are built (see also section 6.8.4.2).

The use of trench-capacitor structures can also reduce the SER of DRAMs, increasing the number of stored electrons per cell without requiring the lateral device or chip to be made larger. A trench cell in which the storage plate is inside the trench provides better SER than one in which the storage plate is on the outer walls. (If there is little or no isolation between closely spaced trench cells and the track of an alpha particle intersects two of the cells, considerable charge can be transferred. The charge flow along the alpha track can cause transient forward-biasing of the cell junctions. This bipolar-like effect becomes more dominant as the cell-to-cell spacing decreases.)[105]

A final method used to reduce SER is to apply a thick coating of a radioactive-contaminant-free polymer on top of the IC. For example, alpha particles with energies of up to 8 MeV are completely absorbed by a 50-μm-thick layer of polyimide. In addition, packaging materials are now manufactured with much lower concentrations of radioactive impurities. Purification of the materials used in wafer fabrication (particularly metallization) is also being pursued, and trace levels of radioactive impurities are now low enough that they can be eliminated them as a significant source of alpha particles.

Even with thick polymer coatings, chips are still vulnerable to strikes by high-energy cosmic rays (in the form of high-atomic-number nuclei). A small fraction of such very energetic particles can penetrate the earth's atmosphere, the package, and the polymer coating, entering the silicon. Under some conditions, these nuclei can also produce soft-errors through the generation of electron-hole pairs. At present, it is estimated that cosmic-ray events produce an SER about an order of magnitude lower than that due to alpha particles from currently available "clean" packaging materials.

8.3.6 The DRAM as a Technology Driver

The DRAM has also been used as a *technology driver* over a large part of its life, since it makes a good test vehicle for advancing silicon integrated-circuit process technology. The regular, repetitive architecture of the DRAM chip requires the least amount of engineering design time to create a circuit with hundreds of thousands, or millions, of devices that can be produced in full-scale production in the factory. Hence, a new MOS technology can be fully tested in the shortest time using the DRAM as the production test vehicle. After the flaws in the process and manufacturing procedures have been ironed out using the technology driver, the process can be transferred to the manufacture of other, more design-intensive circuits.

In the mid-1980s, U.S. semiconductor manufacturers increasingly turned to high-density MOS logic arrays as a replacement technology-driver circuit, since most of them had been forced to abandon the the DRAM market after the price erosion of the 64-kbit

DRAM. The American preference for the logic test chip was also based on the fact that considerable engineering and manufacturing expertise from the production of such dense logic chips as the 80386 and 68000 MPU families. By the late 1980s, however, the world wide manufacturing trend had swung to CMOS SRAMs, since these circuits are easier to design than DRAM circuits (which require intricate clock circuitry). In addition, the 4-Mbit and larger DRAMs use three-dimensional capacitor structures that are unique to the DRAM, and their yield statistics may thus not be as valuable for learning how to manufacture other circuits that do not utilize such structures.

In 1989, however, Osamu of Toshiba presented a case for why DRAMs could still serve as excellent technology drivers.[80] First, the repetitive structure allows failure categories to be easily identified so that efforts can be undertaken to correct processes that lead to the failures. Second, the factors that impact the RAM yield can be analyzed more completely and in a much shorter period of time than those in logic devices (i.e., the tasks of test-vector generation and design for testability are much less onerous for RAMs than for logic circuits). Finally, the volume production of DRAMs is so great that memory-fabrication processing experience is acquired much more rapidly. These lessons can be transferred to other volume IC production lines. For example, DRAM production can reach 10^6 parts per year at a single manufacturer (with 2×10^6 devices per chip in the 1-Mbit DRAM), versus ~50K microprocessor parts per year (with 1×10^6 devices per typical 32-bit microprocessor).

8.4 MASKED READ-ONLY MEMORIES (ROMS)

Masked (or *mask-programmed*) ROMs are nonvolatile memories into which information is permanently stored through the use of custom masks during fabrication. Users thus must provide the ROM manufacturer with the desired bit pattern of the memory. Subsequent changes of stored data are impossible, and only *read* operations can be performed. Since only a single customized mask is required to personalize these ROMs for a specific application, however, many designs can be economically implemented.

Such memory circuits have been implemented in bipolar, NMOS, and CMOS technologies. As of 1989, 4-Mbit CMOS ROMs were the largest available, but parts containing as many as 32 Mbits are envisioned. The fastest high-density MOS-based ROMs have access times of about 80 ns,[81] while less dense (256-kbit) ROMs are faster (50 ns). The faster ROMs are most often used to simplify the interface to microprocessors by eliminating wait states, thereby permitting more rapid program execution. Bipolar ROMs are even faster (e.g., 10 ns access time for a 1-kbit ECL ROM), but they are also much less dense.

In addition to interfacing with microprocessors, ROMs have been used for a variety of applications, including these:

- *Look-Up Tables.* These are used for mathematical calculations in which evaluations of square roots and of trigonometric functions, logarithmic functions, and exponential functions are needed. The procedure would usually be much more time consuming if software subroutines were used to calculate the series expansions of particular functions. Other look-up applications include spell checking and dictionary servicing.

▪ *Character Generators.* All digital systems rely on the display of alphanumeric characters of such input and output devices as CRTs and dot-matrix printers. In most of these applications, the patterns of pixels, segments, or dots used to display the set of characters are stored in a ROM. In Japan, large (1-Mbit and larger) ROMs are used as character generators for the complex *kanji* characters.

▪ *Microcontrol Store.* In microprogrammed digital systems, the execution of an encoded macroinstruction involves the generation of a sequence of signal vectors for gating or control purposes. ROMs are often chosen for such applications because their higher packing density and simpler fabrication procedure make them less costly than static or dynamic RAMs. In addition, ROMs are more stable than SRAMs and DRAMs. Parts of computer programs that are completely debugged and that do not need to be rewritten during computation are thus often stored in ROMs.

8.4.1 Masked-ROM Implementation

Each bit of information in a ROM is stored by the presence or absence of a data path from the word (access) to a bit (sense) line. The data path is eliminated simply by ensuring that no circuit element joins a word and bit line. Thus, when the word line of a ROM is activated, the presence of a signal on the bit line will mean that a *1* is stored, whereas the absence of a signal will indicate that the bit location is storing a *0*. As shown in Fig. 8-31 (which uses an NMOS array as the example), there can be two basic forms of the ROM, with implementation by either the NOR function (Fig. 8-31a), or the NAND function (Fig. 8-31b). Note that programming of masked ROMs in bipolar technology is done by selectively omitting a contact, at the contact mask in the emitter-follower or Schottky-diode ROM arrays shown in Fig. 8-31d.

Although the speed of the ROM depends on the details of the MOS fabrication process, NOR arrays usually have faster access times; in addition, the stored bit pattern can be set by the metal-interconnection layer. Therefore, unprogrammed NOR ROMs can be manufactured up until the metal mask step and can then be stored in inventory. Such almost-completed wafers can be quickly completed (programmed) by using a custom mask that patterns the metal layer.

The NAND-type ROMs, on the other hand, have a longer access time, and they must be programmed through the implantation of dopants into the channel of selected transistors wherever stored zeros are desired (Fig. 8-31c). Such a step must be performed earlier in the manufacturing sequence, which increases the TAT. The advantage of NAND ROMs over the NOR type is that they have a considerably higher density when fabricated using the same process and design rules.

The access time of ROMs is limited by the resistance and capacitance of the word and bit lines, as well as by the currents available to drive these lines. Because of their higher density (i.e., 1-Mbit CMOS ROMs were introduced in 1983), ROMs were the

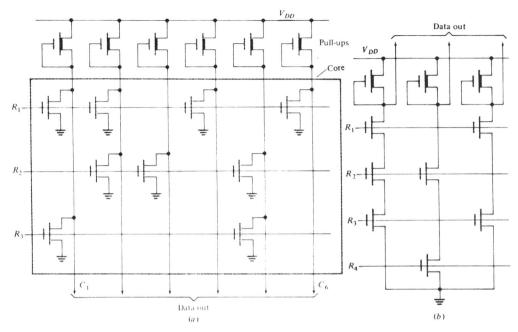

Fig. 8-31 (a) MOS ROM NOR array, (b) MOS ROM NAND array.[1]

first semiconductor memories in which on-chip error-correction circuitry (ECC) was implemented.

8.5 PROGRAMMABLE ROMS (PROMS)

If only a small quantity of ROM circuits is needed for a specific application, custom fabrication of even a single mask layer may be too expensive and/or time consuming. In such cases, it is faster and cheaper for users to program each ROM chip individually. ROMs with such capabilities are referred to as *user* or *field-programmable* memories. Many types of these have been developed.

In this section we describe *programmable read-only memories* (PROMs), a type of ROM into which information can *be programmed only once* and then cannot be erased. Subsequent sections describe ROMs that allow data to be erased after entry.

In PROMs, a data path exists between *every* word and bit line at the completion of chip manufacture (corresponding to a stored *1* in every data position). Storage cells are selectively altered to store a *0* following manufacture by electrically *blowing open* the appropriate word-to-bit connection paths. Since the write operation is destructive, once a *0* has been programmed into a bit location it cannot be erased back to a *1* in a PROM. PROMs were originally implemented in bipolar technology, although MOS PROMs have recently become available as well.

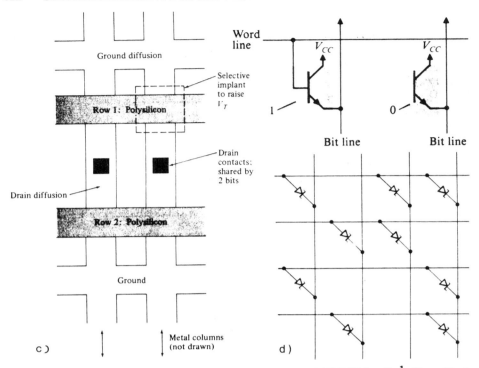

Fig. 8-31 (c) MOS NAND ROM - simplified layout, (d) BJT ROM cells.[1] From D. A. Hodges and H. G. Jackson, *Analysis and Design of Digital Integrated Circuits*, Copyright, 1983 McGraw-Hill Book Co. Reprinted with permission.

In bipolar PROMs, two techniques are used to eradicate the word-to-bit-line paths in desired bit locations. The first involves the use of a bipolar transistor in series with a *fuse*, as shown in Fig. 8-32. In such *emitter-follower* bipolar PROMs, a small fusible link is placed in series with each emitter. Early links were implemented with nichrome thin films, but these exhibited a *growback* problem (in which disconnected fuses became reconnected after some time). PROM fuses are now generally made of polysilicon. Fuse-link PROMs are designed to operate with a 5-V supply for reading data, but higher voltages (10-15 V) are needed to produce the 10-30 mA required to blow open the fuses. Such large voltages may be supplied by off-chip programming devices or by special electronic circuits available on the chip.

The second technique makes use of a *pn* diode that is short-circuited by an avalanching pulse. An example of a 64-kbit bipolar PROM using *pn* diode cells with an access time of 50 ns is described in reference 82.

MOS PROMs have also been introduced, and in 1985 they became available in 1-Mbit size. Such components are actually MOS erasable and programmable read-only memories (EPROMs) housed in inexpensive plastic packages (rather than in the costly, quartz-windowed ceramic packages needed by *erasable* MOS PROMs). Without a quartz window these MOS PROMs can be field-programmed only once. As a result, they are commonly known as *one-time-programmable* (OTP) ROMs. The advantage of these over bipolar PROMs is that they can be fabricated in much higher densities. High-density CMOS OTP ROMs are now being built with access times close to those of bipolar PROMs, but with more bits per chip and much lower power dissipation. For example, a 256-kbit CMOS OTP ROM with an access times of 50 ns has recently been introduced; this approaches the access time (~40 ns) of large [64-kbit] bipolar PROMs.[87] In addition, OTP ROMs are no longer much more expensive than ROMs, and hence they are also expected to increasingly replace masked ROMs.

8.6 ERASABLE PROGRAMMABLE READ-ONLY MEMORIES (EPROMS)

Erasable PROMs depend on the long-term retention of electronic charge as the information-storage mechanism. The charge is stored on a *floating polysilicon gate* of an MOS device (the term *floating* refers to the fact that no electrical connection exists to this gate). The charge is transferred from the silicon substrate through an insulator.

Each of the various mechanisms implemented to transfer (and remove) charge from the floating gate has been the basis of a different erasable-PROM device type. This section describes the so-called *electrically programmable ROM* (EPROM), which also requires that the device be irradiated with ultraviolet (UV) light for removing (or *erasing*) the stored charge from the floating gate.

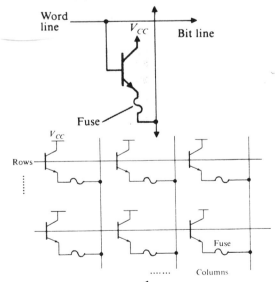

Fig. 8-32 (a) Emitter-follower bipolar PROM.[1] From D. A. Hodges and H. G. Jackson, *Analysis and Design of Digital Integrated Circuits*, Copyright, 1983 McGraw-Hill Book Co. Reprinted with permission.

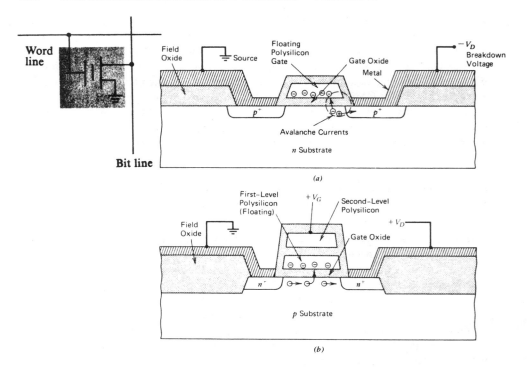

Fig. 8-33 (a) Circuit schematic and cross section showing the mechanism of charge injection into the gate by avalanche in a FAMOS memory element. (b) A FAMOS element made with two layers of polysilicon and suitable for *n*-channel MOS applications. From R. S. Muller and T. I. Kamins, *Device Electronics for Integrated Circuits,* 2nd Ed. Copyright 1986, John Wiley & Sons. Reprinted with permission.

Traditionally, such EPROMs have been used as prototyping vehicles to ensure that no glitches remained in the code. Once the programs were finalized, the code was usually fixed into ROM components. However, the cost of EPROMs is shrinking with advances in technology, and as a result, their use is growing at the expense of ROMs. Another factor in favor of EPROMs is their faster turn around time (which also plays a role in the choice of technology used to implement masked ROMs).

The charge-transfer mechanism is based on the injection of hot electrons into the floating polysilicon gate, which is completely encapsulated by SiO_2. The original EPROM devices were fabricated in PMOS technology and consisted simply of a MOSFET with a floating gate (Fig. 8-33a). If a sufficiently high reverse-bias voltage is applied to the drain, the drain-substrate *pn* junction will experience avalanche breakdown, causing hot electrons to be generated. Some of these will have enough energy to pass over the oxide potential-energy barrier and charge the floating gate (see section 5.6.2). These EPROM devices were thus called *Floating-gate, Avalanche-injection MOS transistors* (FAMOS).

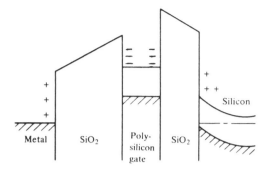

Fig. 8-34 Energy band diagram of a FAMOS device with charge stored in the silicon gate. From E. S. Yang, *Microelectronic Devices*, Copyright 1988, McGraw-Hill Book Company. Reprinted with permission.

Once electrons are transferred to the gate, they are trapped there, as illustrated by the energy-band diagram shown in Fig. 8-34. Since the potential-energy barrier at the oxide-silicon interface is greater than 3 eV, the rate of spontaneous emission of electrons from the oxide over this barrier is negligibly small. The electronic charge on the floating gate can thus be retained for many years.

If the floating gate is charged with a sufficient number of electrons, inversion of the channel under the gate will occur. A conducting channel then forms between the source and the drain, exactly as would occur if an external gate voltage were applied. The presence of a *1* or *0* in each bit location is therefore determined by the presence or absence of a conducting channel in a programmed device.

Subsequent advances in process technology (Fig. 8-33b) made it possible to implement EPROMs with 5 V, *n*-channel devices.[83,84] In such EPROMs the cells can also be laid out in NOR or NAND arrays; we will use the NOR array configuration to describe the operation of these newer cells.

Two layers of polysilicon are used to form a double gate in the transistor, as shown in Fig. 8-33b. Gate #1 is the floating gate and is placed under Gate #2. Cell selection is controlled by Gate #2, which therefore plays the role of the single gate in conventional MOS transistors. Initially, Gate #1 is uncharged; thus, if the drain, source, and Gate #2 of the transistor are grounded, Gate #1 will also be at 0 V. If a voltage (V_2) is subsequently applied to Gate #2, the voltage on Gate #1 (V_1) will be given by:

$$V_1 = [C_2 / (C_1 + C_2)] \ V_2 \qquad (8-3)$$

because the two gates represent a capacitive divider as shown in Fig. 8-35. From the electrical perspective of Gate #2, the transistor appears to have a larger V_T. In order to turn on this transistor, a larger gate voltage must be applied to Gate #2 (typically somewhat more than twice the normal V_T). For example, if a conventional NMOS

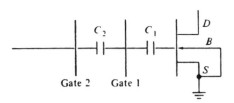

Fig. 8-35 Equivalent capacitive divider of an EPROM structure.

device with a V_T = 1 V was fabricated with the double gate of Fig. 8-33b, a voltage of 2 V would have to be applied to Gate #2 to turn it *ON*; a voltage of 5 V for reading the cell would also cause it to turn *ON* (Fig. 8-36). Such a turned-on device would cause a *positive logic stored zero* to appear at the output of the bit line if the device was used in a NOR array. As a result, the programming of the EPROM begins by discharging all of the floating gates through exposure to UV radiation, so that every cell initially stores a *0*. A *1* is then selectively written into the desired cells.

For a *1* to be written into a cell, both Gate #2 and the drain are raised to about 12 V (for a few hundred microseconds), while the source and substrate are kept grounded (early EPROMs required 30 V programming voltages for several milliseconds). Hot electrons are created near the drain and are attracted to the floating gate (which, due to capacitive coupling, has a more positive potential than the drain). Some fraction of the electrons will traverse the oxide and charge the floating gate. When the voltages on Gate #2 and the drain are returned to zero, these charges remain trapped on Gate #1. The electrons trapped on Gate #1 cause its potential to be at about -5 V. Therefore, if a signal of only 5 V is applied to Gate #2 when the EPROM is being read, no channel will form in the transistor. Under this circumstance, a *1* is stored in the cell. The electron- trapping process is self-limiting, because once electrons are stored on the floating gate they begin to inhibit further electron injection.

In order for the cells to be erased, the stored charge must be removed from the floating gate. This is accomplished by flood exposure of the EPROM with strong ultra-violet light for approximately 20 minutes. The UV light creates electron-hole pairs in the SiO_2, providing a discharge path for the charged floating gate.

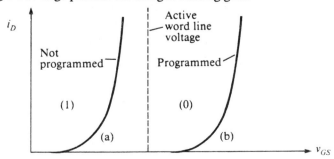

Fig. 8-36 Transfer characteristic of a floating-gate transistor.

One of the advantages of EPROMs is that the cells consist of only one transistor, allowing them to be fabricated with high densities (e.g., a 4-Mbit CMOS EPROM with an access time of 120 ns and 0.8-μm channel-length transistors has been reported).[85] In addition, they cost less to manufacture than electrically erasable PROMs (EEPROMs – see the next section).

A disadvantage of EPROMs is that they require UV light for erasing and must therefore be packaged in an expensive ceramic package with a UV-transparent quartz window. In addition, they must be removed from the circuit board and put into a special UV eraser. (Note that since sunlight and fluorescent lamps contain some UV, one week of sunlight or three years of room-level fluorescent lighting are likely to erase some of the cells. Therefore, except during erasure, the window should be covered at all times with an opaque label.) Another disadvantage is that the high voltage needed to program the EPROM is generally not available on the integrated circuit, so a special programming setup must also be provided. This limitation, combined with the fact that EPROM programming takes a relatively long time, means that these cells are used primarily for reading information and are only occasionally rewritten. (Note, however, that the programming time is decreasing dramatically. In the first 64-kbit EPROMs, it took about 50 ms to program each byte, adding up to almost seven minutes for the entire chip; in the 4-Mbit EPROM,[85] the program time has been reduced to 10 μs/byte, so that the entire chip can be programmed in only five seconds!)

OTP ROMs compete with high-density masked ROMs because they offer the benefit of a significantly shorter TAT (albeit at a somewhat higher cost). OTP ROMs are also less expensive than bipolar PROMs, and offer a PROM capability with much higher density. While bipolar PROMs are generally faster, a three-transistor EPROM cell was recently reported that would allow CMOS EPROMs to be built with the same speed and density as bipolar PROMs, but with much lower power dissipation and 100% testability.[86] Another one-transistor cell, split-gate 256-kbit CMOS EPROM with an access time of 50 ns has also been reported.[87]

Some of the relevant process and circuit-design enhancements used in fabricating the large CMOS EPROMs include the following:[88]

• Use of thin (20 nm) reliable interpoly dielectric materials, often consisting of composite films of $SiO_2/Si_3N_4/SiO_2$, for increased capacitive coupling between Gates #1 and #2.

• Self-aligned contacts to the control gate (as well as a self-aligned floating gate) to achieve the 3.1 x 2.9 μm^2 small cell size.

• Use of a low-resistance polycide gate for the word line to achieve high speed.

• Reduction of the programming voltage to 10.5 V, along with a reduction of the programming time to ~10 μs/byte.

• On-chip test circuits.

A novel self-aligned planar-array EPROM cell has also been proposed.[89,90] This cell appears to make possible the fabrication of 4-Mbit EPROMs with 1-μm design rules because it uses buried n^+ bit lines that are self-aligned to the FAMOS transistor.

8.7 ELECTRICALLY ERASABLE PROMS (EEPROMS)

In some applications it is desirable to erase the contents of a ROM electrically, rather than to use a UV light source. In other circumstances it is useful to be able to change one byte at a time, without having to erase the entire IC. A variety of *electrically erasable PROMs* have been developed to serve these applications. Such EEPROMs are the most sophisticated of the ROM families in terms of the physical operating principles and process complexity. For example, EEPROMs must be fabricated with unique tunnel oxides, as well as with high-voltage transistors (for programming and erasing the devices).

Three technologies have been developed for EEPROM fabrication: (1) MNOS transistors; (2) **Floating-gate Tunnel Oxide (FLOTOX)** MOS transistors; and (3) textured-polysilicon floating-gate MOS transistors. Although MNOS transistor-based devices were among the first EEPROMs to be commercially manufactured, their technology limitations have made them less widely adopted than the others. Therefore, we will devote most of our attention to FLOTOX and textured-poly EEPROMs.

8.7.1 MNOS-Based EEPROMs

The MNOS EEPROM cell consists of a single MOS-like transistor that employs a composite gate-dielectric layer (Si_3N_4, ~50 nm thick, on top of a very thin [~2 nm thick] SiO_2 layer), as shown in Fig. 8-37. (See ref. 118 for more details on MNOS devices.) Unlike in floating-gate MOS devices, the charge is stored in discrete traps in the nitride bulk. The charge transfers from the substrate to the nitride traps (and back, during erasure) by tunneling through the thin oxide layer. Programming is accomplished by applying a high voltage to the top gate; erasing is done by grounding the top gate and raising the well to a high potential. MNOS transistors are built within wells (akin

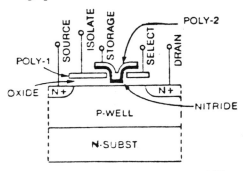

Fig. 8-37. Cross-sectional structure of an MNOS memory cell.[107] (© 1983 IEEE). to those used in CMOS) so that their channel potentials can be controlled.

to those used in CMOS) so that their channel potentials can be controlled.

The manufacturing process of MNOS transistors involves the following modifications to the standard single polysilicon-gate MOS technology: thin oxide growth, nitride deposition, and post-nitride temperature cycles. Mastering the processes used to grow the ultra-thin oxide and deposit the nitride and to control their quality is a challenging task. Furthermore, while the basic transistor is very small and highly scalable, each cell of the memory array requires a select transistor. This requirement, coupled with the need to fabricate wells, produces a relatively large effective cell size. Finally, the charge stored on the nitride traps continually leaks away through the thin oxide by means of tunneling, even when no erase voltage is applied. The charge loss is thus time dependent, making charge retention the main reliability concern with MNOS devices. (With MNOS structures, as the switching speed is increased, the ability to retain stored charge is decreased. Thus, devices with a retention time of tens of years can be fabricated if a slow switching speed can be tolerated.)

Nevertheless, MNOS exhibits higher tolerance to ionizing radiation than do either of the other EEPROM technologies. Thus, MNOS EEPROMs currently find their main use in low-density military applications that need radiation-hardened EEPROMs; this appears to be the niche to which MNOS EEPROMs will be relegated in the future.[91]

8.7.2 FLOTOX EEPROMs

The *floating-gate tunneling oxide (FLOTOX)* transistor, shown in Fig. 8-38a, consists of an MOS transistor with two polysilicon gates. A thin (8-12 nm) gate oxide (or oxynitride) region is formed near the drain. The lower polysilicon layer is the floating-gate while the other is the control gate. The remainder of the floating-gate oxide is typically 50 nm thick, and the interpoly oxide is ~50 nm thick. Programming of this transistor is done by causing electrons to be transferred from the substrate to the floating

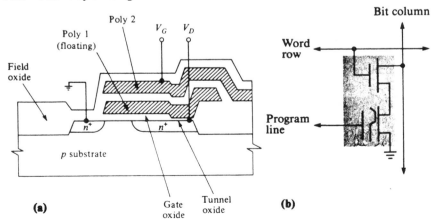

Fig. 8-38 (a) Cell structure of a Flotox transistor structure, (b) Connection in an EEPROM.[91] (© 1986 IEEE).

gate through the thin oxide layer by means of Fowler-Nordheim tunneling.[92]

The control-gate voltage is raised to a sufficiently high value so that tunneling ensues (e.g., 12 V in modern FLOTOX EEPROMs). As electronic charge builds up on the floating gate, the electric field is reduced, which decreases the electron flow. Since the tunneling process is reversible, the floating gate can be erased by grounding the control gate and raising the drain voltage, indicating that tunneling is used both to *program* and *erase* the FLOTOX transistor. Programming and erase times are on the order of 9 ms. Electron transfer by Fowler-Nordheim tunneling, however, requires a minimum electric- field strength of around 10 MV/cm. Thus, for oxides of 10 nm in thickness, such tunneling will be negligible when normal 5-V signals are applied. As a result, FLOTOX transistors can be expected to retain their charges for more than 10 years if the memory is subjected only to normal read cycles.

The FLOTOX transistor must be isolated by a select transistor. Otherwise, the high voltage applied to the drain of the selected cell during erasing would also appear on the drain of the other *unselected* cells in the same memory column. A FLOTOX EEPROM cell must therefore consist of two transistors (Fig. 8-38b). Although this limits the density of such EEPROMs in comparison to EPROMs and flash EEPROMs, it makes it possible to erase and re-program one byte of the memory without having to erase the entire IC. In addition, two cycles are needed to load the correct data into the memory. In the first, all the cells in a byte are programmed (i.e., the floating gates are charged); in the second, selected cells are erased, with the drain used for data control.

The fabrication of FLOTOX EEPROMs involves a modification of the polysilicon-gate MOS process. A double-polysilicon process is used, together with a thin tunnel-oxide growth process. The growth of a high-quality, thin tunneling oxide is, in fact, the critical manufacturing step in this technology. The tunneling dielectric reportedly can be successfully implemented with nitrided oxides, since the barrier between silicon nitride and silicon is lower than that between SiO_2 and Si. As a result, a higher tunneling current can be obtained for the same voltage.[93]

Despite the fact that a reliable process for growing thin tunneling oxides must be developed, FLOTOX-based devices have become the most widely manufactured of the EEPROM types. They are still the easiest to learn to manufacture for companies that have already successfully developed an EPROM process. Since it is desirable to be able to program and erase the EEPROM while it remains in place on a PC board, considerable effort has also been expended to make this memory type fully operational with a 5 V power supply. (This type of operational capability is referred to as *5 V only*.)

FLOTOX-based EEPROMs are best suited for applications in which low-cost, low-density, nonvolatile memories are required – for example, in microcontrollers and programmable logic devices. Another potential application is for smart credit cards; several Japanese companies have announced 64-kbit FLOTOX-based EEPROMs for this market.

On the other hand, scaling and reliability considerations appear to limit the maximum size of FLOTOX EEPROMs to 256 kbits. The need for two transistors, and the relatively large size of the select transistor (due to the large voltages needed for

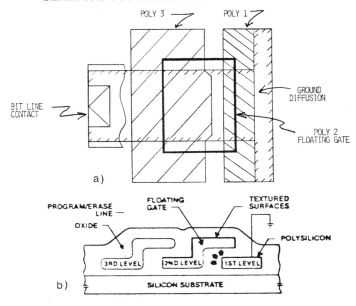

Fig. 8-39 Textured-polysilicon memory cell. (a) Top view. (b) Cross-sectional structure.[91] (© 1986 IEEE).

programming and erasure) are contributing factors. The poly-to-poly area of the sense transistor must also be large, due to the high oxide capacitance of the thin tunnel oxide.[40] Furthermore, the FLOTOX EEPROMs exhibit high failure rates, caused by defect-related oxide-breakdown problems as memory size is increased.[94] Error-detection and correction codes (EDCC) can be used to overcome this limitation, but such solutions impose a penalty of increased die cost. Reference 95 gives an example of a 50-ns access time, 256-kbit FLOTOX-based CMOS EEPROM using a single-bit EDCC.

The reliability of FLOTOX-based EEPROMs compares favorably with that of the other two EEPROM types. As with the others, there is a very low failure rate during 5-V operation; reliability problems occur as a result of the high voltages that must be used during programming and erasing. Random single-bit failures occur in FLOTOX devices due to oxide defects, resulting in leaky oxides that lose charge over time. The number of cycles that most FLOTOX EEPROMs are specified to be able to endure before the thin oxide becomes too leaky to retain data sufficiently, is 10^3–10^4 cycles. However, a process for increasing this endurance level to 10^6 cycles has been reported.[96]

8.7.3 Textured-Polysilicon EEPROMs

Textured-polysilicon EEPROMs, introduced in 1983 as an alternative to the tunneling oxide types of devices, are also based on the floating-gate MOS technology. The cell consists of three layers of polysilicon that partially overlap (Fig. 8-39) to create a cell that behaves like three MOS transistors in series. The floating-gate MOS transistor is formed by the middle polysilicon structure, which is encapsulated with SiO_2 to enable high charge retention. While charge is still transferred to the floating gate by means of Fowler-Nordheim tunneling, tunneling takes place from one polysilicon structure to another rather than from the substrate to the floating gate. The interpoly oxides through which the tunneling takes place can be made significantly thicker than the tunneling oxides in FLOTOX devices (60-100 nm in textured poly devices, versus <12 nm in FLOTOX devices), since the electric field that promotes the tunneling is enhanced by the geometrical effects of the fine texture at the surface of the polysilicon structures.

Textured-poly cells are programmed by causing electrons to tunnel from *poly 1* to the *floating poly*. Erasure is accomplished by causing electrons to tunnel from the floating poly structure to *poly 3*. The voltage of poly 3 is taken high in both the programming and erase operations. The drain voltage, however, determines whether tunneling occurs from *poly 1* to the *floating gate*, or from the *floating gate* to *poly 3*. As a result, the state of the drain voltage determines the final state of the memory cell. This provides an advantage, in that the cell represents a *direct write cell* – there is no need to charge all the cells and then remove the charge from selected cells, as with FLOTOX EEPROMs.

Textured-poly EEPROMs depend on a tunneling process whose physical mechanisms are not as well understood as those of tunneling through thin oxides, and which appears to require tighter control of empirically determined process parameters. In addition, the three poly layers require a more complex (and therefore more costly) fabrication sequence. Furthermore, textured-poly EEPROMs require a higher operating voltage than FLOTOX devices (>20 V). Finally, an intrinsic endurance problem is caused by the very high electron trapping that occurs as a result of tunneling in the poly oxides. This eventually leads to a condition in which the memory can no longer be programmed or erased.

For all of the above reasons, the textured-poly approach has been less widely pursued than the FLOTOX approach. Only one company, Xicor, is heavily involved in manufacturing these devices.[97] However, because the poly cells can be made about one-half the size of FLOTOX cells, it is possible to fabricate them in high-density configurations. In 1989, the largest textured-poly EEPROMs being offered had a 1-Mbit capacity. Although the cell-size advantage gives the textured-poly approach an edge over the FLOTOX EEPROMs for memories larger than 256 kbits, the flash-EEPROM technology described in the next section, provides a way to achieve equally high-density EEPROMs without the need to develop a textured-poly process.

8.8 FLASH EEPROMS

The *flash EEPROM* device is so named because the contents of all of the memory's array cells can be erased simultaneously as with a UV-EPROM, but through the use of an electrical erase signal. The term *flash* refers to the fact that the cells can be erased much more rapidly (1 or 2 seconds, compared to the 20 minutes required to erase a UV-EPROM). Although it was not possible to erase only a single byte in the first generation of flash EEPROMs, by 1989 parts had become available that offered a byte-by-byte erasable (and 64-byte erasable) feature in a 256-kbit memory.[98]

Flash EEPROMs are attractive for the middle of the programmable semiconductor spectrum, where neither EPROMs nor EEPROMs are particularly cost effective. The applications in this range typically require more memory capacity than EEPROMs can provide, but they also need faster and more frequent reprogramming than can be accomplished with EPROMs. Examples include automotive and automated factory equipment applications. As an example, the average EPROM cost about $7 in 1989, and a flash memory about $25. But the differential is wiped out by the expense of single reprogramming. The in-system reprogramming of a flash device may cost as little as $1, whereas pulling an EPROM out of a system to erase it by exposure to 20 minutes of UV light may cost over $80 when equipment, downtime and labor are factored in.

Meanwhile, EEPROMs are likely to remain popular wherever bytes will have to be erased selectively. But flash products, might do better for updating stored logic, when this must done more than once but less often than in main memory, cache memory, or registers. Reprogramming costs are similar, but flash memories are less than half the price of EEPROMs.

The erasing mechanism in flash EEPROMs is Fowler-Nordheim tunneling off the floating gate to the drain region. Programming of the floating gates, however, is carried out in most flash cells by *hot-electron injection into the gate*.* Unlike floating-gate EEPROMs (which incorporate a separate select transistor in each cell to allow individual byte erasure), flash memories forego the select transistor to obtain bulk erasure. Thus, flash-EEPROM cells are roughly two to three times smaller than floating-gate EEPROM cells fabricated with the same design rules.[99] Figure 8-40 shows the cross-section of a CMOS flash-EEPROM cell implemented with triple polysilicon, and a SEM photo of a double-poly flash EEPROM cell.

Most flash-EEPROM cells use a double-poly structure, as shown in Fig. 8-41 (which also shows the Toshiba triple-poly cell, Fig. 8-41b). The upper poly forms the control gate and the word lines of the structure, while the lower poly is the floating gate. The gate oxide is ~10 nm thick,[100] and the interpoly dielectric is an oxide/nitride/oxide composite film ~45 nm thick.[99] In the structure shown in Fig. 8-40 and 8-41c the control-gate poly overlaps the channel region adjacent to the channel under the floating gate. This structure is needed because when the cell is erased, it leaves a positive charge on the floating gate. As a result, the channel under the floating gate becomes inverted. The series enhancement-mode transistor (formed by the control gate over the channel region), is needed in order to prevent current flow from source to drain. A more recently reported flash-EEPROM cell (Fig. 8-41a) does not require the control gate to form a series enhancement-mode transistor, because it uses a special software-controlled erase procedure that prevents the floating gate from being over erased.[100]

---

* The 5-V-only flash memories from Texas Instruments and Amtel depend on tunneling for both write and erase mechanisms.

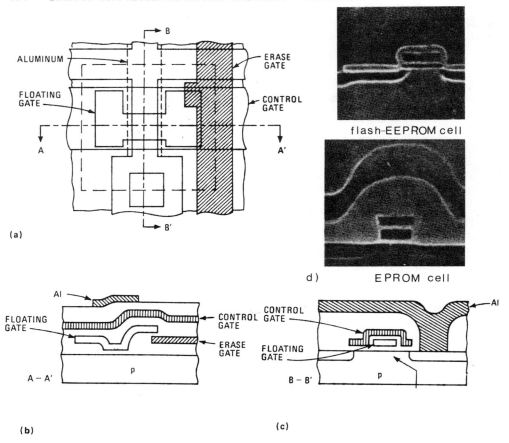

Fig. 8-40 Triple-poly flash-EEPROM cell from Toshiba: (a) Layout; (b) Cross section of the cell; (c) Section at right angles to the section shown in part (c);[101] (d) SEM pictures of double-poly flash-EEPROM and EPROM cells.[99] (© 1988 IEEE).

Flash EEPROMs can be seen to combine the advantages of UV-erasable EPROMs and floating-gate EEPROMs. They offer the high density (Fig. 8-42), small-die size, lower cost, and hot-electron writability of EPROMs, together with the easy erasability, on-board reprogrammability, and electron-tunneling erasure features of EEPROMs. High-density CMOS flash EEPROMs in 1-Mbit sizes are commercially available. It is projected that by the year 2000, 256-Mbit flash EEPROMs will be fabricated with 0.25 μm geometry.

With a memory-cell size of about one-quarter the size of current EEPROM cells, the flash EEPROMs also achieve EPROM die sizes. In addition, there are two types of flash EEPROMs: (1) those that are more akin to the EEPROM (and thus require a 12-V

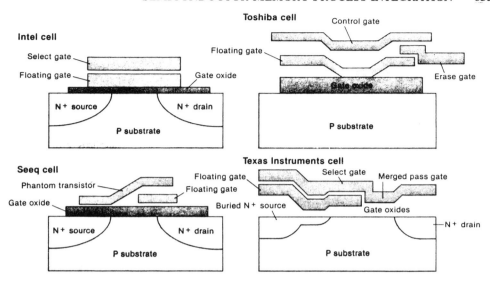

Fig. 8-41 Four approaches to flash memory technology: (a) Intel cell; (b) Toshiba triple-polysilicon cell; (c) SEEQ cell; (4) Texas Instruments 5-V-only cell.[125] (© 1989 IEEE).

external supply for programming and erasure); and (2) those that are closer to EPROMs, and hence need only a 5-V supply. Furthermore, the programming voltage can be applied during read operations, eliminating the need to switch it off when not erasing or programming. Byte-write times are 100 μs, and erasure times are 200 ms. Access times of 110 ns at 30-mA active-current consumption are provided by a 128-kBit CMOS flash EEPROM.[100,101] Endurance (i.e., the number of times a device can be erased and written) is a minimum of 100 cycles, and can be as high as 1000 cycles (note that this is lower than the endurance of EEPROMs, which is typically 1000 – 10,000 cycles).

8.9 NONVOLATILE FERROELECTRIC MOS RAMS

A novel type of nonvolatile MOSFET DRAM memory cell, introduced in late 1987, uses the electrical polarization of a ferroelectric capacitor to store information semipermanently.[102,103] Since ferroelectric polarization retention is nearly perpetual (just as in magnetic core memories), refresh is not needed. The reported write speed is 200 ns in one design,[102] and 60 ns in another,[103] which is much faster than that exhibited by an EEPROM (1 ms) or a UV-PROM (10 ms) without fatigue after 10^{12} write cycles. It is predicted that by 1991 products will be available with operating lifetimes of ~75 years at a cycle time of 100 ns, and 10^{12} read/write cycles. A recent review article has described the latest advances in such nonvolatile RAMs.[124]

The cell contains a ferroelectric capacitor as the charge storage element and an MOS transistor for sensing and writing (Fig. 8-43b). The ferroelectric insulator of the capacitor may be polarized by either a positive or negative voltage, and its polarization state is retained after the voltage is removed (this characteristic is called a *ferroelectric effect*). This effect occurs because in some materials dipoles will align in parallel under the influence of an externally applied electric field, and they will remain aligned (polarized) after the field is removed. Reversal of the field causes polarization in the opposite direction. Thus, a *ferroelectric* is a material which can be permanently polarized by the application of an electric field. (Contrary to the name, the ferroelectric effect has nothing to do with iron.)

A ferroelectric thin-film capacitor exhibits a characteristic hysteresis curve, which describes the amount of charge that the device can store as a function of the applied voltage (Fig. 8-42a). It has two stable polarization states and can be modeled as a bistable capacitor with two distinct polarization thresholds. The *coercive voltage* is the digital switching threshold of the capacitor. For memory applications, it is desirable for the two coercive voltage points to be symmetrical and less than 2.5 V, so that the memory may operate from standard memory power-supply voltages.

Memory arrays of such ferroelectric cells (*FRAMs*) have the potential to replace hard and floppy magnetic disks. Such memory arrays could provide increased reliability compared with the present magnetic disk drives, since the FRAM contains no moving parts. In addition, they could offer much shorter read, write, and access times than

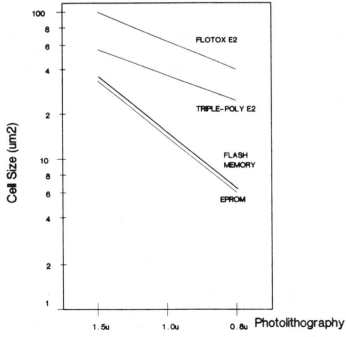

Fig. 8-42 Size of flash-memory cell versus lithographic feature size.[99] (© 1988 IEEE).

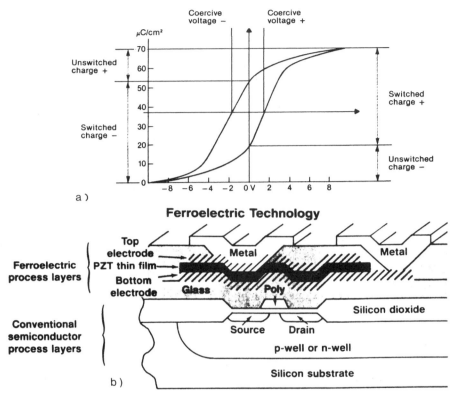

Fig. 8-43 (a) Thin-film ferroelectric capacitors exhibit a stable coercive voltage of less than 2.5 V and a switched charge of more than 20 μcoulombs cm^{-2}.^{125} (© 1989 IEEE). (b) Ferroelectric-dielectric film (lead zirconate titanate, PZT) sandwiched between two metal electrodes to form a non-linear capacitor built above existing circuitry.   Reprinted with permission of Semiconductor International.

current memory disks can.

Ferroelectrics are essentially compatible with conventional wafer processing and memory circuits. The manufacture of MOS transistors involves relatively mature technology, and only the fabrication of ferroelectric thin films on Si and SiO$_2$ substrates still needs to be further developed. In one reported design, a 256-bit demonstration chip uses capacitors fabricated with a thin film of lead zirconate titanate (PZT) ceramic as a dielectric sandwiched between two metal electrodes. This structure forms a "digital memory capacitor" built above existing semiconductor circuitry (Fig. 8-44). Such PZT films remain ferroelectric from -80°C to 350°C, well beyond the operating temperature range of existing silicon circuits. Other ferroelectric materials being studied are lanthanum-doped PZT, and lithium niobate.

FRAMs can be operated and programmed from a single 5 V power supply. In addition, because ferroelectric materials typically exhibit dielectric constants much larger than that of SiO$_2$ (e.g., 1000-1500 versus the 3.8 to 7.0 of current DRAM capacitors),

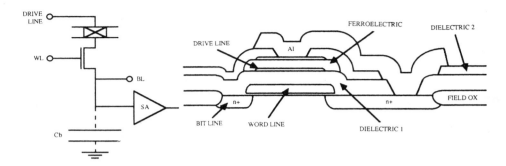

Fig. 8-44 Cross sectional view of a ferroelectric memory cell.[102] (© 1987 IEEE).

a larger charge can be stored on a capacitor of the same size (for example, if a capacitor using a PZT film could store 10 μC/cm^2 of charge, one of the same size using SiO$_2$ could store only 0.1 μC/cm^2). If the smaller ferroelectric capacitor could be used, DRAM manufacturing could thus revert to the simpler planar process.

REFERENCES

1. D. A. Hodges and H. G. Jackson, *Analysis and Design of Digital Integrated Circuits,* McGraw-Hill, New York, 1983.

2. S. Asai, "Semiconductor Memory Trends," *Proceedings of the IEEE,* Dec. 1986, p. 1623.

3. C. T. Sah, "The Evolution of the MOS Transistor," *Proceedings of the IEEE,* October 1988, 1280.

4. O. Minato, *IEEE J. Solid-State Circuits,* October 1982, p. 793.

5. W. C. Holton and R. K. Gavin, "A Perspective on CMOS technology Trends", *Proceedings of the IEEE,* December 1986, p. 1646.

6. R. Hunt, "Memory Design and Technology," in M.J. Howes and D.V. Morgan, Eds., *Large Scale Integration,* Wiley, New York, 1981.

7. S. T. Chu et al., *IEEE J. Solid-State Ckts.,* October 1988, p. 1078.

8. H. Okuyama et al., *IEEE J. Solid-State Ckts.,* October 1988, p. 1054.

9. R. A. Chapman, *Tech. Dig. IEDM,* 1987, p. 362.

10. H. Momose et al., *Tech. Dig. IEDM,* 1984, p. 706.

11. J. S. Fu et al., *Tech. Dig. IEDM,* 1987, p. 540.

12. C. E. Chen et al., "Stacked CMOS SRAM Cell," *IEEE Electron Dev. Letts.,* April, 1983, p. 272

13. L. R. Hite et al., *IEEE Electron Dev. Letts.,* October 1985, p. 548.

14. K. Sakui et al., *Tech. Dig. IEDM,* 1988, p. 44.

15. T. Ohzone et al., *IEEE Trans. Electron Devices,* September 1985, p. 1749.

16. N. C. C. Lu, L. Gerzberg, and J. D. Meindl, "Scaling Limitations of Polysilicon Resistors in VLSI SRAMs & Logic," *IEEE J. Solid-State Ckts,* April 1982, p. 312.

17. R. Saito, Y. Sawahata, and N. Momma, *IEEE Trans. on Electron Dev.*, March 1988, p. 299.
18. S. Voldman et al., *Tech. Dig. IEDM*, 1987, p. 518.
19. N. Hoshi et al., *Tech. Dig. IEDM*, 1986, p. 300.
20. N. Homma et al., *IEEE J. Solid-State Circuits*, October 1986, p. 675.
21. C.-T. Chuang et al., *IEEE J. Solid-State Circuits*, October 1986, p. 670.
22. M. Inadachi et al., *Tech. Dig. IEDM*, 1979, p. 108.
23. Y. Kato, M. Odaka, and K. Ogiuie, "A 16 ns, 16-K Bipolar SRAM," *IEEE Solid- State Circuits Conf.*, 1983, p. 106.
24. K. Ogiue et al., "Technology Improvement for High Speed ECL RAMs," *Tech. Dig. IEDM*, 1986, p. 468.
25. H. V. Tran et al., *IEEE J. Solid-State Circuits*, October 1988, p. 1041.
26. R. A. Kertis, D.D. Smith, and T.L. Bowman, *IEEE J. Solid-State Circuits*, Oct. 1988, p. 1048.
27. Y. Nishioka et al., *IEEE Trans. Electron Dev.*, September 1987, p. 1957.
28. R. H. Dennard, "Field Effect Transistor Memory," U.S. Patent 3,387,286, granted June 4, 1968.
29. R. H. Dennard, "Evolution of the MOSFET Dynamic RAM- a personal view," *IEEE Trans. Electron Dev.*, November 1984, p. 1549.
30. V. L. Rideout, "One-Device Cells for DRAMs: A Tutorial," *IEEE Trans. Electron Dev.*, June 1979, p. 839.
31. C.-T. Chuang et al., *IEEE J. Solid-State Circuits*, October 1988, p. 1265.
32. "Shipment of 1-Mbit SRAMs Ushers in Era of Submicron Linewidths," *Semicond. International*, November 1988, p.28.
33. J. S. Fu et al., "Scaling Studies of CMOS SRAM Soft-Error Tolerances - From 16K to 256K," *Tech. Dig. IEDM*, 1987, p. 540.
34. C. M. Hochstedtler, *Semiconductor International*, April 1989, p. 68.
35. A. F. Tasch, Jr., P. K. Chatterjee, H. S. Fu, and T. C. Holloway, *IEEE Trans. Electron Devices*, **ED-25**, p. 33, (1978).
36. D. K. Schroder, *Advanced MOS Devices*, Addison-Wesley, Redding, MA. 1988.
37. C. Sodini and T. Kamins, "Enhanced Capacitor for One-Transistor Memory Cell," *IEEE Trans. Electron Dev.*, **ED-23**, 1976, p. 1187.
38. S. Chou, *Electronic Design*, October 4, 1984, p. 138.
39. A. F. Tasch, Jr., *IEEE Proceedings*, March, 1989.
40. K. Shimtori et al., *ISSCC Tech. Dig.* 1983, p. 228.
41. H. Sunami, *Tech. Dig. IEDM*, 1985, p. 694.
42. D. S. Yaney et al., *Tech. Dig. IEDM*, 1985, p. 698.
43. L. Risch, W. Mueller, and R. Tielert, *Semiconductor International*, May 1988, p. 246.
44. *Electronics News*, March 6, 1989.
45. B. C. Cole, *Electronics*, February 5, 1987, p. 66.
46. K. V. Rao et al., *Tech. Dig. IEDM*, 1986, p. 140.
47. G. K. Herb, D. J. Riegler, and K. Shields, *Solid-State Technol.*, October 1987, p. 109.
48. K. Yamada et al., *Tech. Dig. IEDM*, 1985, p. 702.
49. K. Yamabe and K. Imai, *IEEE Trans. Electron Dev.*, 1987, p. 1681.

50. S. Rohl et al., *Ext. Abs. Electrochem. Soc. Meeting,* Spring, 1989, p. 194.
51. Y. Miyai et al., *J. Electrochem. Soc.,* January 1989, p. 150.
52. H. Sunami et al., *Tech. Dig. IEDM,* 1982, p. 806.
53. L. Risch, W. Mueller, and R. Tielert, "Four Megabit DRAM Processing," *Semicond. Internatl.,* May 1988, p. 246.
54. D. A. Baglee et al., *Tech. Dig. IEDM,* 1985, p. 384.
55. M. Wada, K. Heida, and S. Watanabe, *Tech. Dig. IEDM,* 1984, p. 244.
56. F. Horiguchi et al., *IEEE J. Solid-State Ckts.,* December 1986, p. 1076.
57. S. Nakajima et al., *Tech. Dig. IEDM,* 1984, p. 240.
58. K. Mashiko et al., *IEEE J. Solid-State Circuits,* October 1987, p. 643.
59. W. F. Richardson et al., *Tech. Dig. IEDM,* 1985, p. 714.
60. S. Sakamoto et al., *Tech. Dig. IEDM,* 1985, p. 710.
61. N. Lu et al., *Tech. Dig. IEDM,* 1985, p. 771.
62. F. Horiguchi et al., *Tech. Dig. IEDM,* 1987, p. 324.
63. A. S. Grove, *Physics and Technology of Semiconductor Devices,* Wiley, New York, 1967, Chapter 10, p. 296.
64. W. F. Noble, A. Bryant, and S. H. Voldman, *Tech. Dig. IEDM,* 1987, p. 340.
65. S. Banerjee et al, *IEEE Trans. Electron Dev.,* January 1988, p. 108.
66. M. Taguchi et al., *Tech. Dig. IEDM,* 1986, p. 136.
67. T. Kaga et al., *Tech. Dig. IEDM,* 1987, p. 332.
68. K. Tsukamoto et al., *Tech. Dig., IEDM,* 1987, p. 328.
69. M. Yanagisawa, K. Nakamura, and M. Kikuchi, *Tech. Dig. IEDM,* 1986, p. 132.
70. W.-H. Lee et al., *IEEE Trans. Electron Dev.,* November 1988, p. 1876.
71. N. C. C. Lu et al., *Tech. Dig. IEDM,* 1988, p. 588.
72. S. Banerjee and M. Bordelon, *IEEE Trans. Electron Dev.,* December 1987, p. 2485.
73. M. Koyanagi, *IEDM Tech. Dig.,* 1978, p. 348.
74. H. Watanabe, K. Kurosawa, and S. Sawada, *Tech. Dig. IEDM,* 1988 p. 602.
75. S. Kimura et al., *Tech. Dig. IEDM,* 1988, p. 596.
76. T. Ema et al., *Tech. Dig. IEDM,* 1988, p. 592.
77. T. May, and M. Woods, "Alpha-Particle-Induced Soft Errors in DRAMs," *IEEE Trans. Electron Dev.,* January 1979, p. 2.
78. G. Sai-Halasz et al., *IEEE Trans. Electron Dev.,* **ED-29,** 1983, p. 725,
79. C. M. Hsieh, P.C. Murley, and R.R. O'Brien, *Int. Reliab. Physics Symposium,* 1981, p. 38.
80. O. Osamu, *Ext. Abs. Electrochem. Soc. Meeting,* Spring, 1989, Abs. No. 130, p. 181.
81. F. Masuoka et al., *Dig. Tech. Papers, IEEE Int.Solid State Cirkts Conf.,* p. 146., 1984.
82. T. Fukushima et al., *IEEE Int. Solid State Circuits Conf.,* p. 14, 1983.
83. P. J. Salsbury et al., *Dig. Tech. Papers, 1977 Inter. Solid-State Circuits Conf.,* 1977, p. 186.
84. D. Frohman-Bentchkowsky, *Solid-State Electronics,* 1974, p. 517.
85. N. Ohtsuka et al., *IEEE J. of Solid-State Circuits,* October 1987, p. 669.
86. S.-S. Lee et al., *Tech. Dig. IEDM,* 1987, p. 588.
87. S. B. Ali et al., J. Solid-State Circuits, February 1988, p. 79.
88. S. Mori et al., Tech. Dig. IEDM, 1987, p. 556.

89. A. T. Mitchell, C. Huffman, and A.L. Esquivel, *Tech. Dig. IEDM,* 1987, p. 548.

90. A. Esquivel et al., *Tech. Dig. IEDM,* 1987, p. 859.

91. S. K. Lai, V. K. Dham, and D. Guterman, "Comparison of Trends in Today's Dominant EEPROM Technologies," *Tech. Dig. IEDM,* 1986, p. 580.

92. E.H. Snow, "Fowler-Nordheim Tunneling in SiO_2 Films," *Solid-State Communications,* **5** (1967), p. 813.

93. D. M. Brown et al., "Properties of $Si_xO_yN_z$ films on Si," *J. Electrochem. Soc.,* **15,** 1986, p. 311.

94. A. Bagles, "Characteristics and Reliability of 10 nm Oxides," *Internat. Reliability Physics Sympos.,* 1983, p. 152.

95. T.-K. J. Ting et al., *IEEE J. Solid-State Circuits,* October 1988, p. 1164.

96. D. Cioaca et al., *IEEE J. Solid-State Circuits,* October 1987, p. 684.

97. D. Guterman et al., *Tech. Dig. IEDM,* 1986, p. 826.

98. B. Santo, *IEEE Spectrum,* Dec. 1989. p. 47.

99. G. Samachisa et al., *J. Solid-State Circuits,* October 1988, p. 676.

100. V. N. Kynett et al., *IEEE J. Solid-State Circuits,* October 1988, p. 1157.

101. F. Masuoka et al., *IEEE J. Solid-state Circuits,* August 1987, p. 548.

102. W. I. Kinney et al., *Tech. Dig. IEDM,* 1987, p. 850.

103. S. Sheffield-Eaton et al., *IEEE Int. Solid State Circuits Conf.,* February 1988, p. 130.

104. A. F. Tasch et al., *Tech. Dig. IEDM,* 1977, p. 287.

105. A. F. Tasch and L. H. Parker, *IEEE Proceedings,* March 1989, p. 374.

106. A. F. Tasch et al., "The Hi-C RAM Cell Concept," *Tech. Dig. IEDM,* 1977, p. 287.

107. A. Lancaster et al., *Tech. Dig. ISSCC.,* 1983, p. 164.

108. C. G. Sodini, S. S. Wong, and P. -K. Ko, *IEEE J. Solid-State Circuits,* February, 1989, p. 118.

109. F. Miyaji et al., "A 25-ns 4-Mbit CMOS SRAM," *IEEE J. Solid State Circuits,* October 1989, p. 1213.

110. N. C.-C. Liu et al., "A 22-ns High-Speed CMOS SRAM with Address Multiplexing," *IEEE J. Solid State Circuits,* October 1989, p. 1198.

111. S. Fujii et al, *IEEE J. Solid State Circuits,* October 1989, p. 1170.

112. K. Rogers, "Bipolar DRAM hits U.S.," *Electronic Engineering Times,* September 11, 1989, p. 2.

113. B.-Y. Hwang et al., *IEEE J. Solid-State Circuits,* April 1989, p. 504.

114. M. Matsui et al., *IEEE J. Solid State Circuits,* October 1989, p. 1226.

115. M. Suzuki et al., *IEEE J. Solid State Circuits,* October 1989, p. 1233.

116. D. M. Brown, M. Ghezzo, and J. M. Pimbley, *Proceedings of the IEEE,* December 1986, p. 1678.

117. D. A. Hodges, "Microelectronic Memories," Scientific American, **237,** p. 130, September 1977.

118. E. S. Yang, *Microelectronic Devices,* McGraw-Hill, New York, 1988, p.342.

119. Y. Numasawa et al., *Tech. Dig. IEDM,* 1989, p. 43.

120. A. Learn et al., *J. Appl. Phys.,* **61,** p. 1898, (1987).

121. T. E. Tang, *Tech. Dig. IEDM,* 1989, p. 39.

122. S. Inoue et al., *Tech. Dig. IEDM, 1989,* p. 31.

123. K. Sunouchi et al., *Tech. Dig. IEDM,* 1989, p. 23.

124. D. Bondurant and F. Gnadinger, *IEEE Spectrum,* July 1989, p. 30.

125. R. Pashley and S. K. Lai, "Flash memories: the best of two worlds," *IEEE Spectrum,* December, 1989, p. 30.

CHAPTER 9

PROCESS SIMULATION

Process simulation is the activity of carrying out processing experiments with the aid of a computer, using mathematical models formulated to describe the phenomena being studied. Specifically, the models are physical equations that are solved for dopant redistribution, oxide growth, and all other deposition and etching steps needed to fabricate complex device structures. If the models can accurately simulate the experimental results, the data obtained from such "paper experiments" are the same as if the experiment had been carried out in the laboratory. Figure 9-1 shows a selected sequence of simulation steps used in the fabrication of a silicon-gate MOS device.[1] The results are incremental with each required process step, and many steps need to be simulated.

In the past several years, process simulators have come into widespread use. Such simulators as SUPREM and SAMPLE are widely distributed, and other proprietary codes have been developed by specific organizations. Silicon VLSI devices have become

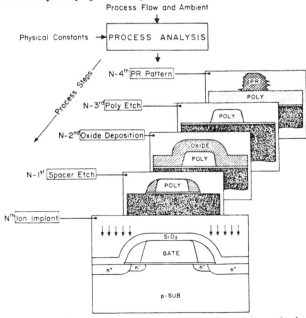

Fig. 9-1 A typical sequence of numerical simulation tools used by a device engineer.

so complicated that it would be prohibitively expensive to optimize process sequences by means of experimental techniques alone. As described in chapter 1, the objective of using these simulators has been to minimize the costly empirical approach. For example, the number of traditional wafer split-runs can be dramatically reduced by using process simulation to obtain target parameters.

This chapter introduces the most widely used process simulators that are publicly available. Enough information is presented to allow readers to understand how the programs function and how to use them. More skill and knowledge can be acquired later through practice and through reference to the user's manual.

9.1 OVERVIEW OF PROCESS SIMULATION

9.1.1 Hierarchy of Simulation Tools for IC Development

Figure 9-2 shows the sequence of simulations that can be performed in integrated-circuit development. Process simulators, the first rung of the ladder of simulation tools, contain models for each of the process steps and allow arbitrary sequences of these steps to be performed. Users thus specify both the sequence of the steps and the process conditions of each step.

There are two types of process simulators: *doping profile* (e.g., SUPREM and SUPRA) and *topography and lithography* (e.g., SAMPLE and PROLITH). The output of the first type is primarily the doping profile in the various layers of the device structure at the end of the sequence being modeled (in one or two dimensions, depending

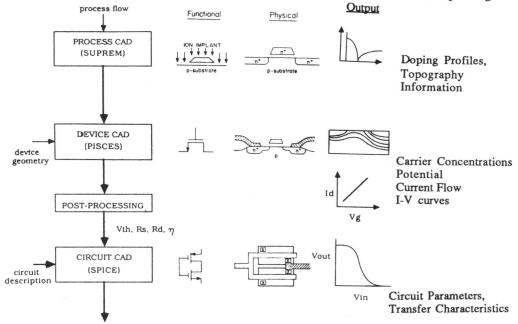

Fig. 9-2 Process analysis – input and output.[1] (© 1986 IEEE).

on the simulator being used). The output of the second type is a two-dimensional cross-section of the device structure (including the surface topography) or the resist feature being modeled.

Another simulation CAD tool related to the process simulator is the *generator of device cross-sections from the layout* (e.g., SIMPL). This type of tool helps designers to visualize the device cross-section generated by a particular combination of mask patterns specified.

The outputs of doping-profile simulators can be used as the inputs to *device simulators* (although user-specified doping profiles can also be used). The outputs of device simulators are such internal device phenomena as potential, charge, and current distributions, as well as such device terminal behavior as threshold voltage, current gain, and current-voltage characteristics. There are MOS (e.g., GEMINI and MINIMOS) and bipolar (e.g., SEDAN and BIPOLE) device simulators, as well as device simulators that can simulate both types of devices (e.g., PISCES and SIFCOD).

Circuit parameters of the appropriate device-circuit models can be extracted from the device-simulator output using *parameter-extractor programs* (e.g., SUXES, TECAP, and TOPEX). Finally, the circuit parameters are entered into the circuit models contained in the *circuit simulators* (e.g., SPICE). Based on the resultant circuit models and connectivity, the circuit simulators produce such outputs as the circuit-switching characteristics (e.g., delay times and rise and fall times), power consumption, and frequency response. Table 9-1 provides a more complete list of the kinds of information that can be obtained from the various types of simulators.

Since this book is concerned with process integration, most of the information in this chapter involves process simulation. However, since process simulation and device modeling are generally closely coupled and exercised iteratively (so that an adequate process can be obtained and the desired device parameters can be reached), a brief description of device and circuit simulators is also provided. Additional information on device simulators can be found in references 2-4, as well as in section 9.8.

9.1.2 Benefits and Limitations of Process Simulation

As noted, process simulation can significantly reduce the cost of developing a new (or modified) IC technology. In addition, some other important benefits can be gained by effectively using process simulation tools, including the following:

- The development cycle of a new technology can made much shorter. It is estimated that without process simulation, submicron-device development effort would take more than a year.

- Simulation allows the analysis of effects that cannot be measured, such as lateral impurity profiles, the location of depletion regions, and the locations of impact ionization and punchthrough.

- Process sensitivities and trade-offs in device design can be analyzed.

Process simulators are capable of simulating many of the individual process steps with a high degree of accuracy. The accuracy of the models is continually being increased. While some examples of the accuracy possible are presented with the descriptions of individual models, interested users of a particular version of a simulator should verify its

---

TABLE 9.1 Applications of Simulation in IC Fabrication

Process Information Obtained by Simulation:

Doping profiles in the silicon substrate
Doping profiles in SiO_2, polysilicon, and silicide layers
Junction depths
Thickness of the material layers
Topography of the silicon substrate surface
Topography of the wafer surface after films have been deposited, etched, flowed, etc.
Lithography: resist profiles after develop (as a function of various processing steps)
Stresses in thin films (e.g., in LOCOS or trench oxides)
Device cross-sections from the layout

Device Characteristics Obtained by Simulation:

Potential variation and electric-field strength in the silicon substrate
Carrier concentrations in the silicon
Current-flow paths
Film sheet resistivity
Threshold voltage (including narrow-width and short-channel effects)
Current-voltage (I-V) device characteristics
Subthreshold currents (including punchthrough currents)
Device-isolation characteristics (e.g., field inversion and punchthrough, both lateral and
    vertical)
Latchup in CMOS
Hot-carrier effects
Soft-error phenomena

Circuit Characteristics Obtained by Simulation:

Device parameters for circuit models
Propagation delay and rise and fall times
Power consumption
Parasitic resistance and capacitance of interconnect lines

---

degree of accuracy for their particular applications, since the evolution of process models is an ongoing effort.

On the other hand, many of the physical processes involved in IC fabrication are still not completely understood. Furthermore, process models must often be simplified so that the computation can be completed in a reasonable time. When many sequential operations are simulated, the small errors introduced in approximating a process for each step can be compounded, so that the final predicted doping profile may be considerably in error. As a result, computer simulations must still be regarded as useful guides, but not yet perfect representations of actual process sequences. Measurements using such techniques as SEM, TEM, and SIMS are still required to confirm computer-simulation results and to point the way to improved models for a given process (see Vol. 1, chap. 17). A symbiotic relationship has developed between experimental process development and simulation, with the simulation results helping to guide the direction of experiments, and experiments establishing the validity of the models.

Despite its limitations, process simulation can provide reasonably accurate quantitative results when used judiciously (and can certainly be cost-effective, especially when compared with silicon-wafer processing costs).

The following are some important guidelines for improving the effectiveness of process simulation tools:[5]

• Everyone using the simulation programs should have a good understanding of the process or system being simulated. This includes familiarity with the models employed in the program, their ranges of applicability, and their limitations.

• The computing environment must be understood by users, who must have the capability to adapt or modify this environment to accommodate the simulation software.

• An understanding of the numerical techniques employed in the simulation program is needed so that problems can be structured or defined to improve the probability of obtaining accurate results. Included in this criteria is an understanding of the numerical algorithms (for correct interpretation of results). This is especially important for questions of convergence.

• The simplest approach should be used in performing a simulation, since it is a waste of computational resources to simulate a device structure with more accuracy than necessary. For example, when only rough estimates are needed, the simplest (and most often the fastest) procedures that will satisfy the relaxed accuracy requirements should be used. Once a concept has been proved to be feasible, slower, more accurate, process simulators can be employed to obtain a more detailed analysis.

9.1.3 Overview of Process Simulators

Table 9-2 lists many of the process simulators that have been developed in the United States up to the time of this writing.* We will limit our discussion to those process simulators available either in the public domain or from commercial vendors; readers wishing more information on the proprietary simulators listed in Table 9-2 (*i.e.* ROMANS,[7] FEDSS,[9] and BICEPS[10]) are directed to the references.

9.1.3.1 Simulator Availability. Table 9-3 lists many of the publicly available process, device, and circuit simulators, together with the organizations from which they can be obtained. While many of the simulators can be procured at nominal cost from the suppliers listed, the others must be purchased from the vendors. While their cost is significantly higher, there are some distinct benefits to acquiring the commercial versions, including the following:

• The commercial versions of process simulators can be integrated by means of proprietary interface software (e.g., the outputs of a process simulator and can be used directly as inputs to device simulators).

* COMPOSITE is a comprehensive process simulator developed in West Germany,[8] and several process simulators developed in Japan are described in a special issue of the *IEEE Transactions on CAD for Integrated Circuits and Systems,* dealing with CAD in Japan.[6]

Table 9-2   PROCESS SIMULATORS

Simulator	Where Originated:	Date Released	Comments
1-D Doping Profile and Oxidation Simulators			
SUPREM I	Stanford Univ.	1977	Simulates doping profiles due to
SUPREM II	"	1979	diffusion and ion implantation, as
SUPREM III	"	1983	well as 1-D thermal-oxide growth. Calculates film resistivities and V_T. (Also available commercially.)
PEPPER[95]	MCC	1985	Handles point-defect interactions and Monte-Carlo-based ion-implant models. Proprietary to MCC members.
PREDICT	MCNC	1987	Designed specially for simulating fabrication of shallow junctions.
2-D Doping Profile and 2-D Oxidation Simulators			
SUPRA	Stanford Univ.	1982	(Also available commercially.)
SUPREM IV	"	1986	Uses point-defect diffusion model, also available commercially.
RECIPE[11]	RPI	1982	
ROMANS	Rockwell	1983	Proprietary simulator.
CREEP	UC Berkeley	1986	Simulates 2-D oxidation, reflow, and oxide growth-related stresses.
2-D Topographical Simulators			
SAMPLE	UC Berkeley	1982	Simulates lithography, deposition, and etching.
PROLITH	Dept. of Defense	1986	Simulates lithography
DEPICT-2	TMA, Inc.	1985	" " " "
SIMPL-1	UC Berkeley	1983	Versions 1 & 2 simulate device
SIMPL-2	"	1985	cross-sections from the layout.
SIMPL/DIX	UC Berkeley	1988	Design Interface CAD tool that provides visual interface for running SIMPL-2, SAMPLE, and CREEP.
2-D Simulators of Doping Profile and Topographical Features			
BICEPS	AT&T Bell Labs	1983	Proprietary simulator.
FEDSS	IBM	1983	Proprietary simulator.

• Program-update and error-correction services are available to purchasers.

• Technical consulting regarding all aspects of the simulators (including interpretation of the results) is provided by the vendors.

• Commercial versions of often contain enhancements not incorporated into the public-domain versions, including additional models, improved accuracy, increased ease of use, and better graphic-output formats.

One of the most significant enhancements to process-simulator usefulness (offered by TMA, Inc., in their version of SUPREM-3; see section 9.2.1), is the capability for the program *to automatically determine the input parameters needed match the target outputs*. Such an *optimization* capability is extremely valuable when a new technology is being developed (i.e., it can find the times, temperatures, and implant doses needed to produce the desired dopant profiles in the devices which are to be fabricated). As an example, the targets might be a CMOS-well doping profile and surface doping concentration (the latter for achieving a desired V_T value). The input parameters that the simulator would determine are the well implant dose and the diffusion time.

TABLE 9-3.   Availability of Process, Device, and Circuit Simulators

---

SUPREM III, SUPREM IV, SUPRA, SOAP, GEMINI, PISCES, SEDAN (Public): Stanford University, Office of Technical Licensing, Stanford University, 105 Encina Hall, Stanford, CA 94305.

DEPICT, SUPREM-3, SUPRA, TSUPREM-4, PISCES, CANDE, SEDAN, TOPEX (Commercial): Technology Modeling Associates, Inc. (TMA), 300 Hamilton Ave., Palo Alto, CA 94301; (415) 327-6300.

SSUPREM3, SSUPREM4, SSAMPL, SSIMPL, S-PISCES2B, SMINIMOS, and DL2000 (Commercial): SILVACO Data Systems, 4701 Patrick Henry Drive, Building 6, Santa Clara, CA 95054;   (408) 988-3482.

SAMPLE, SIMPL-1, SIMPL-2, SIMPL-DIX, CREEP, SPICE (Public): EECS Industrial Liason Program, 457 Cory Hall, University of California, Berkeley, CA 94720.

PROLITH (Commercial): Was formerly available from Chris A. Mack, National Security Agency, Fort Meade, MD. Now being offered by FINLE Technology, Plano TX, 75075.

CADDET (Commercial): Technical Administration, Hitachi, P. O. Box 2, Kokobunji, Tokyo, Japan .

BIPOLE (Commercial): Waterloo Engineering Software, 180 Columbia Street, West, Waterloo, Ontario, Canada N2L 3L3.

SIFCOD (Commercial): Michael Sever Mock, 24 Dubnov St., Tel Aviv 64332, Israel.

TECAP, HPSPICE (Commercial): Hewlett-Packard Laboratories, 3500 Deer Creek Rd., Palo Alto, CA 94304.

FABRICS II (Public): Electrical and Computer Engineering Department, Carnegie-Mellon University, Pittsburgh, PA.

A side benefit of the optimization capability is *sensitivity analysis,* which is the ability of the program to calculate the percentage variation in any output with respect to any input. This capability is very important in centering a design to minimize the influence of processing variations.

9.1.4 General Aspects of Process Simulation

9.1.4.1 Analytical and Numerical Methods of Solving the Equations that Describe Processes.

Many processes can be accurately modeled by differential equations that have analytical expressions as solutions. Examples include the equations describing oxidant transport in the Deal-Grove linear-parabolic model of oxidation, and the Fick's law equation used to model diffusion under low-impurity-concentration conditions in inert ambients (see sections 9.2.4 and 9.2.3.1, respectively). In the case of a drive-in diffusion carried out under the conditions described above, the solution of Fick's equation is given by the well-known Gaussian distribution (note that the boundary conditions in this solution are assumed to be $\partial C/\partial x$ = 0 at x = 0, t, and C [∞, t] = 0, and that the initial condition of the surface concentration is approximated as a delta function; see Vol. 1, chap. 8).

The use of such analytical expressions is advantageous for two reasons. First, such analytical expressions can be rapidly evaluated with a computer, thus allowing simulations to be generated quickly. Second, analytical expressions often provide improved visualization of the process model.

However, simulation of a fabrication process often requires the use of nonlinear differential equations in order for results of sufficient accuracy to be obtained. For example, when diffusion under high impurity-concentration conditions is simulated, the more complex diffusion equation Eq. 9-4a must be used in place of the simpler Fick's-law expression, Eq. 9-4b. In general, such nonlinear equations can be solved only through numerical methods.

In the most advanced models used to simulate diffusion and oxidation, a *set* of coupled nonlinear differential equations must be solved. One of two numerical techniques is typically used: (1) Newton's direct method, or (2) Gummel's iterative method which is applied sequentially to the individual equations (see Fig. 9-3).[12] The direct solution is used for tightly coupled equations, while the iterative method is used for more loosely coupled physics. A more recently used approach combines both Newton's and Gummel's methods.

9.1.4.2 Phenomenological Models versus Physical Models.

The models used to simulate individual processing steps can be based on either phenomeno-logical data (i.e., empirically derived) or physical theory. The first type is therefore referred to as a *phenomenological* (or *empirical*) model, while the second is known as a *physically based* model. The early simulators (SUPREM I-III, SUPRA, and SAMPLE) relied primarily on the first type of models. More advanced simulators for 2-D processes (e.g., SUPREM IV and CREEP) emphasize physically based models more heavily, but phenomenological models are still currently used to some degree in almost all process simulators.

To create phenomenological models (as shown in the right-hand side of Fig. 9-4), it is first necessary to assemble a large set of data concerning the new process. The data

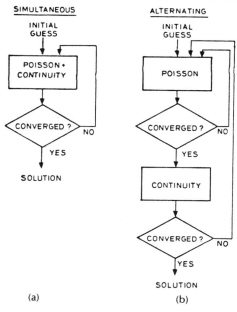

Fig. 9-3 Algorithm flow of simultaneous and alternating methods for solution of the coupled equations. (a) Simultaneous solutions. (b) Iterative solutions.[12] (© 1986 IEEE).

must then be analyzed so that appropriate models can be identified. Additional experimental data may then be used to find the factors that provide the best fit of the experimental data to the analytical functions that simulate the process. For example,

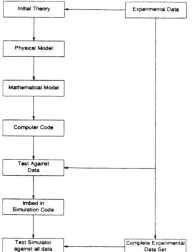

Fig. 9-4 Algorithm for creating new process models and imbedding them in the process simulation code.[13] This paper was originally presented at the Spring 1989 Meeting of The Electrochemical Society, Inc. held in Los Angeles, CA.

experimental data may indicate that some process, such as diffusion or oxidation, appears to depend on particular process parameters (e.g., temperature, pressure, substrate orientation) according to a specific analytical function. As new data is entered, verification must be made that the modeling software will accurately fit it. When a fit does not occur, a new phenomenological model must be created. Phenomenological modeling can thus be seen to be done in hindsight.[13]

Since universal phenomenological models which simulate all conditions have not yet been identified, a process simulator based on models of this type may best be configured with a "decision tree" architecture (e.g., PREDICT). This allows allows models that are valid only over certain ranges of processing parameters to be used for the specific application at hand.

Physically based models are created according to the procedure shown in the left-hand side of Fig. 9-4. In this case, initial experimental data becomes the basis of a new *physical* model. Appropriate *mathematical* models evolved from the physical-model assumptions are then tested against experimental data. If the model accurately simulates the observed phenomena, the mathematical model is included in the simulator code.

The advantage of phenomenological simulators is that they are pragmatic. Many effects that are of great importance in IC processing can be modeled more easily with this approach. Furthermore, when a poor understanding exists of a process's physical nature, a phenomenological model can provide a more accurate simulation. The limitation of phenomenological models is that they cannot be used to discover new processes beyond the parameters of the models. In addition, when a clear understanding of the physical process exists, physically based models will provide more accurate simulations over a wider range of process conditions.

Both types of models therefore find important uses in process simulation. Users of a particular process simulator, however, should be aware of the type of model used in the simulator so that they can determine whether it is appropriate for the task being performed.

9.1.4.3 Gridding. Process and device simulators treat the cross-section of a device structure as a grid of points. At each point, the governing equations are solved for the quantity of interest (e.g., the doping concentration, electrical potential, or current density). In 1-D process simulators (e.g., SUPREM III), the grid spacing is specified only in the vertical direction (i.e., perpendicular to the silicon surface). In 2-D simulators, the grid spacing is specified in both vertical and lateral directions (see Fig. 9-5).

Regions in which the parameter being simulated varies rapidly with distance require a very small grid spacing for accuracy. (This implies that a large number of computations must be made over a short distance.) In regions where the simulated parameter changes less rapidly with distance, the grid spacing can be wider with accuracy and reduced computation time maintained.

Generally, the region just beneath the surface is where the dopant concentration changes rapidly, so tight grid spacing is needed here for high resolution. In the device regions further below the surface, lower resolution may be acceptable because the doping concentration varies less rapidly. In SUPREM III the grid spacing, the depth of the high-resolution region, and the total depth to be simulated are specified by the user.

In 2-D simulators (e.g., SUPRA and SUPREM IV), the grid in both the vertical and horizontal directions is allowed to be more nonuniform, in a manner specified by the

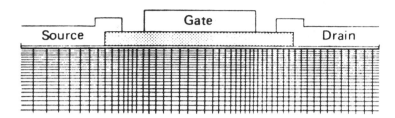

Fig. 9-5 Example of the grid spacing used in 2-D process and device simulators.

user. The calculation of the simulation is performed by finding the solution to the appropriate models at the nodes of the grid (i.e., at each intersection of horizontal and vertical lines).

Note that specification of the grid is an important part of achieving a useful simulation. A suitable trade-off must be established between accuracy and computing costs. In addition, small grid spacing reduces discretization error (i.e., small grid should be selected in regions where large discretization errors would occur if larger, uniform grid had been used).

9.1.4.4 Interfacing of Two Simulators. The output of a process simulation provides the value of impurity concentration at each grid point of the chosen solution mesh. Device simulators must be able to utilize these values to carry out their calculations. Since the device simulation is, in general, performed on a grid that differs from the one chosen for process simulation, some method of interpolating values between the grid points must be provided. Furthermore, since most device simulators are 2-D, some way must be provided to approximately combine several 1-D doping profiles to create a 2-D profile. Finally, some device simulators assume that the wafer surface is planar, when in fact it might be nonplanar (such nonplanarity is represented in the output of 2-D process simulators). A way to handle such gridding differences must also be available.

In an attempt to allow the integration of two process simulators, as well as the interfacing of process simulators with device simulators, a *profile interchange format* (PIF) has been defined.[14] Companies that offer commercial versions of process and device simulators (e.g., TMA, Inc. and SILVACO) generally write the simulators so that they can be interfaced with other programs offered by the same vendor.

9.2 ONE-DIMENSIONAL PROCESS SIMULATORS

In the 1960s and 1970s, the modeling of doping profiles and oxide growth in silicon devices as a result of various process sequences was carried out by solving the analytical equations describing these processes (e.g., by solving the diffusion equation and the Deal-Grove oxidation equation). As process sequences became more complex and device geometries were scaled to the sizes used in the mid-1970s, such secondary effects as the doping dependence of diffusivity and oxidation-enhanced diffusion severely reduced the accuracy of simulations obtained with this simple approach. As a result, models that took such secondary effects into account were developed. Eventually, these individual models were assembled in a single computer program that could simulate the structures obtained from a sequence of fabrication steps.

The first such process simulator that became publicly available (SUPREM I) was developed by Antoniadis and Dutton at Stanford University and was introduced in 1977.[15] SUPREM I is a 1-D process simulator, which means that it calculates doping profiles in the bulk silicon (and in some overlying layers, such as SiO_2) in a direction perpendicular to the wafer surface. In addition, it can determine the thicknesses of the oxide, polysilicon, and epitaxial-Si film layers. The enhanced versions of SUPREM I – released in 1979 and 1983, respectively (SUPREM II and SUPREM III)[16] – are also 1-D simulators.

It is therefore appropriate to apply SUPREM I, II, or III to any region where the impurity distribution or film thickness changes only in the vertical direction. For example, it would be correct to use one of these to obtain simulations of the doping profiles in an MOS device perpendicular to the surface in the region indicated in Fig. 9-6, especially if the minimum lateral device dimension were 2 μm or more. On the other hand, none of the three would be likely to yield an accurate simulation of the doping profile in the device perpendicular to the surface at point X in Fig. 9-6. The reason for this is that the doping profile in this region changes in a lateral as well as in a vertical direction, which is not accounted for by the models used in these simulators.

Figure 9-7 illustrates the evolution of the SUPREM programs. Note that SUPREM I, II, and III are all 1-D simulators, while SUPREM IV is a 2-D process simulator. The first three are generally all based on the same basic models, with enhancements (and a few new models) incorporated into the more recent versions. SUPREM III can handle more layers of device structure than can SUPREM I or II.

SUPREM IV, however, uses many new models, often based on entirely different premises than those used in the first three. In section 9.4.2.5 we will describe the

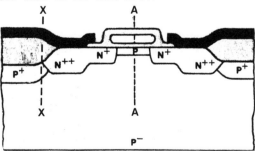

Fig. 9-6 1-D Process simulator can give accurate doping profiles in such regions of the device where the profile only changes in the vertical direction, as along cut line A - A. However, where the doping profile changes in the lateral direction such as in the region penetrated by the cut line X - X, a 1-D simulator will not provide an accurate result.

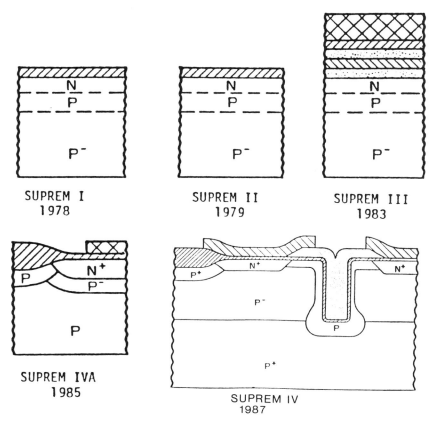

Fig. 9-7 History of the release of SUPREM codes.

capabilities and models of SUPREM IV (as well as its advantages and limitations with respect to SUPRA and SUPREM III). First, however, we will describe the 1-D process simulators. The discussion will be restricted primarily to the latest and most powerful of the SUPREM versions – SUPREM III. However, we will also briefly describe another recently introduced, publicly-available 1-D process simulators, PREDICT. This simulator has been developed especially to treat submicron devices that contain such features as shallow junctions.

9.2.1 SUPREM III (Stanford University PRocess Engineering Model III)

In addition to having been the first process simulator to be publicly introduced, SUPREM has since become the most widely used 1-D process simulator. It has the capability of simulating the following processes in one dimension:

· Ion implantation of impurities.

· Diffusion of impurities in silicon, polysilicon, and SiO_2.
· Oxidation of silicon, polysilicon, and silicon nitride.
· Epitaxial growth of single-crystal silicon layers.
· Deposition.
· Selective etching.

The input file format of the program resembles a process run sheet; the output consists of the 1-D impurity profile in a direction perpendicular to the wafer surface, as well as the thicknesses of the material layers along that direction. The program structure allows process steps to be simulated either individually or sequentially, with the dopant profile obtained after each process step being used as the input for the next step. The profiles can be displayed in various tabular or graphic formats. In the graphic-output format, the impurity concentration (calculated at each point of the grid spacing) is plotted. The individual n-type, p-type, and net-impurity profiles can be calculated and displayed. SUPREM III can also provide such electrical data as charge concentration, sheet resistance, threshold voltage, and channel conductance versus voltage. The program calculates these parameters by numerically solving Poisson's equation after the impurity profiles have been simulated.

The commercially available version from TMA, Inc. (SUPREM-3) also allows the output of the program to be interfaced with various other TMA device simulators (e.g., CANDE, PISCES-2, and SEDAN). That is, it can be formatted so that it can serve as the input file to these programs. Another commercially available version (SSUPREM3 from SILVACO) is also fully interfaced with the SILVACO device simulators, and can be run on PC/AT compatibles and the Apple Macintosh.

9.2.1.1 The Basic Operation and Capabilities of SUPREM III. The SUPREM program has three main parts: (1) an input scanner-supervisor; (2) an output generator; and (3) the part that contains the various process models and performs the actual computations (Fig. 9-8). SUPREM III is also modular so that it can be modified

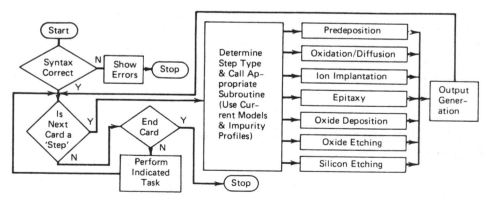

Fig. 9-8 Block flow diagram for the SUPREM II process-modelling computer program.

as new models are developed or as errors in the program are discovered and repaired. A brief introduction to the operation of the program is provided here; more detailed instructions are found in the SUPREM user's manual – cited as reference 17.

The input file, a description of the processing sequence, consists of statements specifying each process step in terms of time, temperature, ambient condition, and other parameters. A single process step can normally be simulated with an input specification of fewer than than 60 alphanumerical characters. Each input specification must also be composed from a set of keywords and numbers, generally taken from the vocabulary of processing jargon. When SUPREM III is initiated, it first lists the entire input file and assigns numbers to each line. Each line of the input file is then checked for syntax validity by comparing the input specifications with a key file containing all of the SUPREM keywords and their related default values.

SUPREM III also provides default values for all physical coefficients through a data file. Users may find occasional discrepancies between measured data and simulation results if they rely only on the default values in the program. In such cases, it is necessary to adjust the parameters used in a model in order to obtain more accurate results. For example, a wafer fab located at an elevation of 5000 feet will have an atmospheric pressure that is less than that at sea level. It is therefore necessary to adjust the appropriate parameters in the oxidation model to obtain correct values of oxide thickness.

The first three lines in the SUPREM III input file are typically a COMMENT line to identify the file, and two initialization lines. The latter two intialize the model parameters by specifying the crystalline orientation of the substrate, the type and concentration of the substrate impurity, the grid spacing, and the total depth to be simulated. There are 500 allowable grid points along the vertical direction.

These lines are followed by the specifications of each process step. Note that while *comment lines* can be inserted for documentation purposes, these lines have no impact on the running of the simulation. SUPREM III is able to handle up to ten material layers, ten different types of materials (silicon, polysilicon, SiO_2, Si_3N_4, aluminum, photoresist, and four user-defined materials), and six impurities (antimony, arsenic, boron, phosphorus, and two user-defined impurities). The models used in SUPREM III to perform such simulations are described in more detail in the upcoming sections.

The final lines of the input file are used to specify the details of the output format for printing or plotting the results. These include the option of providing the electrical parameters that SUPREM is capable of calculating (resistance, threshold voltage, etc.). In addition, process models not included in the basic SUPREM program can be referenced at this point. However, if no call is made for such models, the models that are built into SUPREM are used in the calculations. Note that the simulated doping profiles, etc., can also be saved in data files for future input and continued processing. An example illustrating the specific format of a SUPREM III input is presented in section 9.2.7.

9.2.1.2 Additional Comments on the Use of SUPREM III. As noted earlier, SUPREM III simulations are, strictly speaking, only applicable if planar device

features are much larger than the depth of interest in the silicon substrate. Such dimensional relations still existed in most of the technologies developed up until the late 1980s. For the submicron technologies of the 1990s, however, the depth of interest in the silicon is comparable to the lateral dimensions of the devices. As a result, 2-D simulators will need to be more widely applied in order for adequate simulation accuracy to be achieved.

Nevertheless, it is still advantageous to use SUPREM III in cases in which the program accurately predicts doping profiles, since it requires far less computation time than the 2-D simulators (SUPRA and SUPREM IV). SUPREM III can also be run on a variety of mainframe, minicomputer, and workstation systems (and as noted earlier, with some versions, even on personal computers).

SUPREM III outputs have been used as inputs to 2-D simulators. Some of the issues that need to be resolved when this is attempted were pointed out in section 9.1.4.4. When SUPREM III is used to provide input to an MOS device simulator (e.g., to GEMINI, PISCES, or MINIMOS), at least two 1-D simulations are needed to provide appropriate doping-profile information. In an analysis of NMOS device characteristics, these simulations are the doping distribution in the source/drain region and the doping profile in the channel. The lateral diffusion of the source and drain regions is estimated by multiplying the vertical source/drain profile with Gaussian or complementary-error functions. Reference 5 describes an interface between SUPREM III and the MINIMOS device simulator. Commercially available versions of process simulators (e.g., SUPREM-3 and SSUPREM3) are generally written so that they can generate data files containing 1-D impurity profiles for input into software-compatible device-simulation programs.

9.2.2 SUPREM III Models: Ion Implantation

Ion implantation is a process by which ionized atoms are accelerated directly into a substrate. In VLSI fabrication, this process is primarily used to add dopant atoms (most often selectively) into regions near the surfaces of silicon wafers. The superiority of this method over chemical (diffusion) doping methods has caused it to steadily replace diffusion doping in an increasing number of applications. As a result, it is important to be able to simulate the doping profiles that result when such implantation processes are used in a process sequence. Details of the ion-implantation process are covered in Volume 1, chapter 9.

The mathematical models used to predict ion-implantation concentration profiles are illustrated in order of increasing generality (and complexity) in Fig. 9-9 (see Vol. 1, chap. 9). SUPREM III uses various models from this group, depending on the type of implantation being performed. The *Boltzmann transport equation* (BTE) model is incorporated into SUPREM III and provides superior accuracy in simulations of implantation into multilayer structures. However, range distributions for implanted ions based on the Lindhard, Scharff, and Schiott (LSS) models are retained as the default models. (SUPREM utilizes look-up tables to construct these LSS-based distributions).

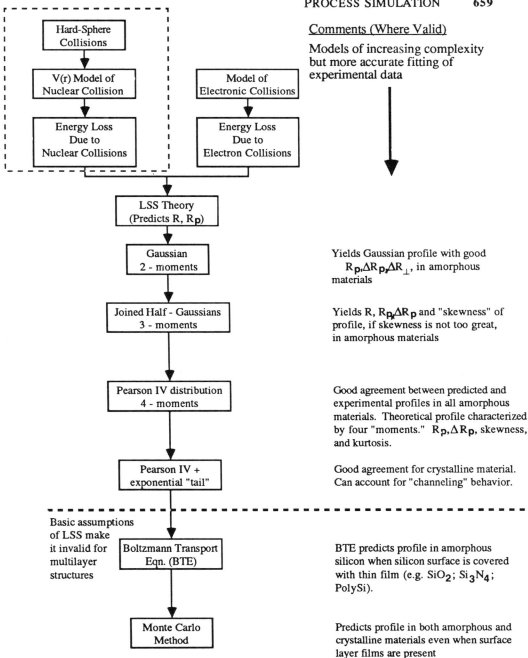

Fig. 9-9 Mathematical models for predicting ion implantation concentration profiles.

The LSS-based provide a sufficiently good fit to experimental data when implants are made directly into silicon, and they require less computation than the BTE approach. When As, P, and Sb implantations are made directly into a Si substrate and the LSS-based model is selected, the user can also specify that the joined half-Gaussian distribution be used to construct the ion distributions. The concentrations n (x) as a function of distance, x, into the substrate on either side of the peak concentration in such joined half-Gaussian profiles are calculated from

$$n_1 \ (x) = n_p \ exp \left[\frac{- (x - R_M)^2}{2 \, \Delta R_{p1}^{\,2}} \right] \qquad x \geq R_M \qquad\qquad (9 - 1)$$

and

$$n_2 \ (x) = n_p \ exp \left[\frac{- (x - R_M)^2}{2 \, \Delta R_{p2}^{\,2}} \right] \qquad 0 \leq x \leq R_M \qquad\qquad (9 - 2)$$

where n_p is the peak concentration. SUPREM calculates the three parameters (R_M, ΔR_{p1}^2 and ΔR_{p2}^2) by using a look up table containing the three moments of the joined half-Gaussian distribution as a function of energy: the projected range, R_p; the standard deviation (or *straggle*), ΔR_p; and the skewness, γ.

The profiles of boron implanted directly into single-crystal silicon substrates, however, exhibit a longer "tail" than is predicted by joined half-Gaussian distributions (due to *channeling effects* – see Vol. 1, chap. 9). As a result, boron implantation profiles into single-crystal silicon are better simulated with the LSS-based models by using a Pearson-IV distribution, together with an added exponential tail with a user-specified decay length (again implemented in SUPREM III with the look-up table approach). It should be noted, however, that if the silicon substrate is preamorphized, a good fit is obtained with a Pearson-IV distribution without the added tail. The accuracy of simulated boron implants is illustrated in Fig. 9-10a,[15] as calculated using SUPREM III, compared to boron implant SIMS data, and in Fig. 9-10b as calculated by TMA's 1989 version of SUPREM-3.

Note that implantation directly into a silicon substrate can be accurately simulated by evaluating the analytical expressions based on the basic theory of ion stopping in solids developed by LSS. However, when the modeling of implantations into multilayer substrates is attempted (e.g., threshold-adjust implants through gate oxides or channel-stop implants through field oxides, as shown in Fig. 9-11), the basic assumptions of the LSS theory do not allow its application. For example, atoms from the surface layers may be knocked into deeper layers by impinging ions (*recoil effect*). Alternative approaches based on numerical solutions rather than on analytical techniques have been developed to treat such cases. These include the solution of the Boltzmann transport equation (BTE)[18] or the use of Monte Carlo (MC) simulation techniques.[19]

In the Monte Carlo approach, ion implantation is simulated by following the history of an energetic ion through successive collisions with target atoms, using the

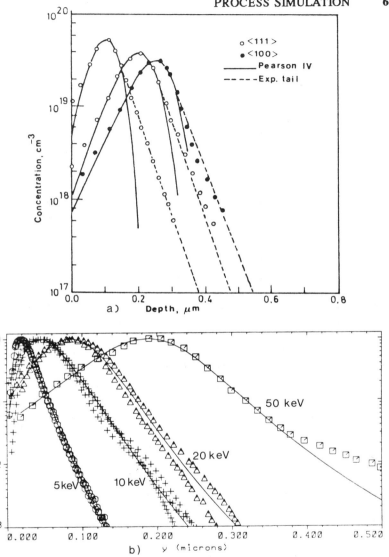

Fig. 9-10 (a) Experimental profiles of boron implanted into <111> (closed circles) and <100> (open circles) silicon. The solid lines represent the Pearson IV distribution and the dashed lines the exponential tail.[15] (© 1979 IEEE). (b) Experimental and simulated curves of boron implanted at 5, 10, 20, and 50 keV using SUPREM-3. Courtesy of TMA, Inc.

binary collision assumption. The calculation of each trajectory begins with a given energy, position, and direction. A large number of ion trajectories is calculated (e.g., 10^3-10^4), and the depth at which each ion stops is determined. The predicted profile is generated by plotting histograms of the number of ions stopped within each interval.

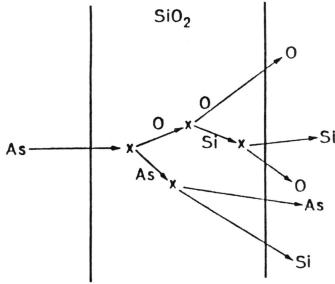

Fig. 9-11 Boltzmann Transport methods can be used to accurately model implantation into multilayer structures. LSS theory is not applicable to multi-layer structures.

MC simulations may be performed for either amorphous or crystalline targets. In amorphous-material simulation the position of the target atoms follows a Poisson distribution, while in crystalline-target simulation the atoms are specified to correspond to the positions they would assume on a lattice.

The MC technique is more general than the BTE approach and it has the advantage of allowing simulated profiles to be calculated in crystalline materials. It is, however, very intensive computationally, generally requiring the use of a supercomputer to produce useful results in reasonable computer times. A recent version of SUPREM-3 available from TMA, Inc. offers the capability of simulating ion implantation using the MC method. This software contains a lookup technique for calculating collision events, which speeds calculation and makes it practical for engineering workstations.*

A simpler analytical technique for predicting implanted profiles in multilayer structures has been developed. This method scales the range of the profile by taking into account the different stopping powers of the various target materials.[20] That is, each layer is scaled in thickness according to the relative ranges of the ions in that layer and in the substrate. This approach is used in the COMPOSITE process simulator.[8]

---

* This Monte-Carlo technique for simulating ion implantation in a 1-D process simulator was developed for the PEPPER process simulator.[95] Although the PEPPER development effort has been discontinued, its Monte-Carlo simulation of ion implantation has been integrated into the SUPREM-3 program offered by TMA, Inc. Other simulation developments provided in PEPPER (such as a physically based model for diffusion, similar to the one used in the 2-D process simulator, SUPREM IV), may be incorporated in the future.

In SUPREM III, either the scaled analytical distribution-function method or the solution of the Boltzmann transport equation can be used to simulate ion implantations into multilayer structures. The BTE numerical technique is selected over the MC technique because it is less computationally intense. With this model, SUPREM III can be used to simulate implantation through up to ten layers. Note, however, that the BTE method does not account for the channeling effects that occur during implantation into single-crystal silicon. Hence, in simulations of direct boron implantations into single-crystal substrates, the Pearson-IV distribution (with an added exponential tail) typically still gives a more accurate fit to observed distributions than does the BTE model.

9.2.3 SUPREM III Models: Diffusion in Silicon and SiO₂, and Segregation Effects at the Si/SiO₂ Interface

It is important to be able to simulate the redistribution of impurities in Si during high-temperature process steps as a result of diffusion and dopant segregation at oxide interfaces. Such dopant redistribution can be accurately modeled in one-dimension; the models used to do this are incorporated in SUPREM III.

9.2.3.1 Diffusion Models Used in SUPREM III. These models are empirically-based models and are derived from observed dopant-diffusion phenomena in silicon. It is assumed that impurity diffusion in silicon can be described by the complete, 1-D continuity equation

$$\frac{\partial C}{\partial t} = \frac{\partial}{\partial x}\left(D\,\frac{\partial C}{\partial x}\right) \pm \frac{q}{kT}\left[\frac{\partial}{\partial x}\left(D\,C_i\frac{\partial \varphi}{\partial x}\right)\right] \qquad (9\text{-}3)$$

where D is the diffusivity and C and C_i are the total and electrically charged impurity concentrations, respectively. The potential (φ) is found from $\varphi = (kT/q)\ln(n/n_i)$, where n and n_i are the electron and intrinsic carrier concentrations at the diffusion temperature. The first term in Eq. 9-3 represents classical gradient-driven diffusion, including nonconstant diffusivity (see Vol. 1, chap. 8, which covers the details of diffusion in silicon). The second term incorporates the electrostatic field-driven flux. The effective electric field-driving force may be established by the concentration gradient of either the impurity under question or another impurity present at a greater concentration.

In many cases of interest, local charge neutrality is maintained, and no electrostatic field exists. Hence, the impurity diffusion in such cases can be described by:

$$\frac{\partial C}{\partial t} = \frac{\partial}{\partial x}\left(D\,\frac{\partial C}{\partial x}\right) \qquad (9\text{-}4a)$$

If D is constant (as is the case with diffusion under low-impurity-concentration conditions), Eq. 9-4a reduces to the simpler Fick's law expression:

$$\frac{\partial C}{\partial t} = D\left(\frac{\partial^2 C}{\partial C^2}\right) \qquad (9 - 4b)$$

If appropriate values of D are used in Eqs. 9-3 and 9-4, the solutions to these equations provide accurate simulations of dopant concentrations in many applications. Therefore, if accurate models of dopant diffusion in Si are to be constructed, correct values of D must be established for the various dopants as a function of temperature and doping concentration.

Two approaches have been used to determine the values of diffusivity needed for these equations. The first is based on extracting the values from measured dopant-concentration profiles. That is, experimentally determined values of D are plotted as a function of temperature and doping concentration. Solutions to Eqs. 9-3 and 9-4 are then fitted fitted to these experimental plots by finding the values of D that cause the equation solutions to match the D values extracted from observed data. The models that result are known as *phenomenological diffusion models*, because they are based on observed phenomena.

The second approach assumes that correct values of D can be established directly from physical principles. In the case of impurity diffusion in Si, the interaction of impurity atoms and point defects in the silicon lattice give rise to the mechanisms that produce diffusion. An understanding of such interactions should allow the development of correct impurity models, which are referred to as *point-defect–based diffusion models.*

Phenomenological models have been implemented in SUPREM III and in other one-dimensional process simulators, such as PREDICT and RECIPE. These models provide a very adequate fit to 1-D diffusion-doping profiles. Since such profiles can be fully characterized, experimentally determined values of D can be accurately determined. The phenomenological approach, however, is not useful for simulating 2-D profiles, for two reasons. First, no measurement techniques have yet been developed to determine 2-D impurity profiles in small devices. Second, the parameter space for a process sensitive to 2-D is greatly expanded. Hence, "shotgun" attempts to optimize device structures become unmanageable. Research work is being actively pursued to improve the capabilities of point-defect–based diffusion models, and such models have already been incorporated into 2-D process simulators (e.g., SUPREM IV; see section 9.4.2).

We now describe the phenomenological models used in SUPREM III for simulating the 1-D profiles due to diffusion of B, As, P, and Sb in silicon. All of the models are based on the *vacancy model under non oxidizing conditions,* proposed by Fair and Tsai.[21] The diffusivity of an ionized impurity species is considered to be the sum of the diffusivities resulting from neutral vacancies and ionized vacancies with an opposite charge. According to this model, there are four charged states for vacancies: doubly negative (2-), singly negative (-), neutral (x), and positive (+). Thus, in general, the *effective diffusivity* under nonoxidizing conditions can be found from

$$D = D^x + D^-\left(\frac{n}{n_i}\right) + D^{2-}\left(\frac{n}{n_i}\right)^2 + D^+\left(\frac{n_i}{n}\right) \qquad (9 - 5)$$

where D^x, D^-, D^{2-}, and D^+ are the diffusivities resulting from interactions between neutral point defects (D^x) and charged point defects (D^-, D^{2-}, D^+). Note, however, that when Eq. 9-5 is substituted into Eq. 9-3 or Eq. 9-4 to fit experimental profiles, however, it does not specify the dominating diffusion mechanisms. The exact mechanisms must still be determined from other experimental evidence or from theoretical considerations. Therefore, Eq. 9-5 can be considered a phenomenological expression of the concentration dependence of the diffusivity.

9.2.3.2 Modeling Low-Impurity-Concentration (Intrinsic) Diffusion in Silicon.
When the impurity concentration is low (i.e., 10^{15}-10^{18} cm^{-3}), n is approximately equal to n_i at the high temperatures needed for diffusion; hence $(n/n_i) \cong 1$. Under such conditions, the effective diffusion coefficient given by Eq. 9-5 reduces to simply the sum of the various diffusion coefficients, independent of concentration, or:

$$D^i = D^x + D^- + D^{2-} + D^+ \qquad (9-6)$$

D^i is therefore referred to as the *intrinsic effective diffusion coefficient*. The intrinsic diffusion coefficients of B, P, As, and Sb are given as default values in SUPREM III.

The vacancy model of *boron diffusion in silicon* assumes that since boron atoms are negatively charged, they diffuse primarily by interaction with neutral and positively charged vacancies, and D^i will thus be essentially due to the sum of D^x and D^+. The intrinsic boron diffusivity, D_B^i, is modeled as:

$$D_B^i = D_B^x + D_B^+ \qquad (9-7a)$$

$$= (0.037) \exp(-3.46 \text{ eV}/kT) + (0.72) \exp(-3.46 \text{ eV}/kT) \qquad (9-7b)$$

$$= 0.76 \exp(-3.46 \text{ eV}/kT) \text{ cm}^2/\text{sec} \qquad (9-7c)$$

The *vacancy model of arsenic diffusion* assumes that since arsenic atoms are donors, they appear to diffuse by interaction with neutral and singly negatively charged vacancies. The intrinsic arsenic diffusivity is given by the sum of D^x and D^-, or:

$$D_{As}^i = D_{As}^x + D_{As}^- \qquad (9-8a)$$

$$= 0.066 \exp(-3.44 \text{ eV}/kT) + 12.0 \exp(-4.05 \text{ eV}/kT) \text{ cm}^2/\text{sec} \qquad (9-8b)$$

For *phosphorus*, the intrinsic diffusivity is dominated by the interaction of impurity atoms with neutral vacancies, and thus the intrinsic phosphorus diffusivity is given by:

$$D_P^i = 3.85 \exp(-3.66 \text{ eV}/kT) \text{ cm}^2/\text{sec} \qquad (9-9)$$

Figure 9-12 plots the intrinsic diffusivities of B, As, and P as a function of temperature. In addition, because the fit between the intrinsic values of diffusivity for B, As, and P is

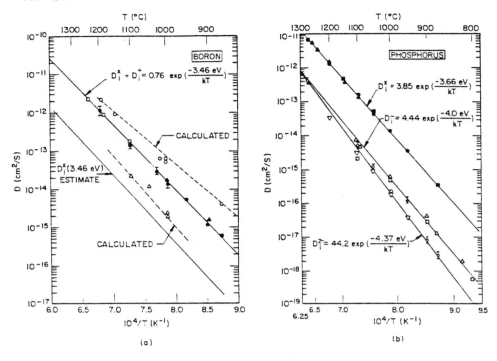

Fig. 9-12 Diffusivity data used in SUPREM III to calculation diffusion profiles.

very good, SUPREM III is capable of providing accurate diffusion profiles for intrinsic diffusion of these impurities.

9.2.3.3 Modeling High-Impurity-Concentration (Extrinsic) Diffusion

in Silicon. At high-impurity concentrations, n or p is likely to exceed the intrinsic carrier concentrations. Under such conditions, the diffusivities of the impurities will be altered from those values observed under intrinsic doping conditions.

In the case of *arsenic*, SUPREM III calculates the effective diffusivity under high-doping concentrations from the following model:

$$D_{As} = D_{As}{}^{X} + D_{As}{}^{-} \qquad (9\text{-}10a)$$

$$= 0.066 \exp(-3.44 \text{ eV}/kT) + 12.0 (n/n_i) \exp(-4.05 \text{ eV}/kT) \qquad (9\text{-}10b)$$

It should also be noted that a *clustering effect* is observed if the doping level of As in silicon approaches the solubility limit. That is, the concentration of ionized As atoms is a fraction of the total As present in the Si, and the difference becomes greater as the diffusion temperature decreases below 900°C. This effect is important because As

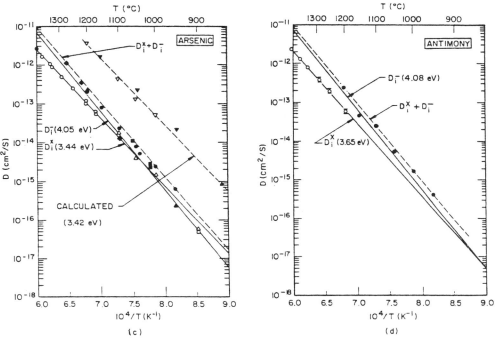

Fig. 9-12 (continued).

clusters are immobile. The latest versions of SUPREM III include this effect in the As diffusion model by determining the concentration of As atoms that do not form clusters and then calculating the diffusion profile due to those atoms. At temperatures higher than 1000°C, the clusters dissociate and then diffuse as separate As species.

The diffusivity of *boron* under high concentrations is calculated by SUPREM III using the following model (noting that $n = n_i^2/p$):

$$D_B = D_B^x + D_B^- \tag{9-11a}$$

$$= [0.037 + 0.72\,(p/n_i)]\,\exp\,(-3.46\,\text{eV/kT})\ \text{cm}^2/\text{sec} \tag{9-11b}$$

It has been observed experimentally that the diffusion of boron in silicon is unique, in that the diffusivity can be reduced by up to a factor of 10 in n^+ silicon but will increase when in p^+ silicon. These variations in diffusivity are accounted for in Eq. 9-11, where the second term decreases when diffusion occurs in heavily n-doped Si (hence decreasing p to a value less than n_i), but increases if the p-doping is high.

The diffusivity of *phosphorus* under high doping concentrations is more complex than that of boron or arsenic. The characteristic doping profile of phosphorus diffusion in silicon consists of three regions: the high-concentration region, the transition region (commonly called the *kink* of the profile), and the low-concentration (or *tail*) region.

According to the Fair-Tsai vacancy model, the diffusivity of phosphorus in the *high-concentration region* is dominated by interaction with neutral, singly ionized, and doubly ionized vacancies (i.e., some of the phosphorus ions, P^+, form pairs with the doubly ionized vacancies, V^{2-}, to form $[PV]^-$). The diffusivity in this region is calculated in SUPREM III as the sum of D^x, D^- and D^{2-}:

$$Dp^{hi} = Dp^x + Dp^- + Dp^{2-} \qquad\qquad (9\text{-}12a)$$

$$= 3.85 \exp(-3.66 \text{ eV/kT}) + 4.44\,(n/n_i)\exp(-4.0 \text{ eV/kT}) \qquad (9\text{-}12b)$$

$$+ 44.2\,(n/n_i)^2 \exp(-4.37 \text{ eV/kT}) \quad \text{cm}^2/\text{sec} \qquad (9\text{-}12c)$$

A band-gap narrowing effect occurs due to lattice-misfit strain when large concentrations (5×10^{20} cm^{-3}) of the smaller-radius phosphorus atoms are present in the silicon lattice. This effect (which is also accounted for in the SUPREM III model), causes the diffusivity to decrease in such heavily-doped phosphorus regions.

In the *tail region,* the Fair-Tsai model proposes that the diffusivity is enhanced relative to the intrinsic value as a result of supersaturation of the silicon lattice by vacancies that arise from the dissociation of the $[PV]^-$ pairs. Hence, the diffusivity in the tail region is calculated from a second expression:

$$D_p^{tail} = Dp^x + Dp^-[\,n_s^3/(n_e^2 n_i)\,]\exp(3\Delta E_g/kT)$$

$$[\,1 + \exp(0.3 \text{ eV/kT})\,]\exp(-\{X - X_e\}/L_v) \qquad (9\text{-}13)$$

where n_s is the surface electron concentration, n_e is the electron concentration at which the Fermi level drops below 0.11 eV from the conduction band edge, X_e is the depth at which n_e is reached, L_v is the diffusion length of the vacancies into the substrate (~25 μm), and ΔE_g is the band-gap narrowing due to the increased doping concentration. L_v is likely to be dependent on such parameters as the oxygen concentration in the silicon wafer, but since these dependencies are not yet well characterized, in SUPREM III $L_v = 25$ μm. This means that essentially all device regions beneath the phosphorus will show diffusivity enhancement.

It might be asked why there is still an interest in phosphorus diffusion under high impurity concentrations, since As is now used as the dopant in the emitters of bipolar transistors and in the source/drain regions of NMOS devices. The reason is that phosphorus is still used as the dopant in the lightly doped extension region of LDD structures in NMOS transistors, while As is used in the heavily doped part of such structures. The diffusivity of phosphorus in the presence of a high concentration of As thus occurs, and the resulting diffusion of phosphorus can impact the depth of shallow source/drain junctions.

9.2.3.4   Oxidation-Enhanced Diffusion Modeling in SUPREM III.

The latest versions of SUPREM III include phenomenological models that account for *oxidation-enhanced diffusion* (OED) and *oxidation-retarded diffusion* (ORD) effects, which are observed when diffusion occurs during oxidation, oxynitridation, or direct nitridation.[22, 23, 24] This topic is the subject of intense research activity, and readers should consult the technical literature for the latest information on it.

9.2.3.5 Dopant-Segregation Effects at the Si-SiO$_2$ Interface, and Diffusion in SiO$_2$.

As described in Volume 1, chapter 7, the dopant impurities (boron, phosphorus, arsenic, and antimony) near the silicon surface become redistributed during thermal oxidation. Such redistribution continues until the chemical potential is the same on each side of the interface, and it typically results in an abrupt change in the impurity concentration across the interface. The process of redistribution is influenced by three factors: the segregation coefficient, m; the diffusivity of the impurity in the oxide; and the rate at which the Si/SiO$_2$ interface moves with respect to the diffusion rate. The value of m is also dependent on temperature. Details of the model that predicts the dopant-redistribution effects are presented in reference 25.

SUPREM III also simulates such effects, based on the model cited in reference 25. This model allows the initial doping concentration to be specified and then solves the differential equation with a moving boundary to describe the redistribution of the initial doping-concentration profile. Since a complicated mathematical function may be needed to accurately describe this doping profile, numerical methods must be used to solve for the final profile when such interface segregation effects are included.

Dopant transport is less well understood and characterized in SiO$_2$ than in silicon. As a result, diffusion in SiO$_2$ is modeled by means of the usual diffusion equation, with a classical diffusion constant, D. Values of effective diffusivities for the different dopant impurities are taken from the literature where available.

9.2.4 SUPREM III Models: Thermal Oxidation of Silicon in One Dimension

Thermal oxidation at high temperatures is used to form SiO$_2$ layers on silicon. In this process (which is an integral process step in IC manufacturing), oxidant from the gas phase (in the form of O$_2$ or H$_2$O molecules) diffuses through the growing SiO$_2$ toward the Si-SiO$_2$ interface. At the interface the oxidant molecules react with the silicon atoms of the substrate, forming new SiO$_2$ material. Details of the thermal oxidation process are presented in Volume 1, chapter 7.

In 1-D cases, the thickness of the SiO$_2$ on top of Si, x_o, can be obtained from the equation derived by Deal and Grove in their linear-parabolic model of oxidation:

$$x_o^2 + Ax_o = B(t + \tau) \tag{9-14}$$

where A and B are rate constants, and τ is the time displacement needed to account for the initial oxide layer, x_i, at $t = 0$:

$$\tau = (x_i^2 + Ax_i)/B \tag{9-15}$$

Solving the quadratic Eq. 9-14 gives the oxide thickness as a function of time:

$$x_o = \frac{A}{2} \left\{ \left[1 + \frac{(t + \tau)}{A^2/4B} \right]^{1/2} - 1 \right\} \qquad (9\text{-}16)$$

This is the basic equation used in all process-modeling programs to calculate SiO_2 thickness. It requires values of the rate constants B and B/A appropriate to the particular process sequence being simulated. *Intrinsic parabolic* and *linear rate constants* (B^i and $[B/A]^i$, respectively) are defined for (100) wafers for the growth conditions of oxygen pressure at one atmosphere and low-impurity concentrations in the silicon substrate.

For dry O_2, both B^i and $(B/A)^i$ may be well represented as the singly activated processes[90]

$$B^i = C_1 \exp - (E_1/kT) \qquad (9\text{-}17a)$$

and

$$(B/A)^i = C_2 \exp - (E_2/kT) \qquad (9\text{-}17b)$$

where $C_1 = 7.72 \times 10^2\ \mu m^2/h$, $E_1 = 1.23$ eV, $C_2 = 6.23 \times 10^6\ \mu m/h$, and $E_2 = 2.0$ eV. The physical process represented by B is oxidant transport through the SiO_2, and the 1.23-eV activation energy corresponds to O_2 diffusion in SiO_2. For (B/A), the 2.0-eV activation energy has historically been connected with Si-Si bond breaking at the Si/SiO_2 interface. However, recent data suggests that other phenomena may also be involved.

For H_2O oxidation, B^i and $(B/A)^i$ have the same form as in Eq. 9-18, but the apparent activation energy of each changes at ~900-950°C (see Fig. 9-13). The values of the two intrinsic rate constants at various pressures and for (100) and (111) Si are stored as default values in SUPREM.[26] [The (100) curve at 1 atm in Fig. 9-13a is a plot of $(B/A)^i$ versus T, and the 1-atm curve in Fig. 9-13b plots B^i versus T, both under pyrogenic steam.] Figures 9-13a and 9-13b also plot values of (B/A) and B in pyrogenic steam for various other oxygen pressures on (100) and (111) wafers.

The values of the the rate constants B and (B/A) will change, however, when the oxide is grown under other than intrinsic conditions. Thus, in order to accurately simulate oxide thicknesses under all practical growth conditions, more general equations must be available to calculate B and (B/A). We will describe these equations in the following sections.

9.2.4.1 High-Dopant-Concentration Cases.

In cases where the concentration of impurities in the Si substrate is high, the oxidation rate is significantly enhanced. (Practical examples include bipolar n^+ emitter regions and NMOS source and drain regions, where oxidation growth rates may be as much as five times higher than they are over lightly doped regions.) Such higher oxidation rates have been attributed to the increase in total silicon substrate vacancy concentrations,[27] with vacancies at the

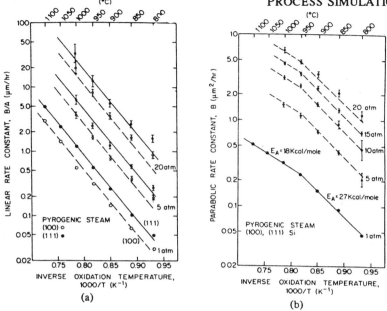

Fig. 9-13 (a) Linear (B/A) and (b) parabolic (B) rate constants versus 1000/T for <100> and <111> silicon wafers oxidized in pyrogenic stem at 1-20 atm.[97] Reprinted by permission of the publisher, The Electrochemical Society, Inc.

Si/SiO$_2$ interface assumed to act as sites for the oxidation reaction. The oxidation-rate enhancement can be included in Eqs. 9-17a and b as follows:

and

$$(B/A) = (B/A)^i \left(1 + \gamma \left(C_V^T - 1 \right) \right) \tag{9-18a}$$

$$B = B^i \left[1 + \delta \left(C_V^T \right)^{0.22} \right] \tag{9-18b}$$

where γ and δ are empirically determined parameters, given as:

and

$$\gamma = 2.62 \times 10^3 \exp \left(-1.1/kT \right) \tag{9-19}$$

$$\delta = 9.63 \times 10^{-16} \exp \left(2.83/kT \right). \tag{9-20}$$

The variable C_T^V is the total normalized vacancy concentration in the substrate at the interface. This concentration depends on doping level and temperature and is given by

$$C_T^V = \frac{\left[1 + C_V^+ \left(\dfrac{n_i}{N} \right) + C_V^- \left(\dfrac{N}{n_i} \right) + C_V^{2-} \left(\dfrac{N}{n_i} \right)^2 \right]}{\left(1 + C_V^+ + C_V^- + C_V^{2-} \right)} \tag{9-21}$$

with the vacancy concentrations C_V^+, C_V^-, and C_V^{2-} calculated from

$$C_V^+ = \exp\left(\left[\, E^+ - E_i \,\right]/kT\right) \qquad\qquad E^+ = 0.35\ \text{eV} \qquad (9\text{-}22a)$$

$$C_V^- = \exp\left(\left[\, E_i - E^- \,\right]/kT\right) \qquad\qquad E^- = E_g - 0.57\ \text{eV} \qquad (9\text{-}22b)$$

$$C_V^{2-} = \exp\left(\left[\, 2E_i - E^- - E^{2-} \,\right]/kT\right) \qquad E^{2-} = E_g - 0.11\ \text{eV} \qquad (9\text{-}22c)$$

where E_i is the position of the intrinsic Fermi level, $E_i \cong E_g/2$, and N is the total dopant concentration. Through C_T^V, (B/A) depends upon the local doping concentration, which accounts for the enhanced oxidation rates of n^+ and p^+ substrates. This high-doping-concentration model is incorporated into SUPREM III. Since the impurity concentration at the Si-SiO$_2$ interface also changes as a result of the diffusion and segregation during oxidation, the enhanced growth rates of SiO$_2$ are recalculated at each time step in SUPREM III.

9.2.4.2 Modeling Other Factors That Impact the Oxide Growth Rate.

Other process conditions that impact the oxide growth rate are the partial pressure of oxygen, the Si-crystal orientation, the presence of HCl, and, under some conditions, the rapid growth of the oxide when it is very thin.

The latter phenomenon has drawn a great deal of attention because the gate oxide in advanced VLSI MOS devices is grown within this thin-oxide regime. Hence, SUPREM III contains an empirical factor that is added to the linear-parabolic model as

$$\frac{dx_o}{dt} = \left(\frac{B}{2x_o + A}\right) + K\exp\left(\frac{-x_o}{L}\right) \qquad\qquad (9\text{-}23)$$

where the decay length, L, is approximately independent of temperature (~7 nm), and K is a singly activated function of temperature with an activation energy of 2.35 eV for <111> and <100> orientations of Si, and 1.8 eV for <110> orientations.[28]

The other conditions that impact the growth rate are included in the equations that calculate B and B/A in the SUPREM III program, as follows:

$$\frac{B}{A} = \left(\frac{B}{A}\right)^i \left(P^n\left[1 + \gamma\left(C_V^T - 1\right)\right]\eta\,\alpha\left[1 + K\exp\left(\frac{-x_o}{L}\right)\right]\right\} \qquad (9\text{-}24)$$

and

$$B = B^i\left\{\, P^m\left[1 + \delta\left(C_V^T\right)^{0.22}\right]\varepsilon\,\right\} \qquad\qquad (9\text{-}25)$$

where P^n and P^m are used to account for the partial pressure of oxygen, η and ε are factors used to account for the presence of HCl, α is the orientation effect, and $(B/A)^i$ and B^i are the low-concentration values of these constants.

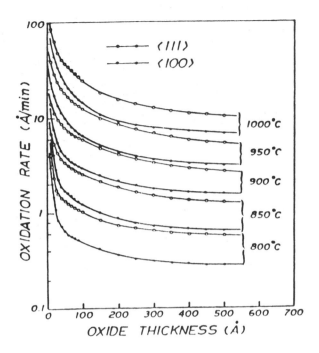

Fig. 9-14 Comparison of simulated and measured oxide thicknesses grown under different temperatures, using SUPREM III.

The addition of chlorine species to the oxidizing ambient has become common in industrial laboratories. The improved threshold stability and more uniform dielectric strength of the SiO_2 films have been well documented (see Vol. 1, chap. 7). Significant increases in the oxidation rate through such chlorine additions have also been observed in dry O_2 ambients. Since there is no quantitative physical model for these effects as of this writing; the enhancements in B and (B/A) due to HCl concentration are stored in a look-up table.

9.2.4.3 Accuracy of Modeling Oxide Growth with SUPREM III.

An example of the accuracy with which the SUPREM III program can simulate thermal-oxide growth is shown in Fig. 9-14, where oxide thicknesses grown at various rates and temperatures in 1-D conditions are compared to the simulated values using SUPREM III. As can be seen, excellent agreement is obtained between the simulated and experimental values. The recent versions of SUPREM-3 (available from Technology Modeling Associates, Inc.) can also simulate the oxidation of polysilicon and silicon nitride, the effects of temperature and pressure ramps on the oxide growth, and oxide growth under rapid thermal processing (RTP) conditions. Figure 9-15 shows the results of simulating oxide growth in dry O_2 (Fig. 9-15a) and wet O_2 (Fig. 9-15b) using

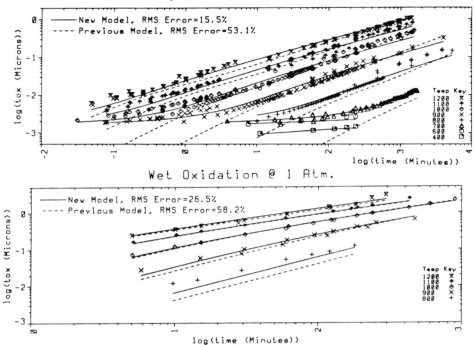

Fig. 9-15 (a) Comparison of simulated an measured growth of SiO_2 under many different temperatures in (a) dry O_2, and (b) wet O_2. Courtesy of TMA, Inc.

SUPREM-3. As a final comment, the modeling of oxide growth is much more complex in two dimensions than in one. This topic will be discussed in section 9.4.3.

9.2.5 SUPREM III Models: Epitaxial Growth

Epitaxy is the process in which a single-crystal film is grown on a single-crystal substrate. The details of epitaxial growth on silicon substrates are described in Volume 1, chapter 5. Epitaxial-film growth is an important process for CMOS, bipolar, and BiCMOS technologies, since it is the only process that can be used to form lightly doped layers on heavily doped substrates. Although it is desirable to have a sharp gradient for the doping concentration between the epitaxial (epi) film and the substrate, impurity diffusion during the high-temperature epi process (typically from 900-1200°C) always smooths out this transition to some degree. That is, impurity atoms in the heavily doped substrate diffuse into the epi film, resulting in a gradual transition between the two regions. It is important to be able to model the degree to which such smoothing occurs, as it can significantly impact device characteristics.

Epitaxial-film deposition processes in SUPREM III are modeled by the one-dimensional model of Reif and Dutton,[29,30] which is capable of simulating epitaxial doping profiles for a variety of growth conditions. This model assumes that silane (SiH_4) is used to grow silicon in a hydrogen ambient in an atmospheric-pressure reactor, and it calculates the epitaxial-film thickness for the specified growth conditions. Phosphorus- and arsenic-doped epitaxial films can be treated by the model. The thermal redistribution of the impurities is simulated by applying Fick's second law during the epitaxial growth.

Vertical autodoping (as discussed in Vol. 1, chap. 5) is also simulated by this model. That is, for cases in which the epi growth rate is more than five times the impurity-diffusion rate, impurity redistribution from the substrate into the growing film can be approximated by complementary-error functions. Such approximations are valid for typical epi growth processes.

The degree of accuracy that the SUPREM III epitaxial-growth model can exhibit is shown in Figs. 9-16a and 9-16b, in which doping profiles simulated with this model are compared with the doping profiles measured in actual epitaxial films (using the spreading resistance technique). In Fig. 9-16a, two epitaxial films were sequentially deposited, the first using an arsine flow rate selected to give an As doping concentration of 10^{17} cm^{-3}, and the second using an As concentration of 10^{15} cm^{-3}. The reactor was also purged for eight minutes with H_2 at 1050°C between the two depositions. Note that excellent agreement is obtained between the simulated and measured impurity profiles, even in the autodoping-produced transition region between the more heavily doped and lighter doped films.

Figure 9-16b illustrates the accuracy with which the model can simulate the deposition of an undoped epitaxial film over a heavily doped region in the Si substrate. In this case, an arsenic layer was implanted into a lightly doped boron (100) Si wafer. Following an anneal step to redistribute the As atoms (so that they had a surface concentration of $\sim 10^{19}$ cm^{-3}), an undoped epitaxial layer of ~ 1.5-μm thickness was deposited. The epitaxial growth rate was ~ 0.27 $\mu m/min$; the deposition time was six minutes, and the temperature was 1050°C. Once again, excellent agreement with the experimental measurements is exhibited by the simulation, including the autodoping profile.

9.2.6 SUPREM III Models: Deposition, Oxidation, and Material Properties of Polysilicon Films

Polycrystalline silicon finds many applications in integrated circuit processing. It is used as a gate and interconnect material in MOS technology, as well as in emitter structures of bipolar transistors. Poly layers in direct contact with the silicon substrate are used as diffusion sources and buried contacts. High-value resistors are realized via the extremely high resistivity of lightly doped polysilicon. In each of these applications the poly structures are exposed to the full range of process technologies, such as oxidation, diffusion, and implantation. Polysilicon deposition and the properties of polysilicon following subsequent processing steps are described in Volume 1, chapter 6.

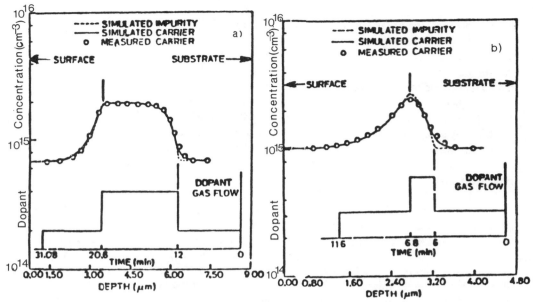

Fig. 9-16 Comparison of simulated and measured doping profiles in a lightly doped epitaxial film on a heavily doped substrate.[30] This paper was originally presented at the Spring 1989 Meeting of The Electrochemical Society, Inc. held in Los Angeles, CA.

To include a modeling capability for such layers in SUPREM III, it is necessary to simulate deposition, oxidation, diffusion, doping segregation across the multiple interfaces, and the resulting electrical resistivity of the poly layers. Such steps are simulated through an iterative approach that updates the properties of the polysilicon as time elapses.

As shown in Fig. 9-17, the properties of the polysilicon are first established after the film has been deposited.[15] The grain growth, dopant distribution, and dopant segregation from the grains to the grain boundaries are then calculated for an incremental time that is part of the cycle used to dope the poly (e.g., if a diffusion step is used, the dopant source, temperature, dopant level, and total doping-cycle time are specified). In addition, any dopant that is lost due to diffusion across any poly/Si-substrate interface is also calculated (e.g., if the doping profile at a buried contact is being simulated). The dopant distribution at the end of the incremental time is then fed back to the model so that the same parameters be calculated for the next time interval.

At the end of the total doping time, the final grain size, doping profile, and doping-segregation conditions are used to calculate the resistivity of the film. Figure 9-18 gives the SUPREM output for the structure and doping profiles in a device section after a 0.5-μm polysilicon film has been deposited and n-doped with POCl$_3$ to the solid solubility of phosphorus at 1000°C.[16] The details of the physical models used here are presented in references 91 and 16.

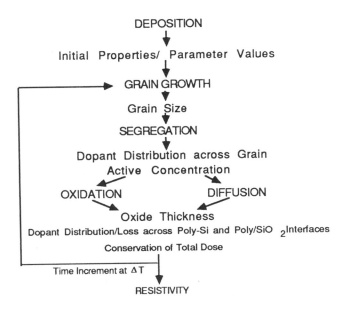

Fig. 9-17 Iterative algorithm used in SUPREM III to simulate polysilicon film characteristics.

The same iterative sequence is used for thermal-oxidation cycles of polysilicon.* In this case, dopant-segregation effects with respect to the growing oxide are also calculated, as is the polysilicon-oxide thickness. Once again, the film resistivity is calculated at the conclusion of the oxidation step.

9.2.7 Creating a SUPREM III Input File

Figure 9-19 illustrates the format used in SUPREM III input files. In this example, the gate region of the n-channel MOS device shown in Fig. 9-6[31] is simulated, and the threshold voltage of the device is calculated.

The first line (TITLE) assigns a title to the file and also serves to identify the analysis. The silicon substrate is defined next, with Lines 2 and 3 (INITIALIZE and +). In this example, the substrate is doped with boron to a concentration of 1×10^{15} cm^{-3}, and it has an <100> crystal orientation. The grid-definition is specified as part of the initialization statement (i.e., the grid spacing is 0.002 μm – note that all the input

---

* Note that the temperature, time, oxidizing ambient, and pressure are the inputs that are specified to the SUPREM III program for a desired thermal-oxidation step. In addition, the effects of high surface-impurity concentrations, oxidant partial pressure, and chlorine can also be simulated.

dimensions in SUPREM are in microns – with 100 grid points calculated, to a depth of 1 μm). The (+) indicates that line 3 is a continuation of the initialization statement. Lines 4-15 then specify the process steps to be simulated. Note that in these process-specification statements the following units are also implied:

time = minutes   concentration = cm^{-3}

thickness = μm   implant energy = keV

temperature = °C   implant dose = cm^{-2}.

COMMENT lines are inserted as desired to provide documentation.

The first process step is a threshold-adjust implant (Line 5). A boron implant with a dose of 2×10^{11} cm^{-2} and an energy of 50 keV is specified. Next, a gate oxide is grown (Line 7) in dry O_2 at 1100°C for 30 minutes. Polysilicon is deposited (Line 9) to a thickness of 0.5 μm, at a temperature of 620°C. Finally, the polysilicon is doped with phosphorus and oxidized (Lines 10 and 11, respectively). The polysilicon doping is done by diffusion at 950°C for 30 minutes to the solid solubility (ss.) of phosphorus. The polysilicon oxidation step is carried out at 1000°C for 20 minutes in dry O_2. The source /drain implant is done at 150 keV and is driven in at 1000°C for 30 minutes in dry O_2. Finally, a layer of phosphorus-doped SiO_2 is deposited.

The remaining lines deal with details of the simulation output. Line 17 specifies how the impurity profile is to be plotted, and Line 19 calls for the program to calculate the threshold voltage. The END card signals the end of the input file.

The doping profile simulated by the sequence after Line 11 is illustrated in Fig. 9-20a, which shows the various doping concentrations in each layer. It can be seen that the phosphorus doping from the gate has penetrated a significant distance into the gate oxide. The boron doping profile in the substrate is plotted. (Note that the simulated

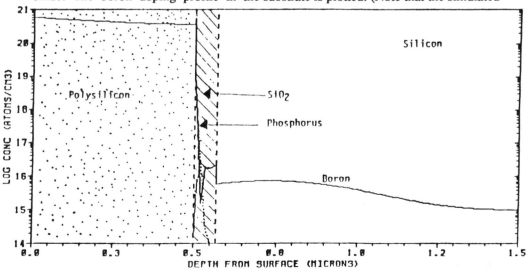

Fig. 9-18 Example of SUPREM III output of a structure containing a heavily doped polysilicon film as the topmost layer.

Fig. 9-19:   Example of a SUPREM III input file.

1 ...	TITLE	MOS Gate Threshold Analysis
2 ...	INITIALIZE	silicon <100> boron = 1e15 thickness = 1
3 ...	+	dx = .002 spaces = 100
4 ...	COMMENT	Implant boron to shift the threshold
5 ...	IMPLANT	boron dose = 2e11 energy = 50
6 ...	COMMENT	oxidize the gate in dryO2 and HCl
7 ...	DIFFUSION	temp = 1100 time = 30 dryO2 pressure = .2 hcl% = 5
8 ...	COMMENT	deposit, dope, and isolate the polysilicon gate
9 ...	DEPOSIT	polysilicon thickness = .5 temp = 620
10 ...	DIFFUSION	temp = 950 time = 30 ss. phosphorus
11 ...	DIFFUSION	temp = 1000 time = 20 dryO2
12	COMMENT	implant source/drain and deposit CVD oxide
13 ...	IMPLANT	arsenic dose = 1e15 energy = 100
14 ...	DIFFUSION	temp = 1000 time = 30 dryO2
15 ...	DEPOSIT	oxide thickness = 0.760 phosphorus concentration = 1e21
16 ...	COMMENT	Plot the impurity profile
17 ...	PLOT	chemical net title = "Net Impurity Profile"
18 ...	COMMENT	Calculate the threshold voltage
19 ...	ELECTRICAL	steps = 11 vth. elec layer = 1 extent = 2 file = elec
20 ...	END	

cross-section in Figs. 9-20b and c show the doping profiles with a layer of Al deposited on the device.)

Figure 9-20b shows the doping profile through the source/drain region and Fig. 9-20c though the field region (the process steps used to form the field oxide were simulated earlier). The heavily doped glass layer is apparent above the thermally grown SiO_2, which contains the As source/drain and boron channel-region implants.

9.2.8   PREDICT

A 1-D process simulator called *PRocess Estimator for Design of Integrated CircuiTs* (PREDICT) has been developed at the Duke University/Microelectronics Center of North Carolina (MCNC) especially for processes used in submicron devices.[32] It includes empirically developed models for low-thermal-budget processing, Ge^+ and Si^+

preamorphization implants, ultra-low-energy implants of B and As, thin oxides, diffusion under rapid thermal processing (RTP), and silicide contacts. Major new process variables include crystal damage produced during implantation and the annealing of this damage, point-defect injection during contacting, and implantation parameters associated with preamorphization. These variables are imbedded in the models used in the PREDICT process-simulation code. The several hundred models in PREDICT include verified models similar to those used in SUPREM III, as well as many new models generated as part of MCNC's shallow-junction submicron-CMOS program. PREDICT uses a decision-tree architecture to call up the appropriate model for the process-step conditions being simulated. One of the unique capabilities of this simulator is that it is able to model the process steps used in fabricating ultra-shallow junctions.

9.3 INTRODUCTION TO TWO-DIMENSIONAL PROCESS SIMULATORS

The output of one-dimensional process simulators is information about the structure of a device in one dimension, perpendicular to the device surface. As noted earlier, such 1-D simulations work well for MOS devices with channel lengths greater than ~2 μm and gate-oxide thicknesses of more than 50 nm, as well as in bipolar transistors with base widths larger than 100 nm. However, even for such "large" devices, between four and eight 1-D simulations must be performed for information about all of the various regions of a device to be obtained. (As an example, Fig. 9-21 shows a cross-section of a BiCMOS device and the eight 1-D profiles that would have to be simulated to cover

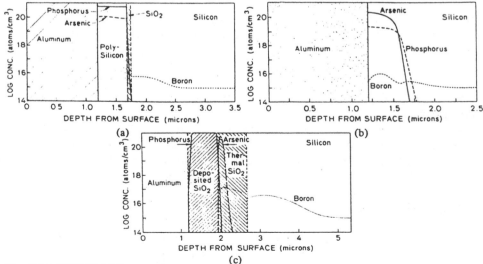

Fig. 9-20 SUPREM III simulations of an NMOS process. Impurity profiles through the (a) channel and (b) source/drain and (c) field regions.[16] (© 1983 IEEE).

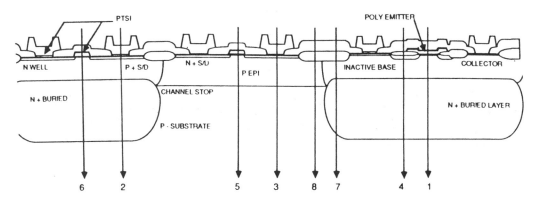

Fig. 9-21 Example of a cross section of BiCMOS structure showing that up to 8 1-D simulations would have to be performed to take into account all of the device regions.

all of the 1-D profiles of interest in the device.) Parametric data can also be extracted from the 1-D simulations to determine some of the device's electrical characteristics, such as sheet resistivity and threshold voltage.

There are many other device parameters, however, that depend on a knowledge of the 2-D distributions of dopants. Such parameters include effective-gate length, drain-bulk sidewall capacitance, and drain-bulk breakdown voltage. If 1-D doping profiles are taken and extended to simulate the 2-D profiles in "small" devices (e.g., those in which minimum gate lengths are less than 2 μm), in many cases the simulation accuracy will be significantly degraded. Since the lateral diffusion, implantation, and oxidation effects significantly change the doping profiles from those predicted by 1-D theory, the magnitude of such errors increases as the device sizes shrink to submicron dimensions. As a result, 2-D simulators are clearly required for such devices.

Figure 9-22 illustrates a cross-section of an NMOS inverter with enhancement-mode and depletion-mode MOS devices (fabricated up to the process step that forms the buried contacts; see chap. 5).[33] Several features of this structure that are pertinent to 2-D process modeling can be identified. First, the wafer topography is not planar, due to the LOCOS structure in the field regions. Second, the polysilicon is used as a gate electrode, an interconnect layer, and a means of forming a contact with the silicon substrate. Hence, there are junctions below the silicon surface in which lateral diffusion reflects the coupled effects of oxidation and diffusion. In addition, the deep n^+-junction results from a multilayer diffusion involving a polysilicon dopant source and the poly-bulk interface. Finally, tight design requirements exist for both the vertical and lateral depth of the junctions in and near the channel regions of both the enhancement-mode and depletion-mode devices.

As shown in Fig. 9-22, the encircled regions require 2-D models for process simulation, while 1-D simulations will suffice in the the regions identified with arrows. Figure 9-23 compares the type of information obtained from a 1-D process simulator (Fig. 9-23a, taken along section A-A of Fig. 9-23b) with that obtained from a 2-D

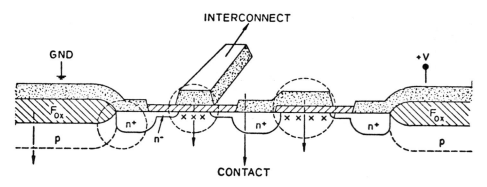

Fig. 9-22 An NMOS cross-section in which the circled regions contain areas where a 1-D simulator would not provide accurate doping profiles.[33] Reprinted with permission of Martinus Nijhoff Publishers.

process simulator (SUPREM IV) when the impurity profiles present in a lightly doped drain region of a MOSFET are simulated.[24]

One example of how 2-D diffusion affects device performance involves an MOS device fabricated with a lightly doped drain (LDD), as shown in Fig. 9-24.[34] In this device, the polysilicon gate is oxidized during the anneal and drive-in step performed following the source/drain implant. The shape of the oxide region grown in the lower corners of the polysilicon gate electrode (i.e., as a result of the lateral penetration and

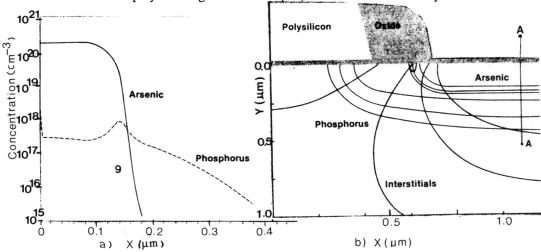

Fig. 9-23 Comparison of (a) process information obtained from the 1-D simulator SUPREM III and (b) SUPREM IV simulation of an LDD MOSFET structure.[24] Reprinted with permission of Solid State Technology.

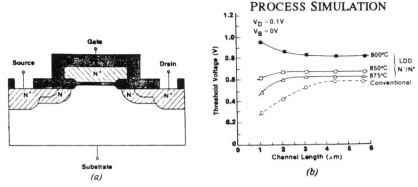

Fig. 9-24 (a) LDD NMOS transistor structure; (b) Device V_T versus channel length with the polysilicon oxidation temperature as a parameter.[24] Reprinted with permission of Solid State Technology.

thickness of the oxide there) will depend significantly on the oxidation temperature. However, the exact shape of this region will be hard to simulate in two dimensions, especially since oxidation will occur in both the heavily doped polysilicon and the lightly doped substrate. Yet, as can be seen in Fig. 9-24, the oxidation temperature has a dramatic effect on the threshold voltage of the device, especially as the channel length is reduced to less than 2 μm. This example illustrates how small process changes in submicron-device fabrication processes can lead to significant 2-D profile changes. Accurate 2-D process models are clearly needed to allow such effects to be studied through simulation-based analysis.

The final impetus for developing such models is that they are needed for device design. It is generally agreed that device-modeling programs do an excellent job of simulating device characteristics, provided that accurate information is available on the device structure (e.g., the doping profiles and layer thicknesses). However, if the doping profiles extended from 1-D simulations are used as the input to the 2-D device simulator, but do not accurately describe the 2-D doping profiles in a submicron device, the output will be incorrect. This is especially true at or near the submicron level, because the extent of point-defect kinetics (which give rise to such effects as oxidation-enhanced diffusion) can extend the diameter of the circled areas shown in Fig. 9-22 by factors of two or three. Thus, without the availability of accurate 2-D process simulators, the rate of progress in submicron device design for VLSI will be significantly reduced.

9.3.1 Classes of Two-Dimensional Process Simulators

At this time, the available 2-D simulators typically fall into two categories: (1) those that simulate 2-D diffusion, ion implantation, and oxidation phenomena (e.g., SUPRA and SUPREM IV), and (2) those that simulate the processes involving the *topography* of the wafer surface, such as deposition, etching, and photolithography processes (e.g., SAMPLE and DEPICT). Although there has been some effort to integrate both types into a single, comprehensive simulator (e.g., BICEPS and COMPOSITE), such simulators are not yet commercially available. Hence, our discussion will treat each of these simulator types separately.

9.4 TWO-DIMENSIONAL DOPING-PROFILE AND OXIDATION-PROCESS SIMULATORS

9.4.1 SUPRA (Stanford University PRocess Analysis Program)

In 1978, Lee developed a 2-D process simulator called BIRD, which utilized Green's function to solve the diffusion equation.[35] The Green's function approach, however, requires constant diffusivity and is therefore valid only under low-doping-concentration conditions. In 1981, Chin and Kump implemented SUPRA, a 2-D process simulator that also uses Green's function when low doping concentrations are encountered, but that handles high-doping-concentration cases with a numerical finite-difference method.[36] (The finite-difference method is described in references 33 and 37.)

SUPRA can handle deposition, etch, ion implantation, oxidation, and epitaxy, as well as diffusion process cycles. Impurity concentrations are calculated only within the thermal-oxide and substrate regions. Mask layers are used as barriers against ion implantation and surface diffusion. Users can specify removal of trapezoidal regions from any material layer to simulate etching. Simulation of LOCOS oxidation in SUPRA is able to account for the variation of oxide shape with silicon-nitride and pad-oxide thicknesses, as well as for oxidation temperature and pressure. In some versions the user can specify arbitrary initial surface and oxide-silicon interface shapes, rather than always having to start the simulation with a flat silicon surface.

SUPRA generates plots for high-resolution graphics devices. Users can choose to have these plots generated in the following forms:

- Cross-section of the device.

- 2-D contour plots of net, total, and individual impurity concentrations.

- 1-D plots of impurity concentrations along arbitrary straight-line paths through the device.

- In some versions of SUPRA, 3-D surface-projection plots.

The process models of SUPRA are divided into two categories: those containing closed forms of analytical equations, and those involving numerical solutions. The analytical models are used to simulate deposition, etch, oxide growth, ion implantation, and low-impurity-concentration diffusions. High-impurity-concentration diffusions are handled with the numerical solution.

9.4.1.1 SUPRA Ion-Implantation Models.   SUPRA generates a 2-D profile for ion-implantation processes by dividing the structure into many slices, each sufficiently narrow that the surface of each slice can be approximated as being flat. The dopant contribution from implantation within each segment is then calculated, and all contributions are superimposed to produce the total implantation profile. Depending on the model used, the vertical distribution function within each layer may be a Gaussian, binormal-Gaussian, or Pearson-IV distribution (with an optional exponential channeling tail). The profile in the lateral direction is a Gaussian profile, multiplied by a complementary-error function in the lateral direction. Implantation through an arbitrary mask can be expressed by a summation of the profiles through all mask segments. Because SUPRA takes into account the stopping powers of different materials, it can therefore simulate implantation into arbitrary combinations of material layers.

9.4.1.2 SUPRA Diffusion Models. If the impurity concentrations are smaller than the intrinsic carrier concentration, n_i, SUPRA solves the linear two-dimensional diffusion equation

$$\frac{\partial n}{\partial t} = D \left(\frac{\partial^2 n}{\partial x^2} + \frac{\partial^2 n}{\partial y^2} \right) \qquad (9\text{-}26)$$

to yield a closed form analytical solution. This gives accurate simulations of diffusion profiles, since at low impurity concentrations D is independent of concentration. Although the analytic solution assumes planar boundaries, structures with nonplanar boundaries are treated by applying the same segmentation and superposition method used for implantation.

In an inert-ambient drive-in diffusion (i.e., diffusion in which no oxidation occurs on the wafer surface), the solution method described above is appropriate. When diffusion is performed simultaneously with thermal oxidation (i.e., an oxidizing-ambient drive-in diffusion), the moving boundary and segregation effects must be accounted for. The one-dimensional diffusion profile following oxidation can be well simulated by adding a correction factor to the inert drive-in concentration. This factor basically subtracts out an appropriate fraction of dopant to account for segregation and out-diffusion from the oxide. This approach is used in SUPRA to model 2-D diffusion during oxidation under low doping-concentration conditions.

When the impurity concentrations are sufficiently large compared to the intrinsic electron concentration at the diffusion temperature, the diffusivity term in the diffusion equation is no longer constant. As a result, the 2-D diffusion equation becomes nonlinear and must be solved numerically. In addition, at high concentrations clustering effects can become significant and electric-field effects may play a role in the diffusion process. SUPRA uses a five-point, finite-difference method to carry out a numerical solution that can handle these complexities. The grid structure chosen for this numerical simulation is rectangular with nonuniform spacing in both spatial directions (Fig. 9-25). The diffusion models used to compute the values of diffusivity in the diffusion equation are the same as those used in SUPREM III. Note that the numerical solution of the diffusion equation in SUPRA has the limitations of being quite CPU intensive and of assuming stationary boundaries.

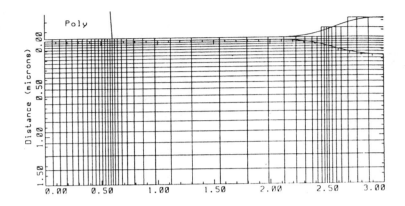

Fig. 9-25 Example of a 2-D rectangular grid used in 2-D process simulators which has non-uniform spacing of grid points in both dimensions.

9.4.1.3 SUPRA Oxidation Models.
SUPRA is designed to simulate both uniform and local oxidation of silicon. The one-dimensional thermal-oxidation processes in SUPRA are handled by the same models used in SUPREM III. Two-dimensional oxidation growth is simulated by SUPRA with the models described in section 9.4.3.2.

9.4.1.4 SUPRA Epitaxial Model.
Epitaxial growth is simulated in SUPRA by specifying the deposition of a silicon layer (with the doping concentration present at the surface of the epi layer), following this with a diffusion step (to approximate the diffusion of dopants from the heavily doped substrate into the lightly doped epi layer).

9.4.1.5 SUPRA Input File.
Figure 9-26 illustrates the format used in SUPRA input files showing the process steps used to produce a lightly doped drain NMOS structure in a *p*-well. The two-dimensional plot of this structure, showing the LDD region, is given in Fig. 9-27.

The input statements specify parameters for each process step, much as is done on a process-run sheet. SUPRA provides default values for all physical coefficients through a data file. Users can, however, modify these coefficients.

In the TMA versions, the parameter names in the input languages of SUPRA and SUPREM-3 are the same, as are the coefficient values for models common to both programs. Thus, users who have gained experience with one program will find the other easy to use. In addition, SUPRA uses the TMA software interface to generate input files to other TMA simulation programs, including DEPICT, CANDE (with which TMA, Inc. has replaced GEMINI), and PISCES.

9.4.2 SUPREM IV

SUPREM IV is another advanced 2-D process simulator that was also developed at
Stanford. First introduced in July 1986,[38] this simulator is quite different from SUPRA

Fig. 9-26 Example of a SUPRA input file.

1 ...	STRUCTURE	<100> concentration phosphorus = 8e14
	+	depth = 3 width = 5 height = 1.5
2 ...	X. GRID	h1 = .2 h2 = .1 width = 5 n.spaces = 25
3 ...	Y. GRID	h1 = .05 h2 = .1 depth = 3 n.spaces = 75
4 ...	END	
5 ...	COMMENT	Form p-well
6 ...	ANALYTIC	
7 ...	DEPOSIT	photoresist thickness = 2
8 ...	ETCH	photoresist x4 = 3
9 ...	IMPLANT	boron dose = 2e12 energy = 100
10 ...	ETCH	photoresist
11 ...	DIFFUSION	time = 175 temp = 1150 dryO2 pressure = .03
12 ...	COMMENT	Implant and oxidize field region
13 ...	DEPOSIT	nitride thickness = 0.07
14 ...	DEPOSIT	photoresist thickness = 2
15 ...	ETCH	photoresist x1 = 4.3
16 ...	ETCH	nitride x1 = 4.3
17 ...	IMPLANT	boron dose = 5e12 energy = 50
18 ...	ETCH	photoresist
19 ...	DIFFUSION	time = 120 temp= 800 steam t.rate = 3
20 ...	ETCH	nitride
21 ...	COMMENT	Define the polysilicon gate and sidewall spacer
22 ...	DEPOSIT	polysilicon thickness = .5
23 ...	ETCH	polysilicon x1 = 0.52 x2 = 0.65
24 ...	IMPLANT	arsenic dose = 1e12 energy = 100
25 ...	DEPOSIT	oxide thickness = .5
26 ...	ETCH	oxide x1 = 1.05
27 ...	ETCH	oxide x4 = 0.52
28 ...	COMMENT	Form n+ source/drain region and remove spacer
29 ...	NUMERICAL	
30 ...	IMPLANT	phosphorus dose = 1e15 energy = 100
31 ...	PRINT	enable summary
32 ...	DIFFUSION	time = 15 temp = 900 print
33 ...	ETCH	oxide

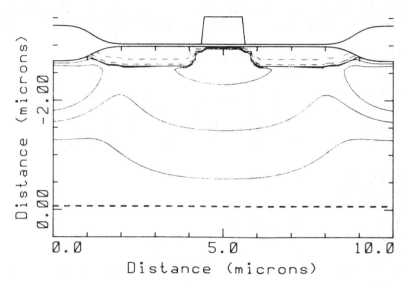

Fig. 9-27 Two-dimensional plot of the device structure for a lightly doped drain MOS device simulated by SUPRA. Courtesy of TMA, Inc.

in that it approaches the simulation of 2-D oxidation and diffusion from an entirely different perspective. The SUPREM IV diffusion and oxidation models are based on the process physics rather than on phenomenological data obtained from measurements.

9.4.2.1 SUPREM IV Models of Diffusion. SUPREM IV was developed primarily to solve coupled oxidation and diffusion problems on the same grid. As a result, grid is used in both the oxide and the silicon. As the oxide grows, the grid is deformed around the interface.

Two-dimensional diffusion is simulated by calculating the local diffusion coefficients based on the concentration of point defects and impurities at that location.[24] Advanced physically based diffusion models are formulated around the belief that the fundamental mechanism of dopant diffusion is involves interactions with point defects.[39] Therefore, if the local point-defect concentrations can be accounted for, a more accurate value of the local diffusivity can be derived. Because these concentrations may be geometry dependent it is important to be able to simulate the movement of the oxide interface as the diffusion proceeds.

In order to accurately model diffusion effects in two dimensions one should know

- the characteristics of the point-defect sources,
- how far the point defects diffuse in the bulk,
- how fast the point defects recombine at nearby surfaces and in the bulk, and
- whether the dopant prefers to move by means of an interstitial or a vacancy.

When these factors are known, it is possible to determine the 2-D distribution of the point defects by solving the diffusion equation for interstitials and vacancies. The dopant diffusivity at each point is then coupled to the local point-defect concentration via the appropriate relationships. Experimental work to determine more data on each of these factors is continuing.

9.4.2.2 SUPREM IV Models of Oxidation.

SUPREM IV can simulate uniform (1-D) as well as 2-D oxidation processes. Empirical models are available for simulating LOCOS structures, as in SUPRA. However, SUPREM IV also allows the user to specify the simulation of 2-D oxidation by means of physically based models. Such models allow simulation at all temperatures of interest, including those temperatures at which the growth of the oxide must be treated according to a nonviscous flow model (e.g., <960°C). The models allow simulation of LOCOS processes and the growth of oxide on the sharp 90° and 270° corners of trenches. More details on these oxidation models will be given in section 9.4.3.

9.4.2.3 SUPREM IV Models of Ion Implantation, Epitaxy, Deposition, and Etching.

SUPREM IV incorporates the same types of ion implantation models as SUPRA and handles the epitaxial process in the same manner. Deposition processes are simulated by assuming that films are deposited in a conformal manner. Etching is simulated by allowing the user to specify the removal of arbitrary regions from any type of layer.

9.4.2.4 SUPREM IV Input-File Format.

An example of the format used in a SUPREM IV input file is shown in Fig. 9-28,[40] which illustrates how process steps can be specified to simulate a LOCOS structure (see also section 9.4.3). The first eleven lines of the file define the structure and the initial grid for the simulation (including the comment lines *#Grid Definition* and *#Structure Definition*). In this example, the six *line* statements create a grid from x_l = -0.75 μm to x_r = 0.75 μm, and from y_s = 0 to y_b = 0.5 μm. The three lines following the *#pad oxide and nitride* comment line specify the deposition and etch parameters for pad oxide and nitride mask. In this case, the nitride mask edge will be at 0.0 μm and will be etched away from x = -0.75 μm to x = 0 μm. The *plot. 2d grid* line generates a plot of the structure to this point, including the grid (Fig. 9-29). Following the *#Numerical Methods* comment line, the *symb* line defines the numerical methods that will be used by SUPREM IV to simulate the thermal oxide. The *diffuse elas* line specifies that the oxide is to be grown via a wet oxidation process for 100 minutes at 1025°C. Figure 9-29b shows the results of this simulation.

9.4.2.5 Comparison of SUPRA and SUPREM IV for 2-D Process Simulation.

Because it relies on analytical solutions wherever possible, SUPRA is relatively fast. Such solutions can be used because they are derived from the phenomenological models that form the basis of the simulations provided. Numerical solutions are used only when necessary (e.g., for simulation of diffusion under high-

```
# Grid Definition

line x loc=-0.75 spacing=0.1 tag=1

line x loc=0 spacing=0.05 tag=m

line x loc=0.75 spacing=0.1 tag=r


line y loc=0 spac=0.05 tag=s

line y loc=0.3 spac=0.05

line y loc=0.5 tag=b


#Structure Definition

region silicon xlo=l xhi=r ylo=s yhi=b

bound exposed xlo=l xhi=r yhi=s ylo=s

initial ori=100


#pad oxide and nitrid

deposit oxide thick=0.02

deposit nitride thick=0.03

etch nitr left pos=0

plot.2d grid


#Numerical Methods

symb lu min.fill symm=f


diffuse elas weto2 time=100 temp=1025 init=1

end
```

Fig. 9-28 SUPREM IV input file for a LOCOS simulation.

impurity-concentration conditions and when electric-field interactions occur during diffusion).

SUPREM IV, on the other hand, emphasizes the use of physically based models for diffusion and oxidation. Because these simulations require numerical solutions, SUPREM IV is slower than SUPRA (by a factor of between 10 and 100). For those cases in which the physics is well understood, however, SUPREM IV provides more accurate 2-D simulations. Intensive research work is being conducted to provide an adequate physical understanding for a wider range of process conditions.

In summary, SUPRA and SUPREM IV are complementary. SUPRA can be used to provide rapid 2-D simulation of device profiles when accurate simulations are not required. It also be used to provide a better simulation than SUPREM IV in cases where the physics of the process is not yet well understood and where the phenomenological models in SUPRA are known to be accurate. On the other hand, SUPREM IV will provide more accurate 2-D impurity profiles in cases where the physics of the process *is* well understood, albeit at the price of much longer computation times.

9.4.3 Two-Dimensional Simulation of Thermal Oxidation

Simulation of oxidation, as described in section 9.2.4, provides accurate predictions of oxide growth under virtually all practical 1-D oxidation conditions. In VLSI technology, however, thermal oxidation can be carried out on such nonplanar surfaces as trench structures and polysilicon lines. Even silicon surfaces that are initially planar can lose their planarity under such oxidation processes as LOCOS. The shapes of the oxide grown on non-planar surfaces are significantly different from those observed on flat surfaces. For example, a "bird's beak" is formed beneath the nitride-mask edge in LOCOS, and the oxide is thinner at the inside and outside corners of trench structures (see chap. 2).

Because of the nonplanar geometries and moving boundaries, it is much more difficult to simulate 2-D than 1-D oxidation processes. That is, the formation of the new oxide at the Si surface involves volume expansion: the newly formed oxide expands and pushes out the old oxide, which rearranges itself through viscous flow.

Both empirical and physically based models have been developed to generate such simulations. The physical models solve the complete set of differential equations governing both the oxide growth and the simultaneous viscous flow. Although such models can give accurate results, they are complex and require long computation times. The empirical models – which are based on relatively simpler analytical expressions – can be evaluated more rapidly, by means of computers.

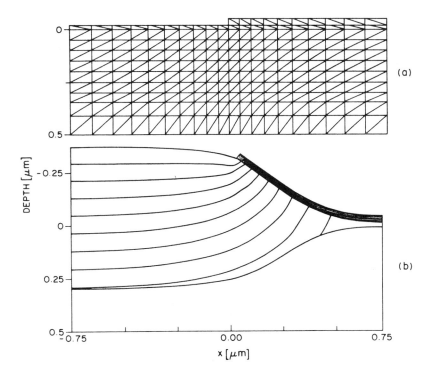

Fig. 9-29 LOCOS simulation with SUPREM IV. (a) Initial grid. (b) Final oxide shape.[40]

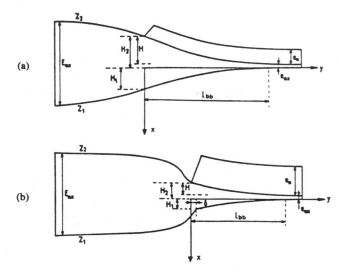

Fig. 9-30 The two shapes of the bird's beak with characteristic lengths (E_{ox}, L_{bb}, and H) and processing parameters (ε_{ox} and ε_n). (a) Shape 1 and (b) shape 2.[41] (© 1987 IEEE).

9.4.3.1 Empirical Models of 2-D Thermal Oxidation.
Empirical models have been implemented in SUPRA and SUPREM IV to simulate the "bird's beak" of LOCOS structures. (SUPREM IV also contains physically based 2-D oxidation models that can be applied to trench-oxidation and LOCOS processes.) The 2-D model in SUPRA for local oxidation uses an empirically determined shape function, based on such input parameters as pad-oxide thickness, silicon-nitride thickness, oxidation temperature, and pressure. That is, the contour of the "bird's beak" is approximated with a complementary-error function whose parameters are defined by the particular LOCOS process conditions. The empirical model in Stanford's SUPREM IV is essentially the same as that used in SUPRA.

Another empirical model that also treats semirecessed LOCOS oxide shapes was developed by Guillemot et al. and has been incorporated in the TSUPREM-4 simulator, available from TMA, Inc.[41] This model again uses a simple parametric relationship that is based on the analysis of experimental data. It classifies the shapes of the bird's beaks that occur in semirecessed LOCOS into two groups, depending on the stress exerted by the nitride masks (Fig. 9-30). If the nitride and pad-oxide layers are both relatively thin, low mask stress exists and smooth birds-beak shapes occur (Case 1 in Fig. 9-30). For such shapes, two analytical, complementary-error functions are used to fit the oxide contours: one for the oxide-silicon interface, and the other for the oxide-nitride interface. In Case 2, strong mask stresses cause the oxide to become "pinched" beneath the nitride edge. Therefore, two functions are used to simulate the bird's beak contours at each interface: one on the left side of the pinch and the other on the right side. Figure 9-31 shows that in each case the bird's beak shape predicted by the simulation closely duplicates the shapes observed in SEM photos.

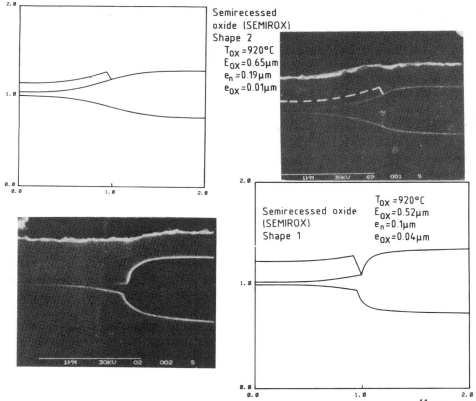

Fig. 9-31 Comparison between experimental and simulated field oxide shapes.[41] (© 1987 IEEE).

9.4.3.2 Physically-Based Models of 2-D Thermal Oxidation.

As mentioned above, the physically based models of 2-D oxidation solve the equations of the Deal-Grove linear-parabolic model (or the equivalent steady-state diffusion model) and at the same time account for the volumetric expansion due to oxide formation and viscous flow. The viscous-flow phenomenon was first quantitatively studied by Eer Nisse, using silicon wafers that had a thermal-oxide film on only one side.[46] Since the thermal-expansion coefficient of SiO_2 differs from that of silicon, the wafer was bowed when the oxidation temperature was less than 960°C. If the temperature was increased above 960°C, however, the wafers remained flat, which suggests that the stress due to thermal mismatch was relieved by the viscous flow of the oxide.

The first physically based model of 2-D oxidation that became available in a public-domain simulator was designed to handle LOCOS processes and was incorporated in the *Stanford Oxidation Analysis Program* (SOAP).[42] It assumed that the growing SiO_2 films behave like viscous fluids that flow when the thermal-oxidation temperature exceeds 960°C, with the viscosity of the oxide determined by the curvature of the wafers.

The oxidation process is modeled by assuming that two mechanisms occur: (1) oxidant diffuses through the SiO_2 to the SiO_2/Si interface, and (2) the oxide flows due to the volume expansion and associated stress buildup. The oxidant diffusion in the SOAP model is described by a generalized diffusion equation (rather than by the flux-conservation equations of the 1-D model given in Volume 1, chapter 7):

$$D \nabla^2 C = \frac{\partial C}{\partial t} \cong 0 \qquad (9 - 27)$$

The boundary conditions for this equation for LOCOS structures are nonhomogeneous, and hence cause the oxidant distribution to be different from place to place along the silicon/oxide interface. The volume-expansion rate therefore also varies with position. The flow of the oxide is described by a simplified Navier-Stokes equation as

$$\mu \nabla^2 V = \nabla P \qquad (9 - 28)$$

where μ, V, and P are the viscosity, the velocity, and the pressure of the oxide, respectively.

The goal of the 2-D oxidation-process simulation is to determine the position of the oxide boundaries at the conclusion of the oxide-growth cycle. Since these boundaries are continuously moving, it is very difficult to obtain this information from the above equations. The SOAP program incorporates a boundary-value numerical technique to solve integral form of Eqs. 9-27 and 9-28 by using Green's function. Details of this method are given in reference 42.

Figure 9-32 shows the simulation of two semirecessed LOCOS field oxides using SOAP. In Fig. 9-32a, the nitride mask is 25 nm thick. Since the stress is small, the oxide moves in a direction normal to the interface throughout its entire growth. In Fig. 9-32b the nitride mask is much thicker (175 nm), and it thus prevents the oxide from moving normal to the interface by forcing it to move laterally during growth. The compressive stress due to the thicker nitride film reduces oxide growth by making the SiO_2 flow toward the open surface.

At temperatures below 960°C, oxides no longer exhibit viscous behavior, but instead deform according to the laws of elasticity. Another model must therefore be used to treat such conditions. Such a model was developed by Kao et al. and has been incorporated into SUPREM IV.[43] This model takes into account the stresses that build up in the growing oxide film and how such stresses impact the growth rate. Experimental evidence to support the model shows that these stresses cause the oxide to grow most slowly at concave corners. The more general nature of this model allows it to simulate the growth of oxide in trenches as well as in LOCOS processes. Figure 9-33 shows a SUPREM IV simulation of an oxide grown in a trench in dry oxygen for 240 minutes at 950°C. Thinning of the oxide in the lower corners is predicted for this structure.

CREEP is the latest physically based 2-D oxidation simulator developed to incorporate the stress effects associated with the growth of thermal SiO_2.[44,45] This simulator incorporates a modified form of Kao's 2-D thermal-oxidation model. The major change from the earlier model is the use of a shear-stress-dependent viscosity

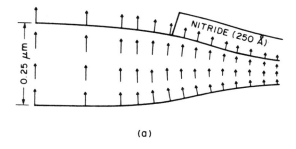

(a)

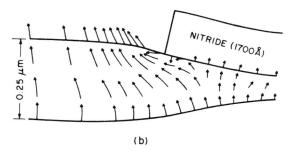

(b)

Fig. 9-32 Two-dimensional oxide growth simulated using SOAP with (a) a thin nitride (25 nm) masking layer, and (b) a thicker (170 nm) nitride masking layer.[42] (© 1983 IEEE).

model in lieu of the hydrostatic-stress-dependent model. CREEP simulations of 2-D oxide thicknesses on both convex and concave cylindrical silicon structures over a wide range of temperatures and curvature conditions show better agreement with experimental measurements than do simulations based on previous models. The model used indicates that the effect of stress during oxidation on the linear growth-rate constant dominates the two-dimensional profile-shape evolution.

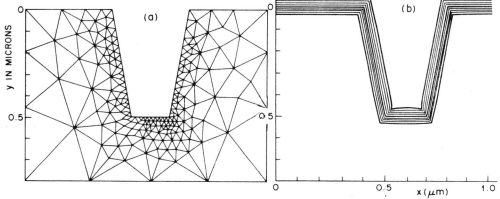

Fig. 9-33 Oxidation of a trench structure using SUPREM IV. (a) Initial grid. (b) Final oxide profile.

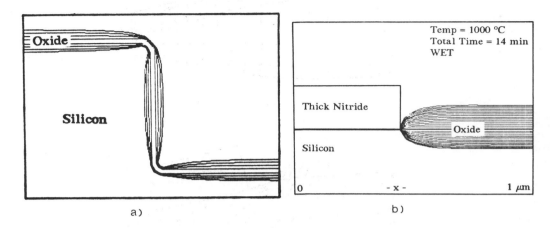

a)   b)

Fig. 9-34 (a) CREEP simulation of an oxide grown on a vertical step in wet oxygen for 3 minutes at 1000°C. (b) CREEP simulation of a SILO process using a 2-nm-thick pad oxide, a thick nitride mask, and an oxide-growth temperature of 1000°C.[44] (© 1986 IEEE).

Figure 9-34 a shows the CREEP simulation of an oxide grown on a vertical step in wet oxygen for 3 minutes at 1000°C. This simulation clearly predicts the thinning of the oxide in the lower and upper corners of the step, as is observed in experimentally grown oxides on trench structures. Figure 9-34b shows a CREEP simulation of a SILO process using a 2-nm-thick pad oxide, a thick nitride mask, and an oxide-growth temperature of 1000°C. CREEP is also capable of calculating the stress along the SiO_2/Si interface as a function of time and to simulate reflow of deposited oxides.

9.5 TWO-DIMENSIONAL TOPOGRAPHY SIMULATORS

Topography growth and process latitude are important issues in VLSI fabrication. New materials and processing techniques are being explored in the attempt to gain greater latitude in the processing of small line widths over topographical features. These novel approaches improve the technology, but they also make process characterization more difficult because of the increased number of process options and parameters that must be considered. Simulation has been pursued as a tool to help designers characterize such topgraphy related processes.

A number of 2-D simulators have been designed especially for process steps in which the wafer surface topography is altered. The most important of these is SAMPLE; we begin the discussion here with a detailed examination of its capabilities. Other topography simulators will be described in later sections.

9.6   SAMPLE (SIMULATION AND MODELING OF PROFILES IN LITHOGRAPHY AND ETCHING)

The SAMPLE program was developed at UC Berkeley to simulate two aspects of IC fabrication: (1) microlithography, and (2) other processes that cause topographical modifications of the wafer surface.[47,48,49] Major applications of SAMPLE in *micro-lithography* involve simulation of the aerial images produced by optical lithography and modeling of exposure and development in resist materials. For *topographical process modeling*, SAMPLE allows the simulation of plasma-etching and deposition processes (by evaporation, sputtering, and lift-off). Arbitrary sequences of these processes can be simulated to allow study of such issues as process latitude and topography control. SAMPLE can also be used on devices that contain multiple layers of many different materials. The outputs are two-dimensional cross-sections of material layers and aerial images of mask features produced at the wafer surface.

Many of the processes simulated by SAMPLE (including resist development, dry etching, and even deposition) can, to first order, be considered as surface-reaction-rate-limited processes. Thus, simulation of such processes can be based on a generalized surface-etching algorithm. The string-advance algorithm commonly used for two-dimensional problems is implemented in SAMPLE to simulate these processes.

SAMPLE consists of approximately 15,000 lines of FORTRAN (F77) and it typically needs a few minutes of VAX11/780 time per process step. It can be run on a microcomputer as small as an IBM-PC-XT with a floating-point chip, a professional quality compiler, and GKS graphics package.

Our discussion on the use of SAMPLE is divided into two parts: the first deals with lithography simulation, and the second with etching and deposition simulation. More information on the actual details of using SAMPLE can be found in reference 50.

9.6.1 Simulating Optical-Lithography Processes with SAMPLE

Microelectronic fabrication relies on the use of lithography to create the very small features and patterns that define the device structures of integrated circuits (see Vol. 1, chaps. 12 and 13). Optical-projection lithography, in which the image of an object (the mask) is projected onto the wafer through a complex, diffraction-limited optical system, is still the dominant technology for creating these patterns. Positive photoresist is used almost exclusively for feature sizes below about 2 μm. The interaction of the incident light and the photoresist material is responsible for the patterns created in the resist films.

We describe here how SAMPLE is used to simulate lithography processes implemented with optical-projection systems and positive photoresists[47] (although it can also be used to simulate electron-beam,[51] and x-ray and ion-beam lithography[52]).

The goal of lithography simulation is to obtain a *simulated profile of a specific resist film at a feature edge, produced after the resist has been exposed and developed through a particular process* (Fig. 9-35). Such a profile contains the following lithographic

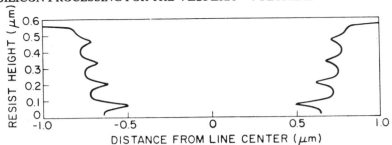

Fig. 9-35 Simulated edge profile of a nominal 1-μm line in AZ 1350 photoresist developed for 85 s in 1:1 AZ developer water.[59] (© 1983 IEEE).

information: the critical dimension of the feature formed by the resist (e.g., line width or space width), the maximum resist thickness, the slope of the resist edge, and any standing-wave effects that may impact the profile.

In some cases, the resist profile will vary with position along a line. One common example involves the crossing of a step by a line of resist. The variation of the resist line width can be caused by nonvertical resist profiles (which make the line width at the substrate surface wider at the bottom of the step; see Fig. 9-36a), as well as by *line notching*; see Fig. 9-36b).[53] Under these circumstances, it may also be necessary to simulate a series of profiles at various points along the line (e.g., as a function of the resist thickness at appropriate locations in proximity to the step). Line width data from these individual profiles can be combined to yield a two-dimensional *plan view* of the resist feature as it crosses the step (Fig. 9-37).

To obtain a resist-film profile, the following sequence of phenomena must be simulated:

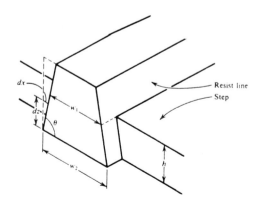

Fig. 9-36 Geometrical effects on the linewidth of a resist line crossing a step.[53] Reprinted with permission of Solid State Technology.

1. The intensity of the (projected) incident light, I, on the surface of the resist film must first be calculated. The distribution of this intensity on the surface of the resist is known as the *aerial image* and is a function of the optical system used to project the image on to the surface. For such features as uniformly wide lines or spaces, only a one-dimensional plot of intensity near the feature edge is needed. However, for such two-dimensional features as contact holes or defects in a uniformly wide line, a 2-D aerial image must be calculated (often plotted in the form of *contours* of equal intensity; see Fig. 9-38).[54] Note that I at each point of the aerial image is assumed to be constant with time for the duration of the resist exposure.

2. The intensity of the light as a function of depth in the resist must then be calculated. The overall intensity decreases with depth due to light absorption by the resist. Next, reflections from the substrate cause a vertical standing-wave pattern in the resist, which modulates the intensity versus depth. Finally, since the resist undergoes chemical changes during the exposure process, the intensity versus depth also varies as a function of time, and so must be calculated at each instant of time for each wavelength present.

3. Next, the normalized concentration of inhibitor (M) in the positive resist as a function of position and time must be calculated. Note that these values must be found for a two-dimensional matrix of points in the resist. The final value of M at each point after the exposure step will be one of the parameters determining the rate at which the resist undergoes development. The value of M versus depth is found using Dill's exposure-bleaching model, in which the resist-exposure parameters A, B, and C must be entered by the user.

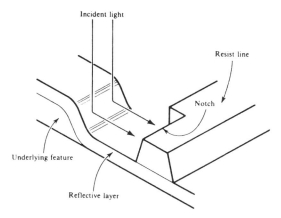

Fig. 9-37 Line notching due to reflection of a nearby topological feature.[53] Reprinted with permission of Solid State Technology.

4. The rate of development at each point in the resist film (R) is calculated next. This rate will depend on such parameters as the local concentration of inhibitor, the prebake process, the developer type and temperature, and the intrinsic resist properties. In some resists a surface-retardation effect occurs, and this effect should also be modeled by the simulation.

5. The development process itself is simulated by using a string advance algorithm (which will be described in more detail in section 9.6.2). A string of nodes, initially parallel to the wafer surface, is allowed to advance according to the local etch rate at each node (as determined from step **4**). The structure of the resist after a specific time of development yields the line-edge profile.

The SAMPLE process simulator has three basic subprograms that are used to model these various aspects of the microlithography process. The *optical-imaging subprogram* simulates the aerial image at the resist surface. The *resist-exposure subprogram* calculates the intensity versus depth in the resist film, as well as the inhibitor concentration as a function of position in the resist after exposure. The *resist-development subprogram* determines R and the final resist profile.

9.6.1.1 Optical-Imaging Subprogram.

The model in the optical imaging subprogram is based on the Hopkins theory of imaging with partially coherent light.[55,56] This image-intensity pattern is calculated by means of the transmission cross-coefficient weighting the Fourier transform of the object transmittance. The cross-coefficient is determined from the pupil function, given the degree of partial coherence, σ, and defocus, μ (see Vol. 1, chap. 13). In the SAMPLE implementation, a combination of analytical and numerical integration is used to evaluate the convolution integrals.[57] Periodic patterns of lines and spaces or isolated lines may be effectively simulated. The user specifies the image pattern (line and space widths), wavelength, numerical aperture and coherence, focus error, and imaging window.

The light intensity of an ideal printed line would be zero within the designed line area and large elsewhere. In real, diffraction-limited optical systems, the printed line is not perfect, even in the focus plane. That is, the intensity falloff near the image edge is more gradual than it would be in the ideal printed line. SAMPLE calculates this falloff for partially coherent imaging systems. Figure 9-39a shows a SAMPLE simulation of the intensity of light as a function of position from the mask edge and focus error in a pattern consisting of 6-μm-wide lines and 2-μm-wide spaces.[56]

The optical-imaging subprogram is also capable of simulating the aerial image of the 2-D patterns used to define such features as rectangular contact holes. Two-dimensional effects at corners in lines and proximity effects between features can impose fundamental design-rule limitations. Figure 9-38 shows calculated diffraction images of a 1 x 1.2-μm contact hole. Each point in the figure is assigned a number that indicates the intensity level; the level 0 indicates all intensities equal to or larger than unity (which is the intensity level required to completely expose a large uniform transparent area at a predetermined location).

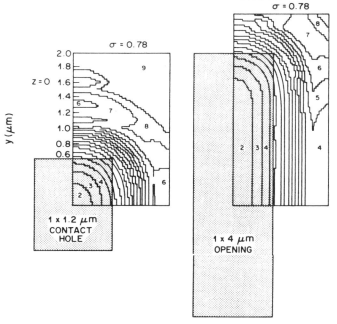

Fig. 9-38 Aerial image of a 1x1.2-μm contact hole and a 1x4-μm rectangular opening. The imaging lens has a numerical aperture of 0.32. The wavelength of the light is 4045 Å.[54]

The optical image of 2-D objects is important in defect inspection as well as in lithography. The rapid falloff in signal intensity with feature size is a key concern in finding small defects. A principal concern is the tendency of a defect in the vicinity of a feature to cause a change in the critical dimension of that feature. Figure 9-40 shows a SAMPLE simulation of how an intrusion into an opaque line tends to reduce the local line width.[58] The intensity contours are shown for a 0.5-μm-square intrusion into a 1.25-μm line at a wavelength of 436 nm, an NA of 0.28, and a partial coherence factor of 0.7. Judging from the optical image, the change in the resist line width would be about 0.25 μm. This would produce a 20% line-width variation, which might be unacceptable if it occurred in a critical device region.

9.6.1.2 Resist-Exposure Subprogram.

Positive photoresist consists of a diazo-type sensitizer in a novolac base resin. Exposure to UV light destroys the sensitizer (which is a photoactive compound, or PAC), enhancing the solubility rate of the novolac resin in basic developers. The exposure process is accompanied by an optical bleaching due to PAC destruction. The *exposure-bleaching model* proposed by Dill et al. is used to simulate this process.[59] The key to obtaining an accurate simulation of the resist exposure with this model is to select accurate values for the exposure parameters (Table 9-4), which are comprised of the *exposure constants A* (the bleachable absorption coefficient), *B* (the nonbleachable absorption coefficient), and *C* (the bleach rate).

The absorption constant, α, of the resist is position dependent and is also dependent on M, according to

$$\alpha = A\ M\ (x,\ y,\ z,\ t)\ +\ B. \qquad (9\text{-}29)$$

The constant C relates the decomposition of the photoactive compound to the local light intensity by

$$\frac{dM}{dt} = -I\ (x,y,z,t)\ M\ (x,y,z,t)\ C \qquad (9\text{-}30)$$

Values of these constants have been measured for specific resists by a variety of workers under various exposure and development conditions.[60,61] The exposure constants for a particular resist are entered by the user via the resist model lines in the SAMPLE program *(resmodel)*.

Table 9-4   Typical Resist-Exposure Parameters

	A	B	C
Shipley 1400-27 (g-line)	$0.601\ \mu m^{-1}$	$0.066\ \mu m^{-1}$	$0.013\ cm^2/mJ$
Shipley 1400-27D1 (i-line)	0.91	0.357	0.019
KTI 820 (g-line)	0.563	0.052	0.130

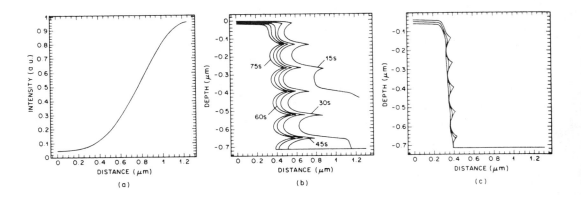

Fig. 9-39 Output of the SAMPLE input file given in Fig. 9-46. (a) Intensity distribution around the mask edge; (b) Development contours; (c) profiles after final descum. After ref. 50.

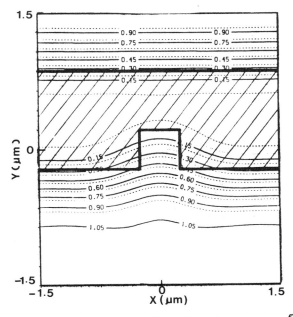

Fig. 9-40 Optical aerial image of an opaque line with a clear intrusion.[58] Reprinted with permission of Solid State Technology.

Dill's model is combined with the light-intensity values calculated from the optical-image and thin-film reflection models to establish the normalized value of M at specific locations in the resist, as a function of exposure process conditions and resist-exposure parameters. The exposure conditions include the optical constants of the substrate and of any overlying layers, as well as the thicknesses of these layers.

These models are used to calculate the intensity of the light as a function of position in the resist layer. This intensity depends on several factors, as follows:

- intensity of the aerial image at the resist surface

- depth below the resist surface (because the intensity decreases with depth due to light absorption by the resist)

- vertical standing-wave effects (caused by reflections from the substrate, which also modulate the intensity versus depth)

- time (as a result of the chemical changes that the resist undergoes during the exposure process).

Consequently, the intensity as a function of position must be calculated at each instant of time for each wavelength.

In SAMPLE, the vertical intensity of the light is first computed at each time step. The resist layer is subdivided into sublayers thin enough to be treated as though they had

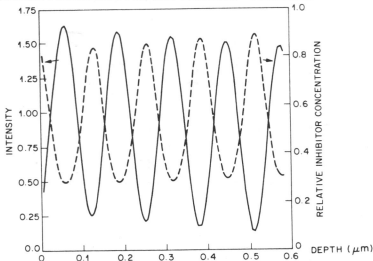

Fig. 9-41 Simulated intensity and inhibitor profile within 0.584 μm AZ1350 photoresist on 600 Å of SiO_2 on Si. Solid line: intensity of exposing light; dashed line: inhibitor concentration after exposure to 57 mJ/cm^2 at λ = 4358 Å.[59] (© 1983 IEEE).

isotropic properties (e.g., <0.03λ). The resist is bleached at each sublayer according to the weighted intensity and sensitivity summed over the wavelengths used. Time is advanced, and the vertical-intensity profiles are recomputed. Figure 9-41 shows a SAMPLE simulation of the intensity of the exposing light versus depth in a resist film on 60 nm of SiO_2 on silicon (normalized to the intensity present at the surface).[50]

The state of the resist is described by the normalized inhibitor concentration, M, at each time step, yielding a vertical profile of the inhibitor concentration (M [z]) (also shown in Fig. 9-41). At appropriate time intervals M (z) is saved, and the exposure is continued until some total energy dose in each sublayer E_{max} is achieved. The result of the exposure simulator is a two-dimensional array describing M (z, E) as a function of depth and energy.

In a separate operation, the horizontal intensity profile (from the aerial image) and the specified exposure (dose) are combined with the M (z, E) array to produce a new array, M (x, z), that describes the state of the resist at each point across the image, x, and at each depth, z.

Post exposure baking prior to development is known to smooth the standing wave pattern by redistributing the inhibitor concentration through diffusion (see Vol. 1, chap. 12). The effect of such bakes on the resist is modeled by modifying the M (x, z) array (that is, it is modeled by means of a simple diffusion in which the standard deviation is specified).

9.6.1.3 Resist-Development Subprogram.
Simulation of the resist-development process is based on the *development-etching model* introduced by Dill in the early 1970s.[62] This model assumes that in positive photoresists the development

process is a simple surface-etching phenomenon, with the resist assumed to dissolve in the developed solution at a rate that is a function of only the local inhibitor concentration. Therefore, the etch rate at each point in the resist (R) is a function of the inhibitor concentration (M) remaining after exposure.

The two-dimensional matrix of inhibitor concentration values M (x, z) is calculated by means of the resist exposure subprogram. A polynomial analytical relationship between the concentration and the etch rate (based on a least-squares fit to experimental data) was used in the early versions of SAMPLE:

$$R(M) = \exp(E_1 + E_2M + E_3M^2) \qquad (9-31)$$

However, a more accurate analytical expression has since been derived and is incorporated in the later versions:

$$R(M) = \frac{1}{\left\{\dfrac{1 - M \exp[R_s(1-M)]}{R_1}\right\} + \left\{\dfrac{M \exp[-R_s(1-M)]}{R_2}\right\}} \qquad (9-32)$$

The parameters R_1 and R_2 are the fully exposed and unexposed development rates, respectively (in units of $\mu m/min$). The sensitivity of the development rate to changes in M is reflected by the dimensionless parameter R_s.

As noted earlier, the etching of any layer in which the kinetics are surface-rate limited may be simulated by the string-advance algorithm. The velocity of the advancing etch front at each point is determined by the local etch rate, R, at that point. These values of R are established by combining the M (x, z) array with either Eq. 9-31 or Eq. 9-32.

The resist profile is simulated by letting the development process proceed according to the model and the algorithm described above. Development starts along the resist surface in contact with the developer. A string of nodes parallel to this surface is allowed to advance according to the local etch rate, R (M [x, z]). Points in the high-exposure area move downward rapidly to clear the center of the exposed region and then push laterally toward the unexposed areas to form the edge profile. Examples of developed profiles simulated by SAMPLE are shown in Figs. 9-39b and 9-39c.[63]

For many positive resists, the development rate is truly a function of M only, and the above description is complete. Other resist materials show an important surface-rate retardation effect, for which R parameterization has been further extended.[63] The more recent versions of SAMPLE can also simulate the contrast-enhancement-layer (CEL) process. An example of the simulated profiles obtained for a CEL process over a highly reflective aluminum layer is shown in Fig. 9-42. It can be seen that the high reflectivity produces very severe vertical standing waves. Even with the CEL process, a 1.25-μm feature is hardly acceptable. However, if a post exposure bake step is also added to

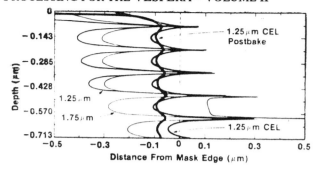

Fig. 9-42 Resist line edge profiles on aluminum.[58] Reprinted with permission of Solid State Technology.

diffuse the standing-wave effects and help reduce the contrast requirement, a usable profile is obtained.

9.6.2 Simulating Etching and Deposition with SAMPLE

A *string-advance algorithm* is implemented in SAMPLE to treat two-dimensional etching and deposition processes as well as the resist-development process.[64] Figure 9-43 illustrates the most important principles of this algorithm. In the case of etching, the advancing etch front is approximated by a series of straight-line segments, each of which moves a specific distance during each time step. To simulate the etch-front position at the next time step, each point along the string is advanced along the angle bisector of the two adjoining segments. The distance moved by each point during each step is determined by the local value of the etch rate. These values are entered by the user. A typical string, composed of between 40-100 line segments, is started on the surface and advances through the layer being etched.

As an example, points 1-5 in Fig. 9-43 show the surface of an oxide at the bottom of a resist-mask window prior to etching. If the oxide is etched with an isotropic-etch process, points 1 and 5 move laterally (in a direction under the resist mask) and points 2, 3, and 4 move vertically. At the next time step, the string connecting points 1' through 5' would simulate the position of the front.

It can be seen, however, that the string segment between points 1' and 2' does not accurately simulate the position of the etch front (the actual position of which is given by the dotted line). In addition, this is much longer than that between 1 and 2. To keep the string segments approximately equal in length, and thus provide a more accurate simulation, the program adds new points when the length of a segment exceeds a maximum limit.

For example, a new point would be added between point 1' and 2' if the string segment there were much longer than that between, say, 2' and 3'. The new point is added by bisecting the angle between 1' and 2' and locating the new point along this bisector at a distance equal to that moved by point 1 or 2 in the time step (Fig. 9-43c). For regions where the segments become shorter, contraction and deletion of small

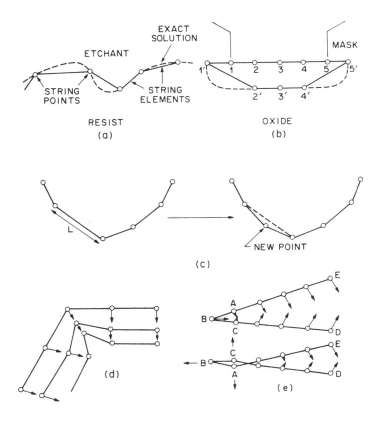

Fig. 9-43 String algorithm. (a) String model approximation to an etch front. (b) Problems with uniform etching of an oxide. (c) Length reduction by bisection. (d) Strings during contraction. (e) Loop formation, budding and expansion. (after Jewett et al., Ref. 50).

segments can occur (Fig. 9-43d). When the SAMPLE program is used to simulate deposition, the same string algorithm is employed, but in the reverse direction.

An example of the use of SAMPLE to simulate a simple plasma-etch process is shown in Fig. 9-44. The isotropic and anisotropic components of the etch (which are used to determine the distance traveled in the horizontal and vertical directions during each time step at each point of the advancing etch front) are obtained from experimentally determined etch rates of actual structures. Although obtaining accurate data on such etch-rate components directly from the physical, electrical, and chemical tool parameters can be a painstaking task, the simulated profile can often track the experimental profiles rather well as a function of power, pressure, and voltage.

In plasma etching, the profile of the etched feature – as well as line-width control and sublayer erosion – depends on a trade-off between the degree of anisotropy and selectivity in the etching process. With multilayer gate materials having dissimilar etching characteristics, the process-design trade-offs are even more complex. Figure

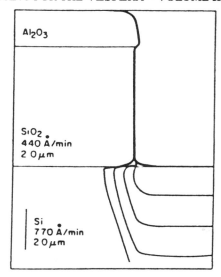

Fig. 9-44 Simulation of reactive ion etching of SiO_2 and Si.[98] (© 1979   Reprinted with permission of AIOP).

9-45 indicates the nature of this problem for a $TaSi_2$-on-polysilicon gate. In this case, etch-rate and profile data from Mattausch et al.[74] have been used to study the profile shape as a function of the mixture of fluorine- and chlorine-containing gases.

9.6.3 Creating Input Files for SAMPLE

The FORTRAN source code of SAMPLE is divided into blocks of subroutines that

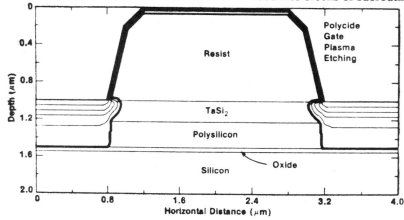

Fig. 9-45 Line edge profile of a polycide ($TaSi_2$ on polysilicon) gate showing the difficulty in obtaining a smooth and nearly vertical profile.[58] Reprinted with permission of Solid State Technology.

simulate various processing steps, such as thin-film deposition and resist development. The user specifies the inputs to the program using a set of keywords. Since this vocabulary is processing jargon, it is not necessary for the user the user to have knowledge of the programming language.

Figure 9.46 presents an example of a SAMPLE input file,[50] in which a diffraction-limited optical system is used to expose a periodic pattern with identical lines and spaces. The first line is the title of the input file. The second and third lines specify the wavelength of the light source (λ = 435.8 nm) and the numerical aperture (NA) of the optical system (NA = 0.28). The fourth line specifies the line and space dimensions, and the fifth specifies partial-coherence and defocus factors. The *resmodel* lines specify the resist parameters (A, B, and C as 0.551, 0.058, and 0.010, respectively), as well as the refractive index **n** (where 1.68 is the real part of **n**, and 0.713 is the magnitude of the imaginary part of **n**). The *layers* lines indicate that the wafer has a thick substrate with a refractive index (4.73, -0.136), and an additional layer whose thickness is 0.741 μm and whose index of refraction is (1.47, 0.0). *Run 3* commands the resist-exposure subprogram to calculate the standing waves in the resist and to determine the inhibitor matrix M (x, z). The *devrate* line sets the constants E_1, E_2, and E_3 in Eq. 9-29. The time of the development process is specified in the *devtime* line. *Run 4* executes the actual development calculation and determines the development contours of the resist layer.

The results of this simulation are shown in Fig. 9-39. Since the line width is 1.25 μm and only half of a mask and space pattern is shown (due to the periodicity of the pattern), the edge of the mask is located at 0.625 μm. The light intensity on the surface

```
# Single wavelength projection lithography (default)   -- samop0

lambda 0.4358                           ; # lambda parameter

proj 0.28                               ; # numerical aperture

linespace 1.25 1.25                     ; # linespace parameters

parcohdef 0 0.7 1.5                     ; # sigma and defocus

run 1                                   ; # run image machine

resmodel ((0.4358))

        (0.551, 0.058, 0.010)

        (1.68, ((-0.02))) (0.7133)      ; # resist exposure parameters

layers (4.73,-0.136)

        (1.47,0.0,0.0741)               ; # layer parameters

dose 150                                ; # dose for exposure in mJcm⁻²

run 3                                   ; # run exposure machine

devrate 1 (5.63, 7.43, -12.6)           ; # resist development parameters

devtime 15 75, 5                        ; # development times

run 4                                   ; # run development machine

descumspec 0.02, 0.04, 3                ; # run descum
```

Fig. 9-46 SAMPLE input file for lithography simulation.

is plotted in Fig. 9-39a, and the resist profile is plotted in Fig. 9-39b as a function of develop time. Finally, the resist profile following a descum step is shown in Fig. 9-39c. It is observed that the line width of the resist layer has decreased to ~0.8 μm.

9.7 OTHER 2-D TOPOGRAPHY SIMULATORS

9.7.1 PROLITH

PROLITH (**p**ositive **r**esist **o**ptical **lith**ography model)[101] is another publically available lithography simulator. It offers several features not available in SAMPLE. In addition to being able to simulate projection lithography (as can SAMPLE), PROLITH can also simulate contact and proximity lithography (a feature unique to PROLITH). The program is designed for both user ease and flexibility, and the current version runs on an IBM PC, AT, PS/2 (or compatible computers). The user can enter multiple values of an input parameter (e.g., focus, NA, exposure energy, or λ), and the program will calculate how the effect of such an input parameter change will impact any desired output parameter (e.g., linewidth or resist sidewall slope). The program was developed by C. A. Mack, and is now available from FINLE Technologies (see Table 9-2).

Some of the other features that PROLITH also offers are the following: (1) The prebake process is modeled as a decomposition of the photoactive compound. The result is a decrease in the value of M. Since the resist parameters A and B in Eq. 9-29 are dependent on the initial photoactive-compound concentration, they can be computed as a function of prebake conditions; (2) Post-exposure bake processes can be modeled as the diffusion of the PAC within the resist; (3) A contrast-enhancement-lithography model is available; (4) Multi-level resists can be modeled with the analytical standing wave expression found in PROLITH; and (5) The three methods used to reduce the standing wave effect (anti-reflective coatings, post-exposure bakes, and dyed photoresists) can all be modeled with PROLITH.

9.7.2 DEPICT

DEPICT is a two-dimensional topography simulator sold by TMA, Inc. (see Table 9-3). While it offers the same type of simulation capabilities as SAMPLE, in some cases it relies on a different set of models. For example, it incorporates three different user-selectable resist-development models; of these, one is similar to the model used in SAMPLE, while another is like the one used in PROLITH.

As of this writing, DEPICT can model structures of to 10 layers, composed of any of 40 user-definable materials (including more than a dozen different photoresists). It has photoresist preexposure and postexposure bake models, as well as a model that simulates bias sputter etching.

Figure 9-47 shows an example of a simulation using DEPICT. For this sequence, a planar silicon surface is specified and this is then covered with a 0.1-μm-thick oxide film. Polysilicon features are then deposited, followed by CVD oxide (assuming a hemispherical deposition source). Finally, an Al film is evaporated, using planetary-

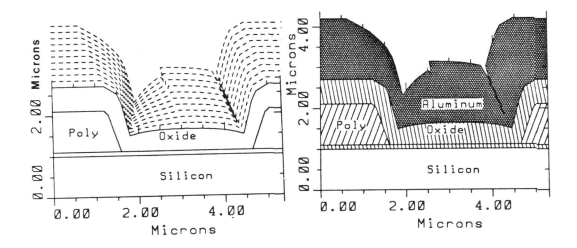

Fig. 9-47 Simulation example using DEPICT. (a) Incremental deposition surfaces. (b) Final structure. Courtesy of TMA, Inc.

source deposition. The incremental deposition surfaces can be plotted (Fig. 9-47a), as can the final overall structure (Fig. 9-47b).

Bidirectional communication allows quick and efficient exchange of information between DEPICT and SUPRA. An example of how these programs can be used together to simulate a MOS device structure with an LDD is shown in Fig. 9-48. First, SUPRA is utilized to simulate the impurity contours up to the implant step that forms the lightly doped extension regions of the source and drain (Fig. 9-48a). That is, SUPRA simulates the boron threshold-adjust and channel-stop implants, field oxidation, polysilicon gate deposition and patterning, and phosphorus source/drain-extension implant. The output of this program is then used as the input to DEPICT, which simulates the oxide-spacer-formation processes of oxide deposition and etching. Finally, the device structure with the oxide spacers (as simulated by DEPICT) is used as the input to SUPRA, which subsequently simulates the heavy source/drain arsenic implant. Figure 9-48b shows the doping profiles after this step (as well as the oxide spacer) as the output of SUPRA.

9.7.3 PROFILE

A dry-etch simulator whose models are based more on the physical principles of the dry-etch process than those used in SAMPLE or DEPICT has been reported by Ulacia and McVittie of Stanford University.[65,66] Although this program also uses the string algorithm to advance the etch front at each time step, it generates appropriate etch rates from physical principles, rather than using the rates extracted from measured data. It separates the simulation problem into the five independent physical and chemical problems:

1. gas-phase kinetics
2. gas electrodynamics
3. particle transport
4. surface kinetics
5. time evolution.

The time-evolution module takes the output of the first four modules and calculates the relative contributions of the different etching mechanisms to the etch rate (i.e., isotropic, ion flux, energy flux, deposition flux, and inhibitor layers).

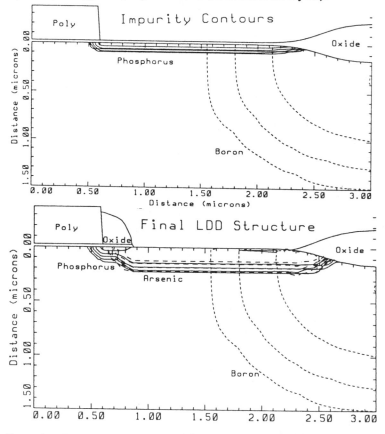

Fig. 9-48 Example of a simulation using both SUPRA and DEPICT. (a) SUPRA is used to calculate the doping contours of the structure up to the lightly doped drain implant step. (b) DEPICT is then used to simulate the formation of the poly sidewall oxide spacer. SURPA is again used to calculate the doping contours after the heavy implant dose performed after spacer formation. Fig. (b) shows this as a SUPRA output. Courtesy of TMA, Inc.

9.7.4 SIMBAD

A simulator called SIMBAD (**SIM**ulation by **BA**llistic **D**eposition) has recently been introduced for modeling the deposition of sputtered and evaporated film over topography.[93,94] Although programs such as SAMPLE and DEPICT have been used to predict the surface profiles of films produced by a variety of techniques, the models used in these simulators do not provide information essential to a full description of the process. They assume that the films are homogeneous, and they provide no information about the microstructure of the films. As feature sizes shrink, knowledge about previously unimportant characteristics (such as density variation and columnar microstructure) will become crucial.

A ballistic-deposition technique is used in SIMBAD to simulate the step-coverage and surface profiles at different stages of growth, as well as to predict local film density and microstructure. Ballistic deposition simulates thin-film growth through the random irreversible deposition of 2-D hard disks launched with linear trajectories toward a surface. The trajectory of each disk does not represent the "path" of one individual atom, but the average path of a large number of atoms that move through very similar trajectories. Figure 9-49 shows a typical simulated film of 30,000 particles over a 1-μm sloped via. Final step coverage is 24 %.

Such simulations are relatively intensive computationally (e.g., one hour on a SUN 3-160 microcomputer). A more enhanced version of the simulator will allow the modeling of 3-D shadowing effects.

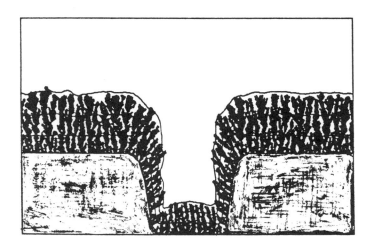

Fig. 9-49 Simulation using SIMBAD of 30,000 particles over a 1-μm sloped via. Final step coverage for a 0.8-μm film is 24 percent.[93] (© 1989 IEEE).

9.7.5 SIMPL (<u>Si</u>mulated <u>P</u>rograms from the <u>L</u>ayout)*

The SIMPL simulator is designed to help circuit and process designers get a physical view of the composite cross-section of the device structure being produced from a given layout. This is done by combining both process and mask information. The mask data is taken directly from a "cut line" drawn on the layout by a designer; SIMPL automatically determines the locations of the intersections of the "cut line" with the mask edges. A separate file of step-by-step process parameters is then used to simulate the cross section along the cut line. Both the layout and the device cross-section are then displayed simultaneously on the designer's terminal.

An example of the layout for a CMOS inverter and its cross-section are shown in Fig. 9-50. The heavy black line at the bottom is the cut line. The various circuit features (i.e., polysilicon, n-well, active area, p-source/drain implant area, contact areas, and metal) are given in different colors (the cover of our book provides an example of the colors – if not a completely correct rendering of either the layout or the cross-section – as they might appear in SIMPL). Note that the scale in the vertical direction of the

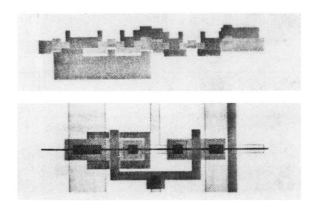

Fig. 9-50 CMOS inverter layout and device cross section simulated by SIMPL-1.[58] Reprinted with permission of Solid State Technology.

* The description of the SIMPL simulator presented here was taken from an article by A. R. Neureuther that appears in *Solid State Technology* (March 1986, p. 71). It is reprinted with the journal's permission.

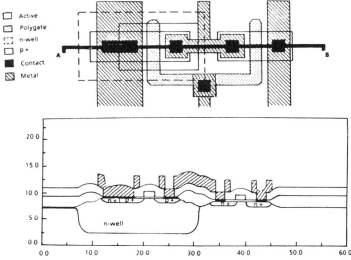

Fig. 9-51 CMOS inverter simulated with SIMPL-2 using polygonal shapes.[58] Reprinted with permission of Solid State Technology.

cross-sectional view in Fig. 9-50 is expanded by a factor of 10 compared to the horizontal scale.

The ability to go from layout to device cross-section is a good test of understanding for a new process engineer. For example, in most processes there is no simple one-to-one correspondence between each device feature and a particular mask. In the n-well CMOS example just given, six masks are used in eight masking steps to fabricate the circuit. While the contact to the n-well does not appear explicitly in the layout, as it is generated by an implicit combination of three masks, it stands out clearly in the cross-sectional view. More subtle device-technology features also appear – for example, the threshold-adjust implant in the channel region and the difference in height of the silicon between p- and n-channel devices. SIMPL is generic in that any set of masks and process-step sequences can be combined. It is also possible to simulate files for other technologies, such as those for bipolar or GaAs devices, by changing the input layout and process sequence.

A second-generation SIMPL-2 has been developed to provide more accurate simulation of such two-dimensional effects as the bird's beak, lateral diffusion, undercutting in etching, and sidewall coverage in deposition. The device profile from a composite process sequence can be generated rapidly using elementary, internal physical-process models. Greater detail can be obtained by interfacing SIMPL with a more rigorous external simulator, such as SAMPLE, for an exchange of profile data and transfer of control.

Figure 9-51 gives an example of the CMOS device cross section generated by SIMPL-2 for the same CMOS process shown earlier using SIMPL-1. In this case the layout and cut line are at the top of the figure. Here, the vertical scale of the device cross-section is three times larger than the horizontal scale. The shape of the bird's beak oxide, the lateral diffusion of the n-well, and the thinning of the metal at steps are depicted (these do not appear in the cross-section generated by SIMPL-1). Even more detail can be obtained by expanding key regions of interest at a larger scale. The ultimate resolution limit is the point-spacing and impurity-doping grid selected by the user for the simulation. Further topography details can be added by including information about resist profiles, plasma-etched polysilicon edges, etched oxide-sidewall shapes, and step coverage of deposited oxides for each process step.

SIMPL offers three modes of operation, depending upon the level of detail desired. In the most primitive of these, *Mode 1*, SIMPL-1 and rectangular shapes are used to generate the cross-section, and the simulation can thus be done very rapidly. For example, the cross-section of the CMOS inverter can be generated in five seconds on a VAX 11/780. *Mode 2* is more elaborate, using SIMPL-2 and arbitrary polygons. With the same computer hardware used, 4 minutes are required to produce the cross-section with Mode 2. The most detailed mode, *Mode 3*, generates cross-sections in which the elementary polygon models are replaced by rigorous external simulation. Each simulator call can take from several minutes to an hour to generate the structure being modeled. For example, calls to SAMPLE in this mode for simulation of lithography, etching, or deposition take several minutes each; for two-dimensional impurity-diffusion or electrical analysis, even longer times are necessary.

Mode 1 is appropriate for on-line viewing of cross-sections from device and test-structure layouts. Mode 2, in which the internal models can be tuned for particular lithography, etching, and deposition technologies, is well suited for exploring worst-case design situations and for assessing the impact of technology innovation. For detailed studies of critical device regions, Mode 3 uses SIMPL-2 to provide a link between layout-based CAD tools and process and development simulators. A common international *Profile Exchange Format* (PIF) has been developed to allow communication among process and device simulators,[14] making it unnecessary to customize each available simulator for use with SIMPL.

The SIMPL programs are written in the "C" language and require a color-display terminal for viewing of the output profiles. SIMPL-1 has 1500 lines of code and generates a standard CIF (Caltech Intermediate Format) file. SIMPL-2 has 12,000 lines of code and uses an additional 3000 lines of the Modified Frame Buffer graphical output interface from the layout system KIC2. Both SIMPL-1 and SIMPL-2 can be run from any standard terminal with the output directed to a file.

9.7.6 SIMPL-DIX

SIMPL-DIX, an interactive computer-aided design tool for running process and device simulators from layout and process specifications,[69] provides a design interface to other process and device simulators through the SIMPL-2 simulator. Device cross-sections can be displayed such on graphics workstations as MicroVAX and SUN systems that run UNIX and Version 10 of the X Window System (hence, the name SIMPL-**D**esign

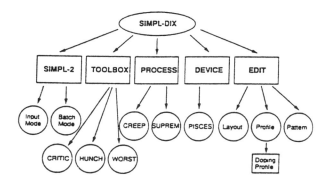

Fig. 9-52 Flowchart of SIMPL-DIX commands.[69] (© 1988 IEEE).

Interface in **X**, or SIMPL-DIX). The SIMPL-DIX CAD tool in effect enhances the capability of other process simulators in a number of ways:

- It allows SIMPL-2 to be interfaced to other 2-D topographical simulators in the following two modes (Fig. 9-52):

    1. As mentioned above, when the basic polygon models of SIMPL-2 are not detailed enough for simulating crucial process steps, the device cross-section can be sent to a more rigorous simulator for a more accurate analysis of topography details. Cross-sections can be sent to SAMPLE for simulation of isotropic or anisotropic etching or deposition (with a unidirectional, dual, hemispherical, conical or planetary source). They can also be sent to CREEP for simulation of 2-D oxidation or reflow.

    SIMPL-DIX sends its current cross-section profile to the given simulator along with a command input file created by the user. The resultant simulation from SAMPLE or CREEP is then automatically stitched back into the SIMPL data structure.

    2. Doping profiles defined from cross-sections and process specifications can be used as initial input-structure files to SUPREM III for further simulation of diffusion and oxidation, and they can also be sent in the form of input-structure files to PISCES II for device simulation.

- SIMPL-DIX contains several interactive design tools to help the designer explore topography-related problems. The first, called HUNCH, highlights combinations of neighboring masks that are likely to generate topography problems. The second, referred to as WORST, allows the effects of mask misalignment on topography features to be examined. The third, CRITIC, extracts significant parameters from a cross-section (e.g., minimum thickness of a particular layer) to aid analysis.

• SIMPL-DIX also has a link to RACPLE, a program that rapidly calculates parasitic resistive and capacitive effects in cross-sections. This capability allows the resistance and capacitance of actual circuit structures to be evaluated in terms of the profiles of metals and insulators passing over topography.[70] Note that the models used in RACPLE determine the resistance of a layer in a cross-section by finding the number of lateral squares in the layer and multiplying this number by the width of the layer perpendicular to the cross-sectional view. Likewise, capacitance is found by multiplying the permittivity of a material by the number of lateral squares and the width perpendicular to the profile. RACPLE calculations have been shown to agree quite closely with exact simulation and experimental data, but they can be computed in a fraction of the time needed for a full solution of the Laplace equation.

9.7.7 Manufacturing-Based Process Simulators

As device sizes decrease, circuit-performance sensitivity to variations in the manufacturing process increases. Because design times and production costs are increasing, designs that will result in low yields should be identified as such before they are committed to production. The circuit designer must therefore be able to predict the effects on circuit performance of statistical variations inherent in the process in order to take corrective action necessary. A new class of simulators known as *manufacturing-based simulators* has thus been developed.[71]

The center of this activity has been Carnegie-Mellon University, where a mixed-level analytical approach for process/device modeling has been developed for manufacturing application. One simulator, named FABRICS II, predicts the parametric yield of a number of processes (NMOS, CMOS, or bipolar).[72, 73] It includes a statistical approach that is applicable to both design centering and on-line manufacturing control. (This simulator will be described in more detail in Volume 3.)

9.8 DEVICE SIMULATORS

The impurity profiles obtained from 1-D or 2-D process simulators do not provide device characteristics. However, *device simulators* can use the doping profiles produced by the process simulators to predict the electronic behavior of the fabricated devices. The doping profiles are used together with the applied bias values as the inputs of a device simulator. The outputs of the programs are the electrostatic potential and free-carrier concentrations in the regions of the wafer where the devices are located. With this data, it is possible to calculate the currents by using the current equations. Several 2-D device simulators have been developed; we will briefly introduce the most popular of those that are publicly available.

9.8.1 Simulation of MOS Device Characteristics under Subthreshold and Linear Operation (GEMINI)

One of the most popular 2-D device simulators is a Poisson equation solver, called GEMINI, which is designed to simulate the device characteristics of MOSFETs.[75] It yields electrostatic solutions, assuming that the potential is not affected by the current. Consequently, the potential, φ, can be solved electrostatically using Poisson's equation alone:

$$\frac{\partial^2 \phi}{\partial x^2} + \frac{\partial^2 \phi}{\partial y^2} = \frac{-\rho\,(x,y)}{\varepsilon_{Si}} \qquad (9 - 32)$$

where ρ is the charge density and ε_{Si} is the dielectric constant of silicon. The approximation that the potential is independent of current is valid only for subthreshold and linear regions of operation in MOS devices, because small currents flow in these regions of operation.

One of the main strengths of GEMINI is that it is extremely computationally efficient, since it solves only Poisson's equation (rather than also solving the current continuity equations, as is done in other device simulators). The dramatically reduced computation time allows GEMINI to provide extensive simulations for device design and process refinements, and makes it possible for the program can be run on a minicomputer. GEMINI is able to take impurity profiles from SUPREM or SUPRA as inputs.

The rectangular gridding can be set up to be variable in two dimensions. The direct output is the potential in two dimensions, which is typically plotted in the form of 2-D equipotential contours. The current can then be calculated using the simulated potential values. GEMINI can also be used to simulate punchthrough currents, electrode and junction capacitance, and avalanche breakdown of reverse-biased junctions. Figure 9-53 shows a two-dimensional plot of potential contours as a function of position generated by GEMINI.

9.8.2 Simulation of MOS Device under All-DC Operating Conditions   (MINIMOS, CADDET, CANDE)

As described above, GEMINI can be used to simulate MOS device operation only at low current levels. When MOSFETs are operated in the saturation region, two-dimensional, current-continuity models coupled with Poisson's equation must be used to simulate transistor-device operation. In MOS devices, however, most current is carried by the majority carrier of the source/drain. Thus, such devices can be simulated by simultaneously solving Poisson's equation and only the majority carrier current-continuity equation.

Several MOS device simulators based on this approach have been developed, including MINIMOS,[76,77] CADDET,[78] and CANDE (the latter being available from TMA, Inc.). These simulators are capable of calculating steady-state (dc) MOS device

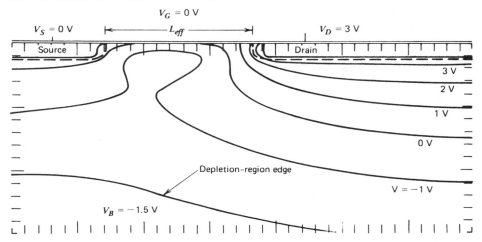

Fig. 9-53 Two dimensional plot of potential as a function of position generated by the GEMINI simulation program. Reprinted with permission of Semiconductor International.

characteristics under any bias conditions (i.e., they are not limited to just the linear and subthreshold regions, as GEMINI is). Since they solve only the majority-carrier current-continuity equation, however, these simulators cannot be used to model the characteristics of bipolar devices.

CADDET and MINIMOS are limited to planar device geometries, while CANDE can simulate most nonplanar MOS devices and can accept general nonplanar 2-D impurity profiles from SUPRA as inputs. Like GEMINI, these simulators employ nonuniform rectangular gridding. All can calculate the complete MOS dc I-V characteristics, I_{Dsat}, punchthrough current, threshold voltage, potential contours, and electric-field contours. Generally speaking, they are easy to use and are not as computationally intensive as the simulators that solve both current-continuity equations (which are described in section 9.8.4).

9.8.3 Bipolar Device Simulators (SEDAN, BIPOLE)

SEDAN is a 1-D simulator for bipolar devices that is available from Stanford University and from TMA, Inc.[92] It solves Poisson's equation and the two current-continuity equations. SEDAN generates I-V characteristics for bipolar transistors and can extract such parameters as bipolar gain, cutoff frequency, Early voltage, parasitic capacitances, and Gummel numbers. It can simulate steady-state and transient operation in bipolar devices. Because it is fast and is easy to use, SEDAN is advantageous to use when the device structure is primarily one-dimensional.

Another bipolar device simulator, BIPOLE,[96] is available commercially from Waterloo Engineering Software. The program is based on a one-dimensional solution of the semiconductor equations in the vertical (emitter to collector) direction using the variable boundary regional approach; this solution is coupled in the quasi-neutral base region to the horizontal solution (direction of base current flow) for majority-carrier drift and diffusion. The program thus includes 2-D effects with very good accuracy and has been tested on numerous devices with f_T values from 10 MHz to 7 GHz and emitter dimensions from 2 x 0.8 μm to 100 μm x 3 cm. The program includes bandgap reduction at heavy doping levels and mobility versus doping. It can run on mainframe computers as well as on PC/AT compatibles.

9.8.4 Combined MOS and Bipolar Device Simulators (PISCES, SIFCOD, PADRE, and FIELDAY)

Several device simulators have been developed that solve Poisson's equation and both current-continuity equations, including PISCES,[79] SIFCOD,[80] PADRE,[81] and FIELDAY.[82,83] Such powerful simulators are able to handle both MOS and bipolar devices, as well as bipolar device effects in CMOS structures. We describe PISCES in most detail because it is the most widely used of the group.

PISCES, developed at Stanford University, is a 2-D, two-carrier device simulator that can calculate device characteristics in either steady-state or transient conditions. It is available both from Stanford and from various commercial vendors (e.g., TMA, Inc., and SILVACO). The program can handle arbitrary physical structures and general doping profiles obtained from SUPREM, SUPRA, or analytic functions. PISCES provides default values for all physical coefficients, such as mobility, work functions, and dielectric permittivities. It can simulate such device characteristics as I-V curves, punchthrough, MOS threshold voltage and I_{Dsat}, electric fields for evaluation of hot-electron effects, bipolar current gain, parasitic capacitances and resistances, and CMOS latchup trigger and holding characteristics. It can also provide printed outputs of such device conditions as potential, carrier concentration, current density, impurity concentration, mobility, and lifetime.

Although two-carrier device simulators can provide accurate simulations of device characteristics under a wide variety of operating conditions, they are computationally complex. This implies that they require long computer times to produce the simulations. In an attempt to reduce the computation time and to allow complicated CMOS device structures to be handled (including such device structure nonplanarities as LOCOS and deep-trench isolation), nonuniform triangular grids are used instead of the rectangular grids employed by GEMINI, MINIMOS, and CANDE. Furthermore, the initial coarse grids (whose definition is based on the structure of a device) can be refined before or during the solution process according to the doping profile or any other physical variable. Figure 9-54 shows an example of the coarse and fine grids defined for an *n*-well CMOS device with trench isolation.

The numerical technique used in PISCES is based on the Gummel and Newton methods, as described in section 9.1.4.4. The Gummel method is preferable in cases in which a MOSFET is operated in zero or reverse bias and low-current conditions (i.e., the subthreshold region). However, under high-current conditions (in which the equations are strongly coupled), convergence is slow when this method is used. The Newton method is more appropriate in such cases, since all the coupling variables are taken into account through a ninefold increase in matrix size. Although such large matrices require large computer memories and long computation times, the Newton method is still more efficient when it is necessary to simulate the MOSFET in saturation, as well as under transient and ac operating conditions.

SIFCOD (**s**imulator **f**or **co**upled **d**evices) is a commercially available simulator that was developed by Mock to handle device structures with arbitrary shapes. PADRE is a proprietary simulator developed at AT&T Bell Labs, and FIELDAY, is a proprietary program developed at IBM. All of these simulators can also perform two-carrier simulation and hence can be used to simulate such transient device behavior as latchup and ionizing radiation effects in CMOS circuits. An example of a soft-error transient analysis in a six-transistor CMOS latch using SIFCOD is given in reference 84.

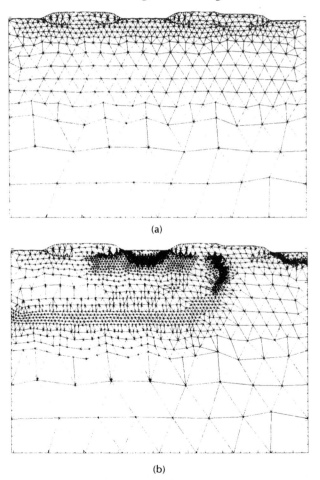

(a)

(b)

Fig. 9-54 Triangular grids for device simulation using PISCES. (a) Initial grid generated from boundary description. (b) Adaptively refined grid during simulation.[12] (© 1986 IEEE).

9.9 CIRCUIT SIMULATORS AND ELECTRICAL PARAMETER EXTRACTORS

The mid-1960s saw the beginning of vigorous efforts in computer-aided circuit design. SPICE (Simulation Program, Integrated Circuit Emphasis), developed at UC Berkeley, has been the generic circuit-simulation tool during the nearly two decades since its introduction.[85] This program combines functional device connections with physical models and parameters (such as device dimensions and junction areas) to produce simulated waveforms of current/voltage versus time, as well as the frequency response for ICs. Efforts have also been made to couple circuit and device analysis in a single simulator environment, as demonstrated with MEDUSA.[86]

In most cases, users must know not only how a single device behaves, but also how to describe it accurately to a circuit simulator by means of a model with the appropriate electrical-device parameters. These parameters can be extracted from the device characteristics calculated by the device simulators and can then be used in circuit simulations. Three examples of such *circuit extraction programs* are SUXES,[87] TECAP (Transistor Electrical Characterization and Analysis Program),[88] and TOPEX. The latter two are commercially available programs offered by Hewlett-Packard and TMA, Inc., respectively. Another program, DL2000, is a SPICE parameter library containing models for more than 2000 commercial discrete devices. It is available from SILVACO.

9.10 FUTURE CHALLENGES IN PROCESS SIMULATION

Some of the issues of future progress in process simulation were addressed by Dutton in two recent articles.[1, 12] and several of the conclusions he drew are paraphrased here.

At MOS channel lengths greater than 2 μm and gate-oxide thicknesses larger than 50 nm, constant bulk parameters for both process and device simulation have been adequate. However, at the much smaller dimensions of state-of-the-art devices, interpretation of physical parameters has become a major bottleneck. For MOS devices, the mobility, avalanche, and oxide-tunneling parameters are all critical, and each presents the challenge of insufficient characterization. For bipolar structures the recombination, band gap, and carrier-transport parameters pose similar problems of insufficient data. Furthermore, such processing steps as thin oxide growth and shallow junction diffusion are reaching atomic dimensions, so that new parameter regimes and even statistical considerations are becoming important. Atomic-scale analytic tools such as transmission and tunneling microscopy are becoming essential.

Although such tools are now available, their use is still rather selective, and the characterization techniques are generally tedious. The problem of materials characterization will therefore be a growing problem and an area of need in VLSI development. (A remark made by Fair in 1989, questioned the accuracy of 2-D simulators, since at that time no one had yet been able to perform measurements to verify the simulation results.[89] Some initial 2-D dopant-profile measurements have in fact, been performed, but the initial results suggest that further work is needed on 2-D process modeling.)

The physical models used to represent the above-mentioned effects are still crude and are thus poorly suited for simulation applications. For example, avalanche models require both field and current-direction information throughout the analysis grid. No elegant means exists for computing the avalanche generation. In the case of gate currents resulting from hot-carrier generation, the physics of how the carriers enter the oxide is poorly understood. Moreover, simulation of these currents requires particle resolutions on the order of $1:10^6$ or better. Finally, much work remains to be done to make it possible to model the physics of high-dopant concentration diffusion and to dramatically reduce CPU costs for simulating coupled-defect phenomena.

In considering the issue of future simulation costs in terms of computational requirements, Dutton concludes that process simulation will be 10-100 times more costly than device analysis for two reasons: (1) multiple transient solutions will be needed to replicate technology cross-sections, and (2) the number of coupled differential equations that must be used to simulate diffusion processes at submicron distances in two dimensions dramatically increases the computational complexity. Substantial computational requirements will thus be needed to support the VLSI research and development efforts.

To overcome some of the difficulties outlined above, Fair has proposed that both phenomenologically- and physically-based 2-D simulators (i.e., point-defect-based process modeling) be used, with the appropriate choice based on the application. One example of a case in which the point-defect-based program must be used involves a moving boundary, where oxidation-enhanced diffusion varies along a curve as the silicon crystalline orientation changes.

On the other hand, in many instances the variability of process conditions makes point-defect-based models qualitative at best. For example, the models available to calculate the point-defect generation and recombination characteristics of silicon regions damaged by ion implantation are still being developed.[99,100] Furthermore, the effects that occur during annealing of end-of-range dislocations depend on the distance from the surface. Hence, many effects involved in diffusion following ion implantation are complex, and the physical processes (as well as the mathematics) are not fully understood. Modeling of processes that occur under RTP, and the formation of silicides (and diffusion in silicides), are also areas for future modelling efforts.

Fair argues that such process steps are more easily simulated through the modeling of their effects on dopant diffusion. With a sufficient data base of 2-D doping profiles, phenomenological models should be adequate for solving many process problems. The level of uncertainty should not be any worse than that occurring with point-defect-based models (which must rely on many unknown coefficients). However, the degree of computational intensity is much less when phenomenological models are used.

REFERENCES

1. R. W. Dutton, *Tech. Dig. IEDM,* 1986, p. 2.

2. K. M. Cham, S.-Y. Oh, D. Chin, and J. L. Moll, *Computer-Aided Design and VLSI Device Development*, 2d Ed., Boston, Klewer Academic, 1989.

3. P. Antognetti et al., *Process and Device Simulation for MOS-VLSI Circuits*, The Hague, Netherlands, Martinus Nijhoff Publishers, 1983.

4. A. Neureuther, *Advances in CAD for VLSI Processes and Device Simulation*, W. Engl, Ed., North-Holland, 1985.

5. R. J. Sokel and D. B. MacMillen, *IEEE Trans. Electron Dev.*, October 1985, p. 2110.

6. *IEEE Trans. CAD for ICs and Systems*, May 1987 (special issue on CAD in Japan).

7. C. D. Maldonado et al., "ROMANS-II," *IEEE Trans. Electron Dev.*, Nov. 1983, p. 1462.

8. J. Lorenz et al., COMPOSITE, *IEEE Trans. Electron Dev.*, October 1985, p. 216.

9. L. Borucki, H. H. Hanse, and K. Varahramyam, *IBM J. Res. Dev.*, vol. 29, no. 3, p. 263, 1985.

10. B. R. Penumalli, BICEPS, *IEEE Trans. Electron Dev.*, Sept. 1983, p. 983.

11. G. Smith and A. J. Steckl, RECIPE, *IEEE Trans. Electron Dev.*, Febuary 1982, p. 216.

12. R. W. Dutton and M. R. Pinto, "The Use of Computer Aids in IC Technology," *Proceedings of the IEEE*, Dec 1986, p. 1730.

13. R. B. Fair, *Ext. Abs. Electrochem. Soc. Mting*, Spring, 1989, Abs. No. 141, May 1989 p. 197.

14. S. G. Duvall, *IEEE Trans. on Computer-Aided Design*, vol. CAD-7, no. 7, July 1988, p. 741.

15. D. A. Antoniadis and R. W. Dutton, "SUPREM I," *IEEE J. Solid-State Circuits*, April 1979, p. 412.

16. C. P. Ho, J. D. Plummer, S. E. Hansen, and R. W. Dutton, "VLSI Process Modeling - SUPREM III," *IEEE Trans. Electron Dev.*, vol. ED-30, (11), p. 1438, (1983).

17. S. E. Hansen, "SUPREM III Users Manual Version 8520," PhD Thesis, Integrated Circuits Laboratory, Department of Electrical Engineering, Stanford University, May 7, 1985.

18. L. A. Christel and J. F. Gibbons, *J. Appl. Phys.*, **51**, 6176 (1980).

19. J. P. Biersack and L. G. Haggmark, *Nucl. Instrum. and Methods*, **174**, 257, (1980).

20. H. Ryssel, J. Lorenz, and K. Hoffmann, "Models of Implantation into Multilayer Targets," *Appl. Phys. A*, **41**, 201, (1986).

21. R. B. Fair, "Concentration Profiles of Diffused Dopants in Silicon," in F. F. Y. Wang, Ed., *Impurity Doping Processes in Silicon*, North-Holland, New York, 1981, Chapter 7.

22. U. Gosele, "Current Understanding of Diffusion Mechanisms in Silicon," *Semiconductor Silicon, 1986*, H. R. Huff, T. Abe, and B. Kolbesen, Eds., Electrochemical Society, Pennington, NJ, 1986, p. 541.,

23. P. Fahey and R. W. Dutton, "Dopant Diffusion Under Conditions of Thermal Nitridation of Si and SiO_2," *Semiconductor Silicon, 1986*, H. R. Huff, T. Abe, and B. Kolbesen, Eds., *Electrochemical Soc.*, Pennington, NJ, 1986, p. 571.

24. P. B. Griffin and J. D. Plummer, "Advanced Diffusion Models for VLSI," *Solid State Technology*, May 1988, p. 171.

25. A. S. Grove, O. Leistiko, and C. T. Sah, *J. Appl. Phys.*, **35**, 2629, (1964).

26. R. Razouk, L. N. Lie, and B. E. Deal, *J. Electrochem. Soc.*, October 1981, p. 2214.

27. W. Shockley and J. L. Moll, *Phys. Rev.*, vol. 119, p. 1480, 1960.

28. H. Z. Massoud et al., *J. Electrochem. Soc.*, **132**, 2685, (1985).

29. R. Reif and R. W. Dutton, "Computer Simulation of Silicon Epitaxy", *J. Electrochem. Soc.*, April, 1981, p. 909.

30. R. Reif, *J. Electrochem. Soc.*, May, 1982, p. 1122.

31. J. D. Plummer, *Solid State Technology*, March 1986, p. 61.

32. R. B. Fair, *IEEE Trans. Electron Dev.*, **ED-35**, March 1988, p. 285.

33. M. Kump and R. W. Dutton, "Two-Dimensional Process Simulation," in *Process and Device Simulation for MOS-VLSI Circuits*, P. Antognetti et al., Martinus Nijhoff Publishers, The Hague, The Netherlands, 1983, p. 235.

34. S. Hsai et al., *IEEE Electron Device Letts.*, Feb. 1982, p. 40.

35. H. G. Lee, TR No. G201-8, Stanford Electronic Laboratories, Stanford University, CA., Aug. 1980.

36. D. Chin, M. Kump, and R. W. Dutton, "SUPRA", Stanford Electronics Laboratories, Stanford University, CA., Oct. 1979.

37. J. A. Greenfield and R. W. Dutton, "Nonplanar VLSI Device Analysis Using the Solution of Poisson's Equation," *IEEE Trans. Electron Devices*, **ED-27**, p. 1520, Aug. 1980.

38. J. D. Plummer et al., "Process Simulators for Silicon VLSI and High Speed GaAs Devices,", Integrated Circuits Laboratory, Stanford University, Stanford, CA, July 1986.

39. P. M. Fahey, "Point Defect and Dopant Diffusion in Silicon," PhD Thesis, Integrated Circuits Laboratory, Department of Electrical Engineering, Stanford University, June, 1985.

40. M. E. Law, C. Rafferty, and R. W. Dutton, "SUPREM IV Users Manual," Technical Report, Integrated Circuits Laboratory, Department of Electrical Engineering, Stanford University, July, 1986.

41. N. Guillemot, G. Pananakakis, and P. Chenevier, *IEEE Trans. Electron Devices*, May, 1987, p. 1033.

42. D. J. Chin et al., *IEEE Trans. Electron Dev.*, ED-30, July, 1983, p. 993.

43. D. Kao et al., *Tech. Dig. IEDM*, 1985, p. 388.

44. P. Sutardja, W. G. Oldham, and D. B. Kao, *Tech. Dig. IEDM*, 1987, p. 264.

45. P. Sutardja, Y. Shacham-Diamond and W. G. Oldham, *Tech. Dig. IEDM*, 1986, p. 526.

46. E. P. EerNisse, "Stress in Thermal SiO_2 during Growth," *Appl. Phys. Lett.*, **35** (1) 1 July, 1979.

47. W. G. Oldham et al., "A General Simulator for VLSI Lithography and Etching: Part I - Application to Projection Lithography," *IEEE Trans. Electron Dev.*, vol. **ED-26** (4), p. 712, Apr. 1979.

48. W. G. Oldham et al., "A General Simulator for VLSI Lithography and Etching: Part I - Application to Deposition and Etching," *IEEE Trans. Electron Dev.*, vol. **ED-27** (8), p. 1455, Aug. 1980.

49. A. R. Neureuther, "IC Process Modeling and Topography," *IEEE Proc.*, Special Issue on VLSI Design,: Problems and Tools, vol. **71** (1), p. 121, Jan. 1983.

50. "SAMPLE Version 1.6a Users Guide," Electronics Research Laboratory, Dept. of Electrical Engineering and Computer Science, UC Berkeley, February 1, 1985.

51. M. G. Rosenfield and A. R. Neureuther, *IEEE Trans. Electron Dev.*, Nov 1981, p. 1289.

52. G. M. Atkinson and A. R. Neureuther, *J. Vac. Sci. Technol., B,* vol. **3** (1), p. 421, Jan./Feb. 1985.

53. W. Arden, H. Keller, and L. Mader, *Solid State Technol.,* July, 1983, p. 143.

54. B. J. Lin, in *Proceedings Conf. Microlith. - Microcircuit Engr 81,* A. Oosenbrug, Ed., Sept. 28, 1979, Lausanne, Switz., p. 47.

55. M. Born and E. Wolf, *Principles of Optics-Electromagnetic Theory of Propagation, Interference, and Diffraction of Light,* Pergamon, New York, 1959.

56. M. M. O'Toole and A. R. Neureuther, SPIE vol. **174,** *Developments in Semiconductor Microlithography IV,* 22, (1979).

57. S. Subramanian, "Rapid Calculation of Defocused Partially Coherent Images," *Applied Optics,* Vol. 20, p. 1854 (1981).

58. A. R. Neureuther, "Topography Simulation Tools," *Solid State Technology,* March 1986, p. 71.

59. F. H. Dill et al., *IEEE Trans. Electron Dev.,* July, 1975, p. 456.

60. A. R. Neureuther and W. G. Oldham, "Simulation in Lithography," in *Process and Device Modeling,* ed. W. Engl, North-Holland, 1986.

61. C. A. Mack, "Absorption and Exposure in Positive Photoresist," *Applied Optics,* **Vol. 27,** p. 4913, December 15, 1988.

62. F. H. Dill et al., *IEEE Trans. Electron Dev.,* July, 1975, p. 445.

63. D. J. Kim, W. G. Oldham, and A. R. Neureuther, *IEEE Trans. Electron Dev.,* Dec. 1984, p. 1730.

64. R. Jewett, Memo UC Berkeley /ERL M 79/68, Univ. of California, Berkeley, 1979.

65. J. I. Ulacia and J. P. McVittie, *Ext Abs. Electrochem. Soc. Meeting, Spring,1988,* Abs. No. 100, p. 144.

66. J. I. Ulacia and J. P. McVittie, *Ext Abs. Electrochem. Soc. Meeting, Spring, 1988,* Abs. No. 101, p. 146.

67. M. A. Grimm, K. Lee, A. R. Neureuther, *IEDM Tech. Dig.,* 1983, p. 255.

68. K. Lee, et al., *1985 Symposium on VLSI Technology, Digest of Technical Papers,* p. 64, May, 1985.

69. H. C. Wu et al., *Tech Dig. IEDM,* 1988, p. 328.

70. E. W. Scheckler, et al., *Proceeding of the Sixth IEEE VMIC Conf.,* 1989, p. 299.

71. S. W. Director, "Manufacturing-Based Simulation" *IEEE Circuits and Devices Magazine,* Sept. 1987, p. 3.

72. W. Maly and A. Strojwas, *IEEE Trans. CAD,* vol. **CAD-1,** no. 3, July, 1981.

73. S. R. Nassif, A. J. Strojwas, and S. W. Director, *IEEE Trans. Computer-Aided Design of Integ. Circuits and Sys.,* Jan. 1984, p. 40.

74. H. J. Mattausch, B. Hasler, and W. Beinvogel, "Reactive Ion Etching of Ta-Silicide /Polysilicon Double Layers," *J. Vac. Sci. Technol. B,* vol. 1 (1), p. 15, Jan.-Mar. 1983.

75. J. A. Greenfield and R. W. Dutton, *IEEE Trans. Electron Dev.,* **ED-27,** 1520 Aug 1980.

76. S. Selberherr, A. Schutz, and H. W. Potzl, *IEEE J. Solid-State Circuits,* vol. SC-15, p. 605, (1980).

77. S. Selberherr, *Analysis and Simulation of Semiconductor Devices,* Springer-Verlag, Vienna, 1984.

78. T. Toyabe and S. Asai, *IEEE Trans. Electron Dev.,* vol. ED-26, p. 453, (1979).

79. M. R. Pinto et al., "PICSES II-B," Stanford Electronics Lab., Stanford University, Tech. Rep., Sept. 1985.
80. M. S. Mock, *Solid-State Electronics,* vol. **24**, p. 959, (1981).
81. M. N. Darwish, M. C. Dolly, and C. A. Goodwin, *Tech. Dig. IEDM,* 1988, p. 508.
82. P. E. Cottrell and E. M. Buturla, *IEDM Tech. Dig.* 1975, p. 51.
83. E. M. Buturla, P. E. Cottrell, B. M. Grossman, K. A. Salsburg, *IBM J. Res. Develop.,* July 1981, p. 218.
84. H. T. Weaver, *Tech. Dig. IEDM*, 1988, p. 512.
85. L. W. Nagel, "SPICE-2, A Computer Program to Simulate Semiconductor Circuits," ERL Memo ERL-M520, U.C. Berkeley, May 1975.
86. W. Engl et al., "MEDUSA -- A Simulator for Modular Circuits," *IEEE Trans. CAD,* April, 1982, p. 85.
87. K. Doganis and D. L. Scharfetter, *IEEE Trans. Electron Dev.,* vol **ED-30**, p. 1219 (1983).
88. E. Khalily, *Hewlett-Packard Journal,* June, 1981.
89. R. B. Fair, *Extended Abs., Spring Meeting Electrochem. Soc.,* May, 1989, Abs. No. 141, p. 197.
90. B. E. Deal, *J. Electrochemical Soc.,* April 1978, p. 576.
91. L. Mei and R. W. Dutton, *IEEE Trans. Electron Dev.,* vol. ED-29, p. 1726, Nov. 1982.
92. Z. Yu, G. Y. Chang, and R. W. Dutton, *Supplementary Report on SEDAN,* Stanford Electronics Laboratory, Stanford University, Technical Report - G201, June 1988.
93. K. L. Westra, T. Smy, M. J. Brett, *IEEE Electron Dev. Letts.,* May 1989, p. 198.
94. T. Smy et al., *Proceedings of 6th Internat. IEEE VLSI Multilevel Interconnect Conf.,* Santa Clara, CA, June 1989, p. 292.
95. B. J. Mulvaney, W. B. Richardson, T. L. Crandle, "PEPPER - A Process Simulator for VLSI," *IEEE Trans. CAD,* vol. 8, No. 4, April, 1989, p. 336.
96. D. J. Roulston, S. G. Chamberlain, and J. Sehgal, *IEEE Trans. Electron Dev.,* **ED-19** p. 809, 1972.
97. B. E. Deal, *J. Electrochem. Soc.,* **125**, (1978) 576.
98. J. L. Reynolds, A. R. Neureuther, and W. G. Oldham, *J. Vac. Sci. Technol.,* **16**, 1772, (1979).
99. T. L. Crandle, W.B. Richardson, and B. J. Mulvaney, *Tech. Dig. IEDM,* 1988, p. 636.
100. R. B. Fair, *Tech. Dig. IEDM,* 1989, p. 691.
101. C. A. Mack, "PROLITH, A Comprehensive Optical Lithography Model," *Optical Microlith. IV, Proc.,* SPIE Vol. 538 (1985), p. 207.

1

APPENDIX A

IC RESISTOR FABRICATION

Resistors in ICs are implemented using either diffused regions in the silicon substrate, or thin-films deposited on the wafer surface. In either case, it is useful to describe their resistance value in terms of the *sheet resistance* parameter, R_s.

Sheet Resistance Definition

Assume that a rectangular block of conducting material with a uniform resistivity (as a function of its depth) is given as shown in Fig. 1. The resistance, R, of this structure is calculated from

$$R = \rho L/A \tag{1}$$

where L and A represent the length and cross-sectional area of the block, respectively, and ρ is the resistivity of the material. (Note that if the resistivity of the block is not constant with depth, ρ in Eq. 1 is considered to be the *average* resistivity of the layer. A technique for accurately calculating ρ in such non-uniformly doped layers is outlined in Muller and Kamins, *Device Electronics for Integrated Circuits,* 2nd Ed., John Wiley & Sons, 1986, p. 112. In this reference, it is pointed out that detailed numerical analysis may be necessary to obtain an accurate value of ρ in regions of doped silicon formed by diffusion, if the diffusion profile deviates from a simple Gaussian distribution.)

Using W as the width of the conducting region, and t as its thickness, the resistance of the block can be written as

$$R = (\rho/t) (L/W) = R_s (L/W) \tag{2}$$

where $R_s = (\rho/t)$ is called the *sheet resistance* of the layer of material. Note that R_s

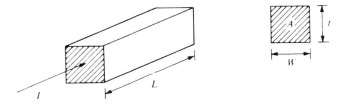

Fig. 1 Resistance of a block of material having uniform resistivity. The ratio of resistivity to thickness is called the *sheet resistance* of the material.

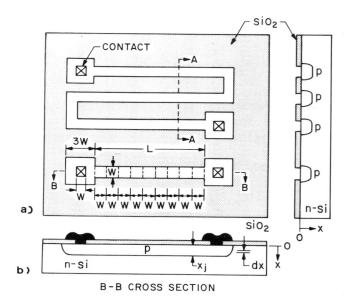

Fig. 2 (a) Top view of a zig-zag and a straight diffused resistor. (b) Side view of the straight diffused resistor.

measured for a given layer is numerically the same for any size square. It should also be noted that the unit for sheet resistance is the ohm, since the ratio of L/W is unitless. To avoid confusion between R and R_s, however, sheet resistance is given the special descriptive unit of ohms per square (Ω/sq).

If the sheet resistance value is known, a designer need only specify the length and width of the resistor to define its value. The ratio L/W can be interpreted as the number of unit squares of material in the resistor. If it is necessary to use a large number of squares to obtain a desired resistance value, a folded or zig-zag structure is normally employed (Fig. 2a).

Figure 2b depicts the side views of the straight resistor structure (in this example, a diffused IC resistor is shown). The layout of the resistor has a dumbell shape, with the contacts made at each end. In this illustration, the resistor body is nine "squares" long. If the sheet resistance of the diffusion were 50 Ω/sq, the resistor body would have a resistance of 450 Ω. Each end of the resistor adds approximately 0.65 squares to the resistor, and the total resistance of the resistor would therefore be approximately 515 Ω. The resistance contributed by a corner square can be estimated by taking it to be approximately 65% of the straight-path value. Figure 3 gives the number of squares contributed by various end and corner configurations.

Diffused Resistors

Various diffused resistor structures can be formed in either bipolar or MOS processes. Diffused IC resistors in *npn bipolar processes* can be formed by using the shallow diffusion for the transistor base and emitter regions, or by using legally doped epitaxial collector regions. In *MOS processes,* diffused resistors are formed using the process steps that produce the source/drain regions. In *CMOS processes*, the diffused regions that form the wells can also be used to form diffused resistors.

Use of diffused resistors is popular because of their compatibility with the remainder of the rest of the planar process. That is, they can be formed at the same time as the other circuit elements, and hence do not add to the fabrication cost. To keep the diffused resistor structure isolated from the rest of the circuit, the isolation scheme is usually the same as that used for the transistors. Proper biasing must also be observed. For example, when the base region of the bipolar process is used to form resistors, the *n*-type region surrounding the *p*-diffused region must be tied the most positive potential in the circuit to keep the resultant *pn* junction from ever becoming forward biased. To conserve space, diffused resistors may therefore be put into a common isolation region, since only one contact to the isolation region is needed to establish the appropriate bias.

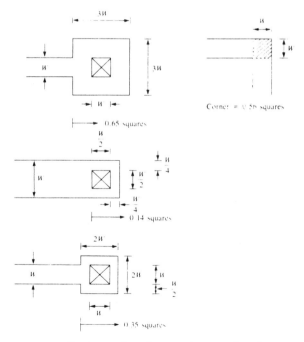

Fig. 3 Effective square contributions of various resistor end and corner configurations. From R. C. Jaeger, *Introduction to Microelectronic Fabrication.* Copyright, 1988, Addison-Wesley. Reprinted with permission.

The resistance of the structure depends on the length, width, depth of the diffusion, as well as on the resistivity of the material. For large-width resistors, the dimension W is simply determined by the mask dimensions. For very narrow resistors, however, the effective width may differ substantially from the mask dimension because impurities diffuse laterally under the oxide as well as vertically. Since minimization of the die area is important, such narrow resistors are frequently encountered. When they are used, the effective width must be determined in order to accurately estimate the resistance value.

The sheet resistance value may vary somewhat from run to run, and it is thus difficult to precisely predict the resistor values in a given circuit. However, it is easy to obtain matched resistors on a given chip, and significantly greater precision can be maintained in the ratios of paired resistors than in the resistance values themselves. For this reason, ICs are often designed so that their critical behavior depends on the ratio of two resistor values rather than on the absolute value of a specific resistor in the circuit.

In addition, since the depletion region of the isolation junction varies with voltage, diffused resistors exhibit more nonlinear I-V characteristics than do thin-film IC resistors. Finally, it should be noted that diffused resistors suffer from two additional limitations: (1) A parasitic junction capacitance is associated with the resistor and the underlying region; and (2) they exhibit a high temperature coefficient.

Bipolar IC Diffused Resistors: Base Resistors. Resistors in bipolar ICs are most commonly formed using the p-type base diffusion of an *npn* transistor, with a contact at each end (Fig. 4a). The resistor structure is created within a separate n-type island so that it is isolated from other devices on the chip. The resistivity of the base diffusion is usually determined during the design of the *npn* transistor (typically ~100- to 200-Ω/sq). Resistors formed by the base diffusion are usually limited to a range from approximately 50 Ω to 100 kΩ. Note that a *pnp* parasitic transistor exists between the resistor and the base resistor and the substrate. The n^+ buried layers, however, help to reduce the gain of this parasitic transistor.

Bipolar IC Diffused Resistors: Emitter Resistors. Diffused resistors can also be fabricated by forming an n-type emitter diffusion entirely within a p-type base region, and then providing contacts at both ends. Since the emitter region has a higher doping concentration than the base, the sheet resistance of emitter resistors will be smaller (~2- to 10-Ω/sq). These resistors can be used to advantage where very low resistance values are required.

Bipolar IC Diffused Resistors: Epitaxial Resistors. By making contact to the lightly doped epitaxial layer, larger resistor values can be obtained than with the base or emitter resistors (~1000-Ω/sq – the buried layer, however, must be omitted in order to obtain such high resistance values.)

Bipolar IC Diffused Pinch Resistors. For even higher values of sheet resistance, a double-diffused structure, known as *pinched resistor,* is sometimes used. In pinched resistors, the n^+-region is formed by the emitter over the p-base region reduces

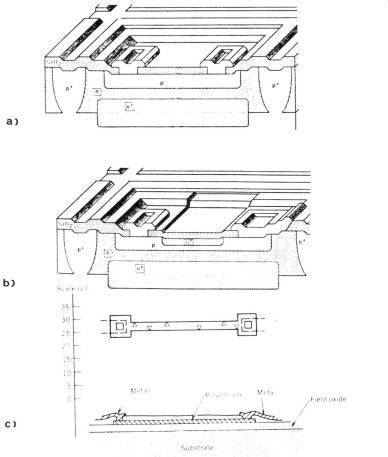

Fig. 4 (a) Base diffused resistor. (b) Base pinched resistor. (c) Thin-film polysilicon resistor. From W. Maly, *Atlas of IC Technologies,* Copyright 1987 by the Benjamin/Cummings Publishing Company. Reprinted with permission.

the vertical dimension of the diffused base resistor (Fig. 4b). This can increase the base resistor sheet resistance to 2-k- to 10-kΩ/sq. The epitaxial layer pinched between the *p*-type base and the *p*-type substrate can also serve as pinch resistor structure (~4-k- to 10-k-Ω/sq). The reproducibility of the sheet resistance of pinched resistors, however, is typically poor (e.g., ±50%). Another significant drawback is that the maximum voltage that can be applied across base pinched resistors is limited to ~6 V because of the breakdown voltage between the emitter-diffused top layer and the base diffusion. The epitaxial pinched resistors can tolerate higher applied voltages.

MOS IC Diffused Resistors: Source/Drain Resistors.   The diffused layer used to form the source and drain of the *n*-channel and *p*-channel devices can be

used to form a diffused resistor. The resulting resistor structure and properties are very similar to the bipolar emitter diffused resistor structures described above.

MOS IC Diffused Resistors: Well Resistors. The well diffused region in CMOS ICs can also be used as the body of a diffused resistor. Since the well is a relatively lightly doped region, the sheet resistance is on the order of 10-kΩ/sq. Its properties are similar to those of the bipolar epitaxial diffused resistors described above.

Thin-Film-Deposited IC Resistors

As noted, diffused resistors have the drawbacks of high temperature coefficient, poor tolerance, and must be junction isolated. The latter implies that a parasitic capacitance exists along with each diffused resistor, and exposure to radiation causes photocurrents to flow across the reverse-biased isolating junction. These disadvantages can be reduced by using thin-film resistors deposited on the top surface of the substrate. After the resistor material is deposited, the individual resistors are patterned in the conventional manner. They are then interconnected to the rest of the circuit using the standard Al interconnect process.

Thin-film resistors exhibit smaller parasitic capacitance values than do diffused resistors, and also do not suffer nonlinear behavior from depletion effects.

Polysilicon Thin-Film Resistors. Polysilicon films are used in silicon-gate MOS processes, as well as in advanced bipolar, polysilicon self-aligned structures, and in BiCMOS technology. Such polysilicon films are often used to form resistors. The nominal sheet resistance of heavily doped polysilicon films is on the order of 20 to 80 Ω/sq. The matching properties of polysilicon resistors are similar to those of diffused resistors. A cross sectional view of a polysilicon resistor is shown in Fig. 4c. High-valued, lightly doped polysilicon resistors (as described in chap. 8) are used as load devices in MOS SRAMs.

Nichrome, Tantalum, and Cermet Thin-Film IC Resistors. Three other materials have found use as thin film resistor materials in IC applications. The value of these resistors can be more reproducibly fabricated than can diffused resistors or polysilicon thin-film resistors, and they exhibit much lower temperature coefficients. These are *nichrome* (80% nickel and 20% chrome), *tantalum,* and *Cermet* (Cr-SiO) films. The sheet resistance ranges of these films are respectively 10 to 1000 Ω/sq, 10 to 1000 Ω/sq, and 30-2500 Ω/sq. The added cost of depositing and patterning another layer of resistive material is the chief drawback of such thin-film resistors.

MOS Devices as Resistors

The MOS transistor biased in the triode region can be used in many circuits to perform the function of a resistor. The effective sheet resistance is a function of the applied gate bias but can be effectively much higher than polysilicon or diffused resistors, allowing large amounts of resistance to be implemented in a small area.

APPENDIX B

PROPERTIES OF SILICON at 300° K

Quantity	Silicon
Atomic weight	28.09
Atoms, total (cm^{-3})	4.995x10^{22}
Crystal structure	Diamond
Density (g /cm^3)	2.33 (solid)
	2.53 (liquid, 1421°C)
Density of Surface atoms (cm^{-2})	
(100)	6.78x10^{14}
(110)	9.59x10^{14}
(111)	7.83x10^{14}
Dielectric constant	11.8
Effective density of states in conduction band, N$_C$ (cm^{-3})	3.22x10^{19}
Effective density of states in valence band, N$_v$, (cm^{-3})	1.83x10^{19}
Energy gap (eV)	1.12
Index of refraction	3.42
Intrinsic carrier concentration (cm^{-3})	1.38x10^{10}
Intrinsic DeBye length (μm)	24
Intrinsic resistivity (Ω-cm)	2.3x10^5
Lattice constant (Å)	5.43
Linear coefficient of thermal expansion, $\Delta L /L\ \Delta T$ (°C^{-1})	2.6x10^{-6}
Melting point (°C)	1421
Mobility (drift) (cm^2 /V s)	
μ_n (electrons)	1500
μ_p (holes)	475
Specific heat (J /g-°C)	0.7
Thermal conductivity (W /cm-°C)	
Solid	1.5
Liquid	4.3
Youngs modulus (kg / mm^2)	10,890

APPENDIX C

FUNDAMENTAL PHYSICAL CONSTANTS

CONSTANT	SYMBOL	SI - UNIT	DEFINITION
Angsrom unit	Å		$1Å = 10^{-1}$ nm $= 10^{-4}$ μm $= 10^{-10}$m
Atomic mass	$m_u = 1$ u	kg	$m_u = 1.6605 \times 10^{-27}$ kg
Avogadro's constant	N_A	mol^{-1}	$N_A = 6.022 \times 10^{23}$ mol^{-1}
Bohr radius	a_B		0.529 Å
Boltzmann constant	k	J °K^{-1}	k $= 1.38066 \times 10^{-23}$ J /°K
Electron rest mass	m_e	kg	$m_e = 9.11 \times 10^{-31}$ kg
Elementary charge	q	C	1.60218×10^{-19} C
Electrom volt	eV		1 eV $= 1.60218 \times 10^{-19}$ J
			$= 23.053$ kcal /mole
Molar gas constant	R	J mol^{-1}°K^{-1}	R $= 1.987$ cal /mole - °K
Permeability in vacuum	ε_o		$\varepsilon_o = 8.854 \times 10^{-14}$ F /cm
Planck constant	h	J s	h $= 6.626 \times 10^{-34}$ J-s.
Proton rest mass	M_p	kg	1.672×10^{-27} kg
Speed of light in vacuum	c	m s^{-1}	2.9979×10^8 m /s
Standard (atmospheric) pressure		N m^{-2}, Pa	1.013×10^5 Pa
Thermal voltage at 300° K	kT /q	V	0.0259 V

$k = 8.62 \times 10^{-5}$ eV/K

INDEX

NOTE: FOR MAJOR SUBJECT HEADINGS ALSO CHECK THE **TABLE OF CONTENETS**.
THE LISTING ARE NOT ALWAYS DUPLICATED IN THE INDEX.

ORDER FORM

LATTICE PRESS
POST OFFICE BOX 340-W
SUNSET BEACH, CA , 90742, U.S.A.

Please send the number of copies indicated below of **SILICON PROCESSING FOR THE VLSI ERA - Volume 1** and/or **Volume 2 (please specify!).** A check or money order for the full amount for the number of copies ordered is enclosed. Make checks payable to: **LATTICE PRESS.** I understand that I may return the book within 30 days for a full refund if not completely satisfied.

_____   _____copies **Vol. 2** @ $ 89.95
NAME   each, plus $4.00 for each
                                      copy, shipping- handling fee.
                                      California Residents: Please add
                                      $5.40 per copy for Sales Tax.

_____   _____ copies **Vol. 1** @ $ 59.95
ADDRESS   each, plus $4.00 for each
                                      copy, shipping- handling fee.
                                      California Residents: Please add
                                      $3.60 per copy for Sales Tax .

_____
CITY

_____
STATE, ZIP CODE   COUNTRY

                              Allow 30 days for delivery
_____
Daytime Phone Number

International Ordering Information:

Overseas Price: Volume 1 - $69.95; Volume 2 - $99.95. Enclose full payment with order in Travelers Checks (US dollars), International Money Order, or by a check payable to Lattice Press in **US dollars drawn on a U.S. bank.** Include a $4.00 shipping /handling fee per copy for overseas orders shipped via surface mail. (Allow 6-10 weeks delivery.) Include a $20 US shipping /handling fee per copy for overseas orders shipped by airmail. (Allow 2 weeks delivery).

ORDER FORM

LATTICE PRESS
POST OFFICE BOX 340-W
SUNSET BEACH, CA , 90742, U.S.A.

Please send the number of copies indicated below of **SILICON PROCESSING FOR THE VLSI ERA - Volume 1** and/or **Volume 2 (please specify!)**.   A check or money order for the full amount for the number of copies ordered is enclosed. Make checks payable to: **LATTICE PRESS.**   I understand that I may return the book within 30 days for a full refund if not completely satisfied.

_____   _____copies **Vol. 2** @ $ 89.95
NAME   each, plus $4.00 for each
                                                                      copy, shipping- handling fee.
                                                                      California Residents: Please add
                                                            $5.40 per copy for Sales Tax.

_____   _____ copies **Vol. 1** @ $ 59.95
ADDRESS   each, plus $4.00 for each
                                                                      copy, shipping- handling fee.
                                                                      California Residents: Please add
                                                            $3.60 per copy for Sales Tax .

_____
CITY

_____
STATE, ZIP CODE   COUNTRY

                                                             Allow 30 days for delivery
_____
Daytime Phone Number

International Ordering Information:
Overseas Price: Volume 1 - $69.95; Volume 2 - $99.95. Enclose full payment with order in Travelers Checks (US dollars), International Money Order, or by a check payable to Lattice Press in **US dollars drawn on a U.S. bank.**   Include a $4.00 shipping /handling fee per copy for overseas orders shipped via surface mail. (Allow 6-10 weeks delivery.) Include a $20 US shipping /handling fee per copy for overseas orders shipped by airmail. (Allow 2 weeks delivery).